RINGS, CLUSTERS AND POLYMERS OF MAIN GROUP AND TRANSITION ELEMENTS

RINGS, CLUSTERS AND POLYMERS OF MAIN GROUP AND TRANSITION ELEMENTS

Edited by

H.W. Roesky

Institut für Anorganische Chemie, Universität Göttingen, Tammannstrasse 4, D-3400 Göttingen, B.R.D.

ELSEVIER

Amsterdam — Oxford — New York — Tokyo 1989

ELSEVIER SCIENCE PUBLISHERS B.V.
Sara Burgerhartstraat 25
P.O. Box 211, 1000 AE Amsterdam, The Netherlands

Distributors for the United States and Canada:

ELSEVIER SCIENCE PUBLISHING COMPANY INC.
655, Avenue of the Americas
New York, NY 10010, U.S.A.

ISBN 0-444-88172-7

Printed in The Netherlands

PREFACE

The chemistry of rings, clusters and polymers is becoming an important branch of inorganic and organometallic chemistry. Although inorganic rings are known for more than 150 years, they have been structurally characterized only within the last fifty years.

Inorganic and organometallic clusters and polymers have recently become the subject of intense interest and study because of their unusual properties and their potential for applications in the areas of catalysis, electronics and ceramics.

Transition metal clusters have been examined as models for heterogeneous catalysts. Both rings and clusters might function as potential building blocks for extended solid state systems. In comparison to organic polymers, however, general and versatile methods for the synthesis of inorganic and organometallic polymers are not readily available. In particular, the preparation of polymers which contain transition metal atoms in the polymer backbone is a challenge to inorganic chemists.

This book is intended to show the rapid development in this field. Due to limited space, the reviews are focused on selected and modern topics. The articles were written by internationally recognized experts, who have intensively contributed to the research of rings, clusters and polymers.

I am grateful to the contributors to this book. Their knowledge and expertise was important in preparing this survey.

H.W. Roesky

CONTENTS

Rings with Phosphorus Carbon Multiple Bonds

by E. Fluck and B. Neumüller

Azaphospholes

by A. Schmidpeter and K. Karaghiosoff

Multiple Bonds between Transition Metals and Main Group Element Atoms
by W.A. Herrmann

Unsaturated Four-, Six- and Eight-membered Metallaheterocycles and Metal-Containing Polymers
by H.W. Roesky

Organometallic π Systems

by G. Huttner and H. Lang

Polynuclear Transition Metal Complexes with Sulfur Ligands

by B. Krebs and G. Henkel

Clusters of Metals and Nonmetals

by K.H. Whitmire

BORON, ALUMINUM, GALLIUM AND INDIUM WITH OXYGEN AND SULFUR

JAMES R. BOWSER[1] and THOMAS P. FEHLNER[2]

[1]Department of Chemistry, SUNY College at Fredonia, Fredonia, NY 14063 (United States of America)

[2]Department of Chemistry, University of Notre Dame, Notre Dame, IN 46556 (United States of America)

1. INTRODUCTION

1.1 APPROACH

This chapter emphasizes rings, clusters and extended bonding arrays ("polymers") of boron plus oxygen and boron plus sulfur. In addition, similar structural types containing aluminum, gallium or indium with oxygen and sulfur as principal components of the structure are briefly reviewed. A search of the literature reveals a wealth of information on compounds of boron and oxygen, a significant amount on boron and sulfur and much less on the other main group 3-heteroatom species. This information can be found in a variety of places including general sources (refs. 1-23), monographs and reviews on boron-oxygen (refs. 24-33), boron-sulfur (refs. 33-39), and aluminum and gallium (refs. 40-42) as well as recent research articles. In the case of inorganic rings, the situation has been particularly well covered in a two volume work by Haiduc (ref. 4) which has been followed by a two volume review monograph edited by Haiduc and Sowerby (ref. 5).

Given the availability of standard review literature on the subject, our approach to this topic has been modified accordingly. Our primary objective is to concisely summarize the most important aspects of the rings, clusters and polymers of the specified element combinations. Although the material itself does not differ significantly from that already present in the literature, the interrelationships between the three structural forms become more evident. In addition, as the nature of the bonding in many of these systems has not been comprehensively reviewed, we also present a summary and interpretation of available geometric and electronic structural information on ring systems. Finally, for boron-oxygen rings sufficient thermochemical information exists so that the complementary component of structure, energetics, can be examined for selected systems. These analyses provide insight into the structural types observed for given element combinations as well as an appreciation of the consequences of reactivity parameters and patterns. Our approach does not permit us to be comprehensive in an encyclopedic manner; however, we have tried to be complete in so far as structural types and bonding modes are concerned. The literature references point those readers in need of additional information to the appropriate review.

1.2 NOMENCLATURE

The question of nomenclature is a frustrating one. Even for the simplest structural type considered here, rings, the editors of the latest review monograph were unable to get contributors to use a consistent method (ref. 5). In his original monograph, Haiduc (ref. 4) proposed and used a systematic method but it has not been universally adopted. The situation with three dimensional clusters is even more formidable although useful, systematic methods have been proposed (ref. 43). For structures containing extended bonding, good structural information has led to a more rational method of writing the empirical formula. A shorthand notation that includes the size of a boron-oxygen ring and the number of trigonal and tetrahedrally coordinated boron atoms has been used as a classification system (refs. 44,45). Because of the brevity of this chapter no systematic nomenclature is used and we rely on structural illustrations to avoid ambiguity.

2. GENERAL PRINCIPLES

In this section we briefly recapitulate the fundamental properties of the elements concerned and indicate how these properties are expressed in the observed bond types and chemical reactivities. Detailed geometric, quantum chemical and thermochemical data are then used to provide a more precise analysis of the bonding in the most prevalent ring system.

2.1 EXPECTATIONS FROM ATOM AND BOND PROPERTIES

Selected atomic properties of B, Al, O and S are given in Table 1 (ref. 46). Considering the valence shell alone, boron and aluminum will normally form three covalent bonds and are

TABLE 1 Atom properties

property		boron	aluminum	oxygen	sulfur
configuration		$2s^22p^1$	$3s^23p^1$	$2s^22p^4$	$3s^23p^4$
ionization energy (eV)		8.3	6.0	15.9	11.7
differential ionization energy	a_o	4.27	3.22	7.54	5.56
	a_1	8.06	5.52	12.15	9.59
electron affinity (eV)		0.24	0.46	1.465	2.077
electronegativity		2.01	1.47	3.50	2.44
covalent radius (pm)		82	118	73	102

electron pair acceptors while oxygen and sulfur normally form two covalent bonds and are electron pair donors. Obviously the complementarity of the properties of the main group 3 and 6 elements creates a strong driving force for the formation of heteronuclear bonds as opposed to homonuclear bonds. Based on ionization energies, aluminum is a better acceptor than boron and sulfur is a better donor than oxygen. The electronegativities of oxygen and sulfur place these elements clearly in the non-metal category while that of aluminum (plus gallium and indium) confirms its metallic character. Boron, however, is a metalloid in terms of electronegativity as well as other properties (ref. 47). Hence, we really have three types of atoms to consider, B vs Al(Ga, In) vs O and S.

The normal "single bond" distances expected for 3-6 bonds can be estimated from the Schomaker-Stevenson relationship (ref. 46). These are: B-O:139.4; B-S:182.7; Al-O:162.4; and Al-S:213.4 pm. All are shorter than the sum of the covalent radii due to the correction term which is proportional to the difference in electronegativities. The shorter bond distances parallel enhanced bond strengths attributed to bond polarity (ref. 48). As this enhanced bonding is also proportional to the square of the difference in electronegativities, the "extra" energy associated with heteronuclear bond formation will follow the order AlO > BO > AlS > BS. A more precise estimate of relative bond polarity can be obtained using the differential ionization energies in

Table 1. For BO, BS, AlO, AlS the calculated charge separations are 0.162, 0.073, 0.244 and 0.155 and thus the polarities follow the order AlO > BO > AlS > BS. Note that the difference in charge separation for oxygen _vs_ sulfur is very large.

2.2 BASIC STRUCTURAL UNITS

For the rest of section 2 we will restrict ourselves to boron, oxygen and sulfur. The boron atom is found in a tri- or tetracoordinate environment while the group 6 atom is either di- or tricoordinate. Boron and sulfur exhibit higher formal coordination numbers in cluster compounds whereas oxygen is only rarely found in such an environment (see 3.2). Because tetracoordination for the group 3 atom and tricoordination for the group 6 atom involve formation of a donor acceptor bond, a distinction between covalent and coordinate interactions in rings has been made (ref. 4). As shown in Fig. 1, the $(RBO)_3$ ring with tricoordinate boron and dicoordinate oxygen can be viewed as a "simple" covalent system. Coordinating one to three X^- ions to the three available acceptor sites on the boron atoms of the $(RBO)_3$ ring leads to the second type of ring, the ionic compound $[(X)_n(RBO)_3]^{n-}$, (n:1 - 3) with one to three tetracoordinate boron atoms and dicoordinate oxygen atoms. The third case with tetracoordinate boron and tricoordinate oxygen can be viewed as the cyclic trimer of X_2BOR or a "coordination heterocycle". Similar interplay between tri- and tetracoordinate boron units,

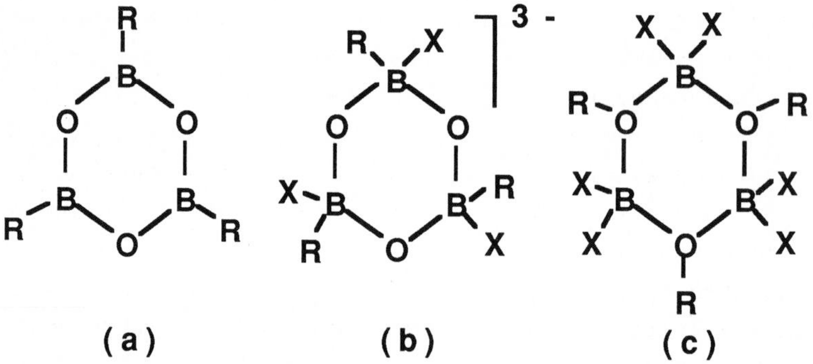

Fig. 1. Schematic representation of (a) "simple" covalent, (b) "ionic" and (c) "internally compensated" six-member boron-oxygen rings.

albeit of a more complex sort, govern the structural behavior of the extended structures in, for example, the borates. These are considered in more detail in 3.3 below. Note that these rings constitute electron precise systems, i.e., there is a bonding electron pair for each line representing an atom-atom interaction. Hence, they have carbon analogues, i.e., C_6H_6 is isoelectronic with $(HBO)_3$ and C_6H_{12} is isoelectronic with $(H_2BOH)_3$.

Heteronuclear boron-group 6 compounds, of which rings are but one example, can be related to heteronuclear polyboranes as follows. Although three dimensional electron precise cages are known, e.g. adamantane analogues, the hallmark of polyboranes is the "electron deficient" cluster structure, i.e., fewer electron pairs than lines indicating nearest neighbor interactions (ref. 49). The driving force for cluster formation is the same as that causing the dimerization of BH_3 to B_2H_6. The formal coordination of two BH bonds with two empty valence orbitals on boron leads to a net reduction in the energy of a system containing two BH_3's of 148 kJ/mol (ref. 50). In the presence of electron pair donors, this "self-coordination" is unraveled, i.e., B_2H_6 is cleaved to form BH_3 base adducts and clusters are opened or fragmented. The relationship of the structures formed from orbitally rich fragments to the number of bonding electrons available from the borane and heteroatom fragments present is described by the Wade, Mingos electron counting rules (refs. 51-54). These rules provide the connecting link between borane clusters and the rings and cages of the group 3-6 systems.

Consider five and six atom systems beginning with the clusters $B_5H_5^{2-}$ and $B_6H_6^{2-}$ shown in Fig. 2. These are closo clusters (6 and 7 skeletal pairs, respectively) with trigonal bipyramidal and octahedral structures. An oxygen atom is equivalent to a BH^{2-} fragment, hence, B_4H_4O, B_5H_5O, $B_3H_3O_2$, $B_4H_4O_2$, $B_2H_2O_3$, and $B_3H_3O_3$ are equivalent to $B_5H_5^{2-}$, $B_6H_6^{2-}$, $B_5H_5^{4-}$, $B_6H_6^{4-}$, $B_5H_5^{6-}$, and $B_6H_6^{6-}$ respectively. Thus, as shown in Fig. 2, one oxygen leads to a closo structure (n + 1 pairs), two to a nido structure (n + 2 pairs) and three to an arachno structure (n + 3 pairs) where n is the number of cluster atoms. At present this constitutes a formal relationship as only $B_5H_5^{2-}$ and $B_6H_6^{2-}$ and $B_2H_2O_3$ and $B_3H_3O_3$ are known. For sulfur, however, neutral cluster compounds analogous to $B_nH_n^{2-}$ are well known, e.g., $B_{11}H_{11}S$, (see 4.2) and give the structural relationships

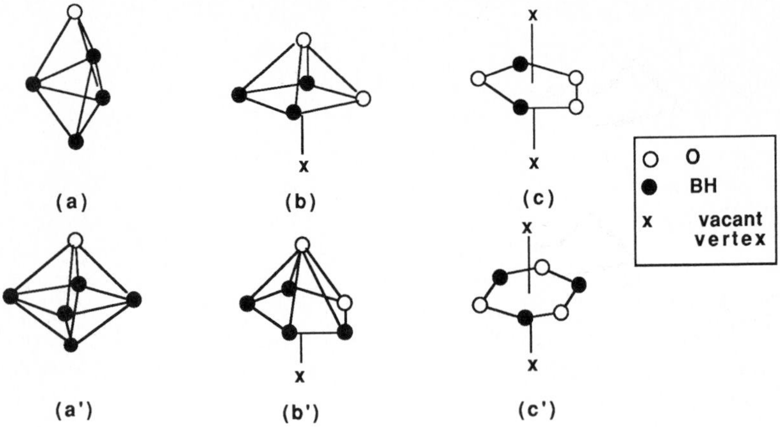

Fig. 2. Predicted structures of (a) B_4H_4O, (a') B_5H_5O, (b) $B_3H_3O_2$, (b') $B_4H_4O_2$, (c) $B_2H_2O_3$, and (c') $B_3H_3O_3$ ($B_5H_5^{2-}$, $B_6H_6^{2-}$, $B_5H_5^{4-}$, $B_6H_6^{4-}$, $B_5H_5^{6-}$ and $B_6H_6^{6-}$) based on the polyhedral skeletal electron counting theory. For (b,b') and (c,c') only one possible isomer is shown.

shown in Fig. 2 more credibility.

As with all structural units there are general ways in which more extended structures can be generated. For monomeric units, chains, rings and cross-linked materials are possible (Fig. 3). For rings, bridged and fused systems are possible, i.e., at least one atom is shared by both ring systems. Likewise, clusters can be joined by bridges or by fusion as schematically shown in Fig. 3. The electron counting rules have been modified to accommodate the latter type of cluster (ref. 55).

2.3 BASIC REACTION PROPERTIES

As pointed out by Van Wazer (ref. 56) much of the variety and richness of organic chemistry is due to the kinetic stability of the compounds. Indeed one of the challenges of modern chemistry is to "activate" the unreactive CC and CH bonds in saturated hydrocarbons. Often the reactive sites in carbon compounds are heteroatoms contained in functional groups (ref. 4). By their nature, compounds containing main group 3 atoms are subject to nucleophilic attack while those containing group 6 atoms are subject to electrophilic attack. It is not, perhaps, unexpected

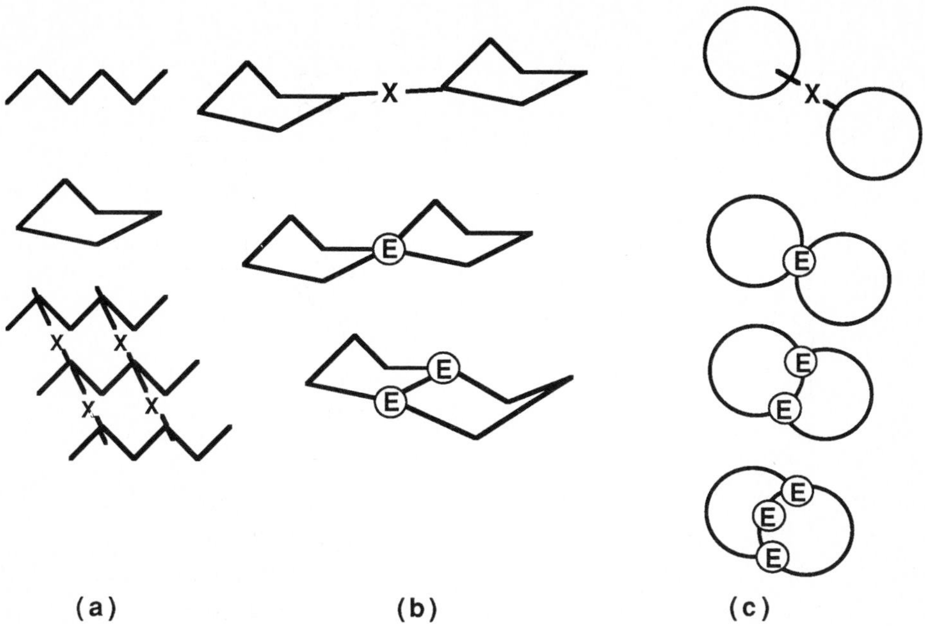

Fig. 3. (a) Chains, rings and networks of monomeric units. (b) Bridged and fused rings. (c) Bridged and fused clusters. "X" refers to a bridging atom and "E" refers to an atom shared by more than one ring or cluster.

that compounds containing both types of atoms will often be kinetically unstable. In these cases net reactivity will be controlled more by reaction thermodynamics than kinetics. Before pointing out the consequences of this observation, it is worthwhile to recall the origin of the low barriers (ref. 57).

The reactions responsible for the lability of these compounds are Lewis acid-base reactions. A thoroughly studied reaction of this type is given in eqn. 1. The structures of reactants and product are known and, from the exceedingly fast reaction rate,

$$BF_3 + NMe_3 \longrightarrow Me_3NBF_3 \qquad (1)$$

the barrier has been estimated to be < 8 kJ/mol (ref. 58). It is the acid, BF_3, that undergoes major structural change on adduct formation. In the free state BF_3 is trigonal planar with short B-F bonds whereas in the adduct it is tetrahedral with longer B-F bonds. As the energy necessary to rearrange BF_3 from the planar to the pyramidal structure is estimated to be about 200 kJ/mol, it is not immediately obvious why the barrier for the reaction is so small. The explanation (ref. 59) involves a characteristic feature of the reaction namely charge transfer. Hence, as shown in Fig. 4, the electronic state of the product correlates with the ionized rather than neutral reactants. If the "interaction" of the two potential energy curves in the vicinity of the crossing point defined by the dashed lines is large enough, then the barrier can be very small.

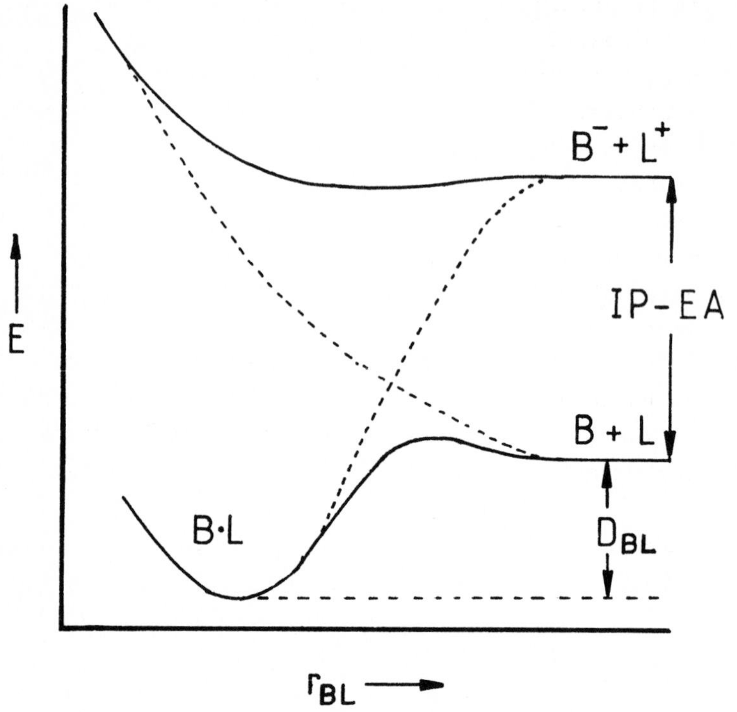

Fig. 4. Schematic drawing of the potential energy surface for the reaction given in eqn. 1 where B represents BF_3 and L represents NMe_3.

Obviously the isolation of free BF_3 in the presence of NMe_3 is impossible. It follows, then, that compounds containing both sites of Lewis acidity and basicity will have a variety of low barrier reaction pathways available to them. Thus, the thermodynamics of a given system will determine the state observed under normal experimental conditions. If several possible products have similar free energies, then concentration can play a significant role in the nature and type of species observed. It follows that many intrinsically stable compounds will not be observed without special efforts to do so. Hence, with low barrier systems, no conclusions concerning bonding can be justified on the basis that certain types of compounds have not been reported.

The consequences of equilibria (thermodynamic control of reactivity) in inorganic systems have been discussed (ref. 56). Because of low activation energies for reaction, most oligomeric compounds readily exchange fragments intra- or intermolecularly. These scrambling reactions complicate the characterization of systems in which they occur in that mixtures of molecules with different structures, but at equilibrium with respect to exchange of fragments, exist. In these cases the ideas of mechanistic chemistry lose their usefulness simply because with low barrier processes a variety of different pathways can lead to identical equilibrium product distributions.

The methyl polyborates constitute an exemplar pertinent to this review (ref. 60). For the system $(MeO)_3B/B_2O_3$ equilibrium is achieved by the exchange of methoxy groups for bridging oxygen atoms around a boron atom. Structures such as that shown in Fig. 5 can be generated and it is possible to set up equilibrium constants for the species that may be in equilibrium with respect

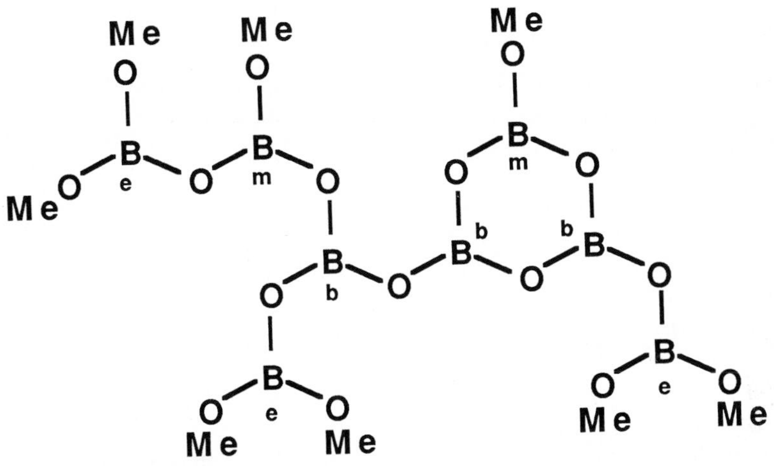

Fig. 5. A possible molecule in the $(MeO)_3B/B_2O_3$ system showing end groups (e), middle groups (m) and branching groups (b).

to exchange of bridging oxygens and methoxy groups. Such a task is a formidable one and a simplification of the problem has been achieved using stochastic graph theory (ref. 61). In this approach the large set of equilibrium constants is replaced with a much smaller one involving "building units". In the example Fig. 5 these units are end groups, $(MeO)_2B(O_{1/2}-)$ (e), middle groups, $(MeO)B(O_{1/2}-)_2$ (m) and branching groups, $B(O_{1/2}-)_3$ (b). ^1H NMR spectroscopy was used to provide the concentration of each group in the system as a function of overall composition (R:MeO/B ratio). The chemical shift data suggested the presence of middle groups in different chemical environments which were attributed to the presence of both rings and chains. The analysis in terms of the three types of groups shows a predominance of middle groups over and above that expected for random sorting of MeO groups and oxygen bridges about the boron atoms. This is expressed in the equilibrium constant: $K = [e]\cdot[b]/[m]^2 \approx 3 \times 10^{-3}$. The gel point (formation of infinite polymers) was shown to occur at R = 1. In addition, this study reported that the half-life for exchange of MeO fragments was 0.04 s at 25°C with a barrier of 19 kJ/mol demonstrating the facile nature of this reaction and the incredible complexity of the mechanistic problem.

Because of the intrinsic kinetic lability of many of these systems, certain types of reactions predominate in the formation and interconversion of group-3/group-6 compounds. The most important is a homo- or heterofunctional condensation reaction in which a small molecule is eliminated, e.g., eqn. 2, thereby driving the equilibrium process in the desired direction. This reaction type leads to the formation of rings, chains, and other

$$X_nB\text{-}OH + HO\text{-}BX_n \longrightarrow X_nB\text{-}O\text{-}BX_n + H_2O \qquad (2a)$$

$$X_nB\text{-}OH + Y\text{-}BX_n \longrightarrow X_nB\text{-}O\text{-}BX_n + HY \qquad (2b)$$

extended structures. The reverse reaction leads to the breakup of such structural types. The reaction also is operative for functional groups other than OH, i.e., Y: halogen or NH_2.

2.4 RING BONDING

A corollary of the observations concerning reactivity is that the most detailed structural information will be available for molecules that can be isolated in a pure state. Of the compound types considered in this chapter, the discrete compounds containing boron-oxygen or boron-sulfur rings are the most thoroughly characterized species. Hence, in this section we make use of this detailed information to explore in a more intimate way the nature of the structure, bonding and energetics. Although not directly transferable, the conclusions are prototypical relative to the other group-3/group-6 systems.

2.4.1 Geometry

The structures of the $H_2B_2O_3$ and $H_3B_3O_3$ molecules have been determined in the gas phase and are shown in Fig. 6 (refs. 62, 63). Both species are planar and the structural data in Table 2 shows that the B-O distance is very close to that estimated for a polar single bond (see 2.1). The planarity of both molecules is difficult to explain on the basis of coordination number as the ring angles in $H_2B_2O_3$ are close to those expected for tetrahedral

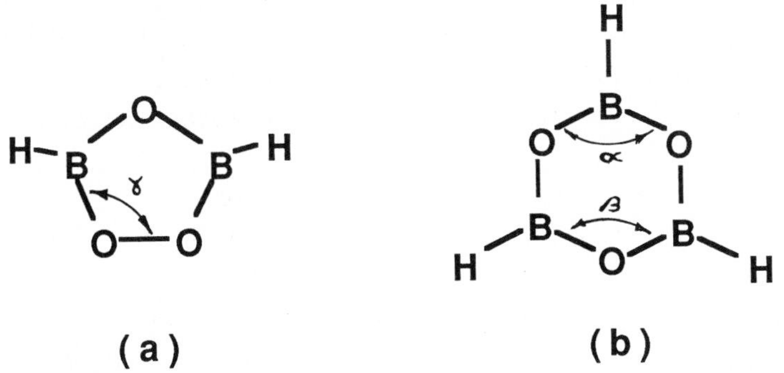

Fig. 6. Structure of (a) $H_2B_2O_3$ and (b) $H_3B_3O_3$.

sp^3 hybridization while those in $H_3B_3O_3$ are those expected for trigonal sp^2 hybridization. On the other hand, if one views these rings as being arachno clusters (see 2.2 above), then in a formal sense the planarity can be accounted for.

As Porter has pointed out, the structural parameters of $H_3B_3O_3$ are more informative when examined in the context of three structurally characterized isoelectronic molecules: C_6H_6, $H_3C_3N_3$, and $H_3B_3N_3H_3$ (ref. 63). Beginning with benzene, one generates $H_3C_3N_3$ by burying the proton of each alternate CH hydrogen in the carbon nucleus. The result is a shrinkage of the ring by 5.9 pm. This effect was attributed to increased electron density in the ring due to the increased nuclear charge. Similarly, the formal conversion of $H_3B_3N_3H_3$ to $H_3B_3O_3$ leads to a ring contraction of 5.9 pm. Note, however, that a redistribution of nuclear charge within the ring has the opposite effect. Going from C_6H_6 to $H_3B_3N_3H_3$ or $H_3C_3N_3$ to $H_3B_3O_3$ leads to a ring expansion of 3.8 pm. This constitutes geometric evidence that increased bond polarity within the rings reduces the π-electron delocalization and the

TABLE 2 Structural parameters for boron-oxygen rings (distances in pm, angles in deg.)

molecule	$B-R_{exo}$	$B-O_O$	$B-O_B$	α	β	γ
$H_2B_2O_3$[a]		136.5(4)	138.0(3)	113.05	104.0	104.95
$H_3B_3O_3$[b]	119.2(17)		137.6(2)	120.0(6)	120.0(6)	
$(HO)_3B_3O_3$[c]	135.5(9)		137.3(7)	120.1	119.9	
$[O_3B_3O_3]K_3$[d]	133.1(10)		139.8(5)	117.3(8)	122.6(8)	
$[O_3B_3O_3]Na_3$[e]	128.0(16)		143.3(9)	114.5	125.5	
$(HS)_3B_3S_3$[f]	181.3(6)		180.3(5)	130.0	109.9	
$[S_3B_3S_3]Na_3$[g]	177		182	125	115	

[a]Ref. 62
[b]Ref. 63
[c]Ref. 64
[d]Ref. 65
[e]Ref. 66
[f]Ref. 67
[g]Ref. 68

bond order. On the other hand, comparison of the structural parameters for $H_3B_3N_3H_3$ with those of $H_3B_3O_3$ indicates that the former is considerably less rigid (ref. 63). The geometric argument above, which is confirmed by calculations discussed below, suggests greater π bonding in $H_3B_3N_3H_3$. Hence, greater rigidity may also be a characteristic of the greater bond polarity in the ring of $H_3B_3O_3$.

Additional structural data from solid state studies on the six-membered B_3O_3 and B_3S_3 ring systems are given in Table 2. Here systematic variations in distances and angles can be noted for related compounds. In comparing the structural parameters for $(RBO)_3$, R:OH, O^-K^+, and O^-Na^+, a relationship between the B-O bond lengths and the O-B-O and B-O-B angles can be seen. Specifically, as the exo-ring B-O distance decreases the endo-ring B-O distance increases while the B-O-B angle increases and the O-B-O angle within the ring decreases. A simple Bent's rule argument suffices to explain this observation (ref. 69). As the B-O exo-bond order increases, the s character increases forcing more p character in the B-O endo-bonds thereby increasing the B-O distance and decreasing the O-B-O angle. To retain planarity, the B-O-B angles open up. The structure of $(HS)_3B_3S_3$ shows that these changes are energetically competitive. As expected, sulfur requires more p character in its bonds than does oxygen and the B-S-B angle has decreased to 110° forcing the S-B-S angle to 130°. Consistent with the expected hybridization at boron, the endo-B-S bond is slightly shorter than the exo-B-S bond. Placing a negative charge on the exo-S in $[S_3B_3S_3]^{3-}$ results in an increase in the B-S bond order and an increase in the B-S-B and decrease in the S-B-S angles as expected. This suggests that the opening of the B-O-B or B-S-B angles is endothermic, i.e., the ring is strained. The thermochemistry (see 2.4.3) corroborates this conclusion.

2.4.2 Electronic structure

Although the geometric data discussed above constitute a first source of information on bonding, a precise model of the intramolecular atom-atom interactions results from experiments that probe the valence electrons. Alternatively, quantum chemical approaches can be used to model the electronic structure for a given nuclear framework.

As $H_3B_3O_3$ is isoelectronic with benzene much of the effort in this regard has been focused on gauging the extent of delocalization of the B-O ring bonding. To illustrate how bond polarity affects a π bond, Fig. 7 presents a comparison of HCN with HBO.

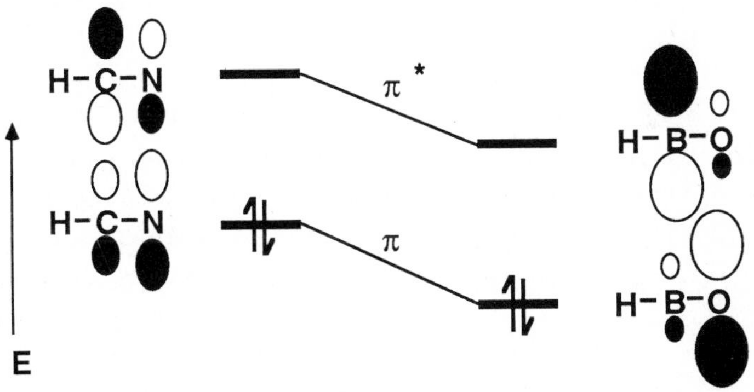

Fig. 7. Schematic representation of the π bonds of HCN and HBO.

The formal transfer of a proton from C to N results in a stabilization of both the π and π^* orbitals coupled with more asymmetry in the B2p and O2p characters of the π and π^* functions (ref. 70). The bonding orbital becomes more like an oxygen lone pair and the antibonding orbital becomes more like an empty orbital centered on boron. Hence, HBO should be a better acid and a poorer base than HCN. The energies of the filled π orbitals can be experimentally compared in the case of HCP and HBS as the photoelectron spectra have been measured. There is a stabilization of 0.33 eV in the π electrons in going from HCP (ionization potential 10.79 eV) to HBS (11.12 eV) (refs. 71,72). This implies a similar stabilization of the lowest unoccupied molecular orbital.

In forming the cyclic trimer from HBO there is a tendency for the σ electrons to "drift" toward oxygen and the π electrons toward boron (ref. 73). Substituents also interact with the B_3O_3 ring in two distinct and opposing ways. Halogen substitution for H leads to a decrease in the σ electron density at boron and an increase at oxygen while the opposite is true for the π densities. Phenyl substitution raises the electron density at boron (ref. 74). Again the nature of $H_3B_3O_3$ is best defined by comparison as the significance of these calculated quantities is difficult to judge in an absolute sense. In comparing $H_3B_3N_3H_3$ and $H_3B_3O_3$ one notes: (a) an increase in first i.p. as expected from the simple argument above, (b) greater BN than BO overlap populations indicating a greater role for π bonding in the former, (c) BH σ donation to N is less than that to O and, (d) B π acceptance from N is greater than that from O (ref. 75). These all suggest that the ring π system in $H_3B_3O_3$ is much less established than that in either $H_3B_3N_3H_3$ or C_6H_6 and it is polarized in the sense of Fig. 7. It is, therefore, less aromatic. Also consistent with Fig. 7 is the trend in first i.p.'s: C_6H_6, 9.5; $H_3B_3N_3H_3$, 10.1(11.2 calc'd); $H_3B_3O_3$, (13.7 calc'd) eV (ref. 75).

On the other hand, the calculations do not rule out all aromaticity and experimental evidence for aromaticity in both $R_3B_3O_3$ and $R_2B_2O_3$ has been cited (ref. 76). For example, downfield shifts in the ^1H NMR spectra relative to model compounds have been attributed to the existence of ring currents. Likewise, the high abundance of borenium ions in the mass spectra of these two ring systems suggests greater stability than that of model compounds for which conjugation is not possible. Finally, the magneto-optical effect observed for $R_3B_3O_3$ and $H_3B_3N_3H_3$ suggests observable ring conjugation in both systems (ref. 77).

2.4.3 Energetics

In contrast to the state of most main group areas there is a significant body of thermochemical information on the B_3O_3 ring system. Hence, for this ring type the partner of structure, energy, can be discussed in an empirical sense. Because of the importance of equilibria in the reactions of these compounds, this type of information is of great practical value.

Finch and Gardner have used thermochemical data to consider the question of ring strain *vs* resonance stabilization in the $R_3B_3O_3$ ring system for R:F, Cl, OH, and Ph (ref. 78). They concluded that this ring is slightly strained when halogen substituted. To consider one aspect of the reactivity of the B_3O_3 ring system we have included R:H and have treated the thermochemical information in a different, but related, manner. Consider the reaction given in eqn. 3 and the pertinent data given in Table 3. The reaction is exothermic for all R for

$$R_3B_3O_3(g) \longrightarrow B_2O_3(c) + BR_3(g), \quad \Delta H_R \qquad (3)$$

TABLE 3 Heats of formation for BR_3 and $(RBO)_3$ and average B-R bond energies

quantity, kJ/mol[a]	F	OH	Cl	Ph	H
$\Delta H_f(BR_3)(g)$	-1135.5(9)	-992(3)	-402(12)	129(9)	96.1[b]
$\Delta H_f[(RBO)_3](g)$	-2362.9(42)	-2270(12)	-1630(8)	-1130[c]	-1216(8)[d]
D(B-R)	642.5	555.1	440.1	443	368
ΔH_R	-45.8	4.2[e]	-45	-15	39[f]

[a]All data from ref. 10 unless indicated otherwise.
[b]Ref. 50
[c]Ref. 78
[d]Ref. 80
[e]The reaction is exothermic for $B(OH)_3(c)$ as the product.
[f]The reaction is exothermic for $B_2H_6(g)$ as the product.

products in their normal states but not exceedingly so. Indeed the reverse of eqn. 3 can be carried out at high temperatures (refs. 81-84) and constitutes a preparative route to the fluoro derivative. The trend in exothermicity of the reaction in eqn. 3 is in accord with qualitative measures of stability, i.e., $(FBO)_3$ disproportionates above -130°C while aryl derivatives melt above 200°C without decomposition (ref. 4).

There is a crude trend of decreasing exothermicity of the reaction in eqn. 3 with decreasing B-R bond energy (Table 3). As always, more than one explanation is possible. The $(RBO)_3$ ring may decrease in aromaticity in going from H to F, the strain energy in the $(RBO)_3$ ring may increase in going from H to F, the B-R bond strength in BR_3 may decrease relative to that in $(RBO)_3$ in going from H to F or some combination of the three. The calculations discussed above show that when possible there is a strong π interaction between exo substituents and the ring. Hence, the first and third explanations are ruled out. That is, substitution of F for H should lower the energy of the B2p orbitals enhancing the π interaction with oxygen (Fig. 7) and there is no reason to believe that the π interactions in BF_3 should be any stronger than those in the B-F bonds of $(FBO)_3$. The explanation based on ring strain is the most likely one and is consistent with the explanation of the geometric effects presented above. As the bond order of the exo B-R bond increases, the O-B-O angle decreases forcing the B-O-B angle to open up to retain planarity. The thermochemical data suggest that the strain energy falls off in the order F,Cl > OH,Ph > H.

3. BORON OXYGEN

3.1 RINGS

3.1.1 Structural types

One or more discrete examples of the four ring structures shown in Fig. 8 are known; however, only the chemistry of the

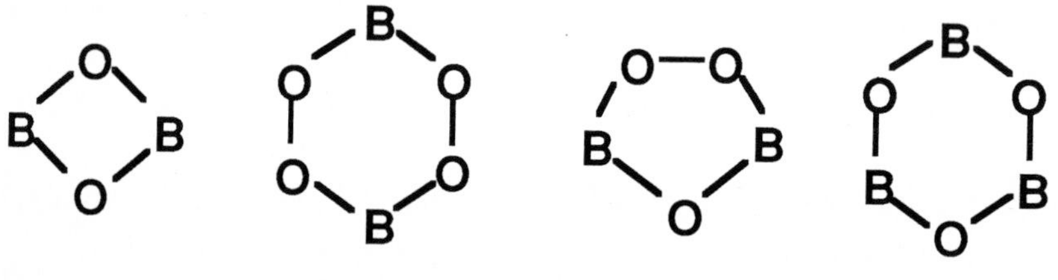

Fig. 8. Known discrete ring types containing only boron and oxygen in the ring. Boron can be tri- or tetracoordinate and oxygen di- or tricoordinate.

six-membered B_3O_3 ring has been systematically explored. The B_2O_4 and B_2O_2 rings are known with four coordinate boron (refs. 5,85,86) and one example of the latter with three coordinate boron is known (ref. 197). Hence, the discussion below is restricted to the B_3O_3 ring system.

3.1.2 Preparative routes

In the following, exemplars of known routes to discrete compounds containing a single B_3O_3 ring are presented. A more complete discussion can be found elsewhere (refs. 4,5).

(i) From sources of the elements. The unsubstituted ring, $(HBO)_3$, can be prepared from the high temperature reaction of elemental boron and water (eqn. 4) (refs. 87,88) or hydrogen, boron and B_2O_3 (refs. 80,89,90). In a similar manner, the

$$3B + 3H_2O \longrightarrow (HBO)_3 + 3/2H_2 \qquad (4)$$

$$B + B_2O_3 + 3/2H_2 \longrightarrow (HBO)_3 \qquad (5)$$

fluoro- and chloro-derivatives can be formed by the reaction of BX_3 with B_2O_3 under forcing conditions (refs. 81-84). Indeed, because of the often favorable thermodynamics of the B_3O_3 rings *vs* $B_2O_3(s)$, see 2.4.3, this ring framework is an ubiquitous product in many reaction systems containing sources of boron and oxygen atoms. In particular, the reaction of "high energy" species such as boron hydrides with oxygen leads to $H_3B_3O_3$ (refs. 91-93). At room temperature in the liquid state, $(HBO)_3$ rapidly disproportionates to B_2O_3 and B_2H_6.

(ii) From the elimination of water. The B-O-B bond can be made by a variety of routes involving the loss of H_2O. For example, B_3O_3 ring formation takes place on the condensation of difunctional organoboron derivatives (eqn. 6) (refs. 94-97). Hence, methods for promoting the dehydration of organoboronic

$$3RB(OH)_2 \longrightarrow (RBO)_3 + 3H_2O \qquad (6)$$

acids serve as useful routes to these rings. A closely related route begins with the reaction of the appropriate Grignard reagent with $B(OR)_3$ which is followed by acid hydrolysis and removal of water (ref. 98). The hydrolysis of organoboranes, e.g., dihalides, can lead to $(RBO)_3$ presumably *via* boronic acids which are never isolated (refs. 99-102).

(iii) Other routes. The reactions of organoboranes with a variety of oxygen atom sources have led to $(RBO)_3$. For example, BR_3 reacts with CO in the presence of BH_4^- as a catalyst to give good yields of $(RBO)_3$ (ref. 103). Further, $PhBI_2$ reacts with Me_2SO to give $(PhBO)_3$ (ref. 104). The thermolysis of many organoboron compounds containing oxygen yields $(RBO)_3$. For example $R_2B{-}O{-}BR_2$ derivatives undergo disproportionation to yield BR_3 and $(RBO)_3$ (ref. 105). Hence, the general reaction (eqn. 7)

$$R_2BOX + YBR_2 \longrightarrow R_2B{-}O{-}Br_2 + XY \qquad (7)$$

in which a small molecule is eliminated and a B-O-B link is formed constitutes a route to $(RBO)_3$. X and Y can be H, OH; C(O)R, OC(O)R; R, Cl; or H, Cl (ref. 4). The reaction of $H_2Os_3(CO)_{10}$ with $BH_3{\cdot}THF$ leads to the elimination of C_4H_{10} and the formation of $[(\mu{-}H)_3Os_3(CO)_9(\mu_3{-}CO)]_3B_3O_3$ where the exo-ring substituent is a transition metal cluster. Three of the oxygen atoms in this compound come from THF (ref. 106). Other specific routes exist (refs. 4,5).

3.1.3 Reactivity

In the following, examples of reactions of compounds containing the B_3O_3 ring are presented. No attempt is made to be comprehensive; however, the basic reaction types are delineated. A more detailed description will be found in refs. 4,5.

(i) Reactions leaving the ring intact. Consistent with our knowledge of the electronic structure of the B_3O_3 (see 2.4.2), Lewis bases add readily to the tricoordinate boron atoms of $(RBO)_3$ (ref. 107). Although in principle three bases can be added, 1:1 adducts are generally observed. Note that OH^- is an exception in this regard. As expected from the behavior of BR_3 compounds, the size of the R group in $(RBO)_3$ affects acceptor ability. Base addition probably precedes some ring cleavage reactions, e.g., H_2O (see below). Some ring types are observed only as the base adducts, i.e., with tetracoordinate boron (see 3.1.1).

Redistribution reactions involving the substituents on boron have been observed as shown, for example, in eqn. 8 (ref. 108).

$$(ROBO)_3 + 3R'OH \longrightarrow (R'OBO)_3 + 3ROH \qquad (8)$$

Some examples of reactions on the ring substituent, R, without cleavage of the B_3O_3 ring are known, e.g., halogenation of the phenyl ring when R:Ph (ref. 109).

(ii) Reactions resulting in ring cleavage. The thermal stability of the B_3O_3 ring varies widely depending on the exo-ring substituents. For example, $(FBO)_3$ decomposes readily as shown in eqn. 9, which is the reverse of a preparative route (refs. 81-84). On the other hand, aryl derivatives are stable to 200°C and above (ref. 4).

$$(FBO)_3 \longrightarrow BF_3 + B_2O_3 \qquad (9)$$

Reaction with bases such as H_2O, ROH, NR_3, etc. can result in ring cleavage, e.g., the $(RBO)_3$ ring is readily cleaved by H_2O forming $RB(OH)_2$. Reaction with more complex bases can lead to the formation of a polycyclic system in which part of the original B_2O_3 ring is retained in the product (refs. 110,111). As shown in Fig. 9 the B-O-B-O-B part of the product results from

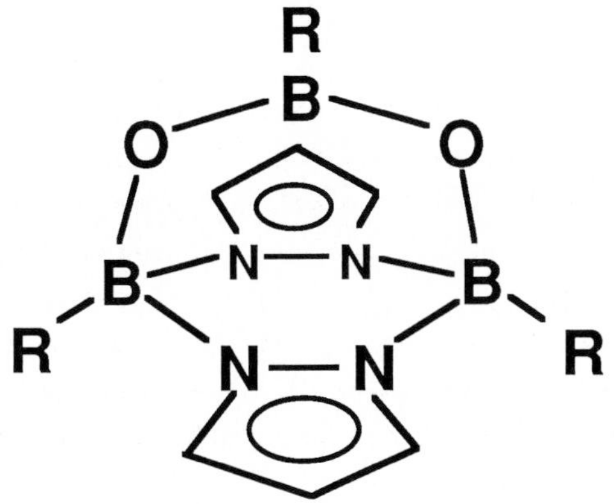

Fig. 9. Structure of $RB(\mu\text{-pz})_2(\mu\text{-OBRO})BR$ where Hpz:pyrazole.

the original ring. It is interesting that this reaction apparently involves the thermally unfavorable amination of a B-O bond. Note that the suggested intermediate is closely related to a bicyclic fragment frequently found in extended systems (Fig. 14b).

Inorganic halides can cleave the B_3O_3 ring. For example, a convenient preparation of organoboron dihalides under mild conditions is given in eqn. 10. Grignard reagents and other organometallic reagents also produce ring cleavage and form

$$(RBO)_3 + 2BX_3 \longrightarrow 3RBX_2 + B_2O_3 \qquad (10)$$

monoboranes with new B-C bonds (ref. 4). The reaction has been recommended (ref. 4) as a convenient route to either R_3B or R_2BOH from $(ROBO)_3$.

3.2 CLUSTERS

Although there are examples of borane cages with exo-cluster oxygen atoms (ref. 112), there are no reported examples with one or more oxygen atoms as contiguous members of the cluster framework. Recently, two metallaboranes have been structurally characterized and shown to contain a framework oxygen atom. The first, $(\eta^6\text{-}C_6Me_3H_3)FeOB_8H_{10}$, Fig. 10, is formed by an unknown mechanism in a metal vapor reactor and has a decaborane-like skeleton with the oxygen atom in the 6-position and the iron atom in the 2-position (ref. 113).

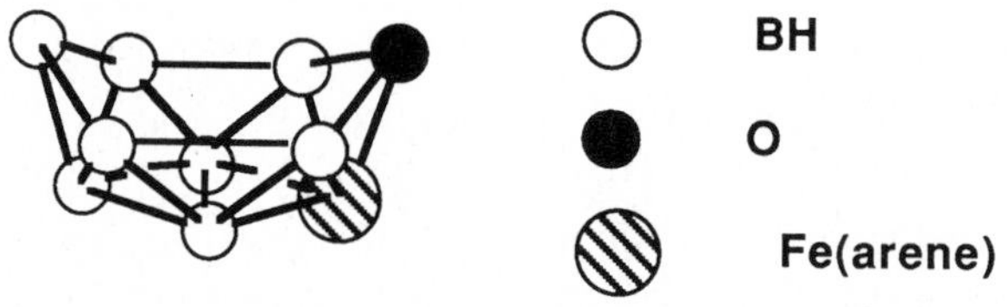

Fig. 10. Idealized representation of the structure of $(\eta^6\text{-}C_6Me_3H_3)FeOB_8H_{10}$.

As such, the oxygen atom behaves as a four-electron donor and the cluster is an example of a ten vertex, 2n + 4, nido cage. The structural parameters provide evidence for a polarization of the cluster bonding related to that discussed in 2.4 where the bonding of the $H_3B_3O_3$ ring was contrasted to that of the C_6H_6 and $H_3B_3N_3$ rings. The average B-O distance (142.5 pm) in the metallaoxaborane cluster is similar to that found in mononuclear compounds, whereas if the oxygen atom were truly involved in multicenter bonding one might well expect an increase in the B-O distance in the cluster. Hence, it was suggested that the bonding is not as delocalized as in the all boron analogue. Support for this view is provided by significant differences in the boron framework of the metallaoxaborane relative to the framework of $B_{10}H_{14}$ as a model.

The second known example of an oxygen atom in a cluster environment is also a metallaoxaborane, $(Cp^*)RhOB_{10}H_9Cl(PMe_2Ph)$ (ref. 114). A representation of the structure is shown in Fig. 11 and demonstrates that this compound is an example of a twelve vertex, 2n + 4, nido cage. Again the oxygen atom appears to act as a four electron donor but, in this case, it is bound solely to boron atoms although still adjacent to the open face. The average B-O distance of 151.1 pm is significantly longer than that reported for $(\eta^6\text{-}C_6Me_3H_3)FeOB_8H_{10}$. In addition no significant geometric changes are observed in the rest of the

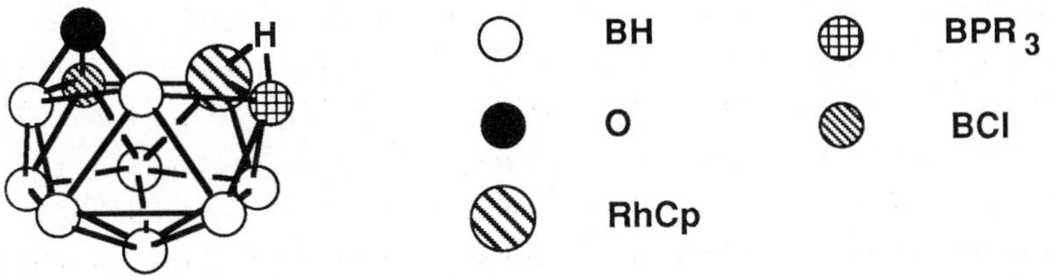

Fig. 11. Representation of the solid state structure $(Cp^*)RhOB_{10}H_9Cl(PMe_2Ph)$, (ref. 114).

borane cage relative to $(Cp^*)RhB_{10}H_{11}Cl(PMe_2Ph)$ (ref. 114). Hence, in this metallaoxaborane, the oxygen atom appears to be participating more fully in the delocalized cluster bonding network. Besides the fact that in $(\eta^6\text{-}C_6Me_3H_3)FeOB_8H_{10}$ the oxygen is bound to two boron atoms and a metal atom while in $(Cp^*)RhOB_{10}H_9Cl(PMe_2Ph)$ it is bound to three boron atoms, the difference in the two cluster bonding systems is not obvious. Note that in $(Cp^*)RhOB_{10}H_9Cl(PMe_2Ph)$ the borons adjacent to the oxygen atoms experience a significant downfield shift in the ^{11}B NMR on formation of the oxo-cage.

The metallaoxaborane, $(Cp^*)RhOB_{10}H_9Cl(PMe_2Ph)$, is produced in 53% yield from the hydrolysis of $(Cp^*)RhB_{10}H_{11}Cl(PMe_2Ph)$. Obviously the H_2O condenses with the B-H-B or B-H-M hydrogens of $(Cp^*)RhB_{10}H_{11}Cl(PMe_2Ph)$ sequentially eliminating two moles of H_2. This suggests that endo-hydrogens having some hydridic character and the proper spatial location are required. As B-H-B protons on large borane cages are usually protonic in character relative to B-H terminal hydrogens, the presence of the metal fragment as well as the exo-cage substituents may play a significant role in facilitating the incorporation of oxygen into the cluster.

In contrast to sulfur, see 4.2, oxygen is not found in a borane cluster environment although the metallaoxaboranes described above demonstrate that such an environment is clearly not an impossible one. In some senses the situation is similar to that in transition metal cluster systems. There are a large number of characterized metal-sulfur clusters in the literature but only a few metal-oxygen clusters (ref. 115). The latter can be very reactive, e.g. $[Fe_3Mn(CO)_{12}O]^-$ must be prepared in the strict absence of CO (ref. 116), thereby decreasing the probability of serendipitous discovery.

There are factors that disfavor oxygen vs sulfur as a cluster atom. A good match of the atomic p functions of the cluster atoms is very important. As noted from the ionization potential data in Table 1, the match between the energies (and size) of B and S is much better than that between B and O. That is, the same factor that promotes strong two center bonding inhibits effective multicenter bonding. However, the very factors that make oxaborane cages difficult to attain, enhance them as synthetic goals in that they may well serve as formal sources of oxygen atoms for controled oxidations.

3.3 EXTENDED STRUCTURES

Because of the acceptor properties of the BO_3 fragment a great variety of extended structural arrangements is possible and many have been characterized (ref. 25). The condensation into fused rings, chains of rings, etc. formally results from the reaction in eqn. 11 combined with Brönsted acid-base reactions.

$$\text{B-OH} + \text{HO-B} \longrightarrow \text{B-O-B} + H_2O \qquad (11)$$

Once one understands the fundamentally simple process whereby extended arrays are built up, then all that is left is the enumeration of the actual observed structures. As the latter is beyond the scope of this article, the following illustrates common structural features with known examples.

The spirocyclic fused ring system is found in many extended systems and can be envisioned as being built up in a stepwise fashion as shown in Fig. 12. Taking into consideration the

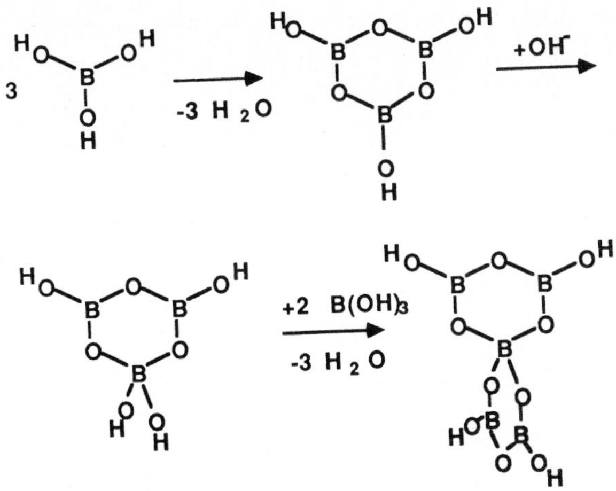

Fig. 12. Formal stepwise build-up of a fused ring system.

importance of equilibria in these systems (2.3), it is obvious that Fig. 12 does not necessarily define an actual route to extended structures. Indeed there may well be a number of operative mechanistic pathways leading to the particular condensed structure stable under the reaction conditions. On the other hand, the two structurally characterized tetraborates in Fig. 13 (ref. 117) clearly demonstrate the overall validity of the formal mechanism.

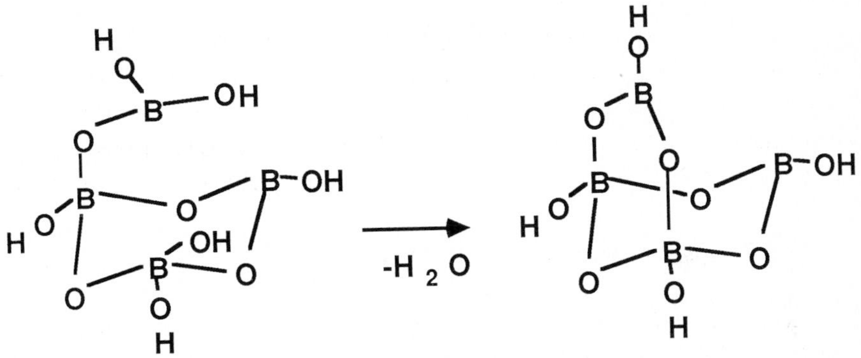

Fig. 13. Structural representations of the $[B_4O_4(OH)_6]^{2-}$ and $[B_4O_5(OH)_4]^{2-}$ anions.

A number of B-O fragments have been found to be useful in visualizing and categorizing the most common structural types (ref. 25). These are shown in Fig. 14(a). An alternate formal route to the spirocyclic ring results from attaching two oxygen

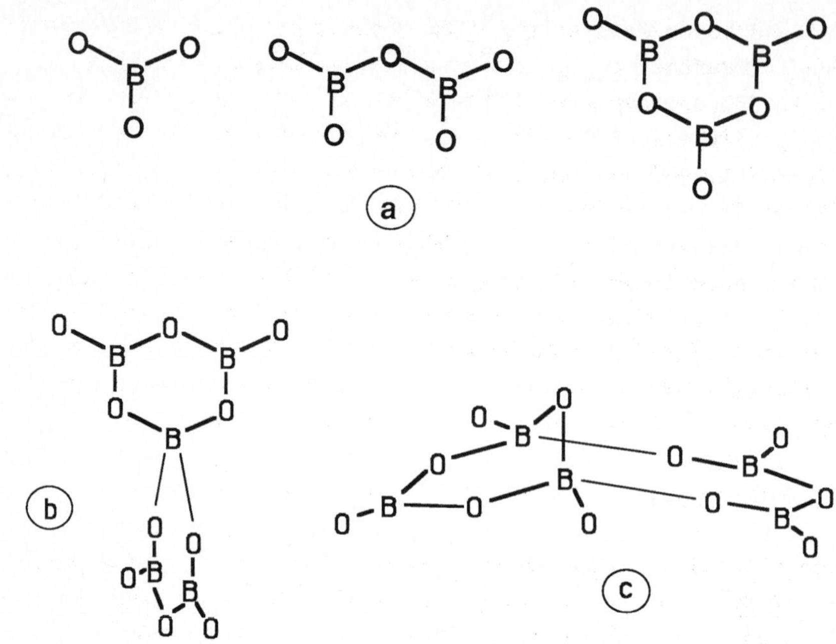

Fig. 14. Fragments for a systemization of extended structural types and the generation of two types of fused rings.

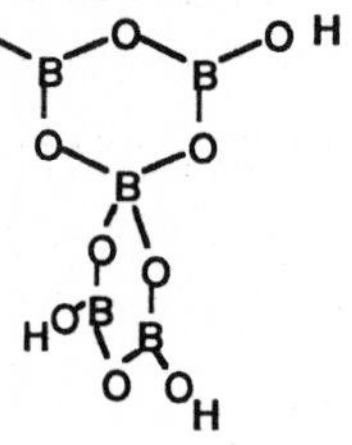

$[B_5O_6(OH)_4^-]$

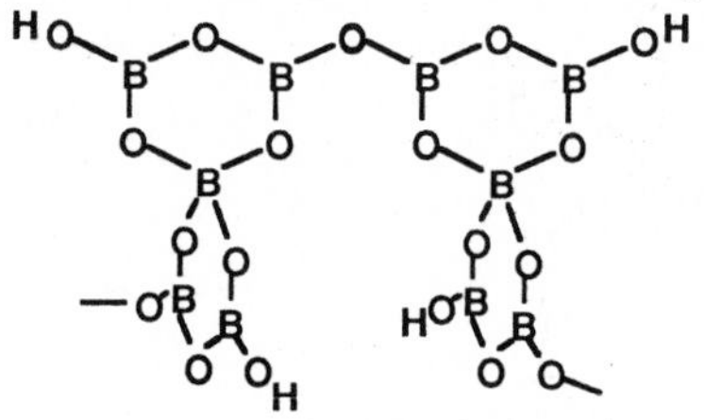

$\{B_5O_7(OH)_2\}_n^{n-}$

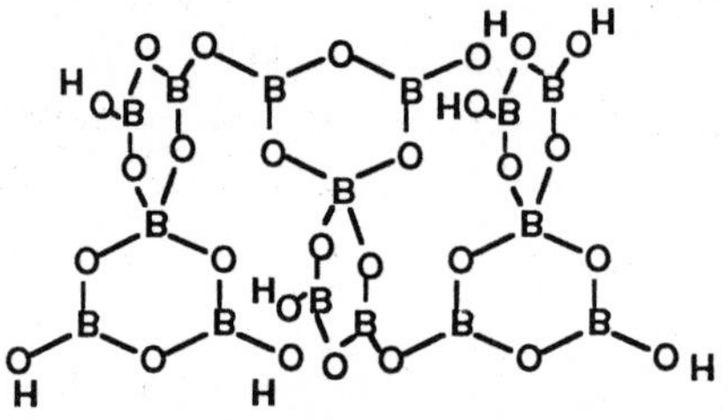

$[B_{15}O_{20}(OH)_8^{3-}]$

Fig. 15. Structures of $[B_5O_6(OH)_4]^-$, $[B_{15}O_{20}(OH)_8]^{3-}$ and $[B_5O_7(OH)_2]_n^{n-}$.

atoms of the diboron fragment to a single boron of the B_3O_3 ring (Fig. 14(b)). Further they can be attached to different boron atoms leading to another common fused ring system (Fig. 14(c)).

Examples of the linking of fused ring systems are shown in Fig. 15 (refs. 118-120). Obviously more complicated extended systems are possible as each spirocyclic ring system has a possible connectivity of four. Hence, layered systems (two dimensional connectivity of units) and networks (three dimensional connectivity of units) can be generated and are observed. Finally in Fig. 16 one example of a complex unit of an extended system is illustrated (ref. 121). This is reminiscent of complex aromatic organic molecules and shows that the structural variety in these systems approaches the richness of those found in organic chemistry.

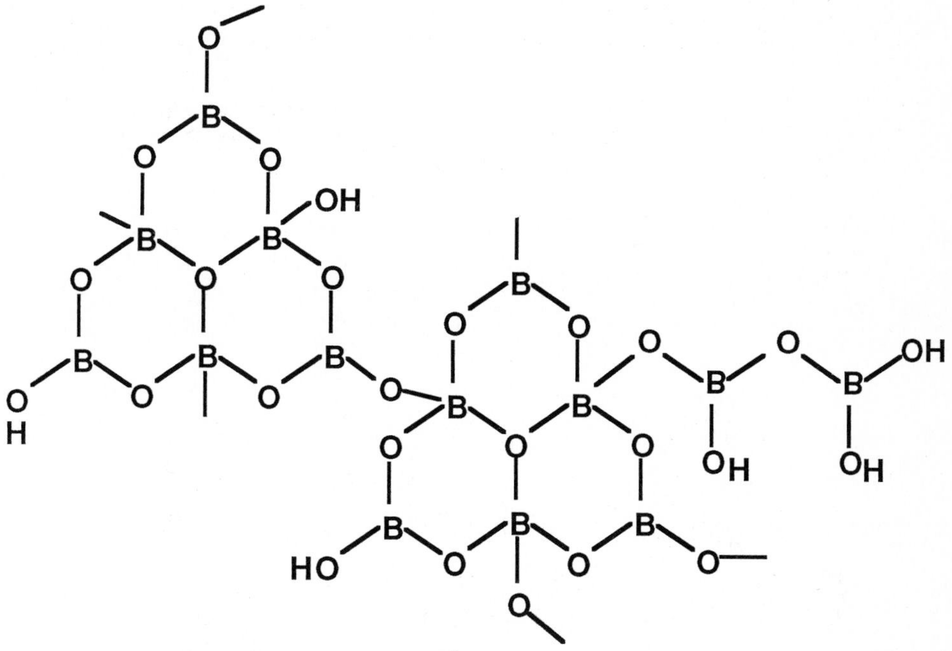

$[B_{14}O_{20}(OH)_6]^{4nm-}$

Fig. 16. Structure of $[B_{14}O_{20}(OH)_6]^{4nm-}$.

4. BORON SULFUR

4.1 RINGS

4.1.1 Structural considerations

The known monocyclic compounds having only boron and sulfur as ring atoms include four- (B_2S_2), five- (B_2S_3 and B_3S_2) and six- (B_3S_3, B_2S_4, and B_4S_2) membered rings. Their structural frameworks are presented in Fig. 17. The two isomers shown for the B_2S_4 system have only been prepared as derivative adducts. Coordination heterocycles (see 2.2 and Fig. 1(c)) of most of these systems are known as well.

All compounds reported to date that have B_xS_x frameworks (that is, equal numbers of boron and sulfur atoms) exhibit regular alternation of the two types of atoms - i.e., only heteronuclear bonds are present. Monocycles with unequal numbers of boron and sulfur (B_xS_y) necessarily contain one or more homonuclear bonds; in such cases the minimum possible number of such bonds has invariably been observed. Most known species of this

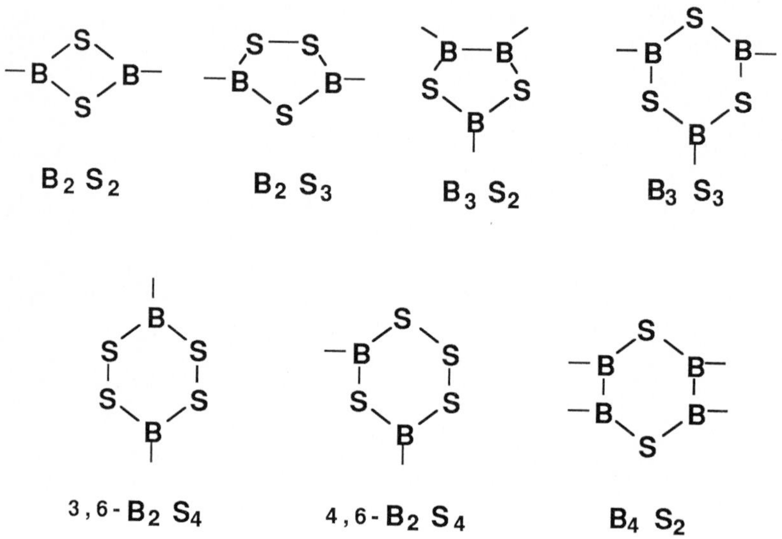

Fig. 17. Skeletal frameworks of monocyclic rings containing only boron and sulfur atoms. The two isomeric B_2S_4 rings have only been isolated as di- or multiple adducts.

type have y > x (more sulfur than boron ring atoms) and therefore contain one or more S-S bonds. These observations are only partially consistent with thermodynamics, as the relative order of bond energy is B-S > B-B > S-S (ref. 69). The paucity of rings containing B-B bonds may be due to their susceptibility to nucleophiles (see 2.3), and it has been suggested that such rings are stabilized by the presence of π-electron-donating substituents on boron (ref. 122). In this regard it is useful to consider the experimental order of stability for $R_3B_3S_3$ rings

$(R_2NBS)_3 > (RSBS)_3 > (ROBS)_3 > (IBS)_3 > (BrBS)_3 > (ClBS)_3$, (ref. 123), which is consistent with the notion that substituents providing electron density and/or steric hindrance enhance the stability of the B_3S_3 skeletal framework.

A number of structures containing bridged rings have been reported. Four- and five-membered units are dominant in such compounds. Two prominent examples are the porphyrin-like

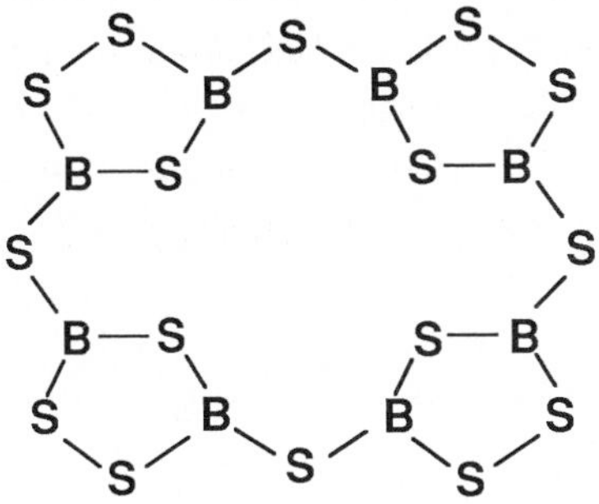

B_8S_{16}

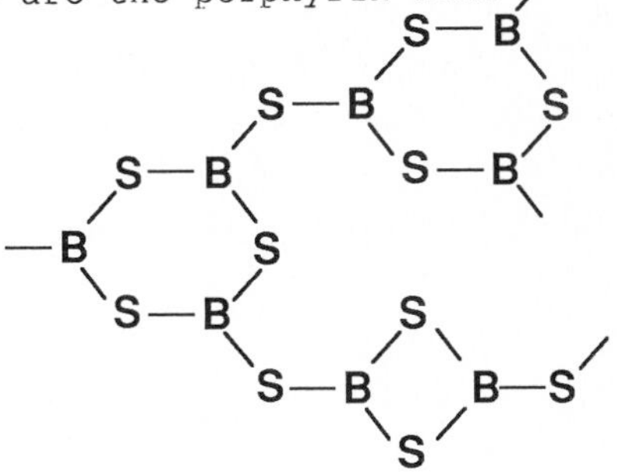

B_2S_3

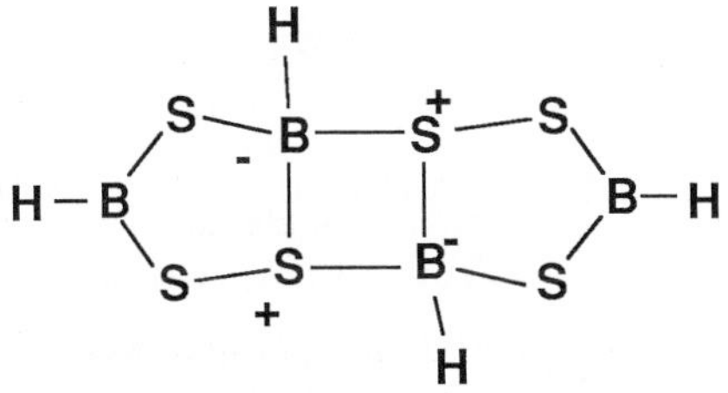

$[H_2B_2S_3]_2$

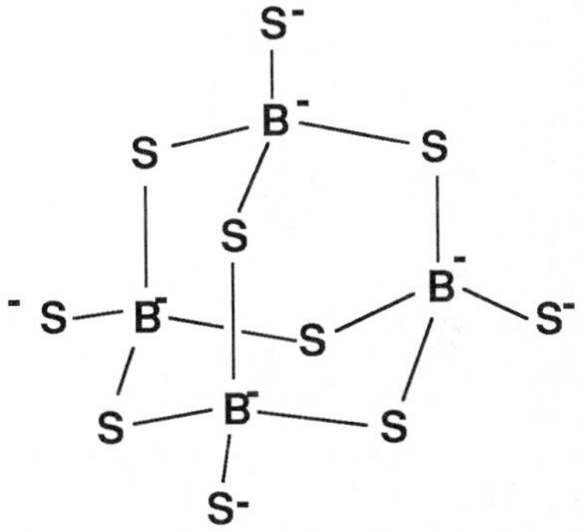

$[B_4S_{10}]^{8-}$

Fig. 18. The structures of some binary boron-sulfur systems containing bridged or fused rings.

molecule B_8S_{16} (ref. 124), with four B_2S_3 rings connected by bridging sulfur atoms, and crystalline B_2S_3. The latter exhibits a layered structure comprising both B_2S_2 and B_2S_3 cyclic units, again joined by sulfur bridges (ref. 125).

Fused rings are also known. A simple example is the dimer of $H_2B_2S_3$, which contains both 3- and 4-coordinate boron and 2- and 3-coordinate sulfur (ref. 126). Multiple fused rings are found in salts that formally contain $B_xS_y^{n-}$ ions. Thus in $Pb_4B_4S_{10}$ the anion is isoelectronic with P_4O_{10} and has an analogous structure, which can be described as four edge-sharing B_3S_3 rings (ref. 68). (A more extensive fused-ring network is found in $Ag_3B_5S_9$ (ref. 127), which will be discussed in Section 4.3.) The structures of the compounds described above are given in Fig. 18.

Structural data for a variety of boron-sulfur ring compounds are compiled in Table 4. Four generalizations consistent with these data are as follows:

(i) The rings tend to be planar or nearly so, but non-planarity can be induced by certain substituents. Thus although $Cl_2B_2S_3$ and $Me_2B_2S_3$ are planar (ref. 128), in $Ph_2B_2S_3$ the unique sulfur lies out of the plane defined by the other four ring atoms (ref. 68); this presumably minimizes the non-bonded interactions with the phenyl groups. $(Me_2N)_4B_4S_2$ is also non-planar, with a geometry that promotes $N \longrightarrow B$ π bonding; this hypothesis is supported by the abnormally long B-S bonds (ref. 122). Cyclic adducts having sp^3-hybridized boron are also non-planar, of course. Thus the compounds $(MeSBX)_3$ (X:Cl and Br) occupy chair conformations (ref. 129).

(ii) B-S bond distances lie in the 178-182 pm range. Exceptions are observed in species having amino substituents at boron (refs. 122,130) and in adducts, where the rehybridization of boron from sp^2 to sp^3 leads to an increase in B-S bond lengths to 190-192 pm (ref. 68).

(iii) For compounds having sulfur-containing substituents, the differences between ring and exocylic B-S bond distances are small. This is in contrast to B-O systems. In $(HSBS)_3$ the exocyclic bond is slightly the longer of the two types, while in $[M^+(SBS)^-]_3$ (M:Na and K) the reverse is true (see 2.4) (refs. 67,68).

(iv) As was noted in 2.4, endo-ring bond angles are generally greater at boron than at sulfur.

TABLE 4 Structural data for compounds containing boron-sulfur rings[a]

ring type	compound	bond lengths (pm) B-S	S-S	angles (°) SBS	BSB	BSS	other	ref.
B_2S_2	B_2S_3	182		104			B-S (exo) 179	(123)
	$R_2B_2S_2$[b]	186		98	82		B-N 140 SBN 132°	(130)
$(B^-S^+)_2$	$(R'BS)_2$[c]	194		95	89		B-S (ring) 183 B-I 213	(131)
B_2S_3	B_8S_{16}	180	208	120	98	101		(124)
	$Cl_2B_2S_3$[d]	179	207	122	97	100	B-Cl 176	(128)
	$Me_2B_2S_3$[d]	180	208	118	102	102	B-C 157	(128)
	$Ph_2B_2S_3$	181	206				B-C 155	(68)
B_3S_3	B_2S_3	180		130			B-S (exo) 183	(125)
	$(BrBS)_3$	181		131	109		B-Br 190	(132)
	$(HSBS)_3$	180		130	110		B-SH 181	(67)
	$(Na^+SBS^-)_3$	182		125	115		B-S 177	(68)
	$(K^+SBS)_3$	181		124	116		B-S 178	(68)
$(B^-S^+)_3$	$(Cl_2BSMe)_3$	195		106	112		B-Cl 180 ClBS 109°	(129)
	$(Br_2BSMe)_3$	194		107	113		B-Br 196 BrBS 109°	(129)
B_4S_2	$(Me_2N)_4B_4S_2$	185					B-B 170 B-N 138	(122)

[a]Structural determinations by X-ray diffraction unless otherwise noted. Bond distances (pm) and bond angles (°) averaged (for symmetrically equivalent positions) and rounded to the nearest integer.

[b]R:-N 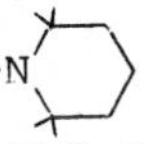 .

[c]R':EtC=C(Et)BI.

[d]By electron diffraction.

4.1.2 Methods of preparation

The most common synthetic route to symmetrical rings (i.e., cyclic dimers and trimers) involves the condensation of di- or trihaloboranes with H_2S.

$$RBX_2 + H_2S \longrightarrow 1/n\ (RBS)_n + 2HX \tag{12}$$

Here R is alkyl or aryl, X, OR, NHR, NR_2, or SR. Many variations of this basic reaction have been reported. For example, the source of sulfur may be Na_2S (ref. 133), HgS (ref. 134), $\underline{t}$-Bu_2S (ref. 135), $(R_2SiS)_3$ (ref. 136), etc. As is the case for B-O, B-N, and several other heterocyclic systems, the value of n in eqn. 12 is most often three. Dimers are occasionally observed, but no well-documented examples of higher cyclic oligomers have yet been reported. The preference for trimers $\underline{vs}$ dimers is probably due to the greater thermodynamic stability of the former (ref. 4). This notion is supported by the irreversible rearrangement of $(HSBS)_2$ upon heating or repeated recrystallization (ref. 123).

$$3(HSBS)_2 \longrightarrow 2(HSBS)_3 \tag{13}$$

Several other routes to B_2S_2 rings are worthy of mention. The first thermally stable dimer to be reported was $(Et_2NBS)_2$ (ref. 137), which was obtained $\underline{via}$ the reaction

$$Et_3N \cdot BH_3 + H_2S \longrightarrow 1/2(Et_2NBS)_2 + C_2H_6 + 2H_2 \tag{14}$$

Recently the coordination dimer of H_2BSH was identified in the product mixture of the reaction of B_2H_6 with H_2S at $-10°$ (ref. 138). SCF calculations indicate that this species exists as a mixture of planar Z and E isomers. ΔH for the reaction

$$B_2H_6 + 2H_2S \longrightarrow (H_2BSH)_2 + 2H_2 \tag{15}$$

was calculated to be about -9 kJ/mol, slightly less favorable than for the competing reaction

$$B_2H_6 + 2H_2S \longrightarrow \begin{array}{c} H \\ \diagdown \\ H \end{array} B \begin{array}{c} SH \\ \diagup \\ \diagdown \\ H \end{array} B \begin{array}{c} H \\ \diagup \\ \diagdown \\ SH \end{array} + 2H_2 \tag{16}$$

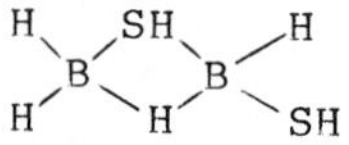

The compounds $(FBS)_n$ (n:2 and 3) have low thermal stabilities, and standard preparative routes give only decomposition products. They were obtained under conditions permitting spectroscopic (UV-PES) study from the reaction of elemental boron with S_2F_2, SF_4 or SF_6 at 1100°. Under these conditions the ratio of monomer to dimer is about 9:1 (ref. 139).

$R_2B_2S_3$ ring compounds are produced by reactions of haloboranes with H_2S_n ($n \geqslant 2$) (ref. 140), Na_2S_2 (ref. 133), or elemental sulfur (ref. 141). A few specific examples are given in eqns. 17-19.

$$2BBr_3 + 2H_2S_n \longrightarrow Br_2B_2S_3 + 4HBr + 2n\text{-}3/8S_8 \qquad (17)$$

$$2PhBBr_2 + 2Na_2S_2 \longrightarrow Ph_2B_2S_3 + 4NaBr + 1/8S_8 \qquad (18)$$

$$2BI_3 + 3/8S_8 \longrightarrow I_2B_2S_3 + 2I_2 \qquad (19)$$

The synthesis of ^{10}B-labeled $Me_2B_2S_3$ is possible <u>via</u> the sequence

$$KBF_4 \xrightarrow{AlBr_3} BBr_3 \xrightarrow{SnMe_4} MeBBr_2 \xrightarrow{H_2S_2} Me_2B_2S_3,$$

which proceeds with an overall yield of 38% (ref. 142).

Both B_3S_2 and B_4S_2 rings have been prepared from the cyclic compound $Me_2SiS_2B_2R_2$ ($R{:}NMe_2$). The Me_2Si unit can be replaced by either a BR or RB-BR fragment depending on the boron-containing reagent (eqn. 20-21) (ref. 122).

Other, less systematic methods for the formation of boron-sulfur rings have been reviewed (refs. 4,34).

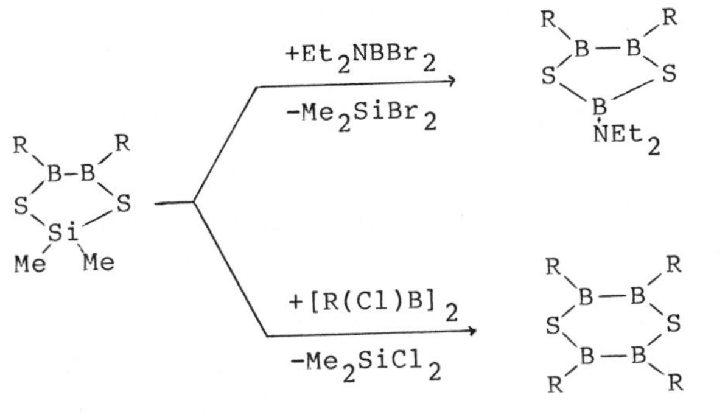

(20)

(21)

4.1.3 <u>Reactions</u>

The reactions of cyclic boron-sulfur compounds can be divided into two types: those in which the ring remains intact (exocyclic substitution and adduct formation) and those in which it is altered or destroyed.

Nucleophilic substitution and addition. The high Lewis acidity of the ring boron atoms makes nucleophilic addition and displacement commonplace in these systems. Substitution reactions are most prevalent for B_2S_3 and B_3S_3 ring compounds, which appear to be the most stable toward fragmentation. Thus it is possible to introduce alkyl, aryl, alkoxy, akylthio, and amino moities through reactions such as

$$(XBS)_3 + BR_3 \longrightarrow (RBS)_3 + BX_3 \qquad (22)$$

X:Br, SH; R:Me, OMe, SEt, NMe_2, etc. (refs. 123,143)

$$Br_2B_2S_3 + 2LiR \longrightarrow R_2B_2S_3 + 2LiBr \qquad (23)$$

R:alkyl, NHR, NR_2, etc. (refs. 144,145)

Certain protic nucleophiles including H_2O and alcohols destroy the framework, with sulfur being extruded as H_2S or H_2S_n.

Addition, rather than displacement, is likely with nucleophiles that lack an appropriate leaving group. Thus the reaction between trimethylamine and $H_2B_2S_3$ yields either the mono- or diadduct depending on the molar ratio (ref. 126). In general, adduct formation appears to stabilize rings of high sulfur content. The two members of the B_2S_4 system shown in Fig. 17 have only been isolated in compounds where both boron atoms are coordinated to nitrogen (ref. 146).

Ring-alteration reactions. Only a few reactions are known in which one type of boron-sulfur ring is converted to another. The $(HSBS)_2$ to $(HSBS)_3$ transformation (eqn. 13) has already been mentioned. B_2S_3 and B_3S_3 frameworks can be interconverted by use of the proper reagents (eqns. 24,25) (refs. 35,140),

$$X_2B_2S_3 + BX_3 \longrightarrow X_3B_3S_3 + X_2 \qquad (24)$$

$$2(XBS)_3 + 3/8S_8 \longrightarrow 3X_2B_2S_3 \qquad (25)$$

while B_8S_{16} is obtained in low yield by the treatment of $Br_2B_2S_3$ with the H_2S-generating agent $S{=}C(SH)_2$ (ref. 125).

The incorporation of heteroatoms, particularly carbon and oxygen, into boron-sulfur rings is often possible; such reactions generally involve addition across multiple bonds. Thus $Br_2B_2S_3$ reacts with disubstituted alkynes, ultimately exchanging a C=C fragment for B-S. In contrast to an initial report (ref. 147), the product contains a sulfur-sulfur bond (ref. 148).

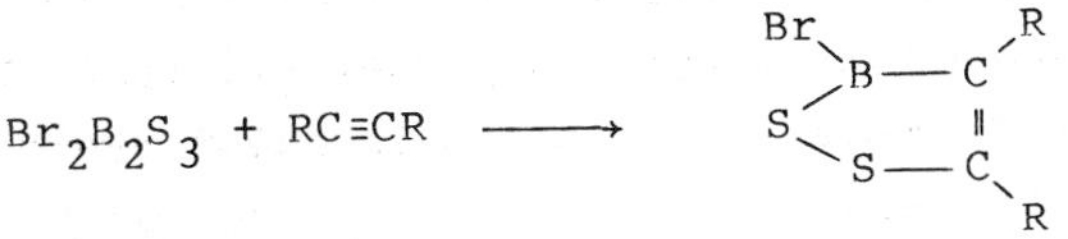

(26)

The boryl-substituted hexene $I_2B(Et)C{=}C(Et)BI_2$ reacts with $(IBS)_3$ to yield a C_2B_2S ring compound (ref. 149).

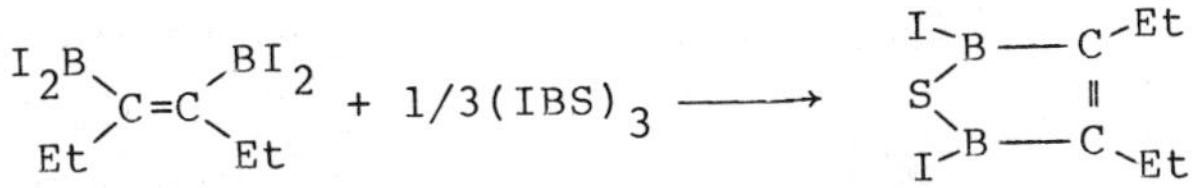

(27)

This product has in turn been used to prepare a number of interesting derivatives (ref. 19).

Hydrazine can be reacted with B_2S_3 rings to replace an S-S unit with N-N, yielding SB_2N_2 frameworks (refs. 150,151). Also, an assortment of reagents containing S=N multiple bonds have been used to convert B_2S_3 to B_2N_2 rings (refs. 152-154).

4.2 POLYHEDRAL CLUSTERS

In contrast to oxygen, a number of polyhedral systems are known which have boron and sulfur in vertex positions. As a vertex atom sulfur typically has no terminal substituents. It is a four-electron donor according to electron-counting rules (refs. 51-54). Thus in terms of framework electrons, compounds conforming to the general formula SB_xH_x are isoelectronic with $B_{x+1}H_{x+1}{}^{2-}$ and $C_2B_{x-1}H_{x+1}$ systems (see 2.2 and Fig. 2). Sources discussing the electronic structures of thiaboranes are available (e.g., refs. 17,155).

4.2.1 Synthesis and structures

Thiaboranes having closo, nido, and arachno frameworks are known. Specific examples (parent compounds or derivatives) include:

closo: SB_9H_9 and $SB_{11}H_{11}$

nido: SB_8H_{10}, SB_9H_{11}, and $SB_{10}H_{12}$
arachno: SB_9H_{13} and $S_2B_7H_9$

The structures of a number of thiaboranes have been determined by X-ray crystallography. These include the tris(triethylphosphine)gold(I) salt of 8-$SB_9H_{12}^-$ (ref. 156), the triethylamine adduct of 6-SB_9H_{11} (ref. 157), and the coupled compound 2,2'-(1-SB_9H_8)$_2$ (ref. 158), whose structures are given in Fig. 19. (X-ray diffraction studies of the adduct $(Me_2S)_2B_{10}H_{12}$ (ref. 159) and of a number of metallathiaboranes have also been reported.)

The first thiaborane reported in the literature had the SB_9 framework (ref. 160). Ammonium polysulfide was used as the

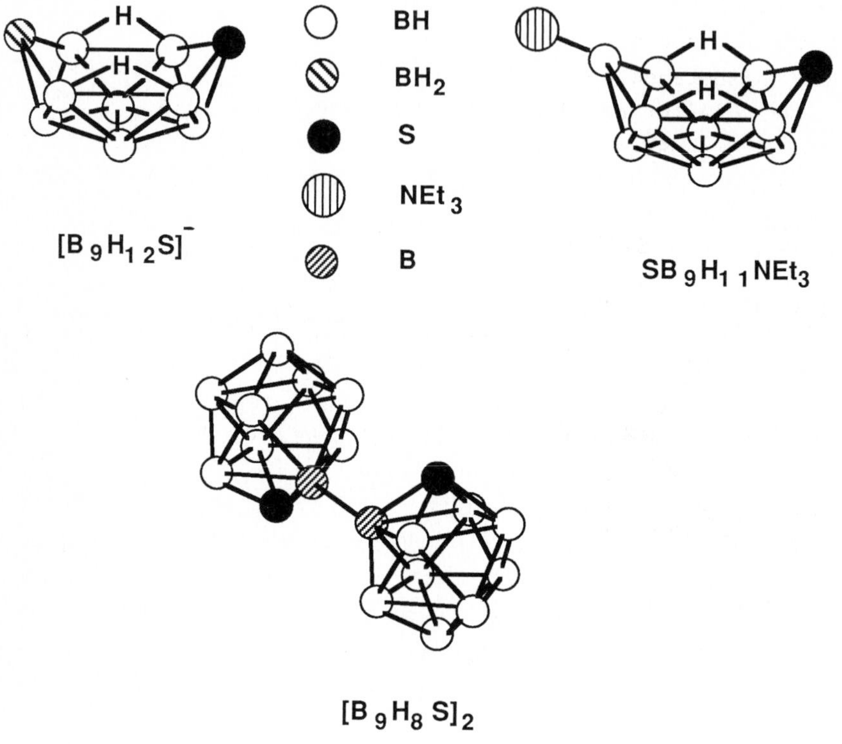

Fig. 19. The solid-state structures of some thiaboranes.

$$B_{10}H_{14} + S^{2-} + 4H_2O \longrightarrow SB_9H_{12}^- + B(OH)_4^- + 3H_2 \qquad (28)$$

source of sulfur. The experimental procedure for this reaction has been optimized (ref. 161).

Other reagents that have been used to incorporate sulfur into polyhedral boranes include t-BuN=S=Nt-Bu (ref. 160), Me_2N-N=S (ref. 160), $K_2S_2O_5/HCl/H_2O$ (ref. 162), $KHSO_3/HCl/H_2O$ (ref. 163), and $Na_2SO_3/HCl/H_2O$ (ref. 164).

4.2.2 Reactions

The chemical reactions of thiaboranes can be classified into those that alter the vertex framework (defined here as cage expansion, degradation, and coupling reactions) and those that do not. Those in which the framework is altered will be considered first. Several methods for the interconversions of thiaboranes are suggested by Fig. 20, which depicts some known reactions of this type.

A general route to cage expansion involves heating nido or arachno thiaboranes to 120-200°C in the presence of $Et_3N{\cdot}BH_3$, causing the evolution of H_2. Specific examples include the conversion of 6-$SB_9H_{12}^-$ to 7-$SB_{10}H_{11}^-$ (or, upon acidification, to

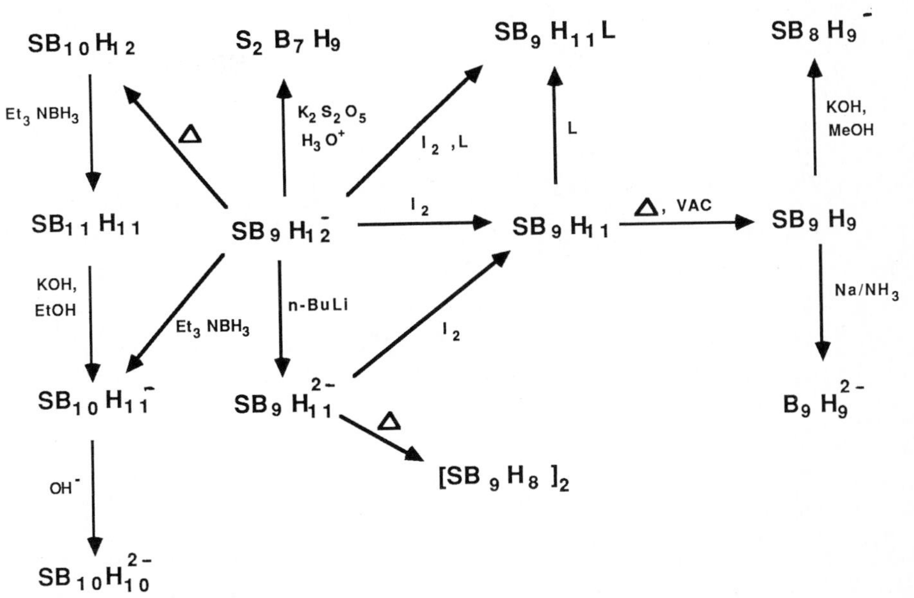

Fig. 20. Some interconversions of thiaboranes.

the conjugate acid 7-$SB_{10}H_{12}$) (ref. 160) and of 7-$SB_{10}H_{12}$ to $SB_{11}H_{11}$ (ref. 165). Dichloro(phenyl)borane can also be used (ref. 160), as is demonstrated by the reaction

$$SB_{10}H_{10}{}^{2-} + PhBCl_2 \longrightarrow 2\text{-}Ph\text{-}1\text{-}SB_{11}H_{10} + 2Cl^- . \qquad (29)$$

A less obvious pathway to cage expansion involves heat (refs. 160,166,167).

$$Cs^+SB_9H_{12}{}^- \xrightarrow{200^0C} Cs^+SB_{10}H_{11}{}^- \qquad (30)$$

$$7\text{-}SB_{10}H_{12} \xrightarrow{450^0C} SB_{11}H_{11} \qquad (31)$$

The coupling of cages through B-B single bonds occasionally occurs during heating. Thus the <u>in vacuo</u> pyrolysis of SB_9H_{11} gave several products, including coupled compounds. The major product was SB_9H_9 (50% yield); the three coupled isomers

$$2SB_9H_{11} \xrightarrow{375^0C} (SB_9H_8)_2 + 3H_2 \qquad (32)$$

(attached at the 2,2', 2,6', and 6,6' positions) were identified as side products in 5% total yield (refs. 158,166).

Hydroxide ion (in either aqueous or alcohol solution) is a standard agent for the degradation of polyhedral boranes and carboranes. The KOH/MeOH reaction system was used to convert the closo compounds SB_9H_9 and $SB_{11}H_{11}$ to $4\text{-}SB_8H_9{}^-$ and $7\text{-}SB_{10}H_{11}{}^-$, respectively. Liquid ammonia can also be used for the degradation of SB_9H_9 to $SB_8H_9{}^-$. However, when metal-ammonia solutions were used the sulfur was lost from the cage and a salt of the $B_9H_9{}^{2-}$ ion was isolated. A similar result was obtained for $SB_{11}H_{11}$ and Na/NH_3 (ref. 167).

Reactions in which the polyhedral framework remains intact include: (a) oxidation/reduction by gain or loss of hydrogen; (b) protonation and deprotonation; (c) addition of Lewis bases; (d) electrophilic substitution; (e) hydroboration; and (f) behavior as ligands towards metals.

For a given framework, simple oxidation/reduction (that is, arachno $\longleftrightarrow$ nido $\longleftrightarrow$ closo conversions) can sometimes be achieved. This is the case for the SB_9 system (refs. 160,161, 167):

$$[SB_9H_{12}\cdot H_2O]^- \xrightarrow{dry} SB_9H_{11} \xrightarrow{375^0} SB_9H_9 \qquad (33)$$

The first step (arachno $\longrightarrow$ nido) can also be accomplished by chemical oxidation of the monoanion by I_2 (eqn. 34) (refs. 161,167) or of the dianion by $SnCl_2$ (eqn. 35) (ref. 167).

$$SB_9H_{12}^- \xrightarrow{I_2} SB_9H_{11} \tag{34}$$

$$SB_9H_{11}^{2-} \xrightarrow{SnCl_2} SB_9H_{11} \tag{35}$$

Thiaboranes with bridging hydrogens are Brönsted acids. SB_9H_{13} deprotonates spontaneously in water to the monoanion, although strong base is needed to produce the dianion. Either or both of the bridge hydrogens of $SB_{10}H_{12}$ can be removed by n-BuLi depending on the reaction conditions (ref. 160).

Like other boron hydrides, thiaboranes show considerable reactivity toward Lewis bases. In addition to commonly used donors (Et_3N, PR_3, Ph_2PH, Me_2S, etc.), adducts have been prepared with the weaker bases Me_2NCHO (ref. 160), CH_3CN (ref. 160), and THF (refs. 167,168).

A variety of electrophilic substitution reactions have been performed on thiaborane substrates. Rudolph and co-workers studied the sequential halogenation and deuteration (refs. 169, 170) and alkylation (ref. 171) reactions of SB_9H_9 and $SB_{11}H_{11}$ using reagents such as Br_2 and I_2, $DCl/AlCl_3$, and $RI/AlCl_3$. (Aluminum chloride acts as a halogenation and deuteration catalyst.) In contrast to the carboranes $C_2B_5H_7$ and $C_2B_{10}H_{12}$, there is little correlation between the site of initial electrophilic attack (found to be the 6 position in both thiaboranes) and calculated charge distribution (ref. 172). Multiple substitutions were achieved, with compounds containing up to nine, six, and five moieties for halogenation, deuteration, and alkylation, respectively, having been identified.

The first hydroborations involving a polyhedral borane were reported for 6-SB_9H_{11}. Alkenes react at the 9 position to give alkyl derivatives. Substituted alkynes react similarly to yield alkenyl-substituted thiaboranes, while acetylene gives the doubly-substituted compound 9,9'-$CH_3CH(SB_9H_{10})_2$ (refs. 173,174). More recently, it was reported that the derivative 1-(3-C_7H_{13})-6-SB_9H_{10} undergoes multiple hydroboration with substrates such as cis-3-hexene. Of the three products (mono-, di-, and trisubstituted), the 1,9-disubstituted species was obtained in greatest yield (59%) under the conditions of the study (ref. 175).

It is not surprising that thiaboranes are known to coordinate to various metals, or that the presence of sulfur appears to create a preference for certain metals. Interactions with "chalcophiles" - in particular, elements of the later d-block families - are most favorable. Thus metallathiaboranes containing elements of the iron (ref. 160), cobalt (refs. 160, 176), nickel (refs. 160,168,177,178) and copper (ref. 156) groups have been reported. It follows that in such compounds the metal and sulfur(s) usually occupy adjacent positions.

Several examples of "sandwich" complexes are known, including $[Fe(SB_{10}H_{10})_2]^{2-}$ and $[Co(SB_{10}H_{10})_2]^-$ (ref. 160) and $[M(SB_9H_{11})_2]^{2-}$ (M:Ni and Pd) (ref. 168). The first thiacarborane reported was 6,8-CSB_7H_{11} (ref. 162), while among the other nonmetals it appears that only arsenic has been incorporated into thiaborane cages (in $As_2SB_8H_8$) (ref. 179).

4.3 EXTENDED STRUCTURES

In comparison to the large number and variety of B-O networks (see 3.3), examples of extended structures having boron-sulfur frameworks are limited. Much of what is known in this area has been summarized by Krebs (ref. 68).

The layer structure of B_2S_3 has already been discussed (see Fig. 18 and Table 4). A second binary compound, BS_2, can be obtained from the reaction of B_2S_3 with sulfur at 120-250°C. It consists of a linear polymer in which B_2S_3 rings are linked by sulfur bridges (ref. 180). This structure is reminiscent of B_8S_{16} (compare Fig. 21 to Fig. 18).

There are two well-characterized examples of extended structures among the metal thioborates. Thallium perthioborate, $Tl^+BS_3^-$, has been prepared by the high-temperature reaction of

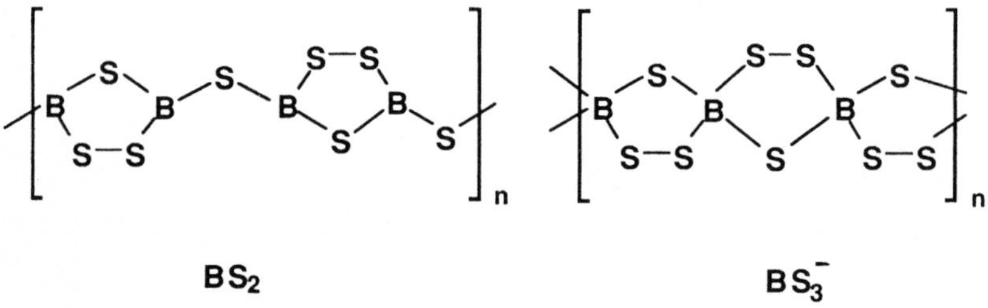

Fig. 21. Portions of the polymeric structures of BS_2 and the anion of $Tl^+BS_3^-$.

Tl_2S with elemental boron and sulfur (ref. 68). Like $(BS_2)_n$, the BS_3^- anion is a polymer chain incorporating B_2S_3 rings. In the anion, however, there are no sulfur bridges; rather, adjacent rings share common boron atoms. Hence all the borons are four-coordinate, and as a result the B-S bonds are long (193 pm). This structure has alternately been described as BS_4 tetrahedra bridged by one S_1 and one S_2 group (ref. 68).

The anionic units of the silver salt $Ag_3B_5S_9$ are "super-tetrahedra" composed of ten boron and 18 sulfur atoms (ref. 127). Each unit comprises ten BS_4 tetrahedral subunits (av B-S 192 pm) linked in an adamantane-like manner. The structure can also be viewed as a fragment of the zinc blende lattice (ref. 68).

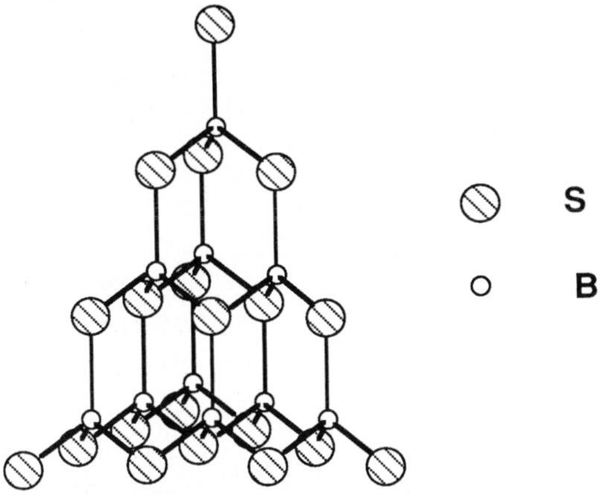

Fig. 22. One of the "supertetrahedral" $B_{10}S_{18}$ units of $Ag_3B_5S_9$.

5 ALUMINUM, GALLIUM AND INDIUM

5.1 GENERAL: CONTRASTS TO BORON

The catenation and cluster chemistry of aluminum, gallium and indium will be treated only briefly. More detailed information is available from other sources (e.g., refs. 4,5,12,40-42).

Compared to boron, the larger covalent radii of these three elements makes a weaker homonuclear bonding; commonly cited bond energies are 293, 113, and 100 kJ/mol for B-B, Ga-Ga, and In-In, respectively (ref. 69). This explains the lack of extensive cluster chemistry for the heavier main group 3 elements. Similarly, the bonds formed by Al, Ga and In to oxygen and sulfur are weaker than are B-O and B-S linkages. This reduces the stability of rings and polymers composed of alternating -M-O-M- and -M-S-M- frameworks, and as a result that chemistry is also comparatively limited.

Another difference is that these M-O and M-S bonds are more polar than those to boron (see 2.1). This has a major effect on solid-state structures, particularly for binary systems. Thus while the lattices of B_2O_3 and B_2S_3 contain four- and six-membered rings, all of the common polymorphs of M_2O_3 and M_2S_3 (M:Al, Ga, In) are best described as ionic-type lattices in which the non-metal atoms are close-packed and the metals occupy either octahedral or tetrahedral interstices (ref. 181).

There are a number of exceptions to the above generalizations, only a few of which will be cited. A variety of ternary Na/Al/O systems have been studied which contain aluminum-oxygen polyanions. Some, such as $[Al_5O_{16}]^{17-}$, are acyclic. However, the anion of $Na_7Al_3O_8$ is a polymeric chain in which twelve-membered (Al_6O_6) rings are linked by bridging oxygens (ref. 182). A somewhat similar structure has been established for the $[Al_6O_{18}]^{18-}$ anion, which is a cyclic hexamer (i.e., $([AlO_3]^{3-})_6$ - see Fig. 23) (ref. 183). The structure of α-$Al(OH)_3$ (bayerite) also contains Al_6O_6 rings. However, in that case the aluminum ions are hexacoordinate (ref. 184).

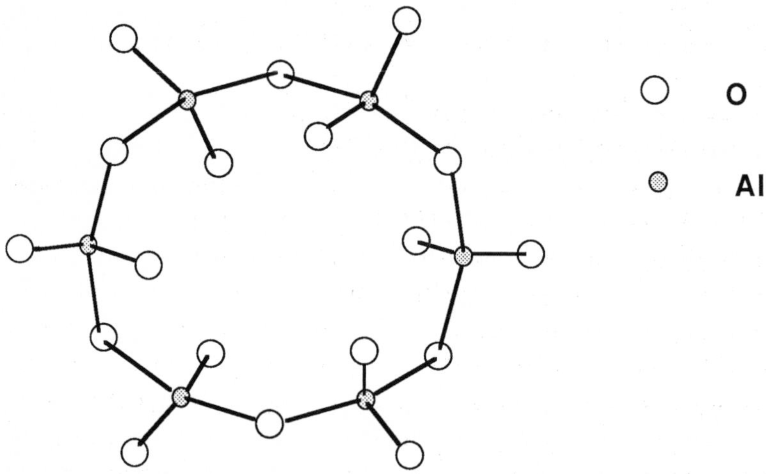

Fig. 23. The structure of the $[Al_6O_{18}]^{18-}$ ion.

5.2 COORDINATION HETEROCYCLES

In contrast to the virtual absence of stable M_xO_y and M_xS_y ring compounds having trivalent aluminum, gallium or indium atoms, cyclic adducts with tetravalent metals are common (ref. 41). Dimers are the most prevalent, but if the substituents are sufficiently small either trimers or tetramers may be observed. Thus Me_2GaOH is tetrameric in the solid state (ref. 185), with alternating gallium and oxygen atoms in an eight-membered ring. Me_2AlOMe is thought to exist as a dimer when freshly prepared from Al_2Me_6 and MeOH, but spontaneously converts to the trimer within minutes (ref. 186); it is also trimeric in the gas phase (ref. 187). The more hindered Et_2AlOt-Bu and t-Bu_2AlOt-Bu are dimeric (ref. 188).

5.2.1 Synthesis

Cyclic adducts can be prepared through reactions of metal alkyls with alcohols or thiols, carboxylic acids, or sulfur oxides. A few representative examples are given in eqns. 36-39 (refs. 186,189-193).

$$Al_2Me_6 + 2MeXH \longrightarrow 1/n[Me_2AlXMe]_n + 2CH_4 \quad \text{X:O and S} \tag{36}$$

$$4Me_3Ga\ OEt_2 + 4H_2O \longrightarrow [Me_2GaOH]_4 + 4CH_4 + 4Et_2O \tag{37}$$

$$2R_3In + 2R'CO_2H \longrightarrow [R_2InO_2CR']_2 + 2RH \tag{38}$$

$$2MeAlCl_2 + 2SO_2 \longrightarrow [Cl_2AlOS(O)Me]_2 \tag{39}$$

An alternate synthesis involves the cleavage of silicon-oxygen bonds by halides (ref. 194):

$$2R_3SiOSiR_3 + 2AlX_3 \longrightarrow [X_2AlOSiR_3]_2 + 2R_3SiX \tag{40}$$

X:Cl, Br, I; R:Me, Et, $SiMe_3$, etc.

5.2.2 Structures

The structures of several of the species mentioned above have been determined by X-ray or electron diffraction. $[Me_2GaOH]_4$ and $[Me_2AlOMe]_3$ contain non-planar eight- and six-membered rings, respectively (refs. 185,187). Among the dimers, $[Br_2AlOSiMe_3]_2$ has a planar ring, and in fact the three bonds about each oxygen all lie in the same plane (ref. 195). The Al-O bond lengths are

181 pm, with O-Al-O angles of 85°.

In several instances the presence of cyclic oligomers has been inferred from spectroscopic data (ref. 41). Such is the case for a number of species believed (ref. 193) to conform to the general structure

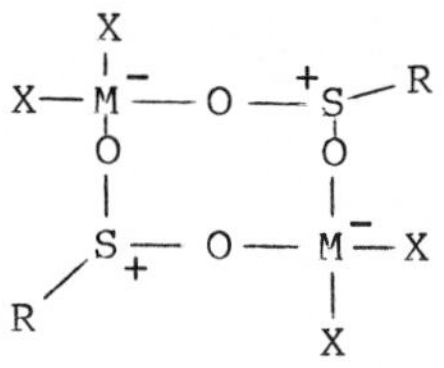

M:Al, X:Cl, Br: R:Me, Et
M:Ga, In: X,R:Me
etc.

as well as for $[Et_2AlSCN]_3$; the latter was predicted on the basis of vibrational spectroscopy to contain the Al_3S_3 ring framework (ref. 196).

Acknowledgment

Our work is supported by the National Science Foundation. J.B. is grateful for the hospitality of the University of Notre Dame while a visiting faculty member during 1987-88.

REFERENCES

1 A.H. Cowley (Editor), Rings, Chains, and Polymers of the Main-Group Elements, American Chemical Society, Washington, 1983.
2 C.E. Carraher, Advances in Organometallic and Inorganic Polymer Science, Marcel Dekker, New York, 1982.
3 N.H. Ray, Inorganic Polymers, Academic, New York, 1978.
4 I. Haiduc, The Chemistry of Inorganic Ring Systems, Part 1, Wiley, New York, 1970.
5 I. Haiduc and D.B. Sowerby (Editors), The Chemistry of Inorganic Homo- and Heterocycles, Vol. 1, Academic, New York, 1987.
6 R. Steudel, Chemistry of the Non-Metals, Walter de Gruyter, New York, 1977.
7 R.W. Parry and G. Kokama (Editors), Boron Chemistry, Pergamon, Oxford, 1980.
8 T.P. Onak, Organoborane Chemistry, Academic, New York, 1975.
9 N.N. Greenwood, Boron, Pergamon, Oxford, 1975.
10 K. Niedenzu, MTP Int. Rev. Sci., Inorg. Chem. Ser. One, 4

(1972) 3.
11 I. Ander, in A. Katrizsky and C.W. Rees (Editors), Comprehensive Heterocyclic Chemistry, Vol. 1, O. Meth-Cohn (Editor), Pergamon, Oxford, 1984.
12 N.N. Greenwood and A. Earnshaw, Chemistry of the Elements, Pergamon, Oxford, 1984.
13 B. Singaram and G.G. Pai, Heterocycles, 18 (1982) 387.
14 B.M. Mikhailov, Sov. Sci. Rev. B, 18 (1980) 283.
15 W. Siebert, Adv. Organomet. Chem., 18 (1980) 301.
16 B.M. Mikhailov, Pure Appl. Chem., 49 (1977) 749.
17 R.W. Rudolph, Acc. Chem. Res., 9 (1976) 446.
18 E.L. Muetterties (Editor), Boron Hydride Chemistry, Academic, New York, 1975.
19 J.H. Morris in G. Wilkinson, F.G.A. Stone and E.W. Abel (Editors), Comprehensive Organometallic Chemistry, Vol. 1, Pergamon, Oxford, 1982, 311.
20 G.E. Herberich in G. Wilkinson, F.G.A. Stone and E.W. Abel (Editors), Comprehensive Organometallic Chemistry, Vol. 1, Pergamon, Oxford, 1982, 381.
21 T. Onak in G. Wilkinson, F.G.A. Stone and E.W. Abel (Editors), Comprehensive Organometallic Chemistry, Vol. 1, Pergamon, Oxford, 1982, 411.
22 L.J. Todd in G. Wilkinson, F.G.A. Stone and E.W. Abel (Editors), Comprehensive Organometallic Chemistry, Vol. 1, Pergamon, Oxford, 1982, 543.
23 K.B. Gilbert, S.K. Boocock, and S.G. Shore in G. Wilkinson, F.G.A. Stone and E.W. Abel (Editors), Comprehensive Organometallic Chemistry, Vol. 6, Pergamon, Oxford, 1982, 879.
24 I. Haiduc in I. Haiduc and D.B. Sowerby (Editors), The Chemistry of Inorganic Homo- and Heterocycles, Vol. 1, Academic, New York, p 109.
25 J.B. Farmer, Adv. Inorg. Chem. Radiochem., 25 (1982) 187.
26 K. Torsell in R.J. Brotherton and H. Steinberg (Editors), Progress in Boron Chemistry, Vol. 1, Pergamon, Oxford, 1964, p 369.
27 P.L. Strong, Mellor's Comprehensive Treatise on Inorganic and Theoretical Chemistry, Supplement, Vol. 5, Longmans, London, 1980, p 795.
28 E.I. Neghishi, J. Organomet. Chem., 108 (1976) 281.
29 Gmelin Handbook of Inorganic Chemistry, Vol. 44, Springer, Berlin, 1977.
30 Gmelin Handbook of Inorganic Chemistry, Vol. 48, Springer, Berlin, 1977.
31 Gmelin Handbook of Inorganic Chemistry, 1st. Suppl., Vol. 1, Springer, Berlin, 1980.
32 Gmelin Handbook of Inorganic Chemistry, 2nd. Suppl., Vol. 1, Springer, Berlin, 1983.
33 Gmelin Handbook of Inorganic Chemistry, 2nd. Suppl., Vol. 2, Springer, Berlin, 1982.
34 W. Siebert in I. Haiduc and D.B. Sowerby (Editors), The Chemistry of Inorganic Homo- and Heterocycles, Vol. 1, Academic, New York, p 143.
35 W. Siebert, Chem.-Ztg., 98 (1974) 478.
36 Gmelin Handbook of Inorganic Chemistry, Vol. 19, Berlin, 1975.
37 Gmelin Handbook of Inorganic Chemistry, Vol. 34, Berlin, 1976.
38 Gmelin Handbook of Inorganic Chemistry, 1st. Suppl., Vol. 3, Springer, Berlin, 1981.

39 Gmelin Handbook of Inorganic Chemistry, 2nd. Suppl., Vol. 2, Springer, Berlin, 1982.
40 K. Wade and A.J. Banister in J.C. Bailar et al. (Editors), Comprehensive Inorganic Chemistry, Vol. 1, 1973, p 993.
41 J. Weidlein, J. Organomet. Chem. 49 (1973) 257.
42 S.K. Boocock and S.G. Shore in G. Wilkinson, F.G.A. Stone and E.W. Abel (Editors), Comprehensive Organometallic Chemistry, Vol. 6, Pergamon, Oxford, 1982, 947.
43 J.B. Casey, W.J. Evans, and W.H. Powell, Inorg. Chem., 22 (1983) 2228.
44 C.L. Christ, Am. Mineral., 45 (1960) 334.
45 C.L. Christ and J.R. Clark, Phys. Chem. Miner., 2 (1977) 59.
46 W.W. Porterfield, Inorganic Chemistry, Addison-Wesley, 1984.
47 P.P. Edwards and M.J. Sienko, Acc. Chem. Res., 15 (1982) 87.
48 R.L. DeKock and H.B. Gray, Chemical Structure and Bonding, Benjamin-Cummings, Menlo Park, 1980, p 90.
49 K. Wade, Electron Deficient Compounds, Nelson, London, 1973.
50 T.P. Fehlner and C.E. Housecroft in J.F. Liebman and A. Greenberg (Editors), Molecular Structure and Energetics, Vol. 1, VCH, Deerfield Beach, 1986, p 149.
51 K. Wade, J. Chem. Soc. Chem. Commun., (1971) 792.
52 K. Wade, Adv. Inorg. Chem. Radiochem., 18 (1976) 1.
53 D.M.P. Mingos, Nature Phys. Sci., 236 (1972) 99.
54 D.M.P. Mingos, Acc. Chem. Res., 17 (1984) 311.
55 D.M.P. Mingos, J. Chem. Soc. Chem. Commun., (1983) 706.
56 J.R. Van Wazer and K. Moedritzer, Angew. Chem., 78 (1966) 401; Angew. Chem. Int. Ed. Engl., 5 (1966) 341.
57 J.C. Lockhart, Inorganic Reaction Mechanisms, Van Nostrand, Princeton, 1966.
58 G.B. Kistiakowsky and R. Williams, J. Chem. Phys., 23 (1955) 334.
59 R.S. Mulliken, J. Am. Chem. Soc., 74 (1952) 811.
60 H.K. Hofmeister and J.R. Van Wazer, J. Inorg. Nucl. Chem., 26 (1964) 1201.
61 D.W. Matula, L.C.D. Groenweghe and J.R. Van Wazer, J. Chem. Phys. 41 (1964) 3105.
62 W.V.F. Brooks, C.C. Costain and R.F. Porter, J. Chem. Phys., 47 (1967) 4186.
63 C.H. Chang, R.F. Porter and S.H. Bauer, Inorg. Chem., 8 (1969) 1689.
64 C.R. Peters and M.E. Milberg, Acta Cryst., 17 (1964) 229.
65 W. Schneider and G.B. Carpenter, Acta Cryst., B26 (1970) 1189.
66 M. Marezio, H.A. Plettinger and W.H. Zachariasen, Acta Cryst., 16 (1963) 594.
67 V.W. Schwarz, H.D. Hausen, H. Hess, J. Mandt, W. Schmelzer and B. Krebs, Acta Cryst., B29 (1973) 2029.
68 See B. Krebs, Angew Chem., 95 (1983) 113; Angew. Chem. Int. Ed. Engl. 22 (1983) 113; and references therein.
69 J.E. Huheey, Inorganic Chemistry, 3rd ed., Harper & Row, New York, 1983.
70 T.A. Albright, J.K. Burdett and M.H. Whangbo, Orbital Interactions in Chemistry, John Wiley, New York, 1985.
71 D.C. Frost, S.T. Lee and C.A. McDowell, Chem. Phys. Lett., 23 (1973) 472.
72 T.P. Fehlner and D.W. Turner, J. Am. Chem. Soc., 95 (1973) 7175.
73 D.T. Haworth and V.M. Scherr, J. Inorg. Nucl. Chem., 37 (1975) 2010.
74 D.R. Armstrong and P.G. Perkins, J. Chem. Soc. A, (1967)

790.
75 A. Serafini and J.F. LaBarre, J. Mol. Struct., 26 (1975) 129.
76 L. Barton, D. Brinza, R.A. Frease and F.L. Longcor, J. Inorg. Nucl. Chem., 39 (1977) 1845.
77 J.-P. Laurent and M. Pasdeloup, Bull. Soc. Chim. (Fr), (1966) 908.
78 A. Finch and P.J. Gardner, Trans. Far. Soc., 62 (1966) 3314.
79 A. Finch and P.J. Gardner, in R.J. Brotherton and H. Steinberg (Editors), Progress in Boron Chemistry, Vol. 3, Pergamon, Oxford, 1970, p 177.
80 R.F. Porter and S.K. Gupta, J. Phys. Chem., 68 (1964) 280.
81 M. Farber, J. Chem. Phys., 36 (1962) 661.
82 M. Farber and J. Blauer, Trans. Far. Soc., 58 (1962) 2090.
83 R.F. Porter, D.R. Bidinosti and K.F. Waterson, J. Chem. Phys., 36 (1962) 2104.
84 D.L. Hildenbrand, L.P. Theard and A.M. Saul, J. Chem. Phys., 39 (1963) 1973.
85 B.M. Mikhailov, L.S. Vasil'ev and V.V. Veselovskii, Izv. Acad. Nauk SSSR, Ser. Khim., (1980) 1106; Chem. Abst. 93 (1980) 114589.
86 M.A.A.F. deC.T. Corrondo and A.C. Skapski, Acta Cryst., B34 (1978) 3551.
87 S.K. Gupta and R.F. Porter, J. Phys. Chem., 67 (1963) 1286.
88 W.P. Sholette and R.F. Porter, J. Phys. Chem., 67 (1963) 177.
89 L. Barton and D. Nicholls, Proc. Chem. Soc., (1964) 242.
90 P.L. Timms and C.S.G. Phillips, Inorg. Chem., 3 (1964) 606.
91 L. Barton, F.A. Grimm and R.F. Porter, Inorg. Chem., 5 (1966) 2076.
92 J.F. Ditter and I. Shapiro, J. Am. Chem. Soc., 81 (1959) 1022.
93 G.H. Lee, W.H. Bauer and S.E. Wiberley, J. Phys. Chem., 67 (1963) 1742.
94 D.W. Wester and L. Barton, Org. Prop. Proc. Int., (1971) 191.
95 C.J.W. Brooks and D.J. Harver, J. Chromatogr., 54 (1971) 193.
96 R.T. Hawkins and D.B. Stroup, J. Org. Chem., 34 (1969) 1173.
97 R.A. Bowie and O.C. Musgrave, J. Chem. Soc. C, (1966) 566.
98 P.A. McCusker, E.C. Ashby and H.S. Makowski, J. Am. Chem. Soc., 79 (1957) 5179.
99 E.W. Abel, S. Dandegaonker, W. Gerrard and M.F. Lappert, J. Chem. Soc., (1956) 4697.
100 E.W. Abel, W. Gerrard and M.F. Lappert, J. Chem. Soc., (1957) 5051.
101 V.F. Gridina, A.L. Klebanskii and V.A. Bartashev, Zh. Organ. Khim., 32 (1962) 323.
102 A. Michaelis, Ann. Chem., 315 (1901) 19.
103 M.W. Rathke and H.C. Brown, J. Am. Chem. Soc., 88 (1966) 2606.
104 R.H. Cragg and J.P.N. Husband, Inorg. Nucl. Chem. Lett., 6 (1970) 773.
105 J.P. Tuchagues, Bull. Soc. Chim. (Fr), (1968) 2009.
106 S.G. Shore, D.-Y. Jan, W.-L. Hsu, L.-Y. Hsu, S. Kennedy, J.C. Huffman, T.C. LinWang, A. Marshall, J. Chem. Soc. Chem. Commum., (1984) 392.
107 V.G. Tsvetkov, V.N. Alyasov, N.V. Balakshina, V.P. Maslennikov and Y. A. Aleksandrov, Zh. Obshch. Khim., 51 (1981) 269.

108 P.A. McCusker and J.H. Bright, J. Org. Chem., 29 (1964) 2093.
109 B. Serafin, H. Duda and M. Makosza, Roczniki Chem., 37 (1963) 765; Chem. Abstr., 60 (1964) 2997.
110 J. Bielawski and K. Niedenzu, Inorg. Chem., 25 (1986) 85.
111 J. Bielawski and K. Niedenzu, Inorg. Chem., 25 (1986) 1771.
112 E.L. Muetterties (Editor), Boron Hydride Chemistry, Academic, New York, 1975.
113 R.P. Micciche, J.J. Briguglio and L.G. Sneddon, Inorg. Chem., 23 (1984) 3992.
114 X.L.R. Fontaine, H. Fowkes, N.N. Greenwood, J.D. Kennedy and M. Thornton-Pett, J. Chem. Soc. Chem. Commun., (1985) 1722.
115 R.D. Adams and I.T. Horvath, Prog. Inorg. Chem., 33 (1985) 127.
116 C.K. Schauer and D.F. Shriver, Angew. Chem., 99 (1987) 275; Angew. Chem. Int. Ed. Engl., 26 (1987) 255.
117 N. Morimoto, J. Mineral (Sapporo) 2 (156) 1.
118 S. Merlino and F. Sartori, Acta Cryst., B28 (1972) 3559.
119 S. Merlino, Atti. Accad. Nazl. Lincei Rend. Casse Sci. Fis. Mat. Nat., 47 (1969) 85.
120 S. Merlino and F. Sartori, Science, 171 (1971) 277.
121 J.A. Konnert, J.R. Clark, and C.L. Christ, Am. Mineral., 55 (1970) 1911.
122 H. Nöth, H. Fusstetter, H. Pommerening, and T. Taeger, Chem. Ber., 113 (1980) 342.
123 E. Wibert and W. Sturm, Angew. Chem., 67 (1955) 483.
124 B. Krebs and H.-U Hürter, Angew. Chem., 92 (1980) 479; Angew. Chem. Int. Ed. Engl., 19 (1980) 482.
125 H. Diercks and B. Krebs, Angew. Chem., 89 (1977) 327; Angew. Chem. Int. Ed. Engl., 16 (1977) 313.
126 H. Nöth and R. Staudigl, Z. Anorg. Allg. Chem., 481 (1981) 41.
127 B. Krebs and H. Diercks, Z. Anorg. Allg. Chem., 518 (1984) 101.
128 H.M. Seip, R. Seip, and W. Siebert, Acta Chem. Scand., 27 (1973) 15.
129 S. Pollitz, F. Zettler, D. Forst, and H. Hess, Z. Naturforsch., 31B (1976) 897.
130 E. Hanecker, H. Nöth and U. Wietelmann, Chem. Ber., 119 (1986) 1904.
131 F. Zettler, H. Hess, W. Siebert, and R. Full, Z. Anorg. Allg. Chem., 420 (1976) 285.
132 W. Schwarz, H.D. Hausen, and H. Hess, Z. Naturforsch., 29B (1974) 596.
133 M. Schmidt and F. Rittig, Z. Naturforsch., 25B (1970) 1062.
134 R.H. Cragg, M.F. Lappert, and B.P. Tilley, J. Chem. Soc. A., (1967) 947.
135 M. Schmidt and R. Rittig, Z. Anorg. Allg. Chem., 394 (1972) 152.
136 E.W. Abel, D.A. Armitage, and R.P. Bush, J. Chem. Soc. A, (1965) 3045.
137 J.A. Forstner and E.L. Muetterties, Inorg. Chem., 5 (1966) 164.
138 H. Binder, A. Ziegler, R. Ahlrichs, and H. Schiffer, Chem. Ber., 120 (1987) 1545.
139 T.A. Cooper, H.W. Kroto, C. Kirby, and N.P.C. Westwood, J. Chem. Soc. Dalton, (1984) 1047.
140 M. Schmidt and W. Siebert, Z. Anorg. Allg. Chem., 345 (1966) 87.
141 M. Schmidt and W. Siebert, Angew. Chem., 76 (1964) 687;

Angew. Chem. Int. Ed. Engl., 3 (1964) 637.
142 H. Nöth and R. Staudigl, Inorg. Synth., 22 (1983) 218.
143 E. Wiberg and W. Sturm, Z. Naturforsch., 8B (1953) 689.
144 H. Nöth and T. Taeger, Z. Naturforsch., 34B (1979) 135.
145 A. Meller and C. Habben, Monatsh. Chem., 113 (1982) 139.
146 H. Nöth and R. Staudigl. Chem. Ber., 115 (1982) 813.
147 C. Habben, W. Maringgele, and A. Meller, Z. Naturforsch., 37B (1982) 43.
148 M. Noltemeyer, G.M. Sheldrick, C. Habben, and A. Meller, Z. Naturforsch., 38B (1983) 1182.
149 W. Siebert, R. Full, J. Edwin, and K. Kinberger, Chem. Ber., 111 (1978) 823.
150 D. Nölle and H. Nöth, Z. Naturforsch., 27B (1972) 1425.
151 H. Nöth and R. Ullmann, Chem. Ber., 108 (1975) 3125.
152 C. Habben and A. Meller, Z. Naturforsch. 39B (1984) 1022.
153 C. Habben, A. Meller, M. Noltemeyer, and G.M. Sheldrick, J. Organomet. Chem., 288 (1985) 1.
154 D. Fest, C. Habben, and A. Meller, Chem. Ber., 119 (1986) 3121.
155 R.W. Rudolph and W.R. Pretzer, Inorg. Chem., 11 (1972) 1974.
156 L.J. Guggenberger, J. Organomet. Chem., 81 (1974) 271; F. Klanberg, E.L. Muetterties, and L.J. Guggenberger, Inorg. Chem., 7 (1968) 2272.
157 T.K. Hilty and R.W. Rudolph, Inorg. Chem., 18 (1979) 1106.
158 W.R. Pretzer, T.K. Hilty, and R.W. Rudolph, Inorg. Chem., 14 (1975) 2459; W.R. Pretzer and R.W. Rudolph, J. Chem. Soc. Chem. Commun., (1974) 629.
159 D.E. Sands and A. Zalkin, Acta Cryst., 15 (1962) 410.
160 W.R. Herter, F. Klanberg, and E.L. Muetterties, Inorg. Chem., 6 (1967) 1696.
161 R.W. Rudolph and W.R. Pretzer, Inorg. Synth., 22 (1983) 226.
162 B. Stibr, K. Base, J. Plesek, S. Harmanek, J. Dolansky, and Z. Janousek, Pure Appl. Chem., 49 (1977) 803; V.A. Brattsev, J.P. Knyazev, G.N. Danilova, and V.I. Stanko, Zh. Obshch. Khim., 45 (1975) 1393.
163 K. Base, Collect. Czech. Chem. Commun., 48 (1983) 2593.
164 K. Base, S. Hermanek, and F. Hanousek, J. Chem. Soc. Chem. Commun., (1984) 299.
165 J. Plesek and S. Hermanek, J. Chem. Soc. Chem. Commun., (1975) 127.
166 W.R. Pretzer and R.W. Rudolph, Inorg. Chem., 15 (1976) 1779.
167 W.R. Pretzer and R.W. Rudolph, J. Am. Chem. Soc., 98 (1976) 1441.
168 A.R. Siedle, D. McDowell, and L.J. Todd, Inorg. Chem., 13 (1974) 2735.
169 W.L. Smith, B.J. Meneghelli, N. McClure, and R.W. Rudolph, J. Am. Chem. Soc., 98 (1976) 624.
170 W.L. Smith, B.J. Meneghelli, D.A. Thompson, P. Klymko, N. McClure, M. Bower, and R.W. Rudolph, Inorg. Chem., 16 (1977) 3008.
171 B.J. Meneghelli and R.W. Rudolph, J. Organomet. Chem., 133 (1977) 139.
172 T.P. Fehlner, M. Wu, B.J. Meneghelli, and R.W. Rudolph, Inorg. Chem., 19 (1980) 49.
173 B.J. Meneghelli, M. Bower, N. Canter, and R.W. Rudolph, J. Am. Chem. Soc., 102 (1980) 4355.
174 B.J. Meneghelli and R.W. Rudolph, J. Am. Chem. Soc., 100 (1978) 4626.
175 N. Canter, G.G. Overberger, and R.W. Rudolph, Organometallics, 2 (1983) 569.

176 D.A. Thompson and R.W. Rudolph, J. Chem. Soc. Chem. Commun., (1976) 770; G.J. Zimmerman and L.G. Sneddon, J. Am. Chem. Soc., 103 (1981) 1102.
177 T.K. Hilty, D.A. Thompson, W.M. Butler, and R.W. Rudolph, Inorg. Chem., 18 (1979) 2642.
178 A.R. Kane, L.J. Guggenberger, and E.L. Muetterties, J. Am. Chem. Soc., 92 (1970) 2571.
179 A.R. Siedle and L.J. Todd, J. Chem. Soc. Chem. Commun., (1973) 914.
180 B. Krebs and H.-U Hürter, Acta Cryst., A37 (1981) C163.
181 A.F. Wells, Structural Inorganic Chemistry, Clarendon, Oxford, 5th ed., 1984.
182 M.G. Barker, P.A. Gadd, and M.J. Begley, J. Chem. Soc. Chem. Commun., (1981) 379.
183 P. Mondal and J.W. Jeffrey, Acta Cryst., B31 (1975) 689.
184 See refs. 12 and 181 and citations therein.
185 G.S. Smith and J.L. Hoard, J. Am. Chem. Soc., 81 (1959) 3907.
186 E.A. Jeffrey and T. Mole, Austral. J. Chem., 21 (1968) 2683; also E.G. Hoffmann, Ann., 629 (1960) 104.
187 D.A. Drew, A. Haaland, and J. Weidlein, Z. Anorg. Allg. Chem., 398 (1973) 241.
188 H. Schmidbaur and F. Schindler, Chem. Ber., 99 (1966) 2178; E.G. Hoffmann and W. Tormau, Angew. Chem., 73 (1961) 578.
189 $[Me_2AlOMe]_{2,3}$ can also be obtained by the oxidation of Al_2Me_6 by O_2. See Y. Sakurada, M.L. Huggins, and W.R. Anderson, J. Phys. Chem., 68 (1964) 1934.
190 N.R. Davidson and H.C. Brown, J. Am. Chem. Soc., 64 (1942) 316.
191 M.E. Kenney and A.W. Laubengayer, J. Am. Chem. Soc., 76 (1954) 4839.
192 H.-D. Hausen, J. Organomet. Chem., 39 (1972) C37.
193 J. Weidlein, J. Organomet. Chem., 24 (1970) 63.
194 H. Schmidbaur and M. Schmidt, Chem. Ber., 94 (1961) 1349.
195 M. Bonamico and G. Dessy, J. Chem. Soc. A, (1967) 1786.
196 K. Dehnicke, Angew. Chem., 79 (1967) 942; Angew. Chem. Int. Ed. Engl., 6 (1967) 947.
197 B. Pachaly and R. West, J. Am. Chem. Soc., 107 (1985) 2987.

BORON HYDRIDE CLUSTERS

N.N. GREENWOOD
School of Chemistry, University of Leeds LS2 9JT, England

1. INTRODUCTION

The spectacular growth of metallaborane and metallacarbaborane cluster chemistry during the past 25 years has sometimes tended to obscure the equally impressive parallel developments in the synthesis and structural characterization of the binary boranes themselves. When Alfred Stock summarized his classic work in this area some 55 years ago (ref. 1) only six boron hydrides were established with any certainty; the structure of none of them was known and no binary borane anions had yet been isolated (or at least recognised as such). Today, nearly one hundred neutral or anionic binary boranes have been characterized, in addition to innumerable fugitive species which have been identified spectroscopically or mass spectrometrically, or postulated as short-lived reaction intermediates. The isolated binary borane species can be classified into five main groups according to their stoichiometry and structure:

closo-borane dianions $B_nH_n^{2-}$;
nido-boranes B_nH_{n+4} and the related anions $B_nH_{n+3}^{-}$ and $B_nH_{n+2}^{2-}$;
arachno-boranes B_nH_{n+6} and related anions $B_nH_{n+5}^{-}$ and $B_nH_{n+4}^{2-}$;
hypho-boranes B_nH_{n+8}, and the related anions $B_nH_{n+7}^{-}$ (both rare);
conjuncto-boranes $B_nH_m^{c-}$ (c = 0 - 5), comprising two or more linked clusters.

In recent years extensive reviews have appeared on the preparation, structure, bonding, and chemical reactions of metallaboranes (refs. 2-8), carbaboranes (refs. 8-11), and metallacarbaboranes (refs. 6,8,9,12). Accordingly, the present review concentrates almost exclusively on the homoatomic cluster chemistry of the binary boranes and their anions, only alluding to the various heteroatom clusters where this illuminates the behaviour of the parent compounds themselves. Extensive documentation is available in Gmelin's Handbook (ref. 13). The known boron hydrides are listed in Table 1, and the anions to be considered are in Table 2. Earlier reviews are in refs. 1,14,25.

TABLE 1. Neutral Binary Boranes*

nido-	*arachno-*	*other*	*conjuncto-*
		$\{BH_3\}$, $BH_4^{\cdot}$	
		B_2H_6	
$\{B_3H_7\}$	$\{B_3H_9\}$		
$\{B_4H_8\}$	B_4H_{10}		
B_5H_9	B_5H_{11}		
B_6H_{10}	B_6H_{12}	B_6H_{14}	
			B_7H_{13} i.e [μ-(2-B_5H_8)B_2H_5]
B_8H_{12}	B_8H_{14}	B_8H_{16}	B_8H_{18} i.e. [1,1'-$(B_4H_9)_2$] and [2,2'-$(B_4H_9)_2$]
$\{B_9H_{13}\}$	*n*-B_9H_{15}		B_9H_{17} i.e. [1,2'-$(B_4H_9)(B_5H_8)$]
	i-B_9H_{15}		
$B_{10}H_{14}$			$B_{10}H_{16}$ i.e. [1,1'-, 1,2'-, 2,2'-$(B_5H_8)_2$]
			$B_{10}H_{18}$?[$(B_5H_8)(B_5H_{10})$]
$B_{11}H_{15}$			
			$B_{12}H_{16}$
$B_{13}H_{19}$, $\{B_{13}H_{21}\}$ i.e. $\{(B_6H_{10})(B_7H_{11})\}$			
$B_{14}H_{18}$, $B_{14}H_{20}$, $B_{14}H_{22}$			
$B_{15}H_{23}$			
$B_{16}H_{20}$			
syn-$B_{18}H_{22}$, *anti*-$B_{18}H_{22}$			
$B_{20}H_{16}$			
$B_{20}H_{26}$ $\geq$ 9 isomers: [1,2'-, 1,5'-, 2.2'-, 2,5'-, 2,6', 5,5'-, 6,6'-$(B_{10}H_{13})_2$]etc			
$B_{30}H_{38}$ i.e. [$(B_{10}H_{13})(B_{10}H_{12})(B_{10}H_{13})$]			

*Brackets {} indicate a fugitive but characterized reaction intermediate; some compounds are only isolable at temperatures below 0 °C

Throughout the chapter, where appropriate, reference will be made to the *styx* topological description of the bonding in individual boranes, where *s* is the number of B-H-B three-centre bonds, *t* is the number of BBB three-centre bonds, *y* the number of two-centre B-B bonds and *x* the number of BH_2 groups in the structure (ref. 14)

2. PREPARATIVE REACTIONS

The classic route to the higher boranes has been the controlled vapour-phase thermolysis of diborane under a variety of conditions of temperature, pressure, and time; the effect of weak Lewis bases as catalysts (e.g. Me_2O), and the efficacy of hot/cold tube reactors for particular boranes has also been

TABLE 2. Anionic Binary Boranes

closo-dianions	radical anions	"*nido*-" anions	*arachno*- and other anions and dianions
	$BH_3^{\cdot-}$	BH_4^-	
	$B_2H_6^{\cdot-}$		$B_2H_7^-$
	$B_3H_7^{\cdot-}$		$B_3H_8^-$
		$B_4H_7^-$	$B_4H_9^-$
		$B_5H_8^-$	$B_5H_{10}^-$, $B_5H_9^{2-}$, *hypho*-$B_5H_{12}^-$
$B_6H_6^{2-}$		$B_6H_9^-$	$B_6H_{11}^-$, *closo*-$B_6H_7^-$
$B_7H_7^{2-}$			$B_7H_{12}^-$
$B_8H_8^{2-}$	$B_8H_8^{\cdot-}$		
$B_9H_9^{2-}$		$B_9H_{12}^-$	$B_9H_{14}^-$, $B_9H_{13}^{2-}$
$B_{10}H_{10}^{2-}$		$B_{10}H_{13}^-$	$B_{10}H_{15}^-$ (three), $B_{10}H_{14}^{2-}$, *nido*-$B_{10}H_{12}^{2-}$
$B_{11}H_{11}^{2-}$		$B_{11}H_{14}^-$	*nido*-$B_{11}H_{13}^{2-}$
$B_{12}H_{12}^{2-}$			*conjuncto*-$B_{12}H_{15}^-$

No B_{14} - B_{17} anions yet characterized though $B_{14}H_{21}^-$ has been prepared

conjuncto-anions:

$B_{13}H_{18}^-$, *syn*-$B_{18}H_{21}^-$, *anti*-$B_{18}H_{21}^-$, *syn*-$B_{18}H_{20}^{2-}$, *anti*-$B_{18}H_{20}^{2-}$

n-$B_{20}H_{18}^{2-}$, *photo*-$B_{20}H_{18}^{2-}$, two isomeric $B_{20}H_{19}^{3-}$

three $B_{20}H_{18}^{4-}$[1,1'-, 1,2'-, 2,2'-]

$B_{24}H_{23}^{3-}$, $B_{24}H_{22}^{4-}$, $B_{48}H_{45}^{5-}$

investigated (refs. 15,18,22b,23,24). More recently protonation of $B_3H_8^-$ salts by HCl or syrupy phosphoric acid has been developed as a laboratory-scale route to B_4H_{10} (and B_5H_{11}); this eliminates the need to handle large quantities of gaseous B_2H_6 at high temperatures and pressures but still requires a tedious fractionation to obtain pure products (refs. 26-28).

A major advance has been the development by S.G. Shore and his group of a new systematic synthesis of boranes: this uses the Lewis acids BX_3 (X = F, Cl, Br) to abstract H^- from the more readily available borane anions such as BH_4^-, $B_3H_8^-$, $B_4H_9^-$, and $B_9H_{14}^-$ (refs. 29,30). This method gives high yields of B_2H_6, B_4H_{10}, and B_5H_{11} without the need for extensive fractionations, and permits the ready conversion of B_5H_9 to $B_{10}H_{14}$ via $B_9H_{14}^-$. In principle, it can be thought of as a technique for generating highly reactive neutral borane intermediates which then dimerize or disproportionate to give the desired product, e.g.

$BX_3 + BH_4^- \longrightarrow HBX_3^- + \{BH_3\} \longrightarrow \frac{1}{2}B_2H_6$

$BX_3 + B_3H_8^- \longrightarrow HBX_3^- + \{B_3H_7\} \longrightarrow \frac{1}{2}B_4H_{10} + \frac{1}{2}$"$B_2H_4$" polymer

$BX_3 + B_4H_9^- \longrightarrow HBX_3^- + \{B_4H_8\} \longrightarrow \frac{1}{2}B_5H_{11} + \frac{1}{2}$"$B_3H_5$" polymer

$BX_3 + B_9H_{14}^- \longrightarrow HBX_3^- + \{B_9H_{13}\} \longrightarrow \frac{1}{2}B_{10}H_{14} + \frac{1}{2}$"$B_8H_{10}$" polymer $+ \frac{1}{2}H_2$

A further elegant though less general route devised by Shore (ref. 31) converts the stable *nido*-B_5H_9 to the less stable *arachno*-B_5H_{11} via a two-electron reduction using alkali metal naphthalide followed by protonation of the dianion so formed:

$$B_5H_9 + 2M/C_{10}H_8 \xrightarrow{thf} [M^+]_2[B_5H_9]^{2-} \xrightarrow{2HX} B_5H_{11} + 2MX$$

The perception by R. Schaeffer (ref. 32) that *nido*-B_6H_{10} and its anion $B_6H_9^-$ could act as Lewis bases towards reactive (vacant orbital) borane radicals has led to several new *conjuncto*-boranes, e.g. (refs. 32-36):

$$B_6H_{10} + \tfrac{1}{2}B_2H_6 \xrightarrow{-H_2} \{B_7H_{11}\} \xrightarrow{B_6H_{10}} B_{13}H_{19} + H_2$$

$$B_6H_{10} + B_8H_{12} \longrightarrow B_{14}H_{22}$$

$$B_6H_9^- + Et_2O.B_8H_{12} \longrightarrow Et_2O + B_{14}H_{21}^- \xrightarrow{HCl} B_{14}H_{20} + H_2 + Cl^-$$

$$B_6H_{10} + iso\text{-}B_9H_{15} \longrightarrow B_{15}H_{23} + H_2$$

A useful route to B-B bonded *conjuncto*-boranes involves the photolysis of parent *nido*-boranes, e.g. (refs. 37-42):

$2B_5H_9 \xrightarrow{(Hg)/h\nu} B_{10}H_{16}$ [i.e. 1,1'-, 1,2'-, and 2,2'-$(B_5H_8)_2$]

$2B_{10}H_{14} \xrightarrow{h\nu} B_{20}H_{26}$ [i.e. 1,2'- and 2,2'-$(B_{10}H_{13})_2$]

High-yield catalytic routes to specific B-B coupled *conjuncto*-boranes have also been developed by L.G. Sneddon and his group, e.g. (refs. 43,44):

$2B_5H_9 \xrightarrow{PtBr_2}$ 1,2'-$(B_5H_8)_2$, [i.e. $B_{10}H_{16}$]

$2B_4H_{10} \xrightarrow{PtBr_2}$ 1,1'-$(B_4H_9)_2$, [i.e. B_8H_{18}]

$B_4H_{10} + B_5H_9 \xrightarrow{PtBr_2}$ 1,2'-$(B_4H_9)(B_5H_8)$, [i.e. B_9H_{17}]

When applied to a mixture of B_5H_9 and B_2H_6 in decane at room temperature the method led to the first well authenticated neutral heptaborane (ref. 45):

$B_2H_6 + B_5H_9 \xrightarrow{PtBr_2}$ 2:1',2'-$(B_5H_8)(B_2H_5)$

The method of metal-induced oxidative fusion of clusters which R.N. Grimes and his group devised and used with such success for carbaboranes has recently been extended to the parent binary boranes. For example, the novel edge-fused *conjuncto*-dodecaborane, $B_{12}H_{16}$ was prepared in good yield by reacting KB_6H_9 with $FeCl_2/FeCl_3$ in dimethyl ether at -78 °C (refs. 46,47):

$$4B_6H_9^- + 4Fe^{3+} \longrightarrow B_{12}H_{16} + 2B_6H_{10} + 4Fe^{2+}$$

Similarly, $B_5H_8^-$ when treated with $FeCl_2/FeCl_3$ in thf gave a mixture of $B_{10}H_{14}$ and 2,2'-$(B_5H_8)_2$, whereas use of $RuCl_3$ in thf yielded $B_{10}H_{14}$ as the only coupled product (ref. 48):

$$4B_5H_8^- + 4Ru^{3+} \longrightarrow B_{10}H_{14} + 2B_5H_9 + 4Ru^{2+}$$

Likewise, $B_{10}H_{13}^-$ on treatment with $RuCl_3$ in thf and subsequent exposure to air gave modest yields of a mixture of *syn*- and *anti*-$B_{18}H_{22}$ (ref. 48). A cleaner route to the single *anti*-isomer involves the coupling of two $B_9H_{12}^-$ anions in the presence of $[Os(CO)_3Cl_2]_2$ in refluxing CH_2Cl_2, followed by protonation of the $B_{18}H_{21}^-$ anion so formed (ref. 49). Even higher yields (68%) of the *anti*-isomer are obtained using $HgBr_2$ in CH_2Cl_2 to effect the oxidative fusion (ref. 50):

$$4B_9H_{12}^- + 3HgBr_2 \longrightarrow 2B_{18}H_{22} + Hg + Hg_2Br_2 + 2H_2 + 4Br^-$$

The synthesis of *closo*-borane dianions, $B_nH_n^{2-}$, still relies principally on methods which were first developed over 25 years ago; for example

(a) the thermolysis of boranes in the presence of either the borohydride ion, BH_4^-, or amine-borane adducts such as Et_3NBH_3 (refs. 19,24,51-53):

$$2BH_4^- + B_xH_y \xrightarrow{180\ °C} [(2 + x)/n]B_nH_n^{2-} + \tfrac{1}{2}(y - x + 6)H_2$$

$$2Et_3NBH_3 + B_{10}H_{14} \xrightarrow{90\ °C} (Et_3NH)_2B_{12}H_{12} + 3H_2$$

A more recent variant is the thermolysis of Et_4NBH_4 in mixtures of decane-dodecane under reflux at 175-190 °C (ref. 54): this yields a mixture of $B_{10}H_{10}^{2-}$, $B_9H_9^{2-}$, $B_{12}H_{12}^{2-}$, and $B_{11}H_{14}^-$ as the only products after 12 hours.

(b) Proton transfer reactions involving *bis*-ligand adducts of decaborane (refs. 55-57):

$$2Et_3N + B_{10}H_{14} \xrightarrow{-H_2} \{(Et_3N)_2B_{10}H_{12}\} \longrightarrow (Et_3NH)_2B_{10}H_{10}$$

Such reactions are rapid, convenient, and of high yield.

(c) Oxidative (air) degradation of hydrated $B_9H_9^{2-}$ salts in EtOH, thf, or 1,2-dimethoxyethane gives low yields of the smaller anions $B_8H_8^{2-}$, $B_7H_7^{2-}$, and $B_6H_6^{2-}$ by stepwise cluster contraction (ref. 58).

Much of the recent work on the synthesis of *closo*-borane dianions has been directed towards improving the yields of specific products by varying the detailed conditions employed. For example, $B_6H_6^{2-}$ can now be reliably and conveniently prepared in 10% yield from $NaBH_4$ by the following sequence of reactions (ref. 59):

$$3NaBH_4 + I_2 \xrightarrow[\text{diglyme}]{100\ °C} NaB_3H_8 + 2NaI + 2H_2$$

$$2NaB_3H_8 \xrightarrow[5\ h]{162\ °C} Na_2B_6H_6 + 5H_2$$

To separate the product from unreacted $NaBH_4$ and NaB_3H_8 plus unwanted $Na_2B_{12}H_{12}$, the mixture is cooled to 120 °C (rather than 25 °C) and hot-filtered, the product being finally isolated by crystallization of $Cs_2B_6H_6$ from cold water.

Specific syntheses of other boranes will be mentioned at appropriate points in the following sections in which the structure and properties of individual boranes and borane anions are discussed. The sequence of presentation will be in order of increasing number of boron atoms and increasing hydrogen content. Although monoboron and diboron species are not strictly clusters, a brief account of them is included for the sake of completeness.

3. MONOBORANES

The species to be considered are the fugitive monomeric borane(3) itself, $\{BH_3\}$, the borane(4) radical, $BH_4^{\bullet}$, the trihydroborate(1-) radical anion, $BH_3^{\bullet -}$, and the very stable tetrahydroborate(1-) ion, BH_4^-.

There is now good evidence for the existence of monoborane(3), $\{BH_3\}$ (refs. 22c,60): it is formed during the thermolysis of diborane(6) and by thermal dissociation of $BH_3.CO$ and $BH_3.PF_3$, this last being at present the best source. The infrared spectrum of matrix isolated $\{BH_3\}$ is reasonably interpreted in terms of the expected planar D_{3h} structure (**1**) (ref. 61), and very recently the out-of-plane bending mode, ν_2, has been observed directly in the vapour phase at 1140.88cm^{-1} (ref. 62). The steady-state concentrations of $\{BH_3\}$ depend on the precise conditions of temperature and pressure, but are always low because of its ready propensity to dimerize to B_2H_6 or to react with B_2H_6 to form $\{B_3H_9\}$ (See Sections 4 and 5). The most recent values computed for the B-H interatomic distance in $\{BH_3\}$ are 118.4 pm (ref. 63) and 119.1 pm (ref. 64).

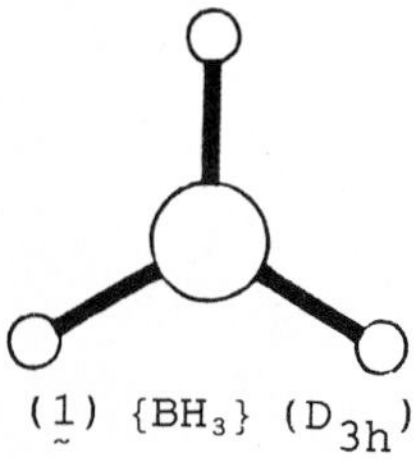

(**1**) $\{BH_3\}$ (D_{3h})

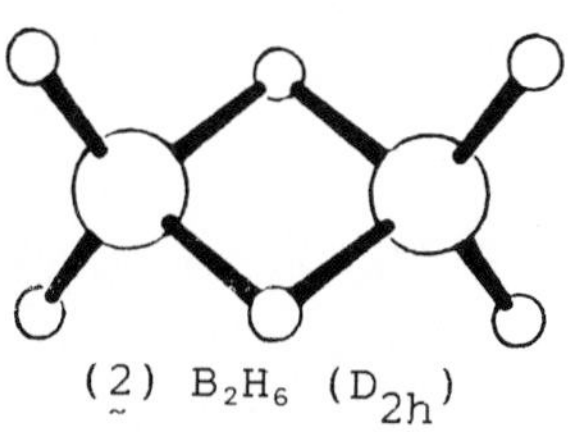

(**2**) B_2H_6 (D_{2h})

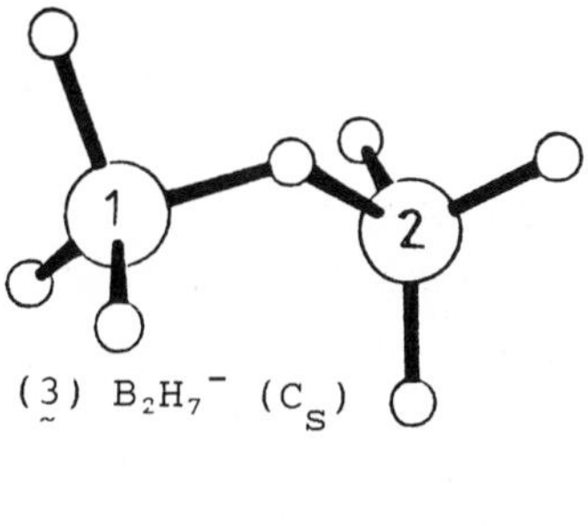

(**3**) $B_2H_7^-$ (C_s)

The borane(3) radical anion, $BH_3^{\cdot-}$, can be generated by photolysis of $Bu^n{}_4NBH_4$ in liquid ammonia solution at 230-290 K (ref. 65):

$$BH_4^-(amm) \xrightarrow[-e^-(amm)]{h\nu} \{BH_4^{\cdot}\} \xrightarrow[-NH_4^+]{+NH_3} BH_3^{\cdot-}(amm)$$

In the mixed solvent ammonia/1,2-dimethoxyethane the half-life of $BH_3^{\cdot-}$ as determined by esr is about 40 ms at 273 K, and this increases to about 108 ms at 250 K. Photolysis of lithium, sodium, or potassium borohydride in liquid ammonia did not give rise to the esr signal for $BH_3^{\cdot-}$, suggesting that the radical interacts with the alkali metal cations under these conditions (ref. 65).

The borane(4) radical, $BH_4^{\cdot}$, has been made by ^{60}Co γ-radiolysis of powdered $NaBH_4$ at 77K (ref. 66). The radical has been shown by esr spectroscopy and high-level MO calculations to have a C_{2v} structure (ref. 66) and the most recent computations indicate two 'short' B-H bonds of 117.7 pm subtending an angle of 129.2° and two longer B-H bonds of 127.2 pm subtending an acute angle of 47.7° (ref. 67). The C_{2v} structure, which is stable at 77K, is 54 kJ mol^{-1} below its dissociation products $\{BH_3\}$ and $H^{\cdot}$ into which it should decompose above 250K (ref. 67). It is interesting to note that the two B-H interatomic distances are close to those in $\{BH_3\}$ (*ca.* 119 pm) and BH_4^- (126 pm).

Metal tetrahydroborates, $M(BH_4)_n$, have been extensively studied (ref. 68) since $LiBH_4$ (ref. 69), $Be(BH_4)_2$ (ref. 70), and $Al(BH_4)_3$ (ref. 71) were discovered nearly 50 years ago. The tetrahedral BH_4^- anion (interatomic distance B-H = 126.0 pm) is isoelectronic with CH_4 and NH_4^+ but is much more reactive and is a versatile reducing agent (refs. 68,72). Though essentially ionic and uncoordinated in the alkali metal tetrahydroborates, BH_4^- is unusual in being able to act as a mono-hapto, di-hapto, or tri-hapto ligand, e.g. in $[Cu(\eta^1\text{-}BH_4)(PMePh_2)_3]$, $[Cu(\eta^2\text{-}BH_4)(PPh_3)_2]$, $(Al(\eta^2\text{-}BH_4)_3]$, and $[Zr(\eta^3\text{-}BH_4)_4]$ (ref. 73). A fuller treatment lies outside the scope of the present chapter.

4. DIBORANES

The species to be considered are B_2H_4(?), B_2H_6, $B_2H_6^{\cdot-}$ and $B_2H_7^-$.

A polymer purporting to be $(B_2H_4)_x$ has been described but it is not well characterized (ref. 74). The hypothetical monomer, B_2H_4, has been the subject of several theoretical studies (refs. 75,76): the most stable topology, (0012), has two BH_2 groups joined by a single B-B bond, thereby implying a vacant p_π orbital on each boron atom. The staggered (D_{2d}) conformer is more stable than the eclipsed (D_{2h}) conformer by some 50 kJ mol^{-1} and the most recent study (ref. 76) leads to a B-B distance of 166.9 pm, B-H 119.5 pm, and angle H-B-H 116.6°. The B-B bond is stable to breakage into two BH_2 units by some 420 kJ mol^{-1}, suggesting that B_2H_4 might be preparable by a suitable synthetic route, though the possibility of dimerization to $\{B_4H_8\}$, or other reactions, might limit its isolability.

Diborane(6), B_2H_6, is one of the most studied and most reviewed of all chemical compounds (see, for example, refs. 1,13-24,72,77), and the elucidation of its structure, bonding and reactivity have had a seminal influence on the development of inorganic chemistry during the past four decades. It is a gas which condenses to a colourless liquid and solid, mp -164.9 °C, bp -92.6 °C.

Diborane(6) has a bridged dimeric structure (2) of D_{2h} symmetry and (2002) topology. The B-H(terminal) distance of 118.4 pm (ref. 78) is very similar to that calculated for monomeric $\{BH_3\}$ whereas B-H_μ (131.4 pm) is some 11% longer. The B...B distance is 174.3 pm, close to that in crystalline boron itself. The bridge angle B-H_μ-B is 83.1° and the terminal H-B-H angle is 121.5° (ref. 78). The enthalpy of dissociation of B_2H_6, which is equal but opposite to the energy of dimerization of $\{BH_3\}$, is a difficult quantity to determine but the most reliable experimental values are in the range 145-165 kJ mol^{-1} (refs. 22c,79). Recent high-level computations give 165 kJ mol^{-1} (ref. 80) and 157 kJ mol^{-1} (ref. 63).

A simple high-yield synthesis of pure B_2H_6 is now available via the hydride-ion abstraction route mentioned in Section 2 (refs. 29,30):

$$MBH_4 + BX_3 \xrightarrow[\text{r.t./3h}]{CH_2Cl_2} \tfrac{1}{2}B_2H_6 + M[HBX_3]$$

Yields of up to 98% are obtained when M is $[NBu^n{}_4]$ or $[PMePh_3]$ and X is Cl or Br. This supplements the numerous other methods already available (refs. 18,23,72).

The thermal decomposition of B_2H_6 in the gas phase was first studied kinetically in 1951 (ref. 81). The reaction is homogeneous and has an order of $^3/_2$, at least in the initial stages, implying that the rate-determining step involves a triboron species:

$$B_2H_6 \rightleftharpoons 2\{BH_3\}$$
$$\{BH_3\} + B_2H_6 \longrightarrow \{B_3H_9\}$$
$$\{B_3H_9\} \longrightarrow \{B_3H_7\} + H_2$$

Cogent arguments have been adduced to suggest that the last of these three reactions is the rate-controlling one (ref. 82) but a recent computational study of the reaction profile at the highest level (ref. 83) suggests that it is the *formation* of $\{B_3H_9\}$ rather than its decomposition which is the slow, rate-determining step. It is interesting to note that experimental values for the activation energy of thermal decomposition of B_2H_6 fall in the range 90-115 kJ mol^{-1} (ref. 84), which is substantially smaller than the enthalpy of dissociation of diborane, namely, *ca.* 160 kJ mol^{-1} (see preceding paragraph). The first isolable compound in the thermolysis is B_4H_{10} (ref. 85) and this is followed by B_5H_{11} and the higher boranes (ref. 84):

$$\{B_3H_7\} + B_2H_6 \longrightarrow B_4H_{10} + \{BH_3\}$$
$$\{B_3H_7\} + \{BH_3\} \longrightarrow B_4H_{10} \text{ etc.}$$

Moving next to anionic diboron species, the radical anion $B_2H_6^{\cdot-}$ has been identified by its esr spectrum in fluid ethereal solution (ref. 86):

$$(Me_3Si)_2N^{\cdot} + B_2H_7^- \xrightarrow[196\text{-}370\ K]{thf} B_2H_6^{\cdot-} + (Me_3Si)_2NH$$

The bis(trimethylsilyl)amino radical was generated by continuous ultraviolet irradiation of the dimer and the heptahydrodiborate(1-) anion (see below) was present as its tetra-n-butylammonium salt; mixed solvents involving tetrahydrofuran, dimethyl ether, and/or 1,2-dimethoxyethane were also successfully used. The spectrum of $B_2H_6^{\cdot-}$ was interpreted in terms of an ethane-like structure with a one-electron B-B σ bond, and semi-empirical intermediate-level calculations on the D_{3d} structure gave r(B-B) 188 pm, r(B-H) 117 pm, and angle HBB 102.9°. Preliminary results suggest that $B_2H_6^{\cdot-}$ may be less reactive chemically than $BH_3^{\cdot-}$.

The $B_2H_7^-$ anion has been known for many years (ref. 87) but only recently has a low-temperature X-ray crystal structure (ref. 88) and an even more precise neutron diffraction study at 80 K (ref. 89) become available. Clear, colourless crystals of the solvate $[N(PPh_3)_2]^+[B_2H_7]^-.CH_2Cl_2$ were grown at room temperature by slow diffusion of Et_2O into a CH_2Cl_2 solution of the salt, which was itself prepared by direct reaction of BH_4^- with the stoichiometric amount of diborane:

$$[N(PPh_3)_2]^+[BH_4]^- + \tfrac{1}{2}B_2H_6 \xrightarrow[-78°C]{CH_2Cl_2} [N(PPh_3)_2]^+[B_2H_7]^-$$

As can be seen from structure (3) the $B_2H_7^-$ anion has C_s symmetry with a bent B-H-B bond (127°) and staggered terminal B-H bonds. [B(1)-H_t 116(2) pm, B(2)-H_t 120(2) pm]. The slight asymmetry of the B-H-B bridge is on the borderline of being significant [B(1)-H_μ 132(2) pm, B(2)-H_μ 121(2) pm] and may suggest a donor-acceptor interaction between the two halves of the anion i.e. $[H_3B\text{-}H]^- \rightarrow BH_3$. The B...B distance is 227 pm, significantly longer than in B_2H_6. The bent-bridge structure is reproduced in the most recent Hartree-Fock calculations (refs. 90,91); the calculated angle is 126.4° (ref. 91) but the energy advantage of the bent over the linear structure is thought to be less than 4 kJ mol^{-1}. The calculated binding energy of the bridge bond $[H_3BH]^- \rightarrow BH_3$ is 148 kJ mol^{-1} (ref. 91) in good agreement with the experimental value of 130 ± 33 kJ mol^{-1} (ref. 92).

5. TRIBORANES

The fugitive species $\{B_3H_7\}$ and $\{B_3H_9\}$, which were briefly alluded to in the preceding section, will be considered in more detail as will the radical anion $B_3H_7^{\overline{\cdot}}$ and the stable octahydrotriborate(1-) anion $B_3H_8^-$.

The great reactivity of $\{B_3H_7\}$ and $\{B_3H_9\}$ precludes their isolation, or indeed their detection by spectroscopic techniques. However, they are amenable to theoretical studies, and the great success of high-level computations in reproducing the known geometrical structures, dimensions, and energies of more stable species, gives confidence that similar methods will yield reliable data on the non-isolable species. The global minimum energy structure of $\{B_3H_7\}$ has C_s symmetry and (2102) topology as shown in (4), though the C_{2v} (staggered) structure (5) with (1103) topology is calculated to lie only 18 kJ mol^{-1} above this (ref. 93). A notable feature of the C_s structure (4) is the asymmetric B-H_μ-B bridge bond, the $H_2B\text{-}H_\mu$ distance being some 12% longer than the HB-H_μ distance. This is now found to be a common feature of boranes which have these structural groups. The less stable C_{2v} structure (5) has an essentially vacant orbital on the unique boron atom and is thought to be the one that is involved in the thermolysis of diborane (ref. 83).

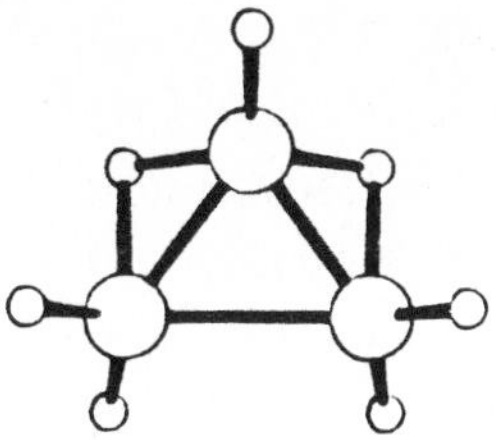

(4) $\{B_3H_7\}$ (C_s)

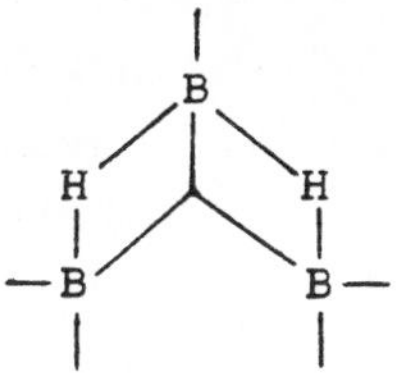

(2102)

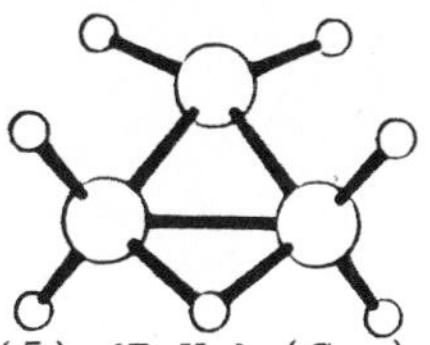

(5) $\{B_3H_7\}$ (C_{2v})

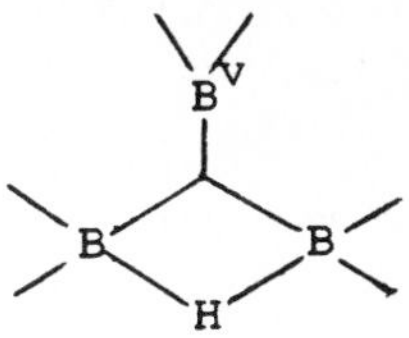

(1103)

Similar calculations on $\{B_3H_9\}$ lead to the D_{3h} (3003) structure (6) (ref. 93). The stabilization per BH_3 unit decreases in the sequence B_2H_6 > $\{B_3H_9\}$ > $\{B_4H_{12}\}$ (D_{4h}, hypothetical species) from 74.5 to 43.5 and 22 kJ mol^{-1}; in the same sequence the B...B distance increases from 175 to 235 and 260 pm, the (symmetrical) $B\text{-}H_\mu$ distance remains essentially constant (130.3, 130.5, 130.6 pm respectively) and the angle $B\text{-}H_\mu\text{-}B$ progressively increases from acute to almost linear (84°, 129°, 172° respectively).

The triborane(7) radical anion $B_3H_7^{\bullet-}$ has been generated in solution by H atom abstraction from $B_3H_8^-$ (ref. 94):

$B_3H_8^- + Bu^tO^\bullet \longrightarrow Bu^tOH + B_3H_7^{\bullet-}$

The t-butoxide radical was generated by ultraviolet photolysis of di-t-butyl peroxide in the presence of $[NBu^n_4][B_3H_8]$ in mixed solvents such as $MeOCH_2CH_2OMe/Me_2O/MeOH$ at *ca*. 250 K. The cyclic C_{2v} (eclipsed) structure (7) was deduced from semi-empirical MO calculations which also furnished the following dimensions: $B\text{-}H_t$ 117 pm, $B\text{-}H_\mu$ 136 pm, B-B(unbridged) 175 pm, B-B(bridged) 181 pm. The $B_3H_7^{\bullet-}$ radical anion appears to be appreciably less reactive than $BH_3^{\bullet-}$ as might be expected for a delocalized electron species of higher ionization energy.

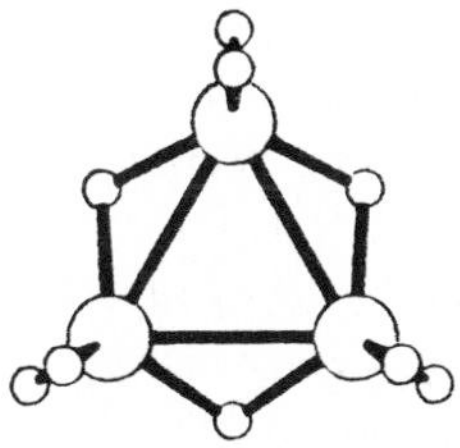

(6) $\{B_3H_9\}$ (D_{3h})

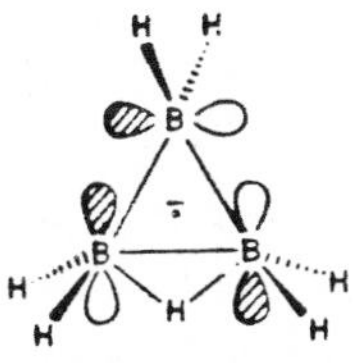

(7) $B_3H_7^{\bullet-}$ (C_{2v})

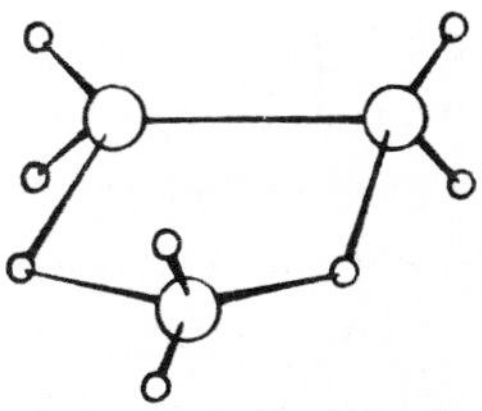

(8) $B_3H_8^-$ (C_{2v})

The *arachno* anion $B_3H_8^-$ is the only binary triboron species that is stable at room temperature and above. It was first made and identified as such in 1957 by the reaction of sodium amalgam with diborane in simple ethers (ref. 95), and was the first polyhedral borane anion to be synthesized:

$2Na + 2B_2H_6 \longrightarrow NaBH_4 + NaB_3H_8$ (80% yield)

Since then numerous synthetic routes have been devised and salts with many different counter cations prepared [ref. 13(d) Part 2]. Convenient high-yield routes which avoid the use of free B_2H_6 include the reaction of $NaBH_4$ and I_2 in diglyme at 100 °C and the reaction between $NaBH_4$ and Et_2OBF_3 in diglyme at 100 °C. Many of the salts are sufficiently robust to withstand being heated to 200-300 °C.

An early X-ray crystal structure determination on $[BH_2(NH_3)_2][B_3H_8]$ at -100 °C (ref. 96) revealed an isosceles triangle of boron atoms structure (**8**) with one B-B distance of ∿180 pm and two BHB distances of ∿177 pm. In addition, $B\text{-}H_t$ distances were in the range 105-120 pm and $B\text{-}H_\mu$ 120-150 pm. More precise dimensions are available from high-level *ab initio* calculations (ref. 93): B-B 192.2 and 181.0 pm, $B\text{-}H_\mu$ 125.5 and 157.6 pm (the very asymmetric $B\text{-}H_\mu\text{-}B$ bond is notable, the longer distance being to the unique boron atom.

The proton and boron-11 nmr spectra of the $B_3H_8^-$ anion have excited considerable attention since their unusual multiplet structure was first observed some thirty years ago. Full documentation of the early work is in ref. 20. The ion is completely fluxional so that, with sufficient resolution, the proton signal is a decet resulting from the splitting of a single proton peak by coupling to three equivalent ^{11}B nuclei each of spin $^3/_2$ [i.e. $\{2 \times (3 \times 1.5) + 1\} = 10$], and the boron-11 signal is a nonet resulting from the splitting of a single boron peak by coupling to eight equivalent protons [i.e. $\{2 \times (8 \times \frac{1}{2}) + 1 = 9$] (ref. 97). The fluxional process is envisaged as the breaking of a $B\text{-}H_\mu\text{-}B$ bond so that the H_μ atom becomes H_t on a BH_3 unit which then rotates before one of its H_t atoms reconverts to H_μ (ref. 99). The most recent calculation of the energy barrier for this process, which permutes all eight H atoms and makes all three B atoms equivalent, is only some 3 kJ mol^{-1} (ref. 93).

The $B_3H_8^-$ anion in powdered crystalline KB_3H_8 is also fluxional on the nmr timescale at temperatures in the range 50-430 K and this has likewise been analysed in terms of BH_3-group rotation with an activation energy of 8.4 ± 0.8 kJ mol^{-1} (ref. 98).

Thermolysis of NaB_3H_8 and its dioxane solvates affords mixtures of $NaBH_4$ and B_5H_9 or $Na_2B_{12}H_{12}$ according to conditions (ref. 100). Thermal decomposition of NaB_3H_8 in the range 80-100 °C first-order kinetics with an activation energy of 115.1 ± 7.5 kJ mol^{-1} and a pre-exponential factor of $10^{12.42\pm1.02}$ (ref. 101); the overall stoichiometry is given by the equation:

$$5NaB_3H_8 \longrightarrow 5NaBH_4 + 2B_5H_9 + H_2$$

The reaction is thought to proceed via the cleavage of $B_3H_8^-$ into BH_4^- and $\{B_2H_4\}$ (see structure 8) followed by polymerization of $\{B_2H_4\}$ according to:

$$5B_2H_4 \longrightarrow 2B_5H_9 + H_2$$

The thermal decomposition of $NaB_3H_8.C_4H_8O_2$ in diglyme at 160 °C yields $NaBH_4$ and $Na_2B_{12}H_{12}$ in a 3:1 molar ratio consistent with the formation of *closo*-$B_{12}H_{12}^{2-}$ from the reaction of the intermediates BH_4^- and B_5H_9: $2BH_4^- + 2B_5H_9 \longrightarrow B_{12}H_{12}^{2-} + 7H_2$.

The $B_3H_8^-$ anion has an extensive reaction chemistry which falls outside the scope of this review, as do the numerous structural variants of the adducts $L.B_3H_7$ of which $B_3H_8^-$ is itself a special example (with L = H^-).

6. TETRABORANES

Tetraborane(10), B_4H_{10}, was the first boron hydride to be isolated and characterized (ref. 102); it is a colourless, mobile, volatile liquid which boils at 18 °C and freezes at -120 °C. Tetraborane(10) is thermally rather unstable (see below) and decomposes rapidly at temperatures above about 40-50 °C. Its *nido* analogue $\{B_4H_8\}$ is even less stable though it has been identified as a reaction intermediate in borane thermolyses. The corresponding anions $B_4H_9^-$ and $B_4H_7^-$ are also known.

The heptahydrotetraborate(1-) anion, $B_4H_7^-$, was prepared by "unsymmetrical cleavage" of B_5H_9 by ammonia in diethyl ether solution at -78 °C over a period of 1-3 weeks (ref. 103):

$$B_5H_9 + 2NH_3 \longrightarrow [BH_2(NH_3)_2]^+[B_4H_7]^-$$

It is notable that there is no BH_2 group in the *nido*-pentaborane source. The compound is a white solid, stable below 0 °C, but it "decomposed spectacularly" on standing at room temperature. Boron-11 nmr spectroscopy suggests a structure in which a tetrahedral cluser $B_4(H_t)_4$ has the three edges of one face bridged by three H_μ.

The nonahydrotetraborate(1-) anion, $B_4H_9^-$, is best prepared by simple deprotonation of B_4H_{10} using KH in Me_2O at -78 °C (ref. 104). Under these conditions KB_4H_9 is stable for months but at room temperature it decomposes after about half an hour. Liquid NH_3 also deprotonates B_4H_{10} (reversibly) to give $[NH_4]^+[B_4H_9]^-$ but there is also a competing reaction which involves unsymmetrical cleavage to $[BH_2(NH_3)_2]^+[B_3H_8]^-$ (ref. 105). Detailed proton and boron-11 nmr spectra of the $B_4H_9^-$ anion have been interpreted on the basis of a C_s structure (9) with two B-H-B bonds and three BH_2 groups i.e. (2013)topology (ref. 104).

The structure of the fugitive *nido*-tetraborane, $\{B_4H_8\}$, has been the subject of several theoretical studies: an extremely high level of computation must be employed to decide whether the ground state configuration corresponds to the loss (from B_4H_{10}) of two H_μ atoms, one H_μ and one H_t, or two H_t atoms, leading respectively to a (2112) toplogy of C_s symmetry, a (3111) C_1 structure, or a (4020) C_{2v} isomer (ref. 93). Several other structures are also close in energy and the molecule is almost certainly fluxional. Dramatic stabilization results from adduct formation with a variety of Lewis bases such as CO, PF_3, PF_2X, PMe_3, $P(NMe_2)_3$ etc. to give LB_4H_8 (note also that $B_4H_9^-$, discussed in the preceding paragraph, is a special case for which L = H^-).

Several routes are now available for the preparation of *arachno*-tetraborane, B_4H_{10} (see Section 2). Most recently (ref. 106) variations of the hydride abstraction reaction (refs. 29,30) have led to almost quantitative yields of B_4H_{10} with no unwanted side reactions or polymer formation. For example, H^- abstraction from an equimolar mixture of BH_4^- and $B_3H_8^-$, using MeI (or I_2) in an inert solvent at room temperature, gives a 96% yield of B_4H_{10}, presumably via $\{BH_3\} + \{B_3H_7\} \longrightarrow B_4H_{10}$

The structure of B_4H_{10} has been established by X-ray crystallography (ref. 107) and by gas-phase electron diffraction (ref. 108) and microwave spectroscopy (ref. 109), this latter giving the most accurate structural parameters. The molecule, structure (10), has a "butterfly" shape of C_{2v} symmetry and (4012) topology, which can be viewed as an octahedral cluster with two adjacent boron atoms removed. The "hinge" distance B(1)-B(3) is 171.8(2) pm and the dihedral angle is 117.4(3)° (ref. 109); this is slightly larger than the dihedral angle between adjacent faces of a regular octahedron which is 109.5°. The electron diffraction data confirm the asymmetry of the bridge bonds: $B(1)-H_{\mu}$ 131.5(9) pm, $B(2)-H_{\mu}$ 148.4(9) pm (ref. 108).

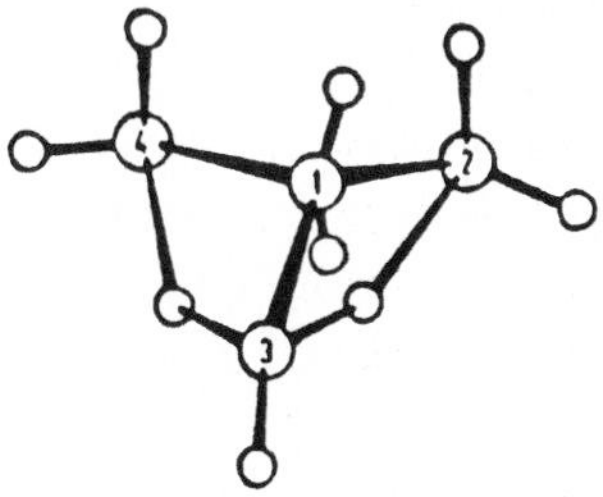

(9) $B_4H_9^-$ (C_s)

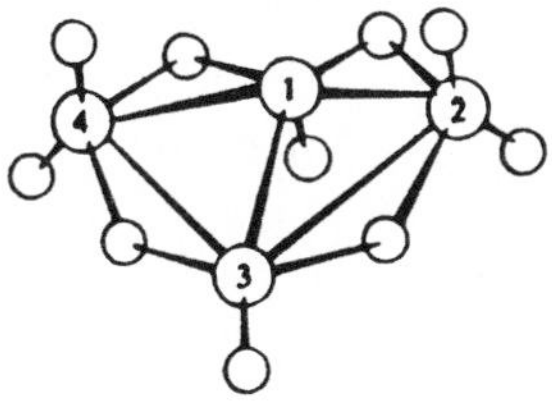

(10) B_4H_{10} (C_{2v})

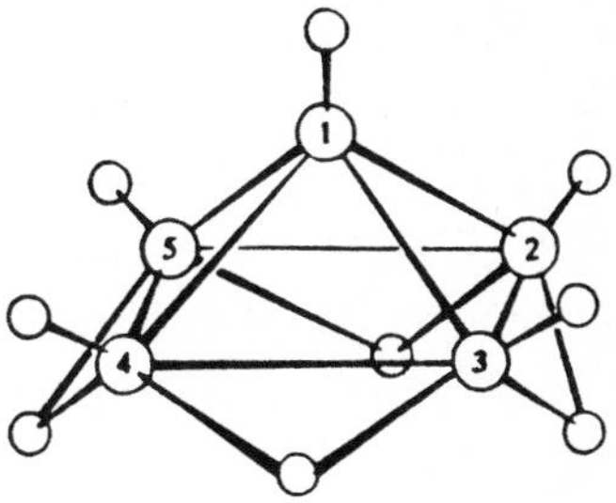

(11) B_5H_9 (C_{4v})

As already noted in Section 4, B_4H_{10} is the first isolable species in the gas-phase thermolysis of B_2H_6. Pure B_4H_{10} itself undergoes ready thermolysis at temperatures above about 40 °C: the initial reaction follows first-order kinetics with an activation energy of 99.2 ± 0.8 kJ mol^{-1} and a pre-exponential factor of *ca.* 6 x 10^{11} s^{-1} (ref. 110). The results can be interpreted in terms of a unimolecular rate-determining dissociation of B_4H_{10} into $\{B_4H_8\}$ and H_2 (refs. 84, 110); consistent with this B_4H_{10} readily undergoes H/D exchange with D_2 (ref. 111). Likewise, thermolysis of B_4H_{10} is considerably repressed in the presence of an excess of H_2, and there is a dramatic change in the distribution of the product boranes which can be explained in detail (refs. 84,112).

Other aspects of the extensive reaction chemistry of B_4H_{10} have recently been summarized (ref. 23). There have also been high-level calculations of the stability of the (unknown) B-B bonded C_2 structural isomer of B_4H_{10}, viz. bis(diboranyl), B_2H_5-B_2H_5, and speculations that it might be detectable or even preparable (ref. 113).

7. PENTABORANES

Both *nido*-B_5H_9 (11) and *arachno*-B_5H_{11} (12) are well known and have been thoroughly studied since the earliest days of boron hydride chemistry (ref. 1). The corresponding mono-anions, *nido*-$B_5H_8^-$ and *arachno*-$B_5H_{10}^-$, are also known as is the *arachno* dianion, $B_5H_9^{2-}$, and the presumably *hypho* anion $B_5H_{12}^-$, structure (13). The *closo* dianion $B_5H_5^{2-}$ has not been isolated or characterized, though the isoelectronic *closo*-dicarbapentaborane(5) 1,5-$C_2B_3H_5$, is well known.

Pentaborane(9) is a colourless, volatile liquid, mp -46.8 °C, bp 60.0 °C; it is the most stable of the lower boranes and resists thermolytic decomposition below about 200 °C (ref. 114), though it is extremely reactive chemically and is spontaneously flammable in air. The structure of B_5H_9 (11) is based on that of a square pyramid of boron atoms formed by removing one vertex from a B_6 octahedron. Each boron atom carries a terminal hydrogen atom and there are four bridging hydrogen atoms around the base, consistent with (4120) topology (ref. 14). The interatomic distances and angles are within the expected ranges, the most precise values being obtained by microwave spectroscopy (ref. 115): B(1)-B(2) 169.0 pm, B(2)-B(3) 180.3 pm, B-H_t 118.4 pm, B-H_μ 135.2 pm, and angle B-H_μ-B 96°. A more recent determination by electron diffraction is in good agreement (ref. 116).

In ether solvents B_5H_9 can be readily bridge-deprotonated at low temperatures by means of NaH, KH, anhydrous KOH, or lithium alkyls to give $B_5H_8^-$ (ref. 117). The stability of crystalline salts of $B_5H_8^-$ depends markedly on the counter cation and is greatest for bulky cations such as PPh_4^+ $AsPh_4^+$ which can be kept for several months at room temperature without decomposition (ref. 118). In solution, however, stability is much less and the anion rapidly affords *arachno*-$B_9H_{14}^-$ (see Section 11). Proton and boron nmr spectroscopy reveal that $B_5H_8^-$ is fluxional, with the three bridge protons moving between the four available basal positions (ref. 119).

If, instead of deprotonating *nido*-B_5H_9 in ether solution, it is treated with an alkali metal naphthalide, a facile two-electron reduction to *arachno*-$B_5H_9^{2-}$ occurs (ref. 120). The salts are sparingly soluble in tetrahydrofuran or 1,2-dimethoxyethane and appear to be quite stable (weeks) under vacuum in both the solid state and in solution. The dianion $B_5H_9^{2-}$ is fluxional in solution. When dissolved in butane at -78 °C the potassium and caesium salts can be protonated by HCl or HBr to give *arachno*-B_5H_{11} in overall yields of up to 38% from the original *nido*-B_5H_9. Since B_5H_9 is commercially available, the method affords the simplest, and safest route to B_5H_{11} yet devised (ref. 120):

$nido$-B_5H_9 + $2M^+[C_{10}H_8]^-$ ⟶ $[M^+]_2[arachno$-$B_5H_9]^{2-}$ + $C_{10}H_8$

$[M^+]_2[B_5H_9]^{2-}$ + 2HX ⟶ B_5H_{11} + $2M^+X^-$

Other syntheses of B_5H_{11} have been mentioned in Section 2.

B_5H_{11} is a colourless, volatile liquid mp -122 °C, bp 65 °C. Its structure (12) is based on a B_5 skeleton formed by removing one apical and one equatorial vertex from a pentagonal bipyramidal *closo*-B_7 cluster. However, the molecule does not have the expected C_s symmetry and, even in the gas phase, there is a definite distortion to C_1 symmetry due to the asymmetric semi-bridging nature of the apical *endo*-hydrogen atom above the open face (ref. 116, and references cited therein). As a result, this H atom is 31 pm closer to B(2) than to B(5) [159.4(9) pm and 189.9(9) pm respectively]. Other key dimensions are B(1)-B(2) 189.2(6), B(1)-B(3) 174.2(8), B(2)-B(3), 181.2(7), and B(3)-B(4) 176.0(12) pm; all B-H_t 119.2(4) pm except for the semi-bridging B(1)-H(1)$_{endo}$ which is 132.7 pm i.e. well within the range of 127-139 pm found for the other B-H_μ distances in the molecule (ref. 116). The proton and boron nmr spectra of the neat liquid show no evidence of the asymmetric structure of B_5H_{11} due to fluxionality (ref. 121) and the most recent theoretical study points to a very low barrier of only 7 kJ mol^{-1} for the C_1-C_s-C_1 fluxional process (ref. 122).

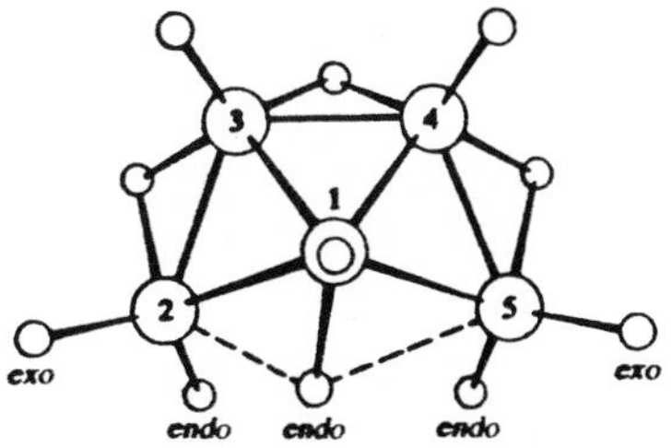

(12) B_5H_{11} (C_1)

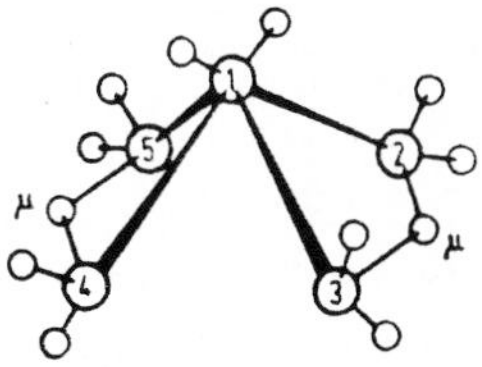

(13) $B_5H_{12}^-$ (C_{2v})

B_5H_{11} is the main volatile product observed in the initial stages of the gas-phase thermolysis of B_2H_6 above 100 °C. It is thought to be formed through the intermediary of B_4H_{10} (see Sections 4 and 6) via the reactions:

B_4H_{10} ⇌ $\{B_4H_8\}$ + H_2

$\{B_4H_8\}$ + B_2H_6 ⟶ B_5H_{11} + $\{BH_3\}$

Recently a detailed kinetic study has also been made of the gas-phase thermolysis of pure B_5H_{11} itself in the temperature range 40-150 °C (refs. 84,112). The reaction is first order in B_5H_{11} with an activation energy of 72.6±2.4 kJ mol^{-1} and a (very low) pre-exponential factor of 1.3×10^7 s^{-1}; the significance of these values will be discussed further when the thermolysis of B_6H_{12} is dealt with in the next Section. The main volatile products per mol of B_5H_{11} consumed are one mol of H_2 plus $\frac{1}{2}B_2H_6$; smaller amounts of B_4H_{10} and the higher boranes are also formed, together with a substantial amount of solid 'polymer' BH_x. These results are consistent with an initial rate-determining dissociation:

$B_5H_{11} \rightleftharpoons \{B_4H_8\} + \{BH_3\}$

The $\{BH_3\}$ then dimerizes to $\frac{1}{2}B_2H_6$ and the $\{B_4H_8\}$ undergoes a complex series of further reactions. In the presence of a large excess of added H_2 the product distribution is dramatically altered (mainly to B_2H_6 and B_4H_{10} and virtually no polymer), but the order of the reaction, the initial rate of thermolysis, and the activation energy are all unaltered . This can be interpreted in terms of the same initial dissociation of B_5H_{11} into $\{BH_3\}$ and $\{B_4H_8\}$, and the dimerization of $\{BH_3\}$ to $\frac{1}{2}B_2H_6$ as above, but with the $\{B_4H_8\}$ now being removed by rapid reaction with H_2 according to the known process $\{B_4H_8\} + H_2 \longrightarrow B_4H_{10}$.

This leads to the observed overall stoichiometry of the long-known 'equilibrium' reaction:

$B_5H_{11} + H_2 \rightleftharpoons B_4H_{10} + \frac{1}{2}B_2H_6$

Treatment of B_5H_{11} with KH results both in proton abstraction and hydride addition (ref. 104):

$B_5H_{11} + KH \longrightarrow K^+[B_5H_{10}]^- + H_2$

$B_5H_{11} + KH \longrightarrow K^+[B_5H_{12}]^-$

The *arachno* anion $[B_5H_{10}]^-$ is also reported as the product of unsymmetrical cleavage of B_6H_{12} with NH_3, though it has not yet been isolated as a pure salt (ref. 104):

$B_6H_{12} + 2NH_3 \longrightarrow [BH_2(NH_3)_2]^+[B_5H_{10}]^-$

An alternative route to the anion $[B_5H_{12}]^-$ is via deprotonation of B_4H_{10} with KH followed by reaction of the resulting $K^+[B_4H_9]^-$ with B_2H_6 in ether at -35° (ref. 104):

$K[B_4H_9] + \frac{1}{2}B_2H_6 \rightleftharpoons K[B_5H_{12}]$

The salt is difficult to isolate from ether since it redissociates to the starting materials on prolonged pumping at -35 °C, but the

$[NBu^n_4]^+$ and $[PMePh_3]^+$ salts are more stable and show no apparent decomposition after 10-15 minutes at room temperature. The proton and boron-11 nmr spectra of $[B_5H_{12}]^-$ imply that it is rare example of a binary *hypho*-borane (i.e. derived from the parent B_nH_{n+8} with n = 5): the proposed structure (refs. 104,123) is shown in (13) from which it can be seen that five BH_2 groups are bonded in a shallow-pyramid by two BHB and two BBB three-centre bonds, i.e. (2205) topology.

8. HEXABORANES

With the hexaboranes we encounter for the first time the complete series of structural types: *closo*-$B_6H_6^{2-}$, *nido*-B_6H_{10}, *arachno*-B_6H_{12} and *hypho*-B_6H_{14}. In addition $B_6H_6^{2-}$ can be protonated to $B_6H_7^-$, and *nido*-B_6H_{10} can (uniquely for a neutral borane) be protonated to the *cation* $B_6H_{11}^+$. Normal deprotonation of B_6H_{10} and B_6H_{12} by loss of bridging protons leads to the anions $B_6H_9^-$ and $B_6H_{11}^-$. Thus there are eight species in all to be considered.

Salts of *closo*-$B_6H_6^{2-}$ are now available in modest yield (see Section 2) and their chemistry has been fairly fully reviewed (refs. 19,24). An early X-ray crystal structure determination on $[NMe_4^+]_2[B_6H_6]^{2-}$ (ref. 124) established the *closo* octahedral structure (14) of the dianion, and a recent redetermination on $K_2B_6H_6$ (ref. 125) led to more precise dimensions: B-B 172.1(3) pm and B-H 107(2) pm.

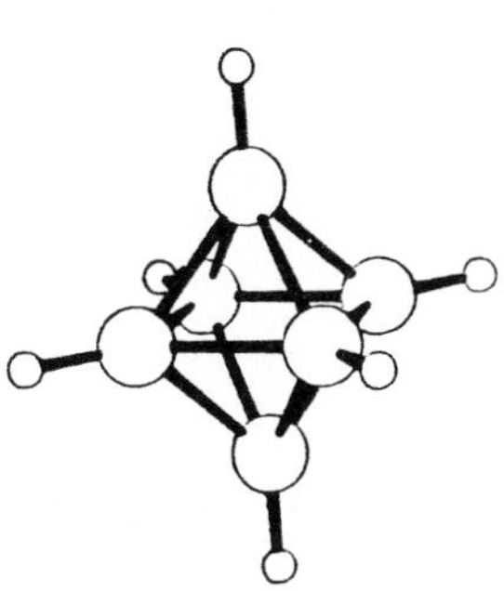

(14) $B_6H_6^{2-}$ (O_h)

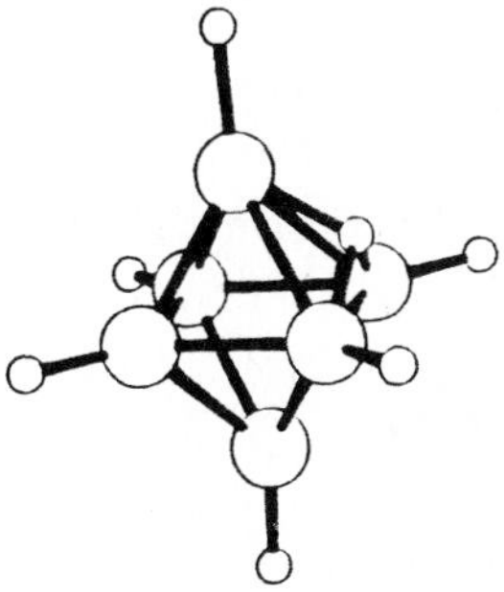

(15) $B_6H_7^-$ C_{3v}

Careful acidification of the alkali metal salts $M_2B_6H_6$ (M = Na, K, Cs) to pH5 using aqueous HCl leads to the mono-anion $B_6H_7^-$ which can be crystallized by use of large counter cations e.g. $[PPh_4]^+[B_6H_7]^-$, $[Ni(bipy)_3]^{2+}[B_6H_7^-]_2$, and $[Ni(phen)_3]^{2+}[B_6H_7^-]_2.Me_2CO$ (refs. 126,127). The X-ray crystal structures (ref. 127) reveal a unique triply bridging H atom above one face of the *closo* cluster thereby imparting C_{3v} symmetry to the anion (15). The three B-B distances in the bridged face, 182(1) pm, are significantly longer than the other nine B-B distances in the cluster, 170(1) pm, but there is no significant difference at the precision of the analysis between $B-H_{\mu}$ [115(6) pm] and $B-H_t$ [112(5) pm]. Proton and boron-11 nmr data show that the anion is fluxional, with proton migration occuring mainly through the B-B edges of the cluster (ref. 128). *Ab initio* calculations reproduce the observed dimensions and lead to a proton affinity of 1732 kJ mol^{-1} for $B_6H_6^{2-}$ (ref. 129). Other calculations agree with the face-protonated structure of $B_6H_7^-$, but there is no consensus on the structure or even the likely stability of the hypothetical doubly protonated neutral species B_6H_8 (refs. 129-132).

Nido-B_6H_{10} is one of the classical boranes (ref. 1) but the development of its chemistry was slow until improved syntheses became available in the 1970's (refs. 133,134) e.g.:

$$LiB_5H_8 + B_2H_6 \xrightarrow[\text{low temp.}]{Me_2O} B_6H_{10} + LiBH_4$$

$$B_5H_9 \xrightarrow[\text{r.t.}]{Br_2} 1\text{-}BrB_5H_8 \xrightarrow[-78\ °C]{KH/Me_2O} K[1\text{-}BrB_5H_7] \xrightarrow[-78\ °C]{\frac{1}{2}B_2H_6} K[B_6H_{10}Br] \xrightarrow{-35\ °C} B_6H_{10} + KBr$$

Hexaborane(10) is a colourless mobile liquid, mp -62.3 °C, bp 108 °C. The molecular structure (16) is based on a pentagonal pyramid of boron atoms each of which has a terminal hydrogen atom attached; in addition there are four bridging hydrogen atoms around the base (ref. 135). Significant features (ref. 136) are the short un-bridged basal B(4)-B(5) bond of 162.6(4) pm and the asymmetric $B-H_{\mu}-B$ bonds, the distances from B(3) [or B(6)] to each H_{μ} being 130 pm whereas those from B(2) to H_{μ} are 116 pm and from B(4) [or B(5)] to H_{μ} 118 pm. The topology is (4220).

When pure, B_6H_{10} is rather stable thermally and there is little decomposition in the gas phase below about 75 °C. Kinetic studies in the temperature range 75-165 °C indicate a second-order process with an activation energy of 79.7 ± 2.7 kJ mol^{-1} and a pre-exponential factor of 4.7×10^6 m^3 mol^{-1} s^{-1} (ref. 137). Subsequent studies suggest that the reaction follows a radical-based mechanism involving $B_6H_9^{\bullet}$ (refs. 84,138). B_6H_{10} is also strongly implicated in the thermolysis of lower boranes *en route* to decaborane (ref. 139) and it has been imaginatively used by

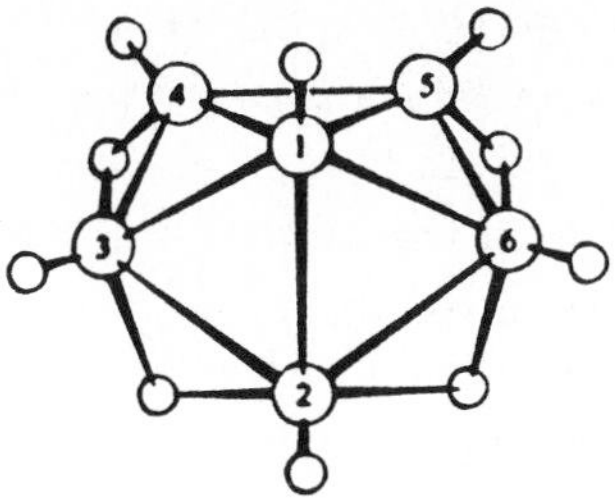

(16) B_6H_{10} (C_s)

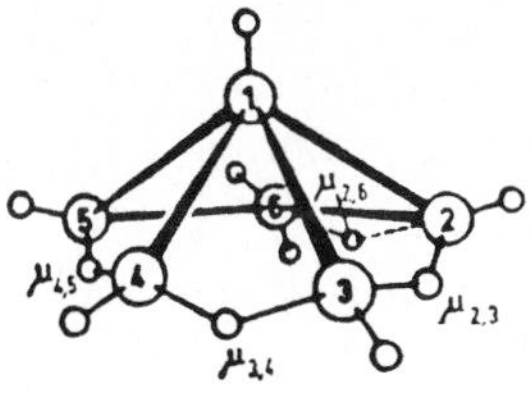

(17) $B_6H_{11}^-$ (C_1)

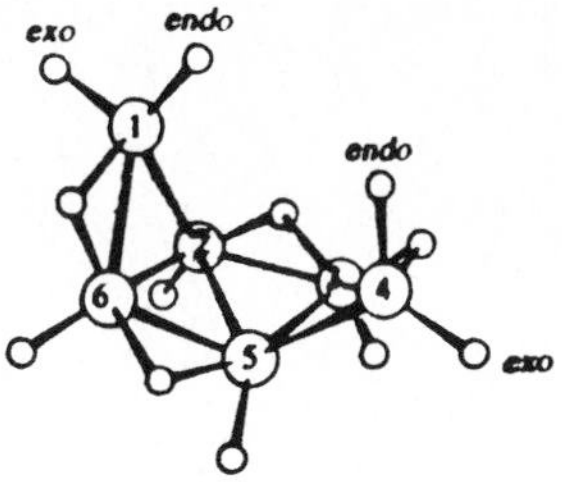

(18) B_6H_{12} (C_2)

R. Schaeffer in the synthesis of macropolyhedral *conjuncto*-boranes such as $B_{13}H_{19}$, $B_{14}H_{20}$, $B_{14}H_{22}$, and $B_{15}H_{23}$ (see Section 2).

B_6H_{10} can be bridge-deprotonated in ethereal solvents by alkali metal hydrides or alkyls to yield salts of $B_6H_9^-$ (ref. 140). The stability of these salts increases with the size of the cation, e.g. $Li^+ < Na^+ < K^+ < [NBu^n{}_4]^+$, and when $[PPh_4]^+$ or $[AsPh_4]^+$ is used the salts are stable for periods of more than one week at room temperature (ref. 118). As expected, $B_6H_9^-$ is more stable than $B_5H_8^-$.

Conversely, protonation of the two-centre B(4)-B(5) bond in *nido*-B_6H_{10} yields the unique cationic species *nido*-$B_6H_{11}^+$ with five bridging protons around the pentagonal base (ref. 141). Thus, treatment of B_6H_{10} with liquid HBr at -80 °C gives a species with pyramidal C_{5v} symmetry (^{11}B nmr) and the solid salt $[B_6H_{11}]^+[BCl_4]^-$ can be isolated at -50 °C.

The corresponding *arachno*-$B_6H_{11}^-$ anion can be prepared either by inserting $\{BH_3\}$ into the basal B-B bond of $B_5H_8^-$ or by deprotonating *arachno*-B_6H_{12} with $B_5H_8^-$ (ref. 104). In ether solutions $Li[B_6H_{11}]$ and $K[B_6H_{11}]$ appear to be stable up to about -15 °C and, with the larger counter cations $[NBu^n{}_4]^+$ and $[PMePh_3]^+$, solutions in CH_2Cl_2 are stable for several hours at room temperature. The two latter salts can also be handled as solids at room temperature. The proton and boron-11 nmr spectra have been interpreted in terms of structure (17) which implies insertion of a BH_3 group into the vacant bridging site of a $B_5H_8^-$ ion; the product of deprotonation of B_6H_{12} — which has structure (18) — would seem to require substantial skeletal and proton rearrangement to achieve this structure.

The next hexaborane to be considered is *arachno*-B_6H_{12} which is best prepared by protonation of $B_6H_{11}^-$ with HCl; the $B_6H_{11}^-$ itself is formed *in situ* by reaction of $\frac{1}{2}B_2H_6$ with $B_5H_8^-$ as outlined in the preceeding paragraph (ref. 104). Yields of up to 70% can be obtained provided the Me_2O solvent is removed as completely as possible before the addition of HCl. B_6H_{12} melts at -82 °C and has a vapour pressure of 10.5 mmHg at 0 °C. The structure (18) has been established by gas-phase electron diffraction (ref. 142) and has C_2 symmetry. The nmr spectra (refs. 143-145) can also be convincingly interpreted on the basis of this structure which implies a (4212) bonding topology. The boron framework is a chiral six-atom fragment of a closed triangular dodecahedron from which two adjacent five-connected vertices have been removed. The structure features two BH_2 groups, two asymmetric B-H-B bonds and two symmetric B-H-B bonds (ref. 142). Typical dimensions are $B(1)-H_\mu$ 142(2), $B(6)-H_\mu$ 120(4), $B-H_t$ 122(1), B-B in the range 170-191 pm. B_6H_{12} is stable both as a liquid and as a gas at room temperature but above about 70 °C it decomposes in the gas phase by a clean first-order reaction which yields B_5H_9 and B_2H_6 in a mole ratio of 2:1 (ref. 146). The activation energy of 75.0 ± 5.8 kJ mol^{-1} and the pre-exponential factor of 3.8 x 10^7 s^{-1} are very similar to those obtained for the thermolysis of B_5H_{11} and have been interpreted, as there, in terms of the rate-determining loss of a $\{BH_3\}$ unit:

$$B_6H_{12} \longrightarrow B_5H_9 + \{BH_3\}; \quad 2\{BH_3\} \longrightarrow B_2H_6$$

This also reflects the structural-geometrical similarity between the two *arachno*-boranes since the notional replacement of H_μ and H_{endo} on B(4) in B_4H_{10} (10) by a BH_3 group yields B_5H_{11} (12) and repetition of this process on the opposite side of the molecule i.e. at B(2) generates the observed structure of B_6H_{12} (18) (ref. 146). These changes are accompanied by systematic trends in the molecular dimensions e.g. the "hinge" B-B distance increases from 170.5(12) pm in B_4H_{10} through 174.2(8) pm in B_5H_{11} to 182.1(13) pm in B_6H_{12}, and the dihedral angle between the two BBB planes at this hinge gradually opens from 117.1(7)° through 138.9° to 167.4(22)° respectively.

The most recently synthesized hexaborane is *hypho*-B_6H_{14}; this is a *conjuncto*-borane formed by the fusion of two $\{B_3H_7\}$ units (ref. 147). The preparative route involves the oxidative extraction of a hydride ion from $B_3H_8^-$ using I_2 or HI in inert solvents at -40 °C, e.g.:

$$B_3H_8^- + I_2 \longrightarrow \{B_3H_7\} + HI + I^-$$

The counter ions are $[NEt_4]^+$ or $[PPh_4]^+$. On the basis of nmr spectra a structure was proposed for B_6H_{14} in which two triangular $\{B_3H_7\}$ units (4) were symmetrically bonded via two bridging hydrogen atoms. However, subsequent calculations (which did not include polarization functions for the H atoms) suggested that a tris(diboranyl) structure with B-B bonds, B_2H_5-B_2H_4-B_2H_5, might be more stable (ref. 148). Clearly, further experimental and theoretical work is required in order to characterize more fully this newest borane.

9. HEPTABORANES

The first neutral heptaborane, *conjuncto*-B_7H_{13} was reported as recently as 1985 (ref. 45); the unrelated *arachno*-$B_7H_{12}^-$ anion (ref. 104) and the dianion *closo*-$B_7H_7^{2-}$ (refs. 19,22d) have been known for rather longer.

The pentagonal bipyramidal structure (19) for *closo*-$B_7H_7^{2-}$ was deduced from its nmr spectrum; no crystal structure is available. As for other intermediate-sized *closo*-borane dianions, $B_7H_7^{2-}$ is prepared by air oxidation of *closo*-$B_9H_9^{2-}$ in 1,2-dimethoxyethane. Yields are low since $B_7H_7^{2-}$ slowly hydrolyses to $B(OH)_3$ and H_2 in neutral solution, making isolation difficult; hydrolysis is very rapid in acid solution and even faster in aqueous $AgNO_3$. Indeed, $B_7H_7^{2-}$ is the least stable member of the *closo*-$B_nH_n^{2-}$ series; the caesium salt begins to decompose at about 400 °C.

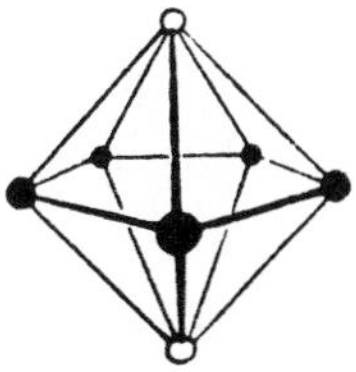

(19) $B_7H_7^{2-}$ (D_{5h})

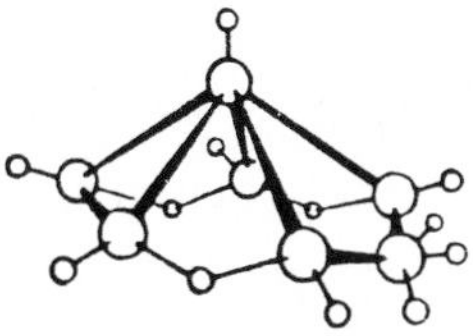

(20) $B_7H_{12}^-$ (proposed)

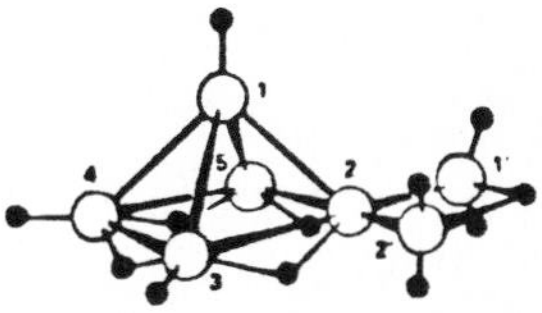

(21) B_7H_{13} (C_s)

The anion $B_7H_{12}^-$ has been made by $\{BH_3\}$ insertion into $B_6H_9^-$ at -78 °C (ref. 104); the synthesis parallels those of $B_5H_{12}^-$ and $B_6H_{11}^-$ above. The structure is unknown but that shown in (20) reflects the method of preparation and is consistent with the nmr spectra.

The *conjuncto*-borane B_7H_{13} was prepared by the $PtBr_2$-promoted dehydro-coupling of B_2H_6 and B_5H_9 in dried decane at room temperature (ref. 45). It was essential to have at least a 5-fold excess of B_2H_6 to prevent the formation of $[1,2'-(B_5H_8)_2]$ (see Section 12). The structure (21) was proposed on the basis of nmr spectroscopy and implies coupling of the two moieties by removal of an H_μ on B_2H_6 and a basal H_t on B_5H_9 i.e. $[\mu-(2-B_5H_8)B_2H_5]$. The compound was a liquid above 0 °C and moderate heating to 40 °C for 2-3 hours caused decomposition to B_2H_6 and B_5H_9.

10. OCTABORANES

Five neutral and two anionic octaboranes are known. The *closo*-$B_8H_8^{2-}$ dianion and *closo*-$B_8H_8^{\cdot-}$ radical anion will be considered first, followed by *nido*-B_8H_{12}, *arachno*-B_8H_{14}, the mysterious B_8H_{16}, and the two B-B *conjuncto*-boranes of formula B_8H_{18} i.e. 1,1'- and $2,2'-(B_4H_9)_2$.

Salts of the *closo*-$B_8H_8^{2-}$ dianion are prepared by air oxidation of *closo*-$B_9H_9^{2-}$ salts in 1,2-dimethoxyethane at 70 °C (refs. 58,148). The caesium salt (which was stable up to 600 °C) was isolated in 40% yield, and other salts e.g. $[Zn(NH_3)_4]^{2+}$, $[Cd(NH_3)_4]^{2+}$, $[NBu^n{}_4]^+$ and $[PMePh_3]^+$ were obtained from the sodium by metathesis in aqueous or ammoniacal solution. The compounds are colourless when pure but the dianion readily oxidizes in solution to the burgundy red radical anion $B_8H_8^{\cdot-}$. Thus, when solutions of the $[NBu^n{}_4]^+$ or $[PMePh_3]^+$ salts in Me_2SO, MeCN, $CHCl_3$, or H_2O are exposed to air,up to 2% of the diamagnetic *closo*-$B_8H_8^{2-}$ is converted to the paramagnetic radical $B_8H_8^{\cdot-}$ which has a strong esr spectrum with more than 300 resolvable lines. The initial stages of the synthesis of *closo*-$B_8H_8^{2-}$ in 1,2-dimethoxyethane at 70 °C (see above) are also accompanied by the formation of the characteristic red colour of $B_8H_8^{\cdot-}$ before the solution turns a milky white after about 2 hours (ref. 58).

An X-ray diffraction study of $[Zn(NH_3)_4][B_8H_8]$ showed (ref. 149) that the dianion had a slightly distorted dodecahedral D_{2d} structure (22); there was a systematic increase in B-B distances as the cluster-connectivity of the boron atoms increased, being 156(3) pm between two 4-connected vertices, 174(2) pm between a 4- and a 5-connected vertex, and 193(2) between two 5-connected vertices. The ^{11}B nmr spectrum of *closo*-$B_8H_8^{2-}$ is temperature-dependent and indicates fluxionality. This was originally interpreted in terms of D_{4d} (square antiprismatic) and C_{2v} (bicapped trigonal prismatic) geometries (ref. 58) but more recent work favours fluxionality involving the ground-state D_{2d} dodecahedral structure (22) and a C_{2v} structure lying some 27 kJ mol^{-1} above this (ref. 150).

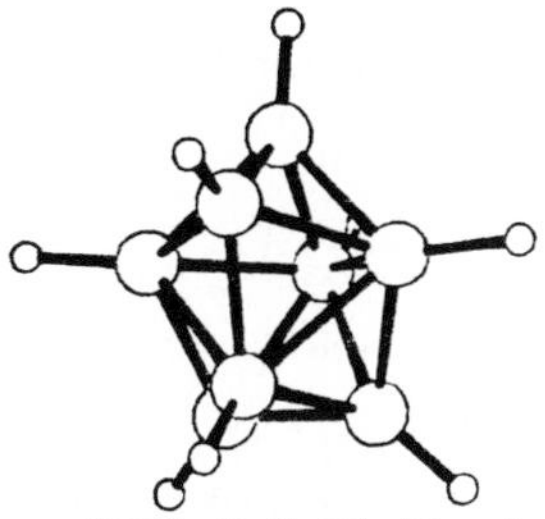

(22) $B_8H_8^{2-}$ (D_{2d})

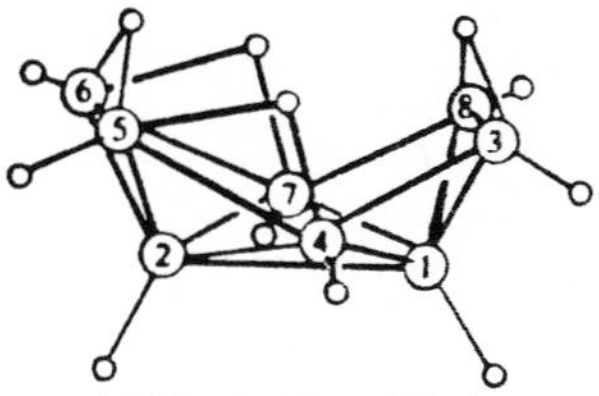

(23) B_8H_{12} (C_s)

Nido-octaborane, B_8H_{12}, was first isolated and characterized in 1964 using a low pressure of H_2 (12 mmHg) to sweep a 2:1 mixture of B_2H_6 and B_5H_9 through an electric discharge. The compound is unstable above its mp of about -20 °C and this greatly complicated its isolation from the product mix (ref. 151). B_8H_{12} is also among the low-presure decomposition products of *n*-B_9H_{15} ($\longrightarrow B_8H_{12} + \frac{1}{2}B_2H_6$) (ref. 152) but the best preparative route is via the low-temperature decomposition of *i*-B_9H_{15} (ref. 153):

$$B_{10}H_{14} \xrightarrow{KOH} B_9H_{14}^- \xrightarrow[-78\,°C]{HCl} i\text{-}B_9H_{15}$$

$$2\ i\text{-}B_9H_{15} \xrightarrow[-45\ \text{to}\ -30\,°C]{\text{pentane}} B_8H_{12} + B_{10}H_{14} + 2H_2$$

Gram amounts can be obtained in this way. The X-ray molecular structure (23) (ref. 151) indicates a boron skeleton which is more open than the (rather crowded) *nido*-framework derived by removal of one 5-connected vertex from a *closo*-B_9 tricapped trigonal prism. Instead, the structure resembles an eight-membered icosahedral fragment with four H_μ symmetrically disposed around the 6-membered open face. The average terminal B-H_t distance is 114(4) pm and there are two symmetrical B-H_μ-B bonds (between B(5)B(6) and B(3)B(8) i.e. across the C_s mirror plane) and two unsymmetrical B-H_μ-B bonds between B(4)B(5) and B(6)B(7) with, for example B(4)-H_μ 127(4) pm and H_μ-B(5) 151(4) pm. As expected, the nmr spectra indicate fluxional behaviour in solution (ref. 154). The thermal stability of *nido*-B_8H_{12} is greatly enchanced by adduct formation to give B_8H_{12}.L where L = Et_2O, Me_3N, or MeCN (ref. 153).

Little is known about *arachno*-octaborane, B_8H_{14}: it is prepared by an obscure reaction (ref. 155) and is very unstable, undergoing 30% decomposition in 30 seconds at 0 °C and 6% decomposition in 1 hour at -30 °C. The nmr data have been interpreted in terms of a structure which has a boron skeleton similar to that in (**23**) but with all six B-B links in the open face carrying bridging H_μ atoms. The compound B_8H_{16} is also of unknown structure. It was obtained as one of several products from the gas-phase co-thermolysis of B_5H_9 with B_2H_6 in a hot/cold reactor at 140°/-20 °C (ref. 156), and was identified by elemental analysis, molecular weight, and mass spectrometry; the nmr was uninformative.

There are three possible ways of joining two *arachno*-B_4 units via a B-B bond to form a *conjuncto*-octaborane(18) *viz.* 1,1'-, 2,2'-, and 1,2'-$(B_4H_9)_2$. Reaction between $[NMe_4][B_3H_8]$ and polyphosphoric acid at 40 °C led (amongst other major products) to trace quantities of 2,2'-$(B_4H_9)_2$ (ref. 157). Yields of up to 30% were obtained when thf.B_3H_7 was treated with an excess of BF_3 at -45 °C (ref. 158). The structure (**24**) was deduced on the basis of low-temperature X-ray and nmr data, and the compound was said to be stable only for very short periods above room temperature. More recently the 1,1'-isomer, structure (**25**) was prepared in essentially quantitative yield by means of the room-temperature $PtBr_2$-catalysed dehydrocoupling of B_4H_{10} (ref. 44). The product was purified by low-temperature distillation at about -55 °C but decomposed in the liquid phase above -30 °C. The third isomer, 1,2'$(B_4H_9)_2$, has not so far been sythesized.

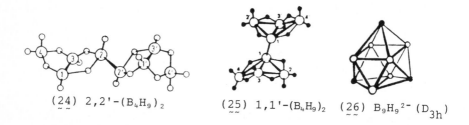

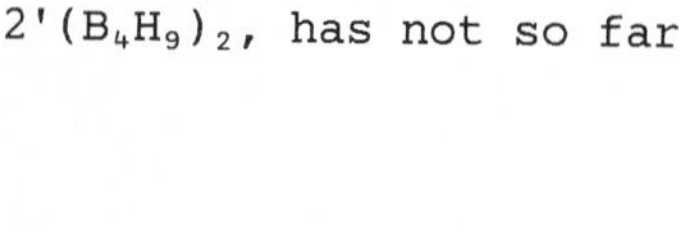

(**24**) 2,2'-$(B_4H_9)_2$

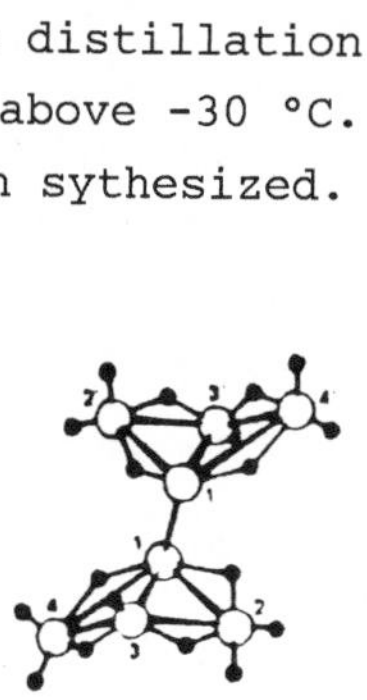

(**25**) 1,1'-$(B_4H_9)_2$

(**26**) $B_9H_9^{2-}$ (D_{3h})

11. NONABORANES

The eight known nonaborane species vary enormously in stability from the robust *closo*-$B_9H_9^{2-}$ (which is stable as its caesium salt at 600 °C in a sealed tube) to the fugitive *nido*-{B_9H_{13}} which has been fleetingly detected but not isolated. The other species are *nido*-$B_9H_{12}^-$, the four *arachno* clusters $B_9H_{13}^{2-}$, $B_9H_{14}^-$, *n*-B_9H_{15}, and *i*-B_9H_{15}, and finally *conjuncto*-B_9H_{17} i.e. [1,2'-(B_4H_9)(B_5H_8)].

The *closo*-dianion $B_9H_9^{2-}$ can be conveniently prepared (though in low yields of 6.8 or 9% respectively), by pyrolysis of either Cs[B_3H_8] or Rb[B_3H_8] at 230 °C for 30 minutes followed by a further 30 minutes at 180 °C (ref. 159). A higher yield (38%) is obtained by pyrolysis of $Na_2[B_{10}H_{12}]$ at 240 °C for 2 hours, followed by addition of RbCl in aqueous ethanol to precipitate $Rb_2[B_9H_9]$ (ref. 160). Metathesis yields salts of other cations e.g. K^+, $[SMe_3]^+$, and $[Zn(NH_3)_4]^{2+}$. X-ray crystal structure analysis of $Rb_2[B_9H_9]$ shows that the anion has a tricapped trigonal prismatic structure (26) of idealized D_{3h} symmetry, slightly distorted to C_{2v} (ref. 161). The B-B distances between a 4-connected and a 5-connected vertex are in the range 168-173 pm, and are substantially shorter than those between two 5-connected vertices which fall in the range 181-185 pm. The ^{11}B nmr spectrum is consistent with the D_{3h} structure (two B-H doublets of relative intensity 2:1) and this remains unmodified even up to 200 °C, implying a high energy barrier for fluxional cluster rearrangement (ref. 162). $Cs_2[B_9H_9]$ is stable when heated in a sealed tube to 600 °C (ref. 159) and the solid is even stable to air oxidation up to 575 °C (ref. 19). However, stability is lower in solution, and stepwise air oxidation of *closo*-$B_9H_9^{2-}$ in ethanol, thf, or 1,2-dimethoxyethane forms a viable route to $B_8H_8^{2-}$, $B_8H_8^{\cdot-}$, $B_7H_7^{2-}$, and $B_6H_6^{2-}$ (see preceding Sections). The *closo*-$B_9H_9^{2-}$ dianion is stable in neutral and alkaline aqueous solution but is rapidly degraded by acid (ref. 159).

The *nido*-anion $B_9H_{12}^-$ is readily formed by the deprotonation of $B_9H_{13}(SMe_2)$ with strong bases such as $[NR_4]OH$, $Ph_3P{=}CH_2$ or $[N(PPh_3)_2][OCN]$ (refs. 163,164). The structure (27) has been determined by X-ray diffractometry and has the expected *nido*-geometry formed by removal of one 5-connected vertex from the parent *closo*-B_{10} cluster (ref. 164). The resulting pentagonal open face is symmetrically bridged by three H_μ (two involving the lowest connected boron atom) thereby conferring effective C_s symmetry on the cluster.

The unstable *nido*-nonaborane(13), B_9H_{13}, cannot be made by direct protonation of *nido*-$B_9H_{12}^-$ but has been identified mass spectrometrically as a transient intermediate in the pyrolysis of the adducts $B_9H_{13}(SMe_2)$ and $B_9H_{13}(SEt_2)$ at 55° and 135 °C respectively (ref. 165). The major products of pyrolysis were $B_{18}H_{22}$, B_8H_{12}, and B_6H_{10}, with small amounts of $B_{10}H_{14}$. The precursor adducts can readily be made via the reaction sequence:

$$B_{10}H_{14} \xrightarrow[-H_2]{2R_2S} B_{10}H_{12}(SR_2)_2 \xrightarrow[\text{reflux}]{3\ \text{EtOH}} B_9H_{13}(SR_2) + B(OEt)_3 + R_2S + H_2$$

The instability of *nido*-B_9H_{13} may be due to the steric crowding of the four supernumerary H atoms in the open face.

The *arachno*-dianion $B_9H_{13}^{2-}$ is highly reactive but can be isolated as its mixed alkali metal salt $LiCs[B_9H_{13}]$ when one equivalent of $LiBu^n$ is added to a solution of $Cs[B_9H_{14}]$ in thf (ref. 166); little is known of its chemical or spectroscopic properties. By contrast, the *arachno*-monoanion $B_9H_{14}^-$ is very stable and readily prepared by a variety of reactions; it is consequently well characterized and much studied. It was originally made (ref. 167) by base degradation of $B_{10}H_{14}$ using two mole equivalents of aqueous KOH, but can be more conveniently prepared (in 85% yield) by a modification which uses one mole equivalent of KOH in methanol as the degrading solvent (ref. 168):

$$B_{10}H_{14} \xrightarrow[(\text{MeOH})]{\text{KOH}} B_{10}H_{13}^- \xrightarrow{3\ \text{MeOH}} B_9H_{14}^- + B(OMe)_3 + H_2$$

The anion is also readily obtainable from B_5H_9 (ref. 169) especially via the room-temperature decomposition of $B_5H_8^-$ salts in thf (ref. 170); the detailed course of this high-yield cluster-aggregation reaction is not clear.

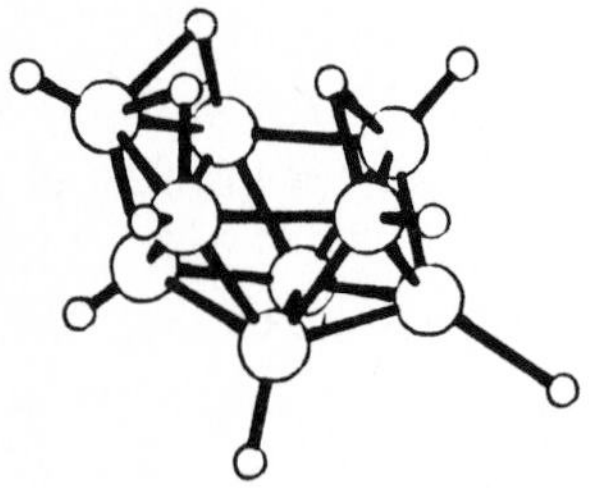

(27) $B_9H_{12}^-$ (C_s)

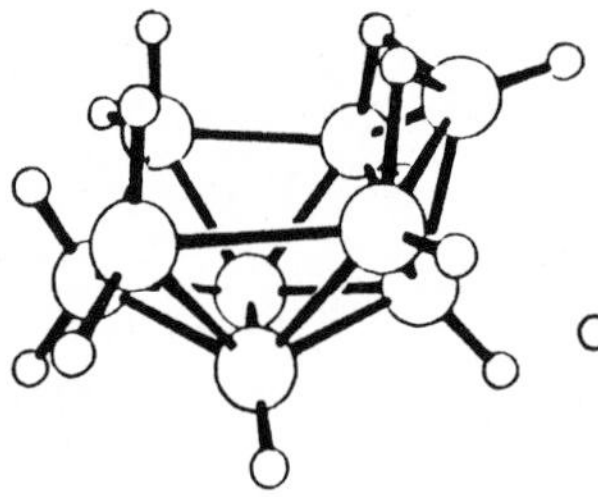

(28) $B_9H_{14}^-$ (C_s)

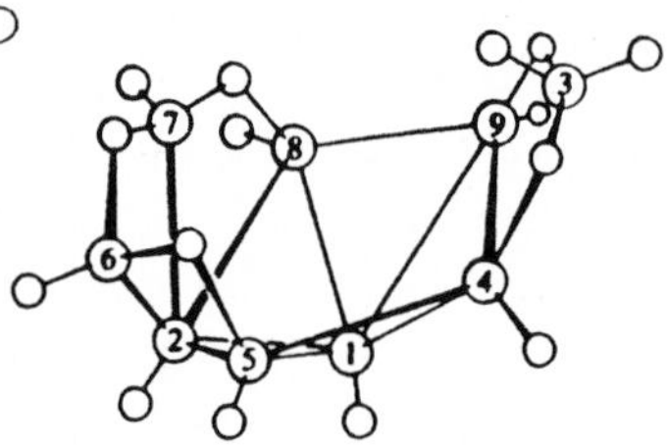

(29) n-B_9H_{15} (C_s)

The X-ray crystal structure of CsB_9H_{14} (ref. 171) reveals that *arachno*-$B_9H_{14}^-$ has C_s symmetry with two B-H-B bonds and three BH_2 groups (28). The ion thus has (2613) topology, but in solution the five *endo*-H atoms are fluxional, thereby giving rise to effective C_{3v} cluster symmetry and three sets of three equivalent boron atoms (refs. 171,172). It will also be noted that $B_9H_{14}^-$ is a special case of the adduct $B_9H_{13}L$ in which the ligand is H^-.

Thermolysis of $B_9H_{14}^-$ salts at 73-153 °C yields *closo*-$B_{12}H_{12}^{2-}$ salts (ref. 173). The anion is also an excellent starter for the synthesis of numerous types of metallaborane (ref. 3).

Two neutral *arachno*-nonaboranes are known: *n*-B_9H_{15} (mp 2.6 °C, vp 0.8 mmHg at 28 °C) and the less stable *i*-B_9H_{15} which decomposes above -30 °C. Together they comprise the first example of geometrical cluster isomers in the boranes, in which the actual arrangement of boron atoms within the cluster differs (as distinct from the position of conjunction of two subclusters). Indeed, *n*-B_9H_{15} was the first neutral binary borane to be added (after a hiatus of nearly 30 years) to Stock's classic suite of six boron hydrides (refs. 1,174). It was originally made in very low yield by passing B_2H_6 through a silent electric discharge (ref. 174) but higher yields can be obtained by allowing B_5H_{11} to interact with the surface of hexamethylenetetramine at 0 °C (ref. 175), or by reacting a 10:1 mixture of B_2H_6 and B_5H_{11} at 25 atm and 25 °C for several days (ref. 176). The detailed mechanisms of these reactions are not yet clear; see however Section 12 for a proposed mechanism for the synthesis of *n*-B_9H_{15} from [2,2'-$(B_5H_8)_2$].

The structure of *n*-B_9H_{15} (29) was established by X-ray diffraction (ref. 177). Formally the *arachno*-cluster of *n*-B_9H_{15} can be derived from the parent *closo*-B_{11} cluster by removal of the unique 6-connected vertex and an adjacent 5-connected vertex. Alternatively it can be thought of as being derived by removal of the B(10) vertex from *nido*-$B_{10}H_{14}$ (see later Sections for these structures). By contrast, the cluster structure of *i*-B_9H_{15} is derived from that of *closo*-B_{11} by removal of the unique 6-connected vertex and an adjacent 4-connected vertex, or by removing the B(9) vertex from *nido*-$B_{10}H_{14}$. It will be noted from structure (29) that *n*-B_9H_{15} features five B-H-B bridges and a single BH_2 group (5421 topology); as usual the B-H-B bridge bonds adjacent to the BH_2 group are unsymmetrical (140 and 118 pm for H_2B-H_μ and HB-H_μ respectively) but the other three are symmetrical, with B-H_μ 128(2) pm in each case (ref. 177).

As might be expected from its structure, *i*-B_9H_{15} can be obtained by the low-temperature protonation of *arachno*-$B_9H_{14}^-$ with liquid HCl at -78 °C (ref. 178). Above -30° it disproportionates rapidly with loss of H_2 into B_8H_{12} and $B_{10}H_{14}$ (see preceding Section). Depending on the conditions respectable yields of *n*-$B_{18}H_{22}$ can also be isolated. It is thought (ref. 178) that decomposition involves initial loss of H_2 followed either by disproportionation or dimerization of the nascent $\{B_9H_{13}\}$ species; ligand complexes of the type $B_9H_{13}L$ could also be isolated. It has not proved possible to obtain an X-ray crystal structure of *i*-B_9H_{15} but the nmr spectra (refs. 178,179) have been interpreted in terms of a C_{3v} structure with six briding H atoms in the open 6-membered face of the B_9 cluster shown in structure (28).

The last nonaborane to be considered is the recently synthesized *conjuncto*-nonaborane(17) i.e. [1,2'-$(B_4H_9)(B_5H_8)$]. The procedure has already been mentioned in Section 2(ref. 44) and can be represented by the subjoined reaction scheme which also indicates the proposed structure of the product (30):

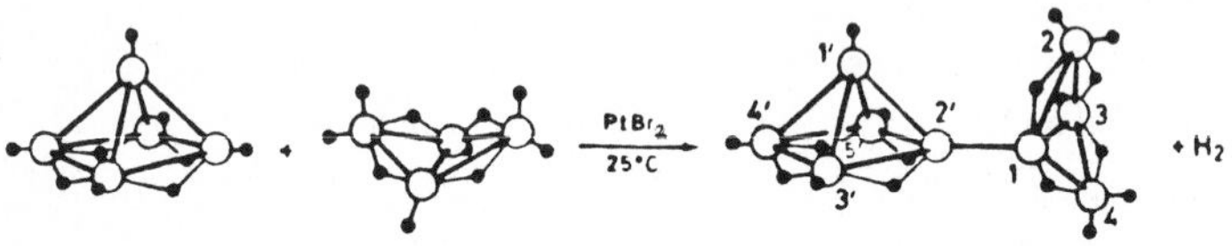

(30) 1,2'- $(B_4H_9)(B_5H_8)$

Low-temperature fractionation affords yields of up to 60% and the structure was assigned on the basis of proton and boron(11) nmr spectra. Other properties were not described except to note that the compound decomposed rapidly in the liquid state above -20 °C.

12. DECABORANES

Twelve (or possibly fourteen) binary decaborane species are known: *closo*-$B_{10}H_{10}^{2-}$; *nido*-$B_{10}H_{14}$ and its anions $B_{10}H_{13}^-$ and $B_{10}H_{12}^{2-}$; *arachno*-$B_{10}H_{14}^{2-}$ and its monoprotonated derivative $B_{10}H_{15}^-$; the three *conjuncto*-boranes 1,1'-, 1,2'-, and 2,2'-$(B_5H_8)_2$, and the monoanions of these last two isomers, *viz.* 1,2'- and 2,2'-$B_{10}H_{15}^-$, which are structural isomers of *arachno*-$B_{10}H_{15}^-$; and finally $B_{10}H_{18}$, which might be a *conjuncto* B-B linked species [B_5H_8)(B_5H_{10})]. Two further species, $B_{10}H_{11}^-$ (i.e. monoprotonated *closo*-$B_{10}H_{10}^{2-}$) and the short-lived purple radical anion $B_{10}H_{14}^{\cdot -}$, have also been briefly reported but little is known apart from their possible existence.

The preparation, structure, bonding, and chemical reactions of *closo*-$B_{10}H_{10}^{2-}$ and *nido*-$B_{10}H_{14}$ have already been extensively reviewed on numerous occasions and little further need be said in detail about these two species (see Section 2 and refs. 14-22,24, 72,73). *Closo*-$B_{10}H_{10}^{2-}$ was the first *closo*-dianion to be discovered (ref. 180) and the original preparation is still the most convenient high-yield route, provided that *nido*-$B_{10}H_{14}$ is available as starting material (ref. 57):

$$B_{10}H_{14} + 2\ NEt_3 \xrightarrow[-H_2]{xylene} B_{10}H_{12}(NEt_3)_2 \xrightarrow[(93\%)]{reflux} [Et_3NH^+]_2[B_{10}H_{10}]^{2-}$$

The thermolysis of Et_4NBH_4 at 185 °C and the reaction of Et_3NBH_3 with diborane at 180 °C are also viable large-scale routes (ref. 181). An early X-ray crystal structure determination (ref. 182) established the bicapped square antiprismatic structure (31) for *closo*-$B_{10}H_{10}^{2-}$ in its copper(I) salt and this was confirmed for the solvated complex $[Cu(PPh_3)_2]_2[B_{10}H_{10}].CHCl_3$ (ref. 183). The most recent structural characterization of *closo*-$B_{10}H_{10}^{2-}$ was done on the two bipyridinium salts $[LH]_2[B_{10}H_{10}]$ and $[LH_2][B_{10}H_{10}]$ where L is 2,2'-bipyridine (ref. 184); as usual the B-B connectivity pattern within the cluster is reflected in the B-B interatomic distances, being 169.0(6) pm between B(apex) and B(antiprism) [e.g. B(1)-B(2)], 182.7(7) pm between B atoms in the square bases [e.g. B(2)-B(3)], and 181.2(8) pm between B atoms in the equatorial triangular faces [e.g. B(2)-B(6)]. The mean B-H distance was 112(3) pm. Most salts of *closo*-$B_{10}H_{10}^{2-}$ are colourless or pale yellow but the diprotonated bipyridinium salt $[LH_2][B_{10}H_{10}]$ is bright red due to change transfer interactions. Many quaternary N, P, or As salts of $B_{10}H_{10}^{2-}$ melt (sometimes with decomposition) in the range 200-230 °C; $Ag_2B_{10}H_{10}$ darkens at 270 °C but is still solid at 360 °C, and the most stable salts, such as those of thallium(I) and caesium, are still solid above 600 °C. They are therefore amongst the most stable polyhedral borane compounds known. The hydrated parent acid $[H_3O]_2B_{10}H_{10}.nH_2O$ is a sirupy liquid which loses water under reduced pressure to yield a crystalline solid of composition approximating to the dihydrate (ref. 185).

Decaborane(14), *nido*-$B_{10}H_{14}$, holds a special place in the hierarchy of polyhedral boranes: it is an air-stable, white crystalline solid, mp 99.6 °C, and was one of the first of the boranes to be isolated by A. Stock (in 1913, see ref. 1, page 80). Even more significantly, it was the first borane to have its polyhedral structure elucidated by X-ray crystallography (ref. 186), and for many years this astonishing "nest-like" or

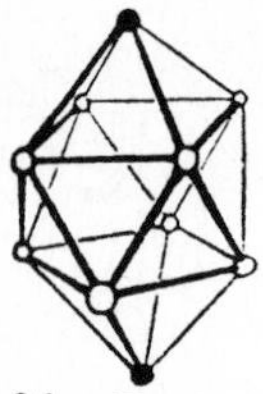

(31) $B_{10}H_{10}^{2-}$ (D_{4d})

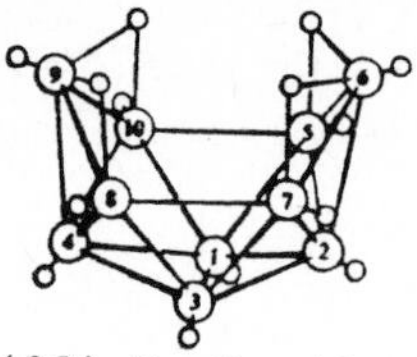

(32) $B_{10}H_{14}$ (C_{2v})

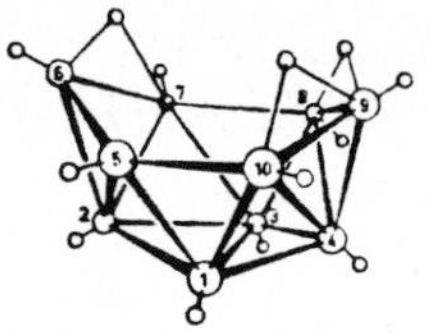

(33) $B_{10}H_{13}^{-}$ (C_1)

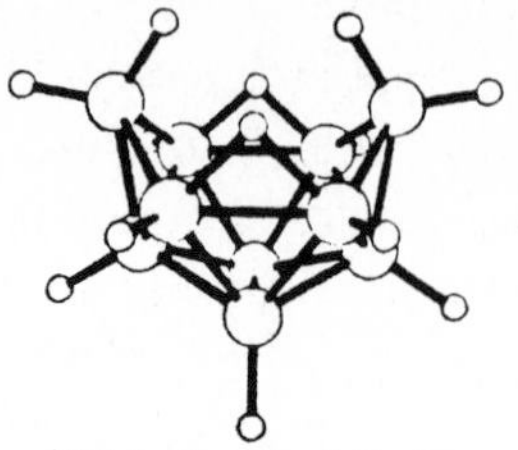

(34) $B_{10}H_{14}^{2-}$ (C_{2v})

nido structure (32) was treated as the exemplar of an icosahedral fragment which dominated the perceptions of polyhedral cluster geometry. In fact, (ref. 187) it is now better thought of as being derived from an eleven-vertex *closo*-B_{11} octadecahedral cluster by removal of the unique 6-connected vertex [see structure (38) in Section 13]. The most precise structural data come from a neutron diffraction study on $^{11}B_{10}D_{14}$ at -160 °C (ref. 188): the B-B distances fall in the range 176-179 pm except for significantly shorter distances of 171.5(4) pm between B(2)-B(6) and B(4)-B(9) pm, and significantly longer distances of 197.3(4) pm between B(5)-B(10) and B(7)-B(8); terminal B-D distances are in the range 117-119 pm and the (slightly asymmetric) bridge bonds are typified by the distances B(5)-D_μ 129.8(5) pm and B(6)-D_μ 135.5(7) pm.

Decaborane(14) can be titrated as a monobasic acid with pK_a 3.5 in methyl cyanide solution (ref. 189) and pK_a 2.7 in ethanol (ref. 190); cf. pK_a 2.85 for $ClCH_2CO_2H$ and 4.75 for CH_3CO_2H. Deprotonation occurs at a bridge site as confirmed by an X-ray crystal structure study on $[NEt_3H]^+[B_{10}H_{13}]^-$ at -170 °C (ref. 191); see structure (33). The interatomic distances are rather similar to those found in *nido*-$B_{10}H_{14}$ itself though there are a few slight differences. Salts of *nido*-$B_{10}H_{13}^-$ tend to be yellow both in solution and as solids and those with large quaternary cations can sometimes be melted with minimal decomposition e.g. $[NEt_3H]^+$ 98 °C, $[PMePh_3]^+$ 127 °C, $[NEt_4]^+$ 250°C (ref. 192).

More forcing conditions are required to bisdeprotonate $B_{10}H_{14}$. For example, $Na_2B_{10}H_{12}$ can be prepared in 75% yield by heating an ethereal solution of $B_{10}H_{14}$ with NaH in a sealed system over a period of days at room temperature (ref. 193). The crystalline disolvate $Na_2B_{10}H_{12}$.2thf can be obtained similarly (ref. 194) and numerous quaternary salts prepared from this by metathesis (refs. 194,195). No X-ray structure is available but nmr spectra (ref. 195) are consistent with the loss of two adjacent H_μ to give a *nido*-$B_{10}H_{12}^{2-}$ dianion of C_s symmetry with four boron atoms B(2), B(4), B(6), and B(9) on the mirror plane. It should be noted however that many metallaborane cluster complexes of $[\eta^4\text{-}B_{10}H_{12}]^{2-}$ are known in which the coordinated dianion adopts the alternative C_s configuration resulting from loss of μ-H(5,6) and μ-H(9,10), which has only B(1) and B(3) on the mirror plane (refs. 3,196).

The *arachno*-dianion $B_{10}H_{14}^{2-}$ is best prepared (in 90% yield) by reacting *nido*-$B_{10}H_{14}$ with an aqueous solution of KBH_4 for half an hour and then precipitating the product by addition of a saturated solution of CsCl (ref. 197). Other salts can be prepared by metathesis. The structure (34) has been determined by X-ray crystallography on $[NMe_4]_2[B_{10}H_{14}]$ (ref. 198) and can be considered as an icosahedral fragment formed by the notional removal of two adjacent vertices from a B_{12} icosahedron. Comparison with structure (32) for *nido*-$B_{10}H_{14}$ is instructive: the two species differ principally in the disposition of the four *endo*-H atoms, the four H_μ in *nido*-$B_{10}H_{14}$ being replaced by $2H_\mu$ and two *endo*-H_t in *arachno*-$B_{10}H_{14}^{2-}$, reflecting the change in topology from (4620) to (2632). Both structures have C_{2v} symmetry and the detailed cluster dimensions are very similar except for a significant contraction in the B(5)-B(10) [and B(7)-B(8)] distance from 197.3(4) pm to 188(1) pm.

Protonation of *arachno*-$B_{10}H_{14}^{2-}$ by aqueous ethanolic HCl affords *arachno*-$B_{10}H_{15}^-$ (ref. 199), and $[PMePh_3][B_{10}H_{15}]$ can be precipitated by addition of an aqueous solution of $[PMePh_3]Cl$ to an acidified solution of $Na_2B_{10}H_{14}$. The $[NMe_4]^+$ and $[NEt_3H]^+$ salts can be prepared similarly. The definitive structure of *arachno*-$B_{10}H_{15}^-$ has not been determined but the ion presumably has the predicted C_s symmetry and (3622) topology (ref. 14), consistent with the nmr spectra of the anion (ref. 200). The parent *arachno*-$B_{10}H_{16}$ has not been isolated, but the three *conjuncto*-boranes having this stoichiometry are well known (ref. 201) and are considered in the next paragraph.

Conjuncto-bipentaboranyls were first made from *nido*-B_5H_9 by proton irradiation (ref. 202) or electric glow discharge in a stream of hydrogen (ref. 203). Yields were low but X-ray crystallography established the eclipsed structure (35) as $1,1'\text{-}(B_5H_8)_2$

(ref. 203). The 1,2'- isomer (36) can be made almost quantitatively by $PtBr_2$ catalysed dehydrocoupling of B_5H_9 at room temperature (refs. 43,44) or (somewhat less conveniently) by heating 2-BrB_5H_8 with a ten-fold excess of B_5H_9 in the presence of catalytic amounts of $AlBr_3$/Al foil at 65 °C (ref. 204). The X-ray crystal structure is available (ref. 205). The third isomer, 2,2'-$(B_5H_8)_2$ (37) can be made in 35% yield by metathesis of KB_5H_8 and 2-BrB_5H_8 in pentane (ref. 204).

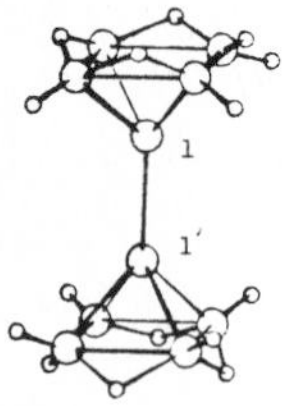

(35) 1,1'-$(B_5H_8)_2$ (D_{4h})

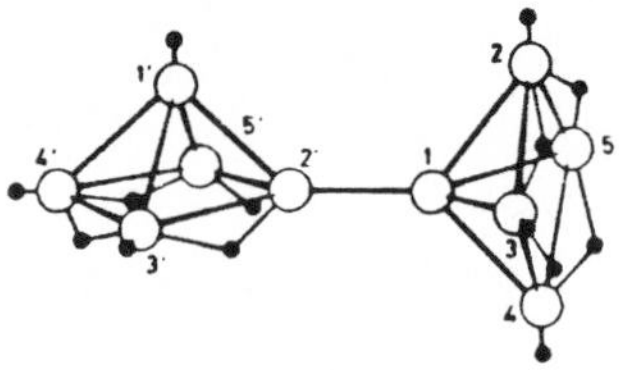

(36) 1,2'-$(B_5H_8)_2$ (C_s)

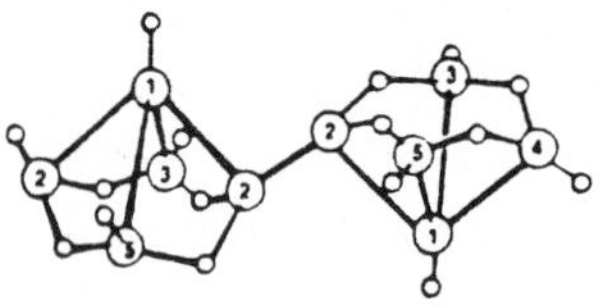

(37) 2,2'-$(B_5H_8)_2$ (C_s)

Properties of the three isomers differ: the 1,1'-isomer is an air stable, readily sublimable white solid at room temperature (ref. 203) with a "poor" mp at *ca.* 81 °C. Treatment of 1,1'-$(B_5H_8)_2$ with HI at room temperature slowly cleaves the central B-B bond to yield B_5H_9 and B_5H_8I; at 150 °C in a sealed tube reaction of 1,1'-$(B_5H_8)_2$ with a stoichiometric amount of I_2 yields $B_{10}H_{14}$ plus 2HI cleanly (ref. 203). The 1,2'-isomer is described as air sensitive colourless crystals mp 18.4 °C (ref. 205). Neither the 1,2'- or 2,2'- isomer shows any tendency to isomerize when heated to 120 °C for several hours under pressure in ether solution (ref. 204). The 2,2'-isomer can be monodeprotonated to $[2,2'\text{-}(B_5H_7)(B_5H_8)]^-$ when treated with KH in Me_2O at -30 °C (ref. 204) or LiMe in Et_2O at low temperature (ref. 205). Likewise LiMe in Et_2O at -35° deprotonates the 1,2'-isomer to $[1,2'\text{-}(B_5H_7)(B_5H_8)]^-$, the bridge proton coming from the pentaboranyl cluster whose apical boron atom is involved in the σ B-B bond (ref. 204). Reaction of an ether solution of $Li[2,2'\text{-}(B_5H_7)(B_5H_8)]$ with B_2H_6 for 6 hours at -40 °C gave $B_{10}H_{14}$, LiB_3H_8, and $LiBH_4$ in low yields. If, however, the addition of B_2H_6 was followed by removal of volatiles and solvent and addition of HCl, then moderate yields of *n*-B_9H_{15} and $B_{10}H_{14}$ were obtained (ref. 205); a mechanism was proposed which involves addition of $\{BH_3\}$ to the basal B-B bond formed by the initial deprotonation followed by proton-induced elimination of a diboron fragment and rearrangement to *n*-B_9H_{15}. By contrast, treatment of the 1,2'-isomer with "superhydride" Li[BEt_3H] in thf at room temperature for two hours, followed by workup, affords [NMe_4][B_9H_{14}] in 94% yield (ref. 205); the suggested mechanism involves hydride ion transfer to give $B_{10}H_{17}^-$ which then loses {BH_3} to give the product. Heating the 1,2'-isomer with Et_3NBH_3 in decane at 100 °C followed by workup gives a 59% yield of $[NEt_3H]_2[B_{12}H_{12}]$ (ref. 206)

Among the many products of the cothermolysis of B_5H_9 and B_2H_6 in a hot/cold reactor at 140°/-20 °C was a new borane characterized as $B_{10}H_{18}$ which came off the low-temperature fractionation column at -15 °C (ref. 156); the yield of $B_{10}H_{18}$ increased somewhat when B_5H_{11} was also initially added to the cothermolysis mixture and it is possible that the product can be formulated as a B-B bonded *conjuncto*-[(B_5H_8)(B_5H_{10})].

13. UNDECABORANES

Members of both the *closo* and *nido* series are known: *closo*-$B_{11}H_{11}^{2-}$, *nido*-$B_{11}H_{13}^{2-}$, *nido*-$B_{11}H_{14}^-$ and the very recently characterized neutral parent borane, *nido*-$B_{11}H_{15}$.

Pyrolysis of $Cs_2B_{10}H_{13}$ (see next paragraph) at 250 °C results in almost quantitative evolution of H_2, and recrystallization affords $Cs_2B_{11}H_{11}$ in up to 70% yield (ref. 159). Other salts can be obtained by metathesis. $Cs_2B_{11}H_{11}$ is stable indefinitely at 400 °C but at 600 °C it quantitively disproportionates in 1 hour to an equimolar mixture of $Cs_2B_{10}H_{10}$ and $Cs_2B_{12}H_{12}$. The *closo*-$B_{11}H_{11}^{2-}$ dianion is thus of comparable thermal stability to $B_6H_6^{2-}$ and is next only to $B_{10}H_{10}^{2-}$ and $B_{12}H_{12}^{2-}$. It is also more stable to hydrolysis than *closo*-$B_nH_n^{2-}$ (n = 6,7,8,9), being unaffected by neutral or aqueous alkali solutions and also by acetic acid or 2N hydrochloric acid, though it degrades in more strongly acid solutions (ref. 159). The structure has not been definitively established but is thought to feature the octadecahedral cluster shown in structure (**38**); this cluster has now been observed in the derivative anion *closo*-[4-(SMe_2)$B_{11}H_{10}$]$^-$ (ref. 207). The parent *closo*-dianion is highly fluxional in solution as revealed by boron-11 nmr.

Salts of the *nido*-$B_{11}H_{13}^{2-}$ dianion can readily be prepared by recrystallization of *nido*-$B_{11}H_{14}^-$ salts from aqueous alkaline solution: $[B_{11}H_{14}]^- + OH^- \longrightarrow [B_{11}H_{13}]^{2-} + H_2O$ (refs. 159,208). The crystal structure of Cs[NMe_4][$B_{11}H_{13}$] has been determined and

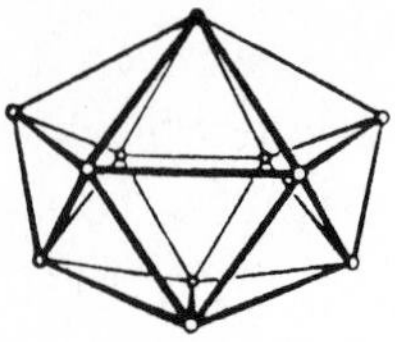

(38) $B_{11}H_{11}^{2-}$ (C_{2v})

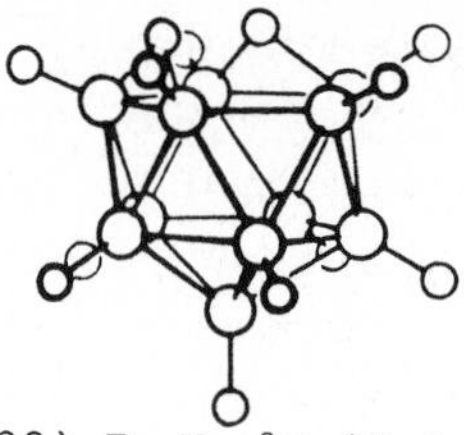

(39) $B_{11}H_{13}^{2-}$ (C_s)

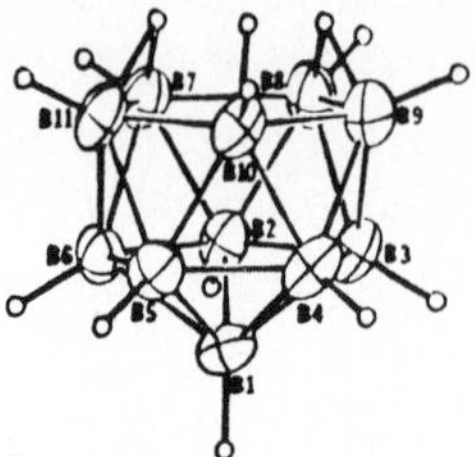

(40) $B_{11}H_{14}^{-}$ (C_s)

the anion has been found to have structure (39) in which the B_{11} cluster consists of a nearly regular icosahedron with one vertex missing (ref. 209). Two H atoms bridge non-adjacent edges of the open pentagonal face [B-H_μ(av) 128(10) pm] and the remainder are terminal [B-H_t(av) 114(10) pm]. The H-bridged B-B distances are 189(1) pm, the other B-B distances in the open face are 181(1) pm, and the remaining B-B distances are in the range 172-183 pm.

The corresponding anion *nido*-$B_{11}H_{14}^{-}$ was first made (ref. 159) by heating *nido*-$B_{10}H_{14}$ with $LiBH_4$ or $NaBH_4$ in 1,2-dimethoxyethane at 90 °C under pressure; (cf. room temperature reaction which yielded $MB_{10}H_{13}$). Several other salts were prepared by metathesis. Subsequently $NaB_{11}H_{14}$ was shown to result from direct reaction of $NaBH_4$ with either B_2H_6 or B_5H_9, or with NaB_3H_8, but the best preparation (ref. 210) comes from the reaction of $NaBH_4$ with $Et_2O.BF_3$ in diglyme at 105 °C using the stoichiometry required by the overall equation:

$$17NaBH_4 + 20Et_2O.BF_3 \longrightarrow 2NaB_{11}H_{14} + 15NaBF_4 + 20H_2 + 20Et_2O$$

Yields can be as high as 70%. The precise structure of *nido*-$B_{11}H_{14}^{-}$ has been the subject of considerable discussion but a very recent X-ray crystal structure analysis of $[PMe_3H][B_{11}H_{14}]$ has established the structure (40) in which the open pentagonal face has two non-adjacent H_μ (as in *nido*-$B_{11}H_{13}^{2-}$) and one BH_2 group thereby retaining approximate C_s symmetry. Interatomic distances are similar to those in *nido*-$B_{11}H_{13}^{2-}$; in particular the B-B distances in the open face [188.6(8) pm] are slightly longer than those within the cluster [176.2(10) pm] and the *exo*-B-H_t distances are slightly shorter than the *endo*-BH_t and the bridging B-H_μ distances [108(5), 113, and 130(9) respectively] (ref. 211).

The elusive parent borane *nido*-$B_{11}H_{15}$ was prepared in the same work (ref. 211) by careful protonation of $K_2B_{11}H_{13}$ or $KB_{11}H_{14}$ using anhydrous HCl at -78 °C; the yield was quantitative after 12 hours. *Nido*-$B_{11}B_{15}$ is not stable above 0 °C and decomposes with evolution of H_2: it is fluxional in solution even at -80 °C but the exact nature of the H atoms associated with the open face is uncertain and it is conceivable that protonation of $B_{11}H_{14}^-$ even results in the opening of an edge in the pentagonal face.

14. DODECABORANES

The existence of the unprecedented icosahedral dianion, *closo*-$B_{12}H_{12}^{2-}$, was predicted by H.C. Longuet-Higgins and M. de V. Roberts in 1955 (ref. 212) some five years before it was serendipidously first prepared in low yield from the obscure reaction between 2-$IB_{10}H_{13}$ and NEt_3 in refluxing benzene (ref. 55). There are now several high-yield routes to salts of *closo*-$B_{12}H_{12}^{2-}$ (see Section 2) and the structure (41) is well established (ref. 213,214). Many salts are known (refs. 13,19,22d,24) and the ion is the stablest of all *closo*-$B_nH_n^{2-}$, the caesium salt being stable up to at least 600 °C. The anion is least distorted in the triethylammonium salt for which the mean icosahedral B-B distance is 178.1(2) pm (ref. 214). The hydrated free acid, $[H_3O]_2[B_{12}H_{12}].4H_2O$ is also known (ref. 215) and is said to be slightly stronger than H_2SO_4.

The only other well characterized dodecaborane is the recently synthesized *conjuncto*-$B_{12}H_{16}$, which was found to be air-stable, mp (decomp.) 64-66 °C (refs. 46,47). The compound was prepared in 43% yield via the metal-promoted oxidative fusion of two *nido*-$B_6H_9^-$ anions using a mixture of $FeCl_2$ and $FeCl_3$ in Me_2O at -78 °C. The structure (42) can be described as resulting from the transoid basal joining of the two pentagonal pyramidal B_6 units by loss of two adjacent H_t from one subcluster; all the other ten B atoms retain their H_t and there are three H_μ on each subcluster. The B_{12} framework thus has C_s symmetry with the two apical B atoms and two basal B atoms on the mirror plane. The structure can also be thought of as resulting from the fusion of a B_8 and a B_6 unit along a common edge (cf. $B_{16}H_{20}$ and the two $B_{18}H_{22}$ species in later Sections). The B-B, B-H_t, and B-H_μ distances are within the expected ranges.

Deprotonation of *conjuncto*-$B_{12}H_{16}$ with KH in thf affords the anion, *conjuncto*-$B_{12}H_{15}^-$, which can be reprotonated with HCl to regenerate the parent borane (ref. 47).

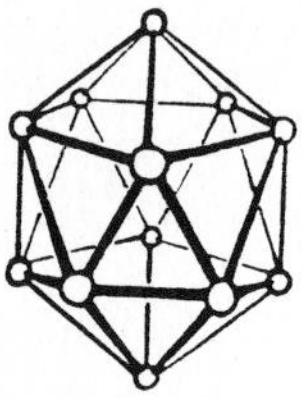

(41) $B_{12}H_{12}^{2-}$ (I_h)

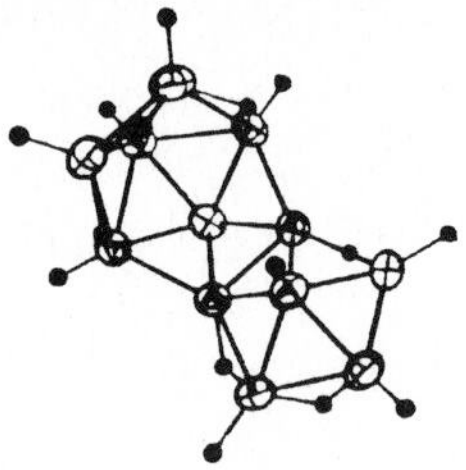

(42) $B_{12}H_{16}$ (C_s)

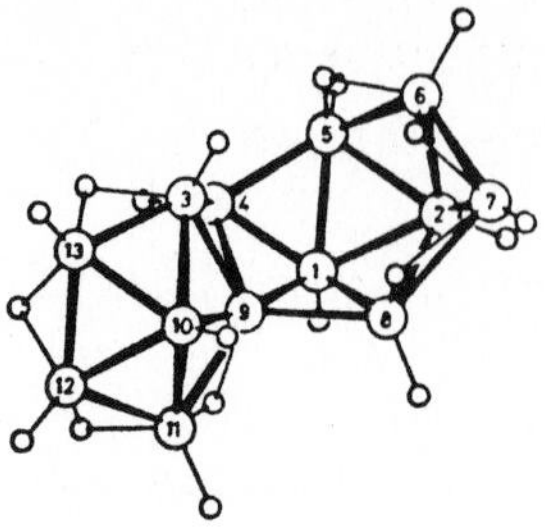

(43) $B_{13}H_{19}$ (C_1)

15. TRIDECA-, TETRADECA-, PENTADECA-, AND HEXADECA-BORANES

No *closo*-borane clusters are known with more than 12 vertices, though 13- and 14-vertex carbaboranes and metallacarbaboranes have been reported (see p.384 of ref. 3 for leading references). Calculations on supra-icosahedral *closo* species $B_nH_n^{c-}$ up to n = 24 suggest that such clusters should have stabilities comparable with that of $B_{10}H_{10}^{2-}$, at least up to n = 22 (ref. 216).

Conjuncto-$B_{13}H_{19}$ was first identified amongst the products of pyrolysis of B_6H_{10} (ref. 35) but is now best prepared (in 32% yield) by heating $K[B_7H_{11}Br]$ with B_6H_{10} in Me_2O at temperatures between -78 °C and ambient (ref. 217):

$$2\text{-}BrB_6H_9 \xrightarrow[Me_2O]{KH/} [2\text{-}BrB_6H_8]^- \xrightarrow[-78\ ^\circ C]{\frac{1}{2}B_2H_6} [B_7H_{11}Br]^- \xrightarrow{B_6H_{10}} B_{13}H_{19} + H_2 + Br^-$$

The product is obtained as yellow crystals, mp 43.5 °C, and the structure (43) was established by X-ray crystallography (refs. 35,218): it can be viewed as a *conjuncto*-borane formed by fusion of either B_6 and B_9 fragments or B_7 and B_8 fragments sharing a common edge. There are 12 terminal and 6 bridging H atoms. A notable feature of the structure is the seven-coordinate B(9) atom which has no terminal H atom attached to it; consistent with expectations, several of the B-B distances involving this atom are rather long - for example, the average distance from B(9) to the four atoms B(3,8,10,11) is 185.3 pm compared with the other B-B distances in the molecule which span the range 170-180 pm.

Deprotonation of *conjuncto*-$B_{13}H_{19}$ by KH in Et_2O at 25 °C affords a quantitative yield of the yellow potassium salt of the anion *conjuncto*-$B_{13}H_{18}^-$ (ref. 217).

Three *conjuncto*-tetradecaboranes are known: $B_{14}H_{18}$, $B_{14}H_{20}$, and $B_{14}H_{22}$. *Conjuncto*-$B_{14}H_{18}$ was first prepared in 1972 by the controlled hydrolysis of $B_{16}H_{20}$ in hexane at 60 °C for 8 hours (ref. 219):

$$B_{16}H_{20} + 6H_2O \longrightarrow B_{14}H_{18} + 2H_3BO_3 + 4H_2$$

The product when purified (30% yield) was a viscous, air-sensitive, pale yellow oil which could not be induced to crystallize even at low temperature. It is stable under nitrogen at room temperature, distils at 10^{-4} mmHg/20 °C, and decomposes above 100 °C. Structure (**44**) was deduced from a detailed analysis of the boron(11) nmr spectrum and is consistant with its mode of formation: the structure can be viewed as being formed by the transoid fusion of *nido*-$B_{10}H_{14}$ and *nido*-B_6H_{10} by the sharing of a common edge [B(5)B(6) in $B_{10}H_{14}$ and B(2)B(3) in B_6H_{10}].

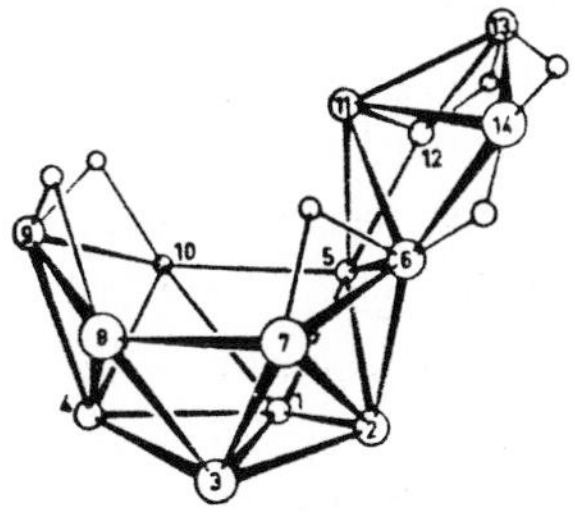

(**44**) $B_{14}H_{18}$ (probable)

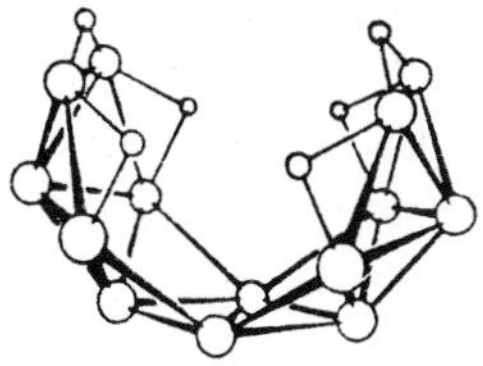

(**45**) $B_{14}H_{20}$ (C_{2v})

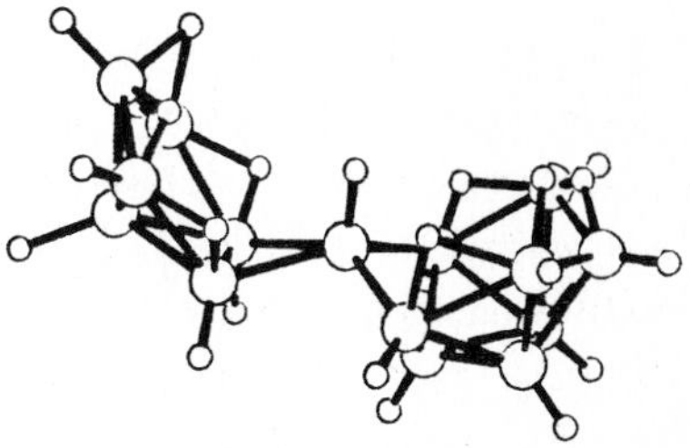

(**46**) $B_{15}H_{23}$ (C_1)

Tetradecaborane(20), *conjuncto*-$B_{14}H_{20}$, is a pale yellow, moderately air-stable crystalline compound made in low yield (2%) by reacting KB_6H_9 with B_8H_{12} in ether solution, followed by removal of solvent and treatment of the $KB_{14}H_{21}$ so formed with HCl at -78 °C; separation from the co-product, $B_{16}H_{20}$, is tedious (ref. 36). An X-ray structure analysis at -164 °C of the colourless (thermochromic) crystals revealed the molecular structure (**45**): two B_8H_{12} fragments fused in cisoid configuration at a common B(3)B(8) edge to give overall C_{2v} symmetry. Each boron atom has a terminal H atom attached and there are 6 bridging H atoms symmetrically disposed on the outer parts of the open face. The inner B(2)B(7)B(3)B(12) lozenge is almost planar (dihedral angle 176.97°). Although not isolated during the course of the synthesis, it is clear that the anion $[B_{14}H_{21}]^-$ is preparable and stable, at least at 0 °C.

Little is known about tetradecaborane(22), *conjuncto*-$B_{14}H_{22}$: it was prepared as a colourless powder in 87% yield by treating B_8H_{12} with a three-fold excess of B_6H_{10} in a sealed tube at -23 °C for 15 min (ref. 34). The compound melts at about 25 °C to give an unstable colourless liquid.

Pentadecaborane(23), *conjuncto*-$B_{15}H_{23}$, has been prepared as colourless crystals in almost quantitative yield by the addition of *nido*-B_6H_{10} to *i*-B_9H_{15} (ref. 33). The structure originally proposed (ref. 34) is now known to be incorrect in some details as a result of subsequent (as yet unpublished) X-ray diffraction experiments (ref. 220). The correct structure (46) features a *commo*-boron atom and is essentially formed by the cisoid addition of B_6H_{10} via its basal B-B bond to form a new 3-centre BBB bond involving the *commo*-B atom on the *i*-B_9 moiety.

The only hexadecaborane species known at the present time is *conjuncto*-$B_{16}H_{20}$ which was first prepared in low yield by thermolysis of $B_9H_{13}.SMe_2$ (ref. 221); using a controlled sequence of temperatures between 80° and 110 °C, an optimum yield of 6.7% was achieved (ref. 222). The compound is also one of the major products of decomposition of B_8H_{12} at 25 °C, either alone or in the presence of a two-fold excess of B_6H_{10} (refs. 34,35). $B_{16}H_{20}$ is a robust compound which forms colourless needles mp 108-112 °C and which can be sublimed at 100 °C (10^{-3} mmHg). The structure (47) has been obtained from single-crystal X-ray diffraction analysis: it can be described as the transoid fusion (without significant distortion) of $B_{10}H_{14}$ and B_8H_{12} along a common edge [B(5)B(6) and B(3)B(8) respectively] (ref. 223). The cluster-connectivity of the two shared boron atoms is six, and neither has a terminal H atom attached.

No 17-vertex neutral or anionic binary borane species are known though some 17-vertex metallaboranes have recently been synthesized (see pp.387-392 of ref. 3 for details). It is also worth noting that molecular orbital studies on hypothetical large *closo*-boranes suggest that the supra-icosahedral *closo*-$B_{17}H_{17}^{2-}$ dianion with C_{2v} symmetry should have a stability which is greater than that of *closo*-$B_{10}H_{10}^{2-}$ and second only to *closo*-$B_{12}H_{12}^{2-}$ itself (ref. 216).

16. OCTADECABORANES

The transoid fusion of two *nido*-$B_{10}H_{14}$ clusters along a common edge can be effected in two ways to give either a centrosymmetric molecule, *anti*-$B_{18}H_{22}$ (48) or a non-centrosymmetric molecule, *syn*-$B_{18}H_{22}$ (49). The corresponding pairs of anions $B_{18}H_{21}^-$ and dianions $B_{18}H_{20}^{2-}$ are also known. The earlier literature usually refers to the *anti*-isomers as *n*- and the *syn*-isomers as *i*- but the structural descriptors are now preferred.

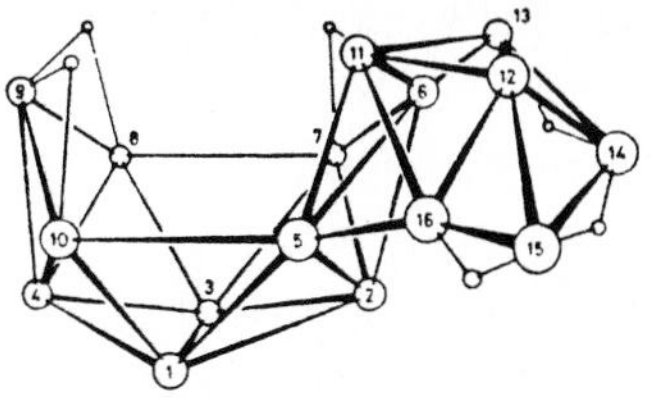

(**47**) $B_{16}H_{20}$ (C_1)

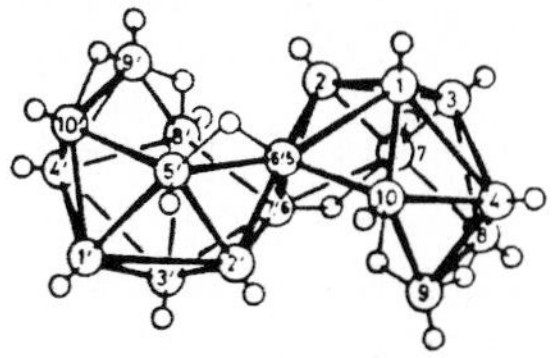

(**48**) *anti*-$B_{18}H_{22}$ (C_i)

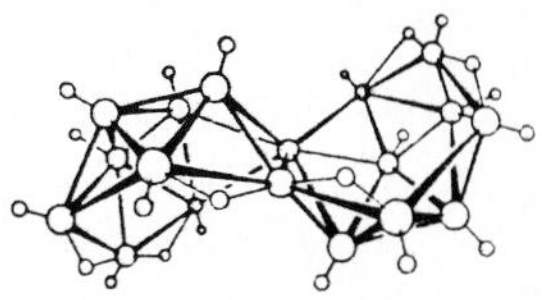

(**49**) *syn*-$B_{18}H_{22}$ (C_2)

A mixture of *anti*- and *syn*-$B_{18}H_{22}$ was first obtained (ref. 224) by hydrolytic degradation of $[H_3O]^+{}_2[B_{20}H_{18}]^{2-}$ (see next Section for synthesis). The total yield was *ca.* 60% but separation of the isomers was tedious (see also below). A better route to *anti*-$B_{18}H_{22}$ is the thermolysis of $B_9H_{13}.SMe_2$ (refs. 221,222) or of similar adducts $B_9H_{13}L$ formed *in situ* by protonation of *arachno*-$B_9H_{14}^-$ in donor solvents such as $Bu^n{}_2O$ (ref. 178). The pure *anti*-isomer can also be obtained by protonating $B_{11}H_{14}^-$ in dioxane solution, removing the solvent and subliming the product at 120 °C under reduced pressure (ref. 225). The oxidative fusion of *nido*-$B_9H_{12}^-$ has already been mentioned in Section 2 as a viable high-yield route to *anti*-$B_{18}H_{22}$ (ref. 49,50). Mixtures of *syn*- and *anti*-isomers result from treatment of $B_{10}H_{13}^-$ with $RuCl_3$ in thf followed by air oxidation (ref. 48). Separation of the two isomers is best effected by partition chromatography on silica gel (ref. 226). The C_2 structure of *syn*-$B_{18}H_{22}$ is chiral and the compound can be resolved into its enantiomers by treating an alkaline aqueous ethanolic solution of the borane with (+)-camphidine followed by repeated crystallization of the yellow needles and work-up via the sodium salt of (+)-*syn*-$B_{18}H_{22}$ (*ca.* 10% yield) and the tetramethylammonium salt of (−)-*syn*-$B_{18}H_{22}$ (3% yield (ref. 226). Optical rotations of the enantiomers are $[\alpha]_D^{20}$ = +2250° and −2500°.

Single-crystal X-ray analysis of the centrosymmetric *anti*-$B_{18}H_{22}$ (ref. 227) and non-centrosymmetric *syn*-$B_{18}H_{22}$ (ref. 228) establish the structures (**48**) and (**49**) respectively. Detailed dimensions are very similar to those in *nido*-$B_{10}H_{14}$ itself (Section 12). Both compounds are air-stable yellow solids mp 179-180 °C (*anti*-$B_{18}H_{22}$) (refs. 224,227) and 125-128 °C (*syn*-$B_{18}H_{22}$)

(refs. 224,226).

Stable salts of the monoanions and dianions of both isomers are readily formed by deprotonation of the parent boranes. Indeed, some of preparative routes described above lead first to the anions which are then protonated to give the neutral boranes. Detailed nmr studies on *anti*-$B_{18}H_{22}$ and its mono- and dianions using [^{11}B-^{11}B]-COSY, [^{1}H-^{1}H]-COSY, and ^{1}H-{^{11}B(selective)} techniques and [^{1}H-^{11}B] shift correlations have shown that *anti*-$B_{18}H_{22}$ deprotonates by loss of one or both of the bridging H atoms associated with the *conjuncto*-boron atoms i.e. H(5,5') and H(6,7) in structure (50) which is a planar projection of structure (48) (ref. 229). This has been confirmed by an X-ray crystal structure study on $[NMe_4]_2$[*anti*-$B_{18}H_{20}$], see structure (51).

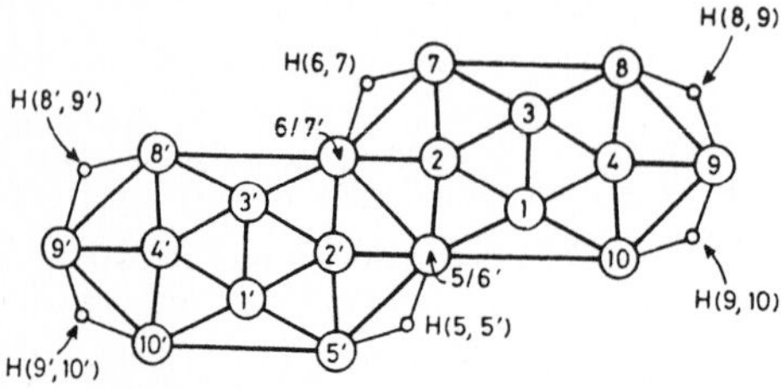

(50) *anti*-$B_{18}H_{22}$ (projection)

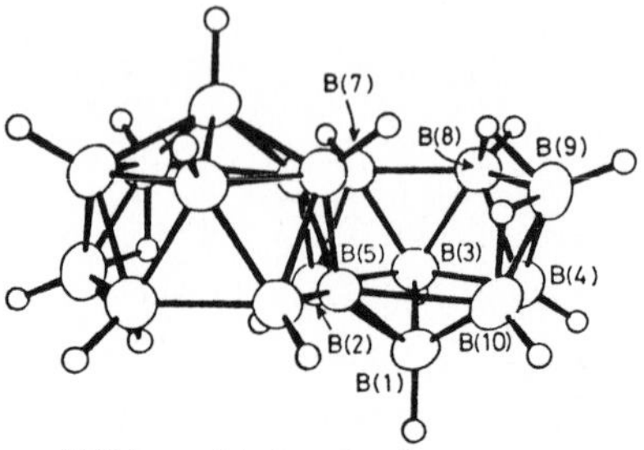

(51) *anti*-$B_{18}H_{20}^{2-}$ (C_i)

The dianion retains its centre of symmetry and detailed interatomic distances are very similar to those in the parent borane except for the B(5)-B(10) and B(7)-B(8) distances, which are almost identical at 197.6(4) and 196.8(4) pm in the neutral borane but are lengthened and shortened to 210.3(5) and 184.2(5) pm respectively in the dianion (ref. 229). The value of 210.3(5) pm for B(5)-B(10) [and B(7')-B(8')] in *anti*-$B_{18}H_{20}^{2-}$ is unusually long for formally bonded B-B distances. The only other notable difference in interatomic distances between the parent borane and the dianion is for the *conjuncto*-link B(5)-B(6) itself which decreases from 183.8(5) to 179.9(5) on removal of the bridging hydrogen atoms attached to these borons.

A rich and fertile metallaborane chemistry based on B_{18} species has been developed (see pp.392-403 of ref. 3).

17. ICOSABORANES

Three main classes of *conjuncto*-icosaborane are known :

(a) the unique pseudo-*closo*-$B_{20}H_{16}$ formed by the notional fusion of two *nido*-$B_{10}H_{14}$ units thereby resulting in the elimination of all eight H_{μ} and all four H_t on the four *commo*-boron atoms: this is the only known binary borane molecule which has fewer H atoms than B atoms in its formula;
(b) a series of seven anionic species formed by fusion of two *closo*-$B_{10}H_{10}^{2-}$ units in various ways; and
(c) a series of at least nine (of the possible eleven) geometrical isomers of the B-B σ-bonded bi(*nido*-decaboranyl), $B_{20}H_{26}$.

Icosaborane(16), *conjuncto*-$B_{20}H_{16}$, can be made either by passing $B_{10}H_{14}$ and H_2 through an A.C. discharge (ref. 230), or by the catalytic pyrolysis of $B_{10}H_{14}$ in the presence of $BMe_2(NHMe)$ at 350 °C (ref. 231); when this latter process is used in a flow reactor yields of up to 10% can be obtained (ref. 232). It is a colourless hygroscopic crystalline solid, mp 196-199 °C (refs. 230, 231). The molecular structure (52) was determined by X-ray diffractometry (refs. 230,233): the D_{2d} symmetry is apparent, as is the absence of terminal atoms on the four central B atoms. The shortest B-B distances are 176.6(6) pm for B(1)-B(2) at the top of the figure and 177.0(4) pm for B(3)-B(9 ("verticals"). The four central *conjuncto* B-B distances at 178.4(4) pm are also shorter than the average of all 52 B-B edges which is 180.5 pm, and this contractional distortion of the two faces of the notional *nido*-B_{10} subcluster precursors has led some to prefer a description of *conjuncto*-$B_{20}H_{16}$ as formed by the fusion of the two four-atom faces obtained by removal of the unique 4-connected vertex from each of two *closo*-B_{13} clusters (ref. 234). The chemistry of $B_{20}H_{16}$ awaits development, though it is known to react with water to give $[H_3O]^+_2[B_{20}H_{16}(OH)_2]^{2-}$ and with ethanol and diethyl ether to give adducts of undetermined structure (ref. 232).

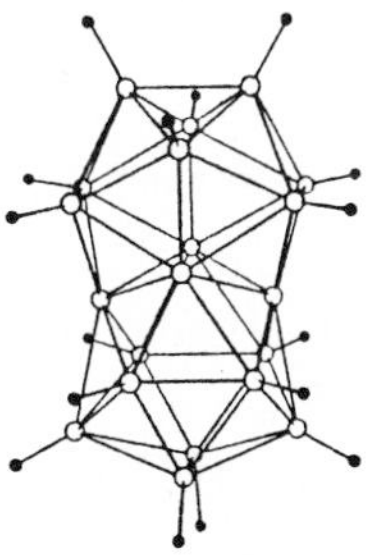

(52) $B_{20}H_{16}$ (D_{2d})

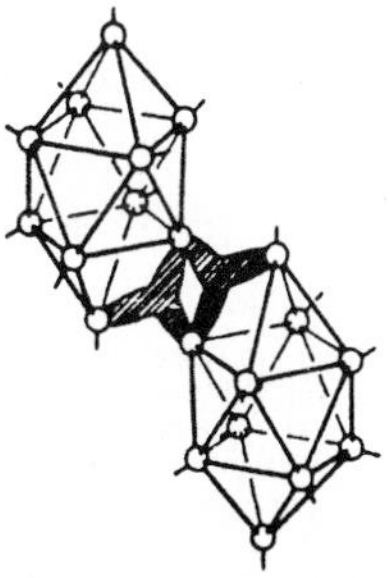

(53) $B_{20}H_{18}^{2-}$ (C_{2h})

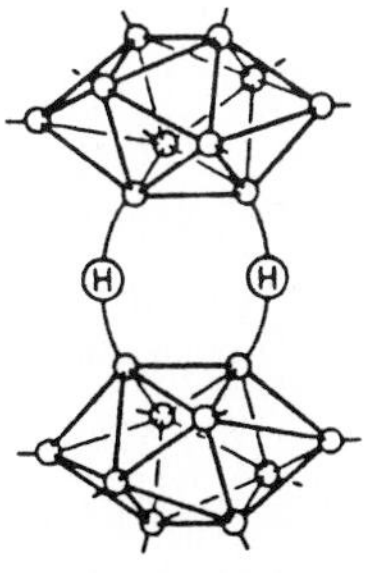

(54) *photo*-$B_{20}H_{18}^{2-}$ (C_{2h})

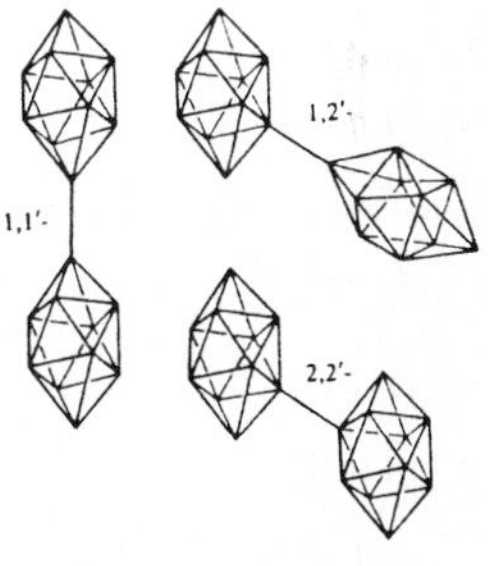

(55) $B_{20}H_{18}^{4-}$ isomers

Numerous icosaborate anions have been characterized. The possible existence of the *conjuncto*-dianion $[B_{20}H_{18}]^{2-}$ was predicted by W.N. Lipscomb in 1961 and it was first prepared by his group via the oxidative coupling of two *closo*-$B_{10}H_{10}{}^{2-}$ ions in aqueous solution (ref. 235):

$$2[B_{10}H_{10}]^{2-} + 4Fe^{3+} \longrightarrow [B_{20}H_{18}]^{2-} + 4Fe^{2+} + 2H^{+}$$

Many other oxidants work well, especially Ce^{4+}, and salts with some 20 different cations have been isolated in high yield (ref. 236), though when $[Et_3NH]_2^+[B_{10}H_{10}]^{2-}$ is oxidized by Ce^{4+} a double-salt is obtained $[Et_3NH]_5^+[B_{20}H_{18}]^{2-}[B_{20}H_{19}]^{3-}$ (ref. 237). In general the salts are stable, yellow, high-melting crystalline solids and an X-ray study on $[Et_3NH]_2[B_{20}H_{18}]$ (mp 173 °C, decomp.) revealed a centrosymmetric structure (53) for the dianion (ref. 238). Each B atom except the inner (*conjuncto*) equatorial atoms B(9,9') has one terminal H atom attached; there are no bridging H atoms in the structure and this, coupled with the close approach of B(9,9') and B(10,10') across the centre of symmetry (172 and 165 pm respectively), suggests the presence of two localized 3-centre BBB bonds between the four atoms. Other dimensions are very close to those in *closo*-$B_{10}H_{10}{}^{2-}$ itself.

When (yellow) solutions of $[Et_3NH]_2[B_{20}H_{18}]$ in MeCN are irradiated with a mercury lamp the solution decolourizes and the $[Et_3NH]^+$ salt of the photo-isomer of $[B_{20}H_{18}]^{2-}$ (mp 150-155 °C) can be isolated in 88% yield (ref. 239). Infrared and nmr spectra indicate the presence of B-H-B bridges and the isomer is formulated as structure (54). The photo-isomer reverts to the original structure (53) when thermolysed in MeCN solution at 100 °C for 36 hours.

The anion $[B_{20}H_{19}]^{3-}$, which was first isolated as the double salt $[NEt_3H]_5[B_{20}H_{18}][B_{20}H_{19}]$ (see above), can be obtained pure as $[NEt_3H]_3[B_{20}H_{19}]$ by appropriate metathetical and ion-exchange techniques (ref. 237) or by direct oxidation of $[NH_4]_2[B_{10}H_{10}]$ using Ce^{4+}/H_2SO_4 at 0 °C followed by workup involving $[NMe_4]Cl$/MeCN to yield the pure white hemihydrate of $[NMe_4]_3[B_{20}H_{19}]$ (ref. 236). Salts with $[NH_4]^+$ and $[H_3O]^+$ are also known (ref. 237), and high-yield syntheses (over 70%) have been devised by careful attention to conditions (ref. 240). Spectroscopic data point to the existence of two isomers, both of which comprise two *closo*-$B_{10}H_{10}{}^{2-}$ ions conjoined by a bridging proton, H_μ^+; the 2,2' isomer is stable only in the solid state and rearranges in solution to the more stable 1,2' isomer (refs. 236,237,241).

If, instead of joining two *closo*-$B_{10}H_{10}^{2-}$ units via a bridging H_μ^+, they are joined via a direct B-B σ bond then three further icosaborane anions are obtained, *viz* 1,1'-, 1,2'-, and 2,2'-$[(closo\text{-}B_{10}H_9)_2]^{4-}$ (55). Salts of these anions were first made by deprotonating $[B_{20}H_{19}]^{3-}$ in aqueous alkali (refs. 236,237) or by reacting $[B_{20}H_{18}]^{2-}$ with Mg. Routes to salts of specific isomers are as follows. $K_4[2,2'\text{-}(B_{10}H_9)_2].2H_2O$ was obtained as white needles in 94% yield by treating $[Et_3NH]_2[B_{20}H_{18}]$ with Na in liquid ammonia at -40 °C and then recrystallizing the crude $Na_4[B_{20}H_{18}]$ using potassium acetate in ethanol (ref. 237). The 2,2'-isomer was also obtained in 98% yield by treating the photo-isomer of $[B_{20}H_{18}]^{2-}$ with Na/liq.NH_3 (ref. 239). $K_4[1,2'\text{-}(B_{10}H_9)_2].2H_2O$ was obtained in 54% yield by isomerizing the 2,2'-isomer in the presence of ice-cold aqueous CF_3CO_2H for 20 seconds before adding aqueous KOH; ethanol was added and, after removing an initial crop of the 1,1'-isomer, the desired 1,2'-isomer was obtained by addition of further EtOH and Et_2O (ref. 237). $K_4[1,1'\text{-}(B_{10}H_9)_2]2H_2O$ can be obtained as colourless crystals in 92-95% yield by treating either the mixed salt $[Et_3NH]_5[B_{20}H_{18}][B_{20}H_{19}]$ (see above) or pure $[Et_3NH]_3[B_{20}H_{19}]$ with aqueous KOH and recrystallizing from aqueous ethanol. The 1,1' isomer can also be obtained by the essentially quantitative isomerization of the other two isomers in boiling water (22 hours) or in 2M hydrochloric acid at 70 °C for 5 minutes (ref. 237). Definitive X-ray structural characterization of the three isomers has not been obtained but the structural assignments implied in (55) are firmly based on boron-11 nmr and infrared spectroscopic data.

The final class of *conjuncto*-icosaboranes comprises the B-B σ-bonded bi(*nido*-decaboranyls), $[n,n'\text{-}(B_{10}H_{13})_2]$. $B_{20}H_{26}$ was first definitely characterized, as an impurity in commercial decaborane(14), by high-resolution mass spectrometry (ref. 242). Eleven geometrical isomers can be envisaged of which four should exist as enantiomeric pairs, making 15 isomers in all; these are shown diagrammatically in the Figure, each $B_{10}H_{13}$ cluster being represented schematically by an inner lozenge [B(1,2,3,4] which is surrounded by an outer hexagon [B(5-10)] and the four $B\text{-}H_\mu\text{-}B$ groups. So far, at least nine isomers have been isolated in pure form and the structure of seven of these has been determined by X-ray crystallography and/or nmr spectroscopy (refs. 40-42,243-246).

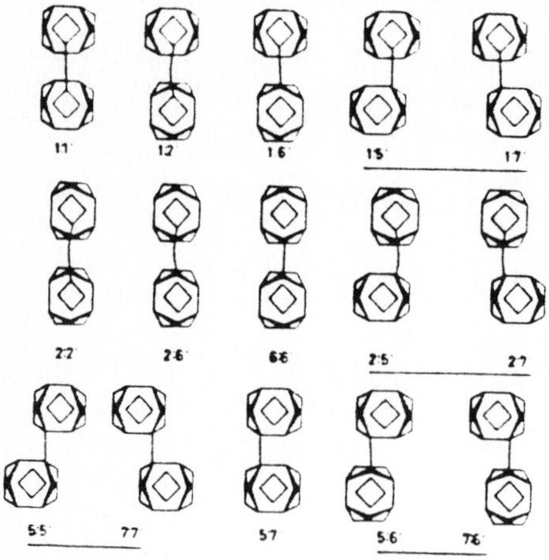

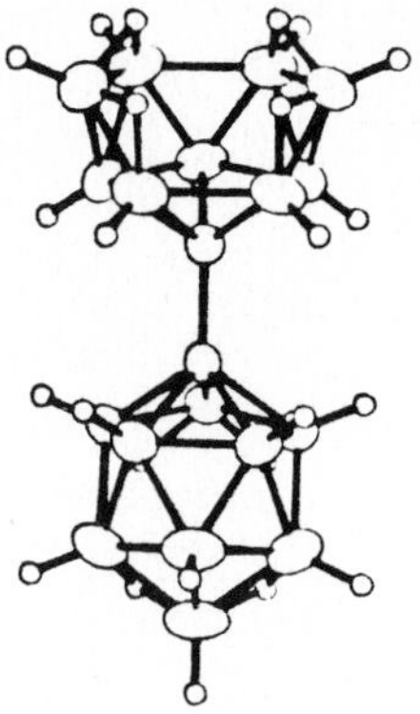

$1,2'-(B_{10}H_{13})_2$
(56)

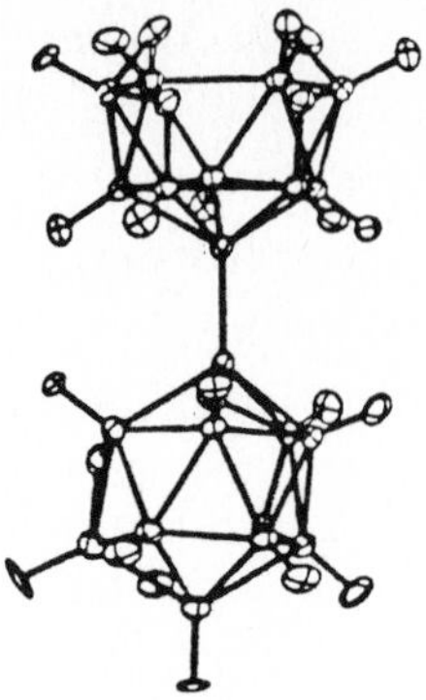

$1,5'-(B_{10}H_{13})_2$
(57)

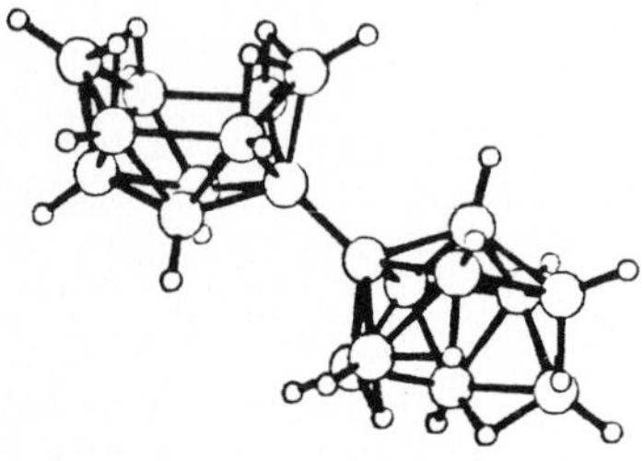

(58) $2,2'-(B_{10}H_{13})_2$

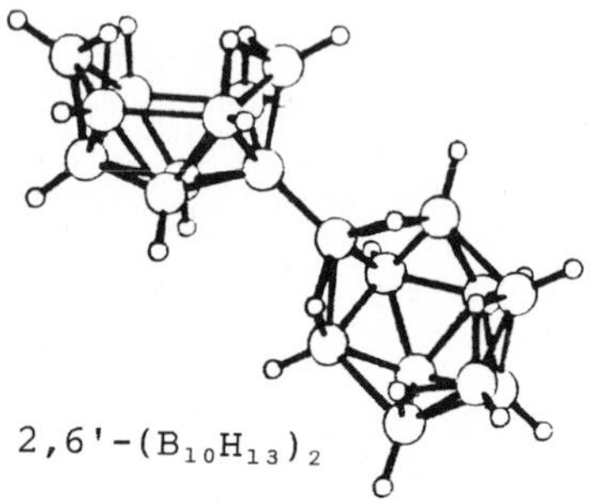

(59) $2,6'-(B_{10}H_{13})_2$

Synthetic routes tend to give mixtures of several isomers in varying proportions and separation can be achieved by preparative-scale thin layer chromatography or by HPLC. All syntheses start from *nido*-$B_{10}H_{14}$, which can be induced to dehydrocouple by (a) thermolysis at 180-185 °C for several hours; (b) thermolysis at 100 °C for 30 min in the presence of "catalysts" such as Me_2S or tetrahydrothiophen; (c) reaction of 6-$ClB_{10}H_{13}$ with $Mg(B_{10}H_{13})I$ at 0 °C; (d) ultravoilet photolysis of $B_{10}H_{14}$ in cyclohexane solution; (e) γ-radiolysis using cobalt-60; (f) irradiation in the beam of a 3MeV van der Graaf generator. The original papers should be consulted for detailed conditions, separatory procedures and yields. For example, ultraviolet photolysis yielded 2,2'-, 2,5'-, 1,2'-, and 1,5'-$(B_{10}H_{13})_2$ in the ratio 40:34:19:7; the overall yield was *ca.* 10% but, allowing for recovered starting material the conversion of reacting $B_{10}H_{14}$ was *ca.* 70% (refs. 41, 245). All the separated isomers were obtained as colourless or off-white solids and their melting points were as follows (cf. $B_{10}H_{14}$ mp 99.6 °C):

isomer	1,2'	1,5'	2,2'	2,5'	2,6'	5,5'	6,6'
mp/°C	139-142	114-115	178-179	109-111	154-155	97-98	198-199

Two further isomers were identified chromatographically and tentatively identified as the 1,6'- and 5,6'-isomers (ref. 245). Molecular structures have been determined by X-ray crystallography as indicated in structures (56), (57), (58), and (59). Detailed dimensions are very similar to those of the parent *nido*-$B_{10}H_{14}$ except for cluster connectivities to the *conjuncto*-boron atoms which are slightly increased by about 1-2 pm. The B-B σ link itself takes on the following values: 169.6(4) (1,2') (ref. 42); 169.8(3) (1,5') (ref. 246); 169.2(3) (2,2') (ref. 244); and 167.9(3) (2,6') (ref. 41). An exploration of the differing chemistries of these last three isomers has begun (ref. 247).

18. HIGHER BORANES

Few neutral or anionic binary boranes having more than 20 boron atoms in the *conjuncto*-cluster have been isolated and identified, though the ubiquitous, intractable "polymeric" boranes that so readily form in many systems are only too well known to all who work in the field. Possible structure motifs for these polymers have been proposed (ref. 248) and the stabilities of supra-icosahedral *closo*-boranes up to $B_{24}H_{24}{}^{c-}$ have been calculated (ref. 216).

The docosaborate dianion, $[B_{22}H_{24}]^{2-}$, is the presumed product formed when an aqueous acidic solution of $NaB_{11}H_{14}$ is treated with C_6H_6/H_2O_2; addition of $[NMe_3H]Cl$ afforded a colourless precipitate of $[NMe_3H]_2^+[B_{22}H_{24}]^{2-}$ (ref. 210). The proposed structure features the linking of *nido*-$B_{10}H_{14}$ and *closo*-$B_{12}H_{12}{}^{2-}$ to give the B-B bonded *conjuncto*-$[(B_{10}H_{13})(B_{12}H_{11})]^{2-}$.

Several salts (Na^+, Cs^+, $NMe_4{}^+$, $NEt_4{}^+$) of the tetracosaborate ion, $[B_{24}H_{23}]^{3-}$, have been made in low yield either by electrochemical oxidation of $Na_2B_{12}H_{12}$ in MeCN (ref. 249) or by chemical oxidation using $FeCl_3$ in MeCN (ref. 250). Decomposition of $[H_3O]_2[B_{12}H_{12}]$ on a steam bath in the presence of oxalic acid followed by workup affords $[NMe_4]_3[B_{24}H_{23}]$ in 4.1% yield (ref. 251). The salts are colourless crystalline solids that behave as 3:1 electrolytes in aqueous solution and spectroscopic data are consistent with the formulation of *conjuncto*-$[B_{24}H_{23}]^{3-}$ as two *closo*-$B_{12}H_{12}{}^{2-}$ units linked by sharing a common bridging H atom,

$[(H_{11}B_{12})\text{-}H_{\mu}\text{-}(B_{12}H_{11})]^{3-}$, *cf.* $[B_{20}H_{19}]^{3-}$ in the preceding section. The major product (59%) from the decomposition of $[H_3O]_2[B_{12}H_{12}]$ mentioned above was identified as $[NMe_4]_5[B_{48}H_{45}].5H_2O$ on the basis of elemental analysis, infrared spectroscopy, and the behaviour of the salt as a 5:1 electrolyte (ref. 251).

The neutral triacontaborane(38), $B_{30}H_{38}$, is probably the largest molecular borane yet characterized. It was first detected mass-spectrometrically as a minor product during the photolytic and thermolytic production of $B_{20}H_{26}$ from $B_{10}H_{14}$ (ref. 40), and was subsequently isolated in larger amounts as a waxy, pale yellow solid from the products of the catalytic thermolysis of $B_{10}H_{14}$ in toluene (ref. 252). Two principal isomers were separated by preparative thin layer chromatography and identified by detailed high-resolution mass-spectrometry and by boron(11) nmr as the bis-*conjuncto*-borane $[(B_{10}H_{13})(B_{10}H_{12})(B_{10}H_{13})]$(*B*-*B*;*B*-*B*). This formulation implies a total of 546 possible isomers consisting of 253 enantiomeric pairs and 40 unique isomers (*i.e.* 293 geometrically distinct species), although a given reaction would probably favour particular families of isomers with related *conjuncto*-features (ref. 252).

ACKNOWLEDGEMENT

I am grateful to Dr Mark Thornton-Pett for help in preparing some of the computer-plotted structures.

REFERENCES

1 A. Stock, Hydrides of Boron and Silicon, Cornell University Press, Ithaca, 1933.
2 J.D. Kennedy, in S. Heřmánek (Editor), Boron Chemistry, World Scientific Publ. Co., Singapore, 1987, pp. 207-243.
3 J.D. Kennedy, Prog. Inorg. Chem., 32 (1984) 519-697; and 34 (1986) 211-434.
4 N.N. Greenwood, Pure Appl. Chem., 55 (1983) 1415-1430.
5 R.N. Grimes (Editor), Metal Interactions with Boron Clusters, Plenum, New York, 1982.
6 R.N. Grimes, in G. Wilkinson, F.G.A. Stone and E.W. Abel (Editors), Comprehensive Organometallic Chemistry, Pergamon Press, Oxford, 1982, pp. 459-542.
7 C.E. Housecroft and T.P. Fehlner, Adv. Organometallic Chem., 21 (1982) 57-112.
8 S. Heřmánek (Editor), Boron Chemistry, World Scientific Publ. Co., Singapore, 1987.
9 G.A. Olah, G.K. Surya Prakash, R.E. Williams, L.D. Field and K. Wade, Hypercarbon Chemistry, Wiley, New York, 1987.
10 T. Onak, in G. Wilkinson, F.G.A. Stone and E.W. Abel (Editors), Comprehensive Organometallic Chemistry, 1982, pp. 411-457.

11 R.W. Jotham in Supplement to Mellor's Comprehensive Treatise on Inorganic and Theoretical Chemistry Vol. V, Part B1, Longman, London, 1981, pp. 450-597.
12 L.J. Todd, in G. Wilkinson, F.G.A. Stone and E.W. Abel (Editors), Comprehensive Organometallic Chemistry, Pergamon Press, Oxford, 1982, pp. 543-553.
13 Gmelin Handbuch der Anorganischen Chemie, Syst. No. 13. (a) Main Volume: Bor, (contains compounds of B with H up to 1925). (b) Supplement Vol. 1, (literature up to 1949). (c) Borverbindungen 8, The Tetrahydroborate Ion and its Derivatives (1950-1974). (d) Borverbindungen 14, Boron-Hydrogen Compounds, Part 1 (1950-1975); Part 2 (1950-1976); Part 3 (1950-1976). (e) Boron Compounds, 3rd Supplement, Vol. 1, Boron and Hydrogen (literature coverage through 1984).
14 W.N. Lipscomb, Boron Hydrides, Benjamin, New York, 1963.
15 R.M. Adams and A.R. Siedle, in R.M. Adams (Editor), Boron, Metallo-Boron Compounds and Boranes, Interscience Publ., New York, 1964, pp. 373-506; R.M. Adams, *ibid*., pp. 507-692.
16 G.W. Campbell, in H. Steinberg and A.L. McCloskey (Editors), Progress in Boron Chemistry, Vol. 1, Pergamon Press, Oxford, 1964, pp. 167-201.
17 M.F. Hawthorne, in E.L. Muetterties (Editor), The Chemistry of Boron and its Compounds, Wiley, New York, 1967, pp. 223-323.
18 R.W. Parry and M.K. Walter, in W.L. Jolly (Editor), Preparative Inorganic Reactions Vol. 5, Interscience Publ., New York, 1968, pp. 45-102.
19 E.L. Muetterties and W.H. Knoth, Polyhedral Boranes, Marcel Dekker, New York, 1968.
20 G.R. Eaton and W.N. Lipscomb, NMR Studies of Boron Hydrides and Related Compounds, W.A. Benjamin, New York, 1969.
21 R.J. Brotherton and H. Steinberg (Editors), Progress in Boron Chemistry, Vol. 2, Pergamon Press, Oxford, especially Chapt. 1 (L.J. Todd) pp. 1-35, and Chapt. 4 (J.D. Odom and R. Schaeffer) pp 141-172.
22 E.L. Muetterties (Editor), Boron Hydride Chemistry, Academic Press, New York, 1975, especially (a) Chapt. 2 (W.N. Lipscomb) pp. 39-78; (b) Chapt. 3 (S.G. Shore) pp. 79-174; (c) Chapt. 4 (T.P. Fehlner) pp. 175-196; and (d) Chapt. 8 (R.L. Middaugh) pp. 273-300.
23 L.H. Long in Supplement to Mellor's Comprehensive Treatise on Inorganic and Theoretical Chemistry Vol. V, Part B1, 1981, pp. 1-185; and (with R.W. Jotham) pp. 186-234.
24 N.N. Greenwood and J.H. Morris, *ibid*., pp. 235-353.
25 L. Barton, Topics in Current Chemistry, Vol. 100, Springer-Verlag, Berlin 1982, pp. 169-206.
26. R. Schaeffer and F. Tebbe, J. Am. Chem. Soc., 84 (1962) 3974-3975.
27 D.F. Gaines and R. Schaeffer, Inorg. Chem., 3 (1964) 438-440.
28 A.C. Bond and M.L. Pinsky, J. Am. Chem. Soc., 92 (1970) 32-36.
29 J.B. Leach, M.A. Toft, F.L. Himpsl and S.G. Shore, J. Am. Chem. Soc., 103 (1981) 988-989.
30 M.A. Toft, J.B. Leach, F.L. Himpsl and S.G. Shore, Inorg. Chem., 21 (1982) 1952-1957.
31 J.R. Wermer and S.G. Shore, Inorg. Chem., 26 (1987) 1644-1645.
32 R. Schaeffer, 24th Internat. Congr. Pure Appl. Chem., Vol. 4, Butterworths, 1974, pp. 1-11.
33 J. Rathke and R. Schaeffer, J. Am. Chem. Soc., 95 (1973) 3402.
34 J. Rathke and R. Schaeffer, Inorg. Chem., 13 (1974) 3008-3011.
35 J. Rathke, D.C. Moody and R. Schaeffer, Inorg. Chem., 13 (1974) 3040-3042.

36 J.C. Huffman, D.C. Moody and R. Schaeffer, J. Am. Chem. Soc., 97 (1975) 1621-1622; Inorg. Chem., 20 (1981) 741-745.
37 J.S. Plotkin and L.G. Sneddon, J. Chem. Soc., Chem. Commun., (1976) 95-96.
38 J.S. Plotkin, R.J. Astheimer and L.G. Sneddon, J. Am. Chem. Soc., 101 (1979) 4155-4163.
39 N.N. Greenwood, J.D. Kennedy, W.S. McDonald, J. Staves and D. Taylorson, J. Chem. Soc., Chem. Commun., (1979) 17.
40 N.N. Greenwood, J.D. Kennedy, T.R. Spalding and D. Taylorson, J. Chem. Soc., Dalton Trans., (1979) 840-846.
41 S.K. Boocock, N.N. Greenwood, J.D. Kennedy, W.S. McDonald and J. Staves, J. Chem. Soc., Dalton Trans., (1980) 790-796.
42 S.A. Barrett, N.N. Greenwood, J.D. Kennedy and M. Thornton-Pett, Polyhedron 4 (1985) 1981-1984.
43 E.W. Corcoran and L.G. Sneddon, Inorg. Chem., 22 (1983) 182.
44 E.W. Corcoran and L.G. Sneddon, J. Am. Chem. Soc., 106 (1984) 7793-7800.
45 E.W. Corcoran and L.G. Sneddon, J. Am. Chem. Soc., 107 (1985) 7446-7450.
46 C.T. Brewer and R.N. Grimes, J. Am. Chem. Soc., 106 (1984) 2722-2723.
47 C.T. Brewer, R.G. Swisher, E. Sinn and R.N. Grimes, J. Am. Chem. Soc., 107 (1985) 3558-3564.
48 C.T. Brewer and R.N. Grimes, J. Am. Chem. Soc., 107 (1985) 3552-3557.
49 J. Bould, N.N. Greenwood and J.D. Kennedy, Polyhedron, 2 (1983) 1401-1402.
50 D.F. Gaines, C.K. Nelson and G.A. Steehler, J. Am. Chem. Soc., 106 (1984) 7266-7267.
51 H.C. Miller, N.E. Miller and E.L. Muetterties, J. Am. Chem. Soc., 85 (1963) 3885-3886; Inorg. Chem., 3 (1964) 1456-1463; see also I.A. Ellis, D.F. Gaines and R. Schaeffer, J. Am. Chem. Soc., 85 (1963) 3885.
52 E.L. Muetterties (ed.), Inorganic Syntheses, Vol. 10, McGraw-Hill, New York, 1967, pp. 81-90.
53 N.N. Greenwood and J.H. Morris, Proc. Chem. Soc., (1963) 338.
54 M. Colombier, J. Atchekzai and H. Mongeot, Inorg. Chim. Acta, 115 (1986) 11-16; H. Mongeot, B. Bonnetot, J. Atchekzai, M. Colombier and C. Vigot-Vieillard, Bull. Soc. Chim. Fr., (1986) 385-389.
55 A.R. Pitochelli and M.F. Hawthorne, J. Am. Chem. Soc., 82 (1960) 3228-3229.
56 M.F. Hawthorne, R.L. Pilling and R.N. Grimes, J. Am. Chem. Soc., 86 (1964) 5338-5339.
57 M.F. Hawthorne and R.L. Pilling, Inorg. Synth., 9 (1967) 16-19.
58 F. Klanberg, D.R. Eaton, L.J. Guggenberger and E.L. Muetterties, Inorg. Chem., 6 (1967) 1271-1281.
59 D.M. Vinitskii, J.V. Rezvova, .A. Solntsev and N.T. Kuznetsov, Russ.J. Inorg. Chem., 33 (1988) 450-452, and references cited therein.
60 B.S. Askins and C. Riley, Inorg. Chem., 16 (1977) 481-484 and references therein.
61 A. Kaldor and R.F. Porter, J. Am. Chem. Soc., 93 (1971) 2140-2145.
62 K. Kawaguchi, J.E. Butler, C. Yamada, S.H. Bauer, T. Minowa, H. Kanamori and E. Hirota, J. Chem. Phys., 87 (1987) 2438-2441.
63 J.F. Stanton, W.N. Lipscomb and R.J. Bartlett, Chem. Phys. Lett. 138 (1987) 525-530.
64 L.A. Curtiss and J.A. Pople, J. Chem. Phys., 89 (1988) 614-615.

65 J.A. Baban, J.C. Brand and B.P. Roberts, J. Chem. Soc., Chem. Commun., (1983) 315-316; see also J.R.M. Giles and B.P. Roberts, J. Chem. Soc., Chem. Commun., (1981) 360-361; J. Chem. Soc., Perkin Trans. 2, (1982) 1699-1711.
66 M.C.R. Symons, T. Chen and C. Glidewell, J. Chem. Soc., Chem. Commun., (1983) 326-328; T.A. Claxton, T. Chen, M.C.R. Symons and C. Glidewell, Faraday Discuss. Chem. Soc., 78 (1984) 121-133.
67 M.N. Paddon-Row and S.S. Wong, J. Mol. Struct. (Theochem), 180 (1988) 353-381.
68 B.D. James and M.G.H. Wallbridge, Prog. Inorg. Chem., 11 (1970) 97-231.
69 H.I. Schlesinger and H.C. Brown, J. Am. Chem. Soc., 62 (1940) 3429-3435.
70 A.B. Burg and H.I. Schlesinger, J. Am. Chem. Soc., 62 (1940) 3425-3429.
71 H.I. Schlesinger, R.T. Sanderson and A.B. Burg, J. Am. Chem. Soc., 61 (1939) 536; and 62 (1941) 3421-3425.
72 N.N. Greenwood, The Chemistry of Boron, Pergamon Press, Oxford, 1973, pp,. 745-763.
73 N.N. Greenwood and A. Earnshaw, Chemistry of the Elements, Pergamon Press, Oxford, 1984, p. 189 and *passim*.
74 R.E. Williams and F.J. Gerhart, Inorg. Nucl. Chem. Lett., 6 (1970) 221-223.
75 W.N. Lipscomb, Pure Appl. Chem., 49 (1977) 691-700.
76 M.A. Vincent and H.F. Schaeffer, J. Am. Chem. Soc., 103 (1981), 5677-5680, and references cited therein.
77 L.A. Long, Progr. Inorg. Chem., 15 (1971) 1-99; Adv. Inorg. Chem. Radiochem., 16 (1974) 201-296.
78 J.L. Duncan and J. Harper, Mol. Phys., 51 (1984) 371-380.
79 T.P. Fehlner and C.E. Housecroft, in J.F. Liebman and A. Greenberg (Editors), Mol. Struct. and Energetics, 1 (1986) 149-207.
80 M. Page, G.F. Adams, J.S. Binkley and C.F. Melius, J. Phys. Chem., 91 (1987) 2675-2678.
81 R.P. Clarke and R.N. Pease, J. Am. Chem. Soc., 73 (1951) 2132-2134; J.K. Bragg, L.V. McCarty and F.J. Norton, *ibid.*, pp. 2134-2140; L.V. McCarty and P.A. DiGiorgio, *ibid.*, pp. 3138-3143.
82 R.E. Enrione and R. Schaeffer, J. Inorg. Nuclear Chem., 18 (1961) 103-107.
83 J.F. Stanton, R.J. Bartlett and W.N. Lipscomb, J. Am. Chem. Soc., 111 (1989) in press.
84 N.N. Greenwood and R. Greatrex, Pure Appl. Chem., 59 (1987) 857-868, and references cited therein.
85 R. Schaeffer, J. Inorg. Nucl. Chem., 15 (1960) 190-193.
86 V.P.J. Marti and B.P. Roberts, J. Chem. Soc., Chem. Commun., (1984) 272-274.
87 H.C. Brown, P.F. Stehle and P.A. Tierney, J. Am. Chem. Soc., 79 (1957) 2020-2021; H.C. Brown and P.A. Tierney, *ibid.*, 80 (1958) 1552-1558; R.K. Hertz, H.D. Johnson and S.G. Shore, Inorg. Synth., 12 (1973) 1875-1877 and 17 (1977) 21-26.
88 S.G. Shore, S.H. Lawrence, M.I. Watkins and R. Bau, J. Am. Chem. Soc., 104 (1982) 7669-7670.
89 S.I. Khan, M.Y. Chiang, R. Bau, T.F. Koetzle, S.G. Shore and S.H. Lawrence, J. Chem. Soc., Dalton Trans., (1986) 1753-1757.
90 A.-M. Sapse and L. Osorio, Inorg. Chem., 23 (1984) 627-628.
91 R. Raghavachari, P. von R. Schleyer and G.W. Spitznagel, J. Am. Chem. Soc., 105 (1983) 5917-5918.

92 W.G. Evans, C.E. Holloway, K. Sukumarabandha and D.H. McDaniel, Inorg. Chem., 7 (1968) 1746-1748.
93 M.L. McKee and W.N. Lipscomb, Inorg. Chem., 21 (1982) 2846-2850.
94 J.R.M. Giles, V.P.J. Marti and B.P. Roberts, J. Chem. Soc., Chem. Commun., (1983) 696-698.
95 W.V. Hough, L.J. Edwards and A.D. McElroy, J. Am. Chem. Soc., 78 (1956) 689.
96 C.R. Peters and C.E. Nordman, J. Am. Chem. Soc. 82 (1960) 5758.
97 W.N. Lipscomb, Adv. Inorg. Chem. Radiochem., 1 (1959) 117-156.
98 E.C. Reynhardt and J.A.J. Lourens, J. Phys., C13 (1980) 2765-2779.
99 I.M. Pepperberg, D.A. Dixon, W.N. Lipscomb and T.A. Halgren, Inorg. Chem., 17 (1978) 587-592.
100 L.V, Titov, E.R. Eremin and V. Ya. Rosolovskii, Russ. J. Inorg. Chem., 27 (1982) 500-502.
101 A.S. Rozenberg, G.N. Nechiporenko, G.N. Titov and M.D. Levicheva, Russ. J. Inorg. Chem., 27 (1982) 1527-1529.
102 A. Stock and C. Massenez, Ber., 45 (1912) 3539-3568.
103 G. Kodama, J. Am. Chem. Soc., 92 (1970) 3482; G. Kodama, V. Engelhardt, C. Lafreze and R.W. Parry, J. Am. Chem. Soc., 94 (1972) 407-412.
104 R.J. Remmel, H.D. Johnson, I.S. Jaworiwski and S.G. Shore, J. Am. Chem. Soc., 97 (1975) 5395-5403.
105 H.D. Johnson and S.G. Shore, J. Am. Chem. Soc. 92 (1970) 7586-7587.
106 B. Brellochs and H. Binder, Z.Naturforsch., 43b (1988) 648-653.
107 C.E. Nordman and W.N. Lipscomb, J. Am. Chem. Soc., 75 (1953) 4116-4117; G.S. Pawley, Acta Cryst., 20 (1966) 631-638 and references therein.
108 C.J. Dain, A.J. Downs, G.S. Laurenson and D.W.H. Rankin, J. Chem. Soc., Dalton Trans. (1981) 472-477.
109 N.P.C. Simmons, A.B. Burg and R.A. Beaudet, Inorg. Chem; 20 (1981) 533-536.
110 R. Greatrex, N.N. Greenwood and C.D. Potter, J. Chem. Soc., Dalton Trans., (1986) 81-89.
111 R. Greatrex, N.N. Greenwood and C.D. Potter, J. Chem. Soc., Dalton Trans., (1984) 2435-2437.
112 M.D. Attwood, R. Greatrex and N.N. Greenwood, J. Chem. Soc., Dalton Trans., (1989) in the press.
113 M.L. McKee, Inorg. Chem. 25 (1986) 2545-2547.
114 A.C. Bond and G. Hairston, Inorg. Chem., 9 (1970) 2610-2611.
115 D. Schwoch, A.B. Burg and R.A. Beaudet, Inorg. Chem., 16 (1977) 3219-3222.
116 R. Greatrex, N.N. Greenwood, D.W.H. Rankin and H.E. Robertson, Polyhedron, 6 (1987) 1849-1858.
117 M.A. Nelson and G. Kodama, Inorg. Chem., 20 (1981) 3579-3980, and references cited therein.
118 N.N. Greenwood and J. Staves, J. Inorg. Nucl. Chem., 40 (1978) 5-7.
119 H.D. Johnson, R.A. Geanangel and S.G. Shore, Inorg. Chem., 9 (1970) 908-912.
120 J.R. Wermer and S.G. Shore, Inorg. Chem., 26 (1987) 1644-1645.
121 A.O. Clouse, D.C. Moody, R.R. Rietz, T. Roseberry and R. Schaeffer, J. Am. Chem. Soc., 95 (1973) 2496-2501, and references cited therein.
122 M.L. McKee and W.N. Lipscomb, Inorg. Chem., 20 (1981) 4442-4444.
123 A.V. Fratini, G.W. Sullivan, M.L. Denniston, R.H. Hertz and S.G. Shore, J. Am. Chem. Soc., 96 (1974) 3013-3015.

124 R. Schaeffer, Q. Johnson and G.S. Smith, Inorg. Chem., 4 (1965) 917-918.
125 I. Yu. Kuznetsov, D.M. Vinitskii, K.A. Solntsev, N.T. Kuznetsov and L.A. Butman, Russ. J. Inorg. Chem., 32 (1987) 1803-1804.
126 D.M. Vinitskii, V.L. Lagun, K.A. Solntsev, N.T. Kuznetsov and I. Yu. Kuznetsov, Koord. Khim., 11 (1985) 1504-1508 [from Chem. Abs., 104 (1986) 61009x].
127 I.Y. Kuznetsov, D.M. Vinitskii, K.A. Solntsev, N.T. Kuznetsov and L.A. Butman, Dokl. Akad. Nauk, SSSR, 283 (1985) 873-877.
128 K.A. Solntsev, Yu.A. Buslaev and N.T. Kuznetsov, Russ. J. Inorg. Chem., 31 (1986) 633-637.
129 I.Yu. Kuznetsov, K.A. Solntsev and N.T. Kuznetsov, Dokl. Akad. Nauk, SSSR, 295 (1987) 138-141.
130 P. Brint, E.F. Healy, T.R. Spalding and T. Whelan, J. Chem. Soc., Dalton Trans., (1981) 2515-2522.
131 T. Whelan and P. Brint, J. Chem. Soc., Faraday Trans. 2, 81 (1985) 267-276.
132 M.A. Cavanaugh, T.P. Fehlner, R. Stramel, M.E. O'Neill and K. Wade, Polyhedron, 4 (1985) 687-695.
133 R.A. Geanangel, H.D. Johnson and S.G. Shore, Inorg. Chem., 10 (1971) 2363-2364.
134 H.D. Johnson, V.T. Brice and S.G. Shore, Inorg. Chem., 12 (1973) 689; R.J. Remmel, H.D. Johnson and S.G. Shore, Inorg. Synth., 19 (1979) 247-253.
135 F.L. Hirshfield, K. Erics, R.E. Dickerson, E.L. Lippert and W.N. Lipscomb, J. Chem. Phys., 28 (1958) 56-61.
136 J.C. Huffman, Thesis, Indiana Univ., 1974.
137 R. Greatrex, N.N. Greenwood and G.A. Jump, J. Chem. Soc., Dalton Trans., (1985) 541-548.
138 P. Davies, PhD Thesis, University of Leeds, 1988.
139 T.C. Gibb, N.N. Greenwood, T.R. Spalding and D. Taylorson, J. Chem. Soc., Dalton Trans., (1979) 1398-1400.
140 H.D. Johnson, R.A. Geanangel and S.G. Shore, Inorg. Chem., 9 (1970) 908-912; V.T. Brice, H.D., Johnson, D.L. Denton and S.G. Shore, Inorg. Chem., 11 (1972) 1135-1137.
141 H.D. Johnson, V.T. Brice, G.L. Brubaker and S.G. Shore, J. Am. Chem. Soc., 94 (1972) 6711-6714.
142 R. Greatrex, N.N. Greenwood, M.B. Millikan, D.W.H. Rankin and H.E. Robinson, J. Chem. Soc., Dalton Trans., (1988) 2335-2339.
143 A.L. Collins and R. Schaeffer, Inorg. Chem., 9 (1970) 2153-2156.
144 J.B. Leach, T. Onak, J. Spielman, R.R. Rietz, R. Schaeffer and L.G. Sneddon, Inorg. Chem., 9 (1970) 2170-2175.
145 A.O. Clouse, D.C. Moody, R.R. Rietz, T. Roseberry and R. Schaeffer, J. Am. Chem. Soc., 95 (1973) 2496-2501.
146 R. Greatrex, N.N. Greenwood and S.D. Waterworth, J. Chem. Soc., Chem. Commun., (1988) 925-926.
147 B. Brellochs and H. Binder, Angew. Chem. Int. Ed. Engl., 27 (1988) 262-263.
148 F. Klanberg and E.L. Muetterties, Inorg. Synth., 11 (1968) 24-33.
149 L.J. Guggenberger, Inorg. Chem., 8 (1969) 2771-2774.
150 D.A. Kleier and W.N. Lipscomb, Inorg. Chem., 18 (1979) 1312-1318.
151 R.E. Enrione, F.P. Boer and W.N. Lipscomb, J. Am. Chem. Soc., 86 (1964) 1451-1452; Inorg. Chem., 3 (1964) 1659-1666.
152 J.F. Ditter, J.R. Spielman and R.E. Williams, Inorg. Chem., 5 (1966) 118-123.
153 J. Dobson and R. Schaeffer, Inorg. Chem., 7 (1968) 402-408.
154 R.R. Rietz, R. Schaeffer and L.G. Sneddon, Inorg. Chem., 11 (1972) 1242-1244.

155 D.C. Moody and R. Schaeffer, Inorg. Chem., 15 (1976) 233-236.
156 J. Dobson, R. Maruca and R. Schaeffer, Inorg. Chem., 9 (1970) 2161-2166.
157 J. Dobson, D.F. Gaines and R. Schaeffer, J. Am. Chem. Soc., 87 (1965) 4072-4074.
158 R.R. Rietz, R. Schaeffer and L.G. Sneddon, Inorg. Chem., 11 (1972) 1242-1244.
159 F. Klanberg and E.L. Muetterties, Inorg. Chem., 5 (1966) 1955-1960; Inorg. Synth., 11 (1968) 24-33.
160 J.C. Carter and P.H. Wilks, Inorg. Chem., 9 (1970) 1777-1779.
161 L.J. Guggenberger, Inorg. Chem., 7 (1968) 2260-2264.
162 E.L. Muetterties, E.L. Hoel, C.G. Salentine and M.F. Hawthorne, Inorg. Chem., 14 (1975) 950-951.
163 B.M. Graybill, J.K. Ruff and M.F. Hawthorne, J. Am. Chem. Soc., 83 (1961) 2669-2670; B.M. Graybill, A.R. Pitochelli and M.F. Hawthorne, Inorg. Chem., 1 (1962) 626-631.
164 G.B. Jacobsen, D.G. Meina, J.H. Morris, C. Thomson, S.J. Andrews, D. Reed, A.J. Welch and D.F. Gaines, J. Chem. Soc., Dalton Trans., (1985) 1645-1654.
165 L.C. Ardini and T.F. Fehlner, Inorg. Chem., 12 (1973) 798-803.
166 A.R. Siedle, D. McDowell and L.J. Todd, Inorg. Chem., 13 (1974) 2735-2739.
167 L.E. Benjamin, S.F. Stafiej and E.A. Takacs, J. Am. Chem. Soc. 85 (1963) 2674-2675.
168 S.K. Boocock, N.N. Greenwood, M.J. Hails, J.D. Kennedy and W.S. McDonald, J. Chem. Soc., Dalton Trans., (1981) 1415-1429.
169 C.G. Savory and M.G.H. Wallbridge, J. Chem. Soc., Chem. Commun., (1970) 1526-1527; Inorg. Chem., 10 (1971) 419-422; J. Chem. Soc., Dalton Trans., (1973) 179-184.
170 V.T. Brice, H.D. Johnson, D.L. Kenton and S.G. Shore, Inorg, Chem., 11 (1972) 1135-1137; D.L. Denton, W.R. Clayton, M. Mangion, S.G. Shore and E.A. Meyers, Inorg. Chem., 15 (1976) 541-548; S.H. Lawrence, J.R. Wermer, S.K. Boocock, M.A. Banks, P.C. Keller and S.G. Shore, Inorg. Chem., 25 (1986) 367-372; see also ref. 30.
171 N.N. Greenwood, H.J. Gysling, J.A. McGinnety and J.D. Owen, J. Chem. Soc., Chem. Commun., (1970) 505-506; N.N. Greenwood, J.A. McGinnety and J.D. Owen, J. Chem. Soc., Dalton Trans., (1972) 986-989.
172 G.B. Jacobsen, J.H. Morris and D. Reed, J. Chem. Soc., Dalton Trans., (1984) 415-421.
173 L.I. Isaenko, K.G. Myakishev, I.S. Posnaya and V.V. Volkov, Izv. Sibirsk Otd. Akad. Nauk SSSR, Ser. Khim. Nauk, (1982) 73-78; from Chem. Abs., 97 (1982) 16161.
174 W.V. Kotlensky and R. Schaeffer, J. Am. Chem. Soc., 80 (1958) 4517-4519.
175 A.B. Burg and R. Kratzer, Inorg. Chem., 1 (1962) 725-730.
176 J.F. Ditter, J.R. Spielman and R.E. Williams, Inorg. Chem., 5 (1966) 118-123.
177 R.E. Dickenson, P.J. Wheatley, P.A. Howell, W.N. Lipscomb and R. Schaeffer, J. Chem. Phys., 25 (1956) 606-607; *ibid.*, 27 (1957) 200-208; P.G. Simpson and W.N. Lipscomb, *ibid.*, 35 (1961) 1340-1343.
178 J. Dobson, P.C. Keller and R. Schaeffer, J. Am. Chem. Soc., 87 (1965) 3522-3523; Inorg. Chem., 7 (1968) 399-402.
179 P.C. Keller, Inorg. Chem., 9 (1970) 75-78; D.C. Moody and R. Schaeffer, Inorg. Chem., 15 (1976) 233-236.
180 M.F. Hawthorne and A.R. Pitochelli, J. Am. Chem. Soc., 81 (1959) 5519.
181 J.M. Makhlouf, W.V. Hough and G.T. Hefferan, Inorg. Chem., 6 (1967) 1196-1198.

182 R.D. Dobrott and W.N. Lipscomb, J. Chem. Phys., 37 (1962) 1779-1783.
183 J.T. Gill and S.J. Lippard, Inorg. Chem., 14 (1975) 751-761.
184 D.J. Fuller, D.L. Keppert, B.W. Skelton and A.H. White, Aust. J. Chem., 40 (1987) 2097-2105.
185 E.L. Muetterties, J.H. Balthis, Y.T. Chia, W.H. Knoth and H.C. Miller, Inorg. Chem., 3 (1964) 444-451.
186 J.S. Kasper, C.M. Lucht and D. Harker, J. Am. Chem. Soc., 70 (1948) 881; Acta Cryst., 3 (1950) 436-455.
187 K. Wade, Adv. Inorg. Chem. Radiochem., 18 (1976) 1-66; R.E. Williams, *ibid.*, 67-142.
188 A. Tippe and W. Hamilton, Inorg. Chem., 8 (1969) 464-470.
189 R.W. Atteberry, J. Phys. Chem., 62 (1958) 1485-1489.
190 S. Heřmánek and H. Polotava, Coll. Czech. Chem. Commun., 36 (1971) 1639-1643.
191 L.G. Sneddon, J.C. Huffman, R.O. Schaeffer and W.E. Streib, J. Chem. Soc., Chem. Commun., (1972) 474-475.
192 M.F. Hawthorne, A.R. Pitochelli, R.D. Strahm and J.J. Miller, J. Am. Chem. Soc., 82 (1960) 1825-1829.
193 P.H. Wilks and J.C. Carter, J. Am. Chem. Soc., 88 (1966) 3441.
194 N.N. Greenwood and D.N. Sharrocks, J. Chem. Soc. A, (1969) 2334-2338.
195 N.N. Greenwood and B. Youll, J. Chem. Soc., Dalton Trans., (1975) 158-162.
196 N.N. Greenwood and J.D. Kennedy, Chapter 2 in ref. 5, pp. 43-118.
197 E.L. Muetterties, Inorg. Chem., 2 (1963) 647-648.
198 D.S. Kendall and W.N. Lipscomb, Inorg. Chem., 12 (1973) 546-551.
199 J.A. Dupont and M.F. Hawthorne, Chem. Ind. (London), 1962, 405.
200 R.R. Rietz, A.R. Siedle, R.O. Schaeffer and L.J. Todd, Inorg. Chem., 12 (1973) 2100-2102.
201 D.F. Gaines, Accounts Chem. Res., 6 (1973) 416-421.
202 L.H. Hall and W.S. Koski, J. Am. Chem. Soc., 84 (1962) 4205-4207.
203 R. Grimes, F.E. Wang, R. Lewin and W.N. Lipscomb, Proc. Natl. Acad. Sci., U.S., 47 (1961) 996-999; R.N. Grimes and W.N. Lipscomb, *ibid.*, 48 (1962) 496-499.
204 D.F. Gaines, M.W. Jorgenson and M.A. Kulzik, J. Chem. Soc., Chem. Commun., (1979) 380-381; J.A. Heppert, M.A. Kulzik and D.F. Gaines, Inorg. Chem., 23 (1984) 14-18.
205 A.M. Barriola, Acta Cient. Venez., 34 (1983) 25-27.
206 J.J. Briguglio, P.J. Carroll, E.W. Corcoran and L.G. Sneddon, Inorg. Chem., 25 (1986) 4618-4622.
207 E.H. Wong, L. Prasad, E.J. Gabe and M.G. Gatter, Inorg. Chem., 22 (1983) 1143-1146.
208 V.D. Aftandilian, H.C. Miller, G.W. Parshall and E.L. Muetterties, Inorg. Chem., 1 (1962) 734-737.
209 C.J. Fritchie, Inorg. Chem., 6 (1967) 1199-1203.
210 G.B. Dunks, K. Barker, E. Hedaya, C. Hefner, K. Palmer-Ordonez and P. Remec, Inorg. Chem., 20 (1981) 1692-1697 and references cited therein; G.B. Dunks, K. Palmer-Ordonez and E. Hedaya, Inorg. Synth. 22 (1983) 202-207.
211 T.D. Getman, J.A. Krause and S.G. Shore, Inorg. Chem., 27 (1988) 2398-2399; T.D. Getman, J.R. Wermer and S.G. Shore in ref. 8, pp. 146-174.
212 H.C. Longuet-Higgins and M. de V. Roberts, Proc. Roy. Soc. (Lond.), A230 (1955) 110-118.
213 J.A. Wunderlich and W.N. Lipscomb, J. Am. Chem. Soc., 82 (1960) 4427-4428; see later volumes of ref. 13 for several other determinations.

214 G. Shoham, D. Schomburg and W.N. Lipscomb, Cryst. Struct. Commun., 9 (1980) 429-434.
215 N.T. Kuznetsov and G.S. Klimchuk, Russ. J. Inorg. Chem., 16 (1971) 645-648.
216 J. Bicerano, D.S. Marynick and W.N. Lipscomb, Inorg. Chem., 17 (1978) 3443-3453.
217 J.C. Huffman, D.C. Moody and R. Schaeffer, Inorg. Chem., 15 (1976) 227-232.
218 J.C. Huffman, D.C. Moody, J.W. Rathke and R. Schaeffer, J. Chem. Soc., Chem. Commun., (1973) 308.
219 S. Heřmánek, K. Fetter and J. Plešek, Chem. Ind. (London), (1972) 606; S. Heřmánek, K. Fetter, J. Plešek, L.J. Todd and A.R. Garber, Inorg. Chem., 14 (1975) 2250-2253.
220 J.C. Huffman, Indiana University Molecular Structure Centre Report No. 81902, Sept. 1981.
221 J. Plešek, S. Heřmánek, B. Stibr and F. Hanousek, Coll. Czech. Chem. Commun., 32 (1967) 1095-1103.
222 J. Plešek, S. Heřmánek and F. Hanousek, Coll. Czech. Chem. Commun., 33 (1968) 699-705.
223 L.B. Friedman, R.E. Cook and M.D. Glick, J. Am. Chem. Soc., 90 (1968) 6802-6803; Inorg. Chem., 9 (1970) 1452-1458.
224 A.R. Pitochelli and M.F. Hawthorne, J. Am. Chem. Soc., 84 (1962) 3218; F.P. Olsen, R.C. Vasavada and M.F. Hawthorne, *ibid.*, 90 (1968) 3940-3951.
225 J.S. McAvoy and M.G.H. Wallbridge, Chem. Commun., (1969) 1378-1379.
226 S. Heřmánek and J. Plešek, Coll. Czech. Chem. Commun., 35 (1970) 2488-2493.
227 P.G. Simpson and W.N. Lipscomb, Proc. Natl. Acad. Sci. U.S., 48 (1962) 1490-1491; J. Chem. Phys., 39 (1963) 26-34.
228 P.G. Simpson, K. Folting and W.N. Lipscomb, J. Am. Chem. Soc., 85 (1963) 1879-1880; P.G. Simpson, K. Folting, R.D. Dobrott and W.N. Lipscomb, J. Chem. Phys., 30 (1963) 2339-2348.
229 X.L.R. Fontaine, N.N. Greenwood, J.D. Kennedy and P. MacKinnon, J. Chem. Soc., Dalton Trans., (1988) 1785-1793.
230 L.B. Friedman, R.D. Dobrott and W.N. Lipscomb, J. Am. Chem. Soc., 85 (1963) 3505.
231 N.E. Miller and E.L. Muetterties, J. Am. Chem. Soc., 85 (1963) 3506.
232 N.E. Miller, J.A. Forstner and E.L. Muetterties, Inorg. Chem., 3 (1964) 1690-1694.
233 R.D. Dobrott, L.B. Friedman and W.N. Lipscomb, J. Chem. Phys., 40 (1964) 866-872.
234 L.D. Brown and W.N. Lipscomb, Inorg. Chem., 16 (1977) 2989-2996.
235 A. Kaczmarczyk, R.D. Dobrott and W.N. Lipscomb, Proc. Natl. Acad. Sci., U.S., 48 (1962) 729-733.
236 B.L. Chamberland and E.L. Muetterties, Inorg. Chem., 3 (1964) 1450-1456.
237 M.F. Hawthorne, R.L. Pilling, P.F. Stokely and P.M. Garrett, J. Am. Chem. Soc., 85 (1963) 3704-3705; 87 (1965) 1893-1899.
238 C.H. Schwalbe and W.N. Lipscomb, J. Am. Chem. Soc., 91 (1969) 194-196; Inorg. Chem., 10 (1971) 151-160.
239 M.F. Hawthorne and R.L. Pilling, J. Am. Chem. Soc., 88 (1966) 3873-3874.
240 A. Kaczmarczyk, Inorg. Chem., 7 (1968) 164-165.
241 R.L. Middaugh and F. Farha, J. Am. Chem. Soc., 88 (1966) 4147-4149.
242 N.N. Greenwood, J.D. Kennedy and D. Taylorson, J. Phys. Chem., 82 (1978) 623-625.

243 S.K. Boocock, N.N. Greenwood, J.D. Kennedy and D. Taylorson, J. Chem. Soc., Chem. Commun, (1979) 16.
244 N.N. Greenwood, J.D. Kennedy, W.S. McDonald, J. Staves and D. Taylorson, J. Chem. Soc., Chem. Commun., (1979) 17.
245 S.K. Boocock, Y.M. Cheek, N.N. Greenwood and J.D. Kennedy, J. Chem. Soc., Dalton Trans., (1981) 1430-1437.
246 G.M. Brown, J.W. Pinson and L.L. Ingram, Inorg. Chem., 18 (1979) 1951-1956.
247 S.K. Boocock, N.N. Greenwood, J.D. Kennedy and W.S. McDonald, J. Chem. Soc., Dalton Trans., (1981) 2573-2584.
248 W.N. Lipscomb, Inorg. Chem., 19 (1980) 1415-1416.
249 R.J. Wiersema and R.L. Middaugh, J. Am. Chem. Soc., 89 (1967) 5078; Inorg. Chem., 8 (1969) 2074-2079.
250 A.H. Norman and A. Kaczmarczyk, Inorg. Chem., 13 (1974) 2316-2321.
251 R. Bechtold and A. Kaczmarczyk, J. Am. Chem. Soc., 96 (1974) 5953-5954.
252 S.K. Boocock, N.N. Greenwood and J.D. Kennedy, J. Chem. Research (S), (1981) 50-51.

POLYSILANES

E. HENGGE and H. STÜGER
Institute of Inorganic Chemistry, Graz University of Technology
Graz, Austria

1. INTRODUCTION

Polysilanes are compounds with more than three Si-atoms, directly bonded together. More than two thousand papers have been published in this field, together with some books and reviews (refs. 1-8).

Polysilanes are known as linear and branched catenated or cyclic structures, but also as higher dimensional frameworks. Because of their technological utility, such as precursors to silicon carbide or as photoresists in microelectronics, especially the polymeric systems attracted particular attention recently. The backbone of all polysilanes is built up by Si-Si linkages, leading to quite different properties compared to carbon based compounds. First, the bond dissociation energy of the Si-Si bond is generally lower than that of the C-C bond, although experimental values vary in a wide range, depending on the compound considered and on the method used for the determination. Therefore, the thermal stability of a Si-Si framework is lower than that of a C-C framework. Second, the Si atom is unsaturated in respect to its coordination number. Thus nucleophilic attacks occur very easily, hydrolysis, halogenations and similar reactions are very common.

2. POLYSILANES SUBSTITUTED WITH HYDROGEN

These facts, mentioned above, are the most important reasons for the considerable differences in the behaviour of hydrocarbons and hydrosilanes. Simple hydrocarbons, as well as their fully or partially halogenated analogues are very stable and well known compounds. Silicon frameworks, however, are much more sensitive and reactive because of the limited stability of the

Si-Si- and Si-H-bonds. Hydrosilane chains only can be isolated up to about eight silicon atoms, chains up to 15 Si atoms have been detected by chromatographic methods (ref.9). Besides the linear, some branched hydrosilanes like i-Si_4H_{10} or the neopentane analogue structure of Si_5H_{12} are also known. First investigations in this field were made by Stock (ref.10) and later on by Fehér (ref.11), both in Germany. Synthesis of hydrosilanes is usually achieved by acid hydrolysis of magnesium-silicide (ref.12).

Cyclic silicon hydrides are not obtained in this way. The first cyclic silicon hydrides were prepared by a completely different synthesis 14 years ago (refs.13,14,15). From the perphenylated silicon cycles, which were already discovered 1921 by Kipping (ref. 73), the perchlorinated cyclosilanes, also completely unknown before, can be obtained by the reaction with hydrogen chloride in presence of aluminiumchloride (Scheme 1).

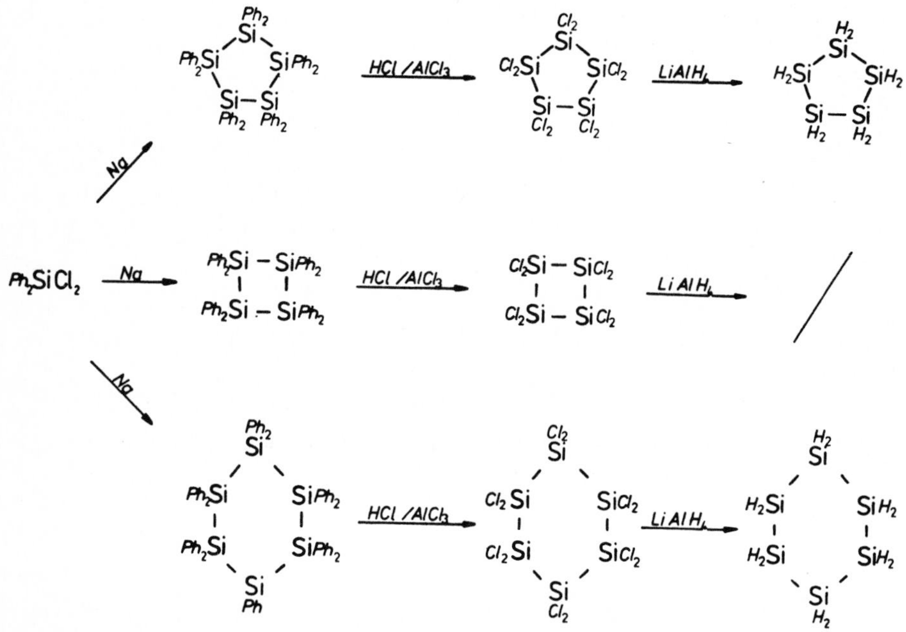

Scheme 1

Hydrogen bromide or hydrogen iodide but not hydrogen fluoride also lead to the corresponding perhalogenated cyclosilanes. Those are reduced by lithium aluminum hydride to the first cyclic hydrosilanes Si_5H_{10} and Si_6H_{12}. The synthesis of cyclotetrasilane failed, probably because of high ring strain. Only the four membered halogenated cycles (Cl,Br,I) (refs.16,17,18) could be isolated.

The cyclic compounds exhibit enhanced thermal stability compared to their open chain analogues, which slowly decompose at room temperature. All silicon hydrides spontaneously ignite or even explode in air. All these compounds show a high decomposition rate at relatively low temperatures, making them prospective precursors for the production of pure silicon for electronic applications. Therefore new pathways to higher hydrosilanes attract considerable attention.

Hydrosilane high polymers are known with defined and non defined structures. Hydrosilanes with non defined structures usually consist of mixtures of $(SiH)_n$, a completely cross-linked polymer, and linear $(SiH_2)_n$ chains. Compositions with lower hydrogen content are also found. Such polymers are obtained by pyrolyzing or photolyzing lower silanes, or by the disproportionation of disilane. Another route employs the reaction of halosilane high polymers with hydrides. In this case, the structure obtained depends on the structure of the halosilane precursor. The reaction of $(SiBr)_n$ with $LiAlH_4$ yields a typical example for a silicon hydride $(SiH)_n$ with defined structure. The silicon monobromide has a layer structure from the starting material $CaSi_2$, which is conserved in the final $(SiH)_n$ polymer. This group of compounds will be discussed later together with other polysilane high polymers (see p.130).

Last not least hydrogen is absorbed on surfaces of crystalline silicon or can be incorporated into amorphous silicon. A series of papers dealing with this interaction has been published, particularly in respect to the semi-conducting properties of these materials. A survey of these aspects is given in refs.12,19.

3. HALOGENATED POLYSILANES

A similar situation is given with polysilanes bearing halogens, mainly chlorine. Usually they are prepared by the reaction of chlorine with silicon or silicides (ref.20). A low reaction temperature is necessary for a good yield of polysilanes. This can be achieved by use of calcium silicide with a very clean surface and catalysts like manganese. The clean surface is obtained by grinding the silicide under inert atmosphere or starting the reaction at higher temperatures as necessary. The separation of the different polysilanes in the reaction mixture is difficult, caused by cracking reactions during distillation and small differences in the boiling points. Therefore, it is impossible, for instance, to separate n- and i-decachlorotetrasilane. Product separation by distillation becomes impossible at all for compounds with more than five silicon atoms in the chain.

Defined perchlorinated polysilanes are of particular interest as precursors of other polysilanes and for industrial application as well. They are, for instance, by-products of the production of trichlorosilane and also of the deposition of pure silicon via trichlorosilane. A large number of per-chloro-polysilanes were recently prepared together with compounds bearing both hydrogen and chlorine substituents. Principal reaction paths are cleavage of cyclic silanes, removal of phenyl groups from Si-phenyl bonds with hydrogen halides, formation of Si-Si bonds by reductive dechlorination or recombination of silyl radicals after photolytic cleavage of (bis)-polysilanyl mercury compounds.

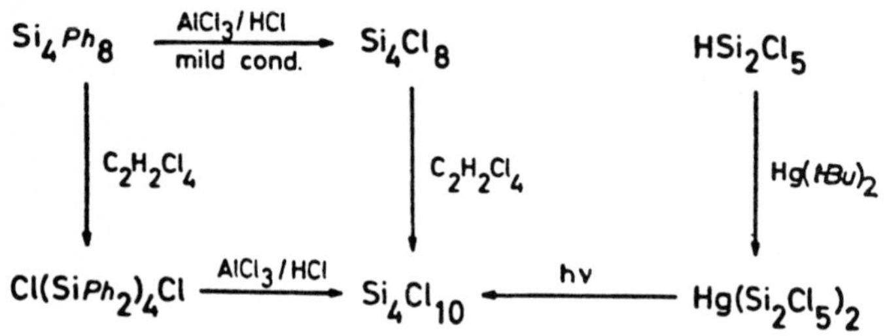

Scheme 2

The synthesis of linear decachlorotetrasilane, unknown for a very long time, is a very illustrative example (Scheme 2). It may be prepared by three different ways. Starting from octaphenylcyclotetrasilane, it is possible under mild reaction conditions to split off the phenyl groups with $HCl/AlCl_3$ without ring cleavage and Si_4Cl_8 is formed ($HBr/AlBr_3$ or HI/AlI_3 lead to the corresponding bromine or iodine derivatives). Subsequent cleavage of one SiSi-bond of the cyclic chloride with tetrachloroethane yields n-Si_4Cl_{10}. Initial ring cleavage followed by removal of the phenyl groups is also possible. A completely different route starts from pentachlorodisilane. After decomposing its mercury derivative by irridiation with light, the linear decachloro-tetrasilane is obtained. All three methods lead to identical samples of the compound (ref.21).

Another example is the preparation of iso-Si_4Cl_{10} achieved by Höfler in Graz in 1975 (ref.22): neo-Si_5Cl_{12} reacts with methanol mainly to the expected neo-$Si_5(OCH_3)_{12}$. In a side reaction, however, methanol is also able to cleave one SiSi-bond by nucleophilic attack and tris(trimethoxysilyl)silane is formed. The course of the reaction strongly depends on the conditions applied. Under suitable conditions the side reaction becomes the favoured one. The Si-H linkage thus formed may be chlorinated with CCl_4 at room temperature and after treatment with BCl_3 the Si-O-C-bonds are split and iso-Si_4Cl_{10} is obtained (Scheme 3).

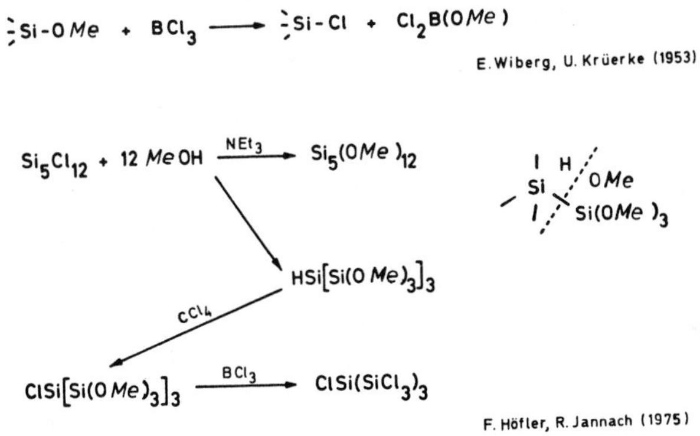

Scheme 3

Chlorinated polysilanes with terminal hydrogen substituents are available from the perphenylated cycles. Ring-cleavage is possible with HCl, PCl_5 or Li. The resulting catenated products subsequently can be reacted to phenylated silicon chains with hydrogens at the end positions. Splitting off the phenyl groups yields the chlorinated products (ref.23). Another interesting reaction, discovered by Gilman (ref.24), is the degradation of 1,5-dihydroxidecaphenylpentasilane by solid aluminum oxide to the corresponding dihydrotrisilane (Scheme 4).

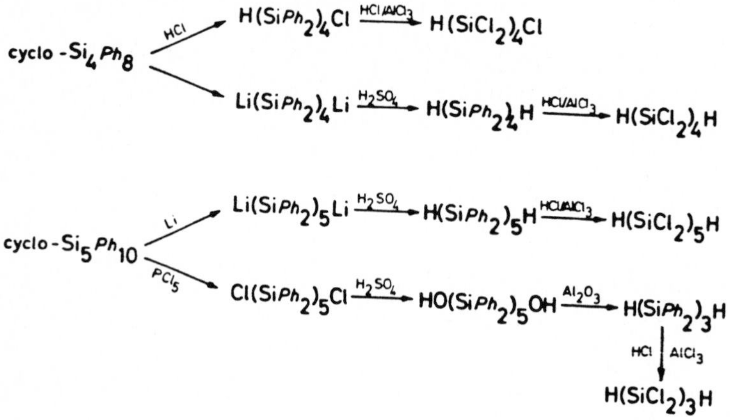

Scheme 4

The synthesis of chlorinated silicon-chains with internal hydrogen substituents by use of silylpotassium compounds represents a third example. Triphenylsilylpotassium reacts with different chlorosilanes to polysilanes. After replacing the phenyl substituents by halogens the desired products are formed (ref.25) (eqn. 1 and 2).

$$Cl-SiH(Ph)-Cl \xrightarrow{+2\,KSiPh_3} Ph_3Si-SiH(Ph)-SiPh_3 \xrightarrow[AlCl_3]{HCl} Cl_3Si-SiH(Cl)-SiCl_3 \quad (1)$$

$$HSiCl_3 \xrightarrow{+3\,KSiPh_3} H-Si(SiPh_3)_3 \xrightarrow[AlCl_3]{HCl} H-Si(SiCl_3)_3 \quad (2)$$

This method may be used for the systematic formation of higher polysilanes increasing the number of silicon atoms step by step. From a polysilane precursor with one hydrogen substituent and n silicon atoms the corresponding silylmercurial is made. Subsequent cleavage of the silicon-mercury bond by sodium/potassium alloy yields the polysilylpotassium compound. This can be reacted with a diorganodichlorosilane to a polysilane with (n+1) silicon atoms in the chain. After restoration of the silicon-hydrogen bond with $LiAlH_4$, this reaction sequence can be restarted. So it was possible to synthesize systematically polysilane chains up to 5 silicon atoms (ref.26).

Silyl-mercury compounds offer a second opportunity for the extension of polysilane frameworks. Photolysis of polysilyl mercurials produces silicon chains twice as long as those observed in the starting material (Scheme 5).

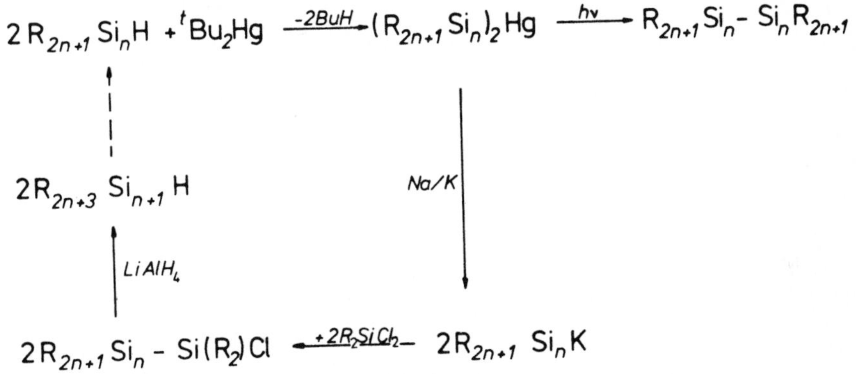

Scheme 5

The chlorinated polysilanes are very sensitive against attacks by nucleophilic reagents like water, caused by the high reactivity of the silicon chlorine bond. The thermal stability is greater than that of hydrogenated polysilanes, but still very low. Extensive decomposition takes place at higher temperatures. Further reduction of stability is observed in strained rings or in chains with increasing number of silicon atoms. Remarkably, however, high polymer silicon chlorides exhibit enhanced stability compared to their low molecular weight analogues. For other polysilane high polymers a similar behaviour is observed.

Physicochemical properties of polysilanes are investigated in numerous publications. It is not possible to give all these details in this short review. Summarizing, from the spectroscopic results on polysilanes, like nmr- and mass spectra, vibrational spectra including normal coordinate analyses, or X-ray structure determinations, the following rules for the SiSi-linkage can be derived:

A) The strength of the SiSi-bond is strongly influenced by substituents. Side groups with free electron pairs and high electronegativity cause an increase of the strength. Ordering different substituents with increasing influence on the SiSi-bond the sequence shown in Table 1 is obtained. The numerical values in Table 1 are taken from measurements of disilane derivatives (ref.27). The influence on larger silicon backbones, however, stays the same.

TABLE 1 Force constants of different disilanes

$X_3Si-SiX_3$

X	f_{SiSi}
H (D)	1,73
Me (CD_3)	1,65
I	1,9
Br	2,1
Cl	2,3
F	2,3
MeO (CD_3O)	2,25
Me_2N	2,2
Ph	2,0

Me < H < I < Ph < Br < MeO - Me_2N < Cl = F

B) The strength of the SiSi-bond also depends on structure and size of the Si-backbone. In strained ring systems, for instance, the SiSi-bonds are weakened, expressed by low force constants and sometimes by longer bond distances (ref.28) (Fig.1).

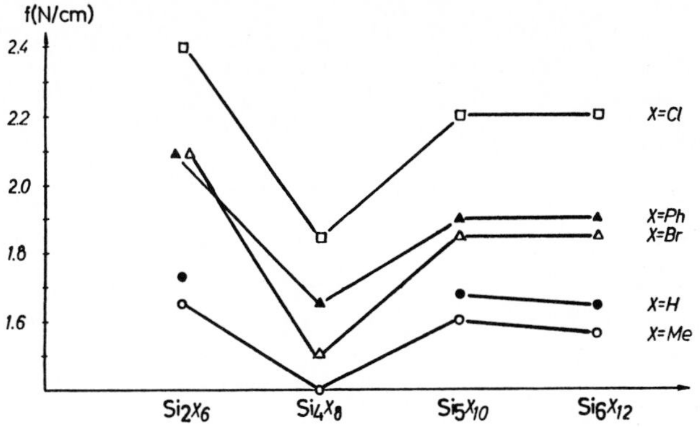

Fig. 1. Si-Si stretching force constants in cyclosilanes

Internal SiSi-bonds in open chained systems are weaker than the terminal ones, documented once more by different force constants, but also by mass fragmentation patterns. The internal SiSi-bond is always split first (ref.29). This fact seems to be responsible for the occurence of branched polysilanes in reaction mixtures obtained after pyrolysis of linear polysilanes (Scheme 6).

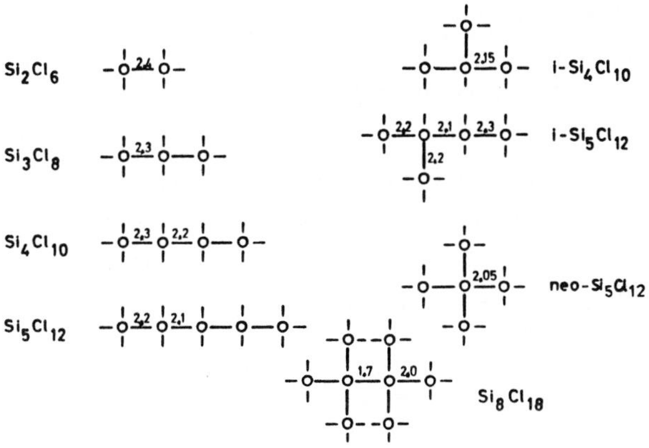

Scheme 6

The strength of the SiSi-bond is not only affected by electronic but also by steric influences. Larger substituent groups, for instance, elongate and weaken SiSi-bonds. This subject, however, will be discussed in more detail later with polysilanes bearing organic substituents.

Similar to corresponding hydropolysilanes halogenated polysilane high polymers are also known having either defined or undefined structures. Two principal methods are available for their preparation. Undefined polymers are formed, when low molecular weight silicon halides are dehalogenated with reactive metals or by pyrolysis. Most papers in this field were published many years ago and have been extensively reviewed already (ref.29). The synthesis of defined halosilane polymers employs halogenation reactions of precursors with preformed polymer structures. More about this second route on p.130.

4. POLYSILANES WITH ORGANIC SUBSTITUENTS

Numerous publications on polysilanes with organic substituents are available. A complete survey of all facts known in this field would go beyond the limits of this article. Therefore only some new aspects should be discussed.

One of these aspects is the influence of the bulkiness of substituent groups on the bond strength and bond length of SiSi-bonds. Wiberg recently found the longest SiSi-distance observed so far in hexakis(t-butyl)-disilane with 269.7 pm (ref.30). Weidenbruch also showed very long SiSi-bonds in a trisilane, substituted with t-butyl groups and iodine (ref.91). Generally, the SiSi-bond turns out to be fairly weak and flexible. Bond lengths are markedly increasing with increasing size of substituents.

Unusual SiSi-distances were also found in cyclotrisilanes. The first successful synthesis of a three-membered silicon ring was achieved by Masamune, followed by other derivatives isolated by Weidenbruch, Watanabe and others. A review of this rapidly growing area has recently been published (ref.31). In all three-membered cyclosilanes, the SiSi-distances are longer than in their unstrained five- and six-membered analogues. The space requirement of the substituents and the ring strain seem to be responsible for this elongation of the SiSi-bond and subsequently for the reduced bond strength.

On the other hand, very short distances between two Si-atoms not bonded directly to each other are sometimes observed. In tetramesitylcyclodisiloxane, for instance, West, Michl et al. found a SiSi-distance of only 231 pm (ref.32). Recent investigations by West suggest the absence of direct bonding interactions between the two Si-atoms (ref.33) (Scheme 7).

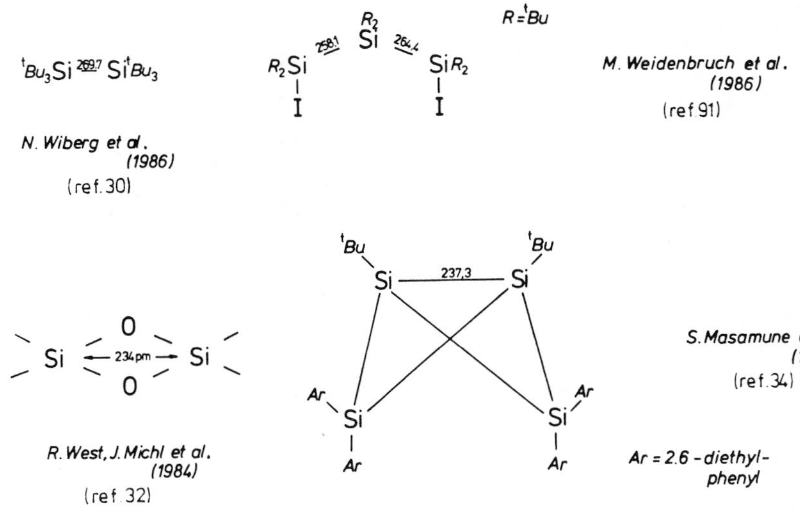

Scheme 7

The unusual flexibility of SiSi-bonds is also demonstrated by tetrasilabicyclo(1.1.0)butane, isolated by Masamune et al. Although an X-ray structure determination affords a SiSi-bond distance of only 237.3 pm, characteristic for strong SiSi-bonds (refs.34,35), this strongly puckered four-membered ring exhibits ring inversion, which must be accompanied by significant stretching of the central SiSi-bond.

In larger ring systems the SiSi-distances approach normal values and are mainly dependent on the substituents attached to the ring. Cartledge has postulated special steric substituent parameters for silicon compounds (ref.36). These Es(Si)-parameters were calculated from kinetic measurements on monosilane derivatives and also found to be valid for cyclic polysilanes by Nagai and Watanabe (refs.37,41). Considering the reductive coupling of dialkyldichlorosilanes with lithium, they observed a preference for the formation of smaller rings with bulky substituents, smaller side groups favour larger ring systems.

Four-membered rings with bulky substituents exhibit relatively long SiSi-bonds. With smaller side groups like the methyl group, normal distances are usually observed.

The folding angle of the four-membered ring also strongly depends on the substituents. The asymmetrically substituted tetramethyl-tetrakis(t-butyl)cyclotetrasilane, for instance, exhibits a strong folding angle of 36.8^{o} (ref.38), the bulky phenyl groups in octaphenylcyclotetrasilane also give rise to a folding angle of about 13 degrees (ref.39). The tetrasilane ring in octakis(trimethylsilyl)cyclotetrasilane, however, is planar in spite of the bulky trimethylsilyl groups (ref.40). Other examples with different folding angles were shown by Watanabe and Nagai (ref.41). The accurate geometry of the permethylated four-membered ring is still controversial. X-ray structure investigations afford a planar ring structure in the crystalline state (ref.42). In the gas phase, however, a slight but definitive deviation from planarity with a puckering angle of about one degree was found by electron diffraction, done by Mastryukov and Rozsondai (ref.43). Thus the folding angle of cyclotetrasilanes not only depends on the size of the substituents but also must be influenced by other parameters. Calculation of potential energy minima seems to be the best way to make correct predictions about this folding angle.

Octamethylcyclotetrasilane now is available in larger amounts for these and other investigations, since a new synthetic route for its preparation has been discovered.

First the phenyl groups in octaphenylcyclotetrasilane are exchanged against chlorine with $HCl/AlCl_3$. Subsequent methylation with dimethyl-zinc affords the desired product in high yield and purity (ref.44) (eqn. 3).

$$Si_4Ph_8 \xrightarrow{AlCl_3/HCl} Si_4Cl_8 \xrightarrow{ZnMe_2} Si_4Me_8 \qquad (3)$$

The chemistry of five- and six-membered silicon rings is well known and many derivatives have been synthesized so far by several authors (refs.45,46,47). The geometries of these rings are consistent with the geometries of their carbon analogues, their properties exhibit no special features except the formation of radical ions, which will be discussed later. In general, silicon cycles are less stable than carbon cycles and show higher flexibility, documented, for instance, by the very facile pseudorotation of Si_5H_{10} or the considerable mobility of Si_6H_{12}. Under the conditions of electron diffraction, Si_6H_{12} shows a perfect equilibrium between chair-, boat- and twist conformation. The energy differences are very low, about 8.4 kJ/mol between chair and twist and only 0.8 kJ/mol between the twist and the boat form. Rearrangements of different stereoisomers therefore occur very easily (refs.48,49) (Fig. 2a and b).

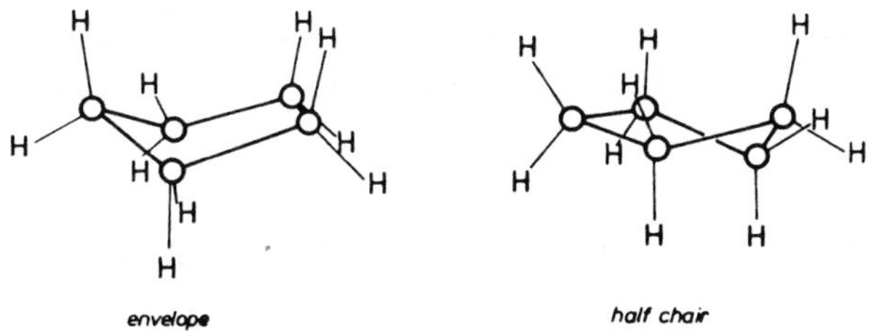

Si_5H_{10}

Fig. 2a. Possible conformations of cyclopentasilane

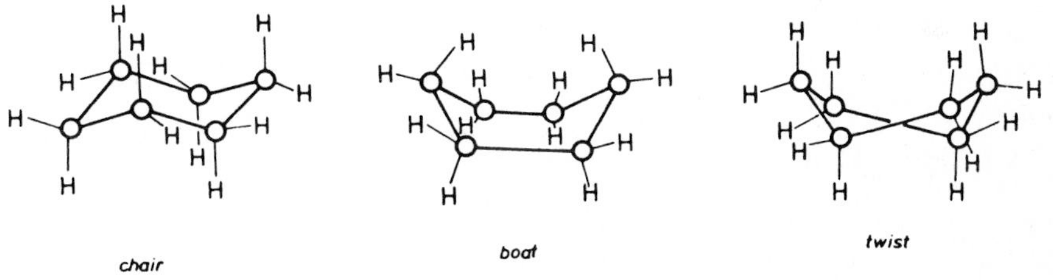

Si_6H_{12}

Fig. 2b. Possible conformations of cyclohexasilane

Only a few polycyclic polysilanes are known so far. Usually they were isolated in small yields as by-products of the reductive dechlorination of dialkyldichlorosilanes or just detected in chromatographic diagrams (refs.50,51).

Recently a way to obtain polycyclic polysilanes by a special synthesis was discovered. The preparation of the starting materials, monofunctional cyclosilanes like monochloro-undecamethylcyclohexasilane was achieved by reacting $HCl/AlCl_3$

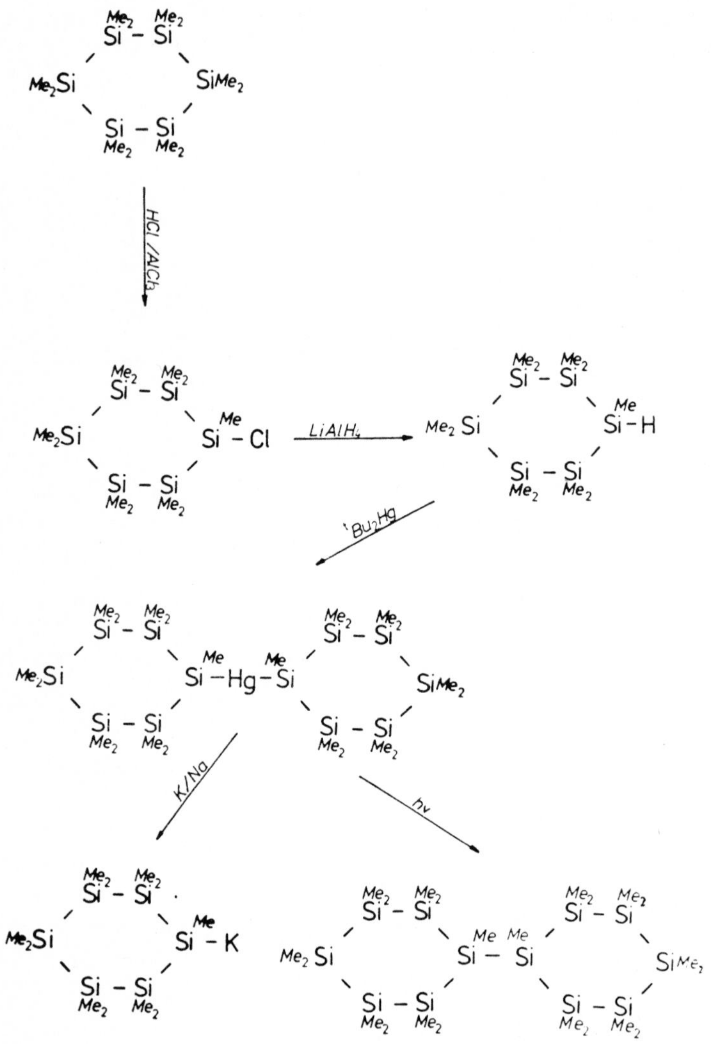

Scheme 8

with dodecamethylcyclohexasilane (ref.52). This reaction, however, is accompanied by side reactions leading to the formation of five-membered rings. The success strongly depends on the quality of $AlCl_3$ used and also has to be controlled by gas chromatography. Other ways to monofunctional cyclosilanes proved to be unsuccessful. West described the synthesis of monophenylundecamethylcyclohexasilane, which easily could be converted into the mono-chloroderivative, but the yields are very low (ref.53) (Scheme 8).

The reduction of the Si-Cl bond in monochloro-undecamethylcyclohexasilane with $LiAlH_4$ results in the formation of the corresponding monohydrocyclosilane, which smoothly reacts with di(t-butyl)-mercury to the new silyl-mercury-compound bis(undecamethylcyclohexasilyl)-mercury, a key substance on the way to polycyclic polysilanes.

Photolysis yields a bicyclic system, sodium-potassium alloy splits the Si-Hg bonds and the first cyclosilylpotassium compound is formed, which further can be reacted with appropriate halosilanes like monobromoundecamethylcyclosilane, dichlorodimethylsilane or 1,2-dichlorotetramethyldisilane to polycyclic silicon compounds (refs.54,55) (Scheme 9).

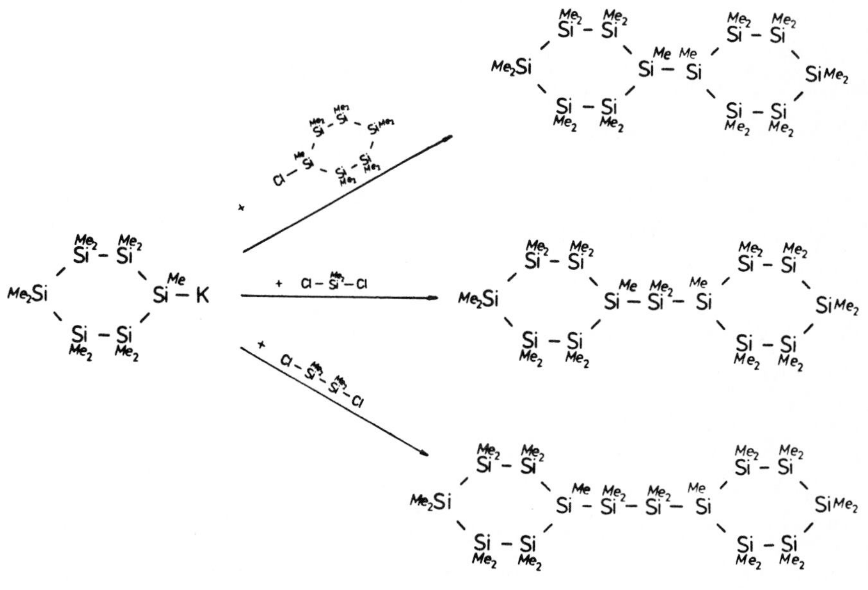

Scheme 9

One of the largest silicon frameworks, made by a special synthesis, is obtained from the reaction of the cyclosilyl-potassium-compound with the isotetrasilane derivative tris-(chlorodimethylsilyl)-methylsilane (Scheme 10).

$$
3\ Me_2Si\underset{\underset{Me_2}{Si}-\underset{Me_2}{Si}}{\overset{\overset{Me_2}{Si}-\overset{Me_2}{Si}}{\diagup\diagdown}}\overset{Me}{Si}-K\ +\ MeSi{\Big\langle}\!\!\begin{array}{l}SiMe_2Cl\\SiMe_2Cl\\SiMe_2Cl\end{array}\ \longrightarrow
$$

$$
MeSi\left(SiMe_2-\overset{Me}{Si}\underset{\underset{Me_2}{Si}-\underset{Me_2}{Si}}{\overset{\overset{Me_2}{Si}-\overset{Me_2}{Si}}{\diagup\diagdown}}SiMe_2\right)_3
$$

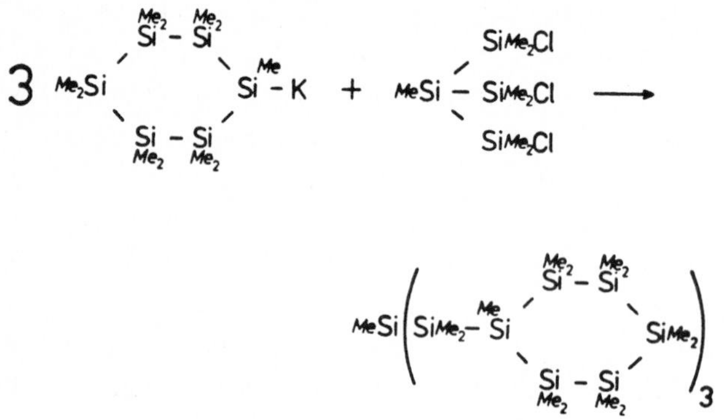

Scheme 10

Remarkable in these systems are the properties of the central SiSi-bonds connecting the cycles. X-ray investigations and force constant calculations exhibit longer and weaker bonds. The differences are not very large, but significant. Mass spectra show primary splitting of these bonds. This is also consistent with the reduced bond strength (ref.56).

$$
2\ Me_2Si\underset{\underset{Me_2}{Si}-\underset{Me_2}{Si}}{\overset{\overset{Me_2}{Si}-\overset{Me_2}{Si}}{\diagup\diagdown}}\overset{Me}{Si}-K\ +\ Br-\overset{Me}{Si}\underset{\underset{Me_2}{Si}-\underset{Me_2}{Si}}{\overset{\overset{Me_2}{Si}-\overset{Me_2}{Si}}{\diagup\diagdown}}\overset{Me}{Si}-Br\ \longrightarrow
$$

$$
\longrightarrow\ Me_2Si\underset{\underset{Me_2}{Si}-\underset{Me_2}{Si}}{\overset{\overset{Me_2}{Si}-\overset{Me_2}{Si}}{\diagup\diagdown}}\overset{Me}{Si}-\overset{Me}{Si}\underset{\underset{Me_2}{Si}-\underset{Me_2}{Si}}{\overset{\overset{Me_2}{Si}-\overset{Me_2}{Si}}{\diagup\diagdown}}\overset{Me}{Si}-\overset{Me}{Si}\underset{\underset{Me_2}{Si}-\underset{Me_2}{Si}}{\overset{\overset{Me_2}{Si}-\overset{Me_2}{Si}}{\diagup\diagdown}}SiMe_2
$$

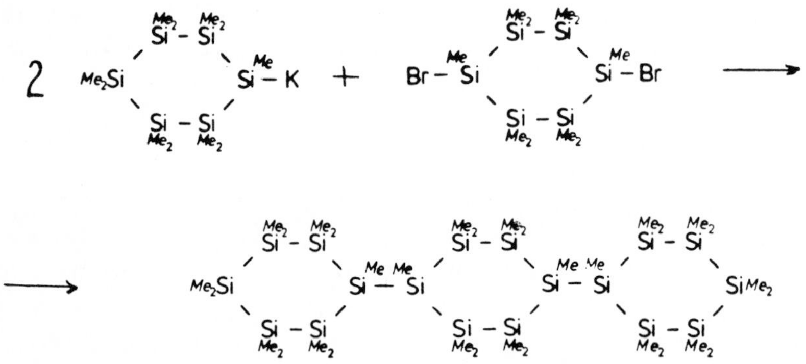

Scheme 11

A tricyclic polysilane, another interesting reaction product, is formed, when the potassium compound is reacted with 1,4-dibromodecamethylcyclohexasilane (Scheme 11).

The 1,4-dibromocyclosilane was prepared similar to the synthesis of its 1,4-dichloro-analogue, recently described as an intermediate on the way to a sulfur bridged cyclohexasilane by Wojnowsky (ref.57). This compound, however, has not been characterized and no preparative details have been reported. Repeating this reaction of antimony pentachloride with dodeca-methylcyclohexasilane, we found the results of Wojnowsky to be correct also under our reaction conditions. Remarkable is the exclusive formation of the 1,4-derivative. Higher substituted products are not observed. Nothing is known about directing effects of substituents in cyclosilanes. Therefore, this observation might be the starting point of further investigations.

Similar studies were done on five-membered rings as well. The preparation of pure permethylcyclopentasilane in larger amounts, however, is not easy to perform. Reductive dehalogenation of dimethyldichlorosilane with alkali metals affords the compound besides other polysilanes in a complicated mixture making the separation very troublesome and considerably lowering the yield. Extensive studies have been performed to find the optimum conditions for the formation of the five-membered ring (ref.58).

Decamethylcyclopentasilane also can be obtained by the method already used for the preparation of octamethylcyclotetrasilane. Decaphenylcyclopentasilane, which easily can be isolated from the reaction of diphenyldichlorosilane with alkali metals, is converted into the perchlorinated five-membered ring with $HCl/AlCl_3$. After treatment with dimethyl zinc complete methylation is achieved. Although this method affords a very pure product in good yields, its applicability is limited by the complicated performance of the reaction in a sealed tube (ref.43). Dodecamethylcyclohexasilane undergoes ring contraction after treatment with catalytic amounts of $AlCl_3$ in presence of tri-methylchlorosilane (ref.59), yielding a five-membered permethy-lated ring with a $SiMe_2Cl$ side group. Although the reaction mechanism is still not understood in detail (ref.60), this reaction can be utilized for the synthesis of polycyclic cyclo-silanes containing five-membered silicon rings. Its course, however, must be controlled by gas chromatography and stopped with acetone at the maximum yield of the desired compound because other products like polychlorosilanyl-cyclopentasilanes are also formed. The reaction sequence to the polycycles is the same as already described for cyclohexasilanes. The silicon-chlorine bond is hydrogenated with $LiAlH_4$, and with bis(t-butyl)-mercury the corresponding mercury compound is formed. Photolytic decomposition once more yields the dimeric structure, treatment with Na/K-alloy affords the K-compound, which further can be reacted with various chlorosilanes to polysilane chains terminated by cyclopentasilane rings (ref.61) (Scheme 12).

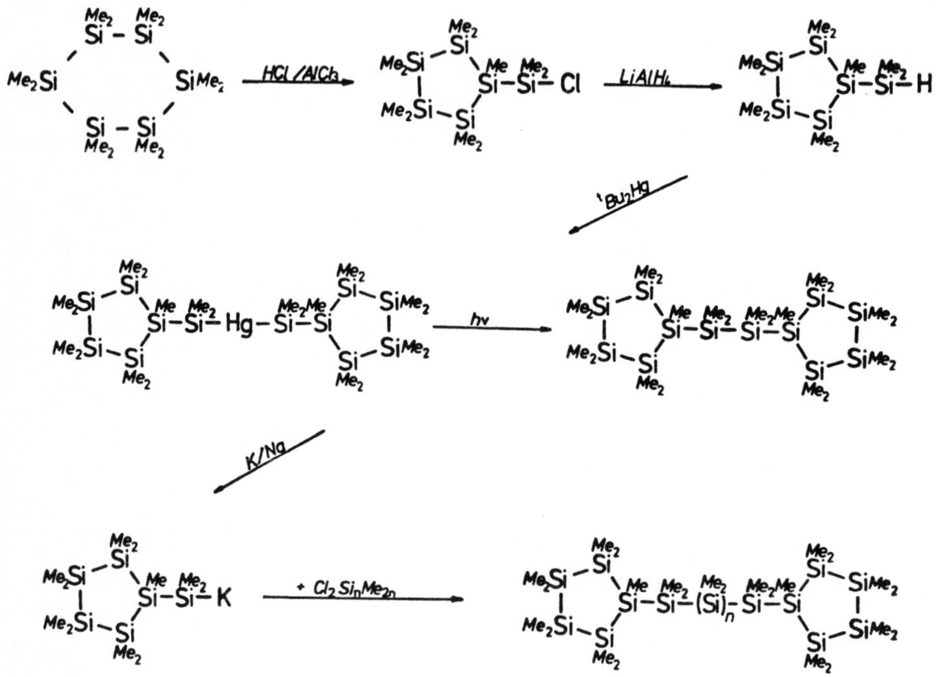

Scheme 12

5. FORMATION OF CATIONS AND ANIONS OF POLYSILANES

One of the outstanding properties of polysilanes is their ability to form radical cations and anions. This was discovered by West on permethylated cyclosilanes (ref.62). He was able to produce polysilanyl-radical anions by adding electrons to the Si-cycles electrochemically or with alkali metals.

As demonstrated by detailed esr investigations and by theoretical studies, the additional charge is completely delocalized within the ring system (refs.63,64,65,66,67,68).

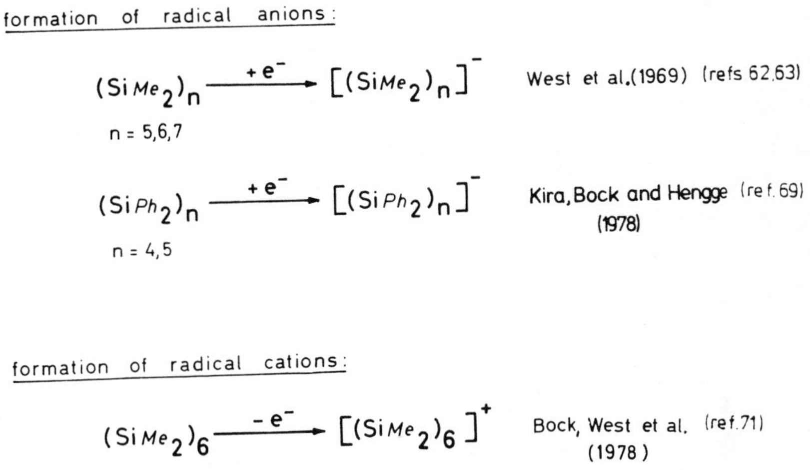

Scheme 13

Perphenylated cyclosilanes also form radical anions. In this case the additional electron is delocalized within the whole molecule, the silicon ring and the phenyl groups (ref.69). Catenated polysilanes also exhibit radical anion formation (ref.70).

Radical cations are generated by treating permethylated cyclosilanes with a $AlCl_3/CH_2Cl_2$ mixture, whose redox potential is sufficient to oxidize compounds with a first ionization potential below 8 eV. ESR studies of these radical cations suggest spin delocalization within the SiSi σ-system (ref.71) (Scheme 13).

6. LINEAR POLYSILANE HIGH POLYMERS

Chemistry and applications of linear polysilane polymers have been extensively reviewed recently (refs.6,72). Therefore this article gives just a brief survey on that class of compounds. High molecular weight polysilane polymers are of particular technological interest. Although they are already known for many years (refs.73,74), they were first considered to be just of minor applicability, because permethyl- or perphenylpolysilanes are highly crystalline, intractable solids with very high melting points. But two important discoveries in the last decade immediately intensified research in that field of chemistry.

In 1975, Yajima and his coworkers found, that polysilane polymers are suitable as precursors for ß-silicon carbide (refs.75,76,77,78). In their process, permethylpolysilane is pyrolyzed above 400°C to produce a polymeric, hexane soluble carbosilane, which can be spun into fibers. Those are X-linked by surface oxidation in air and further pyrolyzed at 1300°C to make black ß-SiC fibers (eqn. 4 and 5).

$$\left[-\overset{CH_3}{\underset{CH_3}{Si}}- \right]_n \xrightarrow{>400^{o}C} \left[-\overset{H}{\underset{CH_3}{Si}} - CH_2 - \right]_n \qquad (4)$$

$$\left[-\overset{H}{\underset{CH_3}{Si}} - CH_2 - \right]_n \xrightarrow[(2)\ N_2,\ 1300^{o}C]{(1)\ air,\ 350^{o}C} \beta - SiC \qquad (5)$$

The second finding, mentioned above, was the discovery, that introduction of side groups other than methyl groups into polysilanes enhances their solubility in organic solvents and aids in their purification and fabrication. Today, more than 50 polysilane homopolymers and copolymers have been prepared with different side groups, among them interesting compounds that contain cycloalkane and unsaturated aliphatic substituents.

6.1 SYNTHESIS

For the synthesis of polysilane high polymers several ways leading to SiSi-bonds are suitable.

Today, the most important one is still the reductive dechlorination of dichlorosilanes with sodium metal leading to a variety of homo and copolymers (eqn. 6 and 7).

$$R^1R^2SiCl_2 \xrightarrow[\text{hydrocarbon}]{\text{Na} \quad 100^{\circ}C} \left[-\underset{R^2}{\overset{R^1}{\underset{|}{\overset{|}{Si}}}}- \right]_n \qquad (6)$$

$$R^1R^2SiCl_2 + R^3R^4SiCl_2 \xrightarrow[\text{hydrocarbon}]{\text{Na} \quad 100^{\circ}C} \left[-\underset{R^2}{\overset{R^1}{\underset{|}{\overset{|}{Si}}}} - \underset{R^4}{\overset{R^3}{\underset{|}{\overset{|}{Si}}}}- \right]_n \qquad (7)$$

However, yields and molecular weights of the reaction products strongly depend on the reaction conditions applied. Therefore numerous investigations about the choice of solvents, metals, temperature and reaction time have been undertaken to optimize this complex reaction. Good progress, for instance, was made by Boudjouk by using supersonic for a better distribution of the alkali metal in the solution (ref.79). The reaction mechanism is also still controversial. Right now, the question whether the reaction follows a radical pathway or rather preformed alkali-silyl-compounds build up the polymer chains cannot be answered without ambiguity.

Major disadvantages of polysilane polymer synthesis by reductive dechlorination are mainly the limited number of side groups suitable for the rough reaction conditions and especially for industrial purposes, the high costs connected with the performance of the reaction.

Therefore other, less expensive ways to polysilane polymers are intensively explored. One of those is the disproportionation of disilane derivatives to monosilanes and polysilanes. Because the chloromethyldisilane starting materials are industrial by-products, numerous patents are dealing with this synthetic route to polysilanes. However, some difficulties are encountered carrying out that reaction. Using the chloromethyldisilane directly for the disproportionation gives rise to certain amounts of chlorine in the final polymer, what strongly effects the properties of the resulting silicon carbide.

This can be avoided by derivatizing the starting material prior to disproportionation by use of amines or ammonia, yielding nitrogen containing products, or with methanol leading to the methoxyderivatives, which can be converted into polymers using Si-H compounds and sodiummethoxylate as a catalyst (ref.80). Starting from methyldisilanes, obtained by reduction of the corresponding chlorodisilanes with $LiAlH_4$, yields polymers with good properties after disproportionation. This route, however, is too expensive for commercial applications once more because of the use of $LiAlH_4$. Pyrolyzing cyano- or cyanatodisilanes might also be a good way to nitrogen containing polysilanes, although the preformation of the starting materials turns out to be rather complicated (ref.81).

Summarizing, it is obvious, that the problem of the inexpensive synthesis of polysilane high polymers in good yields with constant product properties has not been solved so far and therefore remains a major goal in organosilicon chemistry within the next years.

6.2 PROPERTIES AND APPLICATIONS

The properties of polysilane polymers greatly depend on the nature of the organic groups attached to silicon. They appear either as elastomeric, rubbery materials or glassy solids, can be melted and spun into fibers, show strong uv-absorption and are easily degraded by uv-irradiation or heating to high temperatures. Some of them also can be cross-linked by uv-irradiation or by heating them in presence of vinylsilanes and free radial initiators. Results of light scattering experiments show extremely high molecular weights for some polysilane polymers.

Potential applications of polysilane polymers persist in at least three areas:

1. as precursors to silicon carbide fibers or ceramics
2. as photoinitiators for vinyl polymerisation

3. as resist materials in photolithography

Si-C-fibers are among the strongest substances known, stable at high temperatures, chemically highly resistant and show good mechanical properties. Besides other ways, Si-C fibers, moldings or composites can be produced by the process invented by Yajima and coworkers already mentioned, where permethylpolysilane is converted in 2 steps to silicon carbide. Soluble polysilanes in some cases, can be transformed into silicon carbide without the necessity of formation of polycarbosilane, fractionation, or oxidation.

When polysilanes are pyrolyzed, the formation of volatile silicon compounds extremely lowers the yield of SiC. To obtain good ceramic yields and to make sure that preformed fibers retain their shape during conversion, the polysilane chains need to be cross-linked prior to pyrolysis. Such cross-links can be generated by adding multiply unsaturated compounds and initiating free radical formation thermally or photochemically, by oxidizing properly substituted polysilanes or photolyzing polysilane polymers containing unsaturated side groups. Catalytical cross-linking of polysilanes with Si-H-bonds by adding trivinylphenylsilane and chloroplatinic acid is also possible. The potential of polysilane polymers as photoinitiators for vinyl polymerization was discovered by West and his group. Uv light splits SiSi-bonds, forming both radicals and silylenes (eqn.8). These species initiate the polymerization of olefins.

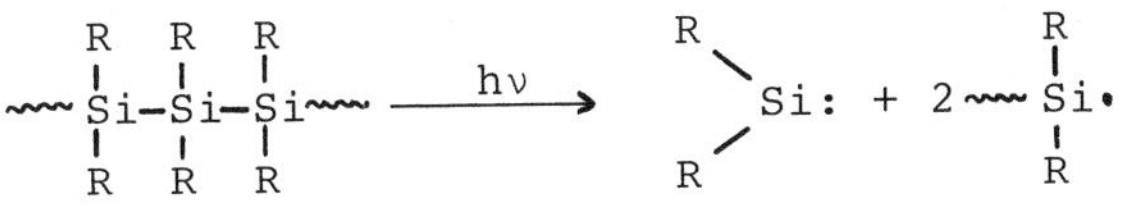

(8)

The unique photochemical properties of polysilane polymers are also the reason for their applicability as photoresists in microlithography. The formation of radicals and silylenes upon uv-irradiation by scission of SiSi-bonds generates reaction products with uv absorptions only shorter than 250 nm, which are generally more soluble than the starting material. Therefore polysilanes can be used as positive photoresists. Photocross-linking of polysilane polymers is also employed in photoresist technology. Cross-linking greatly reduces the solubility making polysilanes suitable to work as negative photoresists as well.

7. NON LINEAR POLYSILANE POLYMERS

Besides linear polysilane chains a number of non linear polysilane polymers exist. Non linear polymers are produced, when $SiCl_4$, $HSiCl_3$ or similar primary silanes are pyrolyzed. They exhibit yellow, red or brown colours (ref.82). The accurate structure of these polysilanes is yet unknown, recent ^{29}Si-nmr investigations indicate the presence of branched polysilane structures (ref.83).

Reductive dechlorination of multifunctional siliconchlorides also affords polysilane polymers with undefined structures. A summary of these mainly older investigations is given in ref. 7.

A second group of non linear polysilane polymers is made from polymeric starting material with already preformed SiSi framework, leading to defined structures in the final product. Two dimensional structures of the general formula $(SiX)_n$ are synthesized from the two-dimensional silicon sheets in the layer lattice of $CaSi_2$ (ref.29). These silicon layers are isolated from

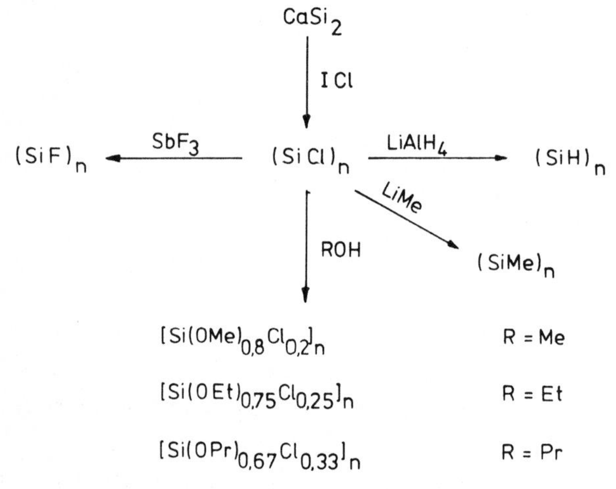

Scheme 14

the lattice with ICl or IBr, yielding yellow $(SiCl)_n$ or $(SiBr)_n$, respectively. The Si-Cl bonds attached to the layers are reactive like all Si-Cl bonds and many derivatives can be formed in solid state reactions. Selected reactions of the polymeric siliconmonochloride are depicted in Scheme 14.

Scheme 15 shows a summary of all derivatives prepared so far.

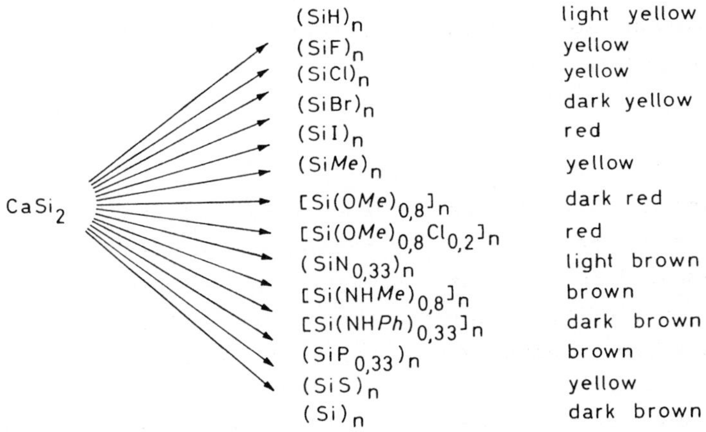

Scheme 15

All these compounds are coloured and the colour depends on the substituents. From the reemission spectra substituents can be ordered with increasing bathochromic shift (Fig.3).

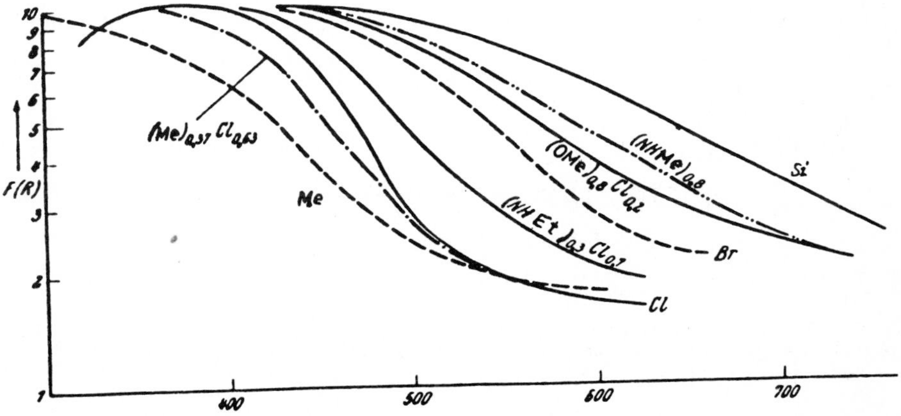

Fig.3. Reemission spectra of $(SiX)_n$

Particularly interesting is the compound of the formula $(Si)_n$. It is silicon in a very reactive state, burning in air and with alcohols and reacting vigorously with water. Esr measurements exhibit a paramagnetic character (ref.84).

All these investigations support the conclusion that electrons are strongly delocalized within the silicon sheets. In all derivatives sometimes radical states are present arising from unsaturated Si valencies, being resonance stabilized in the polymer backbone. More detailed investigations of these phenomena, however, are only possible on smaller SiSi frameworks.

8. ELECTRONIC SPECTRA OF POLYSILANES

The unusual electronic properties of the SiSi-bond have been discovered in the early 1960 s. Today we have a large number of publications dealing with absorption and photoelectron spectra of polysilanes (ref.85). Recently, especially polysilane high polymers have been studied extensively because their electronic characteristics are of particular technological interest (ref.6). In contrast to most carbon based compounds, polysilanes show one or several absorption maxima in the near uv or even

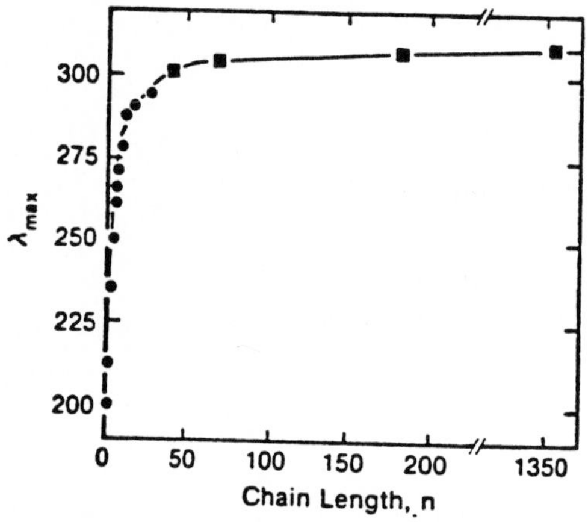

Fig. 4. Plot of uv-absorption maxima versus chain length(n) for poly(alkylsilane). The (●) represent $Me(SiMe_2)_nMe$ and the (■) represent $[(n\text{-dodecyl})(Me)Si]_n$

visible region of the electromagnetic spectrum. Position and intensity of these maxima greatly depend on structure and size of the silicon framework and on the substituents attached to it. Linear permethylated silicon chains, for instance, show a bathochromic shift of the longest wavelength absorption maxima with increasing chain length reaching a constant value of about 305 nm at 25 silicon atoms (Fig. 4).

The absorptivity per SiSi-bond also increases regularly with increasing chain length and finally levels off, when $n \longrightarrow \infty$. Cyclic polysilanes, in general, exhibit absorption maxima at lower energies compared to their open chain analogues. Substituent effects are very well documented by the strong bathochromic influence of phenyl groups.

Theoretical descriptions of the electronic spectra and the bonding situation in polysilanes include treatments by simple HMO (ref.85), semi empirical (refs.86,87) and ab-initio methods at different levels of accuracy (ref.88). They all suggest strong delocalization of σ- electrons in the silicon backbone and electron transitions to unoccupied orbitals of eiter σ- or π- symmetry. The question of silicon 3 d contribution is still controversial, although recent investigations show no necessity to include silicon 3 d-levels for explaining any experimental results.

Just a few publications are available about electronic spectra of polysilanes bearing substituents other than alkyl or aryl groups. Nevertheless, the uv-absorption spectra of perhalopolysilanes seem to be particularly interesting because some of those compounds like octachloro- or octabromocyclotetrasilane or periodocyclosilanes exhibit colours and also might be suitable to serve as model substances to explain the interactions between polymeric silicon frameworks and similar substituents.

Therefore a systematic investigation of the uv-absorption- and He(I)-photoelectron spectra of perhalo- and permethoxycyclopolysilanes has been undertaken in our laboratories (ref.89). The absorption spectra of cyclotetrasilanes are presented in Fig. 5.

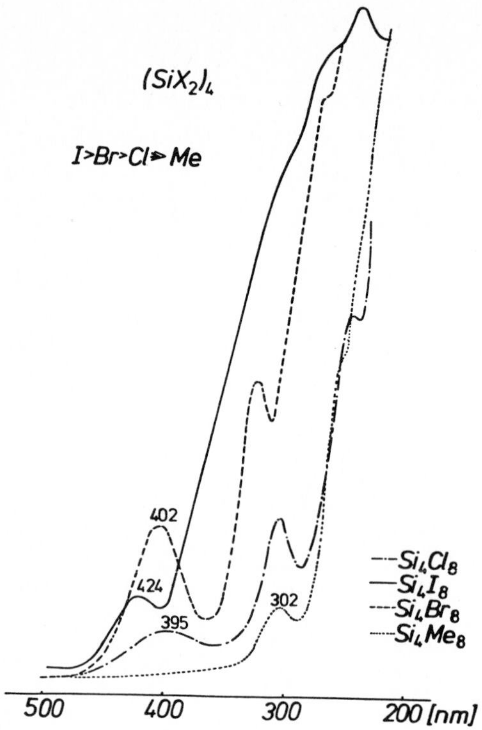

Fig. 5. Uv-vis absorption spectra of cyclotetrasilanes $(SiX_2)_n$ with X = Cl,Br,I,Me

Compared to their permethyl analogues, all compounds exhibit additional low intensity absorption bands with low absorption energies. Those are responsible for the appearance of colours. Position and intensity of the bands depend on the substituents and on the ring size. Iodine substituents cause the most bathochromic shift followed by bromine and chlorine (Fig.5).

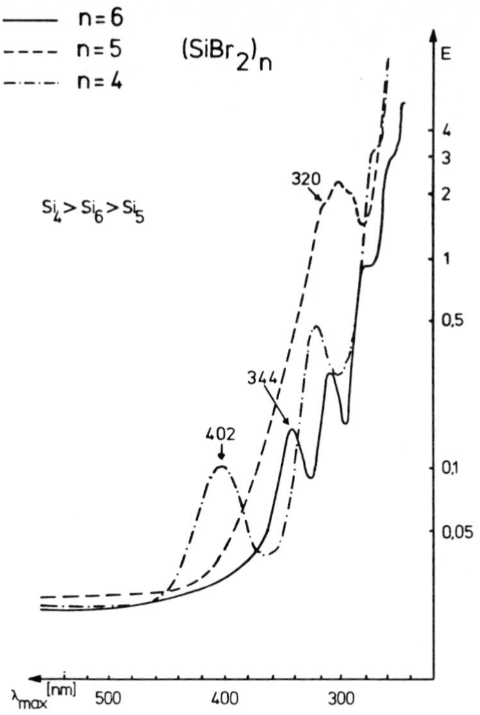

Fig. 6. Uv-vis absorption spectra of bromocyclosilanes $(SiBr_2)_n$ with n = 4,5,6.

Surprisingly we found that the bands in question are shifted to longer wavelengths in the order $Si_4 > Si_6 > Si_5$ (compare Fig.6) and did not follow the ring size, as one would expect from the spectra of permethylcyclosilanes. The same sequence is reflected by the first IP's determined by photoelectron spectroscopy (Table 2), making the corresponding electron transitions highly suggestive of being intramolecular charge transfer transitions between the polysilane framework and the substituents. Similar results and assignments were obtained by Pitt for trisilane derivatives (ref.90).

Comparable absorption bands are also responsible for the colours of selected polysilane polymers. They also appear in the absorption spectra of properly substituted long silicon chains and larger polycyclic ring systems.

TABLE 2 First vertical ionization energies of cyclic silanes Si_nX_{2n}; X = Me, Cl, Br in eV

X	Si_4X_8	Si_5X_{10}	Si_6X_{12}
Me	7.60	7.94	7.79
Cl	8.85	9.50	9.05
Br	8.7	9.25	8.9

REFERENCES

1 E.A.V.Ebsworth in: Volatile Silicon Compounds, p.85, Pergamon Press, London, 1963.
2 E.Hengge in: A.L.Rheingold, Homoatomic Rings, Chains and Macromolecules of Main-Group Elements, p.235, Elsevier Scientific Publishing Comp., Amsterdam, Oxford, New York, 1977.
3 M.Kumada, Pure Appl.Chem., 13 (1966) 167.
4 M.Kumada and K.Tamao, Adv.Organomet.Chem., 6 (1968) 19.
5 E.Hengge, Nova Acta Leopoldina, 264 (1985) 367.
6 R.West, J.Organomet.Chem., 300 (1986) 327.
7 E.Hengge, Fortschr.Chem.Forsch., 51 (1974) 1.
8 E.Hengge, J.Organomet.Chem.Lib., 9 (1980) 261.
9 F.Fehér, D.Schinkitz and J.Schaaf, Z.Anorg.Allg.Chem., 383 (1971) 303.
10 A.Stock, Z.Elektrochem., 32 (1926) 341.
11 F.Fehér, Forschungsber.d.Landes Nordrhein-Westfalen 2632 (1977) 1.
12 Gmelin Handbook of Inorganic Chemistry. Silicon (Syst.-No. 15), Supplement Volume Part B 1, Springer-Verlag, Berlin-Heidelberg-New York, 1982.
13 E.Hengge and G.Bauer, Angew.Chem., 85 (1973) 304; Angew.Chem. Int.Ed.Engl., 12 (1973) 316.
14 E.Hengge and G.Bauer, Monatsh.Chem., 106 (1975) 503.
15 E.Hengge and D.Kovar, Z.Anorg.Allg.Chem., 459 (1979) 123.
16 K.Hassler, E.Hengge and D.Kovar, J.Mol.Struct., 66 (1980) 25.
17 E.Hengge and D.Kovar, Angew.Chem., 93 (1981) 698; Angew.Chem. Int.Ed.Engl., 20 (1981) 678.
18 E.Hengge and D.Kovar, Z.Anorg.Allg.Chem., 458 (1979) 163.
19 J.A.Appelbaum, D.R.Hamann and K.H.Tasso, Phys.Rev.Lett., 39 (1977) 1487.
20 E.Hengge, Rev.Inorg.Chem., 2 (1980) 139.
21 W.Raml and E.Hengge, Z.Naturforsch., 34b (1979) 1457.

22 F.Höfler and R.Jannach, Inorg.Nucl.Chem.Lett., 11 (1975) 743.
23 E.Hengge and G.Miklau, Z.Anorg.Allg.Chem., 508 (1984) 33.
24 H.Gilman and G.L.Schwebke, J.Am.Chem.Soc., 86 (1964) 2693.
25 E.Hengge and F.K.Mitter, Z.Anorg.Allg.Chem., 529 (1985) 22.
26 V.V.Dudorov and A.D.Zorin, Zh.Fiz.Khim., 48 (1974) 721.
27 D.G.White and E.G.Rochow, J.Am.Chem.Soc., 76 (1954) 3897.
28 E.Hengge and G.Miklau, Z.Anorg.Allg.Chem., 508 (1984) 43.
29 E.Hengge in: V.Gutmann, Halogen Chemistry, vol.II, p.169, Academic Press, London, 1967.
30 N.Wiberg, H.Schuster, A.Simon and K.Peters, Angew.Chem., 98 (1986) 100; Angew.Chem.Int.Ed.Engl., 25 (1986) 79.
31 M.Weidenbruch, Comments Inorg.Chem., 5 (1986) 247.
32 M.J.Fink, K.J.Haller, R.West and J.Michl, J.Am.Chem.Soc., 106 (1984) 822.
33 H.B.Yokelson, A.J.Millevolle, B.R.Adams and R.West, VIII. Int.Symp.Organosilicon Chem., St.Louis 1987.
34 S.Masamune, Y.Kabe and S.Collins, J.Am.Chem.Soc., 107 (1985) 5552.
35 R.Jones, D.J.Williams, Y.Kabe and S.Masamune, Angew.Chem., 98 (1986) 176; Angew.Chem.Int.Ed.Engl., 25 (1986) 173.
36 F.K.Cartledge, Organometallics, 2 (1983) 425.
37 H.Watanabe, Y.Nagai in: H.Sakurai, Organosilicon and Bioorganosilicon Chemistry: Structure, Bonding, Reactivity and Synthetic Application, p.107, Ellis Horwood Limited, Chichester, 1985.
38 C.J.Hurt, J.C.Calabrese and R.West, J.Organomet.Chem., 91 (1975) 273.
39 L.Párkányi, K.Sasvári and I.Barta, Acta Crystallogr., B34 (1978) 883.
40 Y.-S.Chen and P.P.Gaspar, Organometallics, 1 (1982) 1410.
41 H.Watanabe, T.Muraoka, M.Kageyama, K.Yoshizumi and Y.Nagai, Organometallics, 3 (1984) 141.
42 C.Kratky, H.G.Schuster and E.Hengge, J.Organomet.Chem., 247 (1983) 253.
43 B.Rozsondai, V.S.Mastryukov and E.Hengge, to be published.
44 E.Hengge, H.G.Schuster and W.Peter, J.Organomet.Chem., 186 (1980) C45.
45 E.Hengge, Phosphorus Sulfur, 28 (1986) 43.
46 H.Gilman and G.L.Schwebke, Adv.Organomet.Chem., 1 (1964) 89.
47 E.Hengge and K.Hassler in: I.Haiduc and D.B.Sowerby, The Chemistry of Inorganic Homo- and Heterocycles, vol.1, p.191, Academic Press Inc.Ltd., London, 1987.
48 Z.Smith, A.Almenningen, E.Hengge and D.Kovar, J.Am.Chem.Soc., 104 (1982) 4362.
49 Z.Smith, H.M.Seip, E.Hengge and G.Bauer, Acta Chem.Scand., A30 (1976) 697.
50 R.West and A.Indriksons, J.Am.Chem.Soc., 94 (1972) 6110.
51 M.Ishikawa, M.Watanabe, J.Iyoda, H.Ikeda and M.Kumada, Organometallics, 1 (1982) 317.
52 M.Ishikawa and M.Kumada, Synth.Inorg.Met.-Org.Chem., 1 (1971) 191.
53 B.J.Helmer and R.West, J.Organomet.Chem., 236 (1982) 21.
54 F.K.Mitter and E.Hengge, J.Organomet.Chem., 332 (1987) 47.
55 F.K.Mitter, G.I.Pollhammer and E.Hengge, J.Organomet.Chem., 314 (1986) 1.
56 K.Hassler, F.K.Mitter, E.Hengge, C.Kratky and U.G.Wagner, J.Organomet.Chem., 333 (1987) 291.

57 W.Wojnowski, B.Dreczewski, A.Herman, K.Peters, E.-M.Peters and H.G. von Schnering, Angew.Chem., 97 (1985) 978; Angew.Chem.Int.Ed.Engl., 24 (1985) 992.
58 L.F.Brough and R.West, J.Organomet.Chem., 194 (1980) 139.
59 M.Ishikawa and M.Kumada, Chem.Commun., (1969) 567.
60 T.A.Blinka and R.West, Organometallics, 5 (1986) 128.
61 E.Hengge and P.K.Jenkner, Z.Anorg.Allg.Chem., in press.
62 G.R.Husk and R.West, J.Am.Chem.Soc., 87 (1965) 3993.
63 E.Carberry, R.West and G.E.Glass, J.Am.Chem.Soc., 91 (1969) 5446.
64 R.West and E.Carberry, Science, 189 (1975) 179.
65 R.West, Pure Appl.Chem., 54 (1982) 1041.
66 R.West and E.S.Kean, J.Organomet.Chem., 96 (1975) 323.
67 C.L.Wadsworth and R.West, Organometallics, 4 (1985) 1659.
68 C.L.Wadsworth, R.West, Y.Nagai, H.Watanabe, H.Matsumoto and T.Muraoka, Chem.Lett., (1985) 1525.
69 M.Kira, H.Bock and E.Hengge, J.Organomet.Chem., 164 (1979) 277.
70 V.V.Bukhtiyarov, S.P.Solodovnikov, O.M.Nefedov and V.I.Shiryaev, Izv.Akad.Nauk SSSR, Ser.Khim., (1968) 1012, Engl.Ed. 967.
71 H.Bock, W.Kaim, M.Kira and R.West, J.Am.Chem.Soc., 101 (1979) 7667.
72 R.West, L'Actualité Chimique, (1986) 64.
73 F.S.Kipping, J.Chem.Soc., (1924) 2291.
74 C.A.Burkhard, J.Am.Chem.Soc., 71 (1949) 963.
75 S.Yajima, J.Hayashi and M.Omori, Chem.Lett., (1975) 931.
76 S.Yajima, K.Okamura and J.Hayashi, Chem.Lett., (1975) 1209.
77 S.Yajima, Y.Hasegawa, J.Hayashi and M.Iimura, J.Mater.Sci., 13 (1978) 2569.
78 Y.Hasegawa, M.Iimura and S.Yajima, J.Mater.Sci., 15 (1980) 720.
79 P.Boudjouk and B.H.Han, Tetrahedron Lett., 22 (1981) 3813.
80 B.Pachaly, V.Frey and N.Zeller (Wacker-Chemie GmbH), DE 35 32 128 Al (Cl.C08G77/60).
81 W.Kalchauer in: Doktor Thesis 1986, T.U. Graz, 1986.
82 R.Schwarz, Angew.Chem., 51 (1938) 328.
83 E.Hengge, unpublished studies.
84 E.Hengge and G.Scheffler, Monatsh.Chem., 95 (1964) 1450.
85 C.G.Pitt in: A.L.Rheingold, Homoatomic Rings, Chains and Macromolecules of Main-Group Elements, p.203, Elsevier Scientific Publishing Comp., Amsterdam - Oxford - New York, 1977.
86 H.Bock, W.Ensslin, F.Fehér and R.Freund, J.Am.Chem.Soc., 98 (1976) 668.
87 T.Veszprémi, M.Fehér, E.Zimonyi and J.Nagy, Acta Chim.Hung., 120 (1985) 153.
88 E.A.Halevi, G.Winkelhofer, M.Meisl and R.Janoschek, J.Organomet.Chem., 294 (1985) 151.
89 E.Hengge and H.Stüger, Monatsh.Chem., in press.
90 C.G.Pitt, J.Am.Chem.Soc., 91 (1969) 6613.
91 M.Weidenbruch, B.Flintjer, K.Peters, H.G. von Schnering, Angew.Chem. 98 (1986) 1090

GERMANIUM-CARBON RINGS

P. MAZEROLLES
Laboratoire des Organométalliques, U.A. 477, Université Paul Sabatier, 118, route de Narbonne 31062 Toulouse-Cedex (France)

INTRODUCTION

Cyclic organogermanium compounds can, like alicyclic compounds (ref. 1) be classified into four series according to ring size :

1. Small rings. These comprise three- and four-membered rings, which are characterized by appreciable deformation of bond angles, causing ring strain and high reactivity.

2. Common rings. These are five to seven-membered rings, for which the saturated derivatives are the easiest to obtain because the spatial proximity of the linear chain ends facilitates cyclization. These molecules are very stable chemically, like their analogous saturated linear derivatives. However, both their thermal stability and chemical reactivity are strongly modified when rings contain vicinal germanium atoms or functional groups, especially in 2-position to a germanium atom. Spirans are also more reactive than the corresponding monocyclic compounds.

3. Medium rings. Those with eight to thirteen atoms in the ring are characterized by crowded conformations, which cause strong steric hindrance. The spatial proximity of opposite sides of such rings allows transannular interactions. This fact, and the large distance between the ends of the linear starting molecules,make special preparative techniques necessary.

4. Large rings. Rings with more than 13 atoms are not strained and their reactivity is similar to that of the corresponding linear derivatives.

1. SMALL RINGS

1.1. THREE-MEMBERED RINGS

1.1.1. Germirenes

The synthesis of germacyclopropenes ("germirenes") was attempted by Vol'pin *et al.* (refs. 2-6) by reaction of germanium diiodide with various alkynes (e.g. acetylene, phenylacetylene, diphenylacetylene) in benzene ; however, the stable products obtained were found afterwards to be the cyclic dimers, e.g. 1,4-digermacyclohexa-2,5-dienes (refs. 7-11).

Two 2,3-substituted germirenes have been isolated recently by addition of a dialkylgermylene to a suitably substituted thiacycloalkyne (refs. 12, 13) :

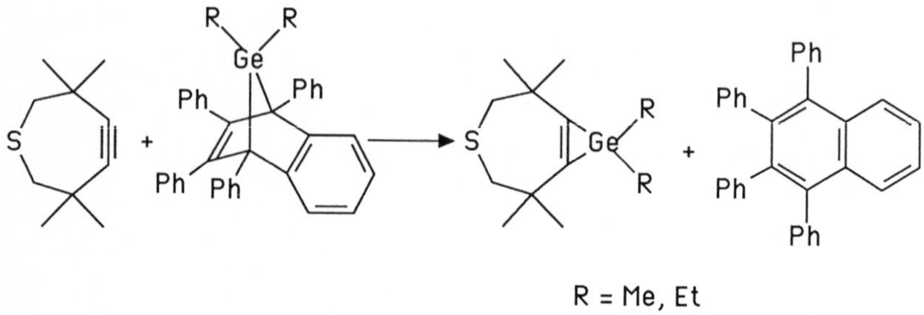

eqn.(1.1)

and the X-ray structure has been determined (refs. 13, 14). These reactive compounds undergo a methanolytic ring cleavage (ref. 12) :

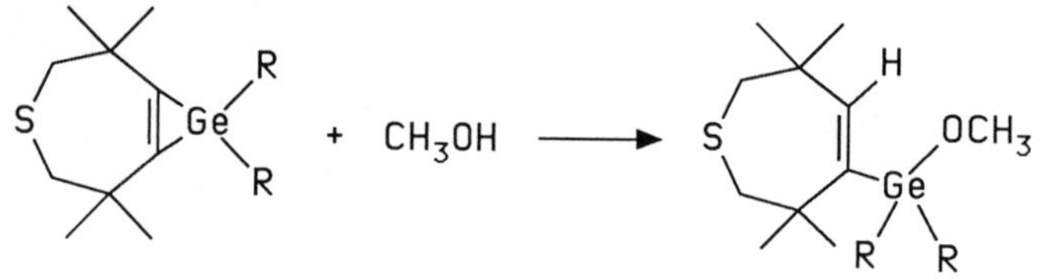

eqn.(1.2)

1.1.2. Digermiranes

A novel and very reactive three-membered ring was recently synthesized by Ando *et al.* by reaction of tetrakis(2,6-diethylphenyl)digermane with diazomethane (ref. 15) :

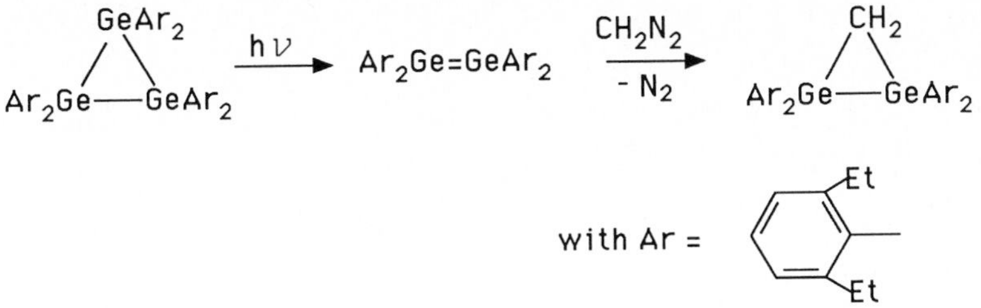

eqn.(1.3)

Photolysis of this digermane gives a mixture of germene and germylene :

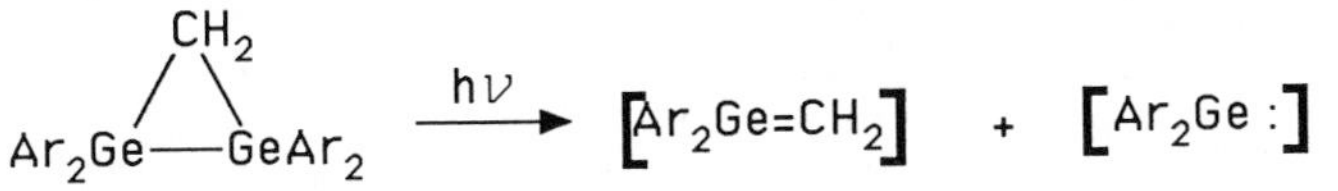

eqn.(1.4)

Ring-expansion reactions are observed by insertion of oxygen, sulfur and selenium into the Ge-Ge bond :

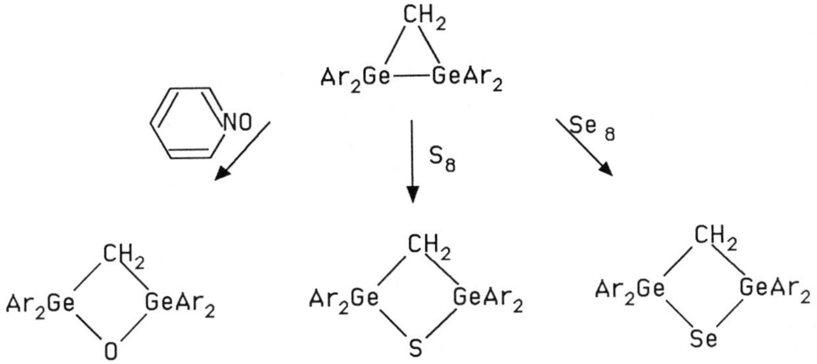

Chart 1. Insertion reactions of digermiranes.

1.2. FOUR-MEMBERED RINGS

1.2.1. Saturated four-membered rings

1.2.1.1. Germacyclobutanes

Synthesis (Chart 2): The first germacyclobutane was prepared by reducing an equimolar mixture of dibutylgermanium dichloride and 1,3-dichloropropane with powdered sodium in boiling xylene (refs. 16,17), reaction (a), but, because of the poor yields (10 %), the intramolecular cyclization of (3-chloropropyl)-germanium chlorides was preferred, reaction (b), (ref. 17). Reduction by an alkali metal in xylene provides good yields of dialkylgermacyclobutanes and 4-germaspiro [3.4] octane (ref. 17). This type of reaction also allows the one-step synthesis of hydrogenogermacyclobutanes (ref. 18). The recent facile preparation of di-Grignard reagents from 1,3-dihalides allowed the synthesis of dialkylgermacyclobutanes and 4-germaspiro[3.3]heptanes in almost quantitative yield (refs. 19-21), reactions (c).

A novel formation of a germacyclobutane ring was observed during the reduction of an unsaturated germanium dichloride, reaction (d) ; the transient dihydride undergoes an intramolecular addition forming a hydrogermacyclobutane (refs. 22, 23) .

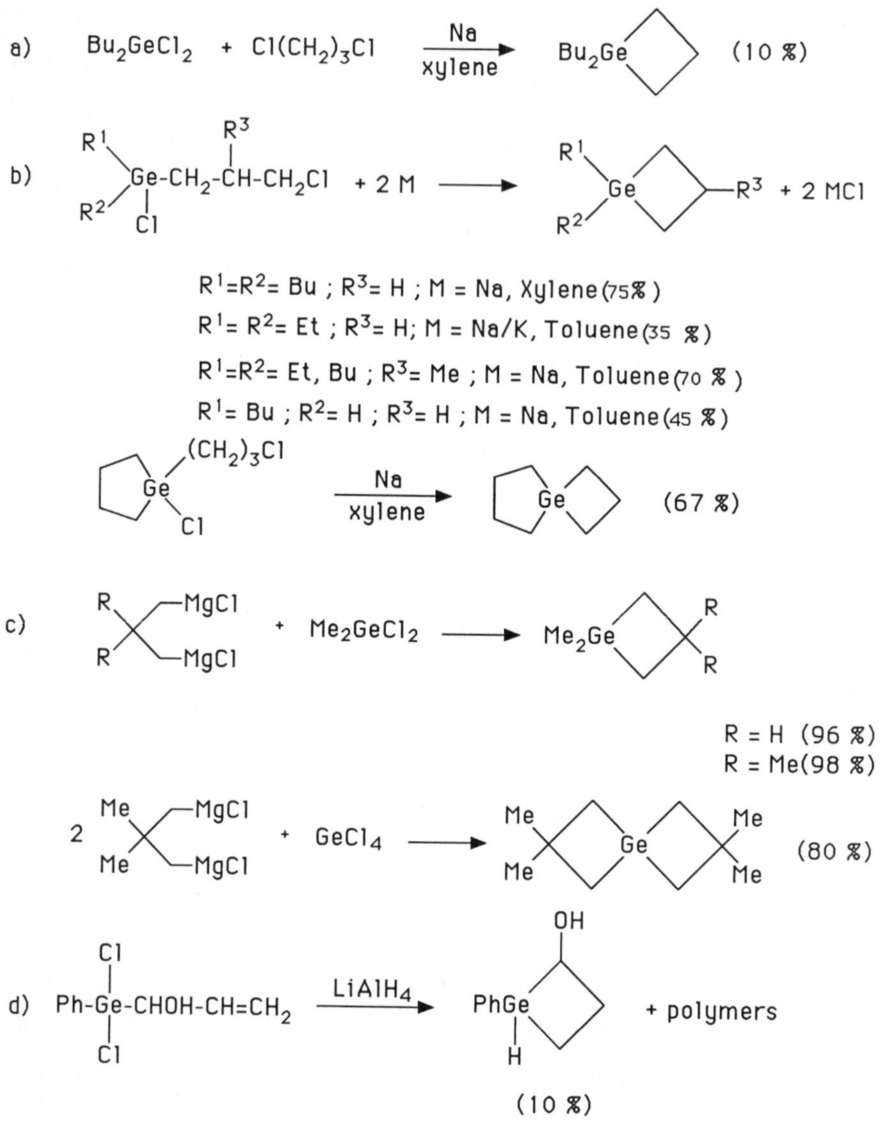

Chart 2. Formation of germacyclobutanes

1.2.1.2. Digermacyclobutanes and metallagermacyclobutanes

1,3-Digermacyclobutanes are formed by dimerization of germaethylenes (ref. 24), reaction (a). The reaction of dimethylgermanium dichloride with the di-Grignard reagent of dibromomethane produces a mixture of 1,3-digermacyclobutane, 1,3,5-trigermacyclohexane and 1,3,5,7-tetragermacyclooctane (refs. 21, 25), reaction (b). Trimethylsilyl-C-substituted-1,3,digermacyclobutanes have been isolated in the reaction of dimethylgermanium dibromide with a suitable lithium reagent at low temperature (ref. 26), reaction (c) and found among the products of a direct synthesis from trimethyl(dichloromethyl)silane (ref. 27), reaction (d). 1,3-Digermacyclobutanes and 1-sila-3-germacyclobutanes, which are prepared in good yield by the reaction of (1-chloromethyl)-3-bromo-1,3-digermane (or silagermane) with magnesium in boiling ether (refs. 28, 29), reaction (e), are used as cocatalysts for the metathesis of olefins (ref. 30). A 1-sila-3-germacyclobutane ring is formed in moderate yield by mixed bimolecular cyclization of a bis(chloromethyl)silane and a germanium dihalide in the presence of magnesium (ref. 31), reaction (f). Lastly, four-membered rings containing germanium and titanium were isolated from the reaction of a polyhalogermane with the appropriate di-Grignard reagent (ref. 21), reactions (g)

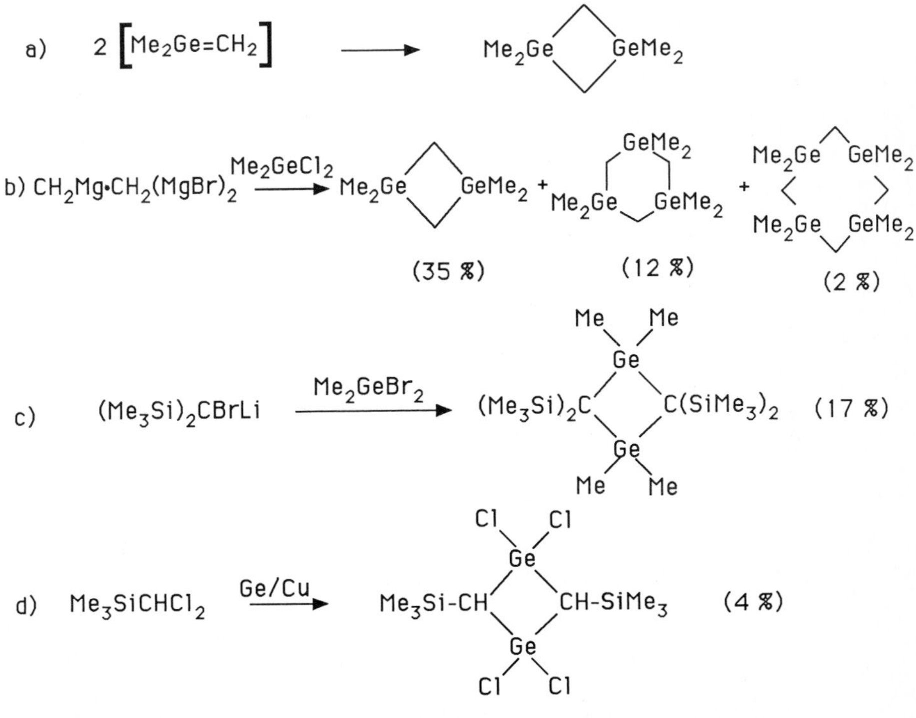

Chart 3. Formation of digerma- and metallagermacyclobutanes.

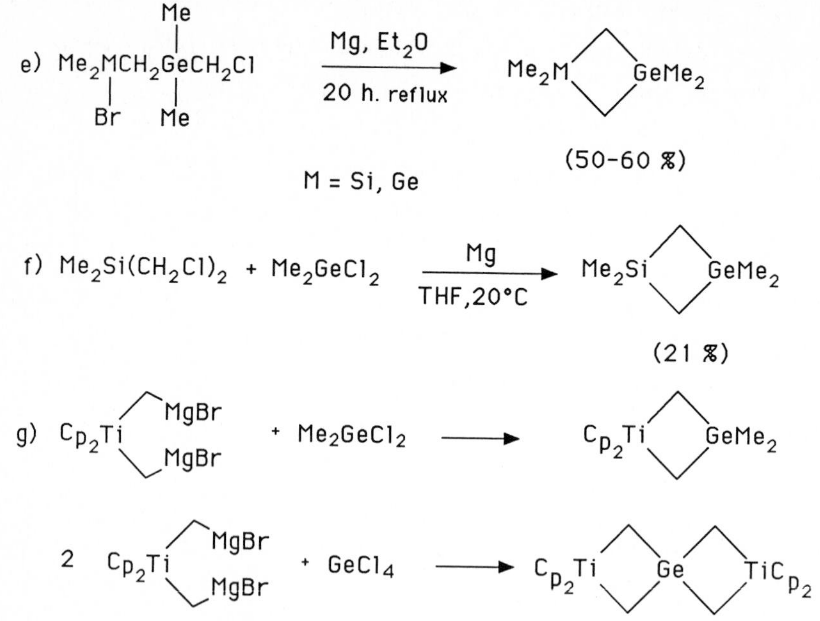

Chart 3 (continued). Formation of digerma- and metallagermacyclobutanes

Reactivity : Unlike silacyclobutanes, which decompose thermally into a mixture of ethylene and silaethylene, germacyclobutanes, when heated to 590° C, form a germylene, which inserts into an intracyclic Ge-C bond of a germacyclobutane ring to yield a 1,2-digermacyclopentane (ref. 32) :

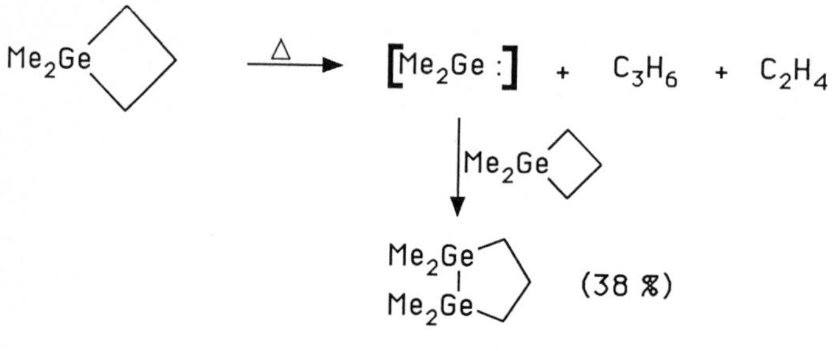

eqn.(1.5)

Because of the high polarizability of intracyclic germanium-carbon bonds resulting from ring strain, germacyclobutanes are very sensitive towards ionic reagents (Chart 4), with ring cleavage occurring easily :

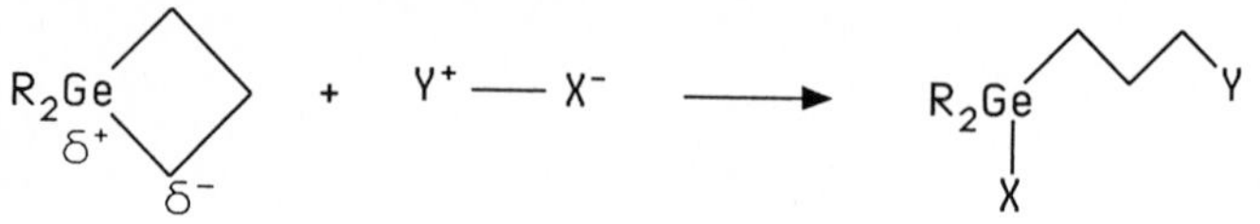

eqn.(1.6)

Cleavage with bromine occurs quantitatively at -78°C in bromoethane, while iodine reacts only at room temperature or above. With iodine monochloride, a germanium-chlorine bond is formed. Ring-opening reactions also occur with compounds such as germanium tetrachloride or sulfuryl chloride that possess a mobile halogen ; the latter compound reacts in two different ways :

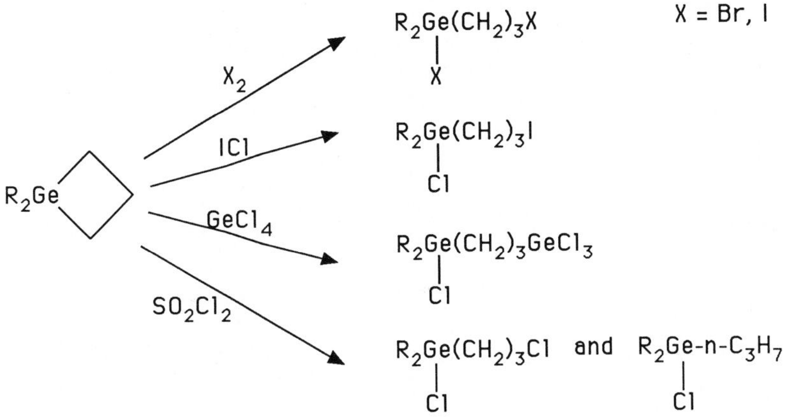

Chart 4. Ring-opening reactions of germacyclobutanes with halogens and inorganic halides. (refs.16, 17).

The acid-catalyzed cleavage of germacyclobutanes results in dialkyl (n-propyl) germanium esters in high yields (Chart 5).

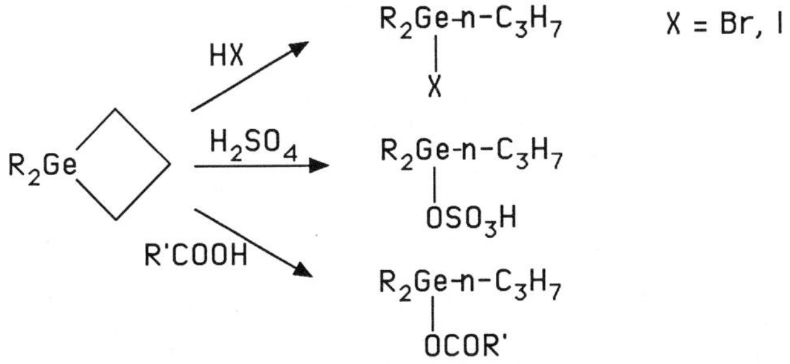

Chart 5. Ring-opening reactions of germacyclobutanes with inorganic and organic acids (ref. 17).

Ring cleavage by nucleophiles (Chart 6) is much more difficult and reactions with organolithiums, sodium alkoxides and lithium aluminum hydride are slow and incomplete :

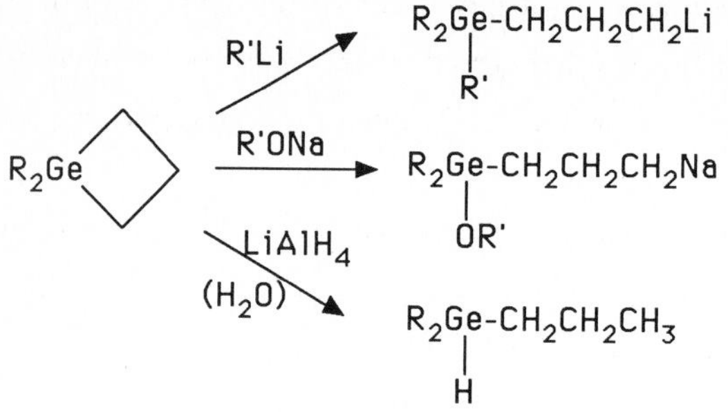

Chart 6. Cleavage of germacyclobutanes by nucleophilic reagents (refs. 16, 18).

In the presence of a suitable catalyst, organosilicon and organogermanium hydrides react with germacyclobutanes giving two different types of ring-cleavage, depending on the nature of the catalyst (Chart 7) :

- in the presence of an ionic catalyst such as chloroplatinic acid, organosilicon and organogermanium hydrides having electron-donor substituents (e.g. trialkylsilanes and germanes), in which the polarity of the metal-hydrogen bond is $M^{\delta+}$-$H^{\delta-}$, add easily upon moderate heating. When trialkylsilane is used, a germanium-hydrogen bond is formed.
- in the presence of a free radical initiator (such as azobis-isobutyronitrile or benzoylperoxide) organosilicon and organogermanium chlorohydrides, having an almost apolar metal-hydrogen bond, produce a dialkyl-(n-propyl)germanium chloride.

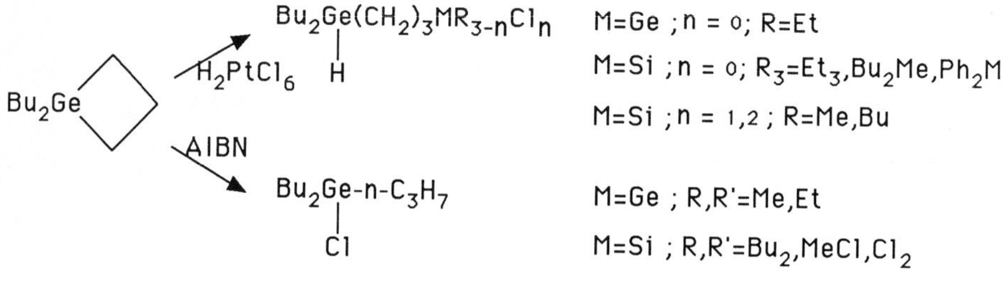

Chart 7. Hydrosilylation and hydrogermylation of germacyclobutanes (refs. 33,34)

Ring-expansion reactions leading to five- or six-membered rings have been described (Chart 8). Thiagermacyclopentanes are formed in quantitative yield by the reaction with elemental sulfur at 250°C. A similar but less ready insertion is observed with selenium. Sulfur dioxide gives quantitatively a germasultine at -10°C while the exothermic addition of sulfur trioxide at low temperature results in the corresponding germasultone. The reaction of phenylbromodichloromethylmercury in boiling benzene, leading also to a five-membered ring, was the first example of dichlorocarbene insertion into a germanium-carbon bond (ref. 35). 1,2-Digermacyclopentanes are formed by thermolysis of germacyclobutanes via the insertion of germylene into the cyclobutane ring (ref. 32).

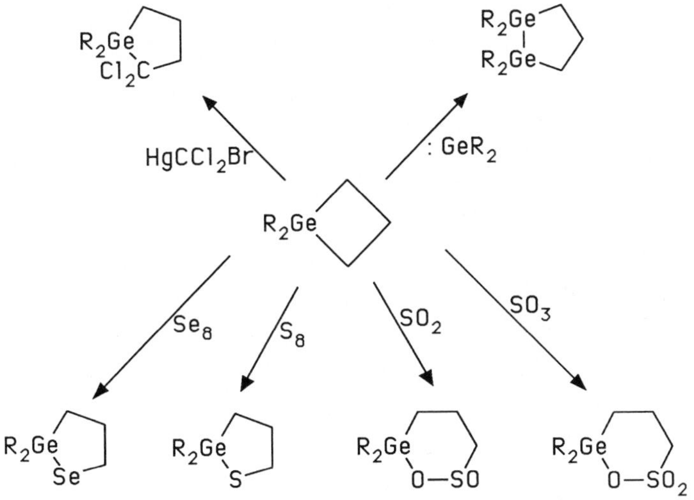

Chart 8. Insertion reactions into an intracyclic bond of germacyclobutanes (refs. 17, 32, 35-39).

1.2.2. Unsaturated four-membered rings

A germacyclobutene was isolated, together with eight-membered isomeric dimers, in the reaction of dimethylgermanium dichloride with 1,2-dihydro-1- magnesiabenzocyclobutene (ref. 40) :

eqn.(1.7)

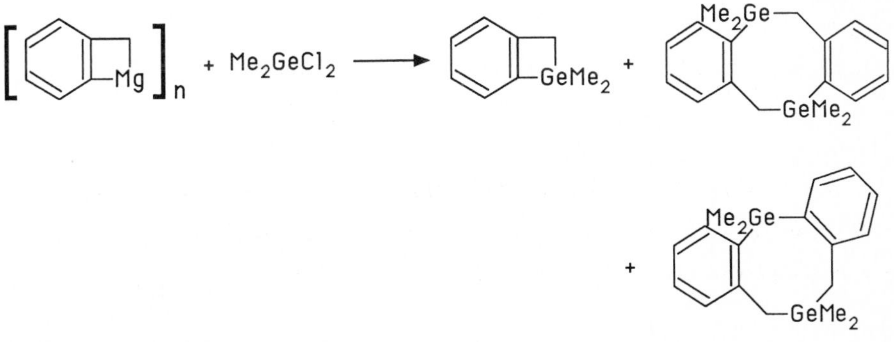

3,4-Substituted digermacyclobutenes are formed by addition of digermenes or dihalogermylenes to a suitable acetylenic compound (refs. 41-45) or, in low yield (3-5 %), by addition of dimethylgermylene to a dimethylgermacyclopropene ring (ref. 45).

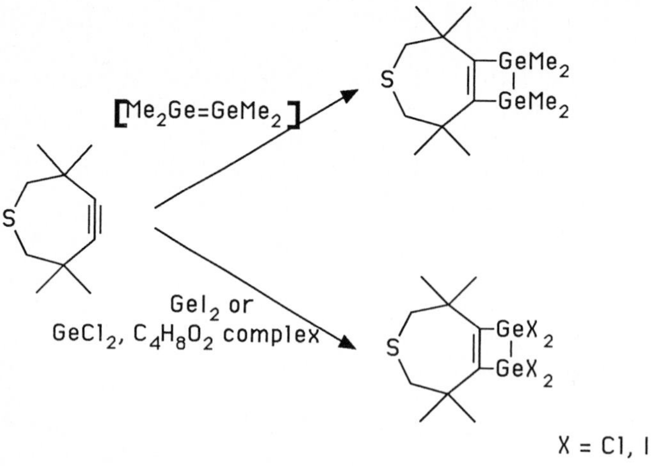

Chart 9. Synthesis of 1,2-digermacyclobutenes

Methylation of the tetrachloride causes the extrusion of the germanium moiety with recovery of the starting acetylene (ref. 45).

2. COMMON RINGS

2.1. SATURATED COMMON RINGS

2.1.1. Monogermanium rings

2.1.1.1. Germacyclopentanes,germacyclohexanes and germacycloheptanes

Synthesis: : 1,1-Dichlorogermacyclopentane is formed with other germanium chlorides by direct synthesis from 1,4-dichlorobutane (ref. 46) :

$$Cl(CH_2)_4Cl \xrightarrow[370-390°C]{Ge/Cu} Cl_2Ge\bigcirc \;(14\ \%) + Cl_3Ge(CH_2)_4Cl \;(18\ \%) \qquad \text{eqn.(2.1)}$$

The reaction of an ethereal solution of the di-Grignard reagent of 1,4-dibromobutane with germanium tetrachloride gives a complex mixture containing 1,1-dichlorogermacyclopentane (refs. 47-49). Even in the presence of a large excess of $GeCl_4$, 5-germaspiro[4.4.]nonane is formed (eqn. 2.2) ;

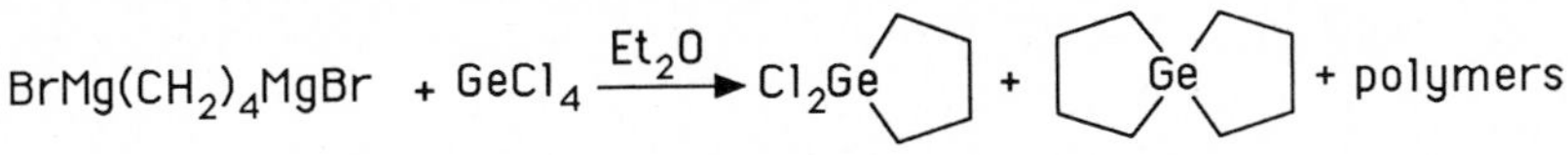

eqn.(2.2)

Reduction of the mixture with lithium aluminium hydride allows the ready isolation of germacyclopentane (b.p. 90°C) from the spiran (b.p. 190°C) and polymers. A similar cyclization is observed with the di-Grignard reagent of 1,5-dibromopentane (eqn. 2.3) :

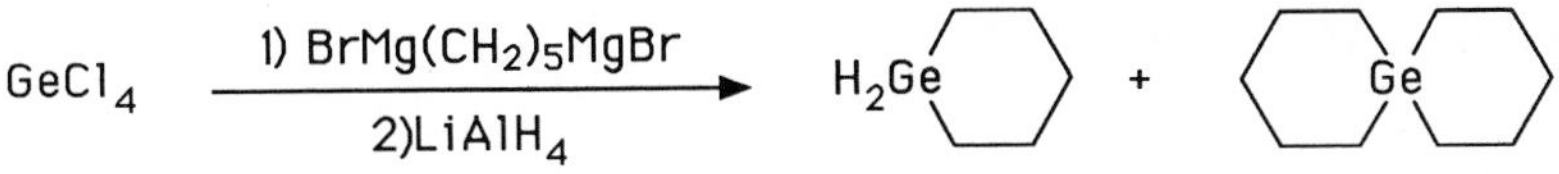

eqn.(2.3)

This reaction is not suitable for the preparation of seven-membered rings (10 % yield). Better yields are obtained by cycloaddition of a dihydrogermane to 1,5-hexadiene (ref. 50) :

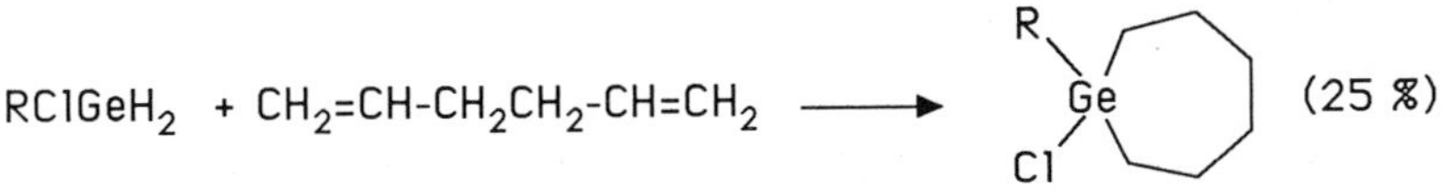

eqn.(2.4)

Analogous hydrogermylation of 1,4-pentadiene leads to chlorogermacyclohexanes in 21-25 % yields (ref. 51) :

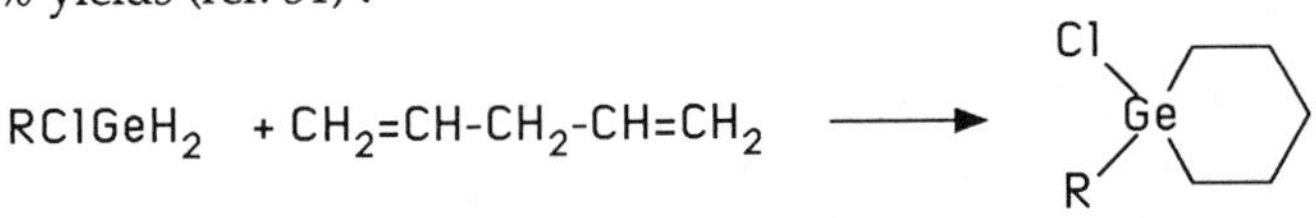

eqn.(2.5)

In the synthesis using a di-Grignard reagent, the yield of germacycloalkanes is increased when diphenylgermanium dihalide is used as the starting germanium halide (refs. 17, 52) :

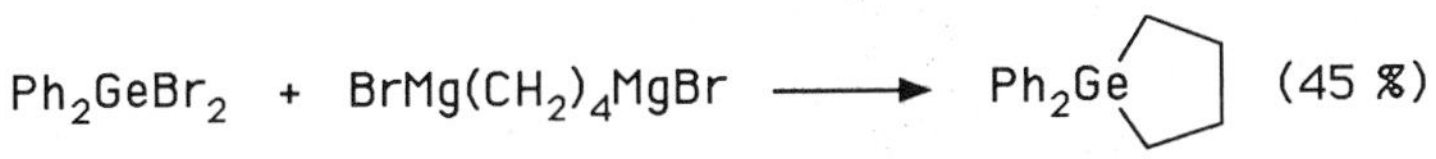

eqn.(2.6)

Diphenylgermacyclopentane and diphenylgermacyclohexane can then be used as starting reagents for the synthesis of other germacyclanes, the cleavage of Ge-C(Ph) bonds by bromine being easier than that of intracyclic Ge-C bonds :

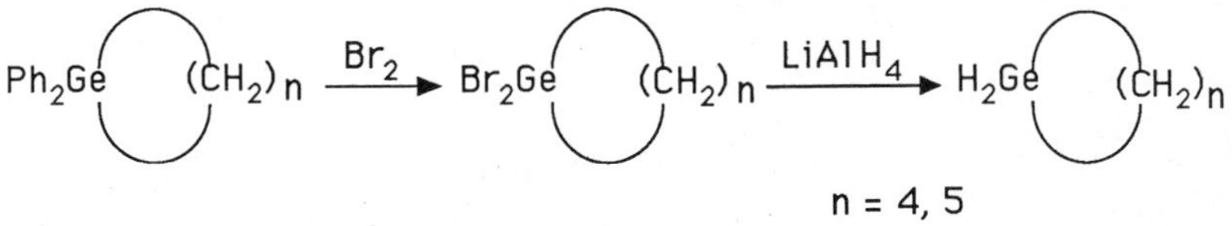

eqn.(2.7)

Other substitution reactions (Chart 10) occur as in aliphatic series and are also suitable for germacyclohexanes (refs.17, 48) :

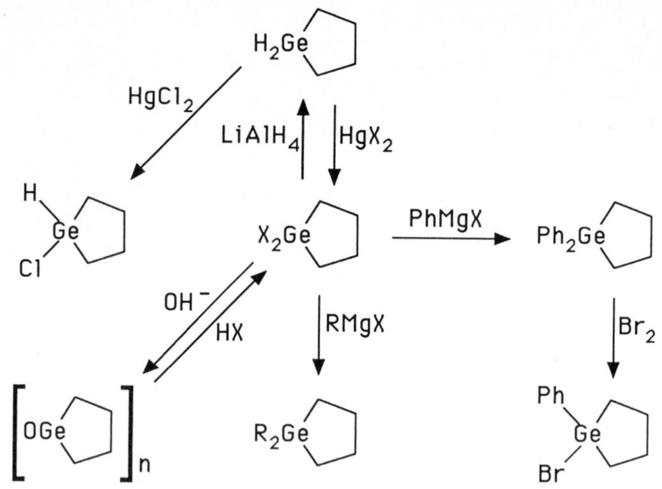

Chart 10. Substitution reactions of germacyclopentanes

The electron-impact fragmentation pattern of germacyclopentanes indicates the formation of ethylene. A study with labelled derivatives (ref. 53) demonstrated that the ethylene (in this case CH_2CD_2) comes from ring carbons in 2- and 3-position with respect to the germanium atom :

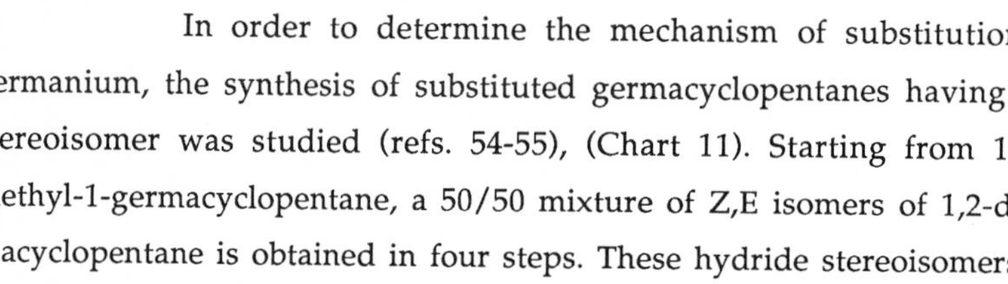

eqn.(2.8)

In order to determine the mechanism of substitution reactions at germanium, the synthesis of substituted germacyclopentanes having a preferential stereoisomer was studied (refs. 54-55), (Chart 11). Starting from 1,1-diphenyl-2-methyl-1-germacyclopentane, a 50/50 mixture of Z,E isomers of 1,2-dimethyl-1-germacyclopentane is obtained in four steps. These hydride stereoisomers can be separated (or at least enriched) by distillation, allowing the study of addition reactions to ethylene and acetylene, reactions (a), substitution by chlorine, and reduction of the chloride by lithium aluminum hydride, reactions (b). These reactions, which occur with retention (addition, chlorination) or inversion (reduction of the chloride) of the configuration are,*stereospecific*. Mixture of Z/E isomers having a predominant

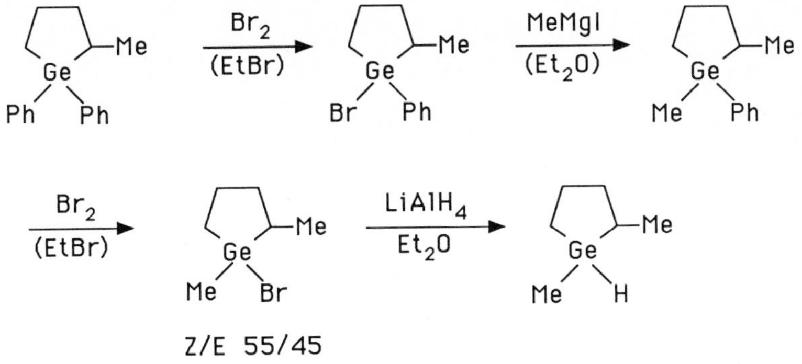

Chart 11. Synthesis of configurationnally stable 1,2-dimethylgermacyclopentane Z,E isomers.

isomer can also be obtained by using *stereoselective* reactions. Thus, nucleophilic substitutions on a mixture containing almost equivalent quantities of Z/E isomers of 1,2-dimethyl-1-bromogermacyclopentane lead to inequivalent mixtures of alkoxy-, alkylthio-, dialkylamino- or dialkylphosphinogermacyclopentanes, allowing the determination of the stereochemistry of the insertion mechanism in Ge-X bonds (X = O, S, N, P) (refs. 54, 56-63), reactions (c), (Chart 12) :

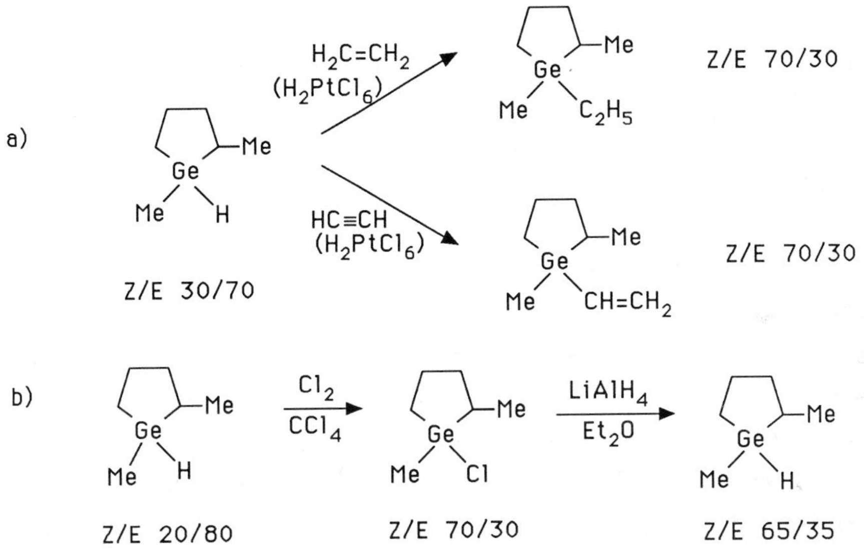

Chart 12. Stereochemistry of addition reactions (a) and substitution reactions (b) at germanium.

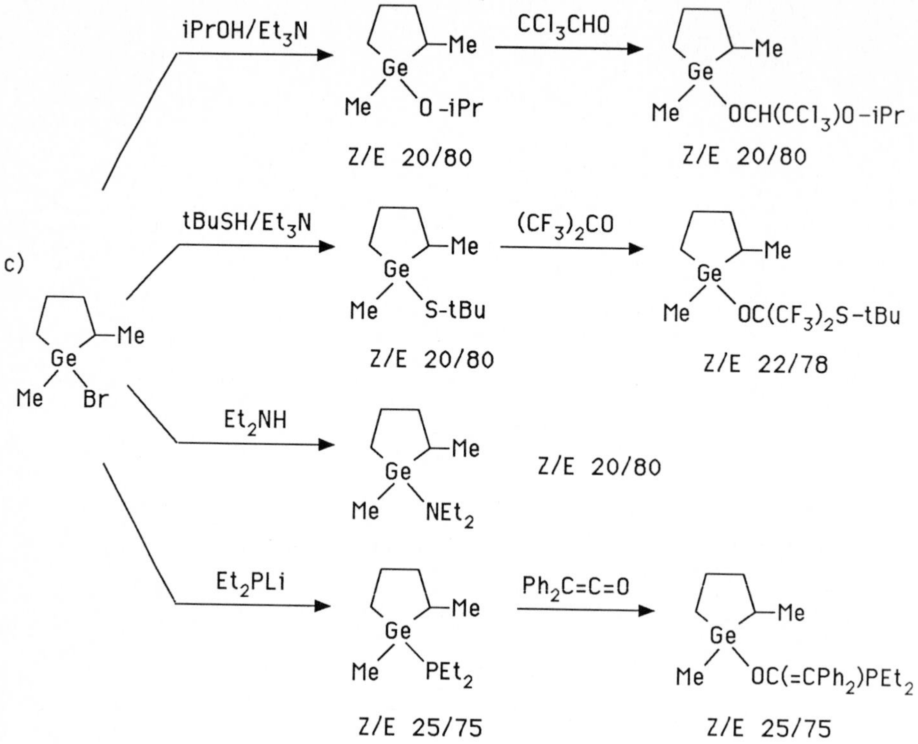

Chart 12 (continued). Stereochemistry of substitution reactions (c) at germanium

2.1.1.2. Functional germacyclopentanes and -hexanes

Germacyclopentan-3-ols, accessible from germacyclopent-3-enes (Chart 19), are very sensitive towards inorganic and organic acids (refs.64, 65) :

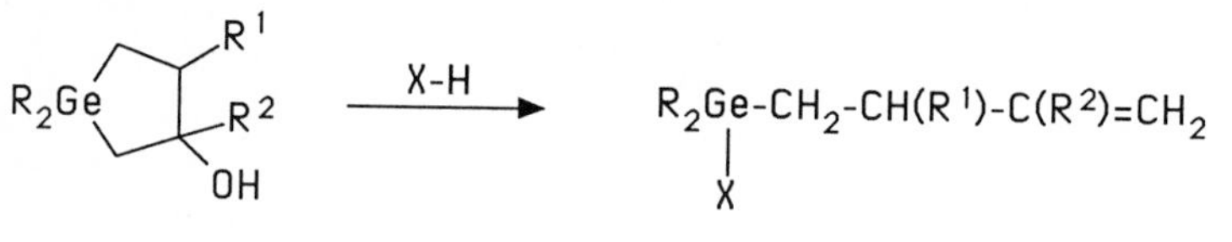

eqn.(2.9)

$R^1, R^2 = H, CH_3$; $X = F, Cl, Br, I, CH_3COO$

Unsubstituted germacyclopentanols give a mixture of germacyclopentanone and oxa-germacyclopentane by reaction with Raney nickel (Chart 13) but with substituted germacyclopentanols the ring is cleaved :

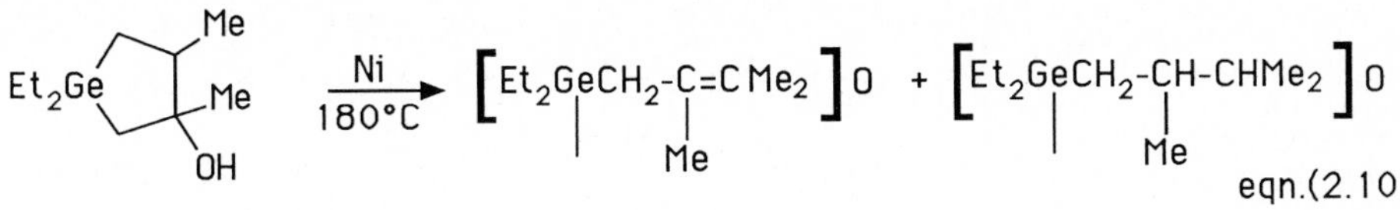

eqn.(2.10)

Germacyclopentan-3-ones are formed, together with an oxagermacyclohexene, by thermolysis of 2-epoxides, reaction (a) and, together with a methyl-substituted oxagermacyclopentane, by reaction of Raney nickel with 3-germacyclopentanols, reaction (b), (refs. 66,67).

Germacyclopentan-2-ones are prepared in three steps from 1-methoxybutadiene, reactions (c), (ref.68).

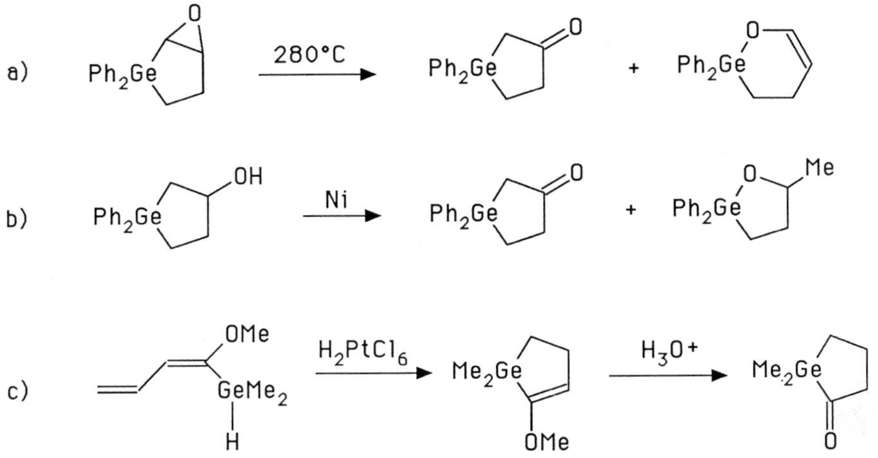

Chart 13. Formation of germacyclopentanones

Germacyclohexan-4-ones are prepared (Chart 14)

- as for the silicon analog (ref. 69) by a modified Dieckmann cyclization of a suitable diester (refs. 70, 71), reaction (a)

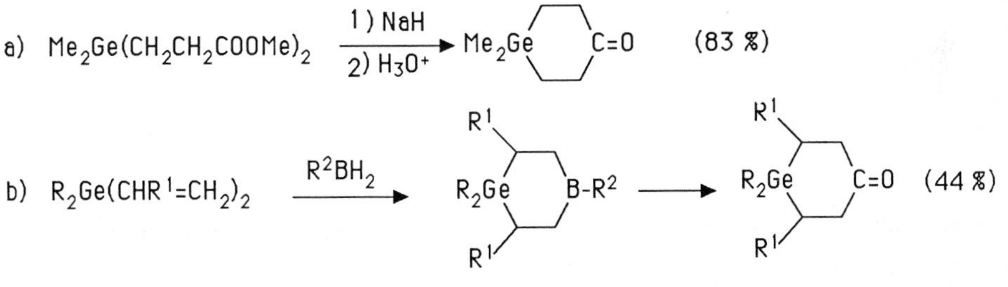

R = Me, R' = Me, Ph ; R = Et, R' = Me

Chart 14. Formation of germacyclohexan-4-ones.

- from dialkyldivinylgermanes (ref. 72), by hydroboration (thexylborane or borane-methyl sulfide complex) followed by cyanoboration (ref. 73) or dichloromethyl methyl ether (ref. 74) process, reaction (b)

6-Oxa-3-germabicyclo[3.1.0]hexanes are prepared in good yield from germacyclopent-3-enes and are reduced to germacyclopentan-3-ols by lithium aluminum hydride (Chart 19). Their base-induced rearrangement with Et_2NLi is a convenient method for the synthesis of germacyclopent-2-en-4-ols (ref. 75) :

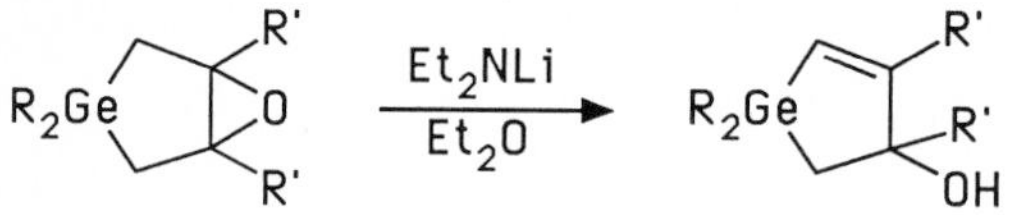

(70-72 %) eqn.(2.11)

R = Me, Ph ; R' = H, Me

Under thermolysis, germaepoxides give a transient germanone, which can be trapped into Si-Cl or Si-O bonds (ref. 76) :

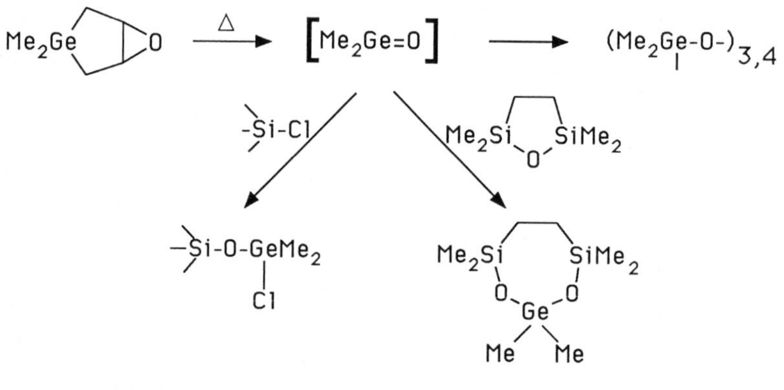

Chart 15. Formation and trapping of dimethylgermanone.

Ion-molecule reactions between F^- and germaepoxides give mainly the Me_2GeFO^- ion (ref. 77).

6-Thia-3-germabicyclo [3.1.0.] hexanes are obtained in good yield from the corresponding epoxides. Their characteristic reactions are summarized in Chart 16 (ref.78)

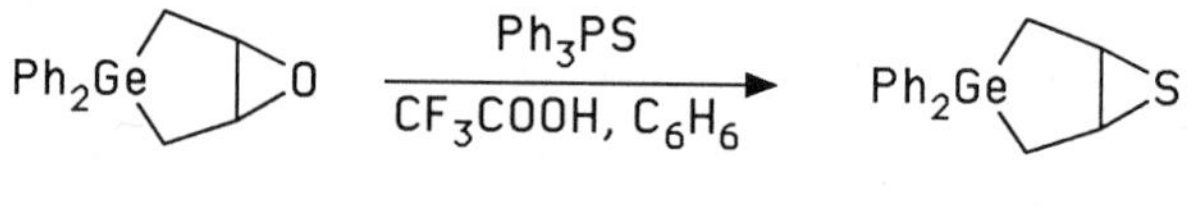

(75 %)

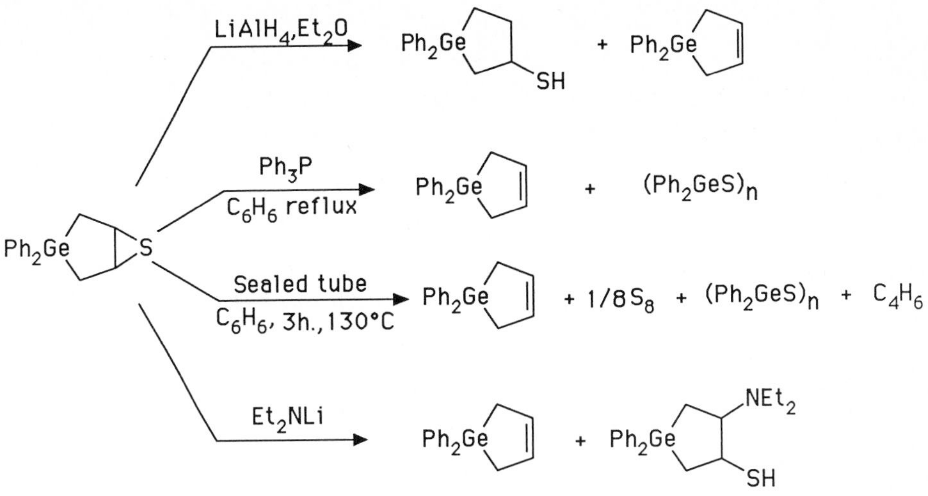

Chart 16. Synthesis and chemical reactivity of 6-thia-3,3-diphenyl-3-germabicyclo [3.1.0] hexane.

2.1.2. Digerma- and germametallacyclopentanes and -hexanes

1,2-Digermacyclopentanes and cyclohexanes are formed in 75% yield by thermal decomposition of the mercuric intermediate resulting from the reaction of diethylmercury with a suitable germanium hydride (ref. 79) :

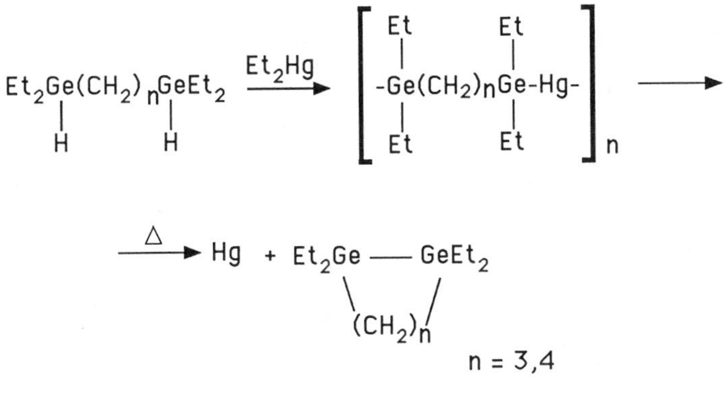

eqn.(2.12)

1,1,2,2-Tetraethyl-1,2-digermacyclopentane has also been obtained in moderate yield (25 %) by ring contraction of a thiadigermacyclohexane :

eqn.(2.13)

whereas only traces are formed by reduction of 1,3-bis(diethylbromogermyl) propane with sodium in boiling xylene. Tetramethyl-1,2-digermacyclopentane is formed in 38 % yield by thermolysis of dimethylgermacyclobutane (ref. 32).

Bromine easily cleaves the germanium-germanium bond at low temperature :

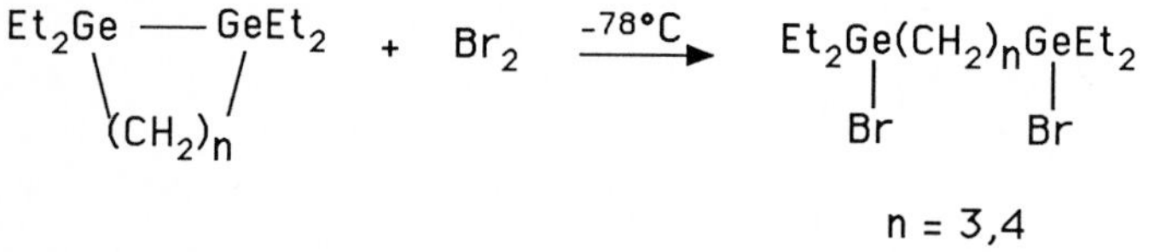

eqn.(2.14)

A competitive reaction showed that the reactivity decreases in the order :

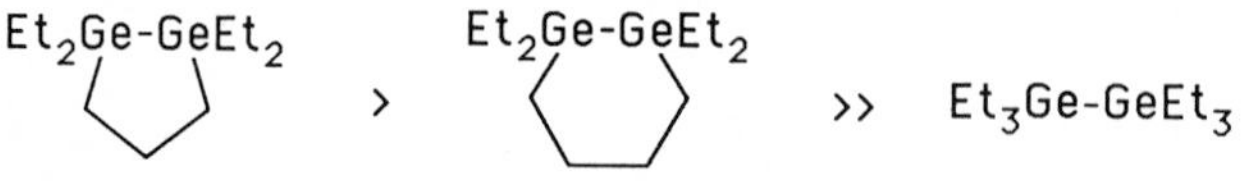

eqn.(2.15)

The five-membered ring derivative reacts easily with oxygen, sulfur and selenium, giving the corresponding six-membered heterocycles by a ring expansion reaction :

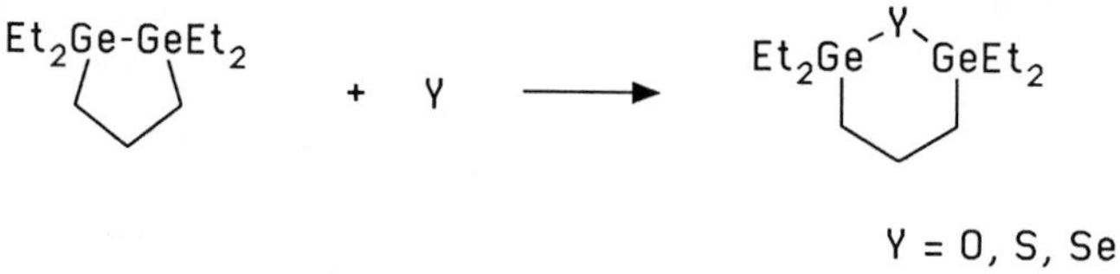

eqn.(2.16)

Tetramethyldigermacyclohexane, which is much less reactive, does not react appreciably under the same experimental conditions.

The phenyl(bromodichloromethyl) mercury-derived dichlorocarbene does not insert into the Ge-Ge bond but into 2- C-H bonds, and thermal decomposition of the adduct causes ring cleavage (ref.80) :

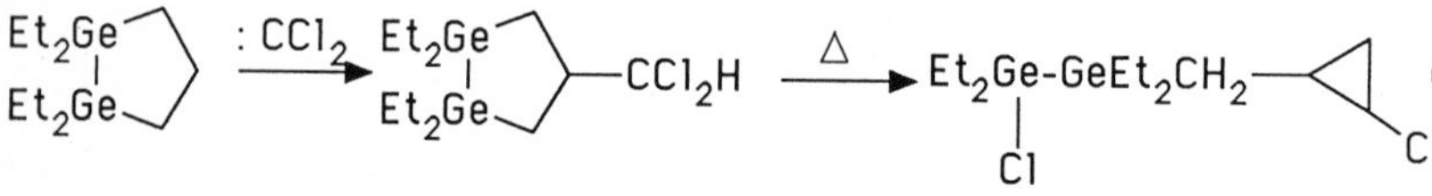

eqn.(2.17)

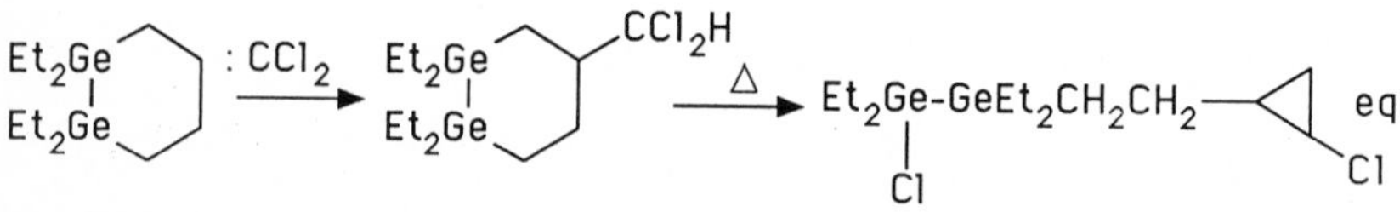

eqn.(2.18)

Radical addition of dihydrodigermanes to 1,4-pentadiene leads to *1,2-digermacycloheptanes* (ref.81) :

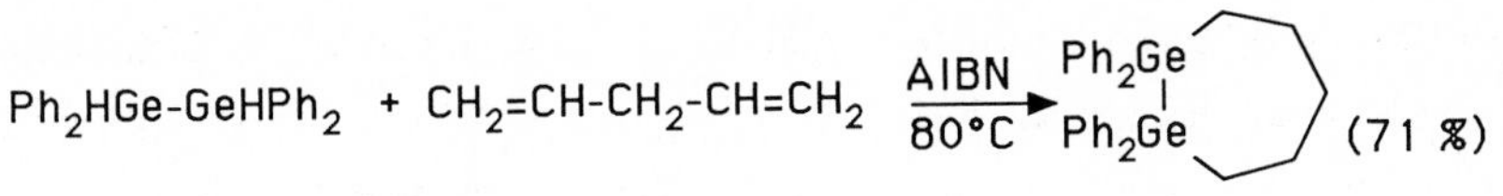

$$Ph_2HGe\text{-}GeHPh_2 \ + \ CH_2\!=\!CH\text{-}CH_2\text{-}CH\!=\!CH_2 \ \xrightarrow[80°C]{AIBN} \ \overset{Ph_2Ge}{\underset{Ph_2Ge}{\Big\langle}} \Big\rangle \quad (71\ \%) \qquad \text{eqn.(2.19)}$$

1,3-Digermacyclopentanes and *1-germa-3-silacyclopentanes* are formed by ring closure of suitable dihalides (refs. 28,82)

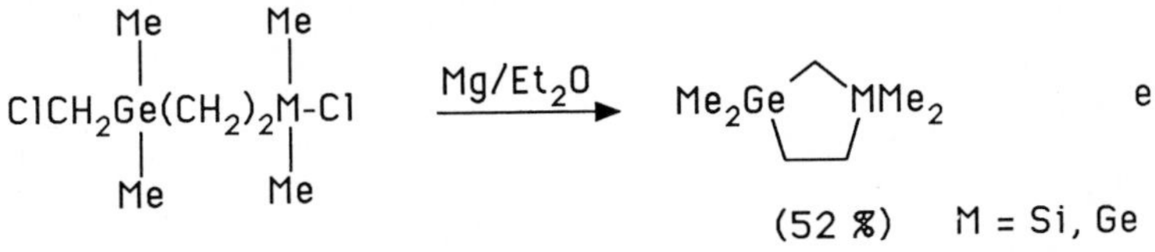

$$\underset{\underset{Me}{|}}{\overset{\overset{Me}{|}}{ClCH_2Ge}}(CH_2)_2\underset{\underset{Me}{|}}{\overset{\overset{Me}{|}}{M}}\text{-}Cl \ \xrightarrow{Mg/Et_2O} \ Me_2Ge\overbrace{\diagdown\underset{\diagup}{}} MMe_2 \qquad \text{eqn.(2.20)}$$

$$(52\ \%) \quad M = Si,\ Ge$$

1-germa-3-silacyclohexanes are obtained similarly in 28 % yield (ref. 28)

1,4-digermacyclohexanes are formed in three ways :

- by catalytic hydrogenation of 1,4-digermacyclohexadienes (ref. 11) :

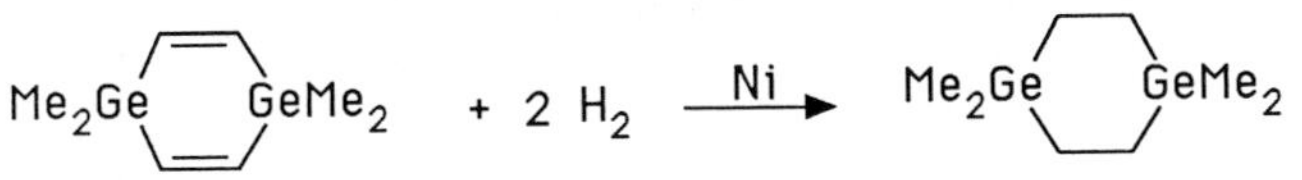

$$Me_2Ge\overbrace{\diagdown\diagup}GeMe_2 \ + \ 2\ H_2 \ \xrightarrow{Ni} \ Me_2Ge\overbrace{\diagdown\diagup}GeMe_2 \qquad \text{eqn.(2.21)}$$

- by addition of an organogermanium (ref. 48) or organotin (ref. 83) dihydride to a dialkyl or diaryldivinylgermane :

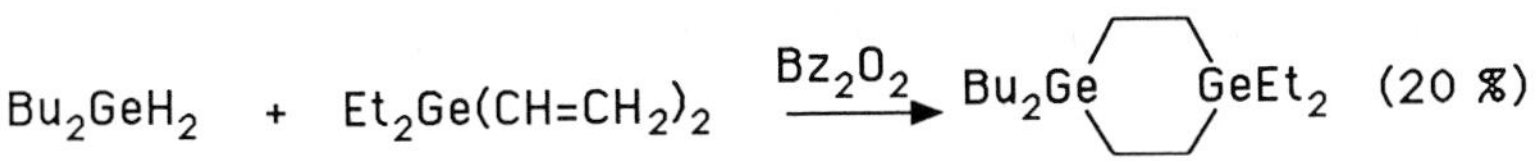

$$Bu_2GeH_2 \ + \ Et_2Ge(CH\!=\!CH_2)_2 \ \xrightarrow{Bz_2O_2} \ Bu_2Ge\overbrace{\diagdown\diagup}GeEt_2 \quad (20\ \%) \qquad \text{eqn.(2.22)}$$

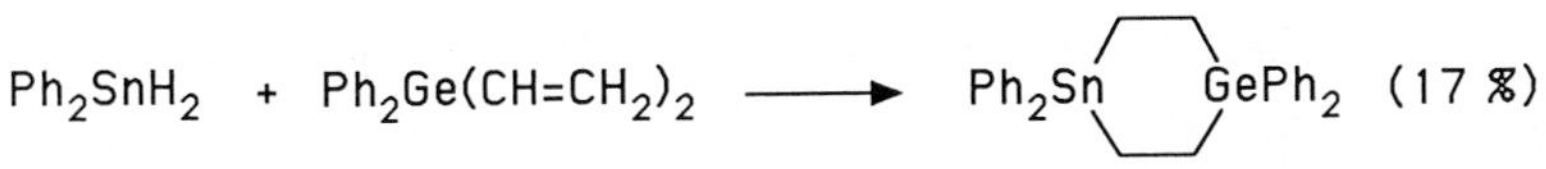

$$Ph_2SnH_2 \ + \ Ph_2Ge(CH\!=\!CH_2)_2 \ \longrightarrow \ Ph_2Sn\overbrace{\diagdown\diagup}GePh_2 \quad (17\ \%) \qquad \text{eqn.(2.23)}$$

- by addition of dimethylgermylene to allenes (ref. 84) :

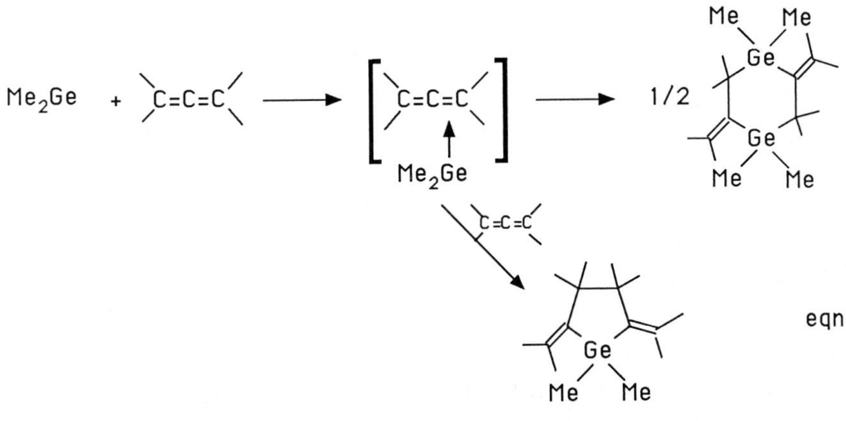

$$\text{eqn.(2.24)}$$

Cyclization of a suitable doubly lithiated silane by dimethylgermanium dichloride affords a [4.1.1] germapropellane (ref. 85)

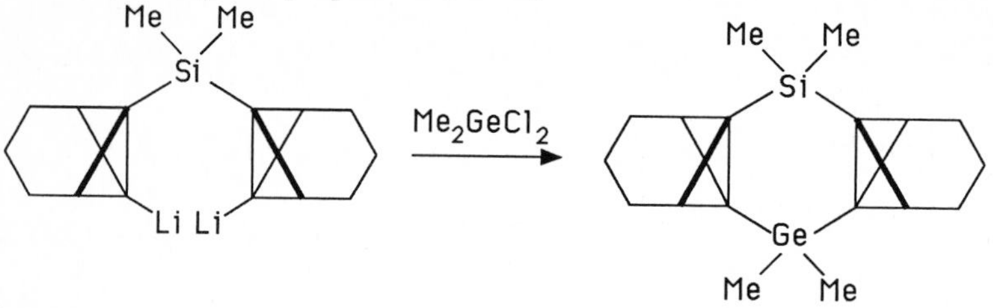

eqn.(2.25)

A series of 1-germa-4-metalla six-membered cyclic compounds, in which the carbon atoms of two o-carborane nuclei are incorporated, have been synthesized by Zakharkin (refs. 86-87) :

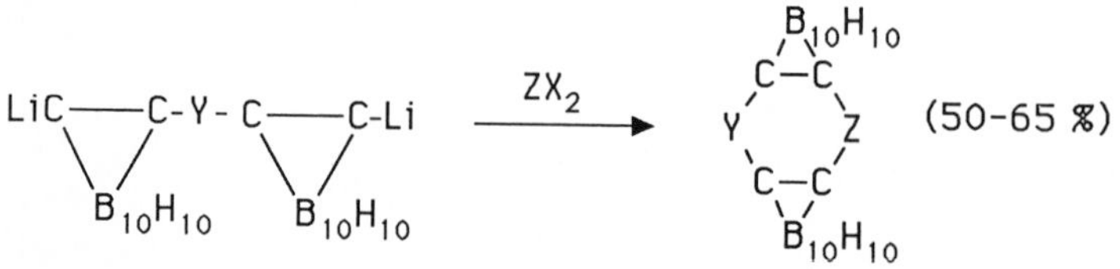

eqn.(2.26)

$Y = CH_2, Me_2Si$; $Z = Me_2Ge$

$Y = Me_2Ge$; $Z = Me_2Ge, Me_2Sn, PhP, MeAs$

2.1.3. Germaadamantanes

1-Methyl-1-germaadamantane is obtained in 37 % yield by aluminum chloride catalyzed redistribution of *cis*-1,3,5-tris(trimethylgermylmethyl)cyclohexane (ref. 88). The methyl group is cleaved by trichlorosilane in the presence of catalytic amounts of chloroplatinic acid :

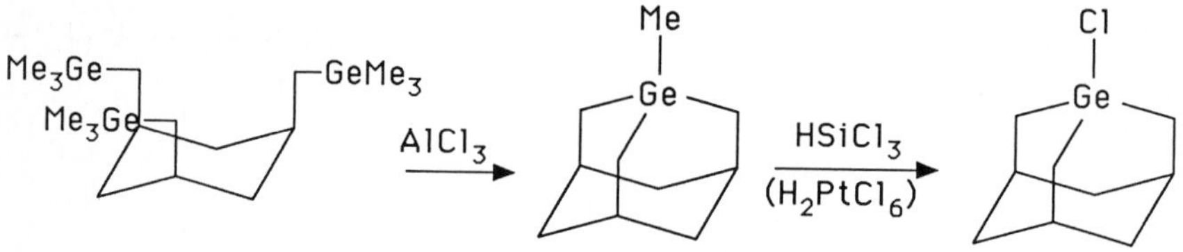

eqn.(2.27)

Tetragermaadamantanes are also formed in 20 % yield when 1,1,3,3,5,5-hexamethyl-1,3,5-trigermacyclohexane is heated with aluminum bromide (ref. 27) :

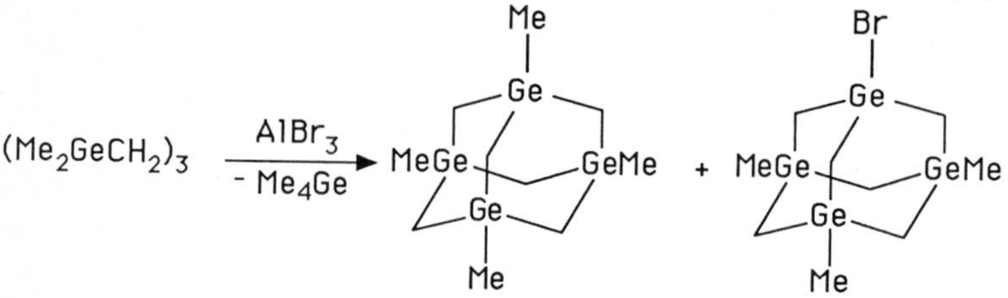

eqn.(2.28)

2.2. UNSATURATED COMMON RINGS

2.2.1. Germacyclopentenes

2.2.1.1. Germacyclopent-3-enes

Synthesis : Dihalogermacyclopent-3-enes were first prepared by reaction of a trihalogermane with a conjugated diene. Because of the equilibrium :

$$HGeCl_3 \rightleftharpoons GeCl_2 + HCl$$

trichlorogermane gives with butadiene a mixture of linear and cyclic compounds (refs. 89-91) :

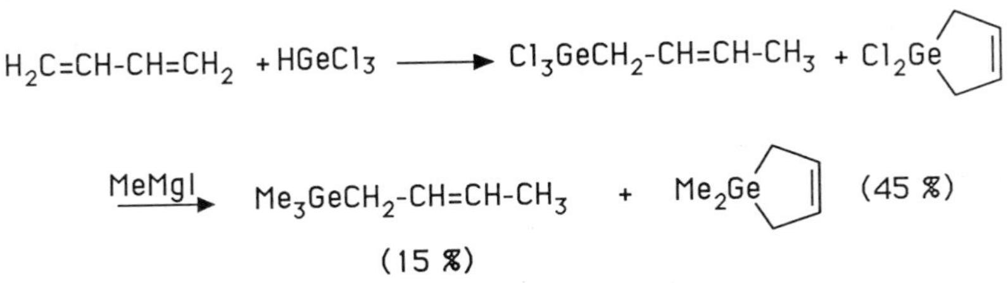

eqn.(2.29)

With isoprene, a similar mixture of linear and cyclic compounds is obtained, but with piperylene, only the linear adduct is formed (refs. 90,92).

An analogous addition of tribromogermane to butadiene produces a mixture in which the linear derivative predominates (refs. 93-94) :

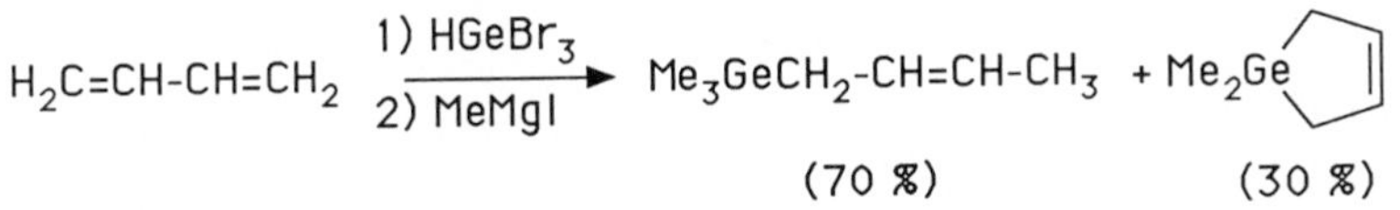

eqn.(2.30)

Nefedov *et al.* (refs. 95-99) showed that the yield of germacyclopentene is markedly increased when trichlorogermane etherate is used. They proposed a mechanism involving the formation of dichlorogermylene by displacement of the following equilibrium :

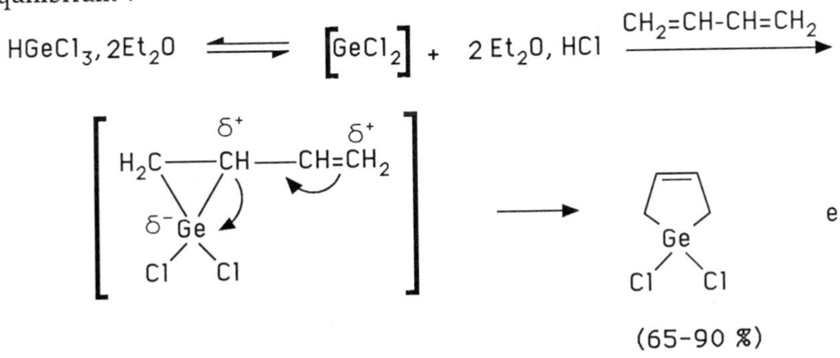

eqn.(2.31)

Germanium diiodide, a stable but reactive germylene, undergoes nearly quantitative 1,4-addition with common conjugated dienes (ref. 100) (Chart 17) :

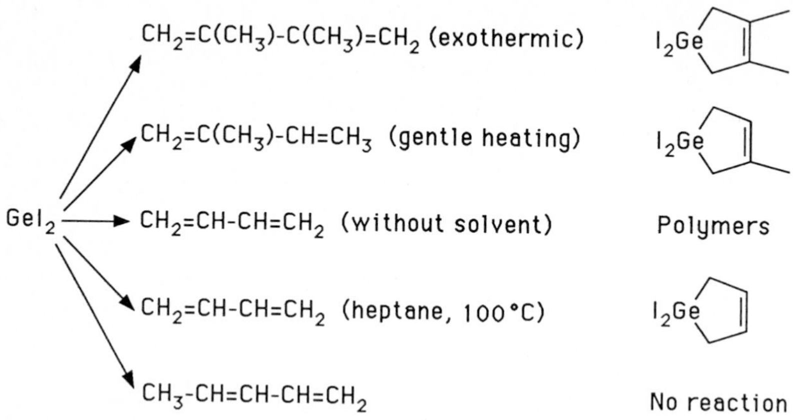

Chart 17. 1,4-Cycloaddition of germanium diiodide to 1,3-dienes

With dimethylbutadiene, the reaction is fast and exothermic. The addition to isoprene is slower and needs gentle heating. In the case of butadiene, polymerization of the diene is observed in the absence of solvent (refs.89,90), but the yield of diiodogermacyclopentene is nearly quantitative when the reaction is carried out in a saturated hydrocarbon at 100°C (ref. 101). However, germanium diiodide does not react with piperylene.

Difluorogermylene adds to dimethylbutadiene to form the expected difluoride in moderate yields (ref. 102) :

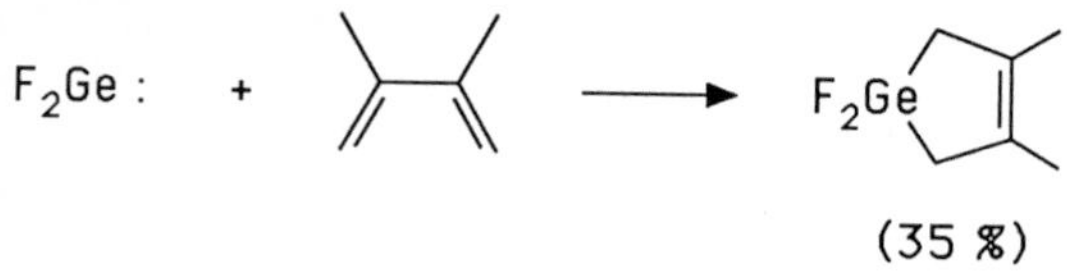

eqn.(2.32)

Dihalogermacyclopent-3-enes are also formed but in low yield by direct synthesis from germanium tetrachloride and 1,3-diene (ref.103) :

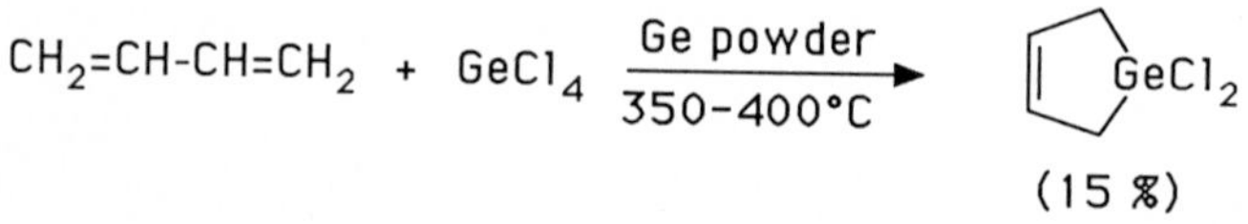

eqn.(2.33)

A number of cycloadditions of mixed (refs. 104-109) and cyclic (refs. 110-112) germylenes to 2,3-dimethylbutadiene have been reported:

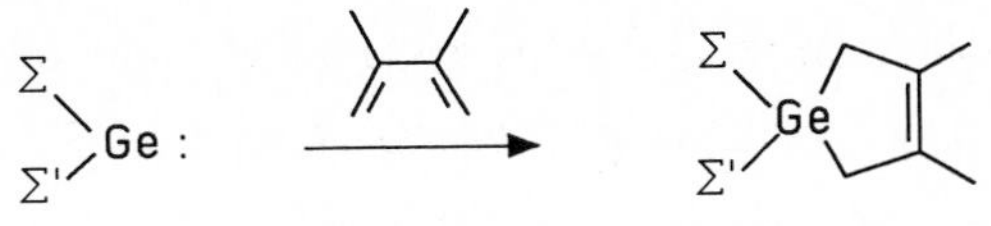

Σ = Et, Ph ; Σ' = F, Cl, Br, I, H, OMe, SMe, Me_2N, Et_2P

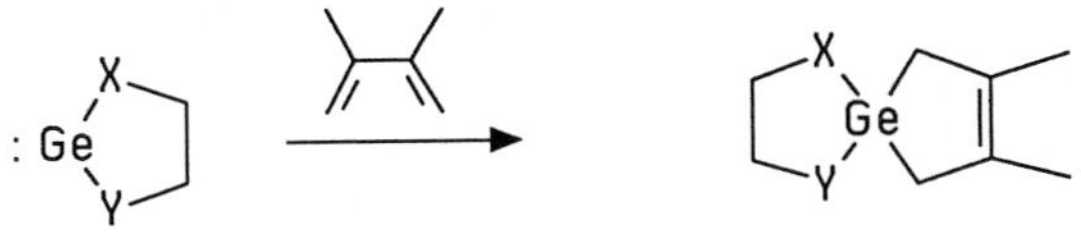

eqn.(2.34)

X = Y = O, S , NMe ; X = O, Y = S

When complexed, germylenes also react with dimethylbutadiene; yields of the cycloadduct are usually low and depend on the nature of the germylene (ref. 113) :

eqn.(2.35)

R=R'=Ph, Et, Mes ; R= Ph, R'= H, Me, Cl

The reactivity was found to decrease in the order :

X_2Ge,Nu > RGeX,Nu > RR'Ge,Nu

with X = halogen ; R = Ph, Mes, ; R' = H, Me, Ph, Mes ; Nu = Et_3N, C_5H_5N

A large number of functional germacyclopentenes were similarly prepared from the corresponding suitably complexed germylenes (ref. 114) :

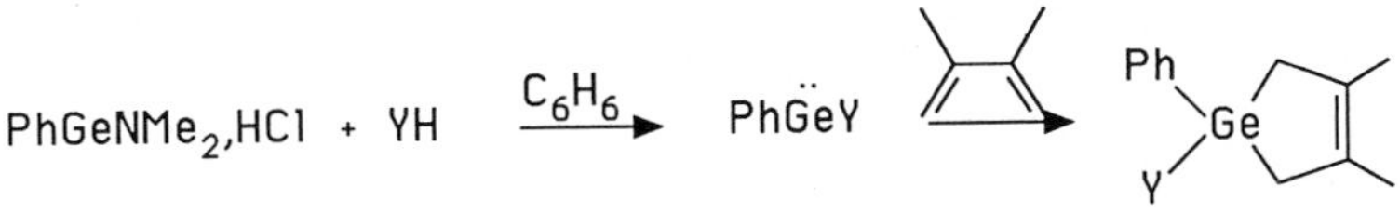

eqn.(2.36)

Y = OR, SR, NR_2, PR_2, R'_3Ge ...

With Y = $OCH(Ph)NMe_2$, the unstable transient adduct decomposes into a germanium oxide (ref. 115) :

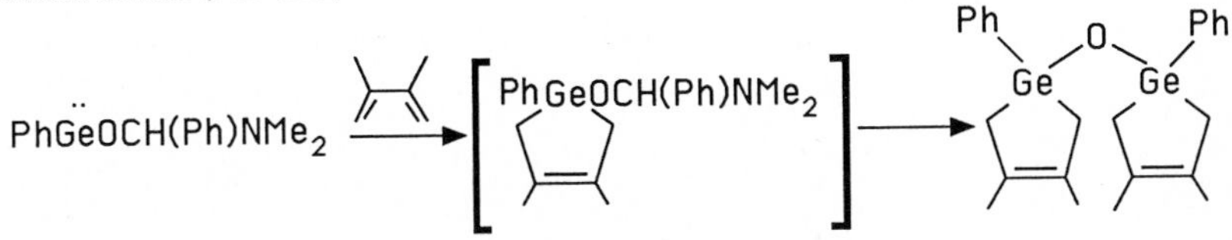

eqn.(2.37)

Cycloaddition of free dimethylgermylene (refs. 116-118) with a series of 1,3-dienes was studied by Neumann *et al* (refs. 84,116, 119) who found that acceptor substituents in 1,(4) position accelerate cycloaddition while donors diminish or suppress it. Usually, besides the expected 1,4-adduct, a 3,4-bisubstituted germacyclopentane (as a mixture of syn/anti isomers), resulting from the reaction of one germylene with two dienes, is isolated. The ratio depends on the nature of the substituents :

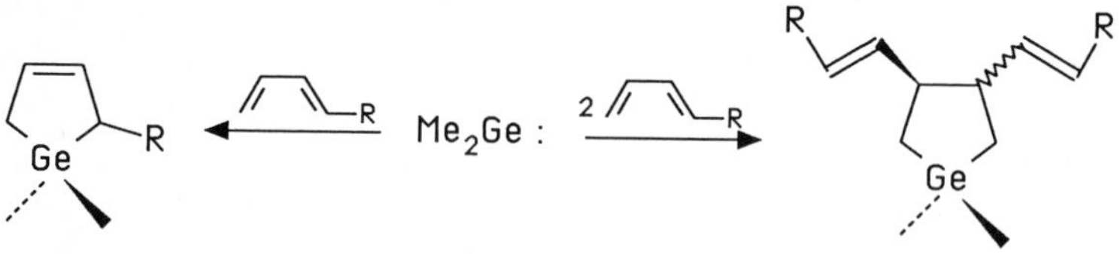

(A) (B)

	A	B
R = H	10 %	50 %
R = Me	11 %	38 %
R = Ph	9 %	55 %
R = t-Bu	58 %	–
R = OCOMe	10 %	–

eqn.(2.38)

Similarly, with 2-substituted butadienes, 1,4-addition competes with 1,2-addition :

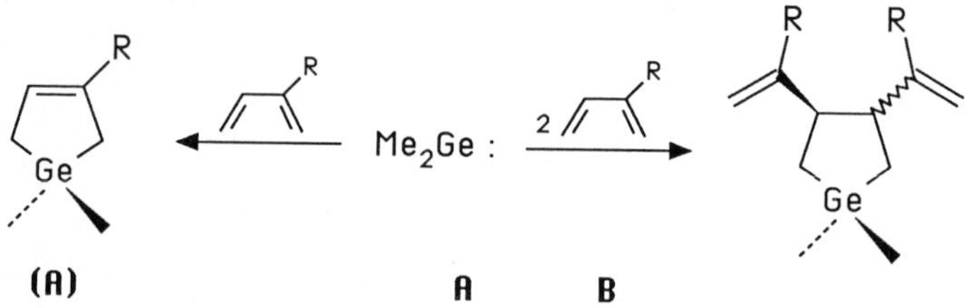

(A) (B)

	A	B
R = Me	13 %	45 %
R = Ph	63 %	17 %
R = t-Bu	74 %	5 %

eqn.(2.39)

I.R. and Raman spectra of a series of sila- and germacyclopent-3-enes have been reported and compared to those of cyclopentenes (ref. 120).

Reactivity : 1,1-Dihalogermacyclopent-3-enes are the best precursors of other germacyclopentenes (ref. 100). Chart 18 shows that :

- Reduction with lithium aluminum hydride gives the corresponding hydrides
- Other halides are easily prepared *via* the oxide in good yields (refs. 121, 122)

- Quantitative dialkylation and diarylation are achieved with Grignard reagents

- Mixed derivatives can be obtained by stepwise cleavage of diallylgermacyclopentenes (ref. 123), the Ge-C(allyl) bond being more reactive towards hydrogen chloride than intracyclic Ge-C bonds.

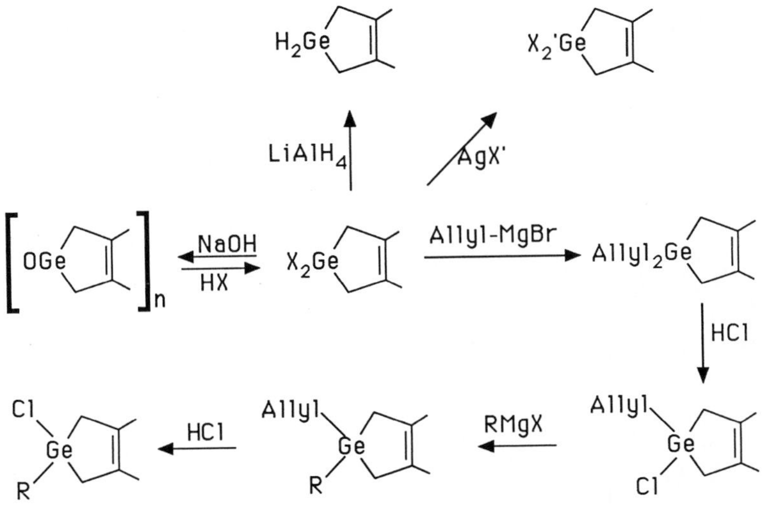

Chart 18. Germacyclopent-3-enes. Substitution reactions at germanium

Germacyclopent-3-enes easily undergo addition reactions at the double bond with retention of the five-membered ring (Chart 19) :

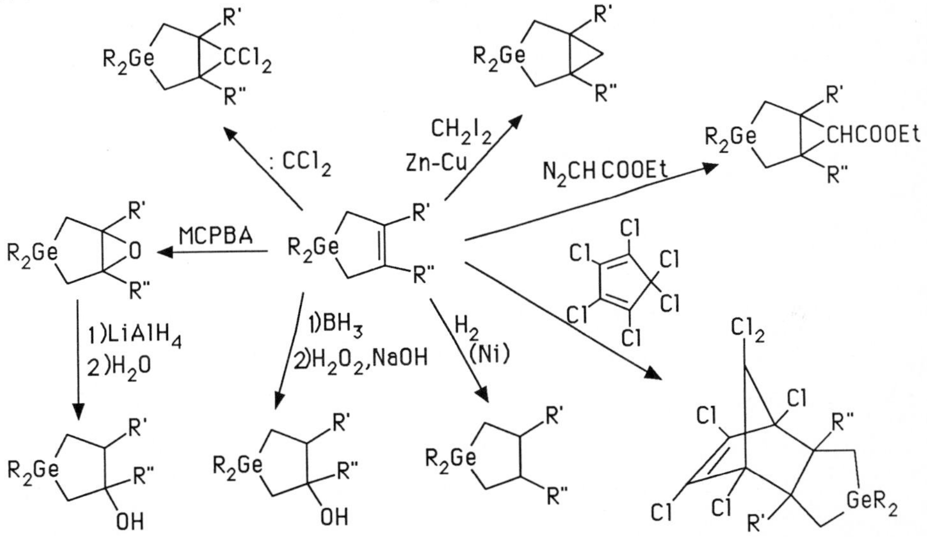

Chart 19. Addition reactions of germacyclopent-3-enes (refs.65,78,101, 124-127)

- hydrogenation occurs quantitatively at room temperature in the presence of Raney nickel
- with carbenes (Simmons-Smith reagent in boiling THF, $PhCCl_2Br$ in boiling benzene and ethyldiazoacetate at 200°C) bicyclo[3.1.0]hexanes are obtained
- addition of diborane (followed by basic oxidation) and epoxidation of the double bond with metachloroperbenzoic acid (followed by reduction with lithium aluminum hydride) are suitable methods of preparing of various 3-germacyclopentanols. Because these reactions have a different regio- and stereospecificity, different isomers (A_2, B_2, A_3, B_3) are obtained in the case of ring-substituted germacyclopentenes :

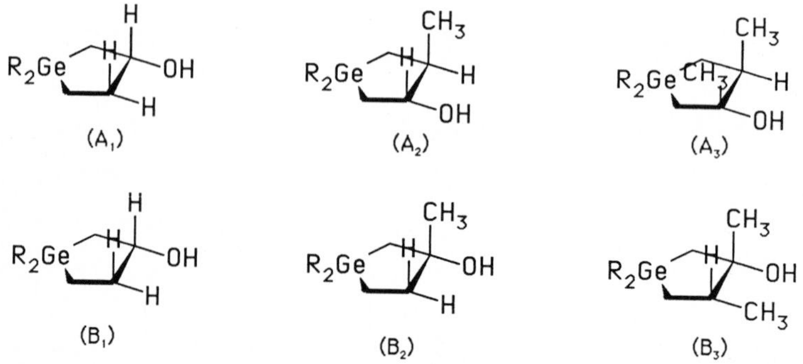

- On the other hand, germacyclopent-3-enes undergo ring-opening polymerization catalyzed by alkyllithium reagents, in the presence of hexamethylphosphoramide or N,N,N',N'-tetramethylene diamine, to yield poly(1,1-dimethyl-1-germa-*cis*-pent-3-ene) (ref. 128) :

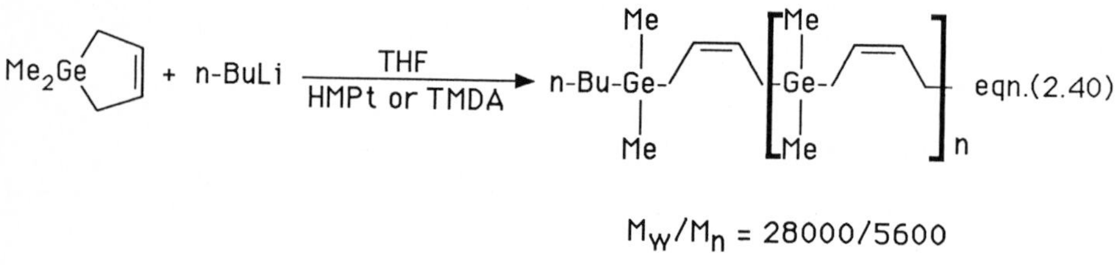

Bicyclic germacyclopent-3-enes

Bis(methylene)cycloalkanes also react with germylenes ; e.g. dichlorogermylene adds to bis(methylene)cyclohexane to give the expected adduct in good yield (ref. 99) :

eqn.(2.41)

but bis(methylene)cyclobutane leads to an unstable compound.

Free dimethylgermylene reacts quantitatively with bis(methylene)cyclopentane and -cyclohexane to form exclusively the 1,4-adduct (refs. 84, 119) :

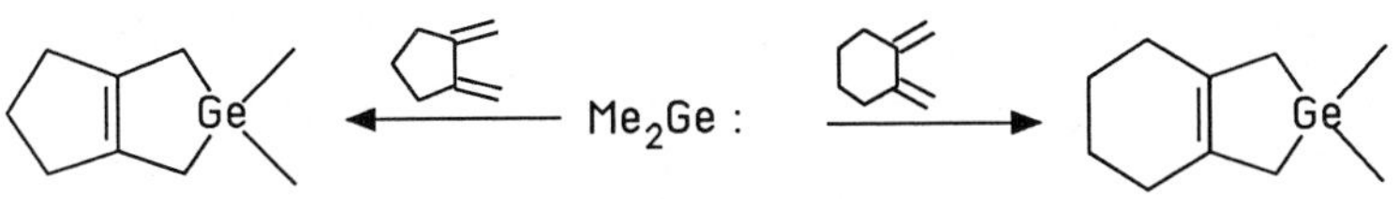

eqn.(2.42)

while with bis(methylene)cyclobutane, the 1,4-addition product is not isolated :

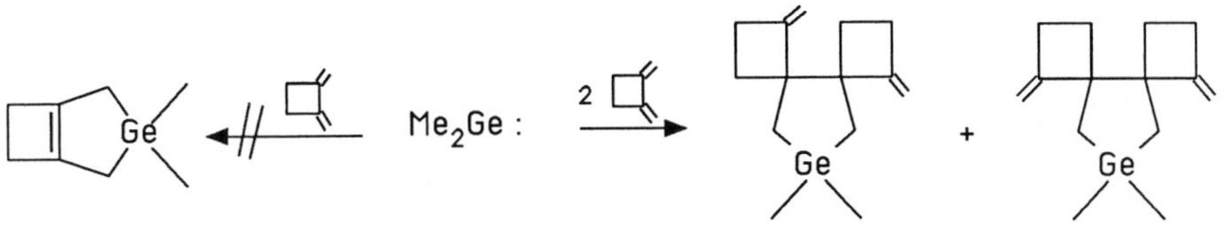

(60 %)

eqn.(2.43)

With rigid s-*cis*-dienes, a mixture of 1,4 and 1,2-adducts is formed :

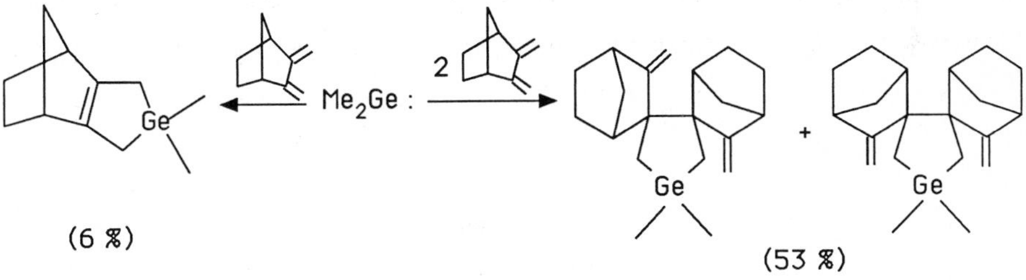

eqn.(2.44)

In these cycloadditions, the reactivity of bis(methylene)cycloalkanes decreases in the order :

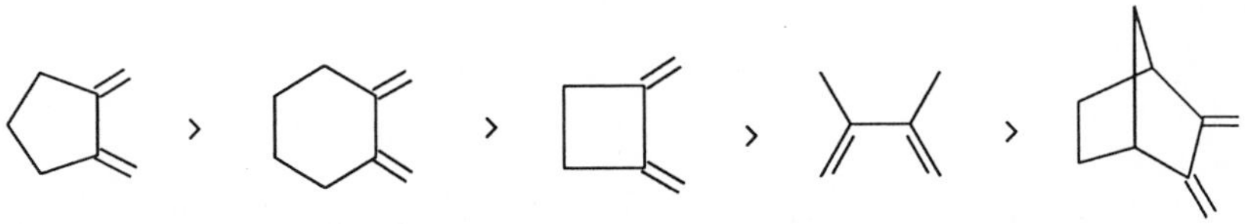

Functional bicyclic organogermanium derivatives are obtained by addition of germanium diiodide to bis(methylene) oxolane, thiolane, selenolane and phenylazolidine (ref. 129), (Chart 20) :

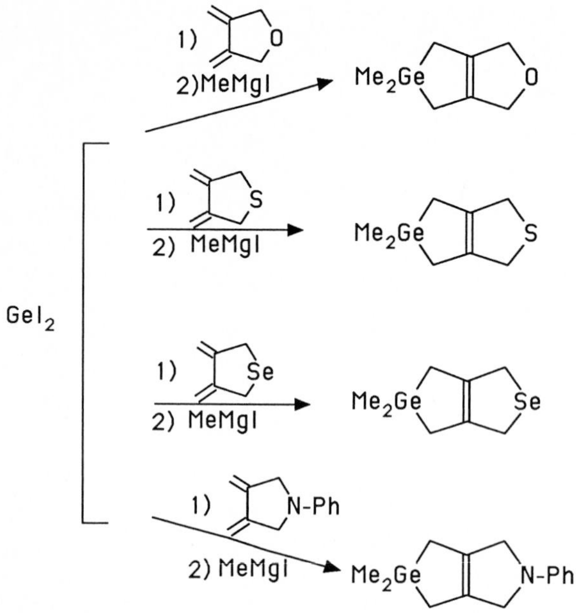

Chart 20. Synthesis of functional bicyclic germanium compounds

Dehydrogenation of the sulfur compound gives the corresponding substituted thiophene, while stepwise oxidation can be observed by the action of metachloro-

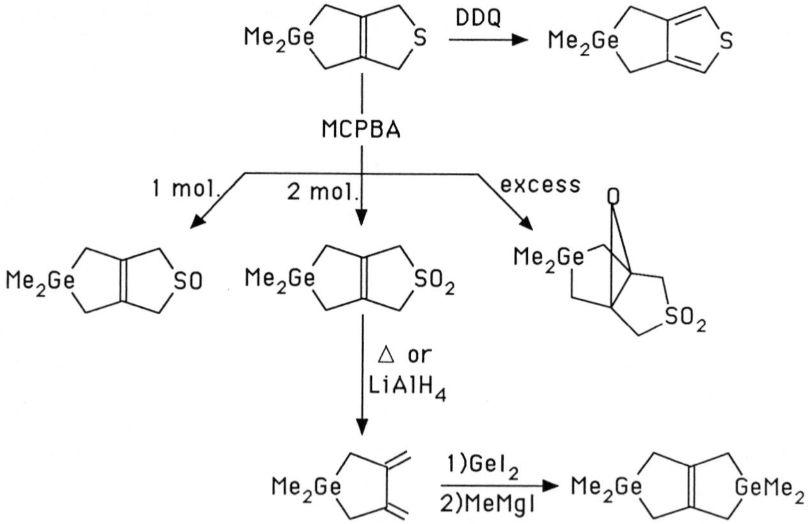

Chart 21. Reactions of 3-thia-7,7-dimethyl-7-germabicyclo [3.3.0] octene

perbenzoic acid. The sulfone is the precursor of bis(methylene)germacyclopentane (Chart 21).

Spirononadienes

A series of 5-germaspiro[4.4]nona-2,7-dienes were obtained by the two following routes :

- Reaction of a dihalogermacyclopentene with an alkali metal in the presence of a conjugated diene (refs. 121, 122) :

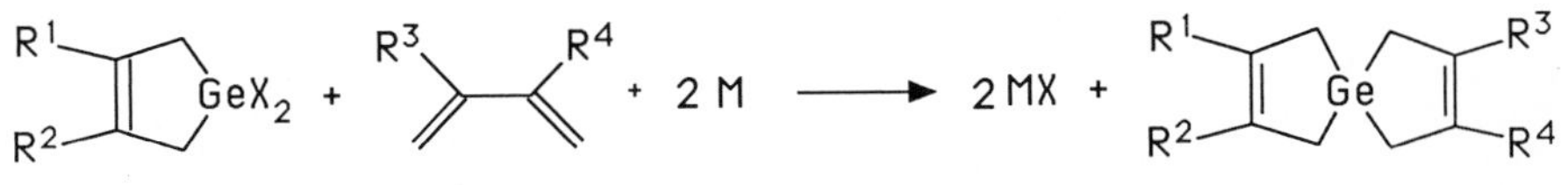

X = F, Cl, Br, I   M = Li, Na R^1, R^2, R^3, R^4 = H, CH_3

eqn.(2.45)

The yield is strongly dependent on the experimental conditions (halide, diene, alkali metal and solvent used). Yields are between 0 and 20 % ; the best results are obtained with chlorides or bromides, lithium and a mixture of ether-THF (5/1) as solvent. The unsubstituted spiran cannot be obtained in this way.

- Reaction of a dihalo- or, better, an alkoxychlorogermacyclopentene with active magnesium (prepared by reduction of $MgBr_2$ with sodium in THF (refs. 130-131)) and a conjugated diene in THF (ref. 132) :

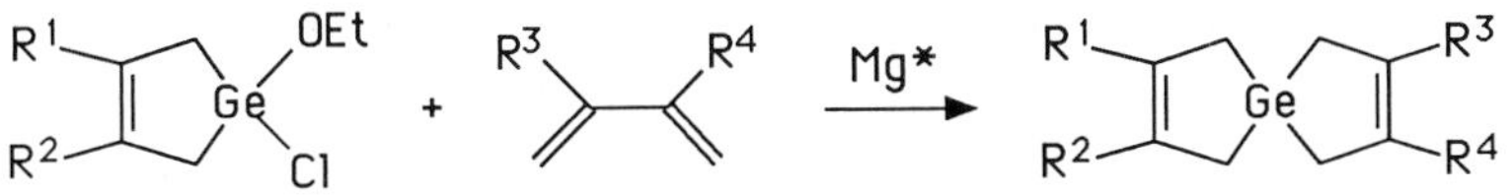

eqn.(2.46)

R^1, R^2, R^3, R^4 = H, CH_3

The yields are usually between 30 and 50 % except for the unsubstituted spiran (12 %). In the mass spectrum of these compounds the principal peak corresponds to the fragment resulting from the loss of one ring in the molecule ; for unsymmetrical spirans, the less-substituted ring is preferentially cleaved. 2,7-Dimethyl-5-germaspiro-[4,4]nona-2,7-diene is one of the best precursors, allowing the formation of amorphous semiconducting thin layers of non-stoichiometric germanium carbide by chemical vapour deposition (refs. 133-134).

2.2.1.2. Germacyclopent-2-enes (Chart 22)

Germacyclopent-2-enes are formed, together with germacyclopent-3-enes, by thermolysis of germacyclopentan-3-ols, reaction (a), desulfuration of germacyclopent-2-ene-3-thiols, reaction (b) or radical bromination of germacyclopent-3-enes, reaction (c). Separation of germacyclopent-2-ene from the mixture is based on the greater chemical reactivity of the 3-isomer.

Ene reactions (d) (refs. 67, 135) provide a general method of synthesis of 3-functional germacyclopent-2-enes. In particular photooxygenation leads, after reduction of the photooxidation product, to germacyclopent-2-en-3-ols, which are precursors of germoles (refs. 136-139).

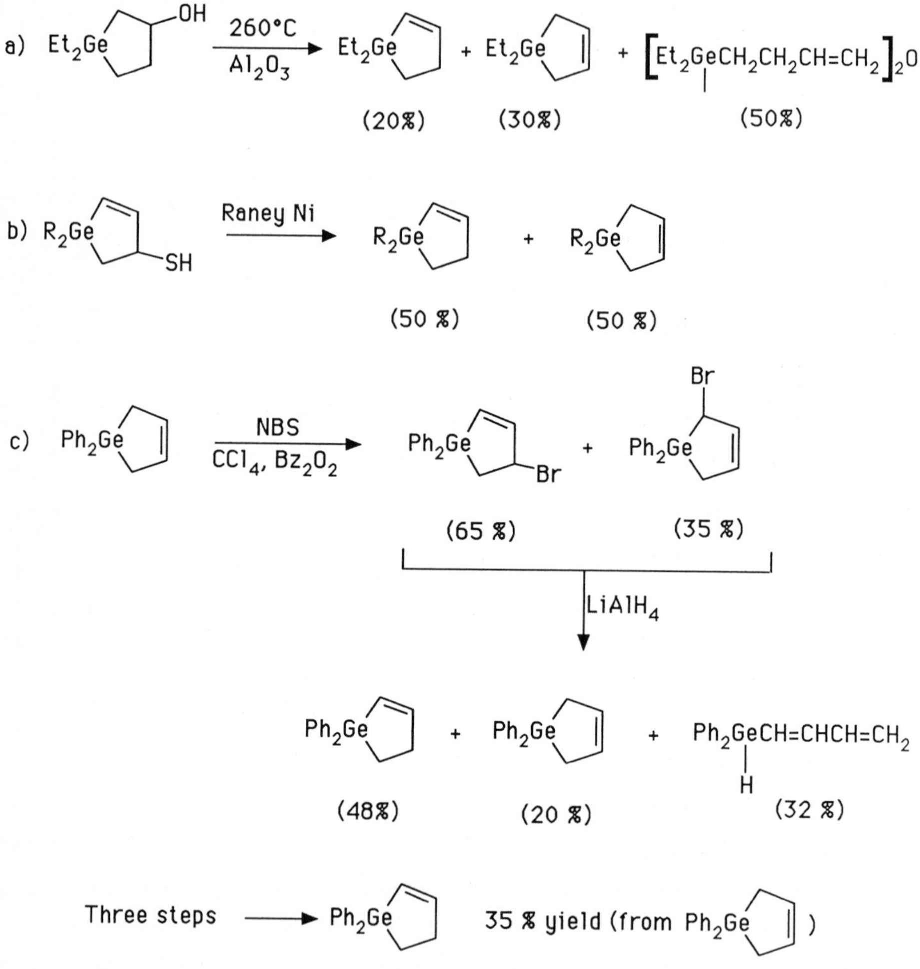

Chart 22. Synthesis of germacyclopent-2-enes.

d) 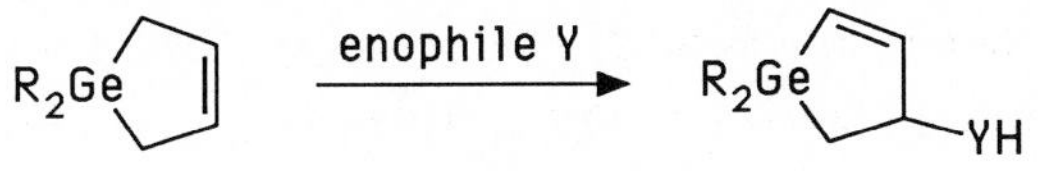

Y = maleic anhydride, formaldehyde, phenyltriazolinedione, benzyne

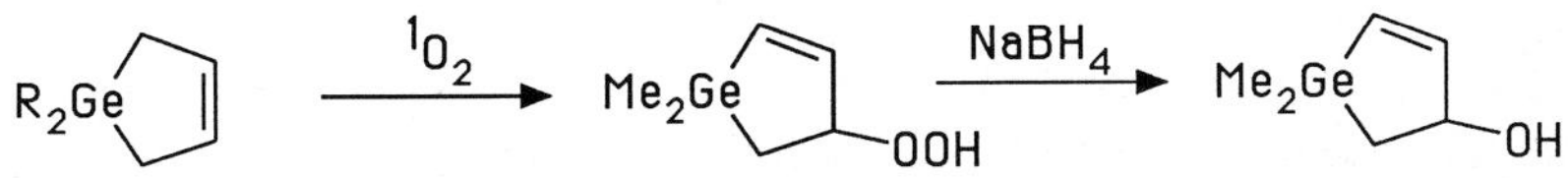

e) 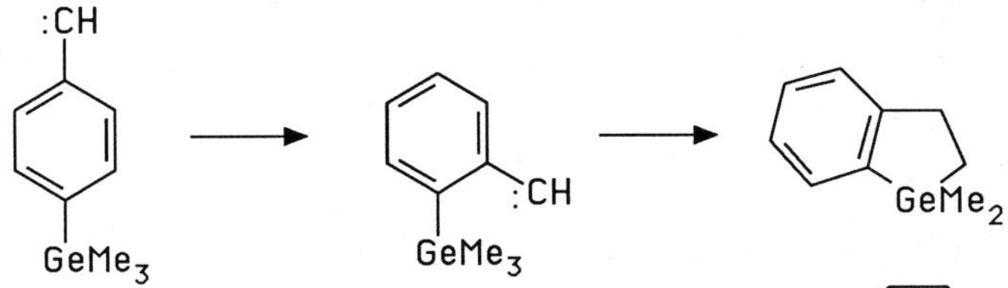

(96 %)

f) 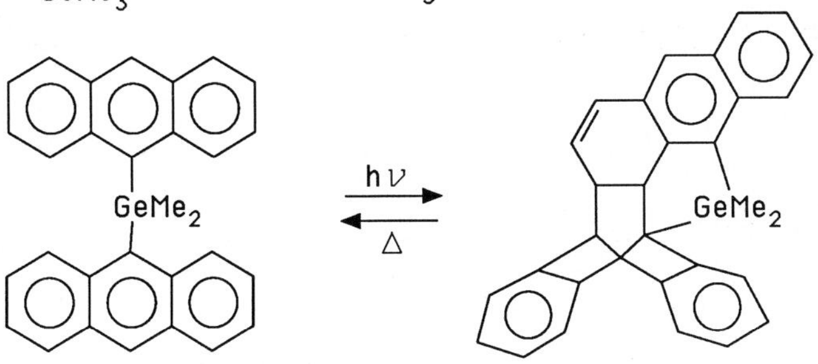

Chart 22 (continued). Synthesis of germacyclopent-2-enes

A novel formation of a germaindane was described by Chambers and Jones (ref. 140). This compound results from the ring closure of a trimethylgermyl-*o*-substituted phenyl carbene by abstraction of hydrogen from a methyl group, reaction (e). Irradiation of a benzene solution of di-(9-anthryl)dimethylgermane with a high-pressure mercury lamp gives in 71 % yield a photoproduct that undergoes a quantitative thermal cycloreversion by heating at 200° C for 1 h. (ref. 141), reaction (f).

2.2.2. Germacyclohexenes and digermacyclohexenes

- *Germacyclohex-3-enes* result from the (2+4) addition of germaethylenes to 2,3-dimethylbutadiene (refs. 24, 142) :

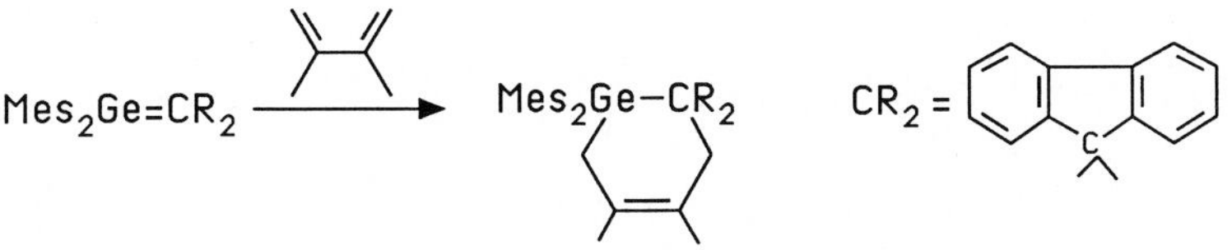

eqn.(2.47)

A 9,10-(germapropano)-bridged anthracene containing germacycloheptene rings was prepared in 50 % yield by Bogdanovic (ref. 143) in two steps from magnesium-anthracene complex :

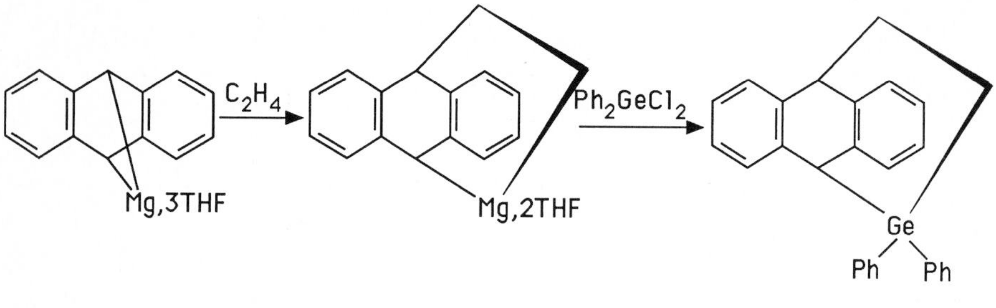

eqn.(2.48)

-1,2-Digermacyclohex-4-enes are prepared in 30-50 % yield by the reaction of lithium with 1,2-dichloro-1,2-digermanes in the presence of a conjugated diene in a suitable solvent (ref. 144) :

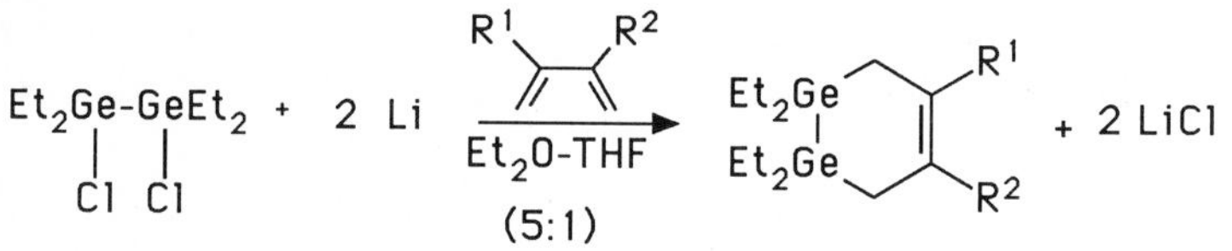

eqn.(2.49)

R^1, R^2 = H, CH_3

Digermacyclohexenes are reactive compounds, giving a cyclic oxidation product on contact with air at room temperature :

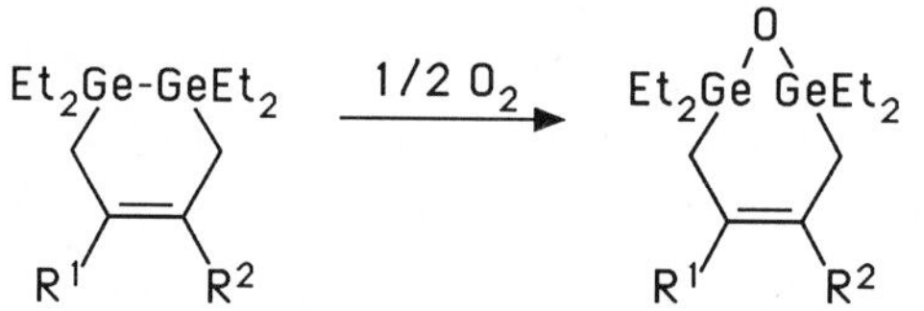

eqn.(2.50)

The principal peak in the mass spectrum of all 1,1,2,2-tetraethyl-1,2- digermacyclohex-4-enes corresponds to the fragment [Et_2Ge=$GeEt_2$]. This intermediate is also formed by vacuum thermolysis and can be trapped with 1,4-diphenylbutadiene (ref. 145), (eqn. 2.51).

^{13}C, I.R., Raman and UV spectra of a series of 1,2-digermacyclohexenes and 1,2-digermacycloalkanes have been studied and compared to those of silicon and carbon analogs (ref. 146).

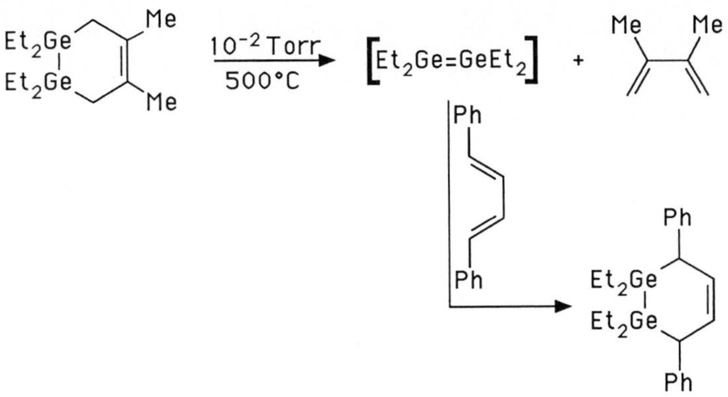

eqn.(2.51)

2.2.3. Germacyclopentadienes (Germoles)

Tetraphenyl substituted germoles are usually prepared by the reaction of a germanium polyhalide with 1,4-dilithiotetraphenylbutadiene (from diphenylacetylene and lithium) (refs. 147-150), (Chart 23) :

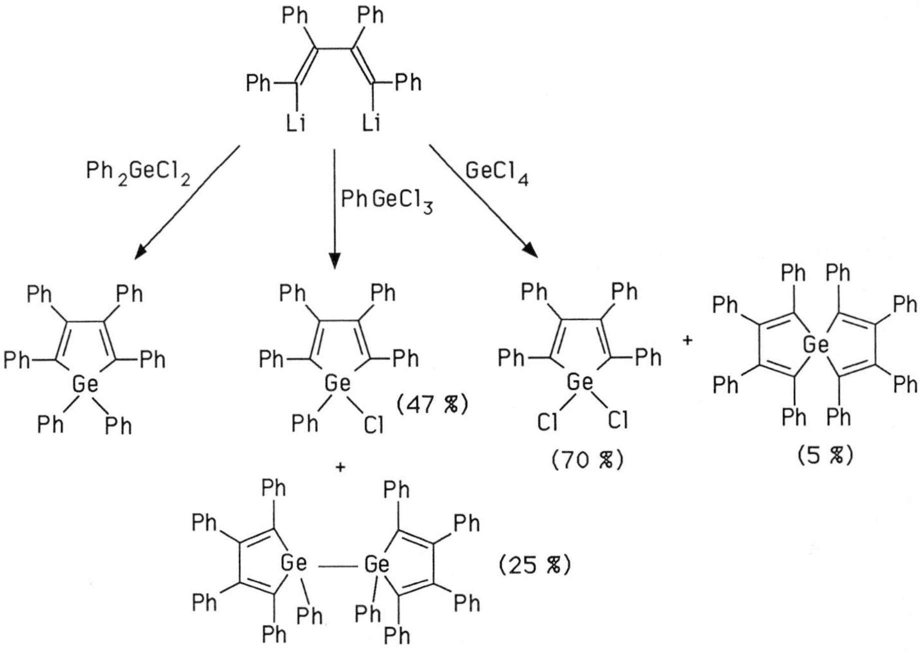

Chart 23. Synthesis of polyphenylated germoles.

From germanium tetrachloride, the spiran can be obtained in 5 to 22 % yield depending on the experimental conditions.

In the reaction with phenylgermanium trichloride, a digermane is formed as a by-product resulting from the side reaction :

eqn.(2.52)

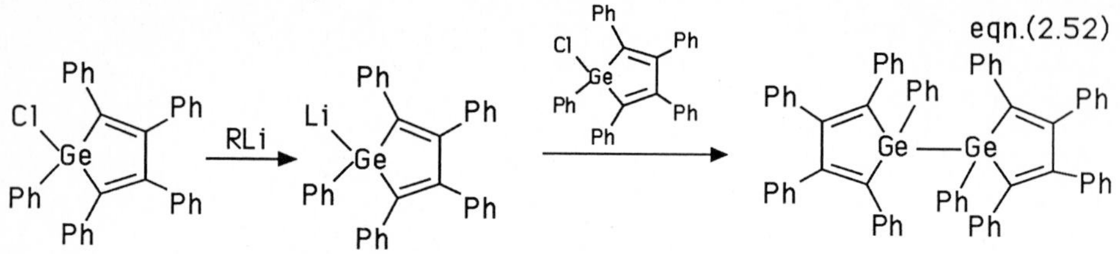

The synthesis of germaindenes and germafluorenes is summarized in Chart 24 :

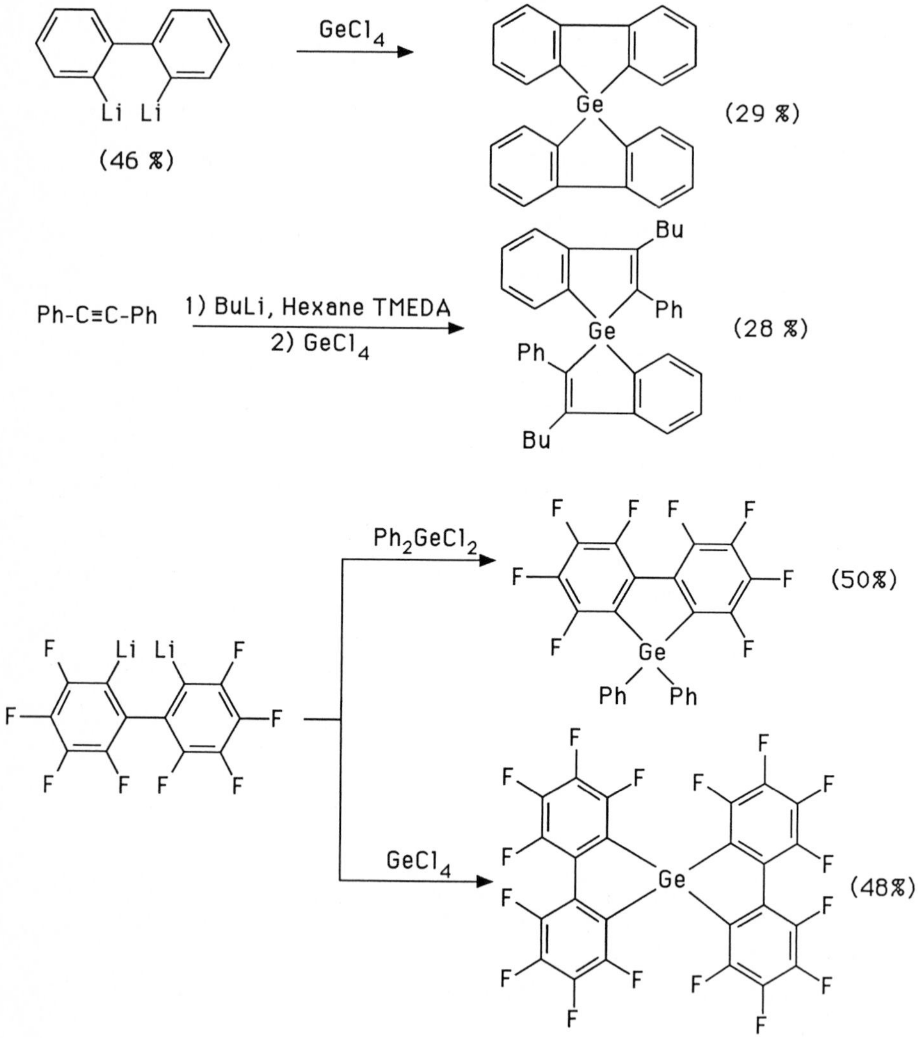

Chart 24. Synthesis of germafluorenes and germaindenes (refs. 151-156).

The synthesis of a methyl-substituted dichlorogermole by transmetalation was recently achieved by Fagan and Nugent (ref. 157) :

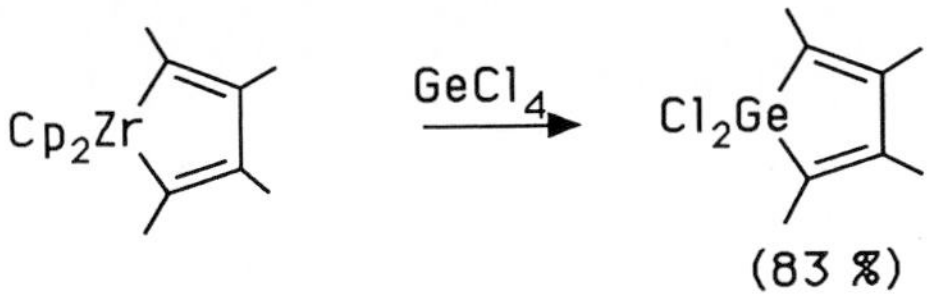

eqn.(2.53)

A novel direct synthesis was reported by Cohen *et al.* (ref. 158) :

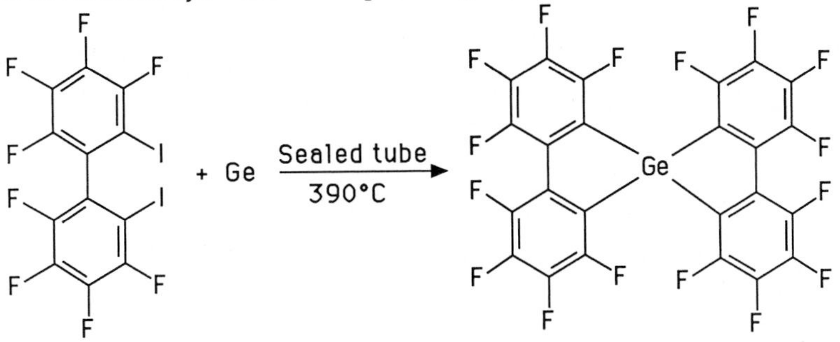

eqn.(2.54)

Dimethyltetraphenylgermacyclopentadiene gives a Diels-Alder adduct with maleic anhydride :

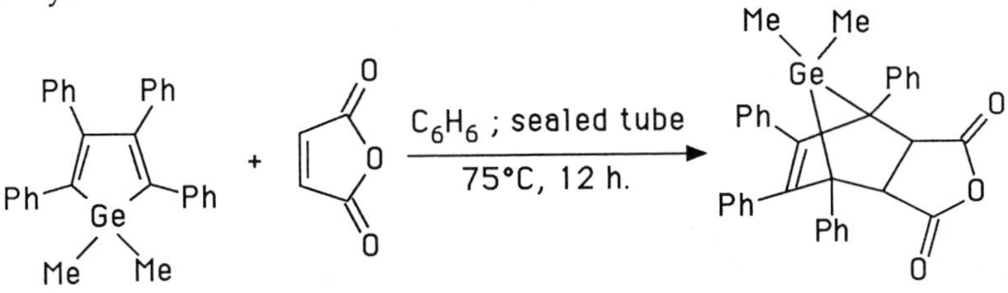

eqn.(2.55)

and with dimethylacetylene dicarboxylate (ref. 159) but with hexafluorobutyne the unstable 7-germanorbornadiene decomposes :

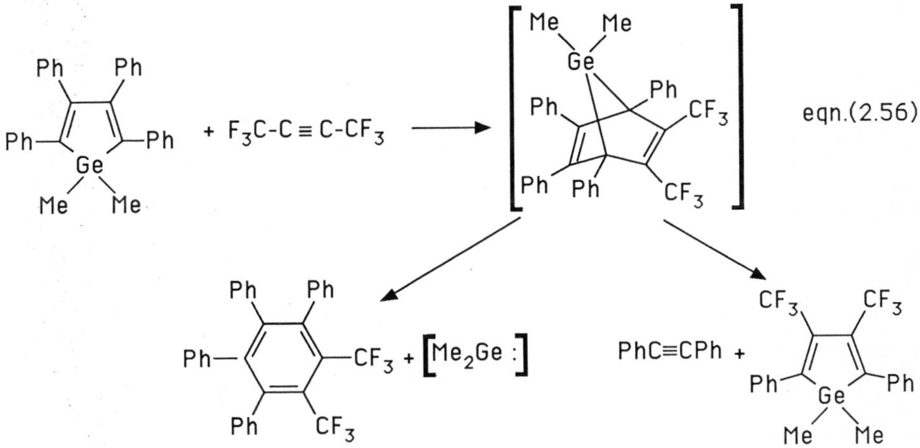

eqn.(2.56)

Photochemical or thermal (70°C) decomposition of the adduct with benzyne is widely used as a source of dimethylgermylene (refs. 12, 13, 45, 116, 118, 160-162) :

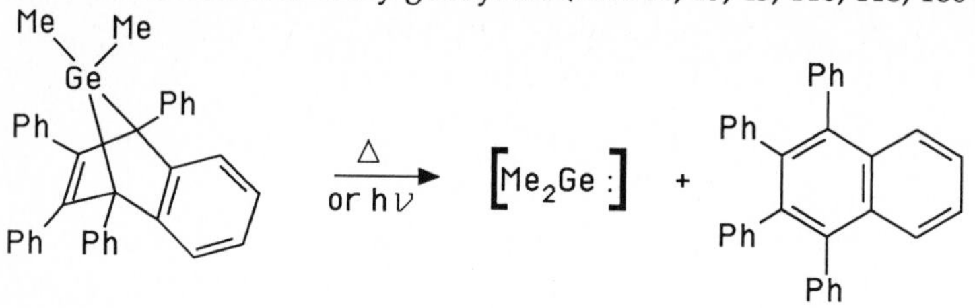

eqn.(2.57)

The corresponding diene-iron tricarbonyl complexes have been prepared by photochemical reaction of iron pentacarbonyl with 2,3,4,5-tetraphenyl-1-germacyclopentadiene (refs. 163-165) :

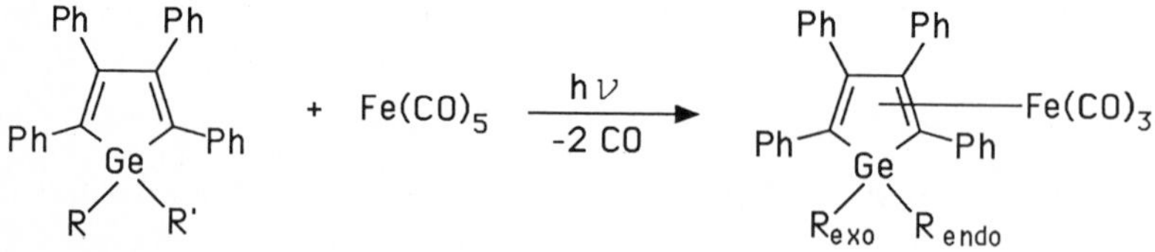

eqn.(2.58)

Cleavage of these complexes by binary halides demonstrates the enhanced reactivity of the *exo* germanium-carbon bond :

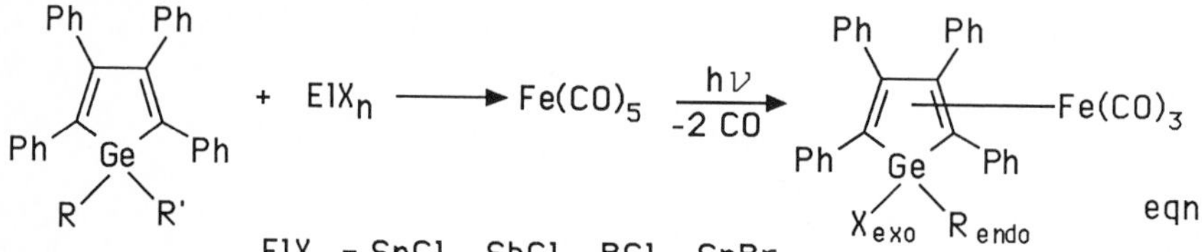

eqn.(2.59)

$ElX_n = SnCl_4, SbCl_5, BCl_3, SnBr_4$

Substitution reactions of these halides are observed with AgF, NaI, NaOMe and $LiAlH_4$, resulting in *exo* functional germacyclopentadiene complexes.

Quantitative removal of the hydride ion (X = H) by Ph_3CBF_4 (in CH_2Cl_2) and chloride ion (X = Cl) by $AgBF_4$ (liq. SO_2) produces salts in which the cation is not an η5-germacyclopentadienyl complex but an η4-diene complex in which the positive charge is localized mainly at the germanium atom (ref. 166) :

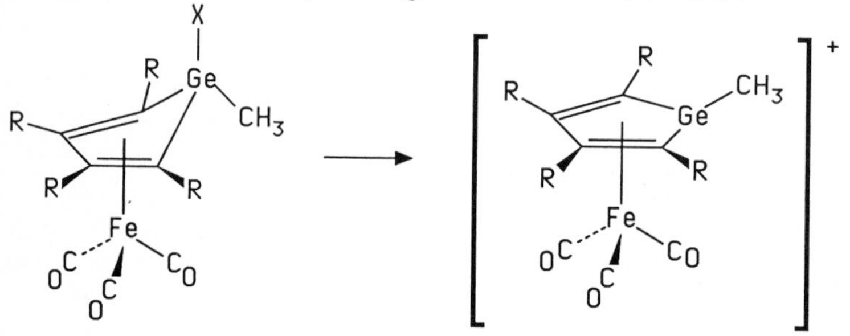

eqn.(2.60)

Chlorogermoles are suitable precursors for the preparation of other germacyclopentadienes. Reduction of pentaphenylchlorogermole gives a germanium hydride that is 10^6 times more acidic than triphenylgermane (ref. 150). With butyllithium at low temperature, a red solution of the anion is formed (ref. 149).

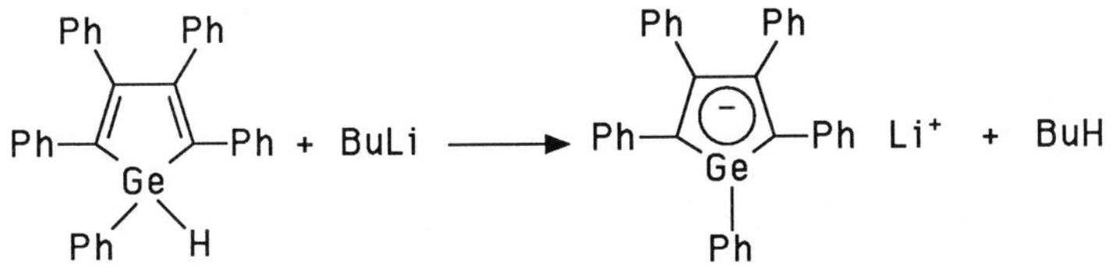

eqn.(2.61)

This lithio derivative reacts easily with water and trimethylchlorosilane, but the expected analog of ferrocene is not formed by reaction with iron (II) chloride (ref. 150). Unlike cyclopentadiene, pentaphenylgermole does not form a π-complex with nonacarbonyldiiron ; "hydroferration" of a Ge-C_σ bond of the germole occurs instead (ref. 167) :

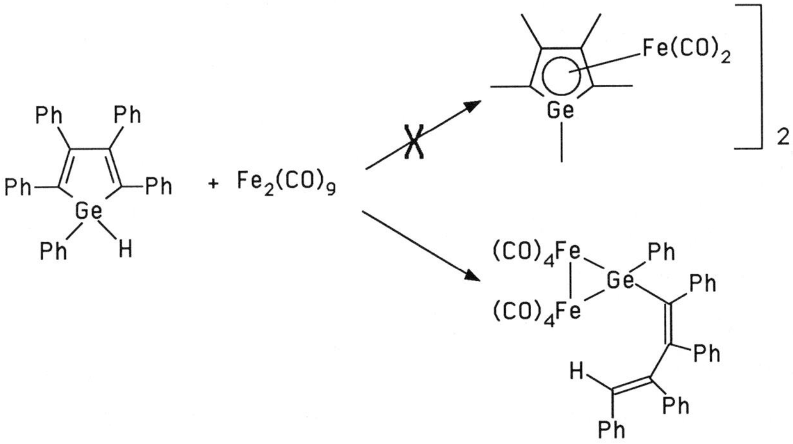

eqn.(2.62)

C-Unsubstituted or C-methylated germoles are prepared either by dehydration of germacyclopentenols on alumina or thoria (200-350°C) or by thermolysis of germacyclopentene phenylcarbamates (refs. 136, 137, 168-171) :

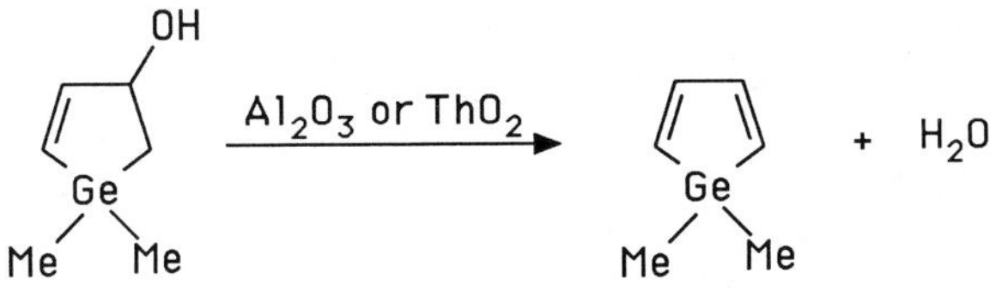

eqn.(2.63)

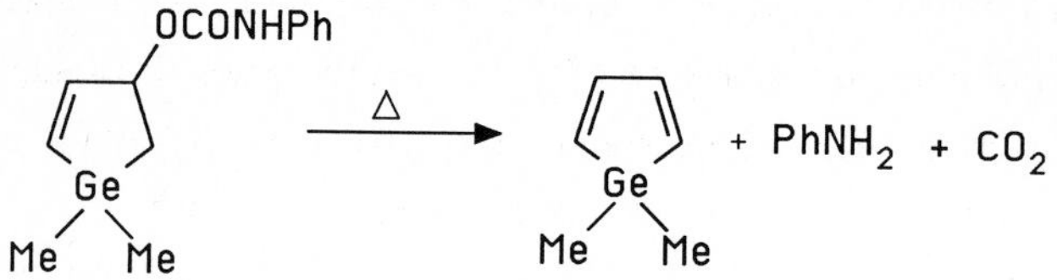

eqn.(2.64)

With 1,1,3-trimethyl- and 1,1,3,4-tetramethylgermacyclopentenols, a mixture of germole and transoid isomer is formed :

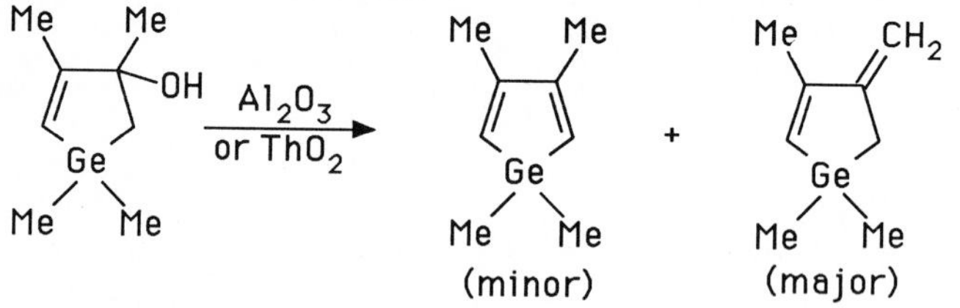

eqn.(2.65)

The tetramethylgermole is easily prepared in high isomeric purity by one-pot synthesis (ref. 168) :

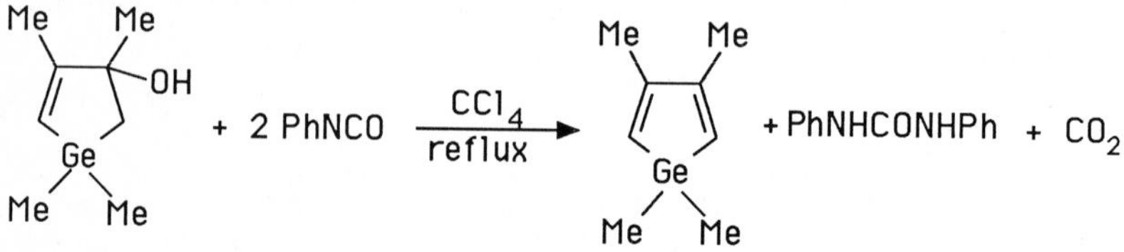

eqn.(2.66)

While tetramethylgermole can be isolated and stored for a few weeks in a refrigerator, dimethyl- and trimethylgermoles are not stable and dimerize rapidly at room temperature (refs. 136, 169, 171) :

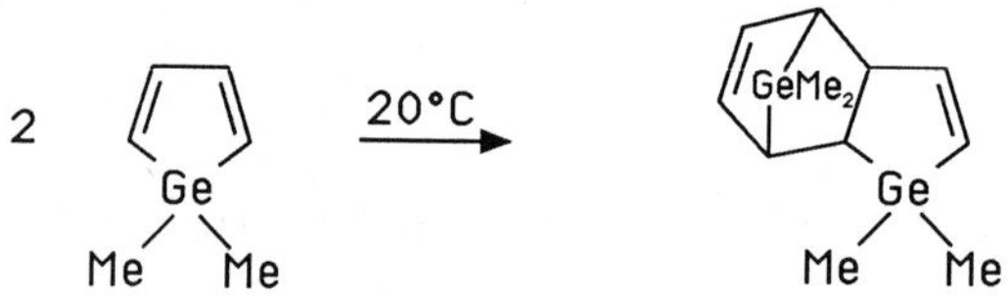

eqn.(2.67)

Thermolysis of 1-allyl-1,3,4-trimethylgermacyclopent-3-ene does not lead to the expected 1,3,4-trimethylgermole. Instead, 2,3-dimethylbutadiene is formed, resulting from a cycloreversion [4+2] reaction (ref. 172).

An unusual cleavage of a Ge-C(exocyclic) bond results, with partial isomerization of the ring, from the reaction of butyllithium with 1,1,3,4-tetramethylgermole (ref. 173) :

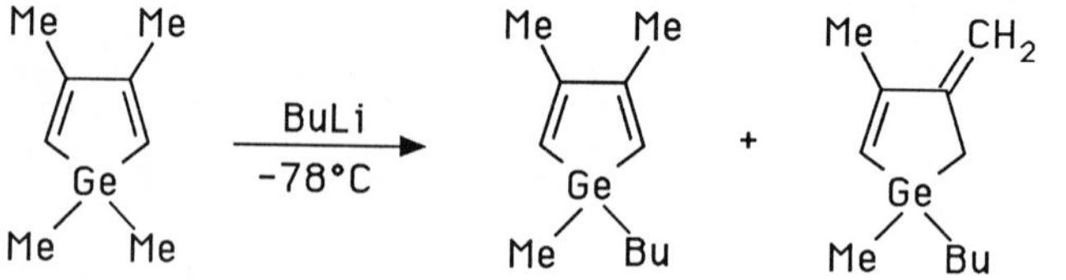

eqn.(2.68)

Molybdenum, iron (ref. 174) and cobalt (refs. 174,175) complexes of tetramethylgermole have been reported. The action of tin tetrachloride on the tricarbonyliron complex causes the cleavage of the *exo* Ge-C(Me) bond (ref. 174) :

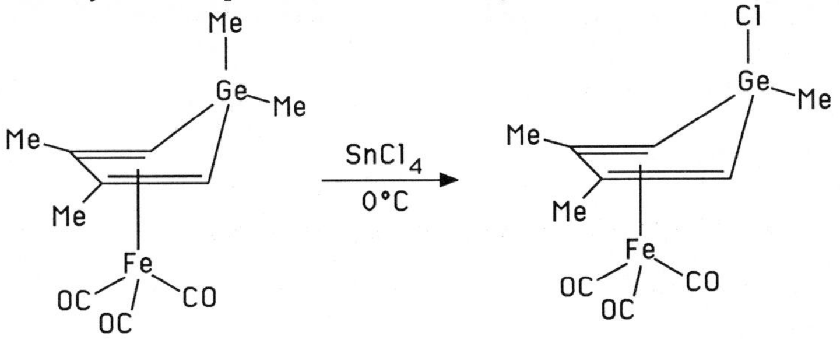

eqn.(2.69)

Spectra of tricarbonyliron complexes of these germoles (refs. 171,176) suggest a strong donation from iron to the metallole ligand giving a partial aromatic character to these compounds.

2.2.4. Germacyclohexadienes and heptadienes

2.2.4.1. Germacyclohexadienes

Diethylgermacyclohexa-2,4-dienes are formed by thermolysis of the bicyclic compounds resulting from the addition of dichlorocarbene to germacyclopent-3-enes (refs. 125, 127) :

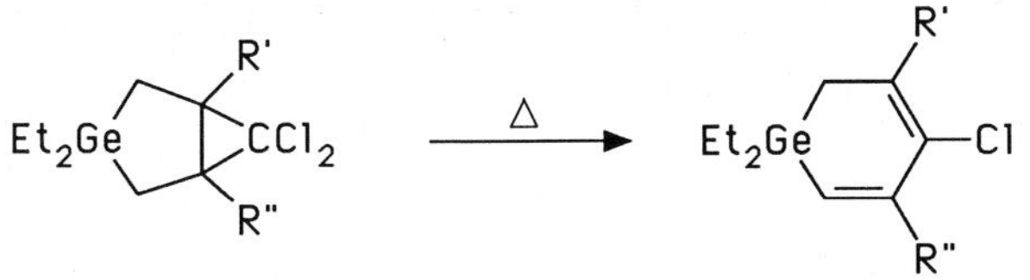

eqn.(2.70)

These conjugated dienes were the precursors of the first double bonded germanium-carbon species (ref. 24) :

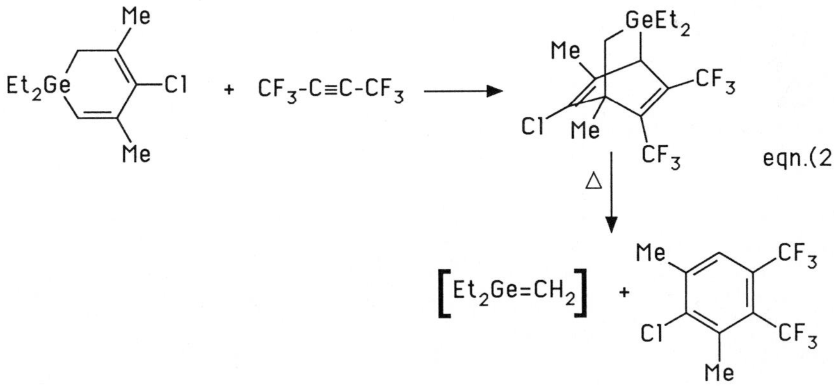

eqn.(2.71)

1-Germacyclohexa-2,5-dienes are prepared by reaction of *cis,cis*-1,5-dilithium-1,4-pentadiene with a germanium dihalide (refs. 177, 178) :

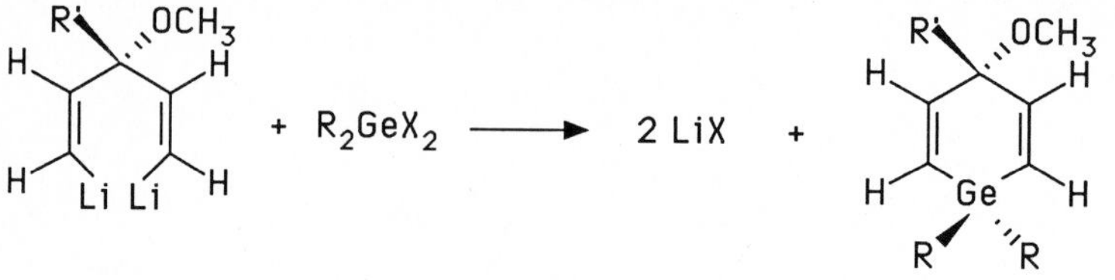

eqn.(2.72)

R = alkyl, aryl ; R' = cyclohexyl, Ph, t-Bu

With an organogermanium trihalide, the two isomeric cyclic germanium monohalides are formed, whereas a mixture of germanium dichloride and spiran is obtained from germanium tetrachloride :

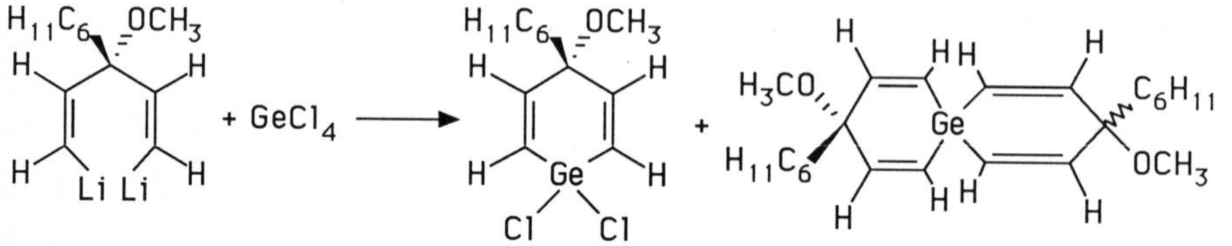

eqn.(2.73)

Reduction of 1-chloro-1-germa-4-methoxycyclohexa-2,5-dienes causes the removal of the methoxy group and the formation of a conjugated germacyclohexadiene hydride, which can be converted into the corresponding halide (ref. 179) :

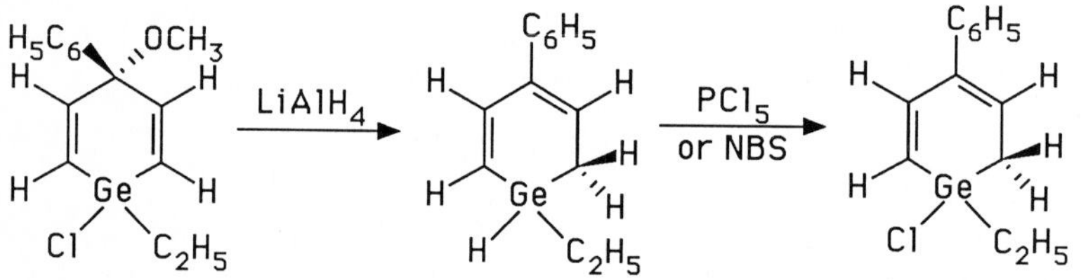

eqn.(2.74)

X = Cl, Br

Elimination of the methoxy group and isomerization of the diene also occurs by reaction with sodium, followed by hydrolysis :

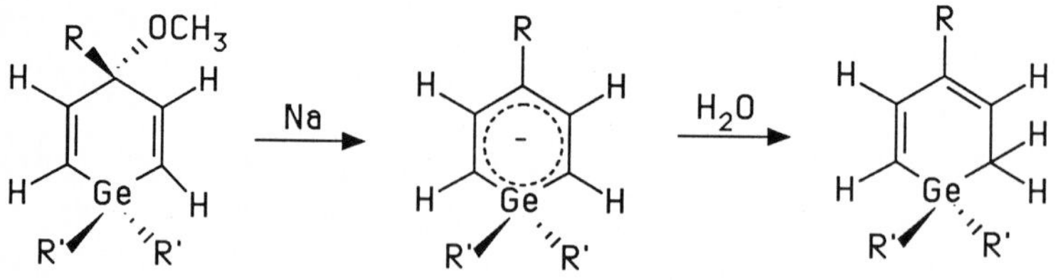

eqn.(2.75)

Dehydrobromination of 1,4-ditert-butyl-1-bromo-1-germacyclohexa-2,4-diene forms 1,4-ditert-butyl-1-germabenzene, which dimerizes or can be trapped by dienes in a Diels-Alder reaction (ref. 180), (Chart 25) :

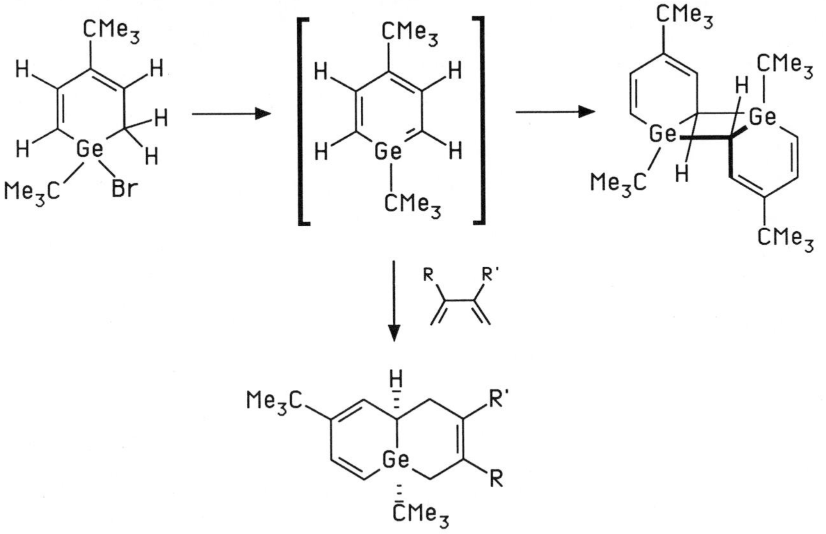

Chart 25. Formation and trapping of 1,4-ditert-butyl-1-germabenzene.

Reaction of methylgermanium trichloride with the di-Grignard reagent of *o,o'*-dichlorodiphenylmethane leads to a 9-germadihydroanthracene (ref. 181) :

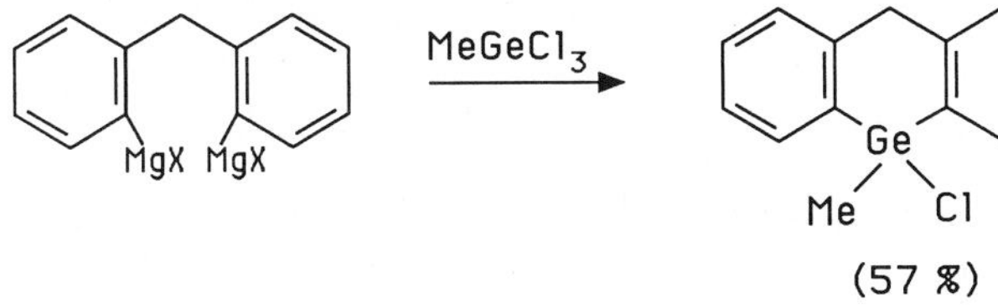

eqn.(2.76)

and the spiro compound is formed by reaction of germanium tetrachloride with the corresponding dilithio compound (ref. 182) :

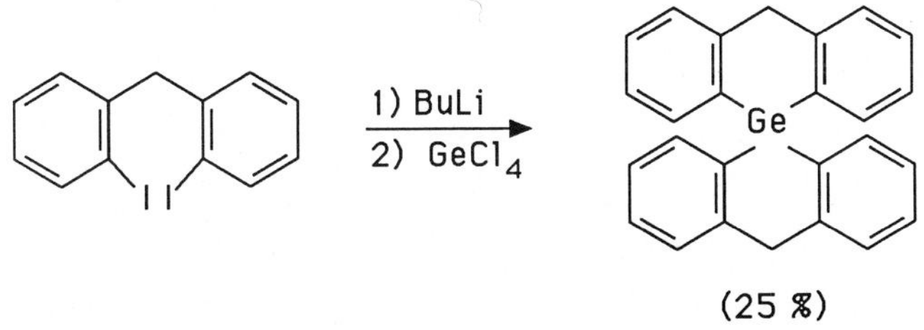

eqn.(2.77)

1,4-Digermacyclohexadienes have been prepared by Vol'pin *et al.* by addition of germanium diiodide to acetylenes (refs. 2, 4, 6) :

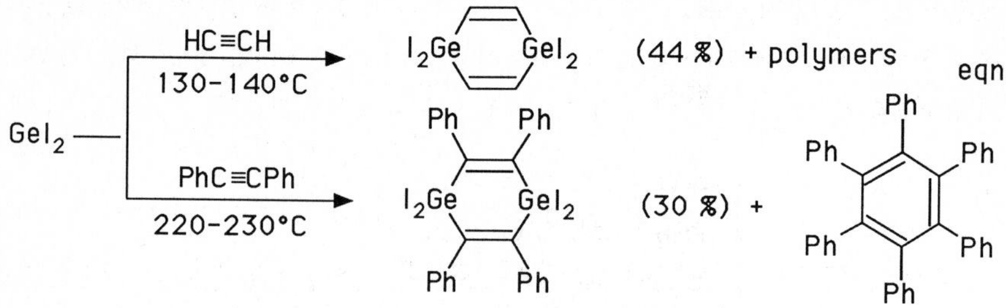

eqn.(2.78)

A similar addition is observed with 2-butyne (ref. 183) but the expected 1,4-digermacyclohexadiene is not formed by addition of dimethylgermylene to dimethylacetylene dicarboxylate (ref. 159).

The stability of the ring is demonstrated by the reactions summarized in Chart 26, (refs. 2, 4, 8, 11, 26) :

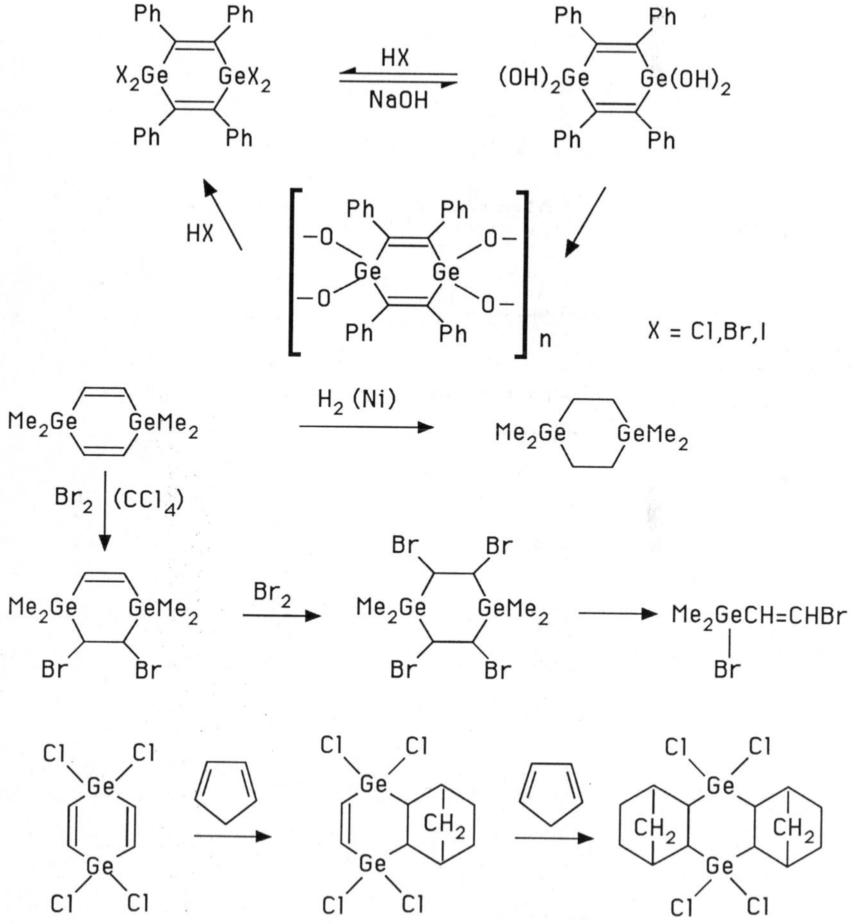

Chart 26. Characteristic reactions of 1,4-digermacyclohexa-2,5-dienes.

Because of the controversy concerning the structure of these compouds, a large number of spectroscopic investigations have been reported. Infrared and Raman spectra (refs. 2-6, 11), NMR spectra (refs. 6-7, 9, 11), mass spectra (refs. 7, 9), UV (refs. 9, 11), electron diffraction studies (ref. 10), and X-ray structures (refs. 8,11) are in agreement with the 1,4-digermacyclohexa-2,5-dienic structure.

2.2.4.2. Germacycloheptadienes

Cycloaddition of dihalogermylenes to 1,3,5-hexatriene yields germacyclohepta-3,5-dienes (refs. 99, 102, 184-185) :

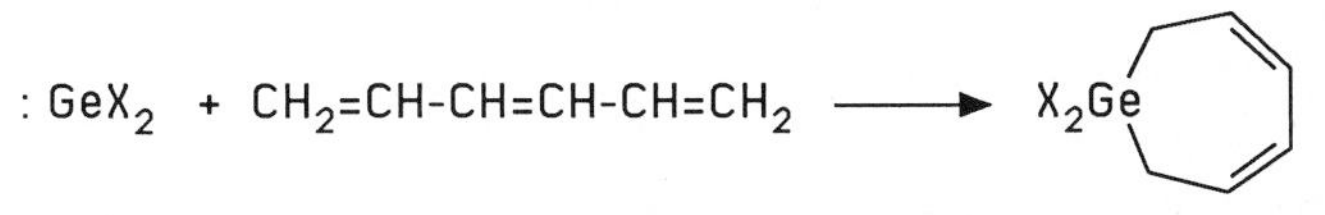

eqn.(2.79)

X = F, Cl (35 %)

Similarly, a bicyclic adduct is obtained with cyclooctatetraene (ref. 99) :

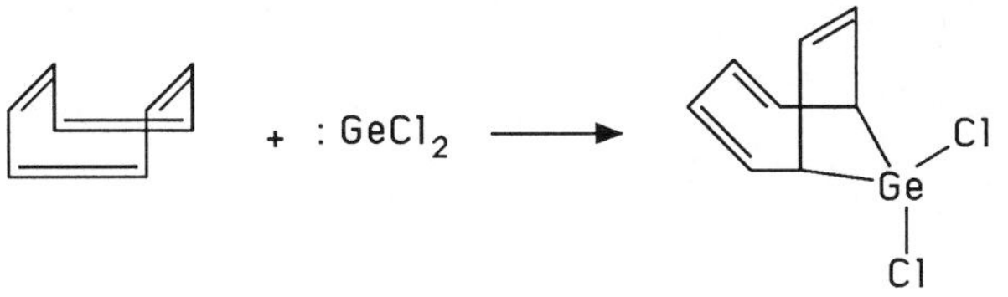

eqn.(2.80)

5,5-Dimethyl (and diphenyl)-10,11-dihydro-5H-dibenzo[b,f]germepins are prepared by action of the appropriate organogermanium dichloride with 2,2'-dilithiobibenzyl (ref. 186) :

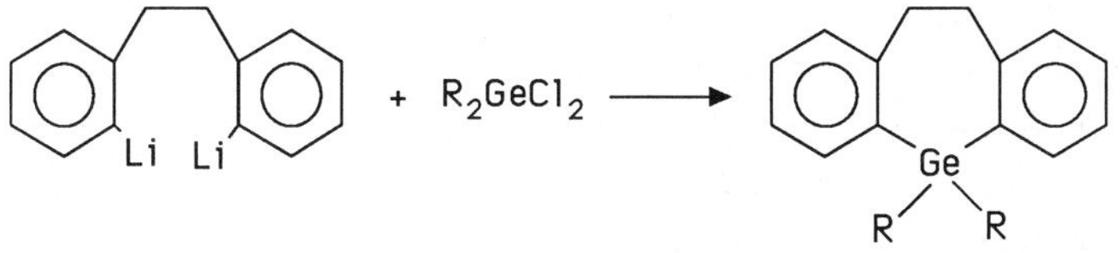

eqn.(2.81)

R = Me, Ph (33-35 %)

These compounds are converted into 5H-dibenzo[b,f]germepins by bromination followed by dehydrobromination.

3- MEDIUM RINGS

3.1. SYNTHESIS

Chart 27 summarizes the different routes leading to the formation of these rings :

- Würtz-coupling of bis (ω-bromoalkyl)germane using sodium in boiling xylene reaction (a) and thermolysis of a suitable thorium salt, reaction (b) give only poor yields of crude product (ref. 187) ; dimethylgermacyclooctanone is obtained by hydroboration of dimethyldiallylgermane by thexylborane, followed by cyanoboration (ref. 72), reaction (c). Germa- and digermacyclanols are formed by radical hydrogermylation of hydrogenogermaenols (refs. 22, 23, 102), reactions (d). Direct synthesis of a eight-membered 1,5-disila-3,7-digermacyclane was effected by Gar *et al*, by passing a stream of dimethylbis(chloromethyl)silane over a mixture of germanium and copper powder (ref. 27), reaction (e).

- The best preparation is the cyclization of a diester in an oxygen-free atmosphere, forming the expected acyloin in good yield (refs. 187-188). With identical ester chains, an odd-membered acyloin is obtained ; even-membered acyloins are similarly isolated from unsymmetrical esters.

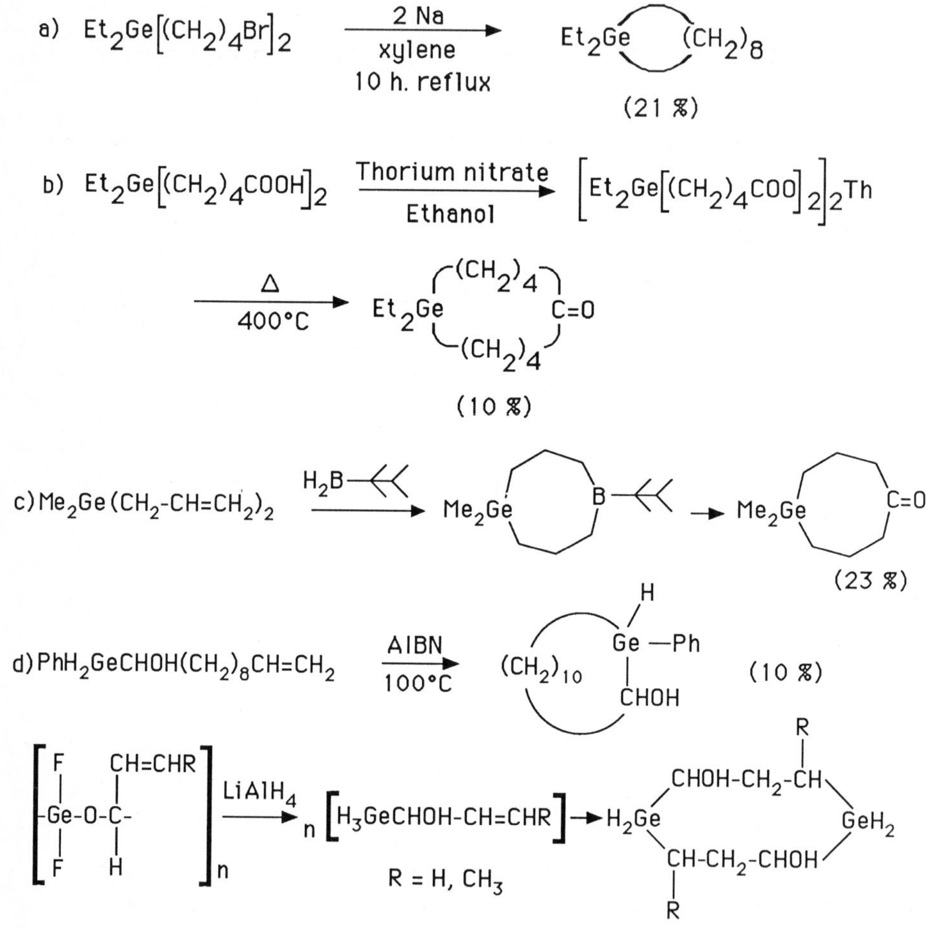

Chart 27. Synthesis of medium-sized germacyclic derivatives.

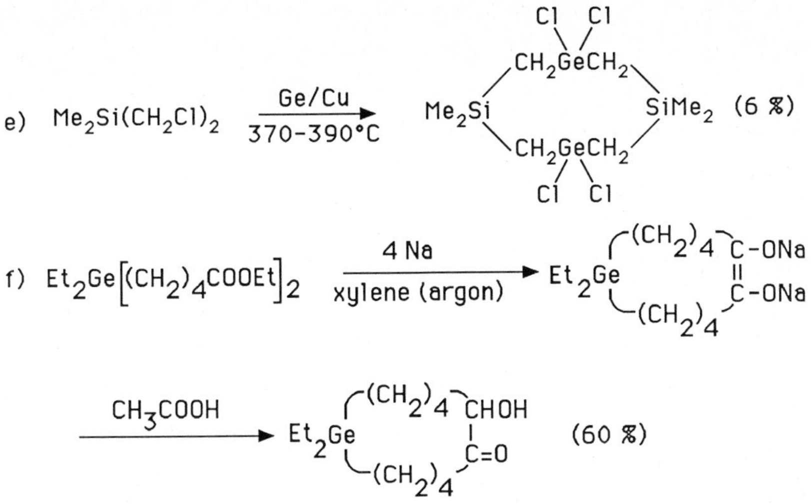

Chart 27 (continued). Synthesis of medium-sized germacyclic derivatives

These acyloins are the precursors of other germanium medium-sized rings (Chart 28), (refs. 187-191) :

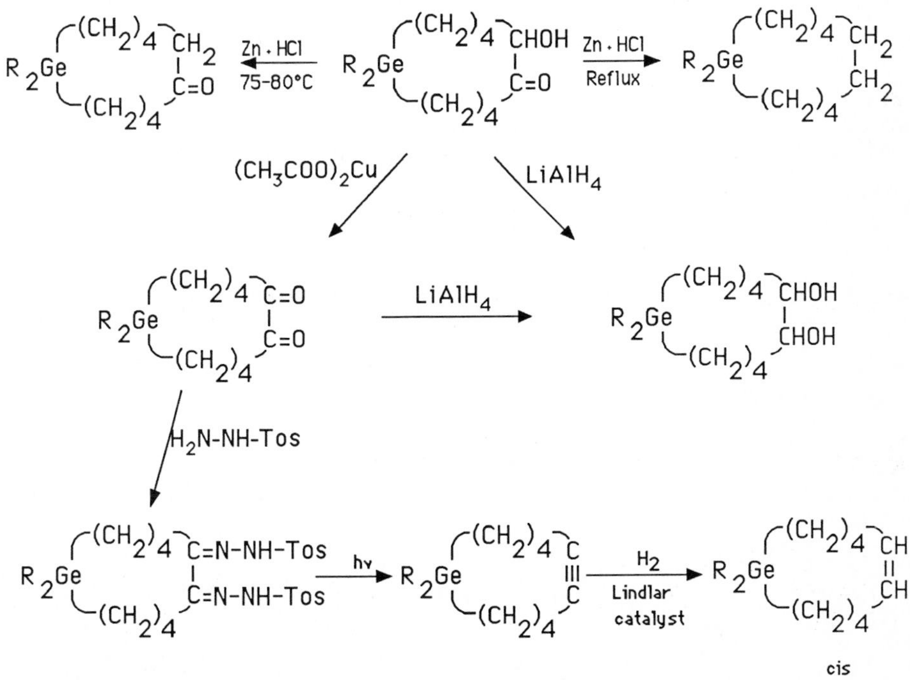

Chart 28. Some reactions of functional medium-sized organogermanium derivatives

- Reduction by zinc in acid medium leads, depending on the temperature, to the ketone or to the cycloalkane, while the α-glycol is formed with lithium aluminum hydride

- Cupric oxidation gives quantitatively a cycloalkanedione, the precursor of germacycloalkyne and cycloalkene.

3.2. TRANSANNULAR INTERACTIONS

Interactions can be detected between the germanium atom and a suitable atom located on the opposite side of the ring. For dimethylgermacycloundecanone, dipole moment, solvent effect in NMR,(ref. 192) intensity of IR absorption bands, (ref. 70), photoelectron spectrum (ref. 193) and magnetooptical measurements (ref. 71) are in agreement with the spatial proximity of the germanium

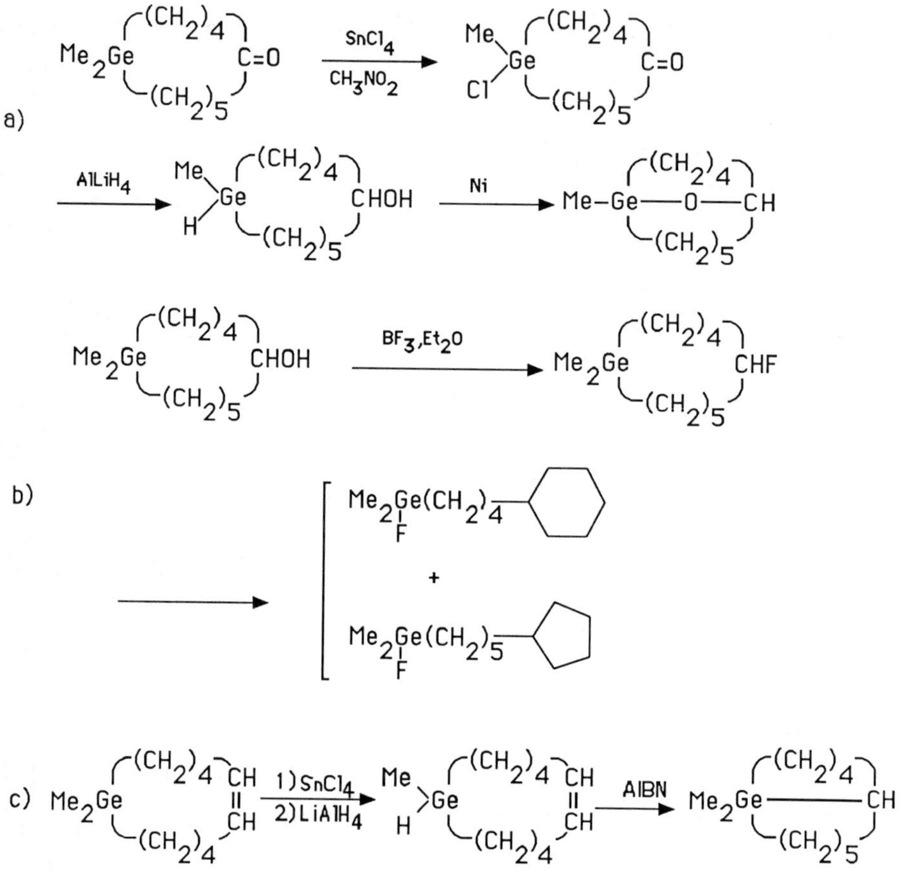

Chart 29. Transannular reactions of functional medium-sized organogermanium rings.

atom to the carbonyl group, in the gas phase or in solution. However, the crystal structure (ref. 194) reveals that at least in the solid state the Ge...CO distance is large enough to prevent any transannular interaction. A study of the IR spectrum at pressures up to 5 GPa showed that only reversible conformational changes occur (ref. 195).

3.3. TRANSANNULAR REACTIONS

The closeness of opposite sides of the ring facilitates intramolecular cyclizations (transannular reactions) (Chart 29) :

- 1-Methylgermacycloundecan-6-ol, which results from the reduction of 1-methyl-1 chlorogermacyclohexan-6-one, undergoes hydrogen elimination in the presence of Raney nickel to give a bicyclic compound having an intracyclic germanium-oxygen bond (ref. 190)
- Intramolecular hydrogermylation of 1-methylgermacycloundec-6-ene in the presence of a radical initiator, leads similarly to a bicyclic germacycloundecane (ref. 191)
- On the other hand, nucleophilic attack of fluorine on germanium causes the cleavage of an intracyclic germanium-carbon bond in 1,1-dimethylgerma-6-fluorocycloundecane (ref. 196).

4. LARGE RINGS

Some 15-membered rings were prepared by using acyloin condensation (ref. 188) :

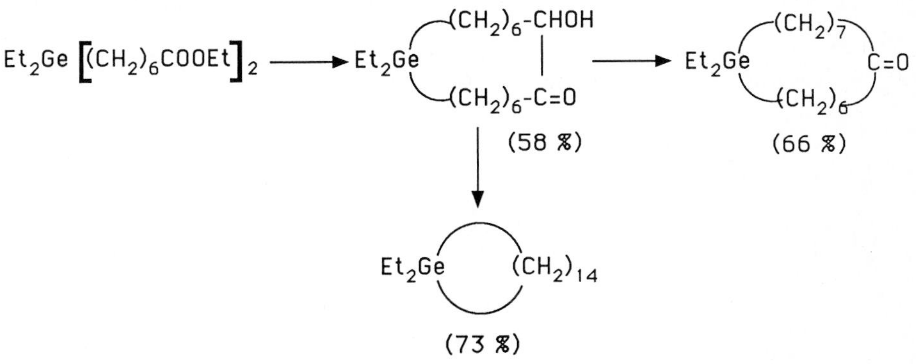

Chart 30. Synthesis of 15-membered organogermanium rings.

Some books and reviews on organogermanium chemistry, including germanium-carbon rings, are listed in references 197 to 205.

REFERENCES

1 D. Lloyd, Alicyclic Compounds, Edward Arnold Ltd., London, 1963.
2 M.E. Vol'pin, Yu.D. Koreshkov, V.G. Dulova, and D.N. Kursanov, Tetrahedron,18, (1962), 107-22.
3 M.E. Vol'pin and D.N. Kursanov, Zh.Obshch.Khim., 32, (1962), 1142-46
4 M.E. Vol'pin and D.N. Kursanov, Zh.Obshch.Khim., 32 (n°5), (1962), 1455-60.
5 L.A. Leites, V.G. Dulova, and M.E. Vol'pin, Izvest. Akad. Nauk SSSR, Ser. Khim., (1963), 731-37.
6 M.E. Vol'pin, V.G.. Dulova, and D.N. Kursanov, Izvest.Akad.Nauk SSR, Ser.Khim., (1963), 727-31.
7 F. Johnson and R.S. Gohlke, Tetrahedron, 26, (1962), 1291-5.
8 M.E. Vol'pin, Yu.T. Struchkov, L.V. Vilkov, V.S. Mastryukov, V.G. Dulova, and D.N. Kursanov, Izvest.Akad.Nauk SSSR, Ser.Khim., n° 11, (1963), 2067.
9 F. Johnson, R.S. Gohlke, and W.H. Nasutavicus, J.Organometal.Chem., 3, (1965), 233-44.
10 L.V. Vilkov and V.S. Mastryukov, Zh.Strukt.Khim., 6, (1965), 811-16.
11 M.E. Vol'pin, V.G. Dulova, Yu.T. Struchkov, N.K. Bokiy, and D.N. Kursanov, J.Organometal.Chem., 8, (1967), 87-96.
12 A. Krebs and J. Berndt, Tetrahedron Lett., 24 (n° 38), (1983), 4083-86.
13 M.P. Egorov, S.P. Kolesnikov, Yu.P. Struchkov, M.Yu. Antipin, S.V. Sereda, and O.M. Nefedov, J.Organometal.Chem., 290, (1985), C27-C30.
14 M.P. Egorov, S.P. Kolesnikov, Yu.T. Struchkov, M.Yu. Antipin, S.V. Sereda, and O.M. Nefedov, Izvest.Akad.Nauk SSSR, Ser.Khim., (4), (1985), 959-60.
15 W. Ando and T. Tsumuraya, Organometallics, 7, (1988), 1882-83.
16 P. Mazerolles, M. Lesbre, and J. Dubac, C.R.Acad.Sci., 260, (1965), 2255-58.
17 P. Mazerolles, J. Dubac, and M. Lesbre, J.Organometal.Chem., 5, (1966), 35-47.
18 P. Mazerolles, J. Dubac, and M. Lesbre, C.R.Acad.Sci., 266, (1968), 1794-96.
19 J.W.F.L. Seetz, B.J.J. Van de Heisteeg, G. Schat, O.S. Akkerman, and F. Bickelhaupt, J.Organometal.Chem., 277, (1984), 319-22.
20 B.J.J. Van de Heisteeg, M.A.G.M. Tinga, Y. Van den Winkel, O.S. Akkerman, and F. Bickelhaupt, J.Organometal.Chem., 316, (1986), 51-55.
21 F.Bickelhaupt, Angew.Chem.Intern.Ed.Engl., 26, (1987), 990-1005.
22 P. Rivière and J. Satgé, Angew.Chem.Intern.Ed.Engl., 10, (1971), 267-68.
23 P. Rivière and J. Satgé, J.Organometal.Chem., 49, (1973), 173-89.
24 T.J. Barton, E.A. Kline, and P.M. Garvey, J.Amer.Chem.Soc., 95, (1973), 3078.
25 J.W. Bruin, G. Schat, O.S. Akkerman, F. Bickelhaupt, J.Organometal.Chem., 288, (1985), 13-25
26 D. Seyferth and J.L. Lefferts, J.Amer.Chem.Soc., 96, (1974), 6237-38.
27 T.K. Gar, A.A. Buyakov, and V.F. Mironov, 42 (n° 7), (1972), 1521-24.
28 V.F. Mironov, T.K. Gar, and S.A. Mikhailyants, Dokl.Akad.Nauk SSSR, 188, (1969), 120-23.
29 V.F. Mironov, S.A. Mikhailyants, and T.K. Gar, Zh.Obshch.Khim, 39, (1969), 2601.
30 E.D. Babich, N.B. Bespalova, V.M. Vdovin, T.K. Gar, and V.F. Mironov, Izvest.Akad.Nauk SSSR, Ser.Khim., (1977), 960-61.

31 D. Seyferth and C.J. Attridge, J.Organometal.Chem., 21, (1970), 103-106.

32 N.S. Nametkin, L.E. Gusel'nikov, R.L. Ushakova, V.Yu. Orlov, O.V. Kuz'min, and V.M. Vdovin, Dokl.Akad.Nauk SSR, 194, (1970), 1096-99.

33 P. Mazerolles, J. Dubac, and M. Lesbre, Tetrahedron Lett., n° 3, (1967), 255-58.

34 P. Mazerolles and J. Dubac, C.R.Acad.Sci., 265, (1967), 403-6.

35 D. Seyferth, S.S. Washburne, T.F. Jula, P. Mazerolles, and J. Dubac, J.Organometal.Chem., 16, (1969), 503-6.

36 P. Mazerolles, J. Dubac, and M. Lesbre, J.Organometal.Chem., 12, (1968), 143-48.

37 J. Dubac and P. Mazerolles, C.R.Acad.Sci., 267, (1968), 411-13.

38 J. Dubac and P. Mazerolles, J.Organometal.Chem., 20, (1969), P5-P6.

39 J. Dubac and P. Mazerolles, Bull.Soc.Chim.Fr., n° 10, (1969), 3608-9.

40 H.J.R. de Boer, O.S. Akkerman, and F. Bickelhaupt, J.Organometal.Chem., 321, (1987), 291-306.

41 S.P. Kolesnikov, A. Krebs, and O.M. Nefedov, Izvest.Akad.Nauk SSSR, Ser.Khim., 9, (1983), 2173-4.

42 A.A. Espenbetov, Yu.T. Strchkov, S.P. Kolesnikov, and O.M. Nefedov, J.Organometal.Chem., 273, (1984), 33-37.

43 S.P. Kolesnikov, M.P. Egorov, A.M. Gal'minas, and A. Krebs, Izvest. Akad.Nauk SSSR, Ser.Khim., 12, (1985), 2832.

44 O.M. Nefedov, S.P. Kolesnikov, M.P. Egorov, A.M. Gal'minas, and A. Krebs, Izvest.Akad.Nauk SSSR, Ser.Khim., (1985), 2834.

45 O.M. Nefedov, M.P. Egorov, A.M. Gal'minas, S.P. Kolesnikov, A. Krebs, and J. Berndt, J.Organometal.Chem., 301, (1986), C21-C22.

46 V.F. Mironov, T.K. Gar, and A.A. Buyakov, Zh.Obshch.Khim., 43 (n° 4), (1973), 798-801.

47 R. Schwarz and W. Reinhardt, Ber., 65, (1932), 1743-46.

48 P. Mazerolles, Bull.Soc.Chim. Fr., (1962), 1907-14.

49 F.J. Bajer and H.W. Post, J.Org.Chem., 27 (n° 4), (1962), 1422-24.

50 K.I. Kobrakov, T.I. Chernisheva, N.S. Nametkin, and A.Ya. Sideridu, Dokl.Akad.Nauk SSSR, 196, (1971), 100-102.

51 K.I. Kobrakov, T.I. Chernysheva, and N.S. Nametkin, Dokl.Akad.Nauk SSSR, 198 (n° 6), (1971), 1340-43.

52 P. Mazerolles and J. Dubac, C.R.Acad.Sci., 257, (1963), 1103-06.

53 A.M. Duffield, C. Djerassi, P. Mazerolles, J. Dubac, and G. Manuel, J.Organometal.Chem., 12, (1968), 123-32.

54 J. Dubac, P. Mazerolles, M. Joly, and F. Piau, J.Organometal.Chem., 127, (1977), C69-C74.

55 J. Dubac, P. Mazerolles, M. Joly, and J. Cavezzan, J.Organometal.Chem., 165, (1979), 163-73.

56 J. Dubac, P. Mazerolles, and J. Cavezzan, Tetrahedron Lett., (1978), 3255-58.

57 J. Dubac, P. Mazerolles, J. Cavezzan, and M. Joly, J.Organometal.Chem., 165, (1979), 175-85.

58 J. Dubac, G. Dousse, J. Barrau, J. Cavezzan, J. Satgé, and P. Mazerolles, Tetrahedron Lett., (1978), 4499-502.

59 J. Dubac, J. Cavezzan, P. Mazerolles, J.Escudié, C. Couret, and J. Satgé, J.Organometal.Chem., 174, (1979), 263-74.

60 J. Escudié, C. Couret, J. Dubac, J. Cavezzan, J. Satgé, and P. Mazerolles, Tetrahedron Lett., (1979), 3507-10.

61 J. Dubac, J. Escudié, C. Couret, J. Cavezzan, J. Satgé, and P. Mazerolles, Tetrahedron, 37, (1981), 1141-51.

62 J. Dubac, J. Cavezzan, A. Laporterie, and P. Mazerolles, J. Organometal.Chem., 197, (1980), C15-C18.

63 J. Dubac, J. Cavezzan, A. Laporterie, and P. Mazerolles, J.Organometal. Chem., 209, (1981), 25-36.

64 P. Mazerolles and G. Manuel, C.R.Acad.Sci., 267, (1968), 1158-61.

65 G. Manuel, P. Mazerolles, M. Lesbre, and J.P. Pradel, J.Organometal.Chem. 61, (1973), 147-65.

66 G. Manuel and P. Mazerolles, J.Organometal.Chem., 19, (1969), 43-51.

67 G. Manuel, G. Bertrand, and P. Mazerolles, J.Organometal.Chem., 146, (1978), 7-16.

68 J.A. Soderquist and A. Hassner, J.Org.Chem., 45, (1980), 541-43.

69 W.P. Weber, R.A. Felix, A.K. Willard, and H.G. Boettger, J.Org.Chem., 36, (1971), 4060-68.

70 R. Mathis, A. Faucher, and P. Mazerolles, Spectrochim.Acta, 35 A, (1979), 1311-14.

71 P. Mazerolles, Organomet.Coord.Chem. Germanium, Tin, Lead,Plen.Lect. Int.Conf., 2nd, (1977), 75-102.

72 J.A. Soderquist and A. Hassner, J.Org.Chem., 48, (1983), 1801-10.

73 A. Pelter, M.G. Hutchings, and K. Smith, J.Chem.Soc., Chem.Commun., (1971), 1048.

74 H.C. Brown and B.A. Carlson, J.Org.Chem., 38, (1973), 2422-24.

75 G. Manuel, G. Bertrand, and F. El Anba, Organometallics, 2, (1983), 391-94.

76 K.T. Kang, G. Manuel, and W.P. Weber, Chem.Lett., (1986), 1685-86.

77 R. Damrauer, W.P. Weber, and G. Manuel, Chem.Lett., (1987), 235-36.

78 G. Manuel, A. Faucher, and P. Mazerolles, J.Organometal.Chem., 327, (1987), C25-C28.

79 P. Mazerolles, M. Lesbre, and M. Joanny, J.Organometal.Chem., 16, (1969), 227-33.

80 D. Seyferth, Houng-Min Shih, P. Mazerolles, M. Lesbre, and M. Joanny, J.Organometal.Chem., 29, (1971), 371-83.

81 P. Rivière, J. Satgé, and D. Soula, C.R.Acad.Sci., Ser. C., 277, (1973), 895-8.

82 V.F. Mironov, S.A. Mikailyants, and T.K. Gar, Zh.Obshch.Khim., 39, (1969), 2281-4.

83 M.C. Henry and J.G. Noltes, J.Amer.Chem.Soc., 82, (1960), 561-3.

84 W.P. Neumann, 31st Intern.Congress of Pure and Appl.Chem., Sofia(Bulgaria) Section 6, (1987), 148-62.

85 Th. Butkowskyi-Walkiw and G. Szeimies, Tetrahedron, 42 (n° 6), (1986), 1845-50.

86 L.I. Zakharkin and N.F. Shemyakin, Zh.Obshch.Khim., 44 (n° 5), (1974), 1085-89.

87 L.I. Zakharkin and N.F. Shemyakin, Izvest.Akad.Nauk SSSR, Ser.Khim., n° 10, (1977), 2350-52.

88 P. Boudjouk and C.A. Kapfer, J.Organometal.Chem., 296 (3), (1985), 339-49.

89 V.F. Mironov and T.K. Gar, Dokl.Akad.Nauk SSSR, 152, (1963), 1111-14.

90 V.F. Mironov and T.K. Gar, Izvest.Akad.Nauk SSSR, Ser. Khim., 3, (1966), 482-89.

91 V.F. Mironov and T.K. Gar, Izvest.Akad.Nauk SSSR, Ser.Khim., (1963), 578.

92 V.F. Mironov and T.K. Gar, Organometal. Chem.Rev.A, 3, (1968), 311-21.
93 V.F. Mironov and T.K. Gar, Izvest.Akad.Nauk SSSR, Ser.Khim., (1965), 755-58.
94 T.K. Gar and V.F. Mironov, Izvest.Akad.Nauk SSSR, Ser.Khim., (1965), 855-62.
95 O.M. Nefedov, S.P. Kolesnikov, and V.I. Sheichenko, Angew.Chem.Int. Ed.Engl., 3 (n°7), (1964), 508-9.
96 O.M. Nefedov and S.P. Kolesnikov, Izvest.Akad.Nauk SSSR, Ser.Khim., (1966), 201-11.
97 S.P. Kolesnikov, V.I. Shiryaev, and O.M. Nefedov, Izvest.Akad.Nauk SSSR, Ser.Khim., (1966), 584.
98 S.P. Kolesnikov, A.I. Ioffe, and O.M. Nefedov, Izvest.Akad.Nauk SSSR, Ser.Khim., 4, (1975), 978.
99 O.M. Nefedov, S.P. Kolesnikov, and A.I. Ioffe, Izvest.Akad.Nauk SSSR, Ser.Khim., 3, (1976), 619-25.
100 P. Mazerolles and G. Manuel, Bull.Soc.Chim.Fr., n° 1, (1966), 327-331.
101 P. Mazerolles, G. Manuel, and F. Thoumas, C.R.Acad.Sci., 267, (1968), 619-22.
102 P. Rivière, J. Satgé, and A. Castel, C.R.Acad.Sci., Ser.C, 284, (1977), 395-98.
103 E.M. Berliner, T.K. Gar, and V.F. Mironov, Zh.Obshch.Khim., 42, (1972), 1172.
104 M. Massol, P. Rivière, J. Barrau, and J. Satgé, C.R.Acad.Sci., Ser.C, 270, (1970), 237-39.
105 M. Massol, J. Satgé, P. Rivière, and J. Barrau, J.Organometal.Chem., 22, (1970), 599-610.
106 P. Rivière, J. Satgé, and D. Soula, J.Organometal.Chem., 63, (1973), 167-74.
107 J. Satgé, R. Rivière, and A. Boy, C.R.Acad.Sci., Ser.C, 278, (1974), 1309-12
108 M. Massol, P. Rivière, J. Barrau, and J. Satgé, C.R.Acad.Sci., Ser. C, 270, (1970), 237-39.
109 P. Rivière, J. Satgé, and D. Soula, J.Organometal.Chem. 72, (1974), 329-38.
110 H. Lavayssière and G. Dousse, J.Organometal.Chem., 297, (1985), C17-C22.
111 H. Lavayssière, G. Dousse, and J. Satgé, Recueil Trav.Chim.Pays-Bas, 107, (1988), 440-8.
112 G. Dousse, Thesis, Toulouse, (France), (1977).
113 P. Rivière, A. Castel, and J. Satgé, J.Organometal.Chem., 232, (1982), 123-35.
114 P. Rivière, M. Rivière-Baudet, and J. Satgé, J.Organometal.Chem., 96, (1975), C7-C10.
115 P. Rivière, M. Rivière-Baudet, and J. Satgé, J. Organometal.Chem., 97, (1975), C37-C40.
116 M. Schriewer and W.P. Neumann, J.Amer.Chem.Soc., 105, (1983), 897-901.
117 D.P. Paquin, R.J. O'Connor, and M.A. Ring, J.Organometal.Chem., 80, (1974), 341-48.
118 W. Ando, T. Tsumuraya, and A. Sekiguchi, Chem.Lett., 2, (1987), 317-18.
119 W.P. Neumann, E. Michels, and J. Köcher, Tetrahedron Lett., 28 (n° 33), (1987), 3783-86.

120 A. Marchand, A. Millan, J. Dunoguès, G. Manuel, and P. Mazerolles, J.Organometal.Chem., 135, (1977), 23-37.

121 P. Mazerolles, F. Grégoire, and G. Manuel, J.Organometal.Chem., 277, (1984), C4-C6.

122 P. Mazerolles and F. Grégoire, J.Organometal.Chem., 301, (1986), 153-60.

123 G. Manuel, G. Bertrand, P. Mazerolles, and J. Ancelle, J.Organometal. Chem., 212, (1981), 311-23.

124 G. Manuel, P. Mazerolles, and J.C. Florence, C.R.Acad.Sci., 269, (1969), 1553-55.

125 D. Seyferth, T.F. Jula, D.C. Mueller, P. Mazerolles, G. Manuel, and F. Thoumas, J.Amer.Chem.Soc., 92, (1970), 657-66.

126 M. Lesbre, G. Manuel, P. Mazerolles, and G. Cauquy, J.Organometal. Chem., 40, (1972), C14-C16.

127 G. Manuel, G. Cauquy, and P. Mazerolles, Synth.React.Inorg. Metal-Org.Chem., 4 (2), (1974), 143-47.

128 X. Zhang, Q. Zhou, W.P. Weber, R.F. Horvath, T.H. Chan, and G. Manuel, Macromol., 21, (1988), 1563-66.

129 P. Mazerolles, C. Laurent, and A. Faucher, (J.Organometal.Chem.), in press.

130 R.D. Rieke and P.M. Hudnall, J.Amer.Chem.Soc., 94, (1972), 7178-79.

131 R.D. Rieke and S.E. Bales, J.Amer.Chem.Soc., 96 (6), (1974), 1775-81.

132 P. Mazerolles and F. Grégoire, Synth.React.Inorg.Metal-Org.Chem., 16 (7), (1986), 905-13.

133 P. Mazerolles, R. Morancho, and A. Reynès, Silicon, Germanium, Tin and Lead Compounds, 9 (n° 2 & 3), (1986), 155-183.

134 P. Mazerolles, F. Grégoire, R. Morancho, A. Reynès, and N. Séfiani, J.Organometal.Chem., 328, (1987), 49-59.

135 M. Lesbre, A. Laporterie, J. Dubac, and G. Manuel, C.R.Acad.Sci., 280, (1975), 787-9.

136 J. Dubac, A. Laporterie, G. Manuel, and H. Iloughmane, Phosphorus and Sulfur, 27, (1986), 191-203.

137 A. Laporterie, G. Manuel, H. Iloughmane, and J. Dubac, Nouv.Journ.Chim., 8, (1984), 437-41.

138 A. Laporterie, J. Dubac, and P. Mazerolles, J.Organometal.Chem., 202, (1980), C89-C92.

139 A. Laporterie, J. Dubac, and M. Lesbre, J.Organometal.Chem., 101, (1975), 187-208.

140 G.R. Chambers and M. Jones Jr., Tetrahedron Lett., (1978), 5193-96.

141 H. Sakuraï, K. Sakamoto, A. Nakamura, and M. Kira, Chem.Lett., (1985), 497-8.

142 C. Couret, J. Escudié, J. Satgé, and M. Lazraq, J.Amer.Chem.Soc., 109, (1987), 4411-12.

143 B. Bogdanovic, N. Janke, C. Krüger, K. Schlichte, and J. Treber, Angew.Chem.Int.Ed.Engl., 26 (n° 10), (1987), 1025-26.

144 P. Mazerolles, M. Joanny, and G. Tourrou, J.Organometal.Chem., 60, (1973), C3-C5.

145 A. Marchand, P. Gerval, F. Duboudin, M. Joanny, and P. Mazerolles, J.Organometal.Chem., 267, (1984), 93-106.

146 A. Marchand, P. Gerval, M. Joanny, and P. Mazerolles, J.Organometal. Chem., 217, (1981), 19-33.

147 F.C. Leavitt, T.A. Manuel, and F. Johnson, J.Amer.Chem.Soc., 81, (1959), 3163-64.

148 F.C. Leavitt, T.A. Manuel, F. Johnson, L.U. Matternas, and D.S. Lehman, J.Amer.Chem.Soc., 82, (1960), 5099-102.

149 D. Curtis, J.Amer.Chem.Soc., 89, (1967), 4241-42.

150 D. Curtis, J.Amer.Chem.Soc., 91, (1969), 6011-18.

151 H. Gilman and R.D. Gorsich, J.Amer.Chem.Soc., 80, (1958), 1883-86.

152 I.M.. Gverdtsiteli, T.M. Doksopulo, M.M. Menteshashvili and I.I. Abkhazava, Soobshch.Akad.Nauk Gruz SSR, 40, (1965), 333-8.

153 S.C. Cohen and A.G. Massey, Tetrahedron Lett., (1966), 4393-94.

154 M.D. Rausch and L.P. Klemann, J.Amer.Chem.Soc., 89, (1967), 5732-33.

155 S.C. Cohen and A.G. Massey, Chem.Commun., (1967), 457-58.

156 S.C. Cohen and A.G. Massey, J.Organometal.Chem., 10, (1967), 471-81.

157 P.J. Fagan and W.A. Nugent, J.Amer.Chem.Soc., 110, (1988), 2310-12.

158 S.C. Cohen, M.L.N. Reddy, and A.G. Massey, Chem.Commun., (1967), 451-53.

159 J.G. Zavitoski and J.J. Zuckerman, J.Amer.Chem.Soc., 90, (1968), 6612-16.

160 W. Ando, T. Tsumuraya, and A. Sekiguchi, Tetrahedron Lett., 26 (37), (1985), 4523-24.

161 M.P. Egorov, A.S. Dvornikov, S.P. Kolesnikov, V.A. Kuzmin, and O.M. Nefedov, Izvest.Akad.Naul SSSR, Ser.Khim., (5), (1987), 1200.

162 J. Köcher, M. Lehnig, and W.P. Neumann, Organometallics, 7 (5), (1988), 1201-07.

163 P. Jutzi and A. Karl, J.Organometal.Chem., 128, (1977), 57-62.

164 P. Jutzi and A. Karl, J.Organometal.Chem., 215, (1981), 19-25.

165 P. Jutzi, A. Karl, and C. Burschka, J.Organometal.Chem., 215, (1981), 27-39.

166 P. Jutzi, A. Karl, and P. Hofmann, Angew.Chem.Int.Ed.Engl., 19, (1980), 484-5.

167 D. Curtis, W.M. Butler, and J. Scibelli, J.Organometal.Chem. 192, (1980), 209-18.

168 J. Dubac, A. Laporterie, and H. Iloughmane, J.Organometal.Chem., 293, (1985), 295-311.

169 A. Laporterie, G. Manuel, J. Dubac, P. Mazerolles, and H. Iloughmane, J.Organometal.Chem., 210, (1981), C33-C36.

170 A. Laporterie, G. Manuel, J. Dubac, and P. Mazerolles, Nouveau Journ.Chim., 6 (n° 2), (1982), 67-69.

171 A. Laporterie, H. Iloughmane, and J. Dubac, J.Organometal.Chem., 244, (1983), C12-C16.

172 J.B. Béteille, G. Manuel, A. Laporterie, H. Iloughmane, and J. Dubac, Organometallics, 5, (1986), 1742-43.

173 J. Dubac, H. Iloughmane, A. Laporterie, and C. Roques, Tetrahedron Lett., 26, (1985), 1315-18.

174 G.T. Burns, E. Colomer, R.J.P. Corriu, M. Lheureux, J. Dubac, A. Laporterie, and H. Iloughmane, Organometallics, 6, (1987), 1398- 1406.

175 L.C. Ananias de Carvalho, M. Dartiguenave, F. Dahan, Y. Dartiguenave, J. Dubac, A. Laporterie, G. Manuel, and H. Iloughmane, Organometallics, 5, (1986), 2205-11.

176 C. Guimon, G. Pfister-Guillouzo, J. Dubac, A. Laporterie, G. Manuel, and H. Iloughmane, Organometallics, 4, (1985), 636-41.

177 G. Märkl and P. Hofmeister, Tetrahedron Lett., n° 38, (1976), 3419-22.
178 G. Märkl and D. Rudnick, J.Organometal.chem., 181, (1979), 305-28.
179 G. Märkl and D. Rudnick, J.Organometal.Chem., 187, (1980), 175-201.
180 G. Märkl and D. Rudnick, Tetrahedron Lett., 21, (1980), 1405-08.
181 P. Jutzi, Chem.Ber., 104, (1971), 1455-67.
182 G.F. Bickelhaupt, C. Jongsma, P. de Koe, R. Lourens, N.R. Mast, G.L. van Mourik, H. Vermeer, and R.J.M. Weustink, Tetrahedron, 32, (1976), 1921-30.
183 J.V. Scibelli and M.D. Curtis, J.Organometal.Chem., 40, (1972), 317-25.
184 S.P. Kolesnikov, A.I. Ioffe, K. Uisassi, I. Tamash, and O.M. Nefedov, Izvest.Akad.Nauk SSSR, Ser.Khim., n° 5, (1977), 1048-51.
185 O.M. Nefedov and S.P. Kolesnikov, Izvest.Akad.Nauk SSSR, Ser.Khim., n° 11, (1971), 2615-16.
186 J.Y. Corey, M. Dueber, and M. Malaidza, J.Organometal.Chem., 36, (1972) 49-60.
187 P. Mazerolles and A. Faucher, Bull.Soc.chim.Fr., n° 6, (1967), 2134-39.
188 P. Mazerolles, A. Faucher, and A. Laporterie, Bull.Soc.Chim.Fr., n° 3, (1969), 887-90.
189 F. Brisse, A. Battat, J.C. Richer, P. Mazerolles, and A. Faucher, Canad.J.Chem., 53 (n° 23), (1975), 3596-8.
190 P. Mazerolles and A. Faucher, J. Organometal.Chem., 63, (1973), 195-203.
191 P. Mazerolles and A. Faucher, J.Organometal.Chem., 85, (1975), 159-163.
192 P. Mazerolles, A. Faucher, P. Mauret, J.P. Fayet, and D. Mermillod-Blardet, J.Chim.Phys., 80 (n° 6), (1983), 553-57.
194 A. Faucher, P. Mazerolles, J. Jaud, and J. Galy, Acta Cryst., B 34, (1978), 442-45.
195 S. Hamann, P. Mazerolles, A. Faucher, and G. Manuel, Austr.J.Chem., 37, (1984), 23-28.
196 P. Mazerolles, A. Faucher, and J.P. Béteille, J.Organometal.Chem., 177, (1979), 163-70.
197 V.I. Davydov, Germanium, Gordon and Breach, New-York, (1967).
198 V.F. Mironov and T.K. Gar, Organogermanium Compounds, Nauka, Moscow, (1967).
199 F. Glockling, The Chemistry of Germanium, Academic Press, New-York, (1969).
200 M. Lesbre, P. Mazerolles, and J. Satgé, The Organic Compounds of Germanium, Wiley, London, (1971).
201 M. Dub, Orgametallic Compounds, Vol 2, Springer Verlag, Berlin, (1967) and first suppl. (1973).
202 E.G. Rochow and E.W. Abel, The Chemistry of Germanium, Tin and Lead, Pergamon Oxford, (1975).
203 F. Glockling, Organogermanium Chemistry, Quart.Rev., 20, (1966), 45-65; B.C. Pant, Cycloalkanes containing Heterocyclic Germanium, Tin and Lead, J.Organometal.Chem., 66, (1974), 321-403; Germanium, Annual Surveys covering the years 1964-78, D. Seyferth, Organometal.Chem.Rev., (B) 1, (1965), 118-124; 2, (1966), 150-161 ; 3, (1967), 214-36 ; E.J. Bulten, Organometal. Chem. Rev.,(B) 4, (1968), 339-358 ; 5, (1969), 663-686 ; 6, (1970),449-485 ; B.C. Pant, Organometal.Chem.Rev., (B) 9, (1972), 205-247 ; J.Organometal.Chem., 48, (1973), 125-181 ; 68, (1974), 221-293 ; 89, (1975), 1-80 ; 119, (1976), 149-228 ;J.Organometal.Chem.Libr., 6, (1978), 251-309 ; D. Quane, J.Organometal. Chem. Libr., 6, (1978), 311-72 ; 8, (1979), 379- 425 ; 10, (1980), 425-483 ; 13, (1982), 487-538.

204 P. Rivière, M. Rivière-Baudet, and J. Satgé, Germanium, vol. 10, in Comprehensive Organometallic Chemistry, G. Wilkinson, Pergamon, Oxford, (1982).
205 F. Glockling, Germanium-Carbon Compounds, in Gmelin Handbook of Inorganic Chemistry, Springer-Verlag, Berlin, in press.

RINGS WITH PHOSPHORUS CARBON MULTIPLE BONDS

E. FLUCK[1] and B. NEUMÜLLER[2]

[1]Gmelin Institute for Inorganic Chemistry and related Areas of the Max Planck Society, P.O. Box 900 467, Varrentrappstr. 40-42, 6000 Frankfurt/Main 90 (West Germany)

[2]Department of Chemistry, Philipps University, Hans-Meerwein-Straße, 3550 Marburg (West Germany)

1. THREE-MEMBERED RINGS

1.1 PHOSPHIRANES, DIPHOSPHIRANES AND 1H-PHOSPHIRENES

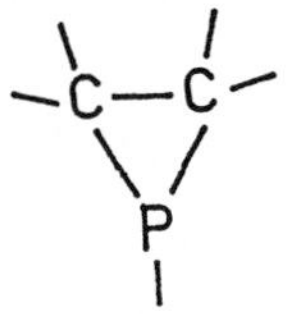
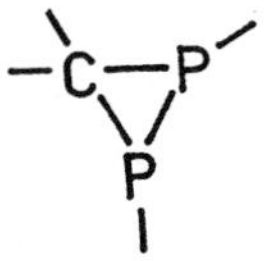
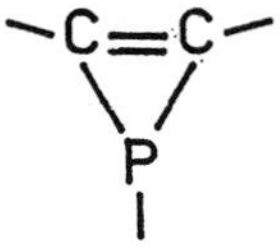

1.2 2H-PHOSPHIRENES

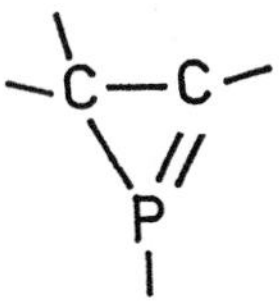

2. FOUR-MEMBERED RINGS

2.1 PHOSPHETANES, 1,2- AND 1,3-DIPHOSPHETANES

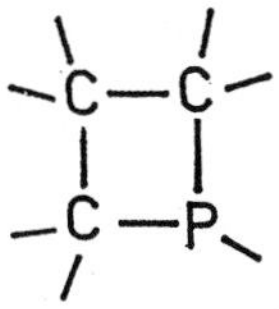
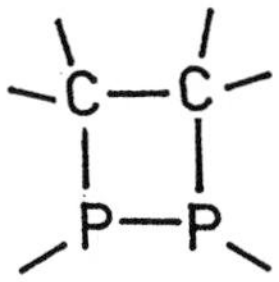
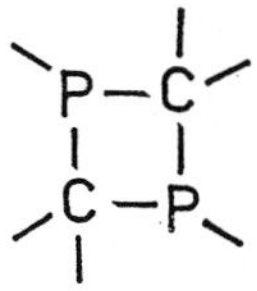

2.2 2-PHOSPHETENES

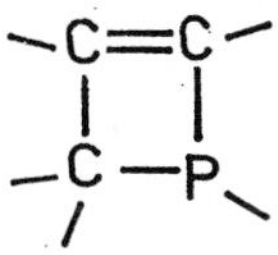
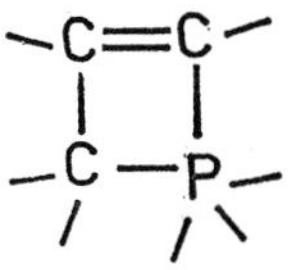

2.3 PHOSPHETES

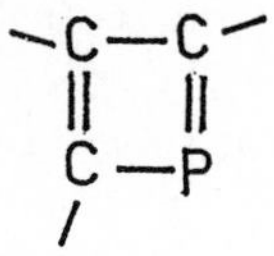

2.4 1,2-DIPHOSPHETENES

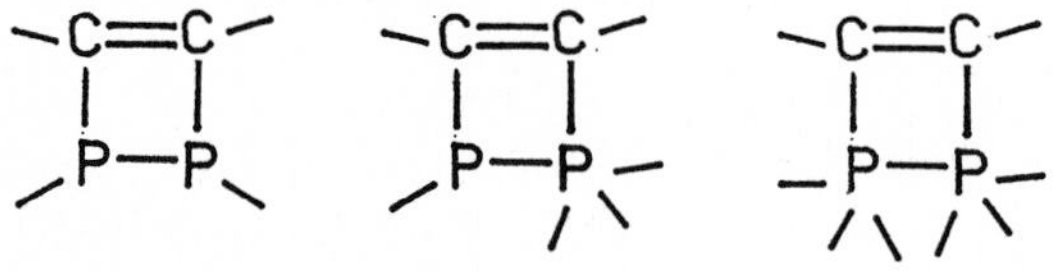

2.5 1,3-DIPHOSPHETENES

2.5.1 $1\lambda^3,3\lambda^3$-Diphosphetenes

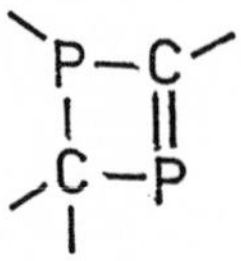

2.6 1,3-DIPHOSPHETES

2.6.1 $1\lambda^3,3\lambda^3$-Diphosphetes

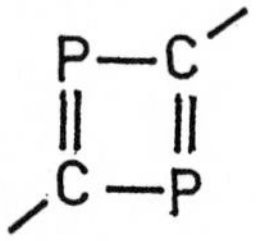

2.6.2 $1\lambda^5,3\lambda^5$-Diphosphetes

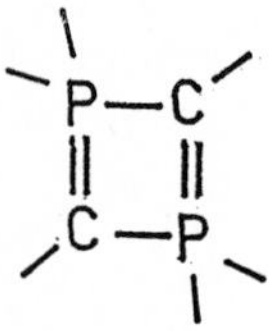

2.7 TRIPHOSPHACYCLOBUTENES

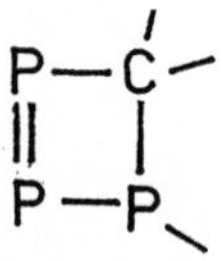

3. FIVE-MEMBERED RINGS

3.1 PHOSPHOLANES, DI-, TRI-, AND TETRAPHOSPHOLANES

3.1.1 Phospholanes

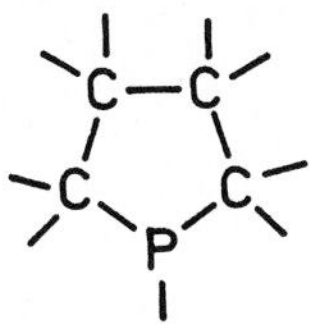
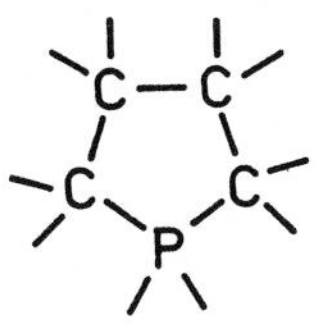
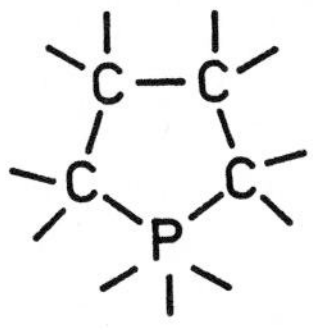

3.1.2 Diphospholanes

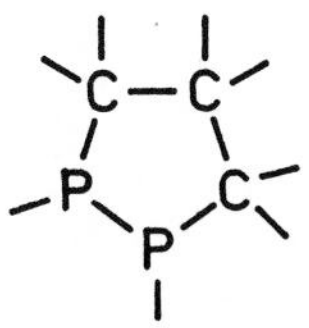
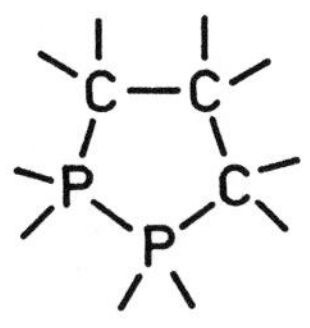
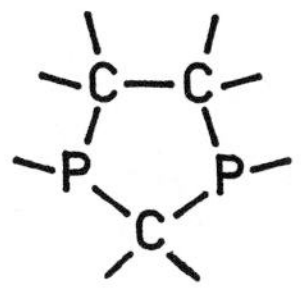
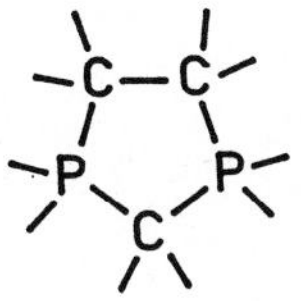

3.1.3 Triphospholanes

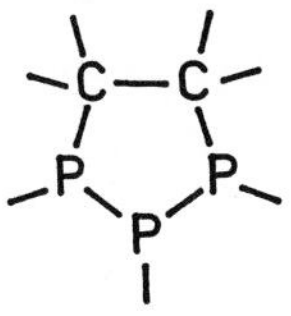
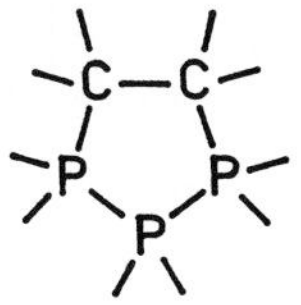
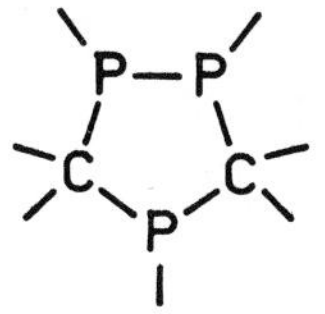

3.1.4 Tetraphospholanes

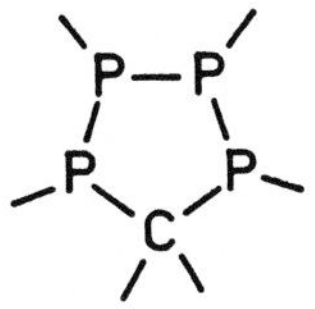
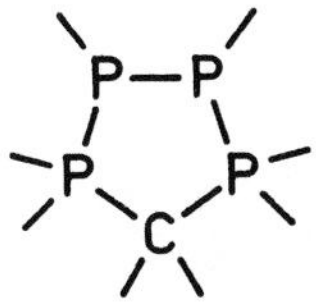

3.2 PHOSPHOLENES AND PHOSPHOLES

3.2.1 Phospholenes

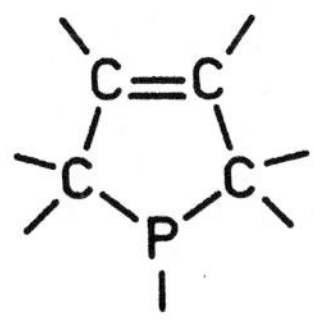
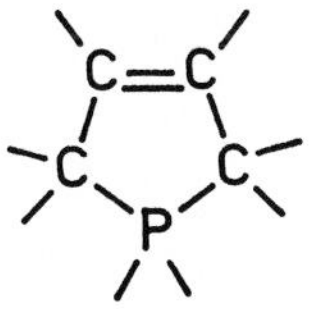
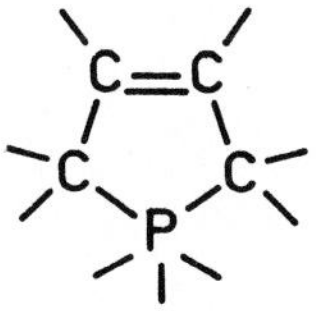

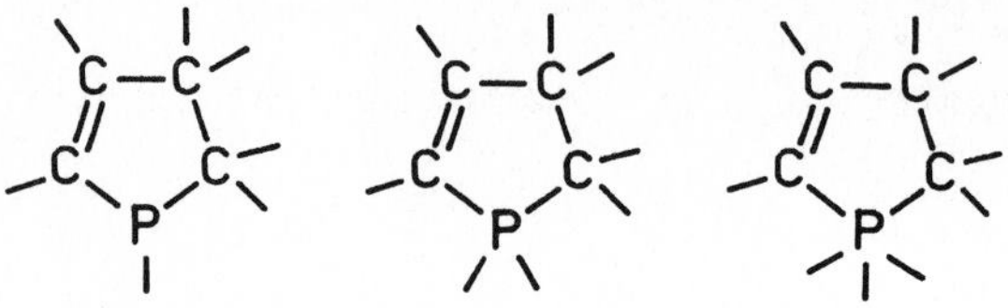

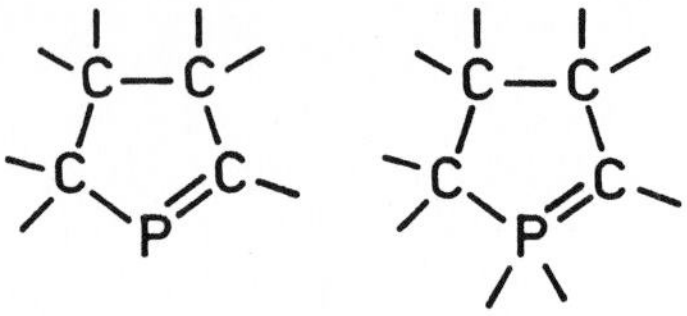

3.2.2 Phospholes

3.2.2.1 $\lambda^3\sigma^3$-, $\lambda^5\sigma^4$-, and $\lambda^5\sigma^5$-Phospholes

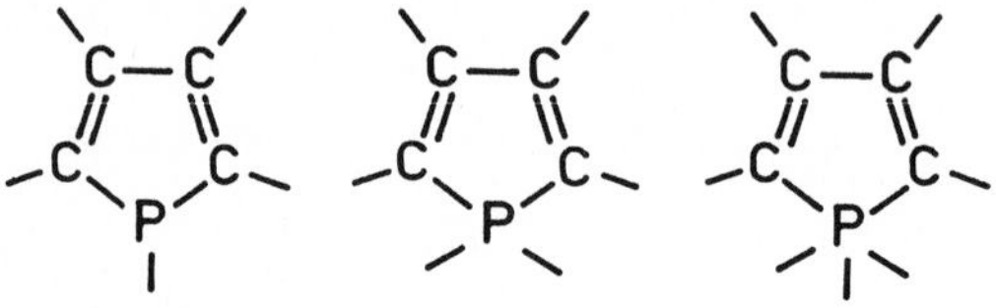

3.2.2.2 $\lambda^3\sigma^2$-Phospholes

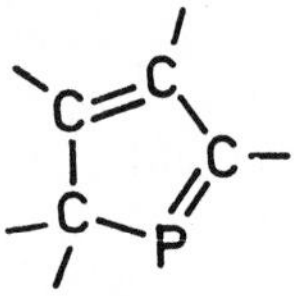

3.2.2.3 $\lambda^5\sigma^4$-Phospholes with P-C-Multiple Bonds

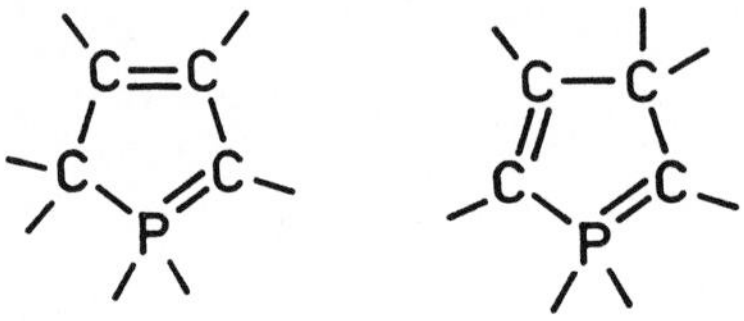

3.2.3 Phospholyl Anions

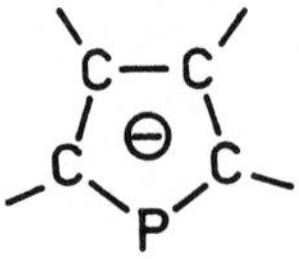

3.3 DIPHOSPHOLENES AND DIPHOSPHOLES

3.3.1 Diphospholenes

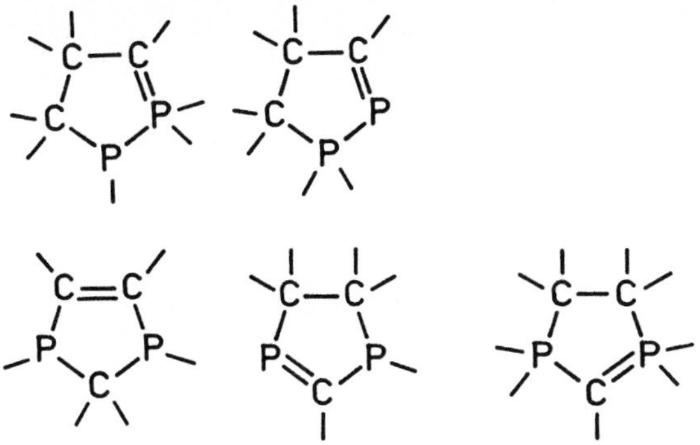

3.3.2 Diphospholes

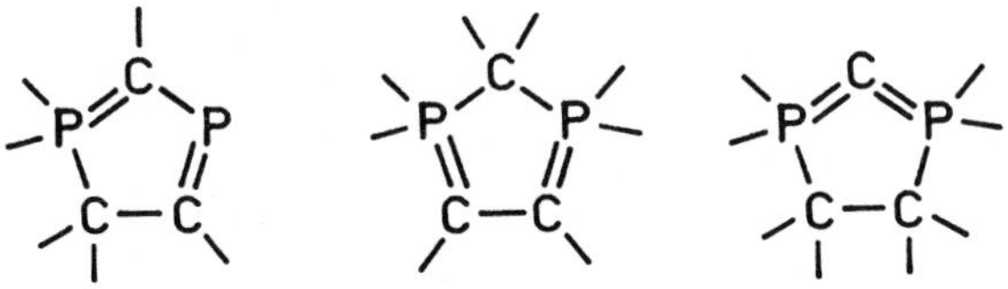

3.3.3 Diphospholyl Anions

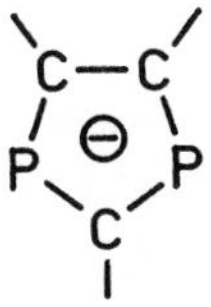

3.4 TRIPHOSPHOLENES AND TRIPHOSPHOLES

3.4.1 Triphospholenes

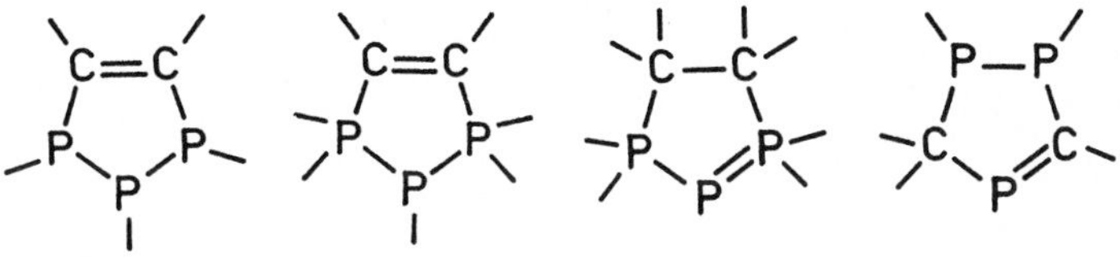

3.4.2 Triphospholes and Triphospholyl Anions

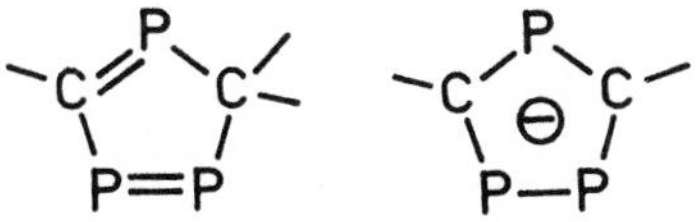

3.5 TETRAPHOSPHOLENES AND TETRAPHOSPHOLES

3.5.1 Tetraphospholyl Anions

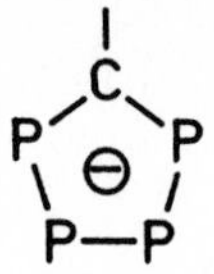

4. SIX-MEMBERED RINGS

4.1 λ^3-PHOSPHORINES

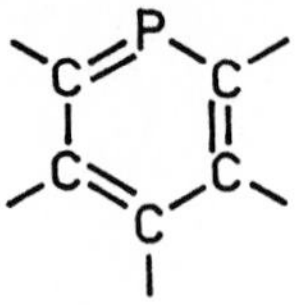

4.2 λ^5-PHOSPHORINES

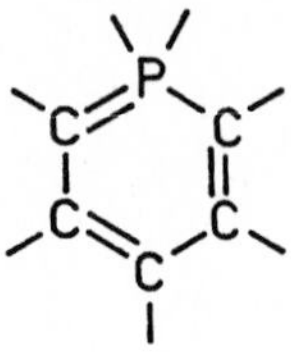

4.3 $1\lambda^3,4\lambda^3$-DIPHOSPHORINES

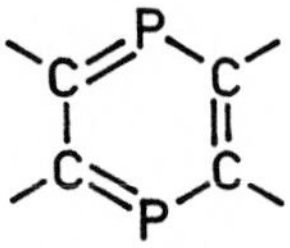

4.4 $1\lambda^5,4\lambda^5$-DIPHOSPHORINES

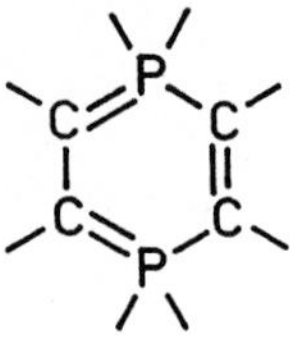

4.5 $1\lambda^5,3\lambda^5$-DIPHOSPHORINES

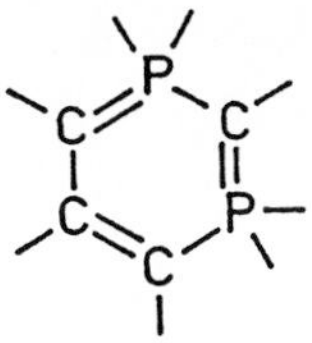

4.6 $1\lambda^5,3\lambda^5,5\lambda^3$-TRIPHOSPHORINES

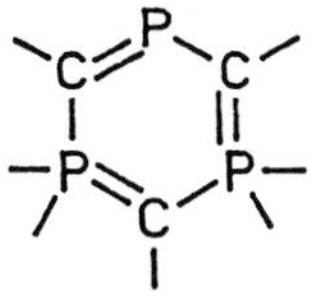

The following chapter reviews all unsaturated phospha-heterocycles in which at least one of the phosphorus atoms is participating in multiple bonds if classical structural formulae are written. In ring systems of this type phosphorus can show oxidation number III and coordination number 2 as well as oxidation number V and coordination number 4. The material is arranged according to increasing ring size, increasing number of phosphorus atoms in the ring and increasing oxidation numbers of phosphorus. For systematic reasons rings, which are not within the scope of the title are mentioned and partly referenced, but not treated in detail.

1. THREE-MEMBERED RINGS

1.1 PHOSPHIRANES, DIPHOSPHIRANES AND 1H-PHOSPHIRENES

Saturated three-membered rings with one or two phosphorus atoms such as phosphiranes with phosphorus of oxidation number III (1a) and V (1b) or diphosphiranes (2) are well known. The chemistry of phosphirenes, however, has only recently been developed. Although the treatment of 1H-phosphirenes is outside of the scope of this article, the various types of compounds known are listed together with corresponding references.

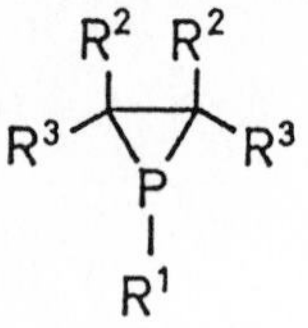

R^1 = H, alkyl, aryl, NR_2
R^2, R^3 = H, alkyl, aryl, $SiMe_3$

1a

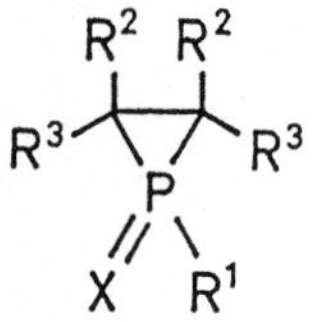

R^1 = alkyl, NR_2
X = O, S, NR
R^2, R^3 = H, alkyl, aryl

1b

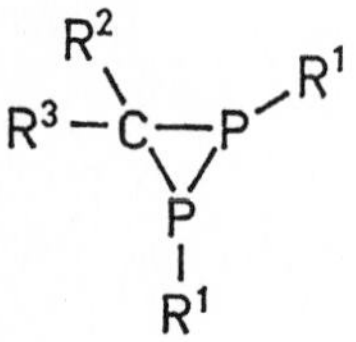

R^1 = alkyl, aryl, NR_2
R^2 = H, alkyl

2

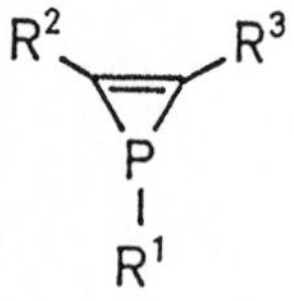

$R^1 = R^2 = R^3$ = Ph (refs. 1,9)
R^1 = Ph; $R^2 = R^3$ = Et (refs. 1,14)
R^1 = Me; $R^2 = R^3$ = Ph (ref. 1)
R^1 = Bu^t; $R^2 = R^3$ = Ph (ref. 1)
R^1 = C(Et)=C(Et)Cl; $R^2 = R^3$ = Et (ref. 1)
R^1 = Me; $R^2 = R^3$ = Et (ref. 6)
R^1 = Bu^t; $R^2 = R^3$ = Et (ref. 6)

3

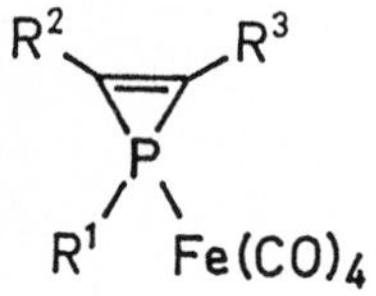

$R^1 = R^2 = R^3$ = Ph (refs. 9,14)

4

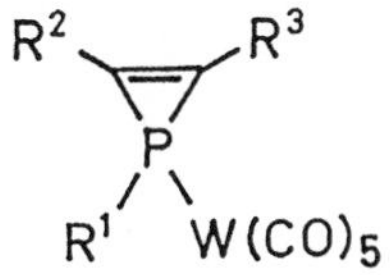

5

$R^1 = R^2 = R^3 = Ph$ (ref. 7)
$R^1 = R^2 = Ph$; $R^3 = COOEt$ (ref. 14)
$R^1 = CH_2CH_2Cl$; $R^2 = R^3 = Ph$ (ref. 4)
$R^1 = CH_2CH_2Cl$; $R^2 = R^3 = Et$ (ref. 4)
$R^1 = CH_2CH_2Cl$; $R^2 = Ph$; $R^3 = H$ (ref. 4)
$R^1 = CH_2\text{-}CH{=}CH_2$; $R^2 = R^3 = Ph$ (ref. 3)
$R^1 = CH_2\text{-}CH{=}CH_2$; $R^2 = R^3 = Et$ (ref. 3)
$R^1 = CH_2\text{-}CH{=}CH_2$; $R^2 = Ph$; $R^3 = Me$ (ref. 3)
$R^1 = CH_2\text{-}CH_2\text{-}CH{=}CH_2$; $R^2 = Ph$; $R^3 = Me$ (ref. 3)
$R^1 = C(Et){=}C(Et)Cl$; $R^2 = R^3 = Et$ (ref. 1)

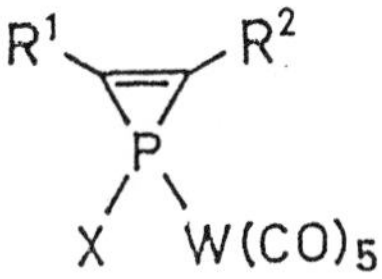

6

$X = NEt_2$; $R^1 = Ph$; $R^2 = H$ (ref. 2)
$X = NEt_2$; $R^1 = R^2 = Ph$ (ref. 2)
$X = Cl$; $R^1 = Ph$; $R^2 = H$ (ref. 2)
$X = OMe$; $R^1 = R^2 = Ph$ (ref. 5)
$X = Cl$; $R^1 = R^2 = Ph$ (ref. 4)
$X = Cl$; $R^1 = R^2 = Et$ (ref. 4)

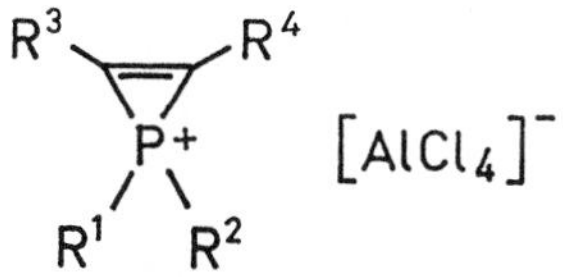

7

$R^1 = R^2 = R^3 = R^4 = Ph$ (ref. 13)
$R^1 = R^2 = R^3 = R^4 = Et$ (ref. 13)
$R^1 = R^2 = Et$; $R^3 = R^4 = Ph$ (ref. 13)

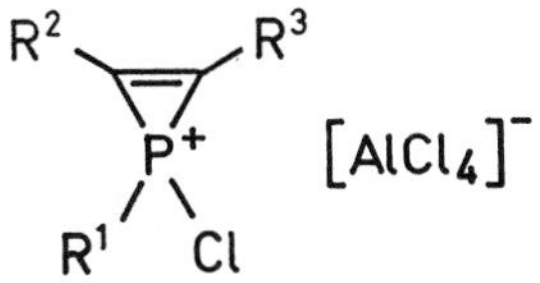

8

$R^1 = R^2 = R^3 = Ph$ (ref. 1)
$R^1 = Ph$; $R^2 = R^3 = Et$ (ref. 1)
$R^1 = Me$; $R^2 = R^3 = Ph$ (ref. 1)
$R^1 = Bu^t$; $R^2 = R^3 = Ph$ (ref. 1)
$R^1 = C(Et){=}C(Et)Cl$; $R^2 = R^3 = Et$ (ref. 1)
$R^1 = Me$; $R^2 = R^3 = H$ (ref. 8)
$R^1 = R^2 = R^3 = Me$ (ref. 8)

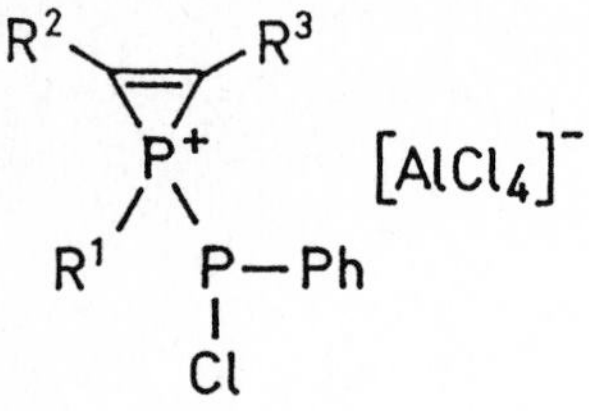

9

R^1 = Me; R^2 = R^3 = Et (ref. 6)
R^1 = Bu^t; R^2 = R^3 = Et (ref. 6)

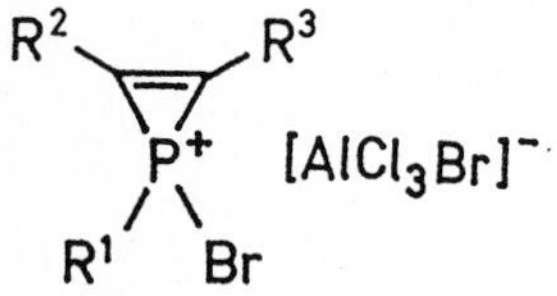

10

R^1 = R^2 = R^3 = Ph (ref. 14)

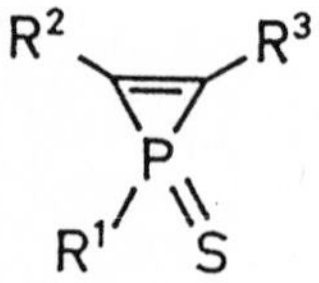

11

R^1 = R^2 = R^3 = Ph (refs. 9, 14)

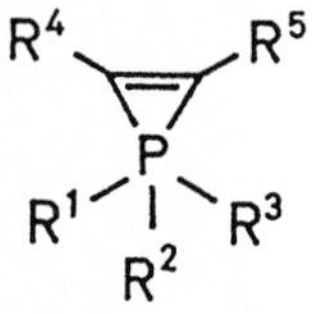

12

R^1 = R^2 = R^3 = Bu; R^4 = OEt; R^5 = H (ref. 10)
R^1 = R^2 = R^3 = Bu^n; R^4 = Ph; R^5 = H (ref. 11)
R^1 = R^2 = R^3 = R^4 = Ph; R^5 = Cl (ref. 12)

It may be pointed out that the compounds of type 12 are not well established and characterized. The material on phosphirenes has been excellently reviewed by Mathey (refs. 15, 16). In another article the relations between phosphirenes and R_2Si  is described (ref. 17). The metal complexes of phosphirenes have extensively been discussed (refs. 18, 19).

1.2.2 2H-Phosphirenes

According to what has been said as yet 1H-phosphirenes A

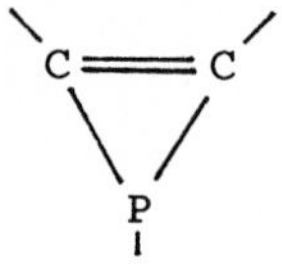

A

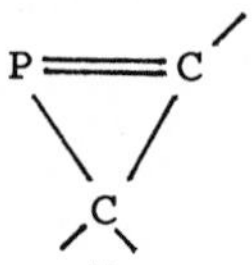

B

can be considered to be a well known class of compound. In contrast, 2H-phosphirenes B were unknown until the first example has been prepared by Regitz et al. in 1987 (ref. 20). 2,2'-Dimethylpropylidynephosphane (t-butylphosphaacetylene) 13 reacts with diazocyclohexane, 14, under [3+2]-cycloaddition. The spirocyclic 3H-1,2,4-diazaphosphol 15 is thermally unstable and is converted already at room temperature into 4H-1,2,4-diazaphosphole 16. Photolysis of 15 results in the formation of a mixture of the 2H-phosphirene 19 and the 1-phospha-1-cyclopentene 20. 19 shows a chemical shift $\delta(^{31}P)$ = 71.7 ppm. With $W(CO)_5$(thf) in tetrahydrofurane (thf) the pentacarbonyltungsten complex of 19 is formed, i.e. compound 21, (m.p. 61°C). Its chemical shift was measured to be $\delta(^{31}P)$ = 123 ppm.

The corresponding (pentacarbonyl)chromium complex was also isolated (m.p. 44°C; $\delta(^{31}P)$ = 107.2 ppm). In 21 the P=C bond distance is very short (163.4 pm) compared with other free and ligated phosphaalkenes. The P-C bond distance in the three-membered ring is about 3 pm longer than the $P-C_{sp^3}$ bond in phosphaalkenes.

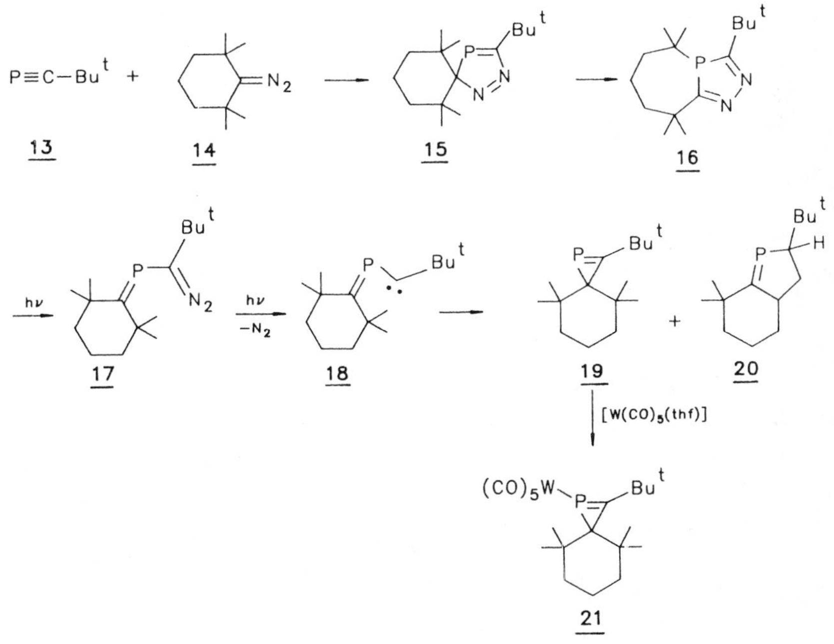

2. FOUR-MEMBERED RINGS

2.1 PHOSPHETANES, 1,2- AND 1,3-DIPHOSPHETANES

Saturated four-membered monophospha and diphospha heterocycles are known in great number. A selection of 1,2- and 1,3-diphosphetanes is shown by the following examples:

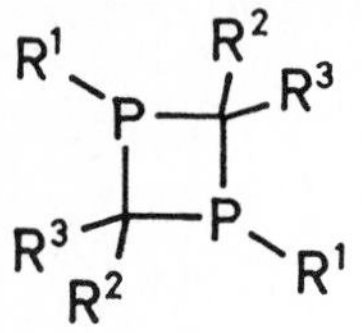

1

R^1 = alkyl, aryl
R^2 = H, alkyl, aryl
R^3 = aryl, NMe_2, $OSiMe_3$

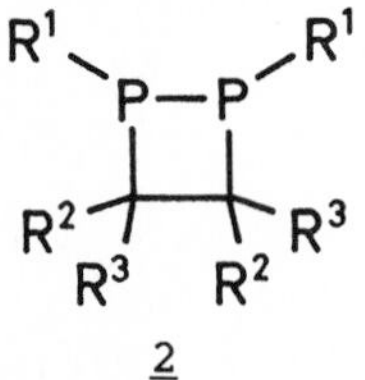

2

R^1 = H, alkyl, aryl, Hal
R^2 = H, alkyl, aryl, Hal
R^3 = aryl, $OSiMe_3$, NMe_2, Hal

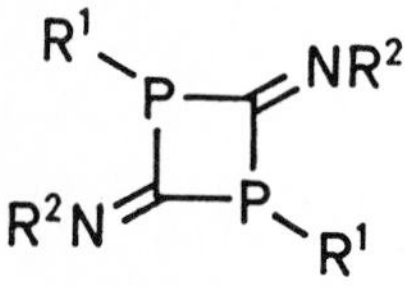

3

R^1 = alkyl, aryl
R^2 = aryl

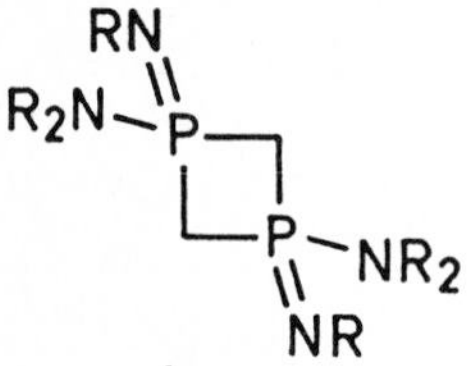

4

R = $SiMe_3$

2.2 2-PHOSPHETENES

The chemistry of 2-phosphetenes has been studied in great detail in recent years. The following list is aimed to allow an easy access to the literature on these compound classes although they are not the subject of this chapter in a strict sense.

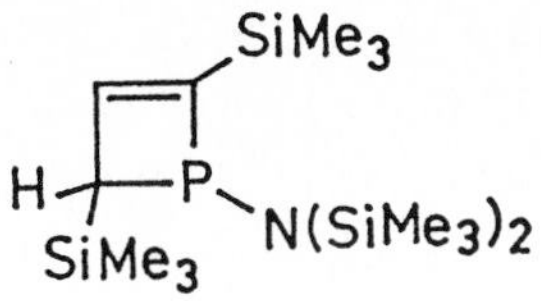

(ref. 22)

5

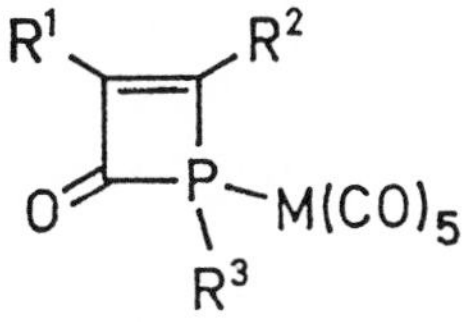

6

$R^1 = R^2 = R^3 = Ph$; M = W (ref. 21)
$R^1 = R^2 = Et$; $R^3 = Ph$; M = W (ref. 21)
$R^1 = R^2 = R^3 = Ph$; M = Cr (ref. 21)
$R^1 = R^2 = Et$; $R^3 = Ph$; M = Cr (ref. 21)
$R^1 = R^2 = R^3 = Ph$; M = Mo (ref. 21)

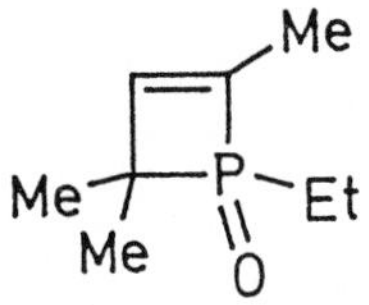

7

(ref. 29)

2.3 PHOSPHETES

Free phosphetes (monophosphacyclobutadienes) are unknown as yet.

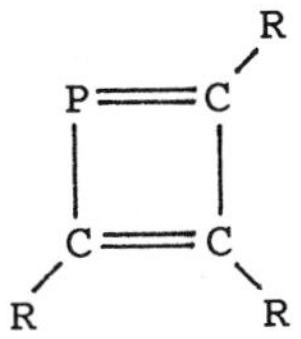

A cobalt(I) complex 8, however, was synthesized by codimerization of t-butylphosphaacetylene with bis(trimethylsilyl)acetylene in the presence of equimolar amounts of η^5-cyclopentadienediethenecobalt(I) (ref. 32). 2,4-Di-t-butyl-1,3-diphosphete-cyclopentadienylcobalt(I) is a by-product besides η^5-cyclopentadienyl-η^4-tetra-(trimethylsilyl)-cyclobutadiene-cobalt(I). It is assumed that the dicobalt-complex is formed as an intermediate for all products. 8 is a red oil, showing a singlet at 3.3 ppm in the ^{31}P-NMR spectrum.

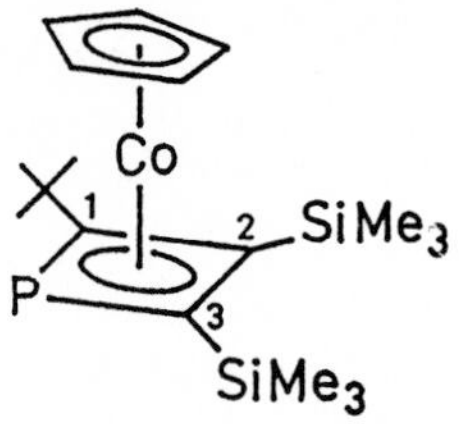

8

In the ^{13}C-NMR spectrum the η^4-bond phosphacyclobutadiene rest shows signals at 116.2 [C1, ^1J(PC) 43.5 Hz], 87.3 [C2, ^2J(PC) 1.5 Hz] and 72.6 ppm [C3, ^1J(PC) 49.8 Hz].

2.4 1,2-DIPHOSPHETENES

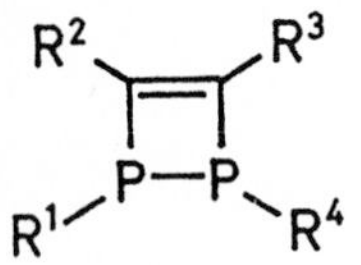

9

$R^1 = R^2 = R^3 = R^4 = CF_3$ (ref. 23)
$R^1 = R^2 = R^3 = R^4 =$ Ph (refs. 24,26)
$R^1 = R^4 = Bu^t$; $R^2 = R^3 = OSiMe_3$ (ref. 25)
$R^1 = R^4 =$ Ph; $R^2 = R^3 = Bu^t$ (ref. 26)
$R^1 = R^4 =$ Ph; $R^2 = R^3 =$ Et (ref. 26)
$R^1 =$ Ph; $R^2 = R^3 = Bu^t$; $R^4 =$ H (ref. 27)
$R^1 =$ Ph; $R^2 = R^3 = Bu^t$; $R^4 = CH_2Ph$ (refs. 27,28)
$R^1 = R^2 = R^4 =$ Ph; $R^3 =$ Me (ref. 28)
$R^1 = R^2 = R^4 =$ Ph; $R^3 = SiMe_3$ (ref. 28)
$R^1 = R^2 = R^4 =$ Ph; $R^3 = Bu^t$ (ref. 28)
$R^1 = R^2 =$ Ph; $R^3 =$ Me; $R^4 = Bu^t$ (ref. 28)
$R^1 =$ Ph; $R^2 = R^3 = Bu^t$; $R^4 =$ Me (ref. 28)
$R^1 =$ Ph; $R^2 = R^3 = Bu^t$; $R^4 = (CH_2)_3Cl$ (ref. 28)
$R^1 =$ Ph; $R^2 = R^3 = Bu^t$; $R^4 = (CH_2)_4Cl$ (ref. 28)
$R^1 =$ Ph; $R^2 = R^3 = Bu^t$; $R^4 = (CH_2)_5Cl$ (ref. 28)
$R^1 =$ Ph; $R^2 = R^3 = Bu^t$; $R^4 = CH(Ph)CH_2Ph$ (ref. 28)
$R^1 =$ Me; $R^2 = R^3 = R^4 =$ Ph (ref. 6)
$R^1 = Bu^t$; $R^2 = R^3 = R^4 =$ Ph (ref. 6)
$R^1 = R^4 =$ Me; $R^2 = R^3 =$ Ph (ref. 6)

R^1 = Me; R^2 = R^3 = Et; R^4 = Ph (ref. 6)
R^1 = Ph; R^2 = R^3 = Et; R^4 = Me (ref. 6)
R^1 = Bu^t; R^2 = R^3 = Et; R^4 = Ph (ref. 6)
R^1 = Ph; R^2 = R^3 = Et; R^4 = Bu^t (ref. 6)
R^1 = Me; R^2 = R^3 = Et; R^4 = Bu^t (ref. 6)

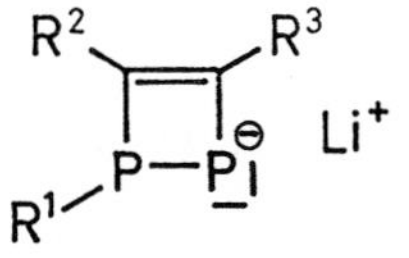

10 — R^1 = Ph; R^2 = R^3 = Bu^t (refs. 27,28)

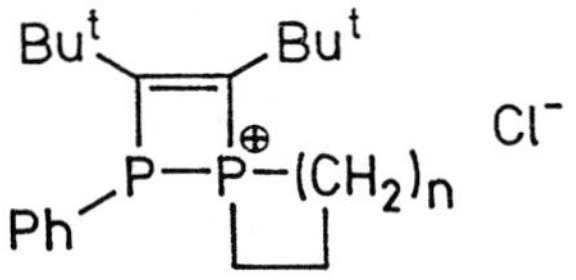

11 — n = 2,3 (ref. 28)

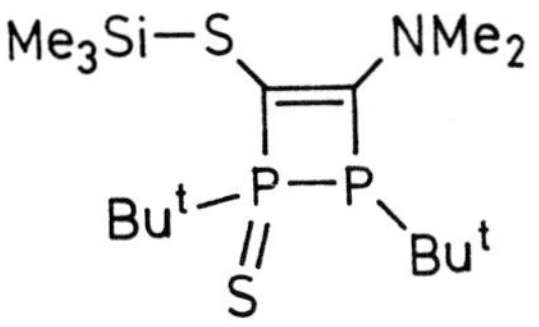

12 — (refs. 30, 31)

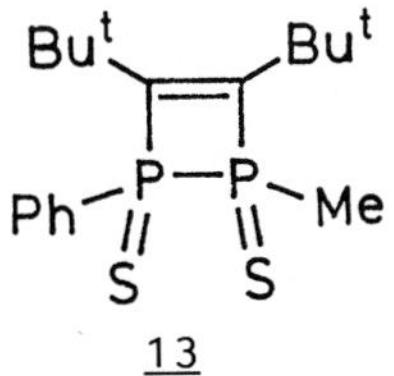

13 — (ref. 28)

2.5 1,3-DIPHOSPHETENES

2.5.1 $1\lambda^3,3\lambda^3$-Diphosphetenes

The formation of 1,3-diphosphetenes was observed in investigations of compounds with low coordination numbers of phosphorus. The reaction of the P-silylated phosphaalkene A in its keto and enol form with phosgen or an isocyaniddichloride yield the diphosphacyclobutenes 14 and 15 (ref. 33).

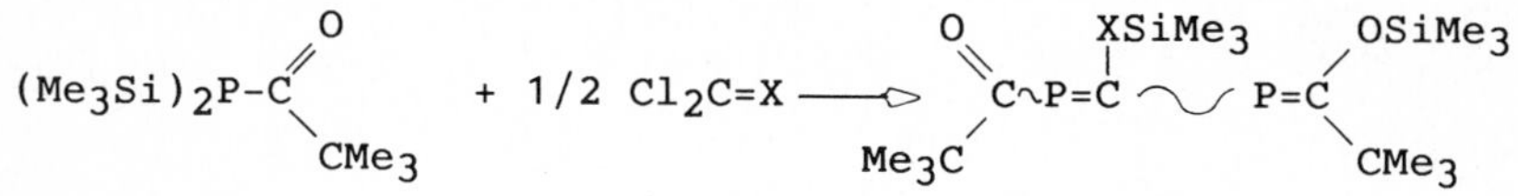

A B

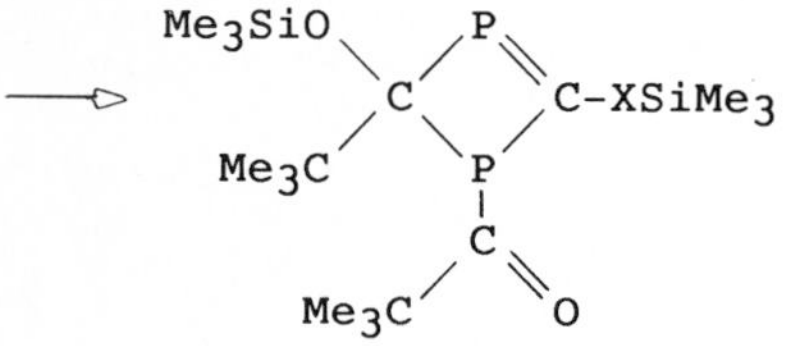

14, 15

14 X = O

15 = $N(C_6H_4CF_3-(2))$

The formation of the diphosphacyclobutenes 14 and 15 can best be understood by assuming an intramolecular [2+2] cycloaddition of the intermediate phosphabutadienes B although it has not been possible to identify the latter by spectroscopic means or addition of dienophiles. Only one of the stereoisomers is formed. 14 has a melting point of 49°C and a boiling point of 88°C (10^{-3} Torr). The chemical shifts $\delta(^{31}P)$ were measured to be 104.1 ppm [d,J(CPC) = 16.2 Hz (C-P-C)] and 279.1 ppm [d,P=C]. The molecular structure of 14 was determined by X-ray analysis and is shown in Fig. 1. The most important distances and angles in the molecule are given in Table 1. The dihedral angle between the planes P2C1P1 and C2P1C1 is 7.5°.

TABLE 1

Bond distances and angles in 14.

distance	[pm]	angle	[°]
P1C1	167.9(7)	C1P1C2	82.4(5)
P2C1	191.2(7)	C1P2C2	80.2(5)
P2C1	179.2(7)	P2C1P1	104.3(5)
P2C2	189.0(7)	P1C2P2	92.3(5)

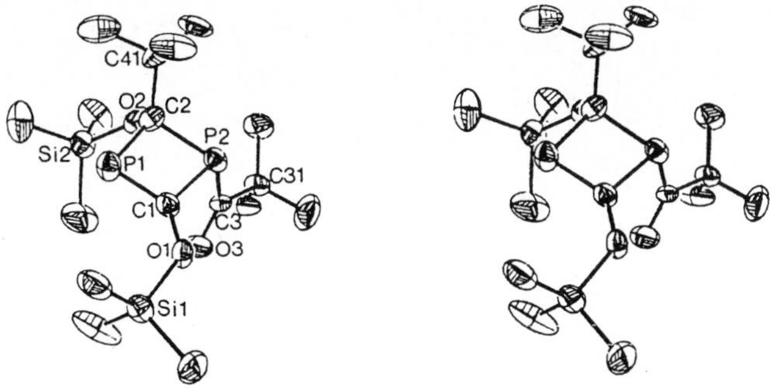

Fig. 1. Molecular structure of <u>14</u>.

The $1\lambda^3,3\lambda^3$-diphosphetene <u>16</u> is formed in the reaction between 2,2-dimethylpropyldynephosphine, $Me_3C\text{-}C{\equiv}P$ and $GeCl_4$ (refs. 34,35). The initial product is a mixture of E/Z-isomeric chloro[1-(trichlorogermyl)alkylidene]-phosphine which reacts with further $Me_3C\text{-}C{\equiv}P$ to give the product <u>16</u>.

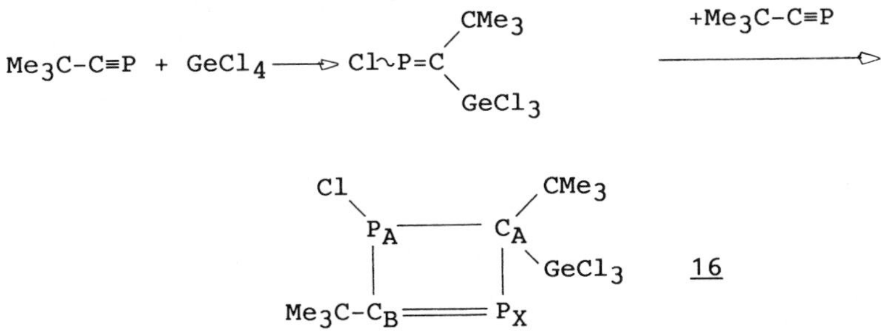

<u>16</u>

The ^{31}P-NMR spectrum is of AX type with chemical shifts of 271 ppm (P_X) and 27.4 ppm (P_A). The ^{13}C-NMR spectrum gives chemical shifts of 245.8 ppm (C_B) and 71 ppm (C_A) and the following coupling constants: $^1J(C_BP_A)$ = 54.5 Hz, $^1J(C_BP_X)$ = 61 Hz, $^1J(C_AP_A)$ = 43 Hz and $^1J(C_AP_X)$ = 42 Hz. Structural data of <u>16</u> are summarized in Table 2.

TABLE 2

Bond distances and angles in 16.

	distance [pm]		angle [°]
P_A-C_B	190	C_B-P_A-C_A	83.8
C_B-P_X	169.4	P_A-C_A-P_X	88.2
P_X-C_A	180	C_A-P_X-C_B	84.4
C_A-P_A	186	P_X-C_B-P_A	97.0
C_A-Ge	197		
P_A-Cl	210		

2.6 1,3-DIPHOSPHETES

While 1,3-diphosphetenes contain only one phosphorus atom with multiple bonding, the class of 1,3-diphosphetes contains two. They have only recently been synthesized.

2.6.1 $1\lambda^3,3\lambda^3$-Diphosphetes

$1\lambda^3,3\lambda^3$-diphosphetes are unknown. They can, however, be

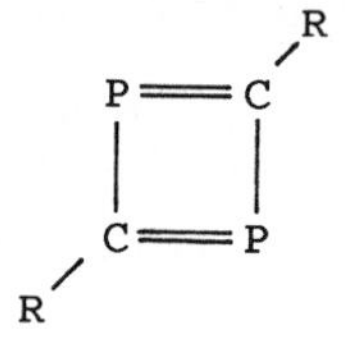

synthesized as ligands in complexes. Displacement of ethylene from cyclopentadienyl(diethene)metal(I) complexes $[M(\eta^5\text{-}C_5H_5)(C_2H_4)_2]$, (R = H, M = Co or Rh; R = Me, M = Co, Rh, or Ir) by treatment with 2,2-dimethylpropylidynephosphane, $Bu^tC{\equiv}P$, affords complexes containing 2,4-di-t-butyl-$1\lambda^3,3\lambda^3$-diphosphete as ligands (ref. 36). The reaction is carried out at room temperature in toluene solution and yields the red-orange complexes 17 - 21. Some of their physical data are given in Table 3.

TABLE 3
Physical Data of Complexes of 1λ^3,3λ^3-diphosphetes.

	δ(^{31}P)[ppm]	^{1}J(PRh)[Hz]	m.p.[°C]	
17	39.0			(ref. 36)
17	38.1			(ref. 37)
18	48.9	31.7		(ref. 36)
19	27.0		92-94 (decomp.)	(ref. 36)
20	38.6	29.3		(ref. 36)
21	32.0			(ref. 36)

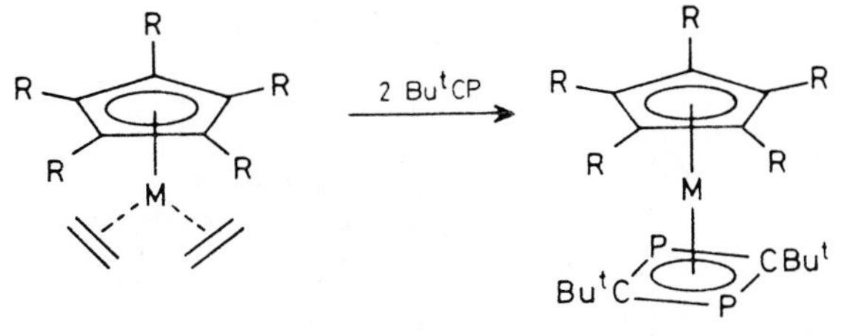

17 R = H, M = Co
18 R = H, M = Rh
19 R = Me, M = Co
20 R = Me, M = Rh
21 R = Me, M = Ir

Fig. 2 shows the molecular structure of [Co(η^5-C_5Me_5)(η^4-($Bu^tCP)_2$)]. The η^4-1λ^3,3λ^3-diphosphete ring is essentially planar and all the P-C bond lengths are equivalent.

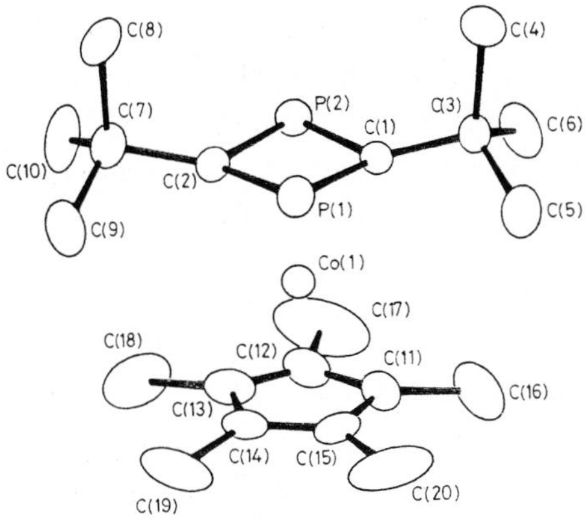

Fig. 2. The molecular structure of 19. Important bond distances (in pm) and angles (in °) are: P(1)-C(1) 180(1), P(1)-C(2) 179(1), P(2)-C(1) 182(1), P(2)-C(2) 180(1), Co(1)-P(1) 224,0(3), Co(1)-P(2) 224.4(4), Co(1)-C(1) 209(1), Co(1)-C(2) 208(1); P(1)-C(1)-P(2) 98.0(5), P(1)-C(2) 98.7(5), C(1)-P(1)-C(2) 82.0(5), C(1)-P(2)-C(2) 81.0(5)

The compound has also been obtained by Binger et al. (ref. 37).

Reacting two moles of 2,2-dimethylpropylidynephosphane with two or three moles of η^5-cyclopentadienyl(diethene)cobalt(I) affords the dicobalt complex 17a and the tricobalt complex 17b as red, crystalline compounds. They also contain $1\lambda^3,3\lambda^3$-diphosphetes as ligands. The crystals of 17a melt at 125°C (decomp.) and those of 17b at 135°C (decomp.). The ^{31}P-NMR spectrum of 17b shows a singulet at 85.9 ppm, while the one of 17a consists of two signals, one at -32.0 ppm for the phosphorus atom with coordination number two and another broadened line at 134.9 ppm.

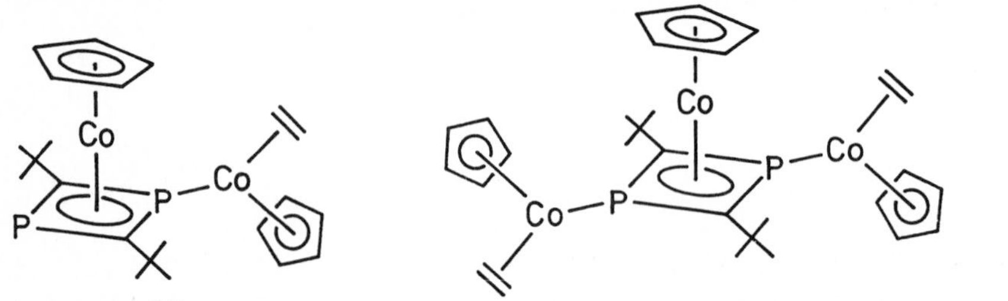

17a   17b

The reaction of 2,2-dimethylpropylidynephosphane (t-butylphosphaacetylene) and bis(ethene)(toluene)iron or (toluene)(1-methylnaphthalene)iron results in the formation of two complexes containing 2,4-di-t-butyl-$1\lambda^3,3\lambda^3$-diphosphete as ligands (ref. 38). Besides these two complexes, (toluene)(2,4-di-t-butyl-$1\lambda^3$, $3\lambda^3$-diphosphete)iron, 22, and (2,4,5-tri-t-butyl-2,3-diphospholyl)(2,4-di-t-butyl-$1\lambda^3,3\lambda^3$-diphosphete)iron, 23, (2,4,5-tri-t-butyl-1,3-diphospholyl)(3,5-di-t-butyl-1,2,4-triphospholyl)-iron, 24, is afforded.

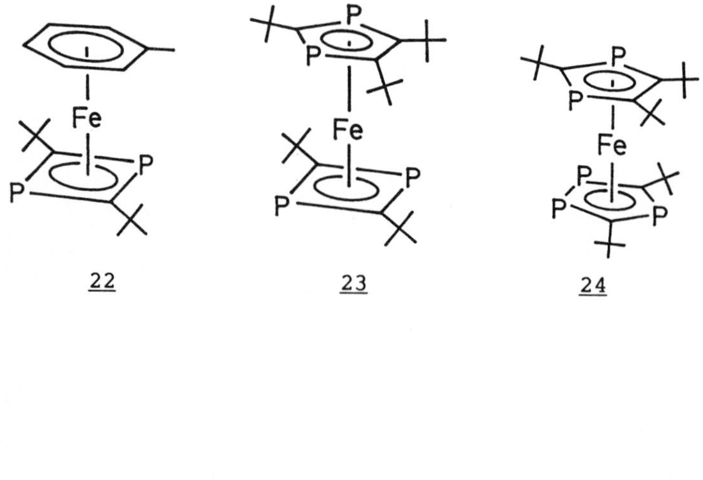

22   23   24

While 22 can be isolated as red crystals, m.p. 57°C, 23 and 24 could not be separated. 22 shows both in CH_2Cl_2 (DCM) and in $MeOCH_2CH_2OMe$ (DME) an unusual electrochemistry. Oxidation occurs reversibly at +0.55 (DCM) and +0.66 V (DME) against an aqueous saturated calomel electrode. Only the second oxidation at +1.50 V (DCM) is irreversible. Reduction at -2.50 V (DME) is irreversible at room temperature but reversible at -60°C. The ions 22^+ and 22^- can be characterized by ESR spectroscopy.

2.6.2 $1\lambda^5,3\lambda^5$-Diphosphetes

The first $1\lambda^5,3\lambda^5$-diphosphete which has been observed can be

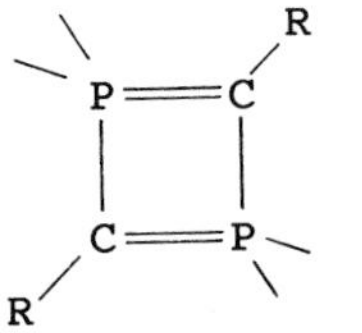

obtained from simple starting materials. Triphenylphosphane and CCl_4 yield (trichloromethyl)triphenylphosphonium chloride which can be reacted with Ph_2PCl to give $[Ph_3P\text{-}C(Cl)\text{-}P(Cl)Ph_2]Cl$ a (ref. 39). With $P(NMe_2)_3$ the thermally unstable intermediate b is formed:

$$\underset{a}{[Ph_3P\text{-}C(Cl)\text{-}P(Cl)Ph_2]Cl} + P(NMe_2)_3 \longrightarrow \underset{b}{Ph_3P{=}C(Cl)\text{-}PPh_2} + (Me_2N)_3PCl_2$$

Upon warming 1,1,3,3-tetraphenyl-2,4-bis(triphenylphosphonio)-$1\lambda^5,3\lambda^5$-diphosphacyclobutadiene dichloride, 25, is the final product:

2 b ⟶ 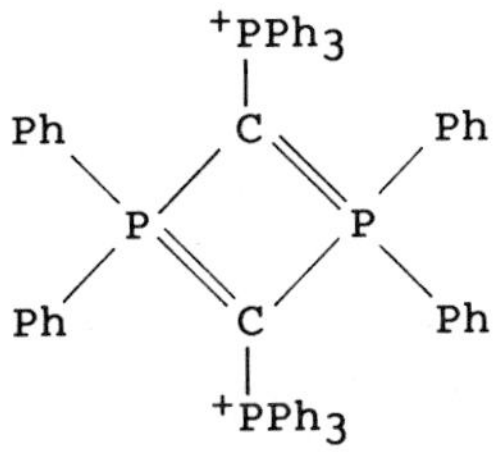

2 Cl⁻ 25

25 melts at 385°C, is soluble in boiling water without decomposition and does not show reactions which are typical for ylids. With elemental bromine the corresponding salt with the

polyhalide ion Br_2Cl is formed, a brilliant yellow solid, m.p. 365°C. The high thermal stability of the cation in 25 and the small polarizing effect on the Br_2Cl^- in spite of its two positive charges indicate that the positive charges are highly delocalized, i.e. the π electron density in the ring is partially removed into the exocyclic P-C bonds.

Comparison with the C-P bond distances for other 1λ^5,3λ^5-diphosphetenes in Table 4 suggests a substantial statistical weight of resonance structures such as Ph_2P^+-C=PPh_3.

In the same way the 1,3-diphosphete 26 was obtained (ref. 40):

$$[(Me_2N)_2]_3P{=}CCl_2 + ClP(NMe_2)_2 \longrightarrow (Me_2N)_3P{=}CCl_2{-}P(NMe_2)_2Cl$$

c

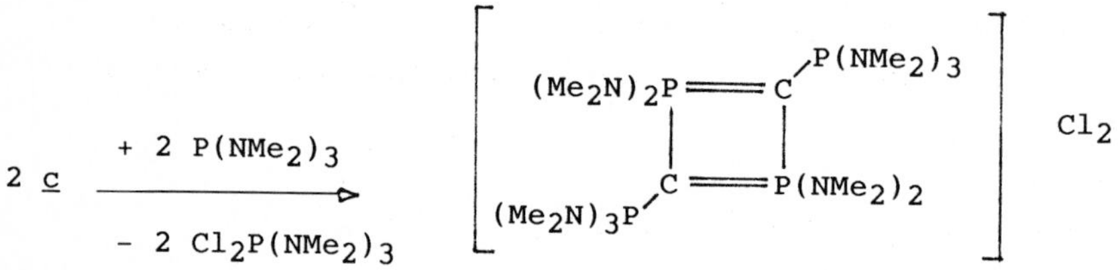

26

Dechlorination of c in CH_2Cl_2 leads to 26, while the reaction in chlorobenzene results in the carbodiphosphorane $(Me_2N)_3P{=}C{=}P(NMe_2)_2Cl$.

The ^{31}P-NMR spectrum of 25 consisting of two triplets at 14.4 and 33.64 ppm [J(PP) = 7.35 Hz] is in agreement with the proposed structural formula which was later confirmed by the results of an X-ray analysis (ref. 41).

An easy access to 1λ^5,3λ^5-diphosphetes was opened by Regitz et al. (ref. 42). Equimolar amounts of chlorophosphanes and diazo compounds of type 28 react in dichloromethane at 20°C to give the 1λ^5,3λ^5-diphosphetes 33a-c.

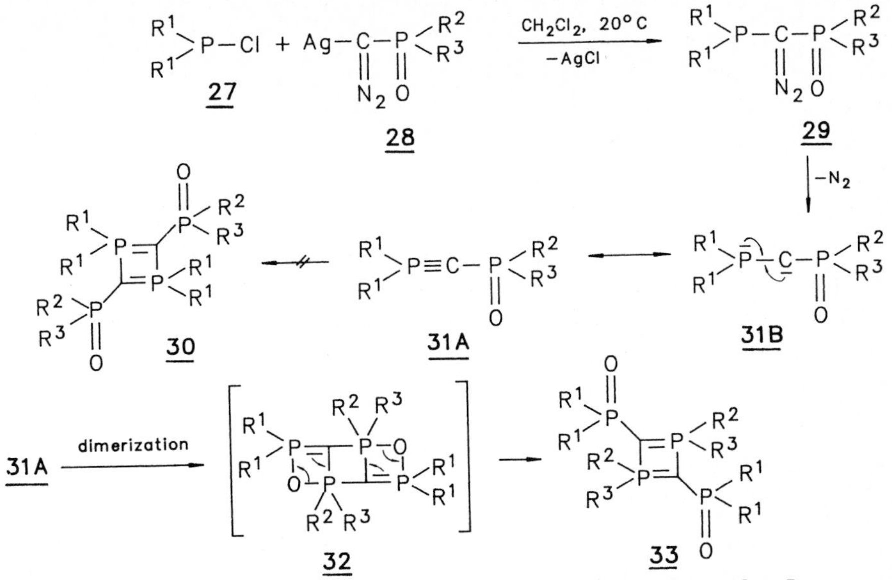

27	R^1
a	NPr^i_2
b	Ph

28	R^2	R^3
a	Ph	Ph
b	Ph	OMe

29–33	R^1	R^2	R^3
a	NPr^i_2	Ph	Ph
b	Ph	Ph	Ph
c	NPr^i_2	Ph	OMe

In the first step of the reaction the phosphino diazo compounds are formed. Spontaneous N_2-elimination yields the phosphaalkynes 31, which dimerize. In this process the ring and exocyclic phosphorus atoms have apparently interchanged their substituents. It can be rationalized by assuming a cycloaddition process involving the P≡C bonds and PO groups of two molecules 31. The formation of the heterocyclic intermediates 32 is followed by ring opening giving rise to the final products.

The ^{31}P-NMR spectra of the exocyclic and ring phosphorus atoms show similar chemical shifts, so that assignments were not made. The data for compounds 33a-c are ($CDCl_3$): 33a δ = 21.1 and 20.8 ppm, 33b δ = 31.6 and 19.3 ppm, 33c δ = 39.6 and 24.3 ($^2J(PP)$ = 6.6 Hz).

The hydrolysis of <u>33b</u> and <u>33c</u> gives

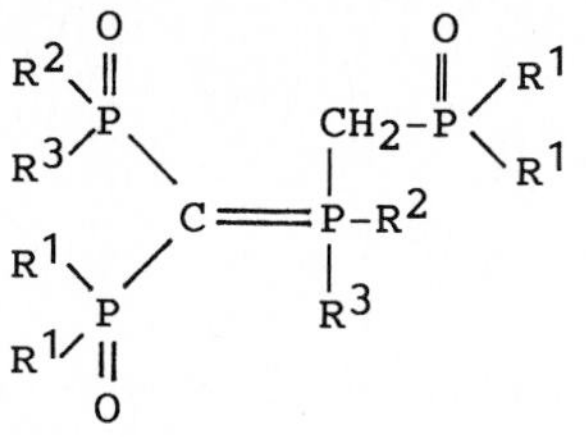

<u>34</u>

Structural data for <u>33a</u> show that the molecule is centrosymmetric. The four-membered ring is practically planar. The exocyclic and endocyclic P-C bond lengths are equal within their standard distances. The virtual identity is considered to be due to contribution from a dipolar resonance structure $Ph_2P\text{-}C\text{=}P^+(O^-)(N\text{-}i\text{-}Pr_2)_2$. Bond distances and angles of the four-membered ring are listed in Table 4.

TABLE 4
Bond distances and angles in $1\lambda^5,3\lambda^5$-diphosphetes in the ring systems (Esd's: 0.2 - 0.3 pm, 0.1 - 0.2°).

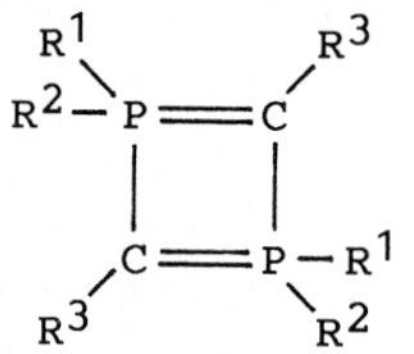

R^1	R^2	R^3	bond distance [pm]		bond angle [°]	
NMe_2	NMe_2	H	P-C	172.5	P-C-P	88.6
					C-P-C	91.4
Ph	Ph	$PPh_3]Cl$	P-C	176.0/176.5	P-C-P	91.7
					C-P-C	88.3
Ph	Ph	$P(O)(NPr^i{}_2)_2$				
			P-C	174.7/175.1	P-C-P	89.5
					C-P-C	90.5
NEt_2	NEt_2	C_6H_5	P-C	174.6/174.1	P-C-P	90.5
					C-P-C	89.5

The first $1\lambda^5,3\lambda^5$-diphosphete unsubstituted at the carbon atoms was synthesized by Svara and Fluck by reaction of methylenefluorobis(dimethylamino)phosphorane with n-butyllithium (ref. 43). Phosphorus-fluoro-ylids had been prepared for the first time in 1985 (ref. 44). In contrast to the phosphorus-fluoro-ylids, the phosphorus-chloro-ylids were previously known (ref. 45) and the reaction between these compounds and alkyllithiums such as methyllithium or butyllithium was studied, although not very exhaustively. The reaction occurs in such a manner that the chlorine atom will be substituted by formation of a phosphorus-carbon bond. It was expected that methylene-fluoro-bis(dimethylamino)phosphorane would behave differently from the phosphorus-chloro-ylids owing to the smaller steric screening of the carbon atom and the greater stability of the phosphorus-fluorine bond in comparison to the phosphorus-chlorine bond. Besides the formation of a phosphorus-carbon bond in the reaction with alkyllithiums which was just mentioned, a metal-halogen exchange could be envisaged, and this in fact occurs very rapidly large, easily polarizable halogens such as bromine or iodine attached to phosphorus in the ylid. Finally an exchange between hydrogen and metal can occur leading to oligo- or polymerization products via intermediates.

If methyl-difluorobis(dimethylamino)phosphorus is reacted at -95°C in n-pentane with butyllithium primarily the phosphorus-ylid

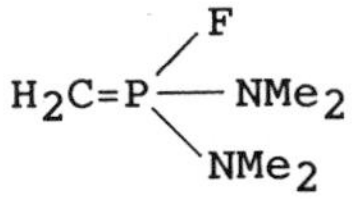

is formed. Further reaction with butyllithium yields 1,1,3,3-tetrakis(dimethylamino)-$1\lambda^5,3\lambda^5$-diphosphete 35. The first step in the reaction occurs by attack of butyllithium on the α-methylene group with formation of the metal-ylid. Then either a double intermolecular fission of lithium fluoride with formation of a second phosphorus-carbon bond, or an intramolecular elimination of lithium fluoride with formation of a second phosphorus-carbon bond, or an intramolecular elimination of lithium fluoride with formation of a system with a phosphorus-carbon triple bond follows, i.e. a σ^3/λ^5 species that subsequently dimerizes.

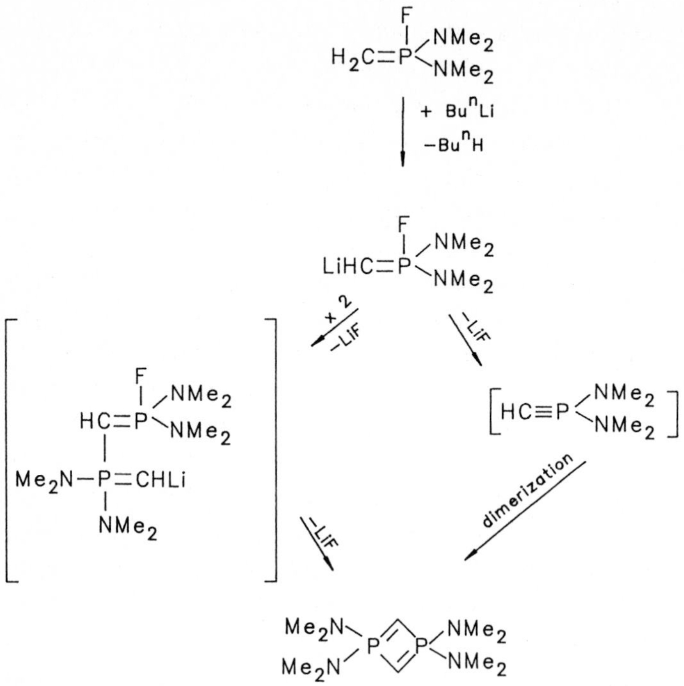

<u>35</u>

The molecular structure of the diphosphete <u>35</u> is shown in Fig. 3. The significant characteristic is the four-membered ring with C_i symmetry. It is completely planar with equal phosphorus-carbon bond length of 172.5 pm. This value lies in the range of phosphorus-carbon double bonds in phosphaalkenes (ref. 46) with typical values 168 and 172 pm or in the range of phosphorus-carbon bond distances in substituted alkylidene-phosphoranes (ref. 47), and is considerably smaller than the bond distances in 1,2-diphosphetenes with sp^2-hybridized carbon atoms, where the distance is about 185 pm. The intramolecular distances between the two carbon atoms and the two phosphorus atoms, 73 and 65 % of the van der Waals radii, respectively, are extremely short and surely associated with a large ring strain.

The planarity of 35 and the fact that the phosphorus-carbon bonds are equal within the error limits, was at first surprising because there is no possiblity of removing electrons from the ring system. The structural analysis does, for instance, not give a clue to a mesomeric exchange between the phosphorus and the exocyclic dimethylamino groups.

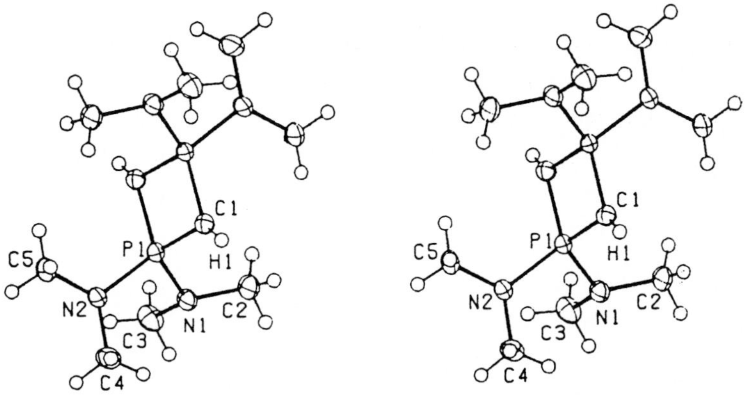

Fig. 3. Molecular structure of λ^5-diphosphete 35

The first results of the calculations, which were made on (ref. 48), or more precisely, on the unsubstituted tetraamino-diphosphete, show that the highest occupied molecular orbital are localized on the carbon atoms. They lie considerably higher on the energy scale than the four linear combinations of the nitrogen atoms. An MNDO calculation shows that the phosphorus atoms carry nearly a full positive charge, whereas the negative charge on the carbon atoms amounts to about -0.73.

Reaction of Bis(diethylamino)fluormethylidenphosphorane, $H_2C{=}PF(NEt_2)_2$, in n-pentane with an equimolar amount of n-butyl-lithium in n-hexane leads in the analogous way to 1,1,3,3-tetrakis(diethylamino)-$1\lambda^5,3\lambda^5$-diphosphete (ref. 49):

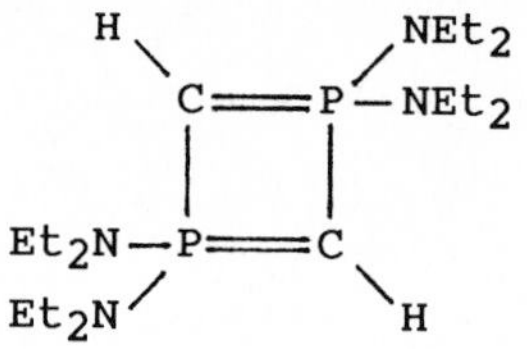

<u>36</u>

forms pale yellow crystals which are sensitive towards oxygen and moisture, m.p. 28°C. The ^{31}P-NMR signal of a solution of <u>36</u> appears at 43 ppm.

2,4-Dimethyl-1,1,3,3-tetrakis(dimethylamino)-1λ^5,3λ^5-diphosphete <u>37</u> was synthesized in a similar way starting with ethylbis(dimethylamino)fluorophosphonium iodide. With the double molar amount of n-butyllithium can be isolated in 19 % yield (ref. 50).

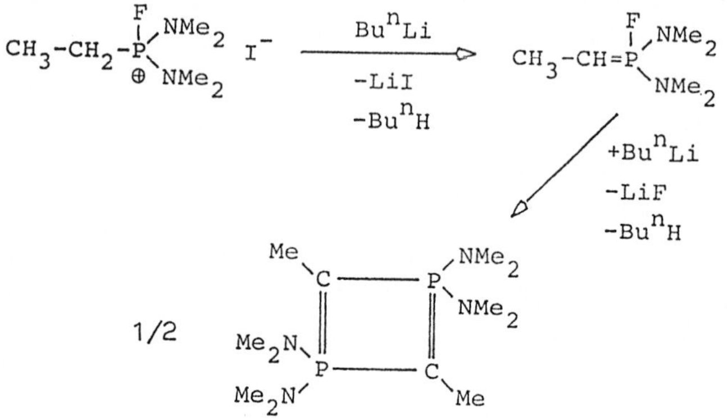

<u>37</u>

<u>37</u> forms ivory colored crystals which melt at 92-95°C which are soluble in n-pentane/ether. In the ^{31}P-NMR spectrum the resonance signal appears at 45 ppm.

The corresponding 2,4-dimethyl-1,1,3,3-tetrakis(diethylamino)-1λ^5,3λ^5-diphosphete <u>38</u> forms, when bis(diethylamino)-fluoroethylidene phosphorane is reacted with the equimolar amount of n-butyllithium (ref. 49):

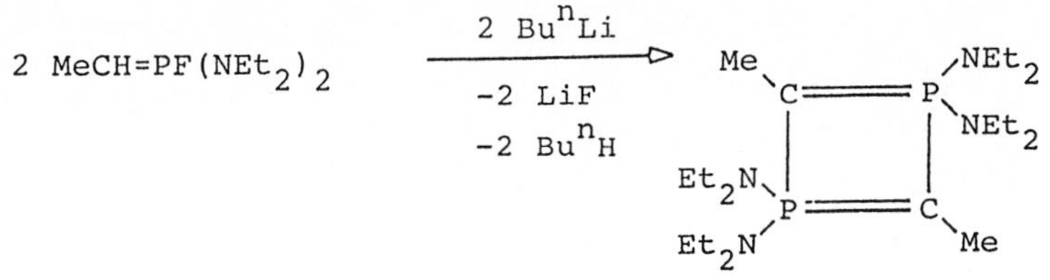

38

The yield is 34 %. The pale yellow crystals can be recrystallized from n-pentane. 38 melts at 31 - 32 °C and has a chemical shift $\delta(^{31}P)$ = 42,6 ppm.

Reacting benzylbis(diethylamino)difluorophosphorane with the double molar amount of n-butyllithium at -70°C in n-pentane leads to 2,4-diphenyl-1,1,3,3-tetrakis(diethylamino)-1λ^5,3λ^5-diphosphete 39 (refs. 49,50), a yellow solid which can be recrystallized from toluene. The yellow crystals are rather stable towards oxygen and moisture and melt at 240°C. The melt is green. The ^{31}P-NMR spectrum of 39 in C_6D_6 shows a singlet at δ = 34,6 ppm.

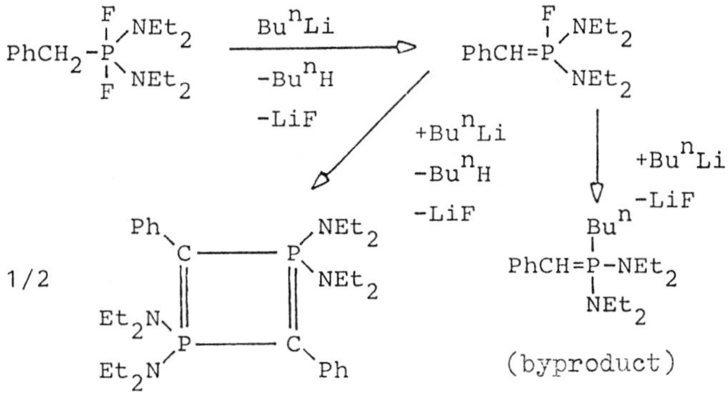

39

The results of an X-ray analysis show 39 to have the molecular symmetry C_i with equal bond distances P1-C1 174.6(2) and P1-C1'174.1(2) pm (ref. 49). The ring is completely planar. The corresponding bond length in the C unsubstituted compound 35 is shorter since no exocyclic substituents are present to accept π electron density. With increasing delocalization of π electron density the P-C bond distance in the rings of diphosphetes increase as is seen from the compilation of bond lengths in Table 4. The largest value is found for 25.

Remarkable in 39 are the intramolecular contact distances between the two carbon atoms and the two phosphorus atoms. With 245.4 pm and 247.7 pm they are extremely short and amount to only 72 and 67 %, respectively, of the van der Waals radii (refs. 51,52). Fig. 4 shows the molecular structure of 39. The phenyl substituents are nearly co-planar with the central ring system. The bond distance between the C atom of the diphosphete and the C atom of the phenyl ring attached to the four-membered ring is signi-ficantly shorter than a C-C single bond.

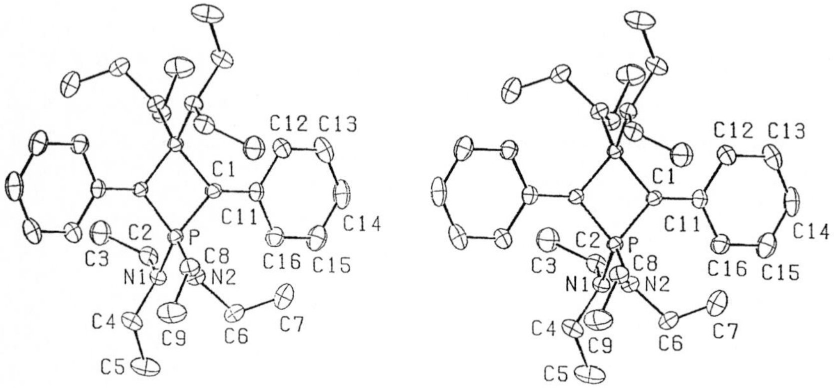

Fig. 4. Molecular structure of 39.

TABLE 5
Bond distances [pm] and angles [°] in 39.

P1-C1	174.6(2)	C1-P1-C1'	89.5(1)
P1-C1'	174.1(2)	P1-C1-P1'	90.5(1)
P1-C11	143.8(2)	P1-C1-C11	134.3(1)
		P1'-C1-C11	135.2(1)

Diethylamino-benzyl-benzylidenfluorophosphorane 40 reacts with lithium bis(trimethylsilyl)amide to form a mixture of compounds the main products being the phosphirane and 1,3-bis(diethylamino)-1,3-dibenzyl-2,4-diphenyl-1λ^5,3λ^5-diphosphete (yield 15.3 %, ref. 251).

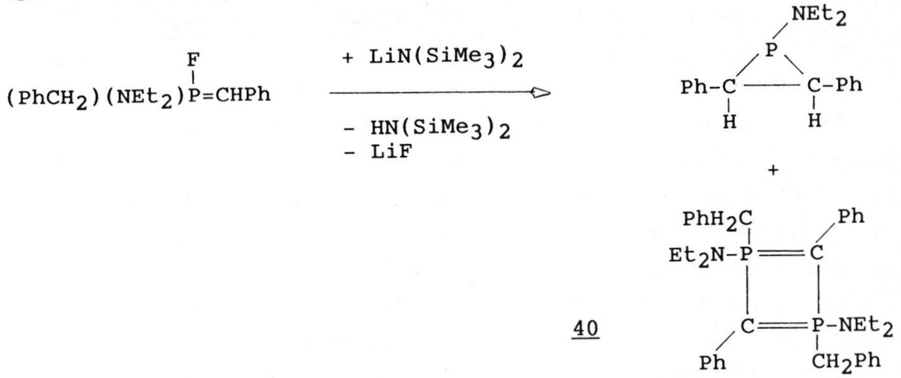

The yellow crystals of 40 melt at 214-217°C and are rather stable in air. The ^{31}P-NMR spectrum of 40 shows only one signal indicating that only one of the two possible isomers

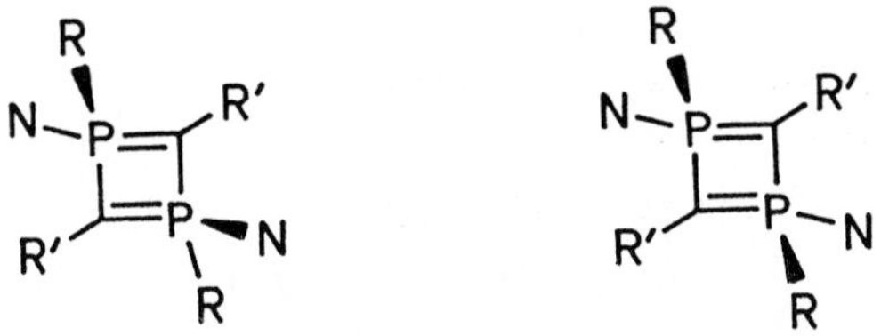

is formed in the reaction described above. The ^{31}P- and ^{13}C-NMR data of 40 is compared with those of other diphosphetes in Table 6 and Table 7.

TABLE 6
NMR data of the ring atoms in λ^5-diphosphetes.

	$\delta(^{31}P)$ [ppm]	$\delta(^{13}C{=}P)$ [ppm]	$^1J(^{13}C{=}P)$ [Hz]	refs.
40	20.6	47.3	99.0	251
39	34.6	50.8	114.9	40,50
35	48.9	8.2	97.3	44
37	44.9	12.2	109.3	50

TABLE 7

^{31}P-NMR chemical shifts of diphosphetes

	Compound	$\delta(^{31}P)$[ppm]	refs.
35	$[HCP(NMe_2)_2]_2$	48,9	43
36	$[HCP(NEt_2)_2]_2$	43,4	49
37	$[MeCP(NMe_2)_2]_2$	44,9	50
38	$[MeCP(NEt_2)_2]_2$	42,6	49
39	$[PhCP(NEt_2)_2]_2$	34,6	49,50
33a	$[Pr^i{}_2N)_2P(O)CPPh_2]_2$	21,2 a)b)	42
33b	$[Ph_2P(O)CPPh_2]_2$	31,6 a)b)	42
33c	$[(Pr^i{}_2N)_2P(O)CP(OMe)Ph]_2$	39,8 a)b)	42
25	$(Ph_3PCPPh_2)_2Cl_2$	14,4 or 33,64 d)	39
26	$[(NMe_2)_3PCP(NMe_2)_2]_2Cl_2$	37,89 or 42,59 d)	40
17	$\{Co(\eta^5-C_5)[\eta^4-(Bu^tCP)_2]\}$	38,1; 39,0 c)	36,37
18	$\{Rh(\eta^5-C_5H_5)[\eta^4-(Bu^tCP)_2]\}$	48,9 c)	36
19	$\{Co(\eta^5-C_5Me_5)[\eta^4-(Bu^tCP)_2]\}$	27,0 c)	36
20	$\{Rh(\eta^5-C_5Me_5)[\eta^4-(Bu^tCP)_2]\}$	38,6 c)	36
21	$\{Ir(\eta^5-C_5Me_5)[\eta^4-(Bu^tCP)_2]\}$	32,0 c)	36

a) chemical shift of the ring phosphorus atom

b) assignment not ascertained

c) calculated from: $\delta[P(OMe)_3]$ = 140 ppm

d) not assigned

2.7 TRIPHOSPHACYCLOBUTENES

Nucleophilic cleavage of white phosphorus with sodium in diglyme results in the formation of the triphosphacyclobutadienide ion (ref. 53)

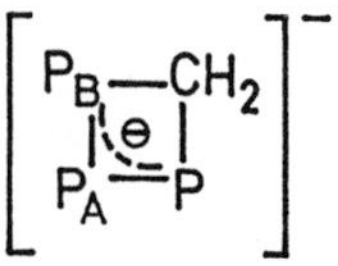

Besides, the pentaphosphacyclopentadienide anion, P_5^-, the tetraphosphacyclopentadienide ion $P_4(CH)^-$ and other polyphosphides are formed. Sodium triphosphacyclobutadienide has not been isolated. The ^{31}P-NMR spectrum of in diglyme consists of an

AB_2-system with the following parameters: $\delta(P_A)$ = 272.0, $\delta(P_B)$ = 262.9 ppm, $J(P_AP_B)$ = -484.7 Hz. These data indicate that the π-bond and negative charge is delocalized over the P_3 fragment of the ring.

3. FIVE-MEMBERED RINGS

3.1 PHOSPHOLANES, DI-, TRI-, AND TETRAPHOSPHOLANES

Although the subject of this article are the unsatured phosphorus-carbon heterocycles, well-known satured ring systems of this kind shall be mentioned. A review until 1981 is given in ref. 2.

3.1.1 Phospholanes

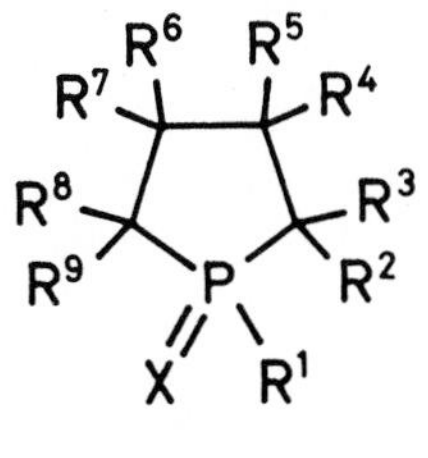

1a

X = lone pair, O, S, $=CH_2$, $W(CO)_5$
R^1 = H, alkyl, aryl, halogen, OR, SiR_3, GeR_3, $P(S)R_2$, $OP(CO)R_2$
R^2, R^3, R^8, R^9 = H, alkyl, aryl, halogen, OR, CO_2R, $P(O)R_2$
R^4, R^5, R^6, R^7 = H, alkyl, halogen, OR, NR_2, NO_2, ONO_2, =O

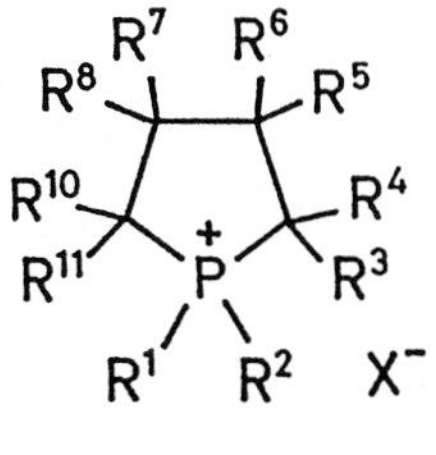

1b

X^- = halogen
R^1, R^2 = alkyl, aryl
R^3, R^4, R^9, R^{10} = H, OH
R^5, R^6, R^7, R^8 = H, alkyl

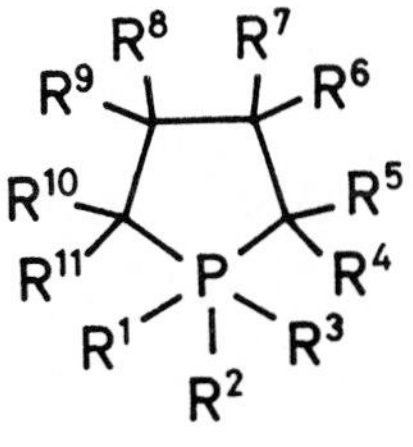

1c

R^1, R^2, R^3 = halogen, alkyl, aryl, OR
R^4, R^5, R^{10}, R^{11} = H
R^6, R^7, R^8, R^9 = H, alkyl

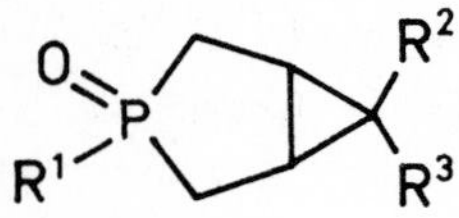

1d

R^1 = alkyl, OH, OR
R^2 = OH
R^3 = CO_2R

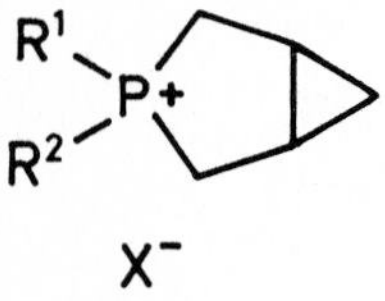

1e

X^- = $AlCl_4^-$
R^1 = Cl
R^2 = NR_2

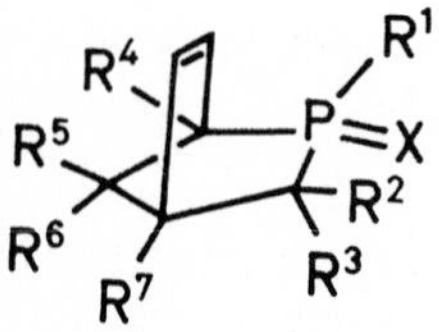

1f

X = lone pair, $W(CO)_5$
R^1 = halogen, alkyl, aryl
R^2 = SiR_3, OR
R^3 = SiR_3, aryl
R^4, R^5, R^6, R^7 = H, alkyl, CO_2R

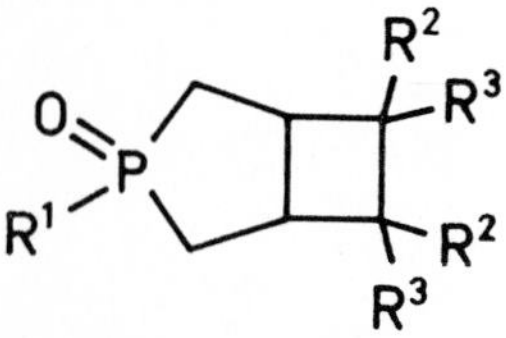

1g

R^1 = alkyl, OR
R^2, R^3 = OR, =O, =N-NPh

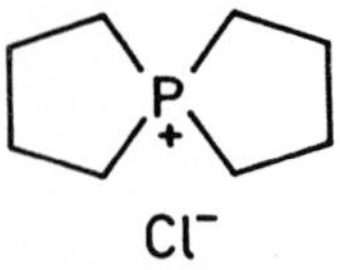

1h

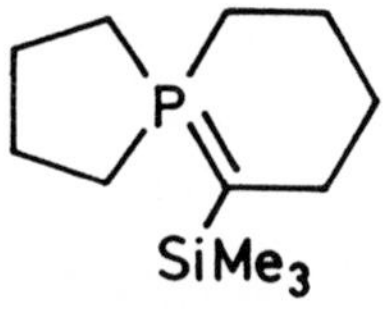

1i

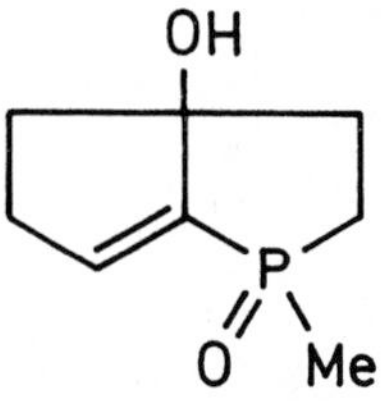

1j

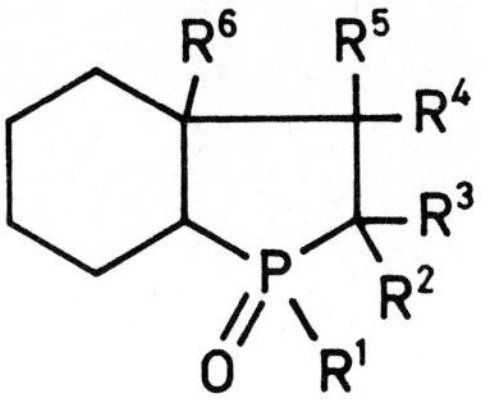

X = lone pair, O
R^1 = alkyl, aryl
R^2, R^3 = H, aryl
R^4, R^5 = aryl, =O
R^6 = H, OH

1k

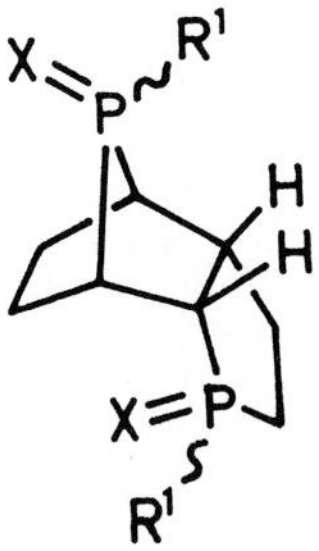

X = lone pair, O, S
R^1 = alkyl, aryl, OR

1l

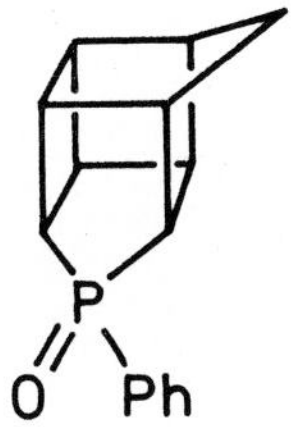

1m

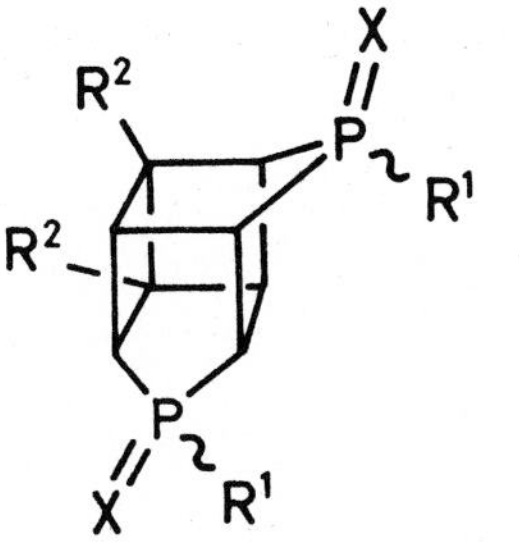

1n

X = lone pair, O
R^1 = H, aryl
R^2 = H, alkyl

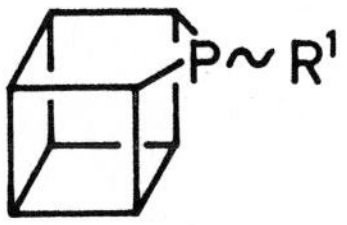

R^1 = alkyl, aryl

1o

3.1.2 Diphospholanes

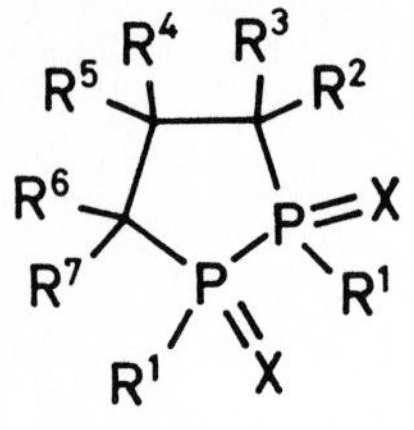

2a

X = lone pair, O
R^1 = H, Li, alkyl, aryl
R^2, R^3, R^6, R^7 = H, CO_2R
R^4, R^5 = H, =O

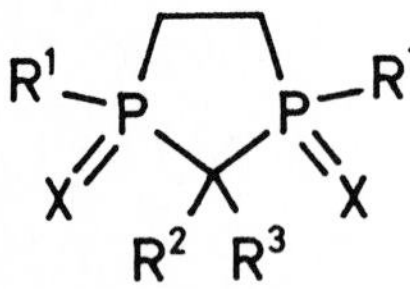

2b

X = lone pair, O
R^1 = alkyl, aryl, OR
R^2, R^3 = H, alkyl, OR, =NR

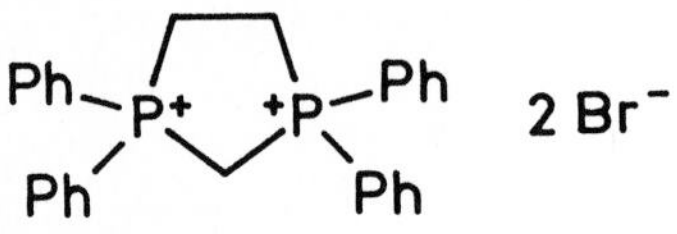

2c

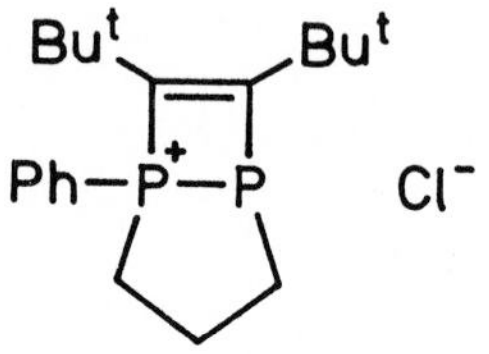

2d

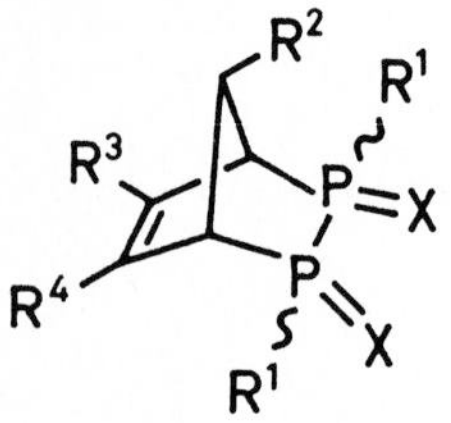

2e

X = lone pair, $Cr(CO)_5$
R^1 = alkyl
R^2, R^3, R^4 = H, alkyl

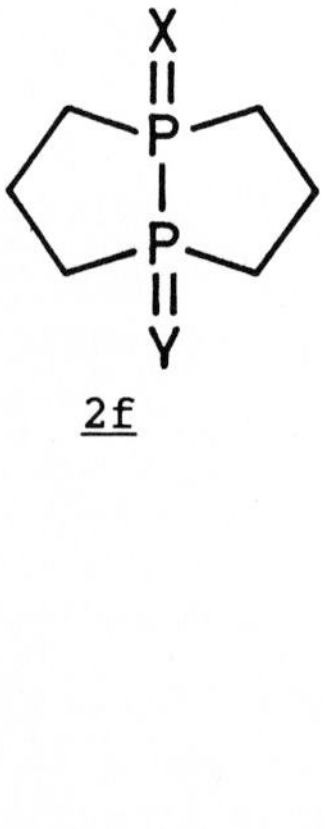

2f

X, Y = lone pair, O, S

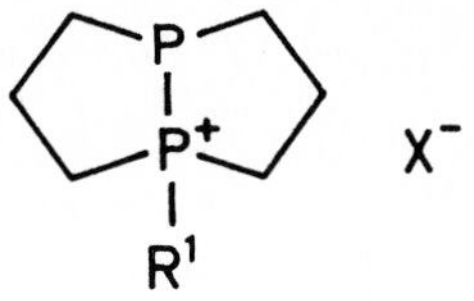

2g

X^- = halogen
R^1 = alkyl, halogen

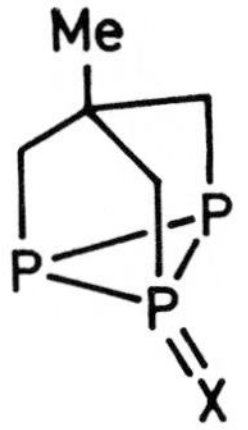

2h

X = lone pair, $Cr(CO)_5$

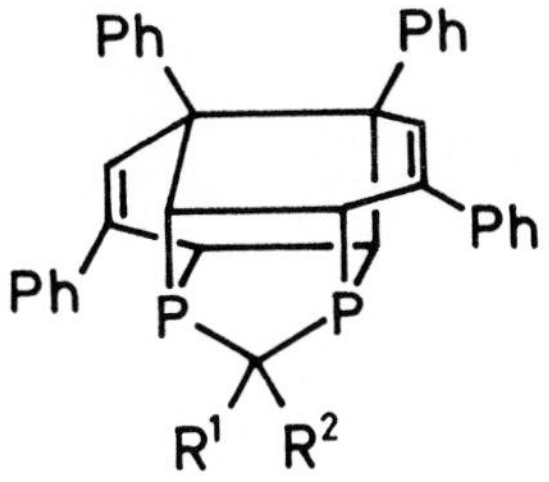

2i

R^1, R^2 = H, alkyl, aryl

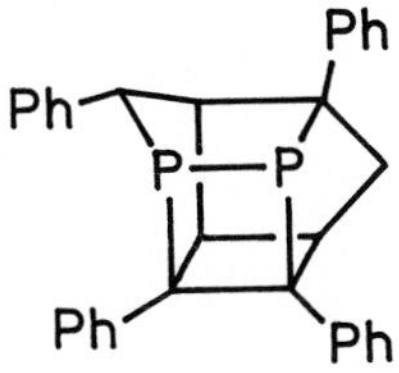

2j

3.1.3 Triphospholanes

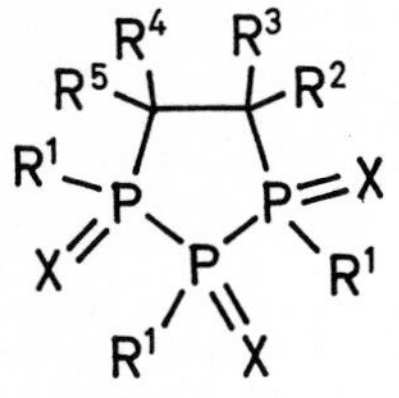

3a

X = lone pair, S
R^1 = H, alkyl, aryl, halogen
R^2, R^3, R^4, R^5 = H, alkyl

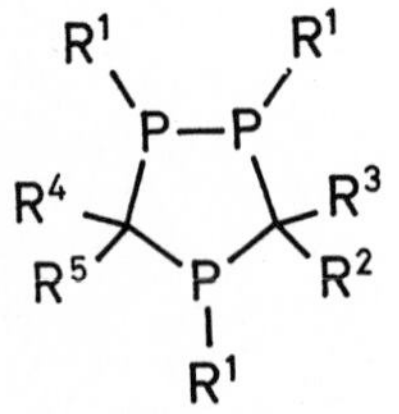

R^1 = aryl
R^2, R^3, R^4, R^5 = alkyl

3b

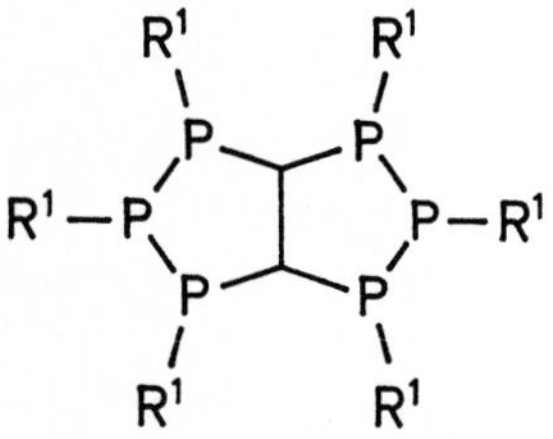

3c R^1 = alkyl

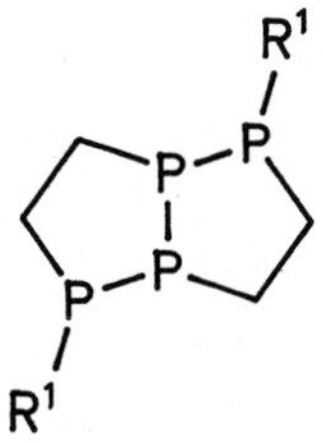

3d R^1 = H, Li

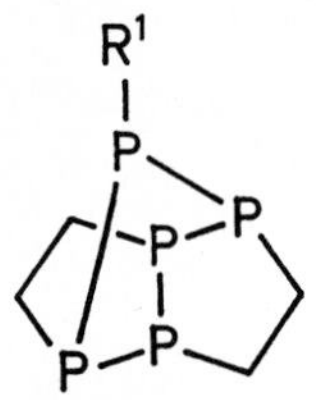

R^1 = alkyl

3e

3.1.4 Tetraphospholanes

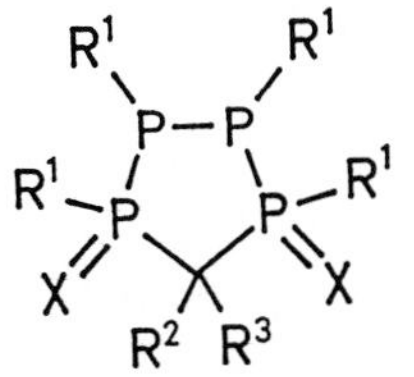

4

X = lone pair, S
R^1 = alkyl, aryl
R^2, R^3 = H, alkyl, =O, =P(2,6-$Bu^t{}_2$-4-Me-Ph)

3.2 PHOSPHOLENES AND PHOSPHOLES

Among the unsatured phosphorus-carbon-heterocycles the five-membered ring is presumably the most common one. Since the synthesis of the first compound belonging to this type - a patent from 1953 - this chemistry has received an impetus (ref. 55). This patent from W.B. McComack is concerned with simple 1,4-cycloadditions of dienes and organophosphorusdihalides, followed by hydrolysis to the phosphole-oxides (eqn. 1, ref. 55).

e.g.: 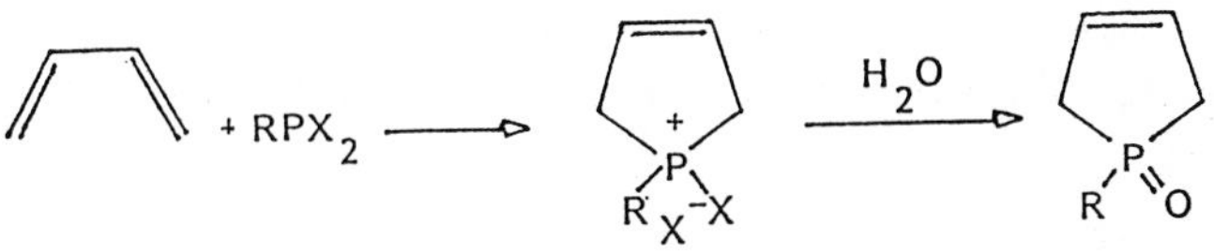 (eqn. 1)

The variety of published syntheses and compounds in the area of phospholenes and phospholes is not surprising, since the first phosphole was described not much later on, in 1959 (eqn. 2, refs. 56-59).

(refs. 56,57)

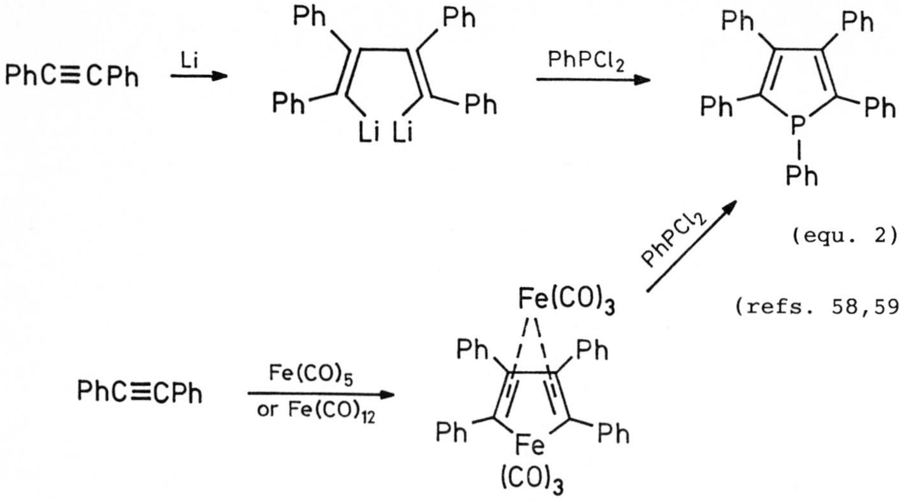

(equ. 2)

(refs. 58,59)

In particular, the groups of L.D. Quin and F. Mathey have contributed to the present comprehension of these classes of compounds by their extensive researches. The fullness of material about phospholenes and phosphides would be beyond the scope of this paper. Therefore, only a selection of the published references and compounds will be mentioned.

3.2.1 Phospholenes

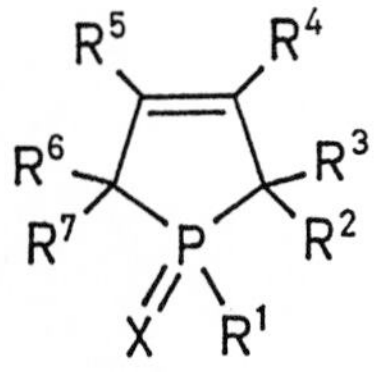

5a

Selected references (sel. refs.):

X = lone pair; R^1 = Me; R^2-R^7 = H (ref. 60)

X = lone pair; R^1 = Me; R^2 = Me; R^3-R^7 = H (ref. 60)

X = lone pair; R^1 = Ph; R^2 = R^7 = Me; R^3-R^6 = H (ref. 61)

X = O; R^1 = Ph; R^2-R^7 = H (ref. 62)

X = S; R^1 = Ph; R^2 = R^3 = R^6 = R^7 = H; R^4 = R^5 = Me (ref. 63)

X = O; R^1 = OMe; R^2-R^7 = H (ref. 64)

X = O; R^1 = NEt_2; R^2 = R^3 = R^5-R^7 = H;

R^4 = Me (ref. 65)

X = lone pair; R^1 = 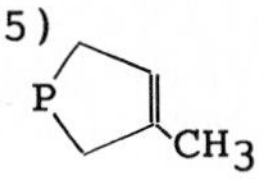 ;

$R^2 = R^3 = R^5$-R^7 = H; R^4 = Me (ref. 65)

X = N(2,4,6-$Bu^t{}_3$Ph); R^1 = Bu^t; $R^2 = R^3 = R^6 = R^7$ = H; $R^4 = R^5$ = Me (ref. 66)

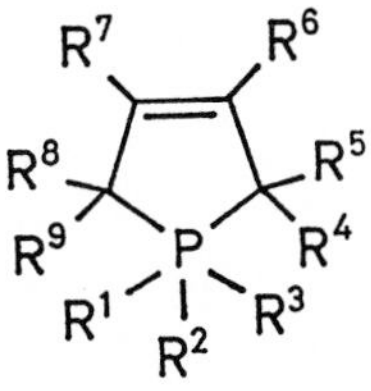

5b

Sel. refs.:

$R^1 = R^2$ = F; R^3 = Me; $R^4 = R^9$ = Me; R^5-R^8 = H; (ref. 67)

R^1-R^3 = Me, R^4-R^9 = F (ref. 68)

R^1 = Et; $R^2...R^3$ = $-(CH_2)_2-$; R^4-R^9 = H (ref. 69)

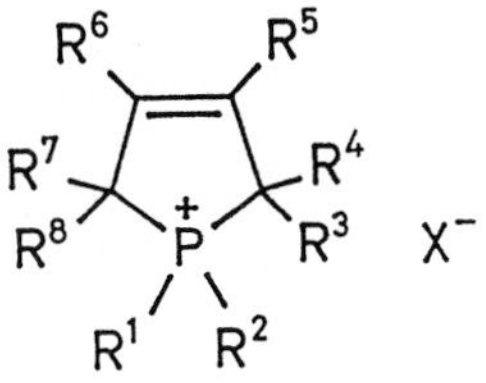

5c

Sel. refs.:

X^- = Cl^-; R^1-$R^3 = R^8$ = Me; R^4-R^7 = H (ref. 70)

X^- = Cl; R^1 = Me; R^2 = Cl; $R^3 = R^8$ = Me; R^4-R^7 = H (ref. 71)

X^- = $AlCl_4^-$; $R^1 = R^2$ = NMe_2, NEt_2, $NPr^i{}_2$; $R^3 = R^4 = R^7 = R^8$ = H; $R^5 = R^6$ = Me (refs. 72,73)

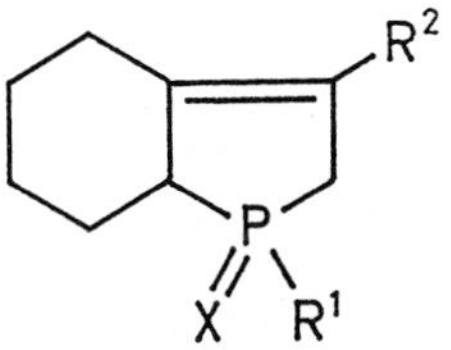

5d

Sel. refs.:

X = lone pair; R^1 = Me; R^2 = H (ref. 74)

X = O; R^1 = Me; R^2 = Br (ref. 74)

X = S; R^1 = Me; R^2 = Br (ref. 75)

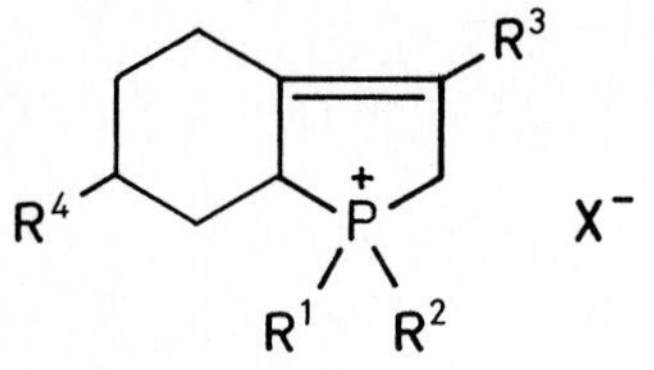

5e

Sel. refs.:

$X^- = Cl^-$; $R^1 = R^3 = R^4 = Me$; $R^2 = Cl$ (ref. 76)

$X^- = Cl^-$; $R^1 = Me$; $R^2 = Cl$; $R^3 = R^4 = H$ (ref. 77)

$X^- = Br^-$; $R^1 = Ph$; $R^2 = Br$; $R^3 = R^4 = H$ (ref. 77)

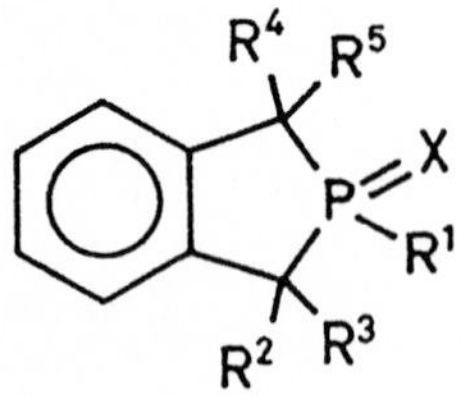

5f

Sel. refs.:

X = O; R^1 = Me, OH; R^2-R^5 = H (ref. 111)

X = lone pair; $R^1 = Bu^t$, Ph; R^2, R^3 = R^4, R^5 = =O (ref. 112)

X = O, S, $Fe(CO)_4$; R^1 = Ph; R^2, R^3 = R^4, R^5 = =O (ref. 112)

X = O; R^1 = Me, Ph; $R^2 = R^4$ = Ph; $R^3 = R^5$ = H (ref. 113)

X = lone pair; R^1 = Ph; R^2, R^3 = R^4, R^5 = =O (refs. 114,115)

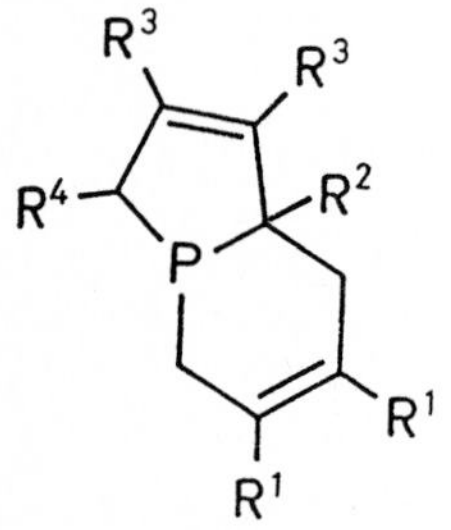

5g

Sel. refs.:

$R^1 = R^3$ = Me; R^2 = Ph; R^4 = H (ref. 78)

R^1 = Me; $R^2 = R^4$ = Ph; R^3 = H (ref. 79)

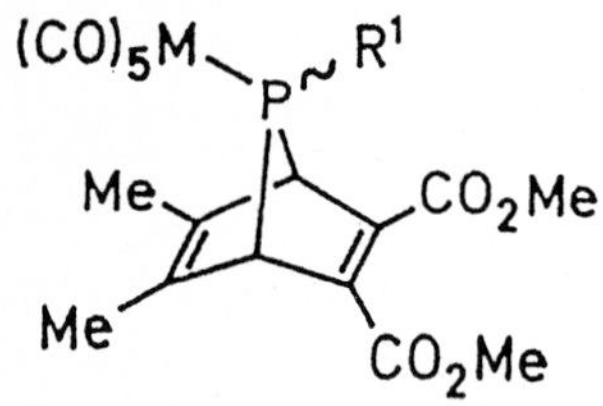

5h

Sel. refs.:

M = Cr, Mo, W; R^1 = Ph, Me (refs. 16,80)

M = W; $R^1 = (CH_2)_nCl$ (n = 1,2) (refs. 4,81)

M = W; R^1 = OMe, F (ref. 5)

M = W; $R^1 = (CH_2)_n$-CH=CH_2 (n = 1-3) (ref. 3)

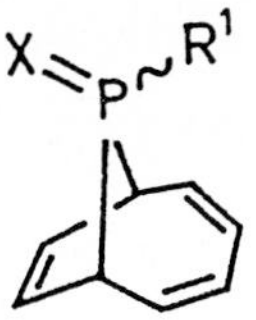

5i

Sel. refs.:
X = lone pair; R^1 = H, Cl, Me, Bu^t, CH_2Ph, NMe_2 (ref. 82)
X = O, lone pair; R^1 = Ph (refs. 83-85)
X = lone pair; R^1 = Bu^t, Ph, c-C_6H_{11}, menthyl, NEt_2, O-menthyl (ref. 86)
X = $Fe(CO)_4$; R^1 = Ph (ref. 87)

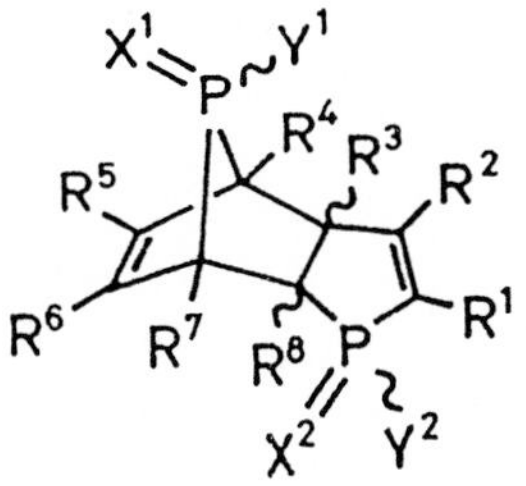

5j

Sel. refs.:
X^1 = X^2 = lone pair; Y^1 = Y^2 = Me, Ph; R^1-R^8 = H (refs. 88,89)
X^1 = X^2 = O; Y^1 = Y^2 = Me, Ph; R^1-R^8 = H (refs. 88,90)
X^1 = X^2 = S; Y^1 = Y^2 = Me; R^1-R^8 = H (ref. 90)
X^1 = X^2 = O; Y^1 = Y^2 = R^2 = R^3 = R^5 = R^6 = Me; R^1 = R^4 = R^7 = R^8 = H (ref. 90)
X^1 = lone pair; X^2 = O; Y^1 = Cl; Y^2 = NEt_2;
R^2 = R^5 = Me; R^1 = R^3 = R^4 = R^6 = R^7 = R^8 = H (ref. 119)
X^1 = X^2 = O; Y^1 = Y^2 = NEt_2; R^2 = R^5 = Me; R^1 = R^3 = R^4 = R^6-R^8 = H (ref. 119)
X^1 = X^2 = lone pair; Y^1 = Y^2 = Cl; R^2 = R^5 = Me; R^1 = R^3 = R^4 = R^6-R^8 = H (ref. 119)
X^1 = X^2 = lone pair; Y^1 = Y^2 = R^2 = R^5= Me; R^1 = R^3 = R^4 = R^6-R^8 = H (ref. 120)

$X^1 = X^2 = Fe(CO)_4$; $Y^1 = Y^2 = Ph$;
R^1–R^8 = H (ref. 87)

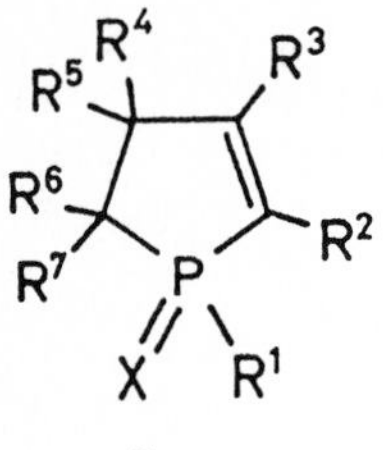

6a

Sel. ref.:
X = O; R^1 = Me; R^2–R^7 = H (ref. 91)
X = S; R^1 = Me; R^2–R^7 = H (ref. 75)
X = lone pair; R^1 = Cl; R^3 = Me;
$R^2 = R^4 = R^5$–R^7 = H (ref. 92)
X = O; R^1 = Cl; R^2–R^7 = H (ref. 93,94)
X = O; R^1 = Ph; R^2 = $P(O)Ph_2$; R^3 = Me;
R^4, R^5 = =O; $R^6 = R^7$ = H (ref. 95)
X = O; R^1 = Me; $R^2 = R^4$–R^7 = H; R^3 =
OMe (ref. 96)
X = O; R^1 = OEt; R^2–R^7 = H (ref. 64)

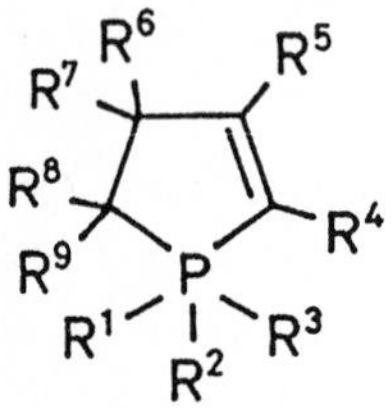

6b

Sel. refs.:
R^1–R^3 = Me; R^4–R^9 = F (ref. 68)

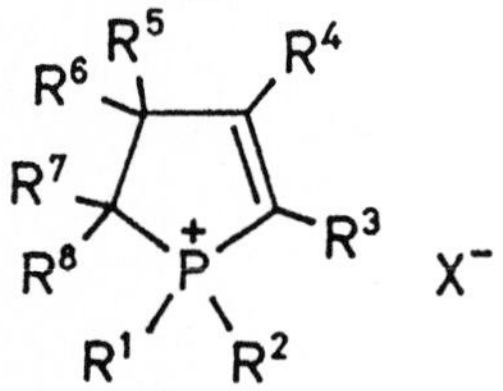

6c

Sel. refs.:
X^- = Cl^-; R^1 = Ph; R^2 = Cl; $R^3 = R^8$ =
Me; R^4–R^7 = H (ref. 61)

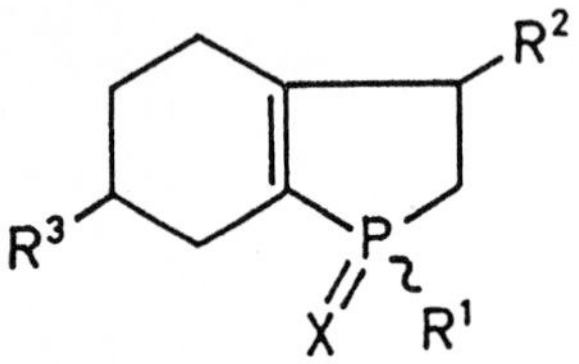

6d

Sel. refs.:
X = lone pair; R^1 = Me; $R^2 = R^3$ = H
(ref. 98)
X = O; R^1 = Me; $R^2 = R^3$ = H (ref. 74)
X = lone pair, O; R^1–R^3 = Me (ref. 76)

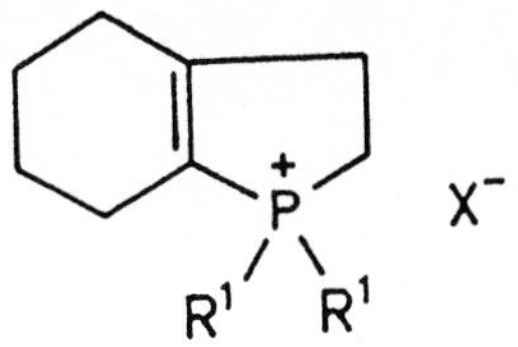

6e

Sel. refs.:
$X^- = I^-$; R^1 = Me (ref. 55)

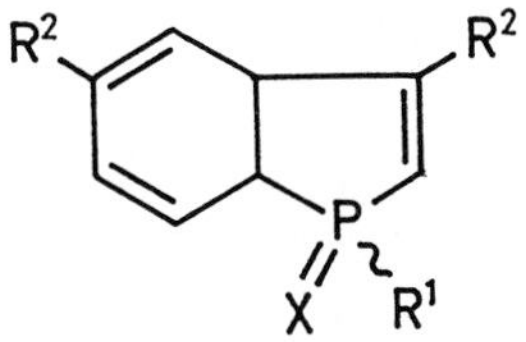

6f

Sel. refs.:
X = lone pair; R^1 = Me; R^2 = H (ref. 88)
X = lone pair; R^1 = Ph; R^2 = H (refs. 84,88)
X = O; R^1 = Ph; R^2 = H (refs. 88,100)
X = lone pair; R^1 = R^2 = Me (refs. 84,99)

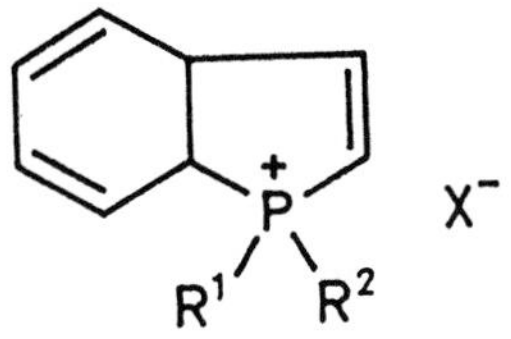

6g

Sel. refs.:
$X^- = I^-$; R^1 = Me; R^2 = Ph (ref. 88)

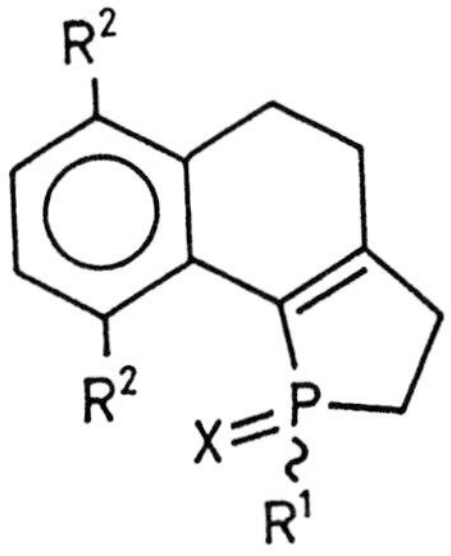

6h

Sel. refs.:
X = lone pair; R^1 = Me; R^2 = H (ref. 98)
X = O; R^1 = Me; R^2 = H (refs. 77,98)

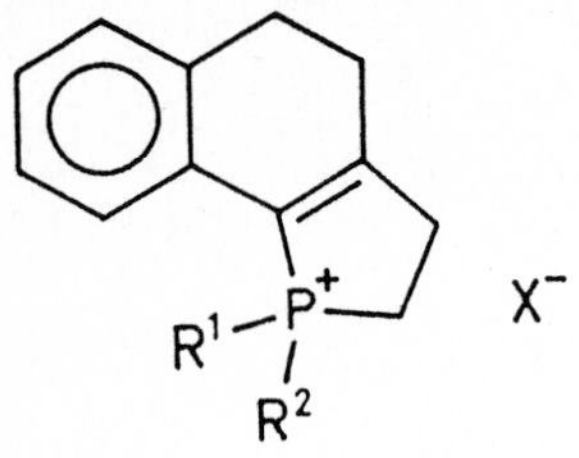

6i

Sel. refs.:
$X^- = I^-$; $R^1 = R^2 = Me$ (ref. 98)
$X^- = Cl^-$; $R^1 = Me$; $R^2 = Cl$ (ref. 77)
$X^- = Br^-$; $R^1 = Ph$; $R^2 = Br$ (ref. 77)

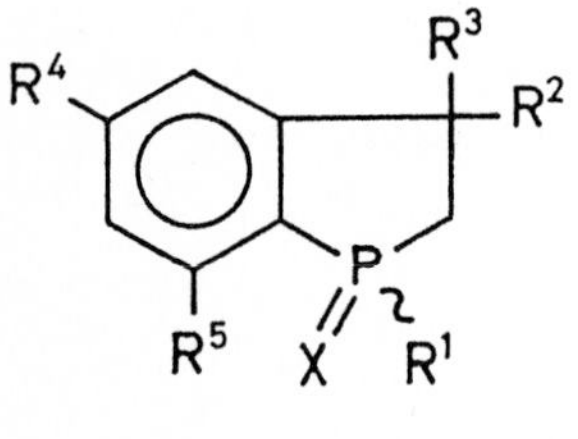

6j

Sel. refs.:
X = lone pair; $R^1 = Et$; R^2–$R^5 = H$ (ref. 101)
X = O; $R^1 = Ph$; $R^2 = R^3 = Me$; $R^4 = R^5 = H$ (ref. 102)
X = O; $R^1 = Me$; $R^2 = R^3 = Me$; $R^4 = R^5 = H$ (ref. 103)
X = lone pair; $R^1 = CH(SiMe_3)_2$; $R^2 = R^3 = Me$; $R^4 = R^5 = Bu^t$ (ref. 104)
X = O; $R^1 = Cl$; $R^2 = R^3 = Me$; $R^4 = R^5 = Bu^t$ (ref. 105)
X = O; $R^1 = OEt$; R^2, R^3 = =O; $R^4 = R^5 = H$ (ref. 106)
X = S; $R^1 = Cl$; $R^2 = R^3 = Me$; $R^4 = R^5 = Bu^t$ (ref. 107)
X = lone pair; $R^1 = HC{=}P(2,4,6\text{-}Bu^t{}_3Ph)$; $R^2 = R^3 = Me$; $R^4 = R^5 = Bu^t$ (ref. 108)

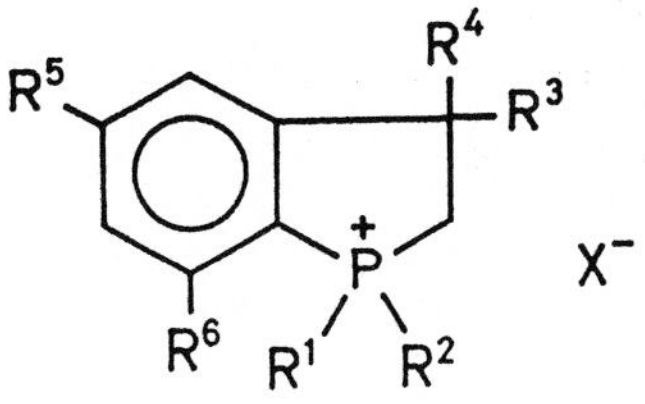

6k

Sel. refs.:
$X^- = Br^-$; $R^1 = R^2 = Et$; R^3-R^6 = H (ref. 101)
$X^- = I^-$; R^1 = Me; R^2 = Ph; R^3-R^6 = H (ref. 88)
$X^- = BF_4^-$; $R^1 = R^2$ = H; $R^3 = R^4$ = Me; $R^5 = R^6 = Bu^t$ (ref. 109)
$X^- = BF_4^-$; R^1 = H; R^2 = HP(2,4,6-$Bu^t{}_3$Ph); $R^3 = R^4$ = Me; $R^5 = R^6 = Bu^t$ (ref. 109)
$X^- = AlCl_4^-$, BF_4^-; R^1= H; R^2 = HC=P(2,4,6-$Bu^t{}_3$Ph); $R^3 = R^4$ = Me; $R^5 = R^6 = Bu^t$ (ref. 110)

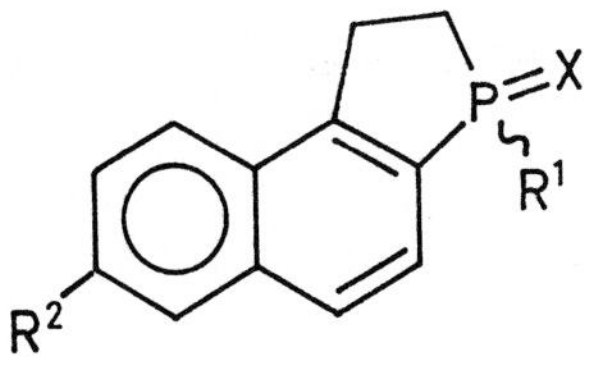

6l

Sel. refs.:
X = O; R^1 = Ph, OH; R^2 = H (ref. 98)
X = lone pair; R^1 = Me; R^2 = H (ref. 77)
X = O; R^1 = Me; R^2 = OMe (ref. 98)

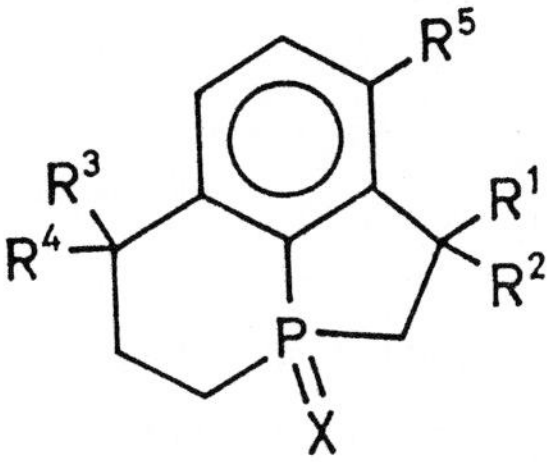

6m

Sel. refs.:
X = O; R^1-R^4 = Me; R^5 = H (ref. 103)
X = lone pair, O; R^1-R^4 = Me; R^5 = Me (ref. 116)

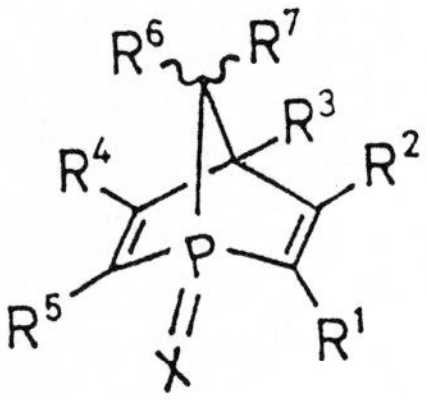

6n

Sel. refs.:

X = lone pair, S; $R^1 = R^2 = R^5 = Ph$; $R^3 = R^4 = Me$; $R^6 = R^7 = H$ (ref. 78)

X = lone pair; $R^1 = R^2 = R^5 = Ph$; $R^3 = R^4 = H$; $R^6 = R^7 = Ph$ (ref. 78)

X = lone pair; $R^1 = R^2 = R^5 = R^7 = Ph$; $R^3 = R^4 = R^6 = H$; (ref. 79)

X = lone pair; $R^1 = H$; $R^2 = R^5 = R^7 = Ph$; $R^3 = R^4 = R^6 = H$ (ref. 79)

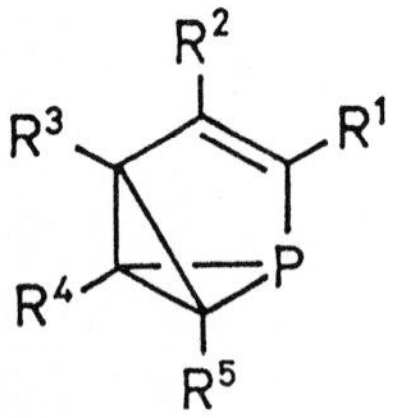

6o

Sel. refs.:

$R^1 = R^3 = R^4 = R^5 = Bu^t$; $R^2 = CO_2Me$ (ref. 118)

$R^1 = R^2 = R^4 = R^5 = Bu^t$; $R^3 = CO_2Me$ (ref. 118)

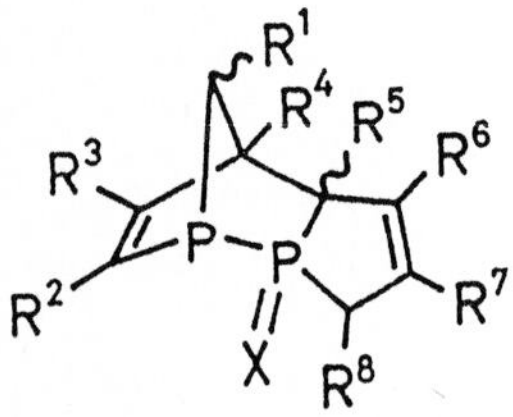

6p

Sel. refs.:

X = lone pair; $R^1 - R^8 = H$ (refs. 79,121)

X = lone pair, S; $R^1 = R^2 = R^5 = R^8 = H$; $R^3 = R^4 = R^6 = R^7 = Me$ (refs. 79,121)

X = lone pair; $R^1 = R^2 = R^5 = R^8 = Ph$; $R^3 = R^4 = R^6 = R^7 = H$ (refs. 79,121)

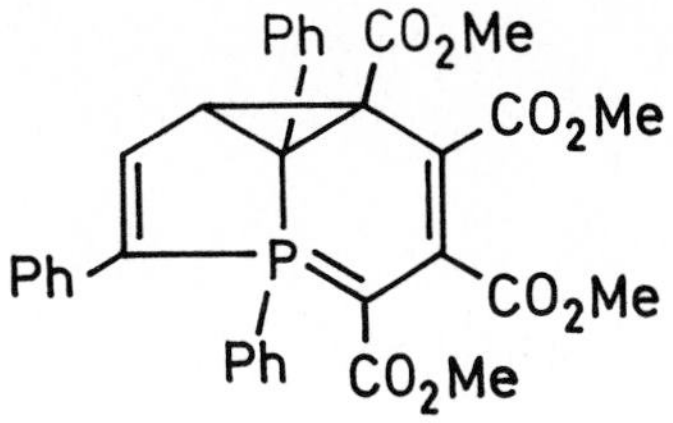

6q (refs. 122,123)

In contrast to the wide variety known in phosphole systems, in which phosphorus does not participate in unsaturation, there are only a few examples known containing C=P-double bonds. In 1986 Appel et al. described reactions of α,ß-dicarboxylic acid chlorides with tris(trimethylsilyl)phosphane (eqn. 3, ref. 124).

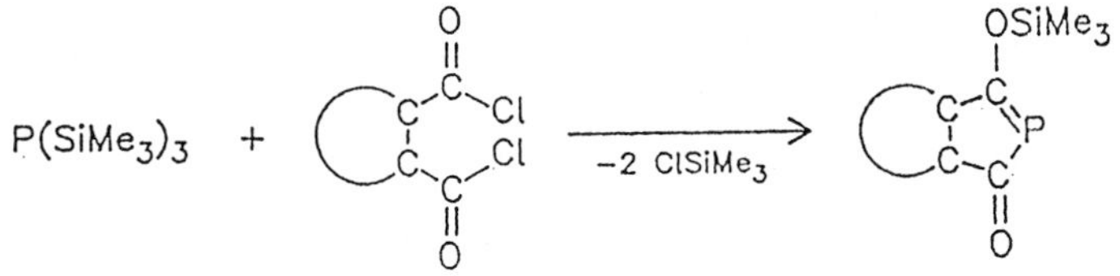

(eqn. 3)

7a

$^{31}P\{^{1}H\}$-NMR			
δ[ppm]:	106.8	109.2	109.7
mp[°C]:	70	77(dec.)	120

Surprisingly those λ^3-phospholenes show no tendency to dimerize, in contrast to

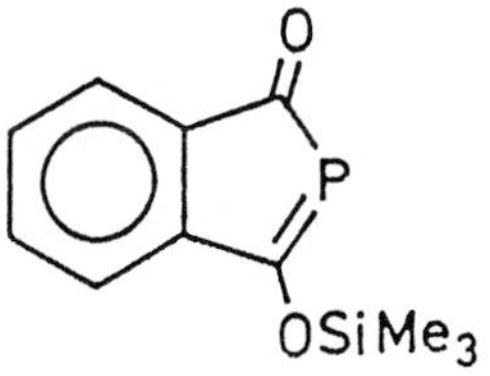

9b (refs. 124,125), which is

synthesized in the same way (see eqn. 3). The thermally unstable 9b dimerises in a [2+2]-cycloaddition above ca. -25°C to the 1,2-diphosphetane 5k (eqn. 4).

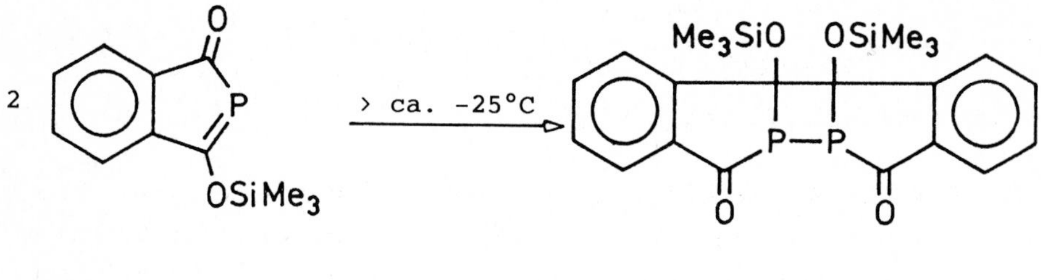

9b

5k (eqn. 4)

Compound 5k is obtainable in good yield (mp. 165°C). In the ^{31}P-NMR spectrum one observes a signal at -42 ppm. The X-ray structure analysis confirms the head-to-head and tail-to-tail connection in 5k (Fig. 5, ref. 124).

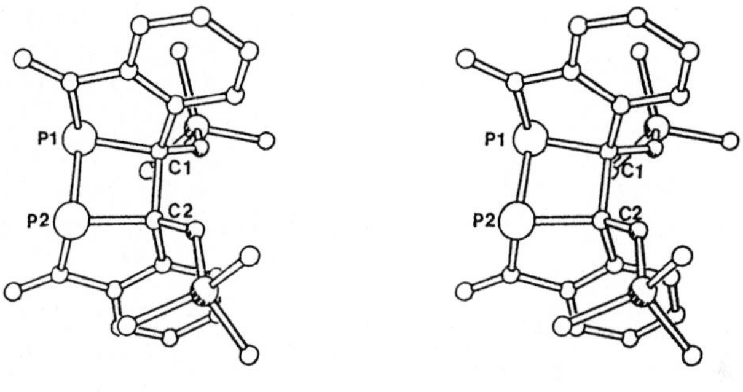

Fig. 5. Stereo plot of 5k

λ^5-2H-phospholenes have not yet been isolated, however, their existence as an intermediate was postulated in a formal [2+2]-cycloaddition between vinylphosphanes and acrylic derivatives (eqn. 5, ref. 126).

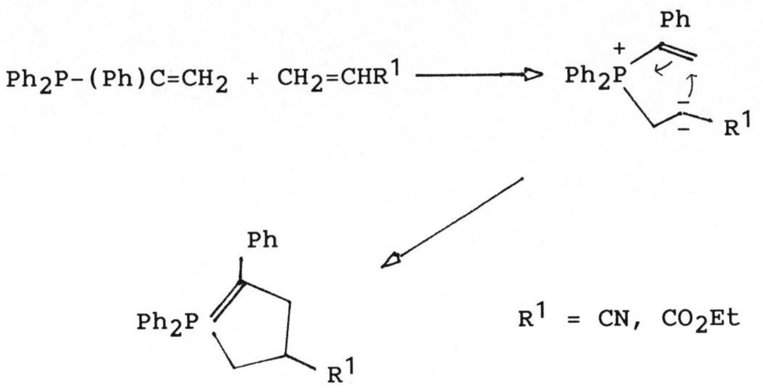

(eqn. 5)

R^1 = CN, CO_2Et

7b

In the presence of aldehydes these compounds reacted to give phosphane oxides, explained by a Wittig reaction of the cyclic ylids 7b with aldehydes resulting in ring opening (ref. 126).

3.2.2 Phospholes

3.2.2.1 $\lambda^3\sigma^3$-, $\lambda^5\sigma^4$-, and $\lambda^5\sigma^5$-Phospholes

The field of phospholes posesses a large amount of literature in comparison with other phosphorus-carbon heterocycles, so that we refer to several review articles (refs. 55,127,128). In spite of the already mentioned fullness of compounds and synthesis methods, the basic compound $(CH)_4PH$ was observed the first time in 1983 (ref. 79).

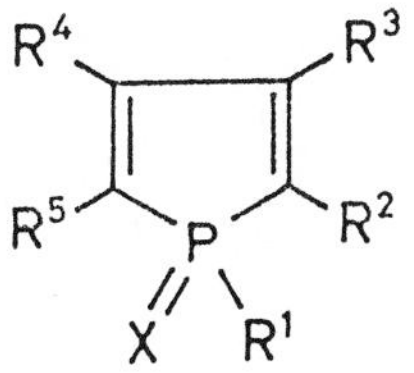

8a

Sel. refs.:

X = lone pair; O, S, Se; $R^1 = R^2 = R^5$ = Ph; $R^3 = R^4$ = H (ref. 129)

X = lone pair; R^1 = Me; R^2 - R^5 = H (refs. 130, 131)

X = lone pair; R^1 = Ph; $R^2 = R^5$ = H; $R^3 = R^4$ = Me (ref. 132)

X = S; R^1 = Ph; $R^2 = R^5$ = H; $R^3 = R^4$ = Me (ref. 133)

X = lone pair; R^1 = Bu^t; R^2 = CO_2Et; $R^3 = R^4$ = Me; R^5 = H (ref. 134)

X = 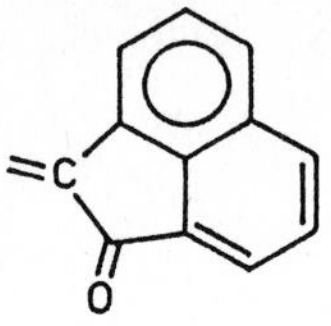 ; R^1 = Ph; R^2 = R^5 = Ph; R^3 = R^4 = H (ref. 135)

X = lone pair; R^1 = 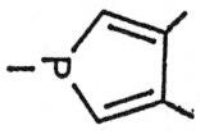 ; R^2 = R^5 = H; R^3 = R^4 = Me (ref. 136)

X = lone pair; R^1 - R^5 = H (refs. 79,137)

X = lone pair; R^1 = H; R^2 - R^5 = Ph (refs. 79,138)

X = $W(CO)_5$; R^1 = Cl, F, OMe; R^2 = R^5 = H; R^3 = R^4 = Me (ref. 5)

X = $W(CO)_5$; R^1 = $(CH_2)_2CH{=}CH_2$; R^2 = R^5 = H; R^3 = R^4 = Me (ref. 3)

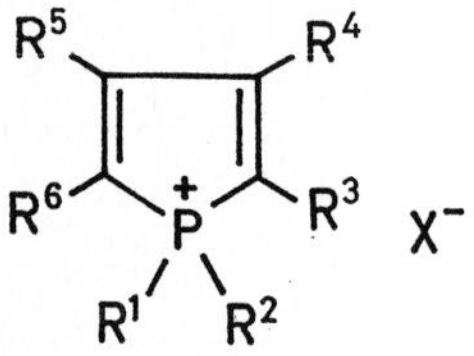

8b

Sel. refs.:

X^- = Br^-; R^1 = R^3 = R^6 = Ph; R^2 = CH_2CO_2Et; R^4 = R^5 = H (ref. 129)

X^- = I^-; R^1 = R^3 = R^6 = Ph; R^2 = Me; R^4 = R^5 = H (ref. 129)

X^- = I^-; R^1 = R^2 = R^4 = R^5 = Me; R^3 = R^6 = H (ref. 97)

X^- = $TaCl_6^-$; R^1 = R^3 = R^6 = Ph; R^2 = R^4 = R^5 = H (ref. 139)

X^- = $AlCl_4^-$; R^1 = Ph; R^2 = Cl; R^3 - R^6 = Me (ref. 140)

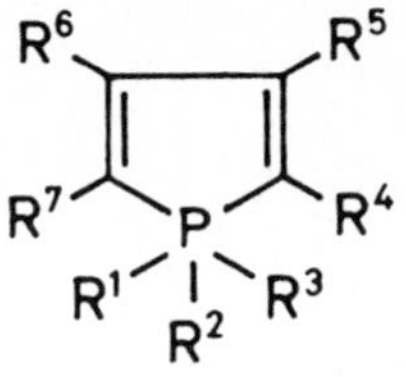

8c

Sel. refs.:
$R^1 - R^3$ = OMe; $R^4 - R^7$ = CO_2Me (refs. 141-143)
$R^1...R^2$ = $OCMe_2CMe_2O$; R^3 = OMe; $R^4 - R^7$ = CO_2Me (refs. 143,144)

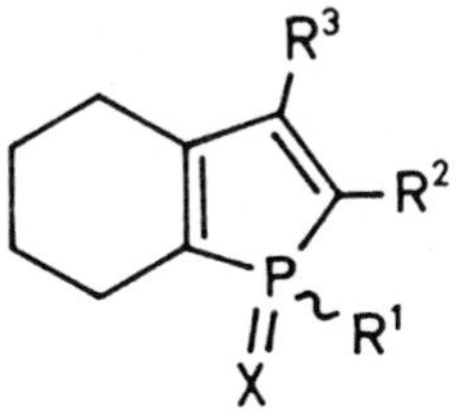

8d

Sel. refs.:
X = lone pair; R^1 = Me, Ph; R^2 = R^3 = H (refs. 77,145)
X = O; R^1 = Ph; R^2 = R^3 = Me (ref. 140)

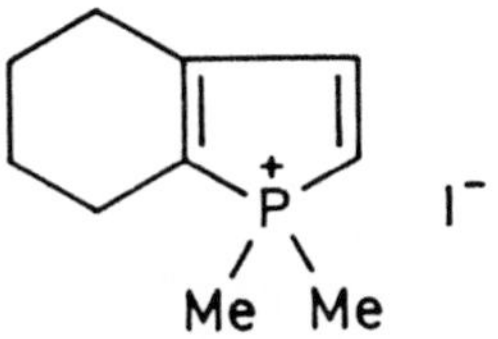

8e (ref. 146)

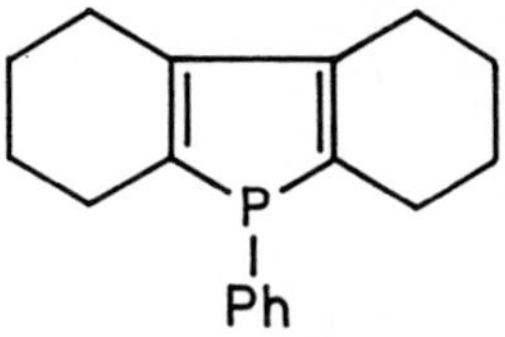

8f (ref. 147)

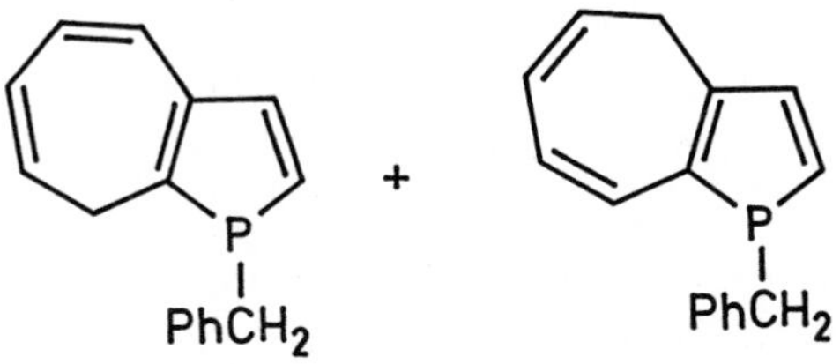

8g (ref. 148)

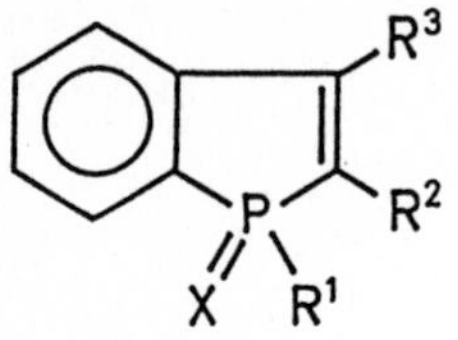

8h

Sel. refs.:

X = lone pair; R^1 = CH_2Ph; R^2 = R^3 = H (ref. 149)

X = lone pair; R^1 = R^2 = Ph; R^3 = Bu^n (refs. 149-151)

X = lone pair; R^1 = Ph, Bu^t; R^2 = R^3 = H (ref. 149)

X = O; R^1 = Ph; R^2 = R^3 = H (ref. 149)

X = S; R^1 = Bu^n; R^2 = R^3 = H (ref. 149)

X = O; R^1 = Ph; R^2 = H; R^3 = Br (ref. 149)

X = O; R^1 = Ph; R^2 = Br; R^3 = H (ref. 149)

X = lone pair; R^1 = R^2 = Ph; R^3 = 2-(C≡CPh)-Ph (refs. 152,153)

X = lone pair; R^1 = R^2 = Ph; R^3 = H, Cl (ref. 154)

X = O, Cp(CO)Co, $Fe(CO)_4$; R^1 = R^2 = Ph; R^3 = Bu^n (ref. 151)

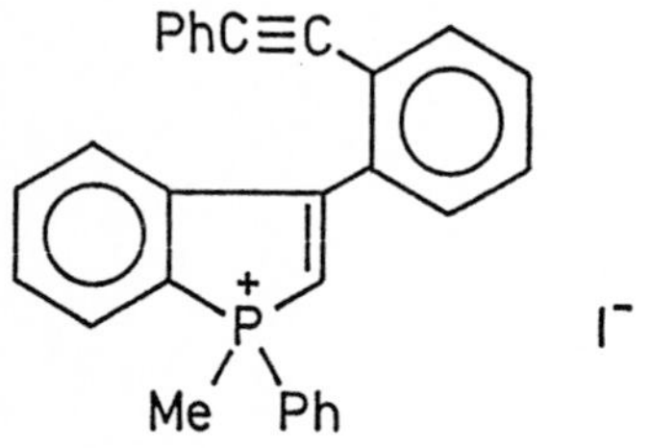

8i (ref. 152)

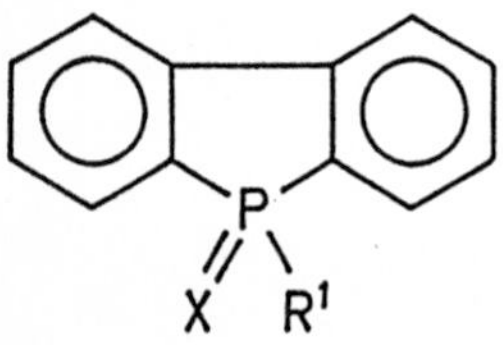

8j

Sel. refs.:

X = lone pair; R^1 = Ph (ref. 252)

X = $=CHCH_2Ph$; R^1 = Ph (ref. 156)

X = lone pair; R^1 = C(O)Me, $C(O)CF_3$ (ref. 157)

X = O; R^1 = OH (refs. 158-160)

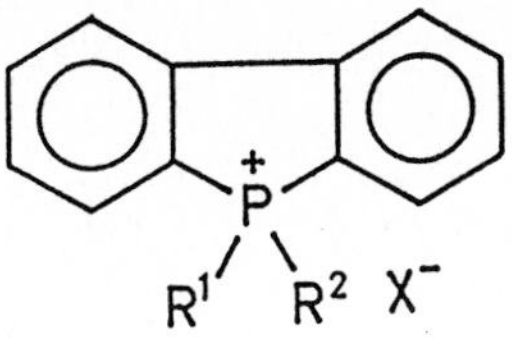

8k

Sel. refs.:
$X^- = I^-$; $R^1 = R^2$ = Me (ref. 161)
$X^- = I^-$; R^1 = Me; R^2 = Ph (ref. 156)

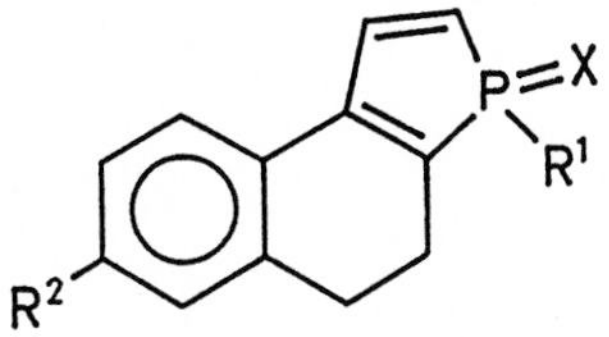

8l

Sel. refs.:
X = O, lone pair; R^1 = Me; R^2 = H, OMe (refs. 77,146)
X = lone pair; R^1 = Me, Ph; R^2 = H (ref. 77)

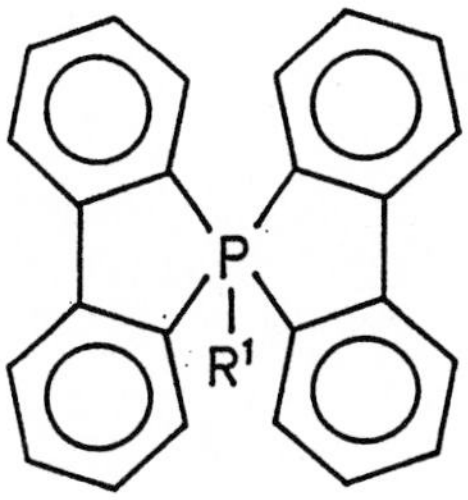

8m

R^1 = Me, Ph (ref. 155)

$X^- = I^-$ (ref. 155)

$X^- =$

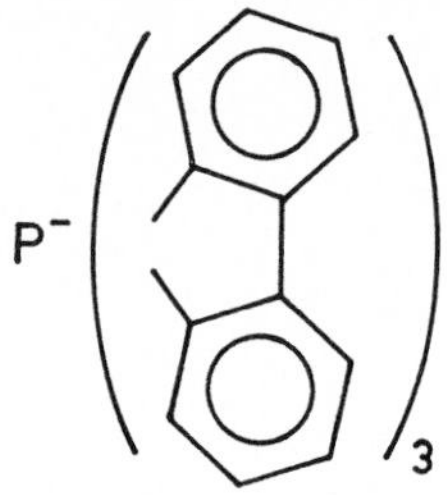

(refs. 155,162)

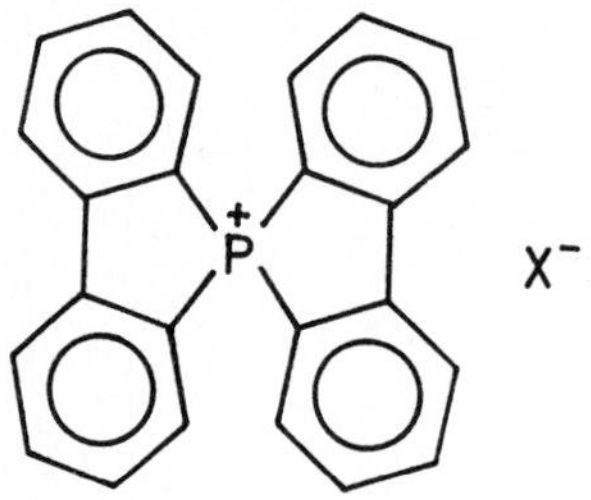

8n

3.2.2.2 $\lambda^3\sigma^2$-Phospholes

λ^3-2H-phospholes, not comprised in condensed systems, were until recently only presumed as intermediates (refs. 79,138,163-165), or observed in metal complexes (ref. 166). In the dimerization of the 1H-phosphole it is supposed that a 1,5-H-shift is the first step. The 1H-phosphole then undergoes a Diels-Alder condensation (eqn. 6, refs. 79,121).

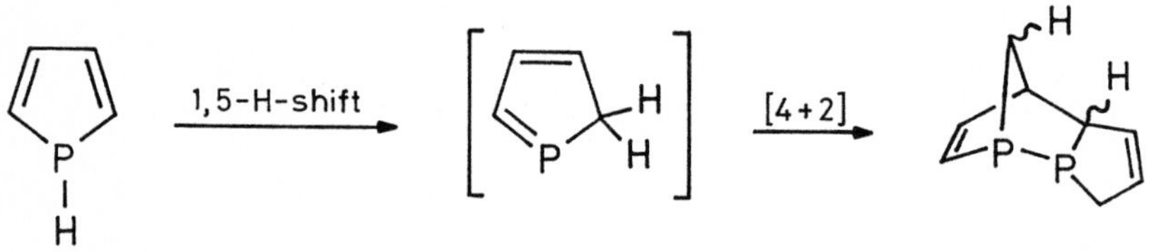

(eqn. 6)

But very recently Regitz et al. were successful in closing the gap. They isolated the λ^3-2H-phospholes 9a by using bulky substituents (eqn. 7, ref. 167).

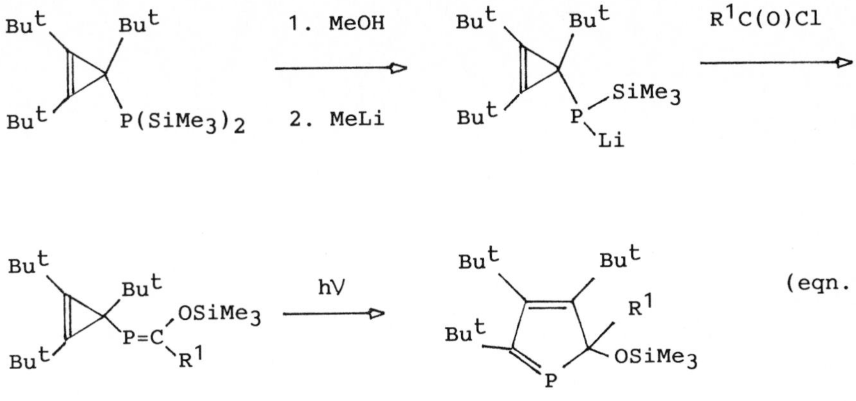

(eqn. 7)

9a

R^1 = Me, Ph, 2,4,6-Me_3Ph, 2,4,6-$(MeO)_3$Ph

The compounds 9a are yellow, thermally very stable oils, which are sensitive to moisture. The ^{31}P-NMR shifts and the yields are shown in Tab. 8.

TABLE 8
Yields and $\delta(^{31}P)$ of 9a.

compound (R^1)	$\delta(^{31}P)$[ppm]	yield [%]
Ph	159	91
2,4,6-Me_3Ph	152	76
2,4,6-$(MeO)_3$Ph	154	74
Me	142	83

The λ^3-benzophosphole 9b can be obtained by the reaction of phthaloylchloride and tris(trimethylsilyl)phosphane (refs. 124, 125). Compound 9b dimerizes at temperatures above ca. -25°C (refs. 124,125) to the 1,2-diphosphetane 5k (eqn. 8, see also 3.2 eqn. 4); the authors of (ref.125) erroneously assumed a 1,3-diphosphetane as product of the dimerization.

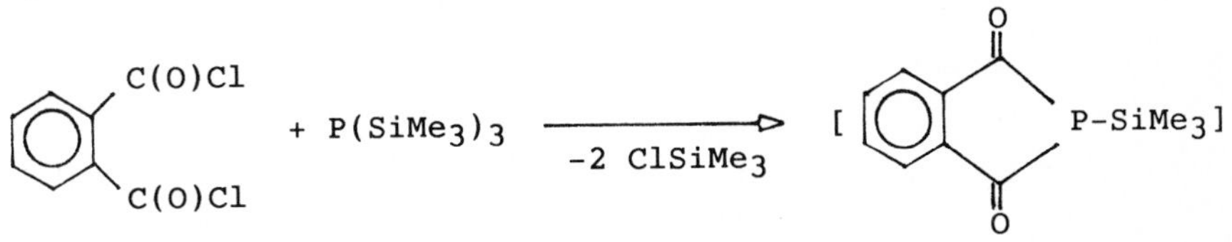

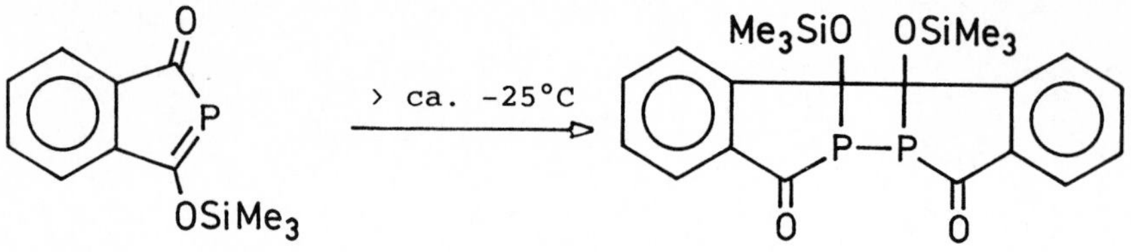

9b 5k (eqn. 8)

The ^{31}P-NMR spectrum of 9b shows a signal at 71.7 ppm (ref. 124) [76.6 ppm (ref. 125)]. The corresponding anion 10 can be synthesized in the reaction of diethylphthalat with KPH_2 in the presence of 18-crown-6 (eqn. 9, ref. 168).

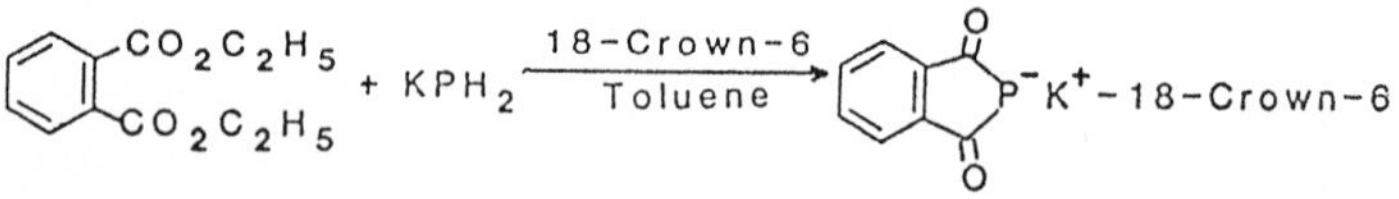

10 (eqn. 9)

The X-ray structure analysis indicates that, surprisingly, the resonance structures α and ß are only minimally involved in charge delocalisation (C-P ca. 180 pm) (Fig. 6).

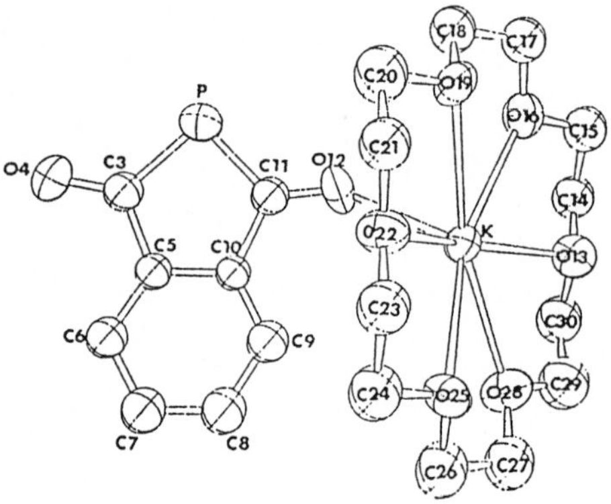

Fig. 6. ORTEP view of 10

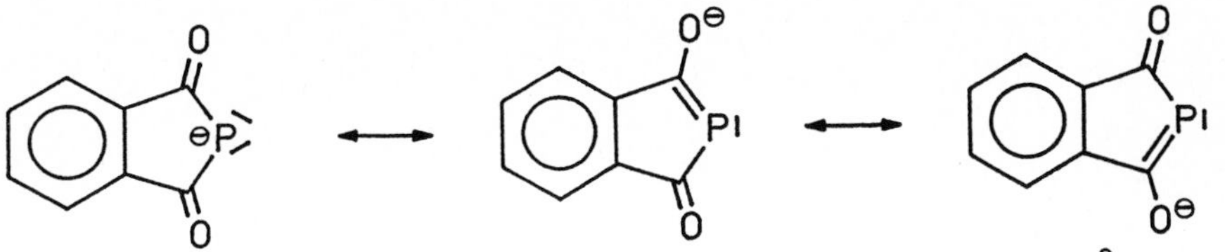

α β

The δ(^{31}P)-NMR value of 43.43 ppm agrees with the increase of the charge density at the phosphorus atom in 10, compared to 9b. One has to point out that 10 is stable in protic solvents. 10 can be dissolved in water or ethanol without decomposition.

In 1983, Märkl et al. synthesized the first phosphaazulenes. In contrast to λ^{3}-2H-phospholes, which can only be stabilized by bulky substituents, phosphaazulenes are thermally stable compounds. A 1-phosphaazulene 9c is formed by thermolysis of the corresponding dihydro-1-phosphaazulenes (1,5-benzyl shift and dehydrogenation) (eqn. 10, ref. 148).

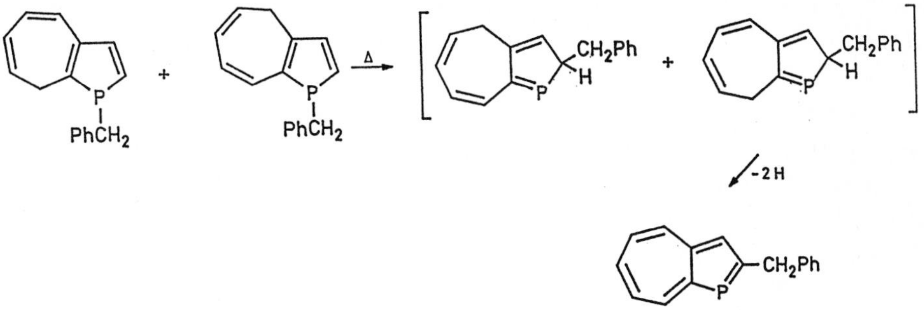

9c (eqn. 10)

Similar is the path of reaction to the first 2-phosphaazulene 9d (ref. 169). 8-Methoxy-heptafulvene in boiling xylene in the presence of KF/18-crown-6 undergoes [8+2]-cycloaddition with Ph-C≡P, followed by dehydrogenation to give 1-methoxy-3-phenyl-2-phosphaazulene 9d (equ. 11).

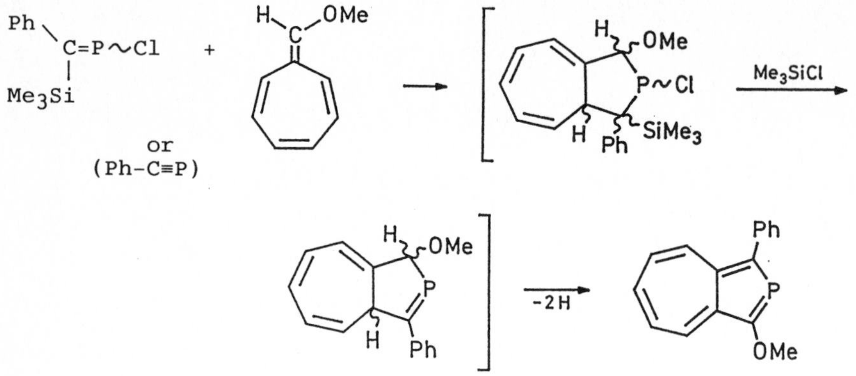

9d (eqn. 11)

[9c: $\delta(^{31}P)$ = 168,3 ppm (ref. 148); 9d: $\delta(^{31}P)$ = 177,9 ppm (ref. 169)].

3.2.2.3 $\lambda^5\sigma^4$-Phospholes with P-C-Multiple Bonds

The first representative of λ^5-2H-phospholes was already prepared in 1961 from Tebby et al. (ref. 170) together with another group (refs. 171,172), but not recognized as the λ^5-2H-phosphole before 1969 (refs. 173,174). Phospholes of this species can be synthesized by the reaction of Ar_3P and acetylenedicarboxylic acid diesters (eqn. 12, refs. 141,170-174). This method has a wide area of application, demonstrated by the reactions with cyclopropyldiphenylphosphane (ref. 175) (see eqn. 12) and some phosphetanes (ref. 176).

$R^1{}_2PR^2$ + 2 $R^3C{\equiv}CR^3$ ⟶ 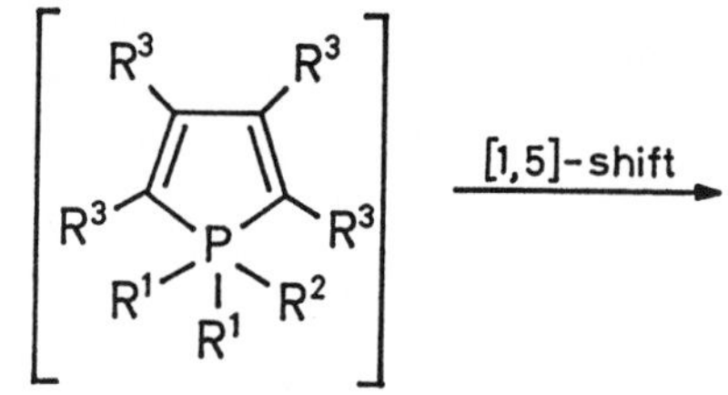 (eqn. 12)

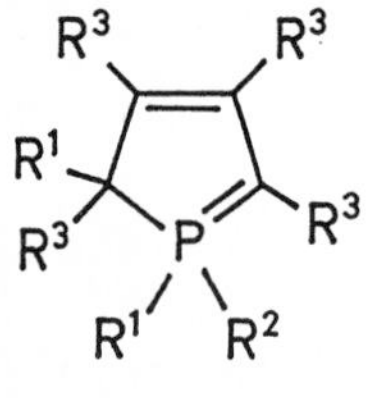

9e

Sel. ref.:
$R^1 = R^2$ = Ph; $R^3 = CO_2Me$ (refs. 170-173)
$R^1 = R^2$ = 4-MePh; $R^3 = CO_2Me$ (ref. 173)
$R^1 = R^2$ = OMe; $R^3 = CO_2Me$ (ref. 142)
$R^1 = R^2$ = OEt; $R^3 = CO_2Et$ (ref. 141)

R^1 = Ph; R^2 = $\overline{CHCH_2CH_2}$; R^3 = CO_2Me
(ref. 175)
R^1 = Ph; R^2 = 4-MePh; R^3 = CO_2Me (ref. 174)

A variation is the reaction of a triphenylphosphanegold complex with hexafluoro-2-butyne (eqn. 13, ref. 177).

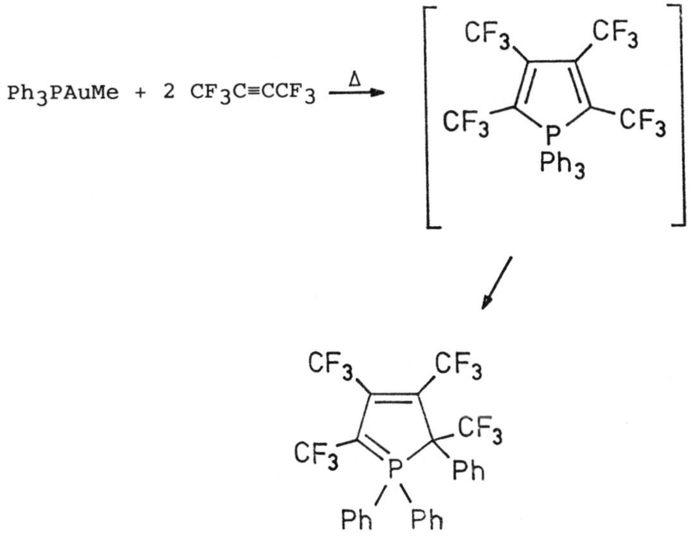

(eqn. 13)

9f

The λ^5-phospholes 9g,h were proposed to be intermediates in cyclisations. Their existence was assumed from the reaction products obtained from aldehydes and 6g,h (refs. 178,179). It is interesting that 6g,h exist as 3H-phospholes, or are at least presumed to be one of their tautomeres (eqn. 14).

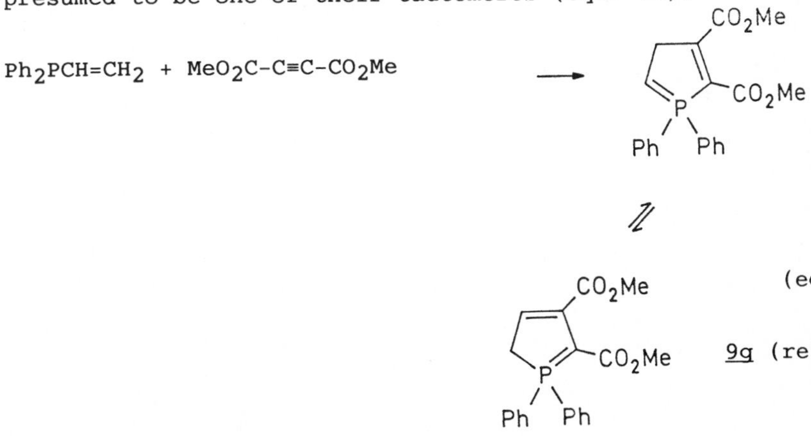

(eqn. 14)

9g (ref. 179)

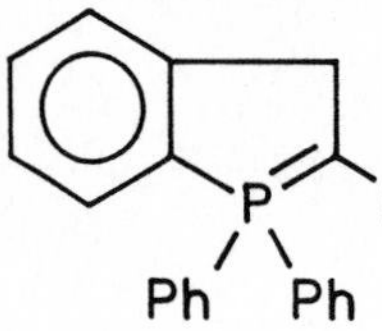

R^1 = H, Ph, CO_2Me (ref. 178)

9h

The 3H-phospholes, described recently by Barluenga et al., are on the other hand much more stable (ref. 180). The compounds 9i are all thermally stable, high-melting solids (eqn. 15).

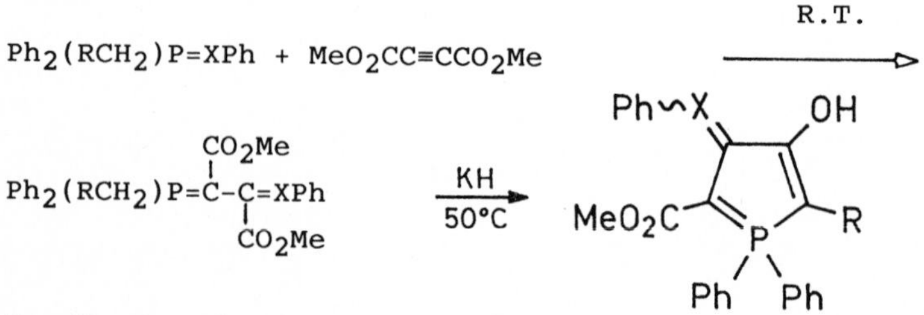

(eqn. 15)

R = H, Me, Ph, $CH{=}CH_2$; X = CH, N

9i

The already mentioned interaction of phosphetanes and acetylenedicarboxylicacid dimethylester leads to the bicyclic 2H-phosphole derivatives 9j (eqn. 16, ref. 176).

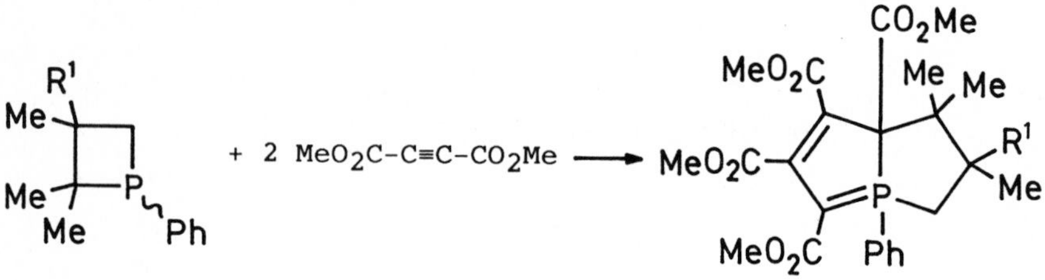

R^1 = H, Me

9j (eqn. 16)

2H-phosphole-containing bicycles are also formed, if the necessary phosphane unit is present in a 1H-phosphole (eqn. 17, ref. 123).

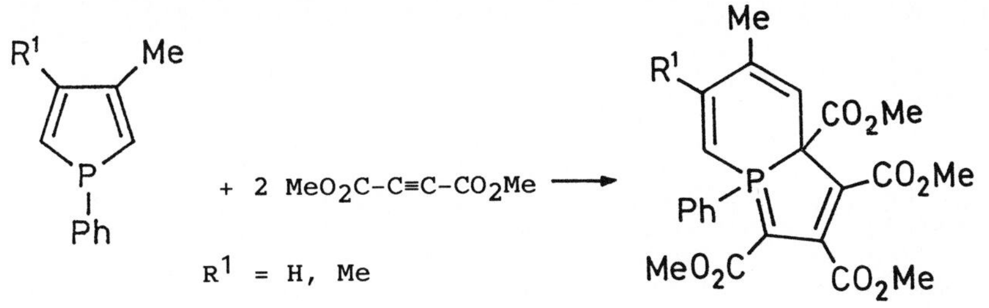

9k (eqn. 17)

3.2.3 Phospholyl Anions

In contrast to the λ^3,σ^3-phospholes, which can be described as "vinylphosphanes", phospholyl anions are truly aromatic compounds (refs. 137,145,181,182,193). The polarization of C-C-double bonds in the α-position of the phospholes are characterized as $^{\sigma+}C_{\beta}$-$C^{\sigma-}{}_{\alpha}$-P (ref. 137). However, one observes an increase in the C_{α}-P bond order in phospholyl anions, compared to phospholes. This is perceptible in the raise of $^1J(C_{\alpha}P)$-coupling constants from phospholes to the anions (e.g.: phosphole $^1J(CP)$ = 6.1 Hz; anion $[(CH)_4P]^-$: $^1J(CP)$ = 46.6 Hz, ref. 137). The delocalization of the charge over the entire ring leads to a partial phosphaalkene character of the C_{α}-P-bonds, observed in the low field shifts of the ^{31}P-NMR signals from the phosphole to the anion (e.g.: phosphole $\delta(^{31}P)$ = -49.2 ppm; $[(CH)_4P]^-$: 77.2 ppm (ref. 137). The most important orbitals of the phospholyl anions are shown in Fig. 7.

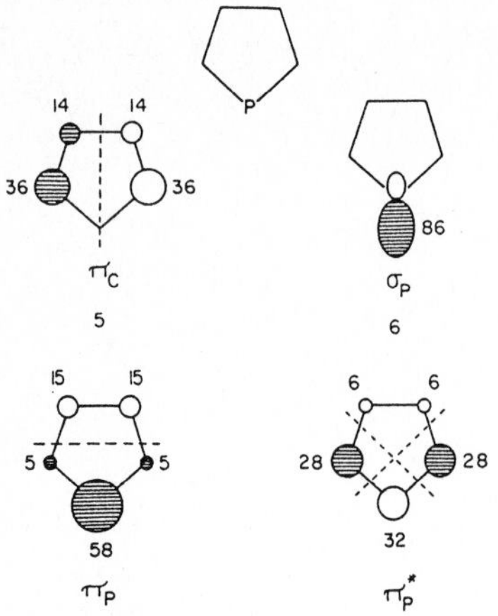

orbital sequence:
$\pi_C < \sigma_P < \pi_P$(HOMO) $<< \pi^*_P$(LUMO)

Fig. 7. Significant orbitals in the phospholyl anion (ref. 182).

For the synthesis of phospholyl anions various technics can be employed, but all depend essentially on a P-C or P-P bond cleavage with alkaline metals (e.g.: eqn. 18, ref. 128).

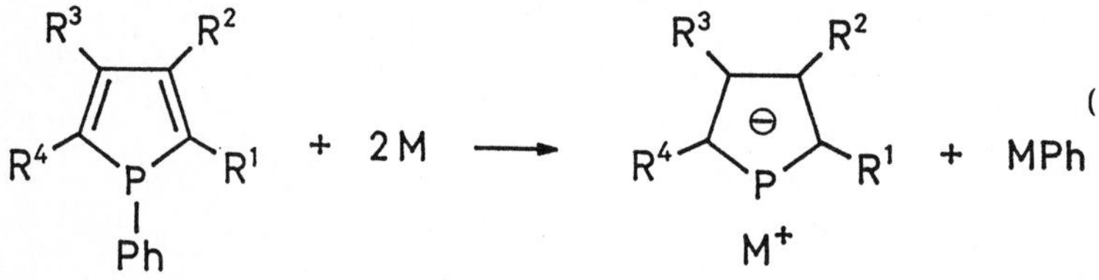

(eqn. 18)

10a

Sel. refs.:
$R^1 - R^4 = H$; $M^+ = Li^+$ (refs. 79,137,138)
$R^1 = R^4 = Ph$; $R^2 = R^3 = H$; $M^+ = Li^+, K^+$ (refs. 79,137)
$R^1 - R^4 = Ph$; $M^+ = Li^+, Na^+, K^+$ (refs. 79,183,184)

Quin et al. developed the synthesis of phospholyl anions in condensed ring systems in 1979: the starting-points were the corresponding P-phenyl-substitued phosphol derivatives which were reacted with potassium in tetrahydrofurane (thf) to give the anions (eqn. 19, ref. 145).

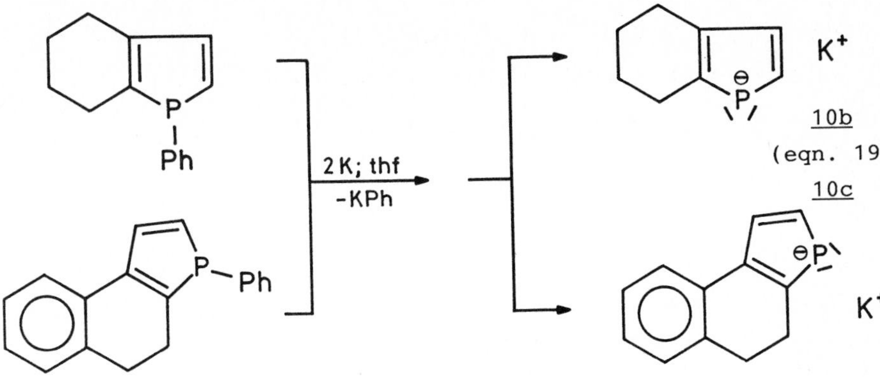

10b
(eqn. 19)
10c

The ^{31}P-NMR shifts 73.3 ppm (10b) and 81.7 ppm (10c) are in the region found for the unsubstituted phospholyl anion $[(CH)_4P]^-$ (72.2 ppm). In contrast to this, direct involvement in an extended aromatic system leads to highfield shifts. The signal for 10d now shows up at 40 ppm (ref. 149). 10d was synthesized by the same procedure as 10a-10c (eqn. 20, ref. 149).

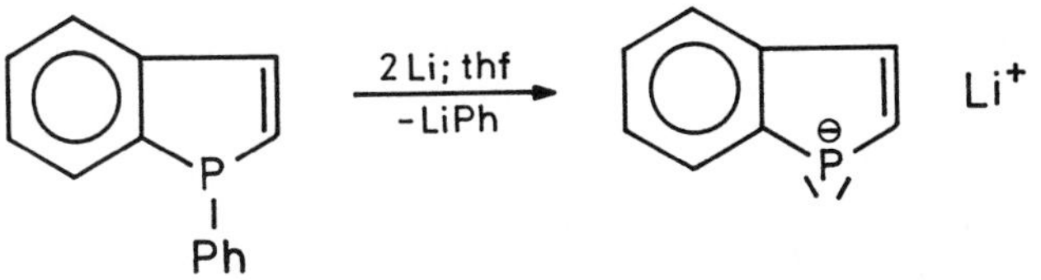

10d (eqn. 20)

Based on the aromatic character of phospholyl anions it is not surprising that the transion metal complexes 11a,b are possessing similarities in structure and properties to the corresponding cyclopentadienyl(Cp)-complexes.

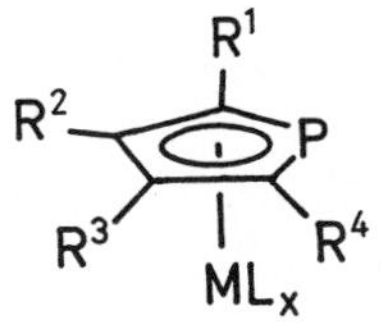

11a

Sel. refs.:

R^1 - R^4 = Ph; ML_x = $Mn(CO)_3$ (ref. 185)

R^1 = R^4 = H; R^2 = R^3 = Me; ML_x = $Mn(CO)_3$ (refs. 186,187)

R^1 - R^4 = Ph; ML_x = FeCp (ref. 185)

R^1 - R^4 = Ph; ML_x = Fe(CpMe) (ref. 188)

R^1 = R^4 = H; R^2 = R^3 = Me; ML_x = FeCp (refs. 117,189)

R^1 = R^4 = H; R^2 = R^3 = Me; ML_x = $Fe(3,4\text{-}Me_2PC_4H_2)$ (ref. 189)

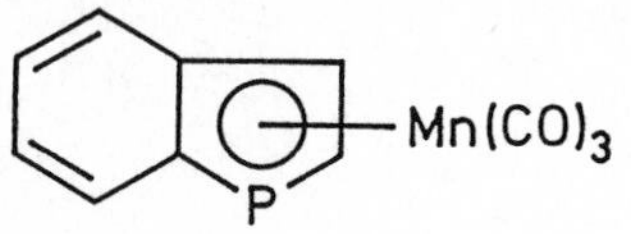

11b (ref. 149)

Particularly the phosphaferrocenes have been investigated. The result of theoretical studies (refs. 182,190) gives an approximate picture, confirmed by ^{57}Fe-NMR (ref. 117) and Mössbauer spectroscopy (ref. 191), as well as by other data (ref. 163).

3.3 DIPHOSPHOLENES AND DIPHOSPHOLES

3.3.1 Diphospholenes

The first $1\lambda^5,2\lambda^3$-diphospholene was found in 1968 by Bergerhoff et al. It was a reaction of malonic acid diesters with aryldichloro-phosphane in the presence of triethylamine (eqn. 21, refs. 192,195).

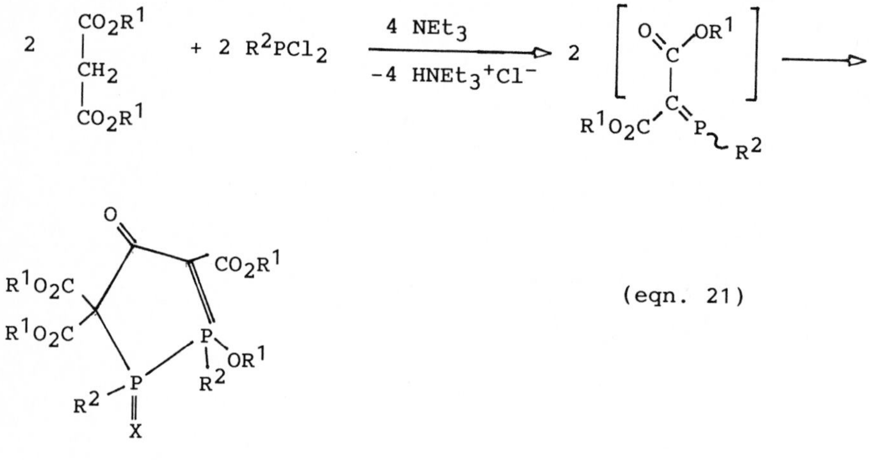

12a

(eqn. 21)

X = lone pair; R^1 = Me; R^2 = Ph (refs. 194,195)
X = $Ni(CO)_3$; R^1 = Me; R^2 = Ph (refs. 194,197)

X = lone pair; R^1 = Et; R^2 = Ph (refs. 194,196)
X = lone pair; R^1 = Me; R^2 = 4-BrPh (ref. 194)
X = lone pair; R^1 = Et; R^2 = 4-ClPh (ref. 195)
X = lone pair; R^1 = Et; R^2 = 4-MePh (ref. 195)
X = lone pair; R^1 = Et; R^2 = 4-NMe_2Ph (ref. 195)
X = lone pair; R^1 = Bu^t; R^2 = Ph (refs. 195,198)
X = lone pair; R^1 = Bu^t; R^2 = 4-NMe_2Ph (ref. 195)
X = lone pair; R^1 = Et; R^2 = Et (ref. 195)
X = lone pair; R^1 = Et; R^2 = Bu^n (ref. 195)

The ^{31}P-NMR spectrum of 12a (X = lone pair; R^1 = Me; R^2 = Ph) shows for P(III) a doublet at -39 and for P(V) a doublet at 86 ppm ($^1J(PP)$ = 198 Hz). From X-ray structure determination data of 12a (x = lone pair; R^1 = Et; R^2 = Ph) one may infer, that the five-membered heterocycle possesses a half-chair conformation. A C=P-double bond distance of 171.4 pm and a P-P-single bond distance of 219.3 pm was found (Fig. 8, ref. 196).
A $1\lambda^3,3\lambda^3$-diphospholene was mentioned in 1975 (ref. 201) and later isolated as a pure compound (ref. 202). The 1,3-diphospholene is formed by the radical catalyzed addition

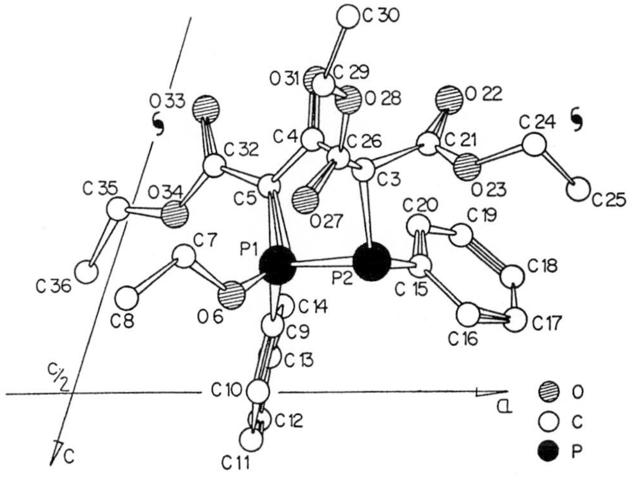

Fig. 8. Molecule structure of 12a (X = lone pair; R^1 = Et; R^2 = Ph)

of monoalkyl(aryl)phosphanes to alkyl(aryl)diethynylphosphanes (eqn. 22).

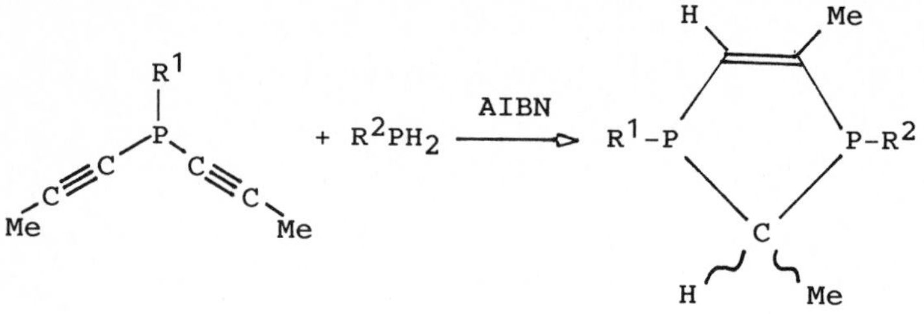

(eqn. 22)

12b

R^1 = Bu^t; R^2 = Ph (refs. 109,110)
R^1 = Ph; R^2 = Ph (ref. 110)
R^1 = CH_2Ph; R^2 = Ph (ref. 110)
R^1 = Bu^t; R^2 = Bu^t (ref. 110)

Diphospholenes containing phosphorus atoms with coordination number 2, were developed very recently. Thus, Issleib et al. were able to generate the 1,3-diphospholene 12c by interaction of 1,2-bis(trimethylsilyl)ethane with $PhN{=}CCl_2$ (eqn. 23, ref. 223).

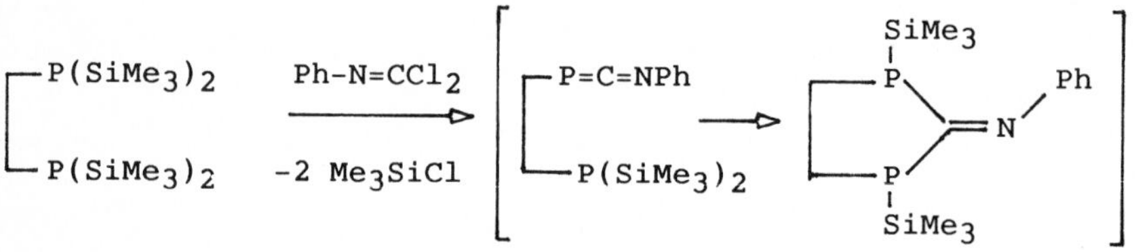

(eqn. 23)

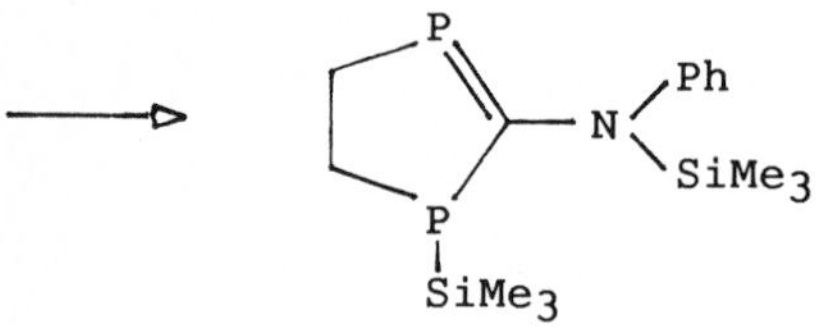

12c

12c could be isolated as a yellow, viscous oil in 25 % yield with a boiling point of 110 - 114°C / 10^{-3} Torr (^{31}P-NMR-spectrum: δ(-P=C) = 162, δ(-P) = -36.5 ppm; $^2J(PP)$ = 24 Hz). Parallel to this, Regitz et al. synthesized oligocyclic systems containing the $1\lambda^3,3\lambda^3$-diphospholene unit (ref. 224).

By reacting cyclohexadiene with phosphaalkynes, the heterocycles 12d,e were produced (eqn. 24).

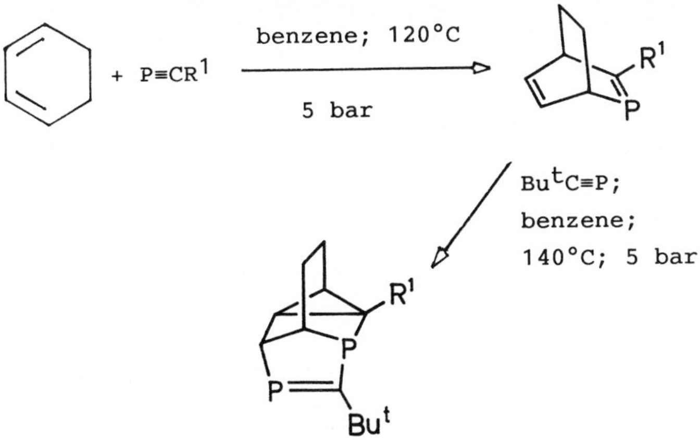

(eqn. 24)

R^1 = Bu^t; CMe_2Et 12d,e

In contrast to (eqn. 24), the corresponding reaction (eqn. 25) of cyclopentadiene with Bu^t-C≡P leads to the phospholene 7c under mild conditions. Only for the additional reaction with a second phosphaalkyne more drastic conditions were required (physical data and yields in Tab. 9).

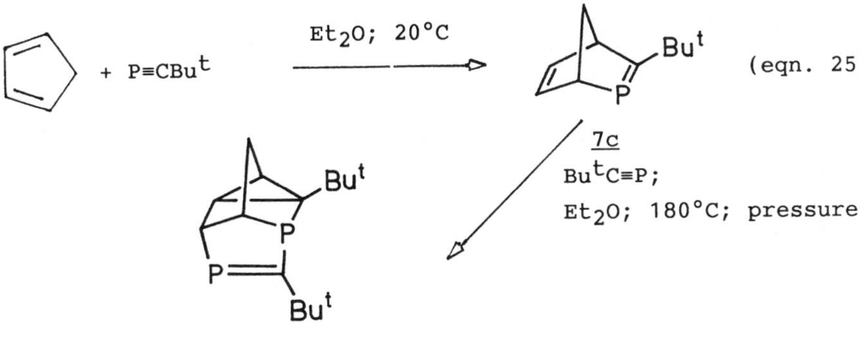

(eqn. 25)

12f

TABLE 9
^{31}P-NMR data and yields of 7c and 12d-f.

	7c	12d	12e	12f
$\delta(^{31}P)$ [ppm]				
-P=C	246	315	315	322
P-	-	-17	-15	5
$^2J(PP)$[Hz]	-	12.6	10.6	18
yield [%]	80[a)]	70	48	80

a) attempts for isolation in pure form failed

A $1\lambda^5,3\lambda^5$-diphospholane unit is included in 12g. 12g can be isolated from the reaction between a diphosphino-substituted (α-carbon atom) ylid and $BrCH_2CH{=}CH_2CH_2Br$ (eqn. 26, ref. 199).

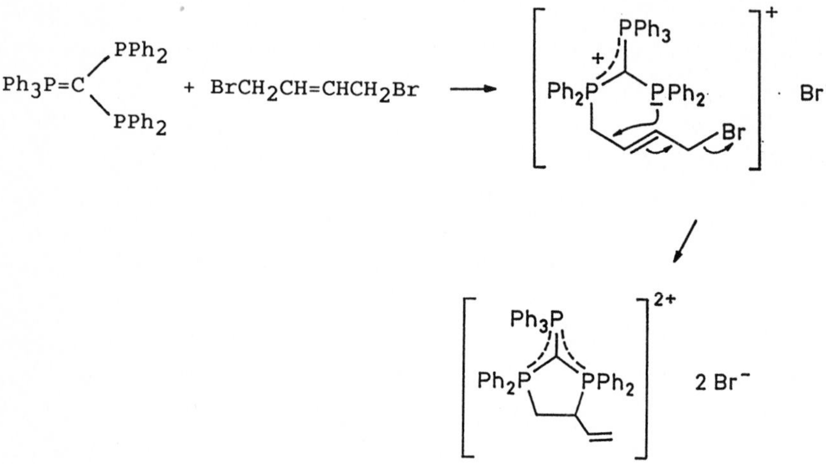

12g (eqn. 26)

12g is thermally a very stable compound (mp. 300°C) and shows in the ^{31}P-NMR spectrum an AMX-spinsystem (11 lines, not correlated; 23.2 ppm - 48.8 ppm).

3.3.2 Diphospholes

Diphospholes were unknown until 1970 when Tebby et al. formed a $1\lambda^5,3\lambda^5$-diphosphole from acetylendicarboxylic acid dimethylester and methylenbis(diphenylphosphane) (eqn. 27, ref. 200).

$Ph_2PCH_2PPh_2 + MeO_2CC{\equiv}CCO_2Me$

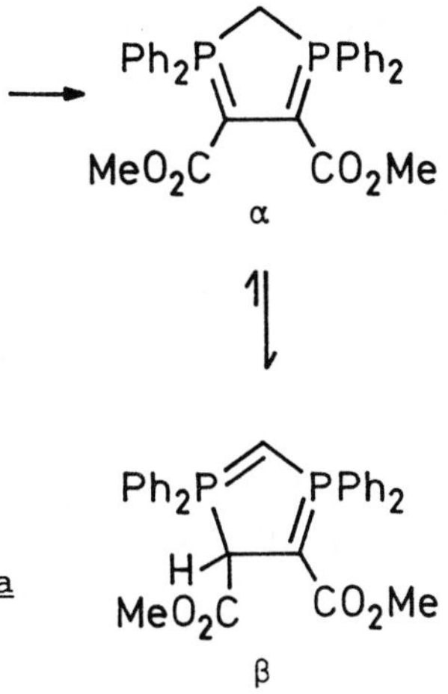

13a (eqn. 27)

The authors substantiated the almost exclusive formation of 13ß (ß/α ≈ 10^4) because of the segment $Ph_2P\cdots CH\cdots PPh_2$, given in ß. This segment usually leads to a favourable delocalization of the charge in ylids. 13a was characterized by ^{1}H-NMR spectroscopy and mass spectrometry (mp. 176 - 177°C). The $1\lambda^5,3\lambda^5$-diphosphole 13b, published by Fluck et al. in 1988 (ref. 203), possesses only ring-C-atoms with sp^2-character. Stirring 1,1,3,3-tetrakis(dimethylamino)-$1\lambda^5,3\lambda^5$-diphosphete (ref. 43) with phenylisonitril 12 h at 70°C in toluene, the ^{31}P-NMR spectrum of the reaction mixture showed a large number of signals between 25 and 100 ppm. By flash chromatography a yellow green oil was isolated from the mixture, which crystallized at -20°C. Bright yellow crystals were obtained in 20 % yield with a melting point of 77 - 79°C. The NMR spectroscopical investigations and data from the mass spectrum allowed an identification: a formal [2+1]-cycloaddition, followed by a spontaneous valence isomerization leads to the first azadiphosphapentafulvene 13b (eqn. 28).

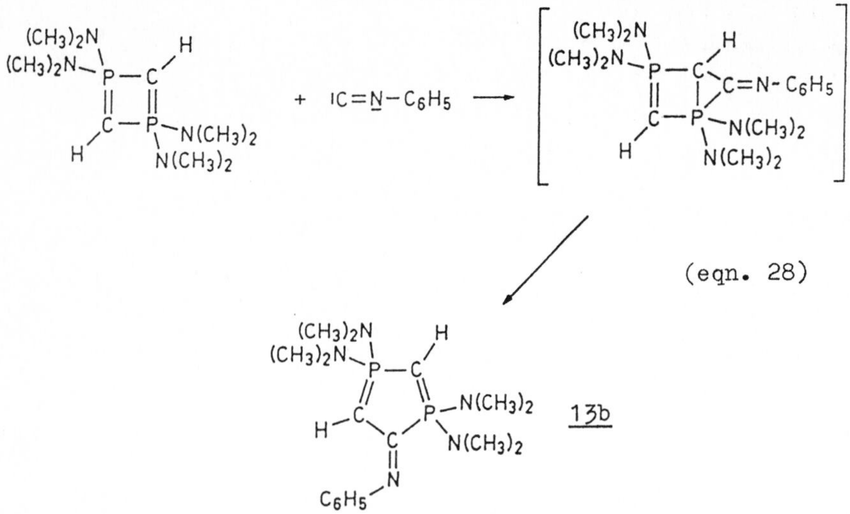

(eqn. 28)

13b

13b turns out to be very stable and is not sensitive to air and moisture. In the mass spectrum (20eV) a molecular peak with a relative intensity of 100 % can be observed. The $^{31}P\{^{1}H\}$-NMR spectrum of 13b shows only one AB-spinsystem ($J(AB)/\nu_0\delta = 0.26$), which means that only one of the imaginable syn/anti-isomeres can be detected at room temperature. By means of ^{13}C-NMR measurements, the anti-form (lone pair of nitrogen and P_A in cis-position; see Table 10 and Fig. 9) could be derived as the one present (ref. 203). By ^{1}H- and ^{13}C-NMR investigations the character of 13b is determined by the conjugated double-ylid-structure ($P_AC^1P_BC^2$) and the imino-carbon atom (C^3) (see Table 10).

TABLE 10

Important NMR-parameters of 13b[α].

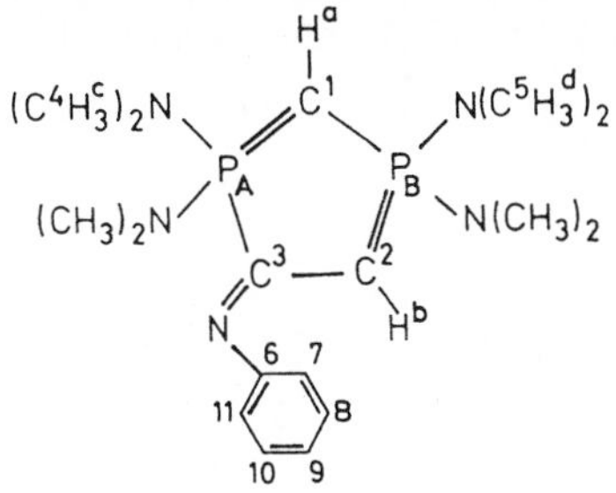

		δ[ppm]	J[Hz]
^{31}P	P_A	56.7	$^2J(P_AP_B)$ = 134.0
	P_B	70.9	
^{13}C	C^1	-2.7	$^1J(C^1P_A;P_B)$ = 124.4; 127.4[ß]
	C^2	54.9	$^1J(C^2P_B)$ = +141.4[γ]
	C^3	159.7	$^1J(C^3P_A)$ = +136.6
^1H	H^a	0.61	$^2J(H^aP_A)$ = $^2J(H^aP_B)$ = 0.8
	H^b	4.21	$^2J(H^bP_B)$ = +14.9
			$^3J(H^bP_A)$ = +47.7
			$^3J(H^aH^b)$ = 2.0

α) measurement at 304 K in C_6D_6

ß) could not be correlated

γ) detailed specification of the NMR-experiments see ref. 203.

The X-ray structure analysis of 13b confirms the structure determination arrived at NMR-data. The molecular structure (Fig. 9) shows a planar five-membered heterocycle. The fragment P2-C1-P1-C3-C2 [d,P2-C1 = 169.3; C1-P1 = 168.7; P1-C3 = 173.2;

C3-C2 = 135.1 ppm] together with the exocyclic imino-nitrogen atom (d,C2-N5 = 133.9 pm) forms a conjugated system (Fig. 10), containing C-P-, C-C-, and C-N-double bond distances. The distance C2-P2 (186.9 pm) corresponds to a C-P-single bond, so that the mesomeric structures, shown in Fig. 10, is dominating.

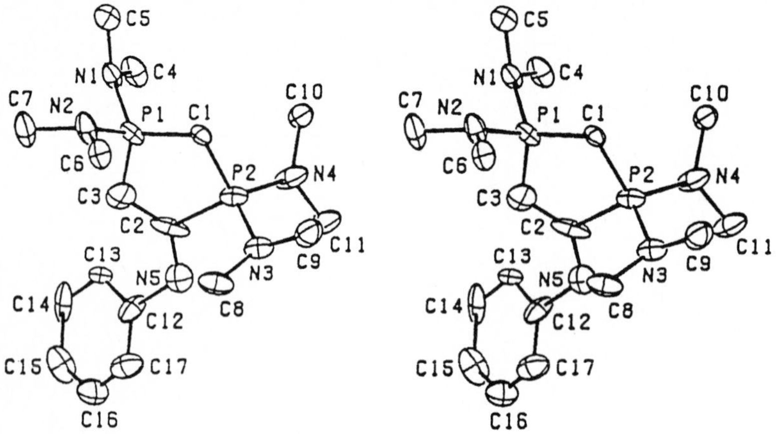

Fig. 9. Molecular structure of 13b.

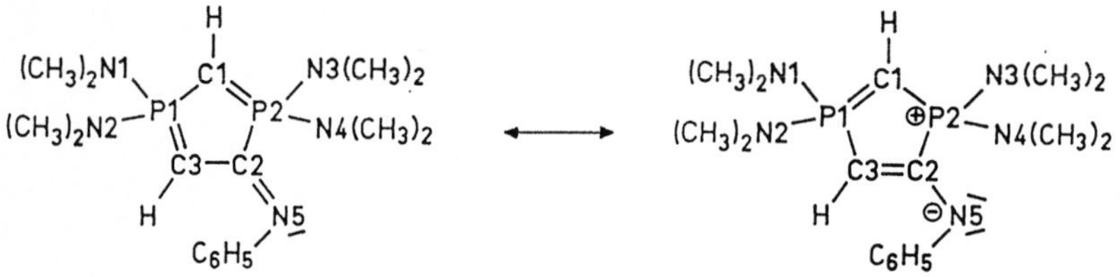

13b

Fig. 10. Mesomeric structures of 13b.

The incorporation of a carbodiphosphorane unit in a five-membered ring leads to "diphospholes" of type 13c (eqn. 29, ref. 204).

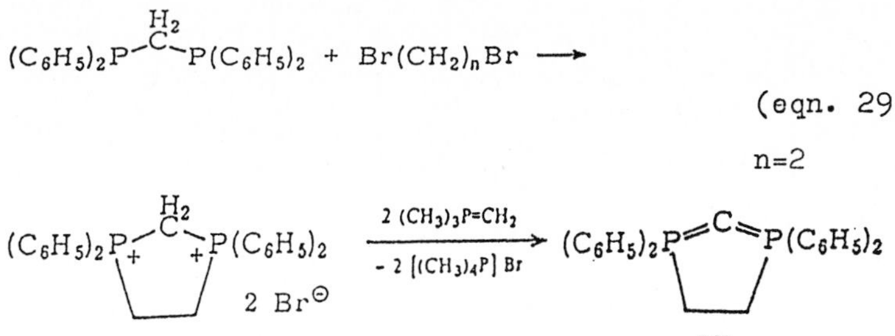

$$(C_6H_5)_2P\overset{\overset{H_2}{C}}{\diagup}\diagdown P(C_6H_5)_2 \;+\; Br(CH_2)_n Br \longrightarrow$$

(eqn. 29)

n=2

$$(C_6H_5)_2\overset{+}{P}\diagup\overset{\overset{H_2}{C}}{}\diagdown\overset{+}{P}(C_6H_5)_2 \quad\underset{-\,2\,[(CH_3)_4P]\,Br}{\overset{2\,(CH_3)_3P=CH_2}{\xrightarrow{\hspace{2cm}}}}\quad (C_6H_5)_2P\!\!=\!\!\overset{C}{\diagup}\diagdown\!\!P(C_6H_5)_2$$

2 Br$^{\ominus}$

13c

Compound **13c** is sparingly soluble in inorganic solvents and thermally unstable, probably because of the ring tension in the P=C=P-sequence. **13c** decomposes at 20°C in several hours [$\delta(^{31}P)$ = -22.45 ppm]. In the reactions of 1,2-diphosphino-substituted benzenes with CH_2Br_2, molecules with 1,3-diphosphainane-frameworks are formed (eqn. 30, ref. 205).

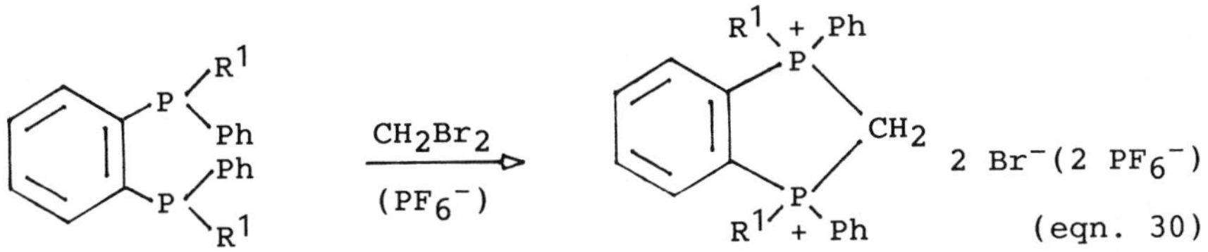

12h,i

12h: R^1 = Me (RR, SS, RS; for Br$^-$ and PF$_6^-$)

12i: R^1 = Ph (Br$^-$)

With ammonia or n-butyllithium **12h,i** can be deprotonated to the semi-ylid-phosphonium salt **13d,e** (eqn. 31, ref. 205).

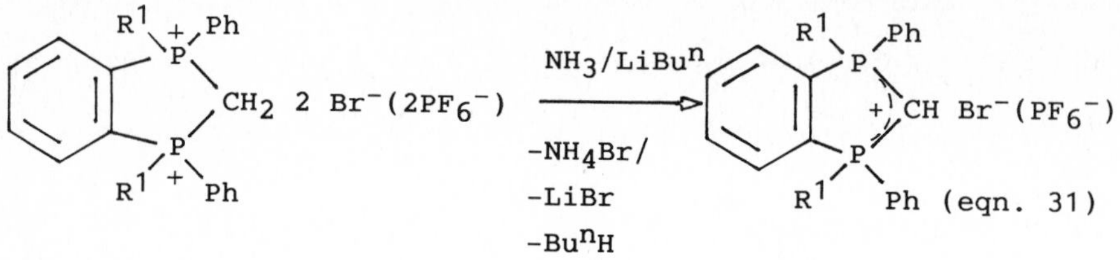

12h,i 13d,e

13d: R^1 = Me (RR, SS, RS; for Br^- and PF_6^-)
13e: R^1 = Ph (Br^-)

By reacting 12h with two equivalents, or 13d with one equivalent of $Et_3P{=}CHMe$, one receives 13f (eqn. 32).

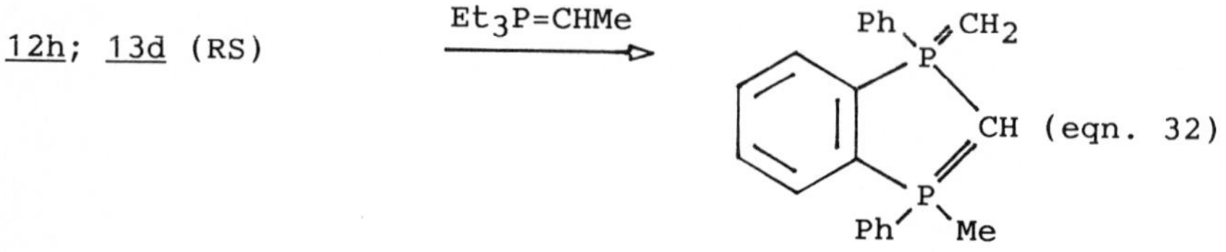

13f (RS)

In contrast to eqn. 32, the lithium-salt adducts 13g can be isolated by treating 12h with 2 equivalents of n-butyllithium (eqn. 33).

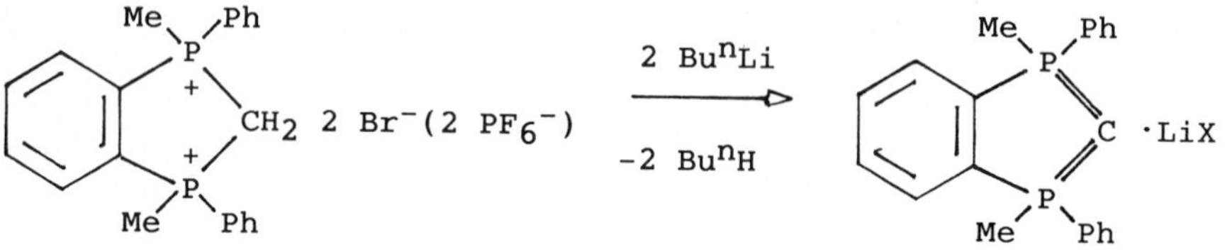

12h (RS) 13g (RS; X^- = Br^-, PF_6^-)
(eqn. 33)

The characterization of 13g is easier by derivatizing 13g to the semi-ylid-phosphonium-salt 13a and to the phosphonium salt 12j (eqn. 34).

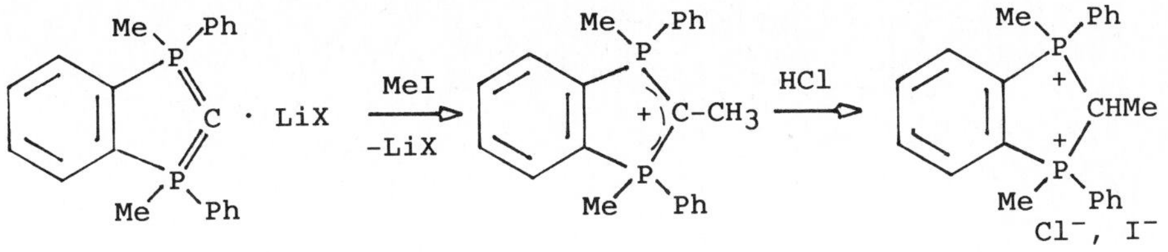

13g (RS; X^- = Br, PF_6) 13h 12j (eqn. 34)

Whereas, according to eqn. 33, only the lithiumsalt-adducts can be isolated, it is possible to synthesize the desired carbodiphosphorane 13i, when the phosphonium-salts 12i and 13e were reacted with a ylid (eqn. 35).

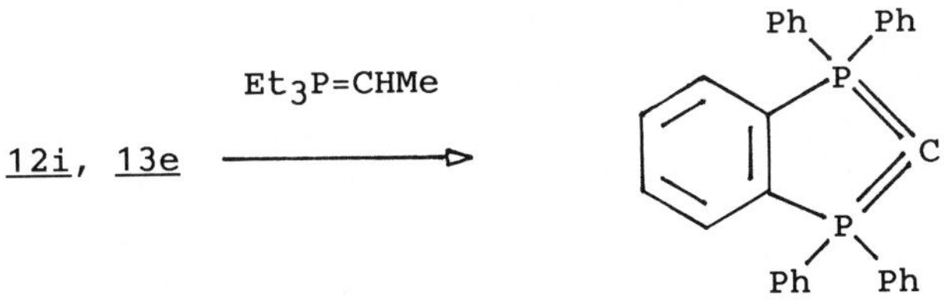

(eqn. 35)

13i

13i is only slightly soluble in organic solvents, and decomposes above -30°C (Tab. 11 shows the ^{31}P-NMR shifts of 12h - 13i).

TABLE 11
$\delta(^{31}P)$-Values of 12h - 13i.

Compound	$\delta(^{31}P)$ [ppm]
12h	41 (RS; X^- = Br^-)$^{a)}$; 39,5 (RR, SS; X^- = Br^-)$^{a)}$
12i	30.7 (X^- = Br^-)$^{b)}$
12j	46.2 (X^- = I^-, Cl^-; RS)$^{b)}$
13d	29.9 (RS; X^- = Br^-)$^{c)}$; 30.7 (RR, SS; X^- = Br^-); 30.9 (RS; X^- = PF_6^-)^{d}; 31.5 (RR, SS; X = PF_6^-)$^{d)}$
13e	31.3 (X^- = Br^-)$^{c)}$
13f	3.9, 26.7 ($^2J(PP)$ = 57 Hz; RS)$^{e)}$; 5.6, 27.0 ($^2J(PP)$ = 57 Hz; RR, SS)$^{e)}$

13g	28.6 (X^- = Br^-; RS)[e]; 28.4 (X^- = Br^-; RR, SS)[e]; 30.8 (X^- = PF_6; RR, SS)[f]
13h	27 (X^- = I^-; RS)[c]
13i	29.5[f]

a) in D_2O; b) in CF_3COOH; c) in $CDCl_3$;
d) in CD_3CN; e) in C_6D_6; f) in THF

3.3.3 Diphospholyl Anions

A $1\lambda^5,2\lambda^3$-and $1\lambda^3,3\lambda^3$-diphosphole, or their diphosphole anions have not yet been isolated as pure compounds. However, in 1987 two groups were independently successful in isolating complexes which contained the $1\lambda^3,3\lambda^3$-diphospholyl unit [in the mixture with $(Bu^tC)_2P_3^-$, $(Bu^tC)_3P_2^-$ 14a could be identified by a ^{31}P-NMR-signal at 185 ppm* (ref. 38)] (eqn. 36, 37, refs. 38,206) (* δ-value was calculated with $\delta(^{31}P[P(OMe_3)_3])$ = 140 ppm). Excellent reviews about reactions of C≡P- and C=P- units in coordination chemistry are available in literature (refs. 249,250).

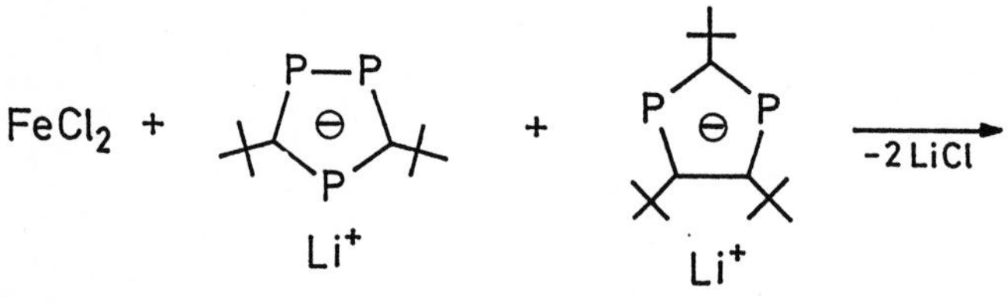

14a

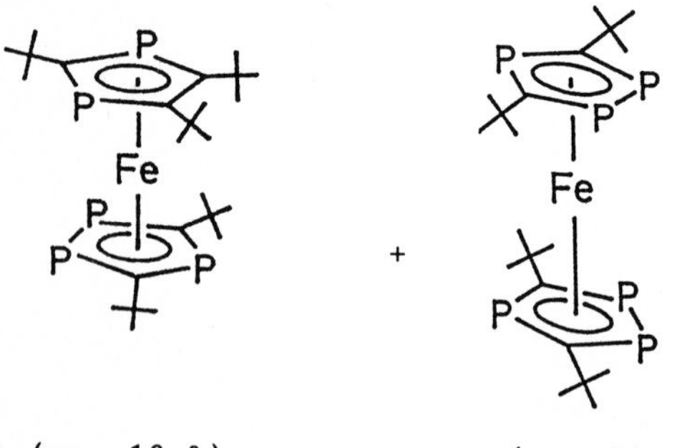

14b (ca. 10 %) (ca. 30 %) (eqn. 36, ref. 206)

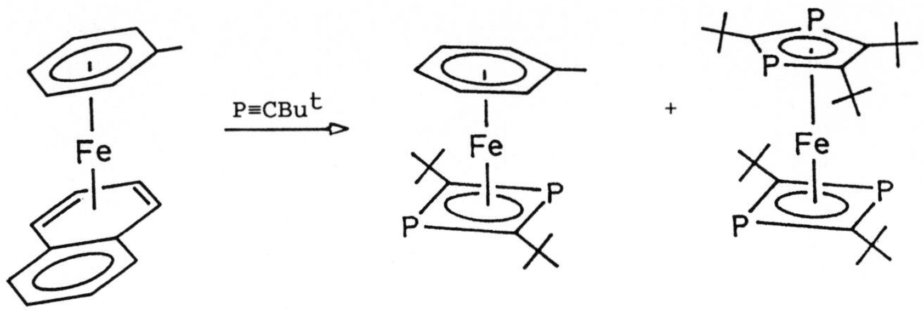

48 %   14c (ca. 3 %)

(eqn. 37)

+ 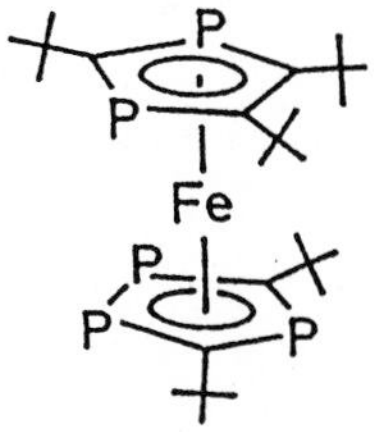

14b (ca. 3 %)

14b can be described as a ferrocene analogue which is verified by an X-ray-structure determination (ref. 206). The P-C-distances in the diphospholyl ligands (the triphospholyl complexes will be discussed in 3.4.2.) are equally long, within the limits of the standard deviations (175 pm), whereas the C-C-distance is measured as ca. 143 pm (see Fig. 11).

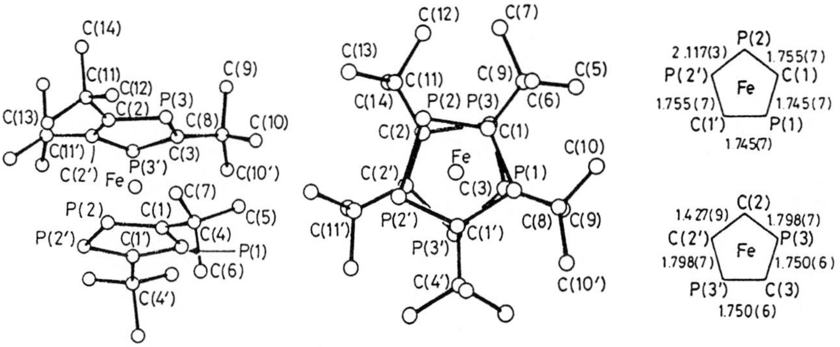

Fig. 11. Two views of the molecular structure and bond lengths of 14b

Although it has been shown my mass spectrometric data that 14b can also be generated by eqn. 37 (ref. 38), the authors of ref. 206 and ref. 38 give contradicting ^{31}P-NMR data for 14b (AB_2-system*: $\delta(P_A)$ = 189; $\delta(P_B)$ = 188.1 ppm; $J(P_AP_B)$ = 44 Hz (ref. 206); $\delta(P^1)$ = 45.9, $\delta(P^2)$ = 32.7 ppm (ref. 38) * see above). 14c is a paramagnetic complex, which was investigated by ESR-spectroscopy (ref. 38, further investigations on diphospholyl and triphospholyl anions see 3.4.2.).

3.4 TRIPHOSPHOLENES AND TRIPHOSPHOLES

3.4.1 Triphospholenes

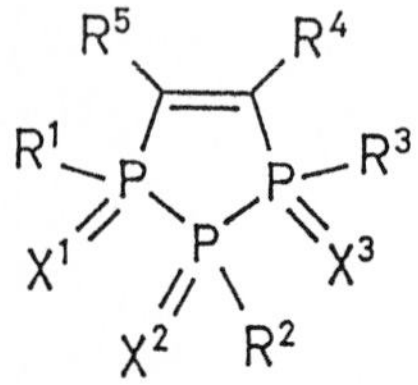

15a

- $X^1 - X^3$ = lone pair; $R^1 - R^5 = CF_3$ (ref. 23)
- $X^1 = X^3$ = S; X^2 = lone pair; $R^1 - R^5$ = Ph (ref. 24)
- $X^1 - X^3$ = lone pair; $R^1 = R^2$ = 4-NMe_2Ph; $R^3 = Bu^t$; $R^4 = R^5$ = H (ref. 207)
- $X^1 - X^3$ = lone pair; $R^1 - R^3$ = Me; $R^4 = R^5$ = H (ref. 208)
- $X^1 - X^3$ = lone pair; $R^1 - R^3$ = Ph; $R^4 = R^5$ = Et (ref. 26)
- $X^1 - X^3$ = lone pair; $R^1 - R^3$ = Et; $R^4 = R^5$ = H (ref. 209)
- $X^1 - X^3$ = lone pair; $R^1 - R^3 = CH_2Ph$; $R^4 = R^5$ = H (ref. 202)
- $X^1 - X^3$ = lone pair; $R^1 - R^3$ = Ph; $R^4 = R^5$ = Me (ref. 28)
- $X^1 - X^3$ = lone pair; $R^1 - R^3$ = Ph; $R^4 = R^5 = SiMe_3$ (ref. 28)
- $X^1 - X^3$ = lone pair; $R^1 - R^3$ = Ph; $R^4 = R^5 = Bu^t$ (ref. 28)
- $X^1 - X^3$ = lone pair; $R^1 - R^3$ = Me; $R^4 = R^5 = Bu^t$ (ref. 28)

In 1964, W. Mahler was able to synthesize the first 1,2,3-triphospholene. Further representatives, or their trapping products, were isolated by using the same procedure, pyrolysis of cyclotetra- or cyclopentaphosphanes in the presence of acetylenes (eqn. 38, refs. 24,26,28).

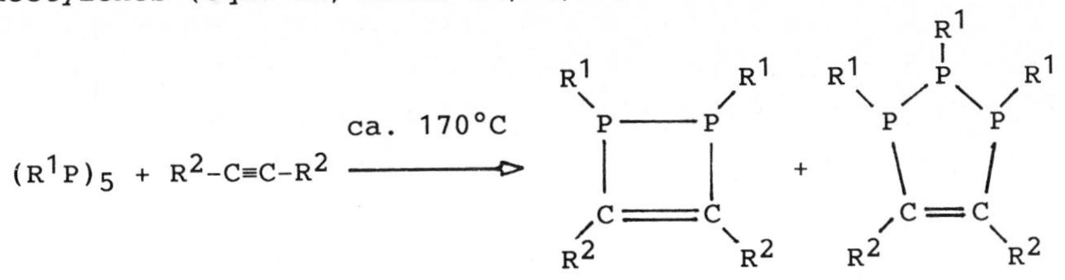

(eqn. 38)

R^1 = CF_3, Me, Ph; R^2 = H, CF_3, Me, Et, $SiMe_3$, Ph, Bu^t 15a

The reaction of $K_2(PR)_4$ with cis-ClHC=CHCl (refs. 208,209) is another possible way to obtain 15a.

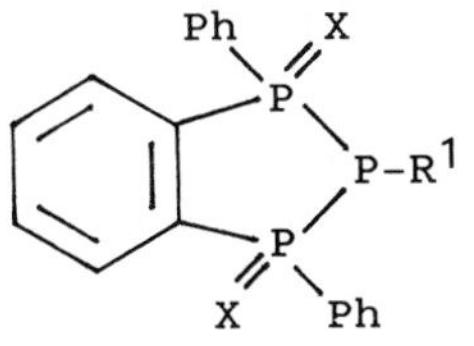

X = lone pair; R^1 = Me (refs. 210,211)
X = S; R^1 = Me (ref. 210)
X = lone pair; R^1 = Ph (refs. 211-214)
X = lone pair; R^1 = H, $SiMe_3$, PPh_2 (ref. 211)
X = lone pair; R^1 = Et (refs. 210,211)

15b

Whereas alkyl and aryl substituted 2,3-dihydro-1H-benzotriphospholes are wellknown (refs. 210-214), a 2-lithio-2,3-dihydro-1H-benzotriphosphole 15c was isolated for the first time in 1985 by Schmidpeter et al. (ref. 211). 15c can be synthesized by reacting o-phenylen-bis(lithiumphenylphosphide) with white phosphorus, crystallizing as thf-adduct (eqn. 39).

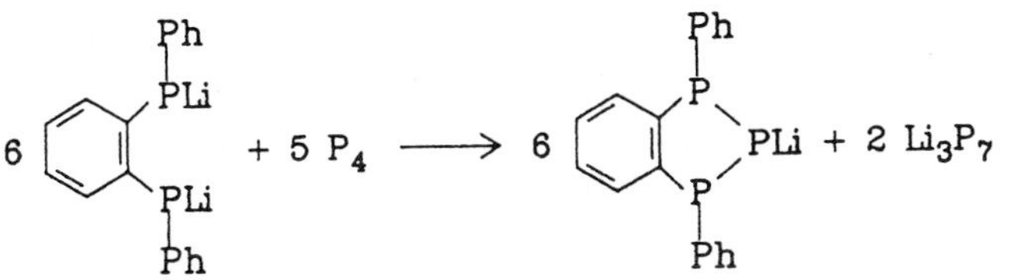

(eqn. 39)

15c

An X-ray structure determination of 15c·$(thf)_3$ shows pyramidal phosphorus atoms with slightly shortened, in comparison to 15b, [X = lone pair; R^1 = Ph; d, P-P = 220.9 ppm (ref. 214)] P-P-distances of 216.8 and 215.3 pm, respectively. This points to an increase in the P-P bond order in the fragment Ph-P-PLi-P-Ph (Fig. 12, ref. 211).

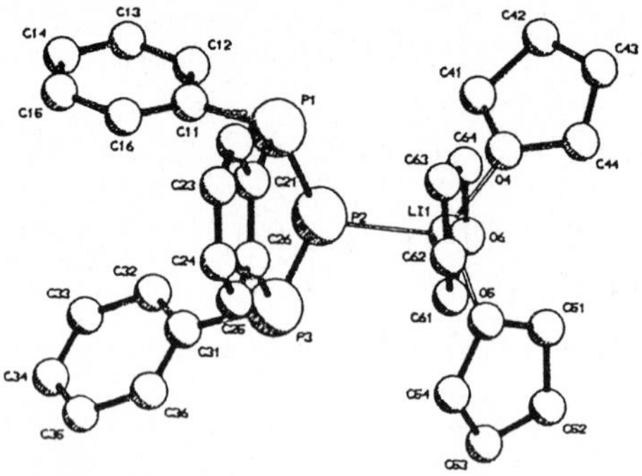

Fig. 12. Molecular structure of 12c·$(thf)_3$

From the ^{31}P-NMR data of 15c one can recognize the significant increase of negative charge at the phosphorus atom P-2 (high field shift of $\delta(^{31}P)$ in comparison to a P-2 substituted benzotriphosphole (Tab. 12).

With 1,4-dipotassium-1,2,3,4-tetramethyl-phosphane and tetrachloracetylene it is possible to build 15d, a bicyclic system with six phosphorus atoms (eqn. 40, ref. 215).

$2\ K_2(PMe)_4 + C_2Cl_4 \longrightarrow$ 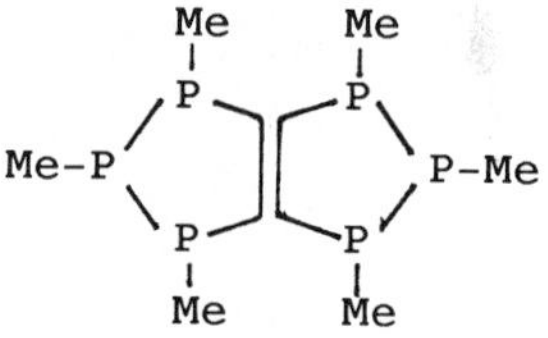 (eqn. 40)

15d

TABLE 12

^{31}P-NMR data of benzotriphospholes (PhP^A-P^BR-P^APh).

compound	R^1	X	$\delta(P^A)$ [ppm]	$\delta(P^B)$ [ppm]	$^1J(P^AP^B)$ [Hz]	solvent	ref.
<u>15c</u>	Li $(thf)_3$	lone pair	31.4	-174.0	368.5	thf	211
<u>15b</u>	H[a)]	lone pair	14.0	-125.9	222.8	thf	211
	Me	lone pair	32.8	- 67.5	252.0	$CHCl_3$	210
	Me	S	60.8	- 77.5	250.0	$CHCl_3$	210
	Ph	lone pair	32.5	- 39.4	265.0	$CHCl_3$	212

a) $^1J(P^BH)$ = 207.2 Hz; $^2J(P^AH)$ = 23.8 Hz

The complex ^{31}P-NMR spectrum shows for $\delta(P^A)$ ca. +12 ppm and $\delta(P^B)$ ca. -4 ppm (MeP^A-P^BMe-P^AMe; chemical shifts are estimated). There exists only a few 1,2,3-triphospholenes with phosphorus-phosphorus multiple bond, whereas such compounds with C-P-multiple bonds are unknown at this time.

The $1\lambda^5,2\lambda^3,3\lambda^5$-triphospholenyl-cation <u>16a</u> is built up from PCl_3, $SnCl_2$ and $Ph_2P(CH_2)_2PPh_2$ (eqn. 41, ref. 216).

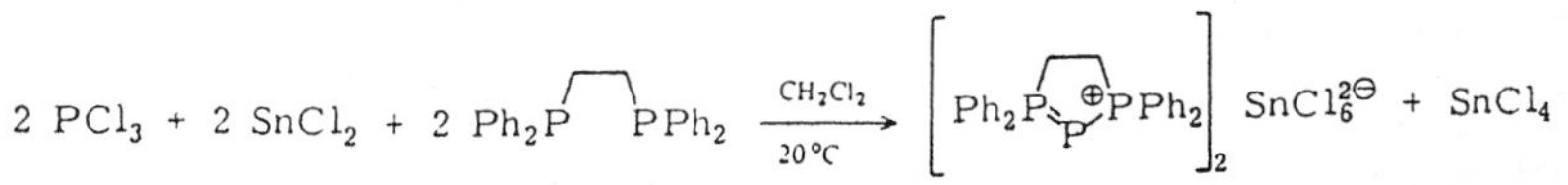

<u>16a</u> (eqn. 41)

In a different way, by a ligand exchange on a formally P^+, one can obtain a similar system (eqn. 42, 43, ref. 217).

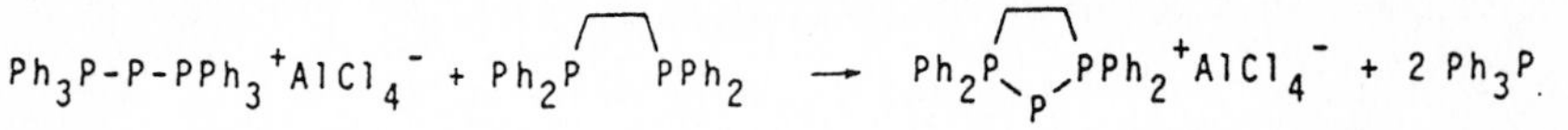

$$Ph_3P-P-PPh_3{}^+AlCl_4{}^- + Ph_2P \underset{\frown}{\quad} PPh_2 \longrightarrow Ph_2P{\underset{P}{\overset{\frown}{\diagdown}}}PPh_2{}^+AlCl_4{}^- + 2\ Ph_3P$$

16b

(eqn. 42)

$$Ph_3P-P-PPh_3{}^+AlCl_4{}^- + \qquad \xrightarrow{\ -\ 2\ Ph_3P\ }$$

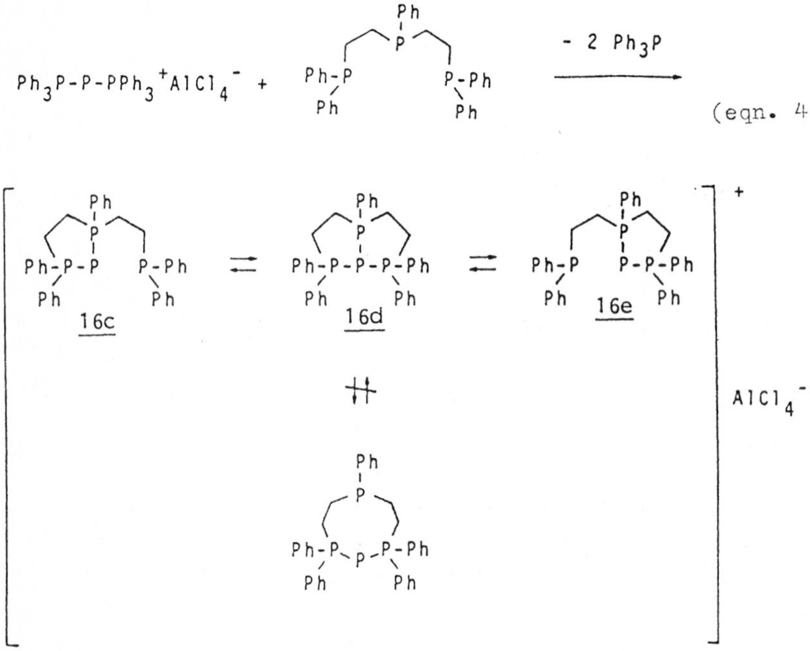

(eqn. 43)

The structure determination of 16a shows as expected a five-membered ring with envelope conformation. The P-P-distances in the P-P-P-subunit are ca. 213 pm, a value between a P-P-single and a P-P-double bond (see Fig. 13, ref. 216).

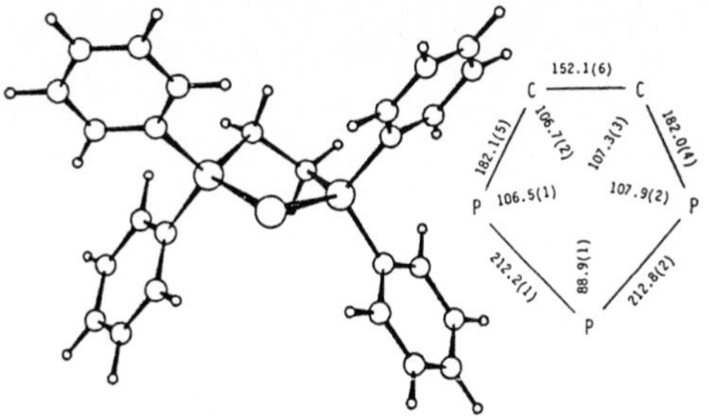

Fig. 13. Structure of the cation in <u>16a</u>·2 CH_2Cl_2 and bond lengths [pm] and angles [°] in the triphospholene ring.

In the $^{31}P\{^1H\}$-NMR spectrum of <u>16a</u> an AB_2-spinsystem with $\delta(P^A)$ = -231.6 and $\delta(P^B)$ = 63.8 ppm ($^1J(P^AP^B)$ = 448.9 Hz) can be observed. <u>16c</u> and <u>16e</u> exist in a dynamic equilibrium, according to ^{31}P-NMR investigations at variable temperatures (ref. 217). <u>16d</u> is supposed to be an intermediate or a transition state in this equilibrium.

3.4.2 Triphospholes and Triphospholyl Anions

The field of triphospholes is very similar to that of the diphospholes. There is no triphosphole reported until 1988, whereas triphospholyl anions are known since 1985. Becker et al. succeeded in isolating a 1,2,4-triphospholyl anion <u>17</u> (ref. 218). The reaction mechanism is not known; however, the intermediates are probably like the triphospholenic anions <u>16e,f</u>. <u>16e,f</u> are presumably formed from phosphaalkenes and an excess of $LiP(SiMe_3)_2$, which is still present in the reaction mixture (eqn. 44).

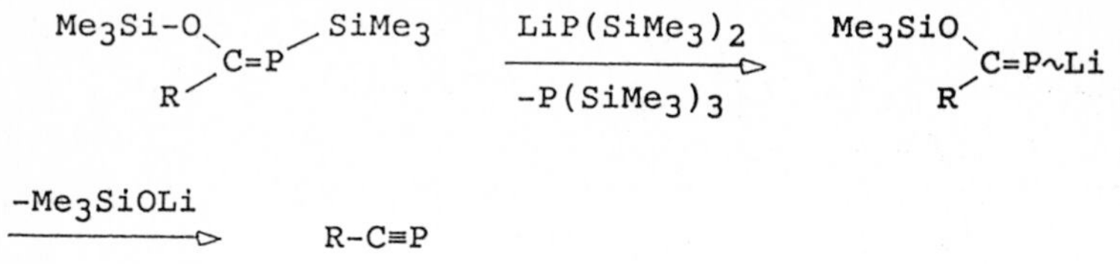

(eqn. 44)

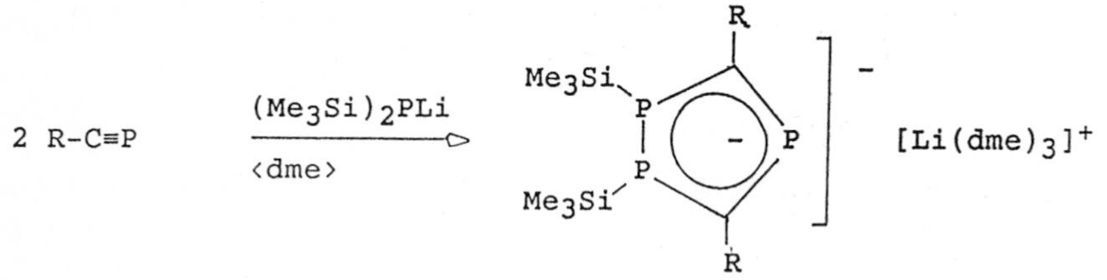

16e: R = Bu^t; 16f: R = Ph (dme: 1,2-dimethoxy-ethane)

16f can also be synthesized in another way. The lithium-3,5-diphenyl-1,2-bis(trimethylsilyl)-1,2,4-triphosphapent-3-en-5-id·3(dme) is obtained as dark green, metallic, shiny needles in 40% yield, from the reaction of ethylbenzoate with $LiP(SiMe_3)_2$ (eqn. 45, ref. 218).

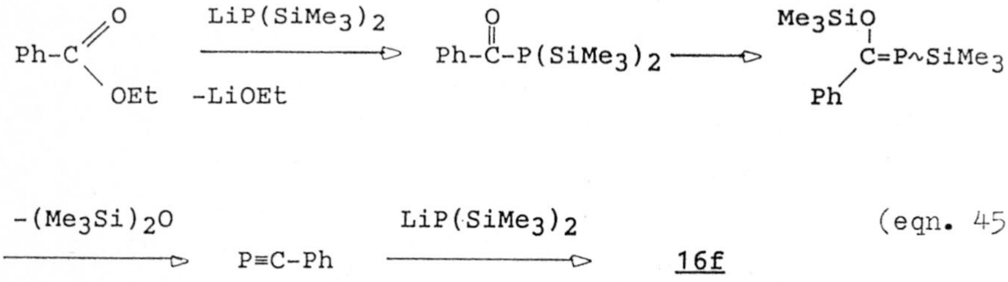

(eqn. 45)

The $^{31}P\{^1H\}$-NMR spectrum of 16f in dme exhibits a characteristic AX_2-spinsystem ($\delta(P^A)$ = 275 ppm; $\delta(P^X)$ = -87.5 ppm; $^2J(P^AP^X)$ = 33 Hz). The bond distances of 172 pm in the fragment Ph-C-P-C-Ph, determined by X-ray structure analysis, are indicating high multiple bond character in these C-P-bonds.

The triphospholenyl anions, generated as shown (eqn. 44) cannot be isolated, but 1,2,4-triphospholyl anions are formed in this reaction by a fragmentation of trimethylsilyl groups (ref. 218). For instance the lithium-3,4-di-tert.-butyl-1,2,4-triphosphacyclopentadienide·3(dme) 17 can be obtained in 25% yield (eqn. 46).

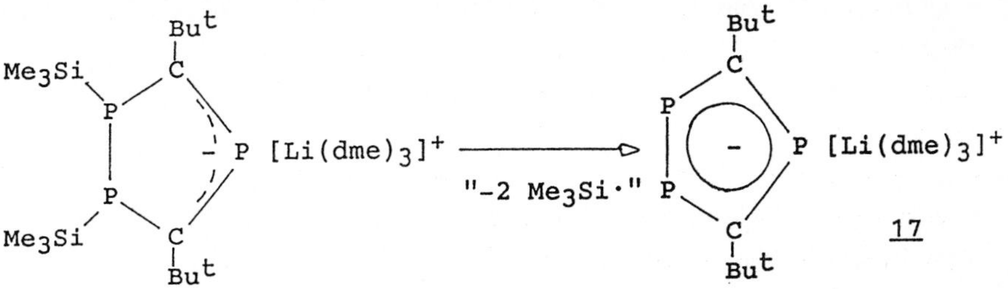

(eqn. 46)

The AB_2-spinsystem of 17 in the $^{31}P\{^1H\}$-NMR spectrum ($\delta(P^A)$ = 252.5, $\delta(P^B)$ = 239 ppm; $^2J(P^AP^B)$ = 51 Hz points to a typical aromatic system, which can be correlated very good with other five-membered phosphorus-carbon-aromatic systems (see 3.5., ref. 53). Reactions of 17 with transition metal complexes were published in 1987 (eqn. 47, ref. 206, see also 3.3.3.).

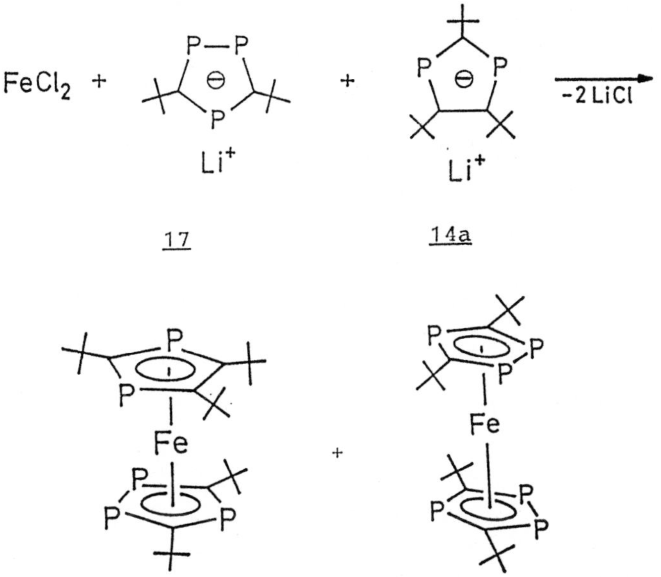

(eqn. 47)

14b (ca. 10 %)

18a (ca. 30 %)

The P-P-distances (triphospholyl ring) in 14b and 18a are 211.5 ppm, while the P-C-distances were determined to be 175-177 pm, which means that all bond distances are between the respective single and double bond distances, but lengthened compared to the bond distances of 17 (Fig. 14, refs. 206,218).

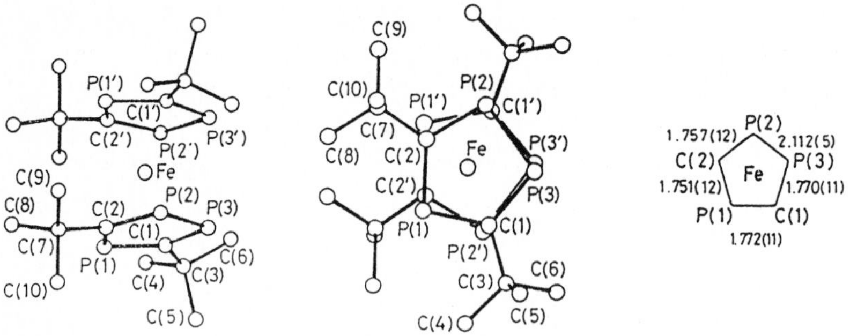

Fig. 14. Two views of the molecular structure and bond lengths of 18a.

The reaction of 17 with $MnBr(CO)_5$ leads, as expected, to complex 18b (eqn. 48, ref. 206).

17 + $Mn(CO)_5Br$ ——▷ (−LiBr)

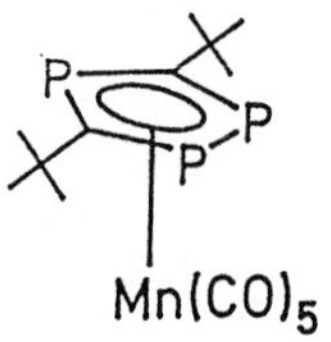

(eqn. 48)

18b

The $^{31}P\{^1H\}$-NMR spectrum of 18a can be described as an AA'BB'CC'- spin system, with $\delta(P^A) = 219$, $\delta(P^B) = 211$, and $\delta(P^C) = 172$ ppm [$^1J(P^BP^C) = 411.2$, $^2J(P^AP^B) = 37.9$, $^2J(P^AP^C) = 42$,

$^2J(P^CP^{C'}) \approx 53$ Hz; other interring coupling constants are insignificant]. The authors assumed, that since η^6-P_6- (ref. 219), η^5-P_5- (refs. 220,221), η^4-[$(CR)_2P_2$]- (refs. 36-38), and η^4-[$(CR)_3P$]- (ref. 32) ring systems could be isolated as their metal complexes, a series of planar rings [$(RC)_mP_n$] (n = 4-0, m = 0-4; n = 5-0, m = 0-5; n = 6-0, m = 0-6) should also exist (ref. 206, see also 3.5. Tetraphospholes).

14b was also isolated together with 14c and a third product in a metal catalyzed cyclization reaction [see 3.3.3., eqn. 37] (ref. 38).

The similarities between the diphospholyl and triphospholyl anions on the one hand and the cyclopentadienyl anion on the other hand in their behavior as ligands in transition metal complexes were shown very recently (ref. 225). A mixture of LiC_5H_5 and $Li(dme)_3(Bu^tC)_2P_3$/$Li(dme)_3(Bu^tC)_3P_2$ (17/14a) in dme was treated with $FeCl_2$. The result was a mixture of 14b, 18a, [$Fe(\eta^5$-$C_5H_5)(\eta^5$-$(Bu^tC)_2P_3)$] 18c, and [$Fe(\eta^5$-$C_5H_5)(\eta^5$-$(Bu^tC)_3P_2)$] 14d. The compounds 14d and 18c, which can be separated as a mixture from the other products by column chromatography, were obtained as orange oils [14d: $\delta(^{31}P)$ = 17.2 ppm; 18c (AB_2-spinsystem): $\delta(P^A)$ = 38.9, $\delta(P^B)$ = 37.9 ppm; $^2(P^AP^B)$ = 45 Hz]. Reaction of 14d/18c with [$W(CO)_5thf$] leads exclusively to 18d (eqn. 49, ref. 225), because of the bulky Bu^t-groups.

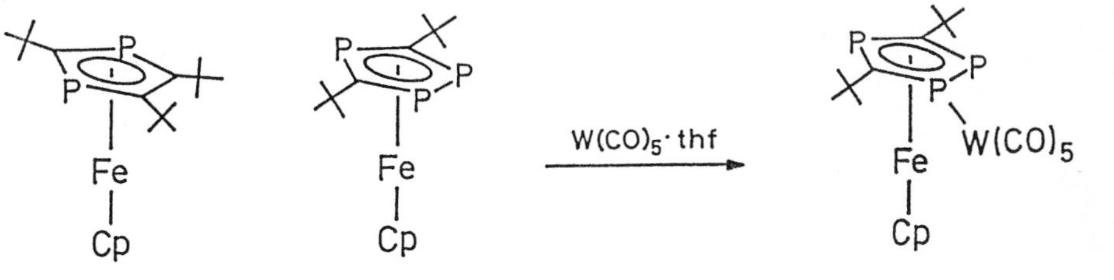

(eqn. 49)

14d 18c 18d

18d can be isolated as red-organge, air-stable crystals [$\delta(^{31}P)$(ABC-spinsystem): $\delta(P^A)$ = 30, $\delta(P^B)$ = 23.4, $\delta(P^C)$ = 13.3 ppm; $^2J(P^AP^B)$ = 47.7, $^2J(P^AP^C)$ = 38.8, $^1J(P^BP^C)$ = 411.6 Hz; $^1J(P^BW)$ = 224 Hz]. The structure of 18d was elucidated from its ^1H- and ^{31}P-NMR spectra and was confirmed by a single crystal X-ray diffraction study (molecular structure and details about bond lengths are given in Fig. 15).

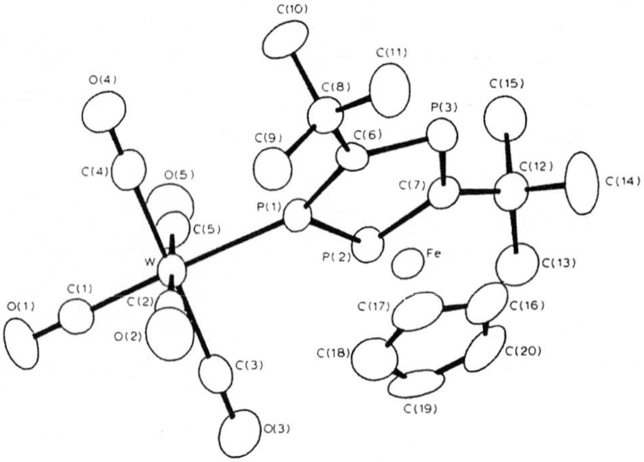

Fig. 15. Molecular structure of 18d with selected bond lengths W-P 251.0(2), P(1)-P(2) 209.8(2), P(1)-C(6) 175.3(7), P(2)-C(7) 176.4(8), P(3)-C(6) 177.7(7), P(3)-C(7) 176.3(6), Fe-P(1) 229.9(2), Fe-P(2) 234.4(3), Fe-P(3) 229.4(2), Fe-C(6) 214.1(4), Fe-C(7) 213.3(8) pm.

Very recently Nixon et al. described the first example of a 1,2,4-triphosphabuta-1,3-diene complex (ref. 226). Treatment of $Li(dme)_3[(Bu^tC)_2P_3]$ 17 (ref. 218) in dme with $CoCl_2$ gave surprisingly the red diamagnetic complex $[Co(\eta^5\text{-}(Bu^tC)_2P_3)(\eta^4\text{-}(Bu^tC)_2HP_3)]$ 18e in low yield (ca. 5%). The proton of the triphosphole unit was probably supplied from the solvent

(Fig. 16). The authors described the $^{31}P\{^1H\}$-NMR spectrum of 18e as two ABX patterns for the two sets of three nonequivalent phosphorus nuclei in the two rings [$\delta(P^1) = 131.2$, $\delta(P^2) = 109.5$, $\delta(P^3) = 93.1$, $\delta(P^4) = 0.2$, $\delta(P^5) = -20.1$, $\delta(P^6) = -64.7$ ppm, $^1J(P^1P^2) = 419.7$, $^2J(P^1P^3) = 41.6$, $^2J(P^2P^3) = 40.2$, $^1J(P^4P^6) = 360.6$, $^2J(P^5P^6) = 24.5$, $^2(P^4P^5) = 24.5$, $^2J(P^2P^4) = 39$, $^2J(P^3P^6) = 24.2$ Hz; see Fig. 16].

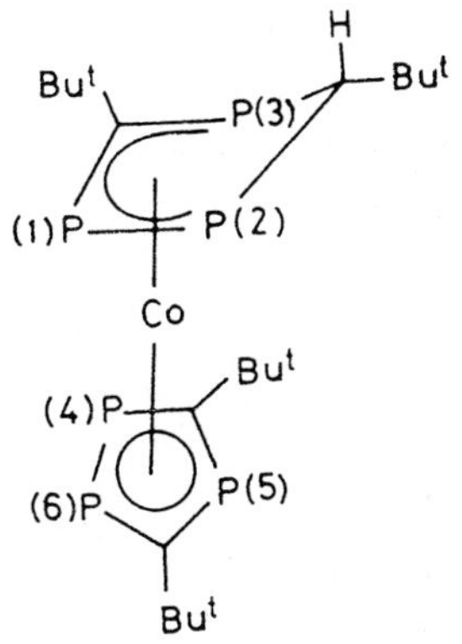

Fig. 16. 18e with labels for the ^{31}P-NMR spectrum

A single crystal X-ray structural determination of 18e revealed a planar η^5-1,3,5-triphosphacyclopentadienyl unit, very similar to the compounds 14b and 18a. The P(2)P(1)CP(3) fragment of the η^4-ligated $[(Bu^tC)_2HP_3]$-ring (see Fig. 17) is planar and the C-P and P-P bond lengths (d, P4-P5 = 213.6, P5-C4 = 179, C4-P6 = 178 pm) lie in the expected range for this unsaturated system.

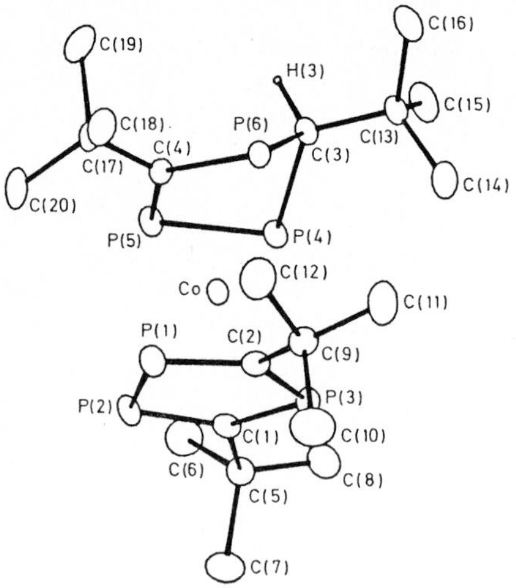

Fig. 17. Molecular structure of 18e.

3.5 TETRAPHOSPHOLENES AND TETRAPHOSPHOLES

3.5.1 Tetraphospholyl Anions

Tetraphospholenes and tetraphospholes are the last link in the series of unsatured five-membered phosphorus-carbon heterocycles. However, as of now no representative of these species has been synthesized. The only exception is the tetraphospholid anion 19, which was published by Baudler et al. in 1987. 19 arises together with P_5^-- and $P_3CH_2^-$-anions as well as other polyphosphides in the reaction of white phosphorus and sodium powder in diglyme (diethyleneglycol-dimethylester). The CH-units were supplied by the solvent. 19 could be identified in the mixture by means of ^{31}P-NMR spectroscopy (Tab. 13, ref. 53).

TABLE 13
^{31}P-NMR parameters[a)] of <u>19</u> (AA'BB'-spinsystem).

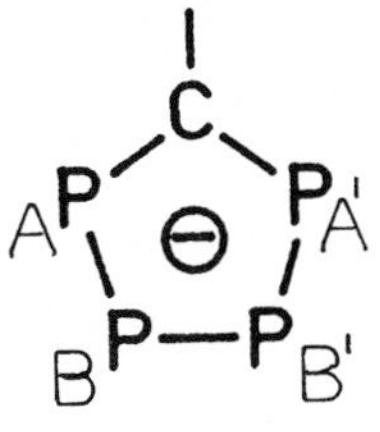

<u>19</u>

$\delta(P^A) = \delta(P^{A'}) = 362.1$ ppm
$\delta(P^B) = \delta(P^{B'}) = 255.1$ ppm
$J(P^AP^B) = J(P^{A'}P^{B'}) = -484.0$ Hz
$J(P^AP^{B'}) = J(P^{A'}P^B) = -3.6$ Hz
$J(P^AP^{A'}) = -53.3$ Hz
$J(P^BP^{B'}) = -505.4$ Hz

a) by assumption $J(PP) < 0$

The center of the AA'BB'-system of <u>19</u> lies at δ = 358.6 ppm and can be correlated with a 6 π-electron species such as P_5^- (refs. 53,222) and $(Bu^tC)_2P_3^-$ <u>17</u> (ref. 218). This value for <u>19</u> is in the middle of the ^{31}P-NMR shifts of P_5^- (470.2 ppm) and <u>17</u> ($\delta(P^A)$ = 252.5, $\delta(P^B)$ = 239 ppm; weighted average, ref. 53). As a comparison, other compounds of the series $[(RC)_nP_m]^-$ (n = 0-4; m = 1-4), can be added as well (Fig. 18).

4. SIX-MEMBERED RINGS

4.1 λ^3-PHOSPHORINES

The first λ^3-phosphorine which became known was the 2,4,6-triphenyl-λ^3-phosphorine, a solid which melts at 171-172°C. It was obtained by heating 2,4,6-triphenylpyrylium-tetrafluoroborate with tris-hydroxymethyl-phosphane (ref. 227).

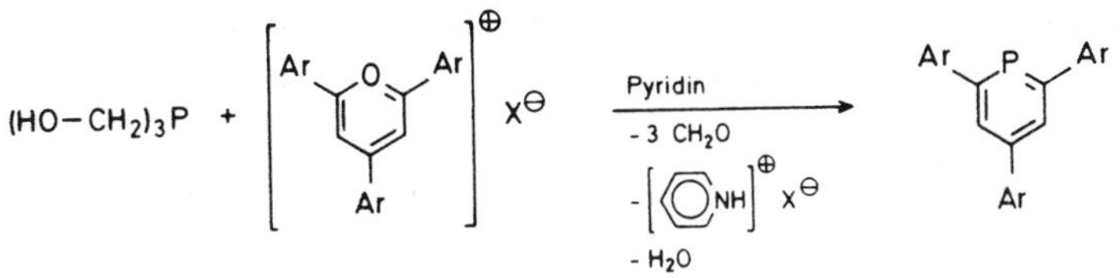

The unsubstituted λ^3-phosphorine was obtained in an elegant synthesis by Ashe III in the following way (refs. 228,229):

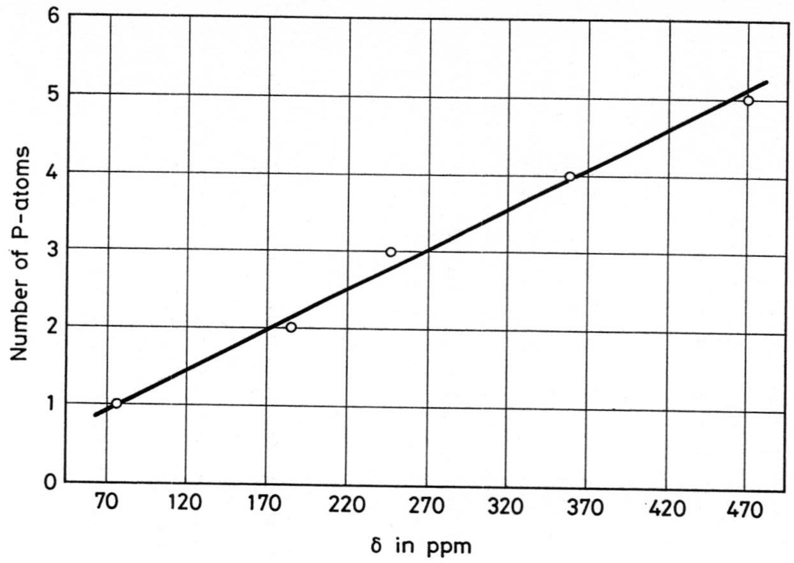

Fig. 18. Comparison of number of P-atoms and $\delta(^{31}P)$-values in $[(RC)_nP_m]^-$ (n = 0-4, m = 1-4; compounds <u>17</u>, <u>19</u> weighted average).

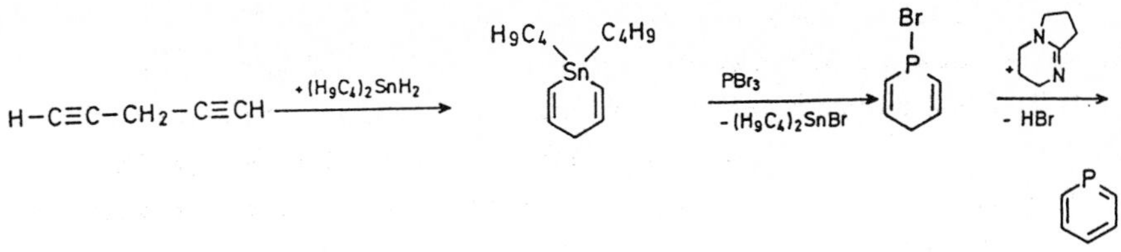

λ3-phosphorines are planar and have with their conjugated 6π electron system aromatic character as was shown by structural investigations and spectroscopic measurements. In addition to the unsubstituted λ3-phosphorine many of its mono-, di-, tri-, tetra- and penta substituted derivatives are known. Their chemical reactivity is determined both by the phosphorus atom and the aromatic character. The latter, e.g., is demonstrated by its ability to form "sandwich"-complexes.

During the last twenty years the chemistry of λ3-phosphorines has been studied in great detail. Excellent reviews are available in the literature (refs. 55, 230-233).

4.2 λ5-PHOSPHORINES

λ5-phosphorines are known in great number although it may be mentioned that only two members of this class of compound have been synthesized in which there are no substituents attached to carbon atoms of the ring, i.e. 1,1-dimethyl-λ5-phosphorine (ref. 234) and 1,1-diphenyl-λ5-phosphorine (ref. 235).

1) Primary phosphanes or alkyl(aryl)bis(hydroxymethyl)phosphanes react with 2,4,6-triphenylpyrylium salts to give λ4-phosphorin cations, which are not stable. With alcohols, phenols or thioalcohols the latter form immediately 2,4,6-triphenyl-λ5-phosphorines:

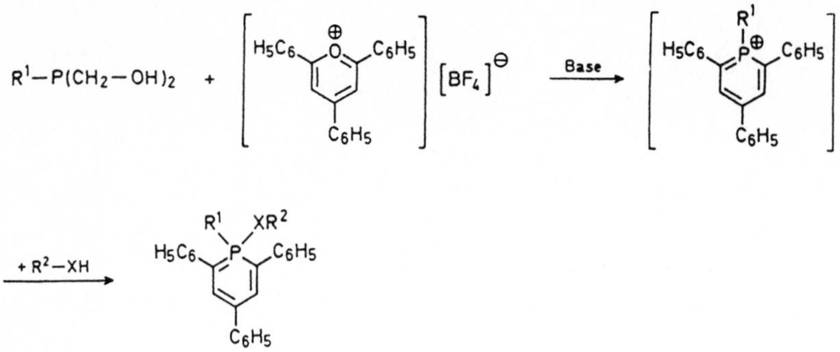

2) 2,4,6-trisubstituted phosphorines add strong nucleophiles such as lithium or Grignard compounds to give the salts of type II, which add electrophiles to give either λ^5-phosphorines or 1,2-dihydro-phosphorines, depending on the solvent and other conditions (refs. 230,236).

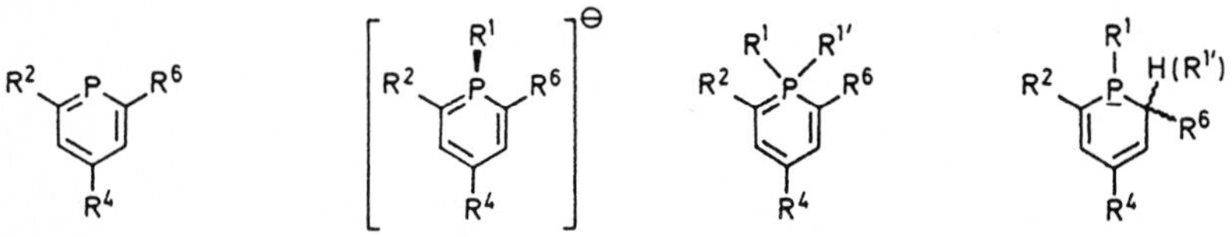

I II III IV

λ^5-phosphorines behave as cyclic phosphonium ylides. The phosphorum atom carries a positive charge, while the negative charge is delocalized over the ring carbon atoms as is shown in formular VI. 1,4,5-Trisubstituted λ^5-phosphorines are stable in air. The six-membered ring of all known λ^5-phosphorines is practically planar.

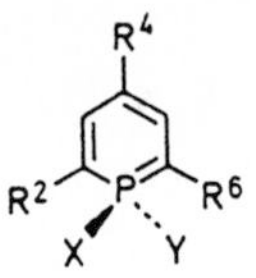

V

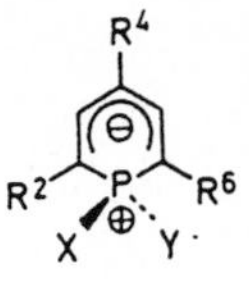

VI

The chemistry of λ^5-phosphorines is described in detail in various recent articles as well as in a handbook chapter.

4.3 $1\lambda^3,4\lambda^3$-DIPHOSPHORINES

The first and only $1\lambda^3,4\lambda^3$-diphosphorine was described by Kobayashi et al. (ref. 237). 2,3,5,6,7,8-Hexakis-[trifluoro-methyl]-7-methoxy-$1\lambda^3,4\lambda^3$-diphospha-bicyclo[2.2.2]octa-2,5-diene 1, (refs. 237,238) eliminates in boiling hexane trans-1,1,1,4,4,4-hexafluoro-2-methoxy-2-butene 2, forming 2,3,5,6-tetrakis-[trifluoro-methyl]-$1\lambda^3,4\lambda^3$-diphosphorine 3. It was not possible to isolate the compound but its intermediate existence was proved by further reactions: With alkines, 3, reacts smoothly in hexane to give the corresponding 1,4-diphospha-bicyclo[2.2.2]octatrienes 4.

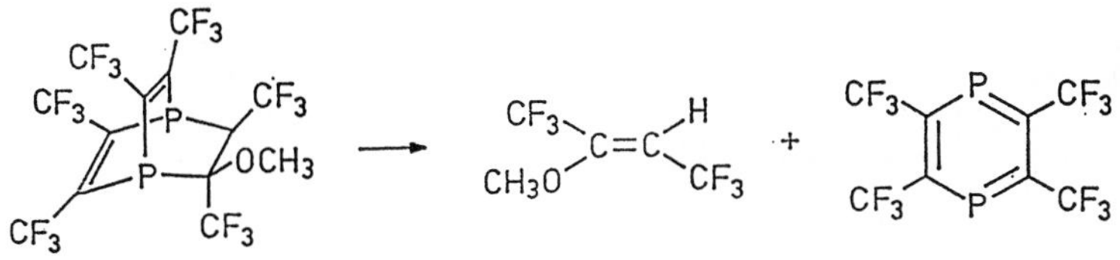

1 2 3

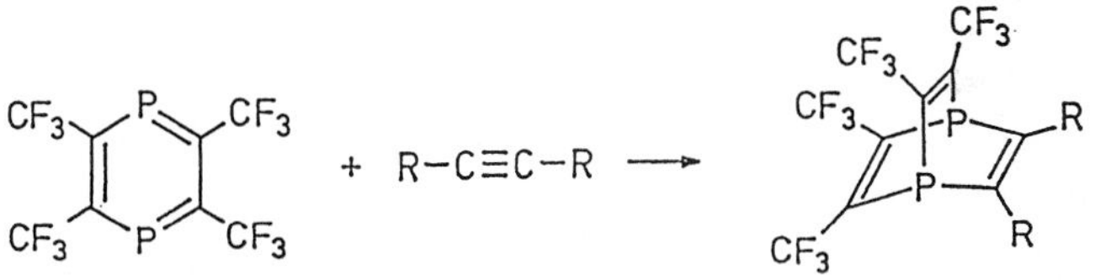

3 4

4.4 $1\lambda^5,4\lambda^5$-DIPHOSPHORINES

Until today there is only one $1\lambda^5,4\lambda^5$-diphosphorine mentioned in the literature. Cis-1,2-bis(diphenylphosphino)ethylene (ref. 239) reacts with dimethyl acetylene dicarboxylate to give dimethyl-1,1,4,4-tetraphenyl-$1\lambda^5,4\lambda^5$-diphosphorine-,2,3-dicarboxylate 5, which forms colorless crystals (m.p. 244°C).

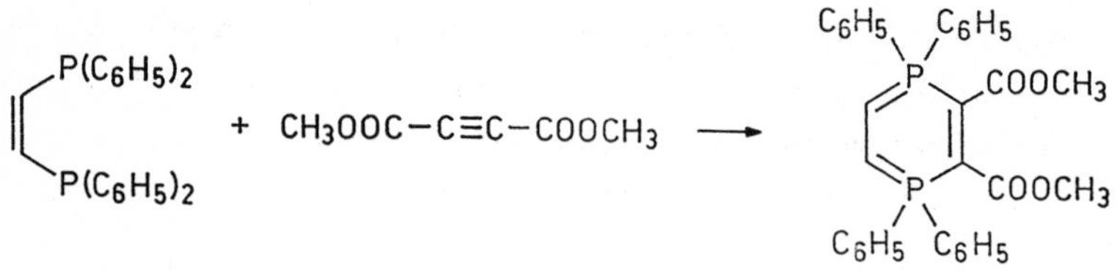

5

The reaction was first reported by Tebby et al. (ref. 200). He characterized the product mainly from the mass spectrum and the IR spectrum which shows that both ester groups are attached to ylidic carbon atoms, and the behavior of 5 on hydrolysis, which leads to the dioxide $Ph_2(O)P\text{-}CH\text{=}CH\text{-}P(O)Ph_2$. Later Davies et al. (ref. 240) described the reaction in more detail and reported the ^1H-NMR spectrum, which confirms the bis-ylide nature of 5. Davies et al. also prepared the bisperchlorate of 5 (ref. 240).

4.5 $1\lambda^5,3\lambda^5$-DIPHOSPHORINES

Diphosphorines of the type

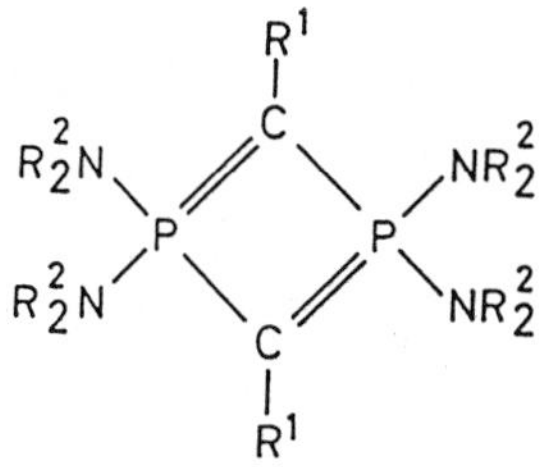

6: R^1 = H, R^2 = Me
7: R^1 = R^2 = Me
8: R^1 = Ph, R^2 = Et

can be conceived as twofold cyclic ylides with a highly carbanionic character of the carbon atoms, as is also purported by their photoelectron spectra (ref. 48). Their chemical behavior can best be interpreted on this basis. The diphosphete 6 reacts smoothly with acetylene dimethoxycarboxylate to give 4,5-dimethoxycarboxylato-1,1,3,3-tetrakis(dimethylamino)-1λ^5, 3λ^5-diphosphorine (ref. 241). Its formation can formally be considered to be an insertion reaction. Mechanistically the following way of formation can be assumed: The anionic carbon atom of the diphosphete 6 attacks a carbon atom of the acetylene forming a P-C bond. In this way the cycloaddition process is induced leading to the intermediate 9. By spontaneous valence isomerization finally the diphosphorine 10 is formed.

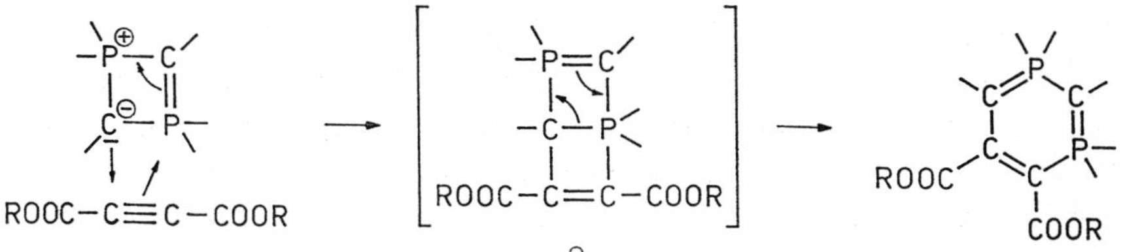

9 10

4.6 1λ^5,3λ^5,5λ^3-TRIPHOSPHORINES

The class of triphosphorines was unknown until the first member was synthesized by Fluck et al. in 1986 (refs. 242,243). 1,1,3,3-tetrakis(dimethylamino)-1λ^5,3λ^5-diphosphete reacts with 2,2-dimethylpropylidynephosphane in a uniform reaction to give 4-t-butyl-1,1,3,3-tetrakis(dimethylamino)-1λ^5,3λ^5,5λ^3-triphosphorine 11. Formally the process can be regarded as an insertion reaction. Mechanistically the reaction proceeds in the following way:

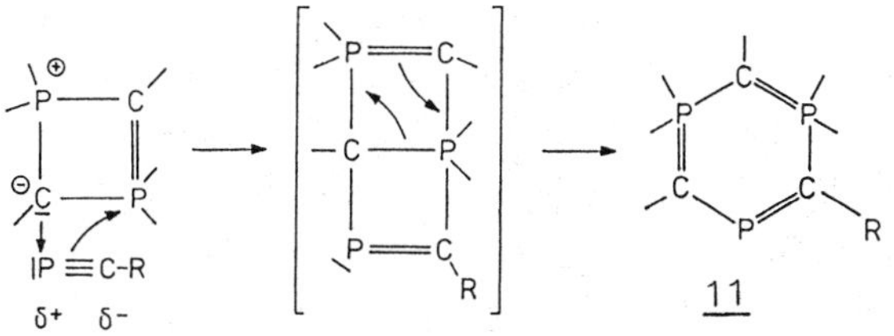

a 11

Compound 11 forms air- and moisture-sensitive, pale yellow crystals, m.p. = 50-55°C, which are soluble in benzene and n-pentane. The anionic carbon atom of the diphosphete is able to attack the P-atom of the phosphaalkyne. The resulting intermediate a rearranges in a spontaneous valence isomerization to 11 (ref. 243). NMR-investigations showed that P^A and P^B are typical ylidic phosphorus atoms, while P^X has a relatively high phosphenium character γ (see Fig. 19, important NMR-data in Tab. 14).

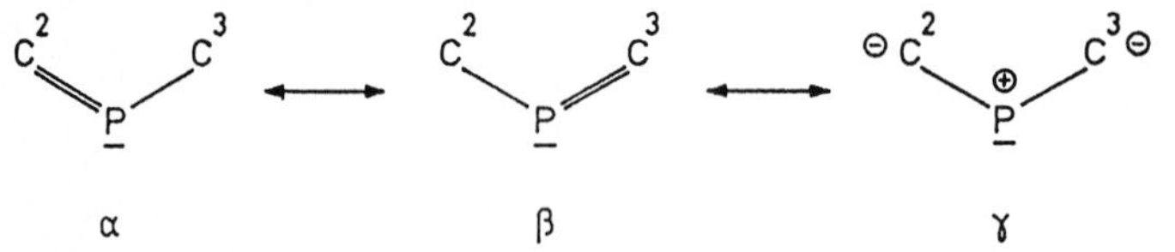

Fig. 19.

The dominating factor for the ^{13}C-NMR shifts of the ring C-atoms is the ylidic character in C^1, C^2, and C^3 (see Tab. 14). The crystal structure determination of 11 gave C-P-distances in the six-membered ring between 168.8 and 171.5 pm. The exception is a distance of 175.8 pm (P^3-C^4), which is forced by the bulkiness of the t-butyl group [Fig. 20; d; P1-C2 168.8(2), P3-C2 169.7(2), P5-C4 170.9(2), P1-C6 171.5(2), P3-C4 175.8(2), P5-C6 170.0(2) pm].

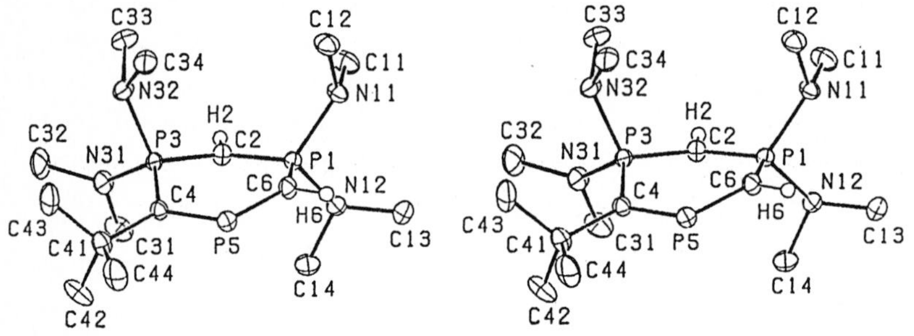

Fig. 20. Molecular structure of 11.

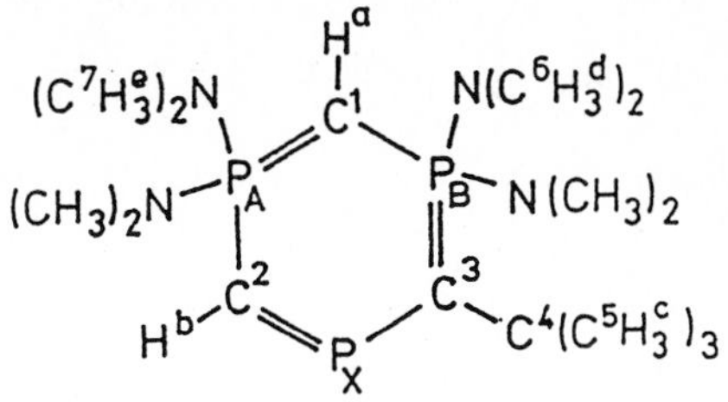

TABLE 14
Significant NMR-parameter of 11[α)].

δ[ppm]		J[Hz]
P^A	57.5	$^3J(P^AP^B)$ = 30.0; $^3J(P^AP^X)$ = 45.0; $^3J(P^BP^X)$ = 49.2
P^B	59.0	
P^X	295.0	
C^1	8.3	$^1J(P^AC^1)$ = 142; $^1J(C^1P^B)$ = 142; $^1J(C^2P^A)$ = 100; $^1J(C^2P^X)$ = 52.5
C^2	62.8	$^1J(C^3P^B)$ = 93; $^1J(C^3P^X)$ = 60; $^3J(C^1P^X)$ = 3.5; $^3J(C^2P^B)$ = 13.5;
C^3	100.8	$^3J(C^3P^A)$ = 8.0
H^a	0.63	$^2J(H^aP^A)$ = 11.6; $^2J(H^aP^B)$ = 11.6; $^2J(H^bP^X)$ = 34.5; $^2J(H^bP^A,P^B)$ = 2.7/3.8[ß)]; $^4J(H^aH^b)$ = 3.8

α) for more details see ref. 243; ß) $^2J(PH)$ and $^4J(PH)$ not distinguishable

The short C-P-bond lengths are caused by polar bond shortening in addition to the endocyclic π-system. The is leads to a formal bond order of 1.5. The six-membered ring is slightly puckered with a largest distance of 19.3 pm from the average plane through the ring atoms.

In 1987 Cowley et al. were successful in the synthesis of a $1\lambda^3,3\lambda^3,5\lambda^3$-triphosphinine derivative, coordinated on a metal fragment (ref. 244) [a recent independent study by Binger et al. has not confirmed the formation of compound 12 (ref. 248)]. Cycloheptatrienmolybdenumtricarbonyl was treated with $Bu^tC{\equiv}P$ in thf at 65°C (eqn. 50).

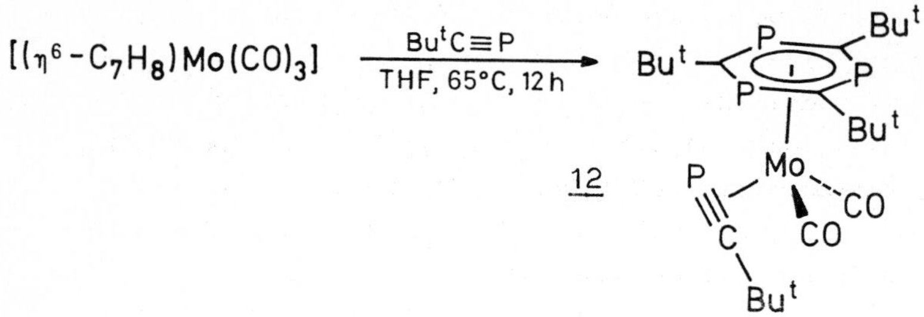

(eqn. 50)

12 was isolated in 59% yield as orange solid with a melting point of 83°C (dec.). The composition was proved by high resolution mass spectrometry. The $^{31}P\{^{1}H\}$-NMR spectrum shows two singlets with a ratio 3:1 at 25.2 and -228.2 ppm. The high field signal at -228.2 ppm belongs to the η^2-phosphaalkyne ligand, while the resonance at 25.2 ppm is caused by the η^6-triphosphabenzene derivative. The suggested mechanism of the reaction is supported by the results obtained in investigations by Regitz et al.

Kinetically stabilized cyclobutadienes react with phosphaalkynes to give $2\lambda^3$-Dewar phosphinines which are thermally stable and can be distilled without decomposition (refs. 245,246). Addition of activated nitriles to kinetically stabilized cyclobutadienes leads to 2-Dewar pyridines. These, however, can be thermally isomerized to give pyridines (ref. 247).

REFERENCES

1 S. Lochschmidt, F. Mathey and A. Schmidpeter, Tetrahedron Lett. 27 (1986) 2635.
2 F. Mercier and F. Mathey, Tetrahedron Lett., 27 (1986) 1323.
3 J. Svara, A. Marinetti and F. Mathey, Organometallics, 5 (1986) 1161.
4 B. Deschamps and F. Mathey, Tetrahedron Lett., 26 (1985) 4595.
5 J.-M. Alcaraz, J. Svara and F. Mathey, Nouv. J. Chim., 10 (1986) 321.

6 L. Ricard, N. Maigrot, C. Charrier and F. Mathey, Angew. Chem., 99 (1987) 590; Angew. Chem., Int. Ed. Engl., 26 (1987) 548.
7 A. Marinetti and F. Mathey, J. Fischer and A. Mitschler, J. Am. Chem. Soc., 104 (1982) 4484.
8 K.S. Fongers, H. Hogeveen and R.F. Kingma, Tetrahedron Lett., 24 (1983) 643.
9 A. Marinetti, F. Mathey, J. Fischer and A. Mitschler, J. Chem. Soc., Chem. Commun., (1984) 45.
10 A.M. Torgomyan, M.Z. Ovakimyan and M.G. Indzhikyan, Arm. Khim. Zh., 32 (1979) 288.
11 G.T. Gasparyan, G.G. Minasyan, A.M. Torgomyan, M.Z. Ovakimyan and M.G. Indzhikyan, Arm. Khim. Zh., 36 (1983) 456.
12 N.M. Kausal, S. Verma, R.S. Mishra and M.M. Bokadia, Indian J. Chem., 19B (1980) 610; N.M. Kausal, S. Verma and M.M. Bokadia, Indian J. Chem., 19B (1980) 611; S. Verma, N.M. Kausal, R.S. Mishra and M.M. Bokadia, Heterocycles, 16 (1981) 1537.
13 R. Breslow and L.A. Deuring, Tetrahedron Lett., 25 (1984) 1345.
14 A. Marinetti and F. Mathey, J. Am. Chem. Soc., 107 (1985) 4700.
15 F. Mathey and A. Marinetti, Bull. Soc. Chim. Belg., 93 (1984) 533.
16 F. Mathey, Angew. Chem., 99 (1987) 285; Angew. Chem., Int. Ed. Engl., 26 (1987) 275.
17 F. Mathey, Pure and Applied Chem., 59 (1987) 993.
18 D. Gonbeau and G. Pfister-Guillouzo, Nouv. J. Chim., 9 (1985) 71.
19 D. Gonbeau, G. Pfister-Guillouzo, A. Marinetti and F. Mathey, Inorg. Chem., 24 (1985) 4133.
20 O. Wagner, G. Maas and M. Regitz, Angew. Chem., 99 (1987) 1328; Angew. Chem., Int. Ed. Engl., 26 (1987) 1257.
21 A. Marinetti, J. Fischer and F. Mathey, J. Am. Chem. Soc., 107 (1985) 5001.
22 B.A. Boyd, R.J. Thoma, W.H. Watson and R.H. Neilson, Organometallics, 7 (1988) 572.
23 W. Mahler, J. Am. Chem. Soc., 86 (1964) 2306.
24 A. Ecker and U. Schmidt, Chem. Ber., 106 (1973) 1453.
25 R. Appel and V. Barth, Tetrahedron Lett., 21 (1980) 1923.
26 C. Charrier, J. Guilhem and F. Mathey, J. Org. Chem., 46 (1981) 3.
27 C. Charrier, F. Mathey, F. Robert and Y. Jeannin, J. Chem. Soc., Chem. Commun., (1984) 1707.
28 C. Charrier, N. Maigrot, F. Mathey, F. Robert and Y. Jeannin, Organometallics, 5 (1986) 623.
29 S.K. Nurtdinov, N.M. Ismagilova, R.A. Fakhrutdinova and T.V. Zykova, Z. Obshch. Khim., 53 (1983) 1045.
30 G. Becker, W. Becker and G. Uhl, Z. Anorg. Allg. Chem., 518 (1984) 21.
31 G. Becker, W. Becker and G. Uhl, Z. Anorg. Allg. Chem., 519 (1984) 31.
32 P. Binger, R. Milczarek, R. Mynott and M. Regitz, J. Organomet. Chem., 323 (1987) C35.

33 R. Appel, V. Barth and F. Knoch, Chem. Ber., 116 (1983) 938.
34 G. Becker, W. Becker, R. Knebl, H. Schmidt, U. Hildenbrand and M. Westerhausen, Phosphorus and Sulfur, 30 (1987) 349; the structure reported in this paper is incorrect (private communication).
35 R. Knebl, Ph.D. thesis, Universität Stuttgart, 1988.
36 P.B. Hitchcock, M.J. Maah and J.F. Nixon, J. Chem. Soc., Chem. Commun., (1986) 737.
37 P. Binger, R. Milczarek, R. Mynott, M. Regitz and W. Rösch, Angew. Chem., 98 (1986) 645; Angew. Chem., Int. Ed. Engl., 25 (1986) 644.
38 M. Driess, D. Hu, H. Pritzkow, H. Schäufele, U. Zenneck, M. Regitz and W. Rösch, J. Organomet. Chem., 334 (1987) C35.
39 R. Appel, F. Knoll and H.-D. Wihler, Angew. Chem., 89 (1977) 415; Angew. Chem., Int. Ed. Engl., 16 (1977) 402.
40 R. Appel, U. Baumeister and F. Knoch, Chem. Ber., 116 (1983) 2275.
41 J. Weiss and B. Nuber, Z. Anorg. Allg. Chem., 473 (1981) 101.
42 H. Keller, G. Maas and M. Regitz, Tetrahedron Lett., 27 (1986) 1903.
43 J. Svara, E. Fluck and H. Riffel, Z. Naturforsch., 40b (1985) 1258.
44 J. Svara and E. Fluck, Phosphorus and Sulfur, 25 (1985) 129.
45 O.I. Kolodiazhnyi and V.P. Kukhar, Usp. Khim., 52 (1983) 1903; Russ. Chem. Rev., 52 (1983) 1096.
46 R. Appel, F. Knoll and I. Ruppert, Angew. Chem., 93 (1981) 771; Angew. Chem., Int. Ed. Engl., 20 (1981) 731.
47 J.C.J. Bart, J. Chem. Soc. B, (1969) 350.
48 T. Veszprémi, R. Gleiter, E. Fluck, J. Svara and B. Neumüller, Chem. Ber., 121 (1988) 2071.
49 E. Fluck, B. Neumüller and H. Riffel, Z. Anorg. Allg. Chem., in press.
50 B. Neumüller and E. Fluck, Phosphorus and Sulfur, 29 (1986) 23.
51 L. Pauling, Natur der Chemischen Bindung, 3. Aufl., Verlag Chemie, Weinheim, 1976.
52 J.E. Huheey, Inorganic Chemistry, Harper and Row, New York, 1972.
53 M. Baudler, D. Düster and D. Ouzounis, Z. Anorg. Allg. Chem., 554 (1987) 87.
54 W.B. McCormack, U.S.P. 2,663,736 (1955)
55 L.D. Quin, The Heterocyclic Chemistry of Phosphorus, John Wiley and Sons, New York, Chichester, Brisbane, Toronto, 1981, and cit. lit. therein.
56 F.C. Leavitt, T.A. Manuel and F. Johnson, J. Am. Chem. Soc., 81 (1959) 3163.
57 F.C. Leavitt, T.A. Manuel, F. Johnson, L.U. Matternas and D.S. Lehman, J. Am. Chem. Soc., 82 (1960) 5099.
58 E.H. Braye and W. Hübel, Chem. Ind., (1959) 1250.
59 E.H. Braye, W. Hübel and I. Caplier, J. Am. Chem. Soc., 83 (1961) 4406.

60 J.J. Breen, J.F. Engel, D.K. Myers and L.D. Quin, Phosphorus, 2 (1972) 55.
61 L.D. Quin and R.C. Stocks, Phosphorus and Sulfur, 3 (1977) 151.
62 K. Moedritzer and R.E. Miller, Synth. React. Inorg. Met.-Org. Chem., 7 (1977) 311.
63 L.L. Chang, D.Z. Denney, D.B. Denney and Y.F. Hsu, Phosphorus, 4 (1974) 265.
64 P. Haake, R.D. Cook, W. Schwarz and D.R. McCoy, Tetrahedron Lett., (1968), 5251.
65 L.D. Quin and J. Szewczyk, Phosphorus and Sulfur, 21 (1984) 161.
66 E. Niecke and M. Lysek, Tetrahedron Lett., 29 (1988) 605.
67 N.J. De'ath, D.Z. Denney, D.B. Denney and C.D. Hall, Phosphorus, 3 (1974) 205.
68 D.B. Denney, D.Z. Denney and Y.F. Hsu, Phosphorus, 4 (1974) 217.
69 N.A. Razumova, N.A. Kurshakova, Z.L. Evtikhov and A.A. Petrov, Zh. Obshch. Khim., 44 (1974) 1866.
70 A. Bond, M. Green and S.C. Pearson, J. Chem. Soc. B, (1968) 929.
71 L.D. Quin and T.P. Barket, J. Am. Chem. Soc., 92 (1970) 4303.
72 C.K. SooHoo and S.G. Baxter, J. Am. Chem. Soc., 105 (1983) 7443.
73 A.H. Cowley, R.A. Kemp, J.G. Lasch, N.C. Norman and C.A. Stewart, J. Am. Chem. Soc., 105 (1983) 7444.
74 C. Symmes, Jr. and L.D. Quin, J. Org. Chem., 41 (1976) 238.
75 C. Symmes, Jr. and L.D. Quin, J. Org. Chem., 41 (1976) 1548.
76 L.D. Quin, K.C. Caster and S.B. Soloway, J. Org. Chem., 49 (1984) 2627.
77 L.D. Quin, K.A. Mesch and W.L. Orton, Phosphorus and Sulfur, 12 (1982) 161.
78 F. Mathey, F. Mercier, C. Charrier, J. Fischer and A. Mitschler, J. Am. Chem. Soc., 103 (1981) 4595.
79 C. Charrier, H. Bonnard, G. de Lauzon and F. Mathey, J. Am. Soc., 105 (1983) 6871.
80 A. Marinetti, F. Mathey, J. Fischer and A. Mitschler, J. Chem. Soc., Chem. Commun., (1982) 667.
81 B. Deschamps and F. Mathey, J. Chem. Soc., Chem. Commun., (1985) 1010.
82 G. Märkl and B. Alig, Tetrahedron Lett., 23 (1982) 4915.
83 T.J. Katz, C.R. Nicholson and C.A. Reilly, J. Am. Chem. Soc., 88 (1966) 3832.
84 L.D. Quin, K.C. Caster and J.C. Kisalus, Phosphorus and Sulfur, 18 (1983) 105.
85 L.D. Quin, N.S. Rao, R.J. Topping and A.T. McPhail, J. Am. Chem. Soc., 108 (1986) 4519.
86 W.J. Richter, Chem. Ber., 118 (1985) 97.
87 A.L. Crumbliss, R.J. Topping and L.D. Quin, Tetrahedron Lett., 27 (1986) 889.
88 L.D. Quin and N.S. Rao, J. Org. Chem., 48 (1983) 3754.
89 L.D. Quin, K.C. Caster, J.C. Kisalus and K.A. Mesch, J. Am. Chem. Soc., 106 (1984) 7021.
90 L.D. Quin, K.A. Mesch, R. Bodalski and K.M. Pietrusiewicz, Org. Magn. Res., 20 (1982) 83.

91 K. Moedritzer, Synth. React. Inorg. Met.-Org. Chem., 5 (1975) 299.
92 D.K. Myers and L.D. Quin, J. Org. Chem., 36 (1971) 1285.
93 V.A. Naumov and V.N. Semashko, Dok. Akad. Nauk. SSR, 193 (1970) 348.
94 K. Moedritzer, Synth. React. Inorg. Met.-Org. Chem., 5 (1975) 45.
95 M. Amin, D.G. Holah, A.N. Hughes and T. Rukachaisirikul, J. Heterocycl. Chem., 22 (1985) 513.
96 L.D. Quin and J.A. Caputo, J. Chem. Soc., Chem. Commun., (1968) 1463.
97 L.D. Quin, S.G. Borleske and J.F. Engel, J. Org. Chem., 38 (1973) 1954.
98 W.L. Orton, K.A. Mesch and L.D. Quin, Phosphorus and Sulfur, 5 (1979) 349.
99 K.C. Caster and L.D. Quin, Tetrahedron Lett., 24 (1983) 5831.
100 N.S. Rao and L.D. Quin, J. Am. Chem. Soc., 105 (1983) 5960.
101 F.G. Mann and I.T. Millar, J. Chem. Soc., (1951) 2205, and references therein.
102 M. El-Deek, G.D. Macdonell, S.D. Venkataramu and K.D. Berlin, J. Org. Chem., 41 (1976) 1403.
103 J.I. Grayson, H.K. Norrish and S. Warren, J. Chem. Soc., Perkin Trans. I, (1976) 2556.
104 B. Cetinkaya, P.B. Hitchcock, M.F. Lappert, A.J. Thorne and H. Goldwhite, J. Chem. Soc., Chem. Commun., (1982) 691.
105 A. Baceiredo, G. Bertrand, P. Mazerolles and J.-P. Majoral, J. Chem. Soc., Chem. Commun., (1981) 1197.
106 B.R. Stults, F.L. May and T.M. Bathazor, Cryst. Struct. Comm., 11 (1982) 1179.
107 M. Yoshifuji, I. Shima, K. Ando and N. Inamato, Tetrahedron Lett., 24 (1983) 933.
108 H.H. Karsch, F.H. Köhler and H.-U. Reisacher, Tetrahedron Lett., 25 (1984) 3687.
109 A.H. Cowley, J.E. Kilduff, N.C. Norman, M. Pakulski, J.L. Atwood and W.E. Hunter, J. Am. Chem. Soc., 105 (1983) 4845.
110 H.H. Karsch, H.-U. Reisacher and G. Müller, Angew. Chem., 98 (1986) 467; Angew. Chem., Int. Ed. Engl., 25 (1986) 454.
111 E.D. Middlemas and L.D. Quin, J. Org. Chem., 44 (1979) 2587.
112 A.R. Barron, S.W. Hall and A.H. Cowley, J. Chem. Soc., Chem. Commun., (1987) 1753.
113 J.M. Holland and D.W. Jones, J. Chem. Soc., Perkin Trans. I, (1973) 927.
114 K. Issleib, K. Mohr and H. Sonnenschein, Z. Anorg. Allg. Chem., 408 (1974) 266.
115 D. Fenske, E. Langer, M. Heymann and H.J. Becher, Chem. Ber., 109 (1976) 359.
116 C.H. Chen, K.E. Brighty and F.M. Michaels, J. Org. Chem., 46 (1981) 361.
117 F. Mathey, A. Mitschler and R. Weiss, J. Am. Chem. Soc., 99 (1977) 3537.
118 K. Blatter, W. Rösch, U.-J. Vogelbacher, J. Fink and M. Regitz, Angew. Chem., 99 (1987) 67; Angew. Chem., Int. Ed. Engl., 26 (1987) 85.

119 L.D. Quin and J. Szweczyk, J. Chem. Soc., Chem. Commun., (1984) 1551.
120 L.D. Quin and K.C. Caster, Phosphorus and Sulfur, 25 (1985) 117.
121 C. Charrier, H. Bonnard, G. de Lauzon, S. Holand and F. Mathey, Phosphorus and Sulfur, 18 (1983) 51.
122 N.E. Waite and J.C. Tebby, J. Chem. Soc. C, (1970) 386.
123 D.G. Holah, A.N. Hughes and D. Kleemola, J. Heterocycl. Chem., 15 (1978) 1319.
124 R. Appel, C. Casser, F. Knoch and B. Niemann, Chem. Ber., 119 (1986) 2915.
125 L.N. Markovskii, V.D. Romanenko, A.V. Ruban and S.V. Iksanova, Zh. Obshch. Khim., 52 (1982) 2796.
126 M.P. Savage and S. Trippett, J. Chem. Soc. C, (1968) 591.
127 F. Mathey, Top. Phosphorus Chem., 10 (1980) 1.
128 F. Mathey, Chem. Rev., 88 (1988) 429, and cit. lit. therein.
129 I.G.M. Campbell, R.C. Cookson, M.B. Hocking and A.N. Hughes, J. Chem. Soc., (1965) 2184.
130 L.D. Quin and J.G. Bryson, J. Am. Chem. Soc., 89 (1967) 5984.
131 L.D. Quin, J.G. Bryson and C.G. Moreland, J. Am. Chem. Soc., 91 (1969) 3308.
132 F. Mathey, Compt. rend., 269c (1969) 1066.
133 F. Mathey, Tetrahedron Lett., (1973) 3255.
134 F. Mathey, Tetrahedron, 32 (1976) 2395.
135 J.I.G. Cadogan, R.J. Scott and N.H. Wilson, J. Chem. Soc., Chem. Commun., (1974) 902.
136 S. Holand, F. Mathey, J. Fischer and A. Mitschler, Organometallics, 2 (1983) 1234.
137 C. Charrier and F. Mathey, Tetrahedron Lett., 28 (1987) 5025.
138 G. de Lauzon, C. Charrier, H. Bonnard and F. Mathey, Tetrahedron Lett., 23 (1982) 511.
139 R. Chuchman, D.G. Holah, A.N. Hughes and B.C. Hui, J. Heterocycl. Chem., 8 (1971) 877.
140 K.S. Fongers, A. Hogeveen and R.F. Kingma, Tetrahedron Lett., 24 (1983) 1423.
141 J.C. Tebby, S.E. Willetts and D.V. Griffiths, J. Chem. Soc., Chem. Commun., (1981) 420.
142 J.C. Caesar, D.V. Griffiths, J.C. Tebby and S.E. Willetts, J. Chem. Soc., Perkin Trans. I, (1984) 1627.
143 R. Burgada, Y. Leroux, Y.O. El Khoshnieh, Tetrahedron Lett., 22 (1981) 3533.
144 P.I. Bkouche-Waksman, P. L'Haridon, Y. Leroux and R. Burgada, Acta Cryst., 38b (1982) 3024.
145 L.D. Quin and W.L. Orton, J. Chem. Soc., Chem. Commun., (1979) 401.
146 L.D. Quin and K.A. Mesch, Org. Magns. Res., 12 (1979) 442.
147 F. Mathey and D. Thavard, Can. J. Chem., 56 (1978) 1952.
148 G. Märkl and E. Seidl, Angew. Chem., 95 (1983) 58; Angew. Chem., Int. Ed. Engl. 22 (1983) 57.
149 F. Nief, C. Charrier, F. Mathey and M. Simalty, Phosphorus and Sulfur, 13 (1982) 259.
150 M.D. Rausch and L.P. Klemann, J. Am. Chem. Soc., 89 (1967) 5732.
151 M.D. Rausch, L.P. Klemann and W.H. Boon, Synth. React. Inorg. Met.-Org. Chem., 15 (1985) 923.

152 W. Winter, Chem. Ber., 110 (1977) 2168.
153 W. Winter and T. Butters, Chem. Ber., 116 (1983) 3271.
154 T. Butters and W. Winter, Chem. Ber., 117 (1984) 990.
155 D. Hellwinkel, Chem. Ber., 98 (1965) 576.
156 I.F. Wilson and J.C. Tebby, J. Chem. Soc., Perkin Trans. I, (1972) 2713.
157 E. Lindner and G. Frey, Z. Naturforsch., 35b (1980), 1150.
158 L.D. Freedman and G.O. Doak, J. Org. Chem., 21 (1956) 238.
159 S. Bracher, J.I.G. Cadogan, I. Gosney and S. Yaslak, J. Chem. Soc., Chem. Commun., (1983) 857.
160 J.I.G. Cadogan, Phosphorus and Sulfur, 18 (1983) 229.
161 D.W. Allen and J.C. Tebby, J. Chem. Soc. B, (1970) 1527.
162 R. Rothuis, T.K.J. Luderer and H.M. Buck, Rev. Trav. Chim., 91 (1972) 836.
163 F. Mathey, Nouv. J. Chim., 11 (1987) 585, and cit. lit. therein.
164 G. de Lauzon, C. Charrier, H. Bonnard, F. Mathey, J. Fischer and A. Mitschler, J. Chem. Soc., Chem. Commun., (1982) 1272.
165 F. Mercier and F. Mathey, J. Organomet. Chem., 263 (1984) 55.
166 S. Holand, C. Charrier, F. Mathey, J. Fischer and A. Mitschler, J. Am. Chem. Soc., 106 (1984) 826.
167 F. Zurmühlen and M. Regitz, J. Organomet. Chem., 332 (1987) C1.
168 C.L. Liotta, M.L. McLaughlin, D.G. van Derveer and B.A. O'Brien, Tetrahedron Lett., 25 (1984) 1665.
169 G. Märkl, E. Seidl and I. Trötsch, Angew. Chem., 95 (1983) 891; Angew. Chem., Int. Ed. Engl., 22 (1989) 879.
170 A.W. Johnson and J.C. Tebby, J. Chem. Soc., (1961) 2126.
171 J.B. Hendrickson, R.E. Spenger and J.J. Sims, Tetrahedron Lett., (1961) 477.
172 J.B. Hendrickson, R.E. Spenger and J.J. Sims, Tetrahedron, 19 (1963) 707.
173 N.E. Waite, J.C. Tebby, R.S. Ward and D.H. Williams, J. Chem. Soc. C, (1969) 1100.
174 N.E. Waite, D.W. Allen and J.C. Tebby, Phosphorus 1 (1971) 139.
175 A.N. Hughes, Heterocycles, 15 (1981) 637.
176 J.R. Corfield, M.J.P. Harger, J.R. Shutt and S. Trippett, J. Chem. Soc. C, (1970), 1855.
177 C.M. Mitchell and F.G.A. Stone, J. Chem. Soc., Dalton Trans., (1972) 102.
178 G. Märkl, Z. Naturforsch., 18b (1963) 84.
179 A.N. Hughes and M. Davies, Chem. Ind., (1969) 138.
180 J. Barluenga, F. López and F. Palacios, J. Chem. Soc., Chem. Commun., (1986) 1574.
181 G. Kaufmann and F. Mathey, Phosphorus, 4 (1974) 231.
182 N.M. Kostić and R.F. Fenske, Organometallics, 2 (1983) 1008.
183 E.H. Braye, I. Caplier and R. Saussez, Tetrahedron, 27 (1971) 5523.
184 L.D. Freedman, B.R. Ezzell, R.N. Jenkins and R.M. Harris, Phosphorus, 4 (1974) 199.
185 E.W. Abel, N. Clark and C. Towers, J. Chem. Soc., Dalton Trans., (1979) 1552.
186 O. Poizat and C. Sourisseau, J. Organomet. Chem., 213 (1981) 461.

187 A. Breque, F. Mathey and C. Santini, J. Organomet. Chem., 165 (1979) 129.
188 R.M.G. Roberts and A.S. Wells, Inorg. Chim. Acta, 120 (1986) 53.
189 G. de Lauzon, B. Deschamps, J. Fischer, F. Mathey and A. Mitschler, J. Am. Chem. Soc., 102 (1980) 994.
190 C. Guimon, D. Gonbeau, G. Pfister-Guillouzo, G. de Lauzon and F. Mathey, Chem. Phys. Lett., 104 (1984) 560.
191 B. Lukas, R.M.G. Roberts, J. Silver and A.S. Wells, J. Organomet. Chem., 256 (1983) 103.
192 G. Bergerhoff, J. Falbe, B. Tihanyi, J. Weber and W. Weisheit, Austrian Patent, (1968) 282,644.
193 G. Bergerhoff, B. Tihanyi, J. Falbe, J. Weber and W. Weisheit, S. African Patent, (1969) 6,806,540.
194 G. Bergerhoff, O. Hammes, J. Falbe, B. Tihanyi, J. Weber and W. Weisheit, Tetrahedron, 27 (1971) 3593.
195 J. Falbe, H. Tummes, J. Weber and W. Weisheit, Tetrahedron, 27 (1971) 3603.
196 W. Saenger, J. Org. Chem., 38 (1973) 253.
197 G. Bergerhoff, O. Hammes and D. Haas, Acta Cryst., 35b (1979) 181.
198 H.J. Padberg, J. Lindner and G. Bergerhoff, J. Chem. Res. (S), (1978) 445.
199 H. Schmidbaur, S. Strunk and C.E. Zybill, Chem. Ber., 116 (1983) 3559.
200 M.A. Shaw, J.C. Tebby, R.S. Ward and D.H. Williams, J. Chem. Soc. C, (1970) 504.
201 G. Märkl, D. Matthes, A. Donaubauer and H. Baier, Tetrahedron Lett., (1975) 3171.
202 G. Märkl, W. Weber and W. Weiß, Chem. Ber., 118 (1985) 2365.
203 E. Fluck, B. Neumüller, G. Heckmann and H. Riffel, Phosphorus and Sulfur, 37 (1988) 159.
204 H. Schmidbaur, T. Costa, B. Milewski-Mahrla and U. Schubert, Angew. Chem., 92 (1980) 557; Angew. Chem., Int. Ed. Engl., 19 (1980) 555.
205 G.A. Bowmaker, R. Herr and H. Schmidbaur, Chem. Ber., 116 (1983) 3567.
206 R. Bartsch, P.B. Hitchcock and J.F. Nixon, J. Chem. Soc., Chem. Commun., (1987) 1146.
207 G. Märkl and D. Matthes, Tetrahedron Lett., (1976) 2599.
208 M. Baudler and E. Tolls, Z. Naturforsch., 33b (1978) 691.
209 M. Baudler, J. Hahn and E. Clef, Z. Naturforsch., 39b (1984) 438.
210 F.G. Mann and A.J.H. Mercer, J. Chem. Soc., Perkin Trans I, (1972) 2548.
211 A. Schmidpeter, G. Burget and W.S. Sheldrick, Chem. Ber., 118 (1985) 3849.
212 F.G. Mann and A.J.H. Mercer, J. Chem. Soc., Perkin Trans. I, (1972) 1631.
213 F.G. Mann and J. Pragnell, J. Chem. Soc. C, (1966) 916.
214 J.J. Daly, J. Chem. Soc. A, (1966) 1020.
215 M. Baudler and E. Tolls, Z. Chem., 19 (1979) 418.
216 A. Schmidpeter, S. Lochschmidt and W.S. Sheldrick, Angew. Chem., 94 (1982) 72; Angew. Chem., Int. Ed. Engl., 21 (1982) 63.

217 S. Lochschmidt and A. Schmidpeter, Z. Naturforsch., 40b (1985) 765.
218 G. Becker, W. Becker, R. Knebl, H. Schmidt, U. Weeber and M. Westerhausen, Nova Acta Leopoldina, N.F. (Nr. 264), 59 (1985) 55.
219 O.J. Scherer, H. Sitzmann and G. Wolmershäuser, Angew. Chem., 97 (1985) 358; Angew. Chem., Int. Ed. Engl., 24 (1985) 351.
220 O.J. Scherer and T. Brück, Angew. Chem., 99 (1987) 59; Angew. Chem., Int. Ed. Engl., 26 (1987) 59.
221 O.J. Scherer, J. Schwalb, G. Wolmershäuser, W. Kaim and R. Groß, Angew. Chem., 98 (1986) 349; Angew. Chem., Int. Ed. Engl., 25 (1986) 363.
222 M. Baudler, S. Akpapoglou, D. Ouzounis, F. Wasgestian, B. Meinigke, H. Budzikiewicz and H. Münster, Angew. Chem., 100 (1988) 288; Angew. Chem., Int. Ed. Engl., 27 (1988) 280.
223 K. Issleib, H. Schmidt and E. Leißring, Synth. React. Inorg. Met.-Org. Chem., 18 (1988) 215.
224 U. Annen and M. Regitz, Tetrahedron Lett., 29 (1988) 1681.
225 R. Bartsch, P.B. Hitchcock and J.F. Nixon, J. Organomet. Chem., 340 (1988) C37.
226 R. Bartsch, P.B. Hitchcock and J.F. Nixon, J. Chem. Soc., Chem. Commun., (1988) 819.
227 G. Märkl, Angew. Chem., 78 (1966) 907; Angew. Chem., Int. Ed. Engl., 5 (1986) 846.
228 A.J. Ashe III and P. Shu, J. Am. Chem. Soc., 93 (1971) 1804.
229 A.J. Ashe III, J. Am. Chem. Soc., 93 (1971) 3293.
230 K. Dimroth, Topics in Current Chemistry, 38 (1971) 1.
231 N.J. Svetsov-Shilowskii, R.B. Bobkova, N.P. Bobkova, N.P. Ignatova and N.N Melnikov, Usp. Khim., 46 (1977) 967.
232 E. Fluck, Topics in Phosphorus Chemistry, 10 (1980) 193.
233 G. Märkl, in Methoden der Organischen Chemie (Houben-Weyl), Vol. E1, Georg Thieme Verlag, Stuttgart, 1982.
234 J. Ashe III, J. Am. Chem. Soc., 98 (1976) 7861.
235 G. Märkl, Angew. Chem., 75 (1963) 669; Angew. Chem., Int. Ed. Engl., 2 (1963) 479.
236 K. Dimroth, in Methoden der Organischen Chemie (Houben-Weyl), Vol. E1, Georg Thieme Verlag, Stuttgart, 1982, pp. 783-821.
237 Y. Kobayashi, I. Kumadaki, A. Ohsawa and H. Hamana, Tetrahedron Lett., (1976) 3715.
238 C.G. Krespan, B.C. McKusick and T.L. Cairns, J. Am. Chem. Soc., 82 (1960) 1515; C.G. Krespan, J. Am. Chem. Soc., 83 (1961) 3432.
239 A.M. Aguiar and D. Daigle, J. Am. Chem. Soc., 86 (1964) 2299
240 M. Davies, A.N. Hughes and S.W.S. Jafry, Can. J. Chem., 50 (1972) 3625.
241 E. Fluck, B. Neumüller and G. Heckmann, Chem.-Ztg., 111 (1987) 309.
242 E. Fluck, G. Becker, B. Neumüller, R. Knebl, G. Heckmann and H. Riffel, Angew. Chem., 98 (1986) 1018; Angew. Chem., Int. Ed. Engl., 25 (1986) 1002.
243 E. Fluck, G. Becker, B. Neumüller, R. Knebl, G. Heckmann and H. Riffel, Z. Naturforsch., 42b (1987) 1213.

244 A.R. Barron and A.H. Cowley, Angew. Chem., 99 (1987) 956; Angew. Chem., Int. Ed. Engl., 26 (1987) 907.
245 J. Fink, W. Rösch, U.-J. Vogelbacher and M. Regitz, Angew. Chem., 98 (1986) 265; Angew. Chem., Int. Ed. Engl., 25 (1986) 280.
246 W. Rösch and M. Regitz, Z. Naturforsch., 41b (1986) 931.
247 J. Fink and M. Regitz, Bull. Soc. Chim. France, (1986) 239.
248 Note in: R. Bartsch, P.B. Hitchcock, T.J. Madden, M.F. Meidine, J.F. Nixon and H. Wang, J. Chem. Soc., Chem. Commun., (1988) 1475; Binger et al., paper presented to Euchem. conference on phosphorus, silicon, boron, and related elements in low coordination states (PSi-BLOCS), Paris-Palaiseau, August 1988.
249 M. Regitz and P. Binger, Angew. Chem., 100 (1988) 1541; Angew. Chem., Int. Ed. Engl., 27 (1988) 1484.
250 J.F. Nixon, Chem. Rev., 88 (1988) 1327.
251 E. Fluck and R. Braun, Phosphorus and Sulfur, in press.
252 G. Wittig and G. Geissler, Liebigs Ann. Chem., 580 (1953) 44.

AZAPHOSPHOLES

ALFRED SCHMIDPETER and KONSTANTIN KARAGHIOSOFF

Institut für Anorganische Chemie der Universität München
Meiserstrasse 1, 8000 München 2, Federal Republic of Germany

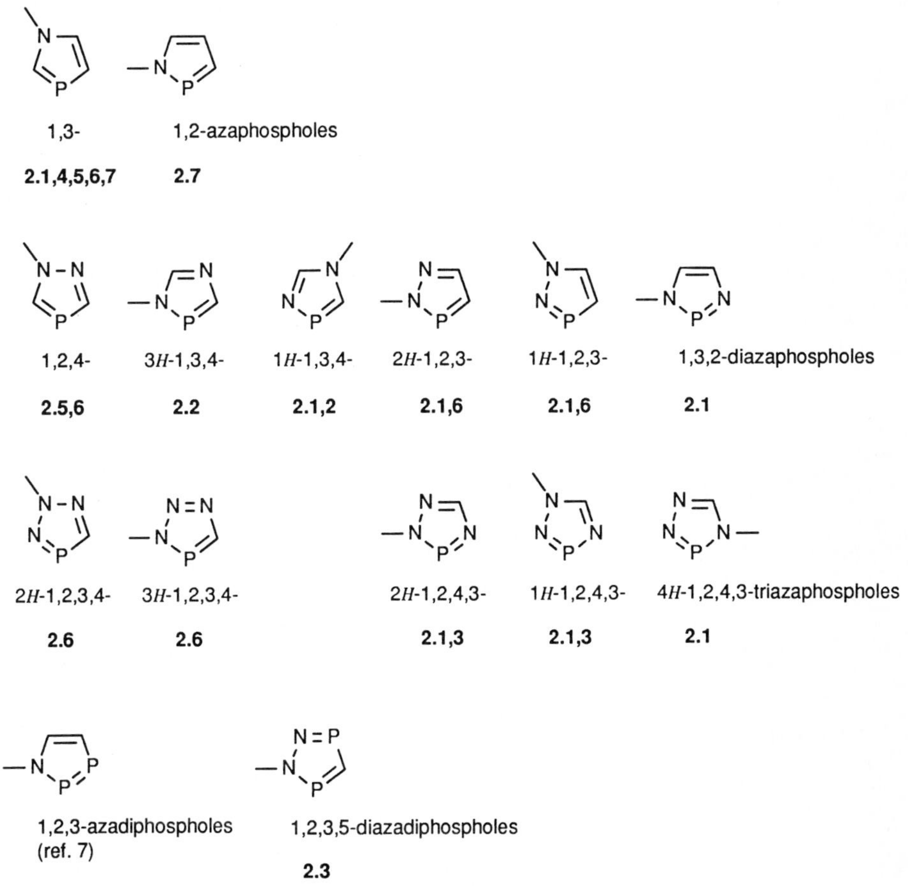

Fig. 1. The known azaphosphole ring systems. The numbers refer to the synthetic routes by which representatives have been prepared. (No representatives are known for 1*H*-1,2,3,4-triazaphospholes and for 1*H*- and 2*H*-tetraazaphospholes.)

1. INTRODUCTION

The synthesis of phosphinines (phosphabenzenes) some twenty years ago (ref. 1) demonstrated that two-coordinate tervalent phosphorus can well participate in a cyclic delocalization. It also demonstrated how effectively a 6π system can stabilize phosphorus two-coordination. This proved even more true for azaphospholes somewhat later (ref. 2). Azaphospholes are five-membered 6π heterocycles which besides the two-coordinate phosphorus contain one three-coordinate nitrogen in the ring. Three-coordinate carbon, two-coordinate nitrogen and eventually another two-coordinate phosphorus constitute the rest of the ring. Representatives of 15 such ring systems are known so far (Fig. 1). Five-membered 6π phosphorus heterocycles containing other elements such as oxa-, thia-, oxaza- and thiazaphospholes are also known but not covered here. The same applies to the phosphole anions $P_n(CR)_{5-n}^-$ which are 6π systems with no other heteroelement than phosphorus and of which the complete series (n = 1 to 5) has become known recently (refs. 3 - 6). Even the azaphospholes alone have become so numerous that not all individual representatives can be mentioned here. A list complete up to summer 1987 is provided however by the compilation of ^{31}P-NMR data of two-coordinate phosphorus in ref. 8.

2. SYNTHESES

In almost all azaphosphole syntheses the five-membered ring is closed by a reaction at phosphorus. The condensations **2.1**, **2.2** and **2.3** are of the type PX + HN, HC, the condensations **2.4** and **2.5** of the inverse type PH, $PSiMe_3$ + XC (X is Cl or some other electronegative substituent). The former are particularly useful for the synthesis of azaphospholes with N adjacent to P, but even certain azaphospholes with C,C-neighboured phosphorus are also prepared this way. The latter condensations only give azaphospholes with C adjacent to P. The cycloadditions **2.6** use phosphaalkenes or phosphaalkynes as dipolarophiles and result in azaphospholes with at least one C adjacent to P. The conversions **2.7** involve a [3+2]-cycloaddition and -elimination in this or the reverse sequence and always result in a reduction of N in the ring. Fig. 1 shows how the different routes contribute to the synthesis of the individual ring systems.

2.1 Cyclocondensations using Trichloro- and Triaminophosphines

An important and inexpensive route to azaphospholes is the condensation of PX_3, X being mostly Cl and NMe_2, to a suitable four-membered chain. After the loss of 2 HX a cyclic intermediate is reached which often can be isolated. In most cases it contains the phosphorus still three-coordinate, sometimes however already two-coordinate. The third HX is lost more or less easily (and reversibly) to yield the azaphosphole.

PCl_3-condensation of N^1- and N^2-methylamidrazones gives 2- and 1-methyl-1,2,4,3-triazaphospholes, R = Me, Ph, according to:

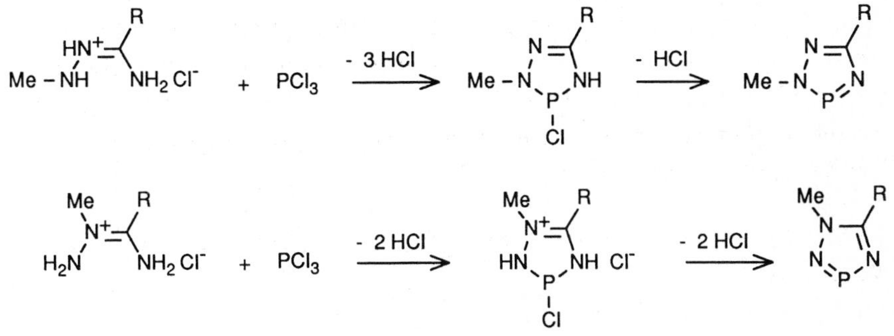

The first step is achieved in refluxing benzene, the second with the help of Et_3N (refs. 9, 10). From N^3-substituted amidrazones result in the same way 4*H*-1,2,4,3-triazaphospholes (ref. 11); they are oligomeric however and no monomeric representative has yet been prepared (refs. 12, 13; see **4.4**).

In place of PCl_3 phosphorus triamides can be used for the condensation and in place of an amidrazone hydrohalide its benzaldehyde derivative (refs. 9, 12 - 14). No intermediates are found in these cases:

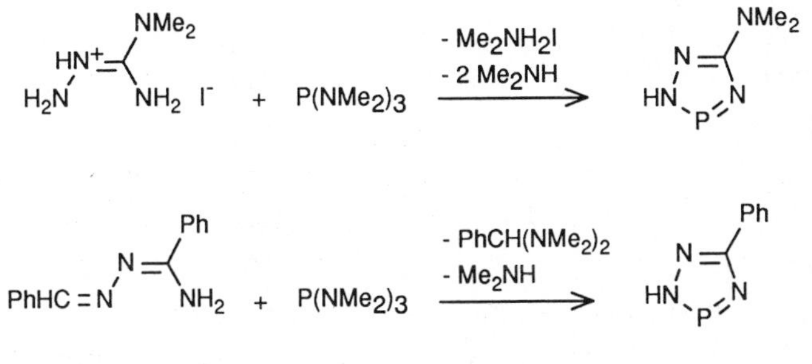

N-unsubstituted triazaphospholes have been obtained this way (refs. 12, 13). In solution they constitute a prototropic equilibrium of all three NH-forms.

By the same condensation also a bicyclic triazaphosphole (R = Me, Ph) has become available (ref. 15):

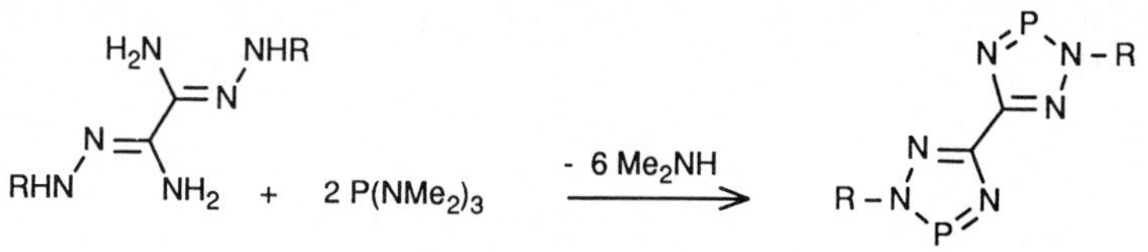

P(V)-compounds have also been used for the amidrazone condensation. PCl_5 yields a dimeric 1,2,4,3-triazaphosphole 3,3-dichloride (ref. 16) which can be reduced by reaction with 1,3-propanedithiol (ref. 17) and a 2-amino-2-imino-1,3,2-dithiaphospholane yields the triazaphosphole directly (ref. 18).

A cationic triazaphosphole analogue with carbon in the ring being replaced by boron has been obtained from the corresponding cyclic chlorophosphine and aluminum chloride (ref. 19):

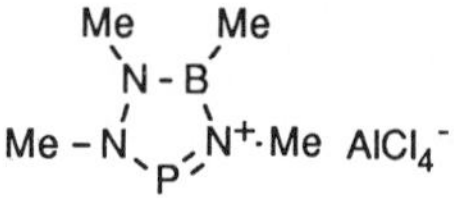

1,3,2-Diazaphospholes require enediamines for their preparation by PX_3-condensation. Diaminomaleodinitrile and PCl_3 in refluxing acetonitrile yield a cyclic chlorophosphine which gives the diazaphospholes by treating with a tertiary base and then with an alkylating agent, R = Me, Et, $CH_2C_2H_3$, CH_2Ph, CH_2COPh, CH_2CO_2Et (refs. 20, 21).

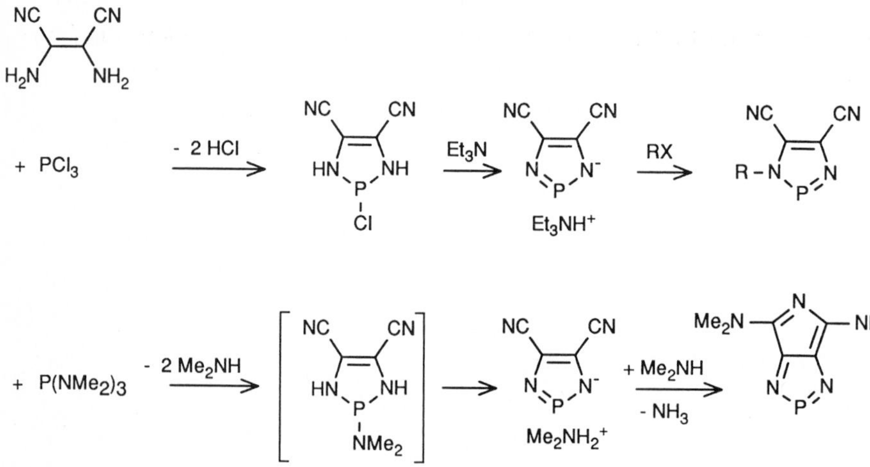

Condensation with $P(NMe_2)_3$ gives a product which is not the cyclic aminophosphine but already the isomeric dimethylammonium diazaphospholate (ref. 22). Further heating converts it to the 4,6-bis(dimethylamino)-1,3,5-triaza-2-phosphapentalene; other secondary amines react analogously (ref. 23). The benzaldehyde derivative of DAMN may also be used for condensation (ref. 13).

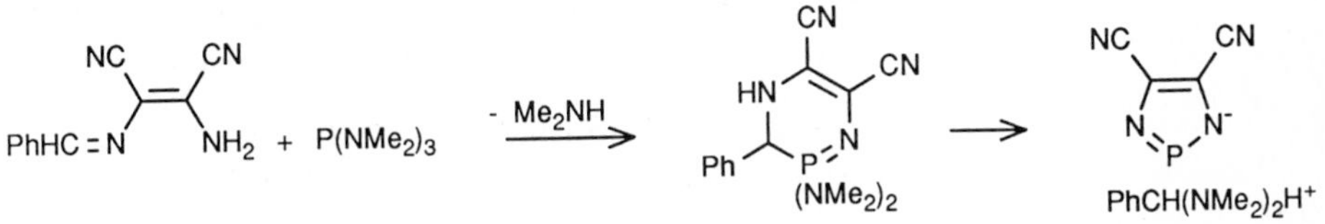

Benzo-1,3,2-diazaphospholes from $P(NMe_2)_3$-condensation of o-phenylene diamines prove to be oligomeric, they monomerize however on heating or as BF_3-complexes (refs. 24, 25; see **4.4**). These results correct an earlier report (ref. 26). Using $P(OPh)_3$ the condensation can be extended to diaminopyridines and -pyrimidines (ref. 27).

The transamination product of sarcosine nitrile and $P(NMe_2)_3$ rearranges to a cyclic aminophosphine which on pyrolysis yields the diazaphosphole (refs. 10, 21).

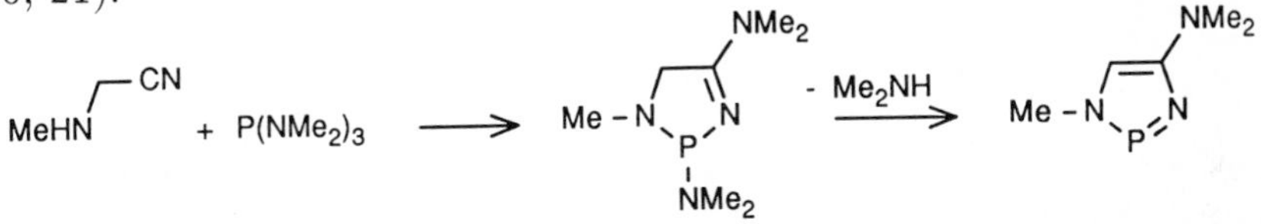

Azaphospholes with just one nitrogen adjacent to phosphorus may also be obtained by a PCl_3-condensation provided the four-membered chain to start with has a sufficiently activated methylene endgroup. This is inherently true for ketone hydrazones. A wide variety of them has been used for the synthesis of 2*H*-1,2,3-diazaphospholes: R^1, R^2, R^3 = H, alkyl, aryl and in addition R^1 = acyl, R^3 = Ph_2PO (refs. 28 - 37).

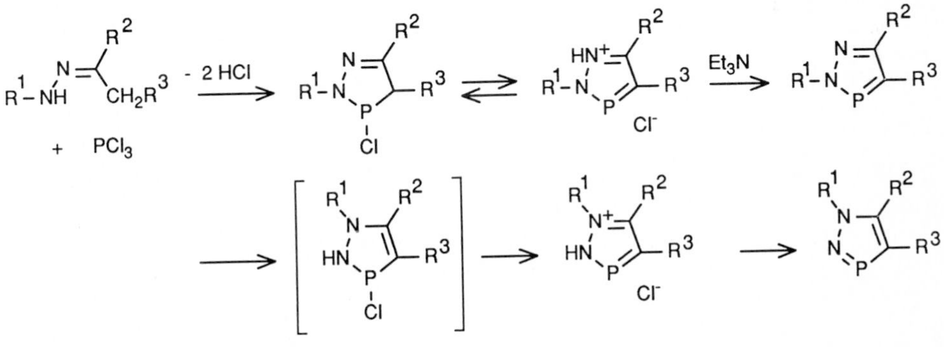

Condensation may be achieved by refluxing in benzene. The intermediate after loss of 2 HCl is depending on the nature of R^1 covalent (R^1 = MeCO, Ph, pyridyl; refs. 29, 36) or ionic (R^1 = H, Me; refs. 34, 36). The third HCl is lost on further refluxing or with the help of a base. From ketone hydrazones with two different alkyl groups of similar reactivity (methyl ethyl ketone as an example) two isomeric products (R^2, R^3 = Me and R^2 = Et, R^3 = H) are obtained (refs. 31, 33). In some cases (e.g. R^1, R^2 = Me, R^3 = H) besides the 2*H*- also the 1*H*-1,2,3-diazaphosphole was found as second product (ref. 35).

A cyclic silyl hydrazone derivative (ref. 38) as well as azo-compounds isomeric to hydrazones (ref. 39) have also been used for the synthesis of 2*H*-1,2,3-diazaphospholes by PCl_3-condensation. The triphenyl-diazaphosphole was observed when the condensation product of phenylazostilbene and PCl_2OMe was thermolyzed (ref. 40).

In special cases also 1,3,4-diazaphospholes have been obtained from a PCl_3-condensation. The necessary four-membered chain with a sufficiently activated methylene group is prepared from 2-amino thiazole or pyridine and bromoacetic ester. Triethylamine is necessary for the condensation (ref. 21).

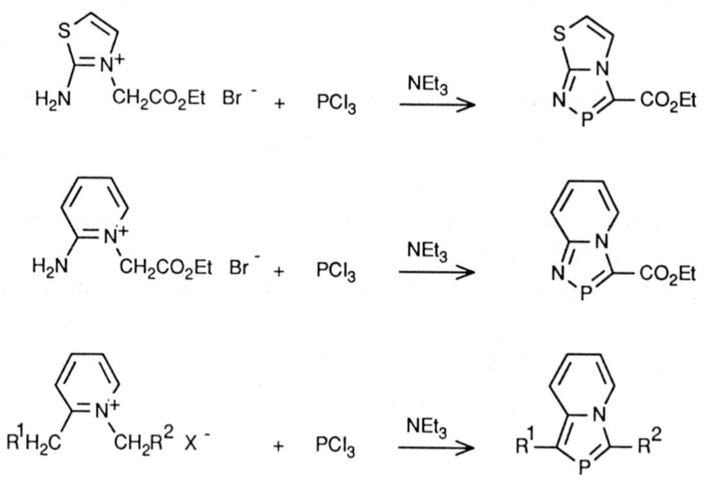

PCl_3 may even react with two methylene groups resulting in rings with a C,C-neighboured phosphorus. Starting from 2-alkylpyridines in a similar way as above bicyclic 1,3-azaphospholes (R^1 = H, Me, Ph; R^2 = COPh, CO_2Et, CN) are obtained (ref. 41).

2.2 Cyclocondensations using Chloromethyl Dichlorophosphine

This reagent can supply the fragment HCP for an azaphosphole ring and has been used for the synthesis of 1,3,4-diazaphospholes with an unsubstituted 5-position. Amidines or similar NCN-compounds can be used as partners in the condensation which has to be induced by Et_3N (refs. 10, 42).

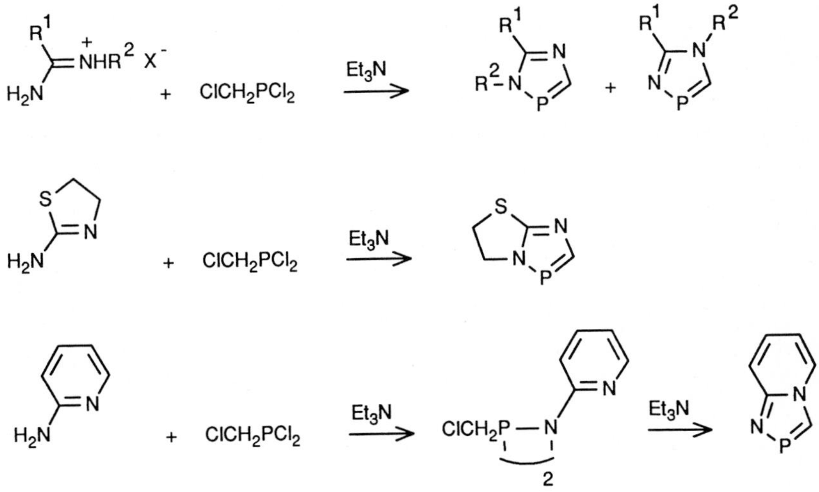

From amidines (R^1/R^2 = Me/H, Ph/H, Ph/Ph, SMe/Me) mixtures of 2*H*- and 1*H*-1,3,4-diazaphospholes are obtained; for R^2 = H they equilibrate by proton exchange. From 2- and 4-aminopyrimidines like from 2-aminopyridine only 1*H*-isomers, from 2*H*-aminothiazole predominantly the 1*H*-isomer are formed. In some cases the diazadiphosphetidine has been isolated as an intermediate.

2.3 Cyclocondensations of Three-membered Chlorophosphines with Hydrazines

A CNP- or PCP-chain with electronegative functions at both ends can be reacted with hydrazines to yield a five-membered ring. An example of the first type gives the synthesis of 1,2,4,3-triazaphospholes (R^1 = Me, Ph; R^2 = *n*Bu, CH_2Ph, Ph) from imidato-chlorophosphines (ref. 14):

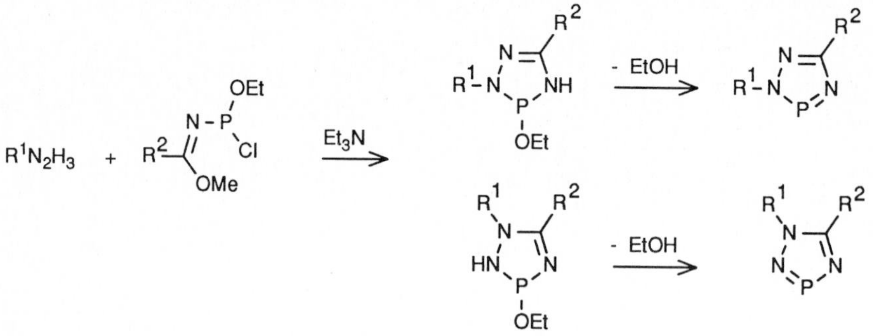

Rather different conditions are necessary to eliminate ethanol from the cyclic ethoxyphosphine intermediates: 1*H*-1,2,4,3-triazaphospholes form easily at room temperature while the 2*H*-isomers need 100°C or OEt/Cl-exchange and subsequent elimination of HCl by base.

1,2,3,5-Diazadiphospholes (R = Me, Ph) result from the condensation of bis(dichlorophosphino)methane with primary hydrazines (ref. 43).

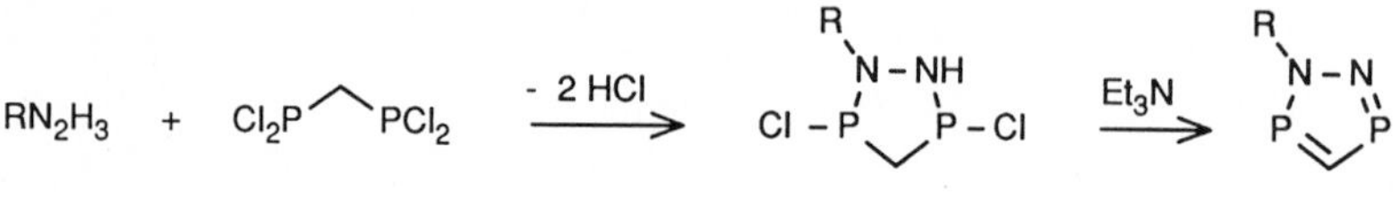

2.4 Cyclocondensation of 2-Aminophosphines

While chlorophosphines are used in most of the aforementioned routes phosphines are used here, actually 2-aminoethyl and 2-aminophenyl phosphines. They are condensed mostly with carboxylic acid derivatives to yield 1,3-azaphospholes and their benzo derivatives. Condensation of iminoesters and 2-aminoethylphosphines gives cyclic butyl-, benzyl- and phenylphosphines (ref. 44). Flash pyrolysis at 700 - 730°C converts them to 1,3-azaphospholes, R^1 = H, Me, Et, Ph (ref. 45).

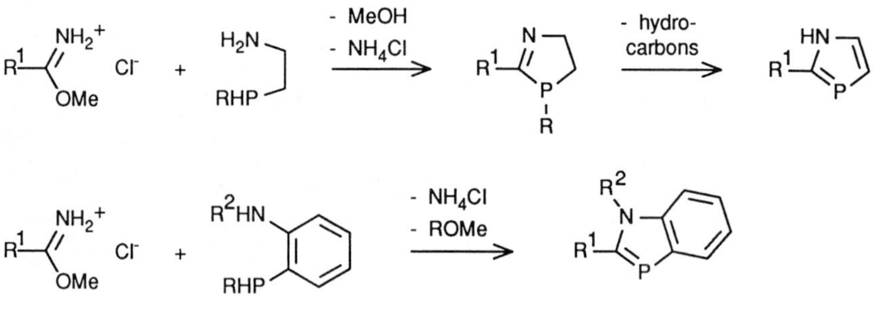

The route has first been demonstrated for 2-aminophenylphosphines R = H, Me, Ph (ref. 46). In place of the imino esters also acyl chlorides, imide chlorides, ortho esters, ortho ester amides, phosgene immonium chloride, benzaldehyde and benzil have been used. The reactions proceed at room temperature and give directly the benzo-1,3-azaphospholes with R^1 = H, R^2 = H, Me, Ph, NMe_2 (refs. 46, 47) and R^1 = Me, R^2 = H, Me, Ph, *t*Bu (ref. 48).

2.5 Cyclocondensations using Trisilylphosphine

Starting from $P(SiMe_3)_3$ 1,3-azaphospholes and 1,2,4-diazaphospholes, i.e. rings with C,C-neighboured phosphorus can be prepared. A two-step route involves the reaction with chloroiminium cations to give 1,3-diamino-2-phosphaallyl cations and their subsequent condensation with hydrazines (ref. 49). The parent 1,2,4-diazaphosphole as well as substituted representatives (R = H, Me, Ph; R^1 = H, *t*Bu, Ph) are accessible this way.

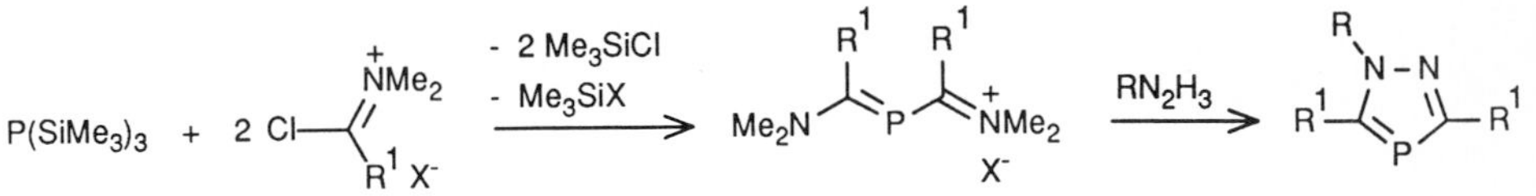

Reaction with oxadiazolium cations (R = Me, Et, Ph) and potassium fluoride also gives 1,2,4-diazaphospholes. Various substituted (R^1, R^2 = H, alkyl, Ph; ref. 50) as well as bicyclic representatives (ref. 51) become accessible this way.

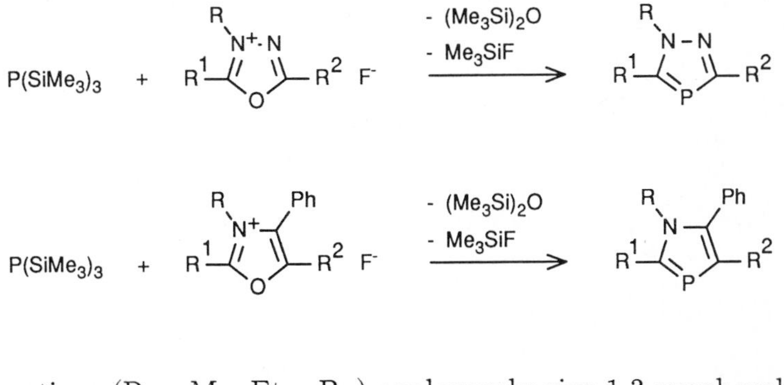

Oxazolium cations (R = Me, Et, *n*Bu) analogously give 1,3-azaphospholes. Again a variety of different substituted (R^1 = H, Me; R^2 = H, Me, Et, Ph;

ref. 52) as well as bicyclic representatives (ref. 51) have been prepared. In place of the oxazolium ion an acyclic equivalent (ref. 52) and in place of $P(SiMe_3)_3$ in some cases $LiP(SiMe_3)_2$ were used (ref. 51).

From zwitterionic oxadiazolium and oxazolium olates (Münchnones) result diaza- and azaphospholes with a trimethylsiloxy group next to the phosphorus. Methanolysis converts them into the corresponding hydroxy derivatives which are stable in the phenolic form (refs. 53, 54).

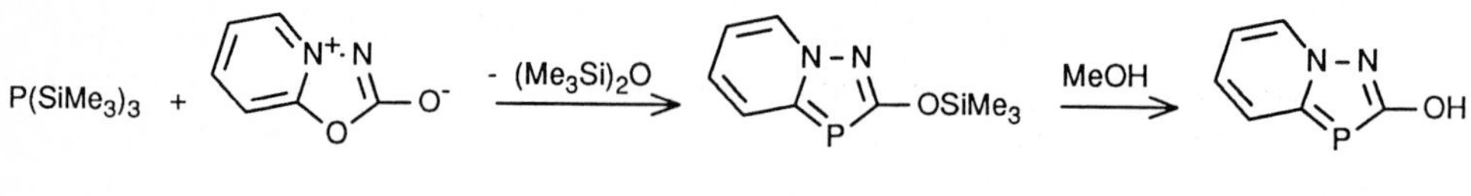

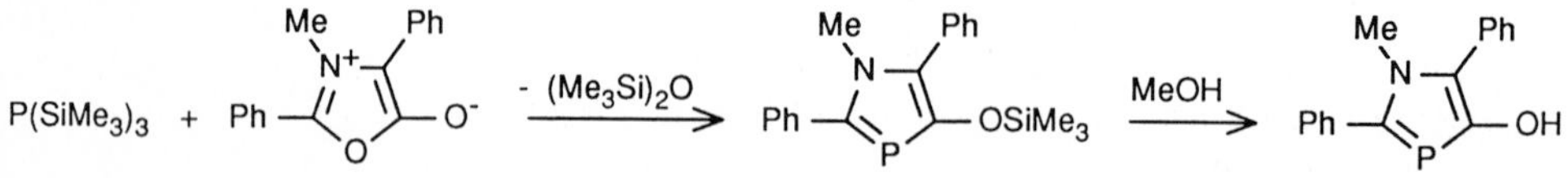

2.6 Cycloadditions of Phosphaalkynes and Phosphaalkenes

In the last five years 1,3-dipolar cycloaddition of phosphaalkynes $R^1C{\equiv}P$ has opened another useful route to azaphospholes with C,C or C,N adjacent to phosphorus, namely 1,2,3,4-triazaphospholes, 1,2,4- and 1,2,3-diazaphospholes and 1,3-azaphospholes (ref. 55). Mostly the phosphaalkyne with R^1 = *t*Bu has been used, but also those with other tertiary alkyl groups, *i*Pr and H. In place of phosphaalkynes phosphaalkenes $R^1(Me_3SiO)C{=}PSiMe_3$ can be used. The cycloaddition is here accompanied by the elimination of $(Me_3Si)_2O$. Cycloaddition of the phosphaalkenes $R^1(Me_3Si)C{=}PCl$ along with the elimination of Me_3SiCl provides access to azaphospholes with R^1 = Ph, CO_2Et, $SiMe_3$. Just the overall reactions are given below; for intermediates and mechanism see the referenced literature (e.g. ref. 55).

Cycloaddition of azides gives 3*H*-1,2,3,4-triazaphospholes with a wide variety of R (refs. 56 - 60 and 62, 63); in case of R = H (ref. 55) and $SiMe_3$ (ref. 60) they rearrange to give the 2*H*-isomers.

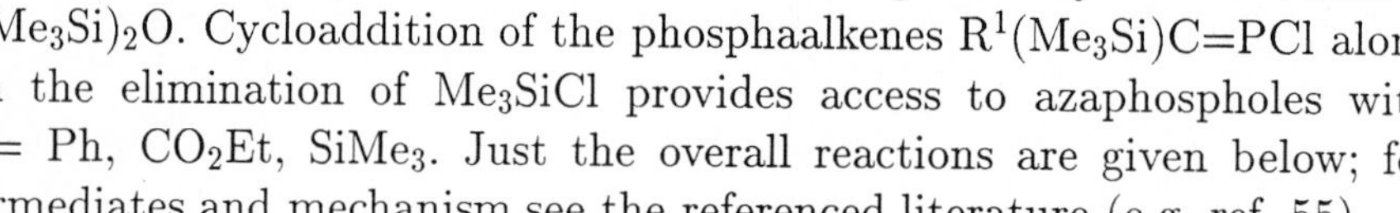

Diazoalkanes add to phosphaalkynes regiospecificly (except to HCP, ref. 61) with the formation of a PC-bond. The cycloaddition products of primary (R = H, refs. 56 - 59, 61, 64), silyl (R = $SiMe_3$, ref. 65), acyl (R = COMe, COPh etc., ref. 64) and phosphoryl diazoalkanes (R = $POPh_2$, $PO(OMe)_2$; ref. 64) rearrange to give 1,2,4-diazaphospholes. Always the isomer resulting from an 1,5-shift of R is formed, for R = acyl in addition a minor amount of the other isomer.

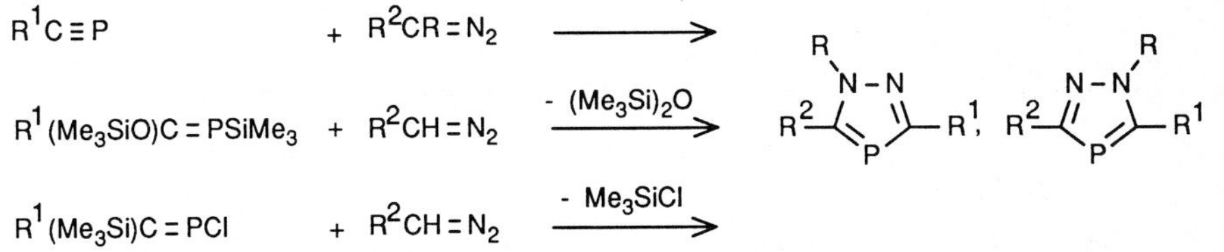

More 1,2,4-diazaphospholes (R = H) result from primary diazoalkanes and the above mentioned phosphaalkenes of the first (refs. 58, 66) or the second type (R^1 = Ph, refs. 63, 67; R^1 = $SiMe_3$, refs. 61, 62, 68; R^1 = CO_2Et, ref. 69).

Diphenyl nitrile imine and a phosphaalkyne (refs. 55, 56, 70) or a phosphaalkene of the second type (ref. 63) also give 1,2,4-diazaphospholes:

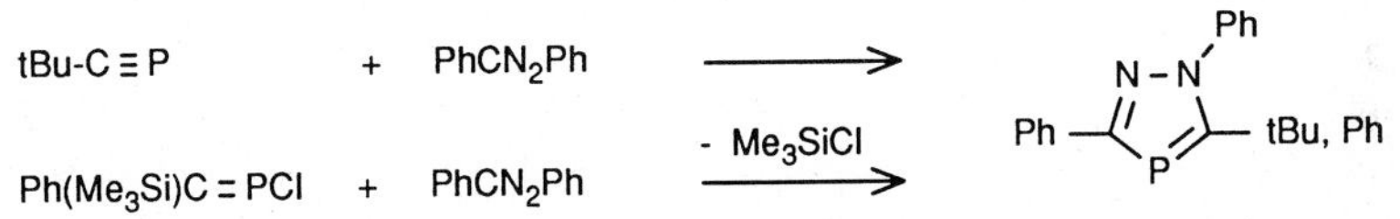

If the nitrile imine has a methyl instead of the one or the other phenyl group 2% of the isomeric 2*H*-1,2,3-diazaphosphole are formed along with the 1,2,4-diazaphosphole (ref. 55).

Sydnones provide another source for the CNN fragment. With a phosphaalkyne (ref. 55) and with both types of phosphaalkenes (ref. 71 and ref. 50) they also give 1,2,4-diazaphospholes (R = H, Me, CH_2Ph, Ph; R^1 = H, Me, Ph). Only in one case a minor amount of 1*H*-1,2,3-diazaphosphole is formed (ref. 55).

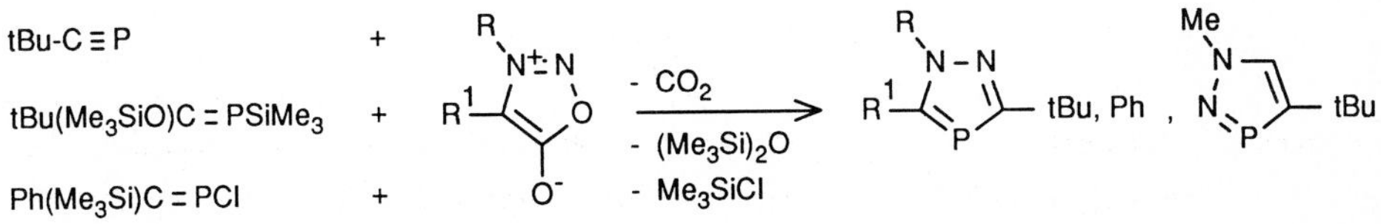

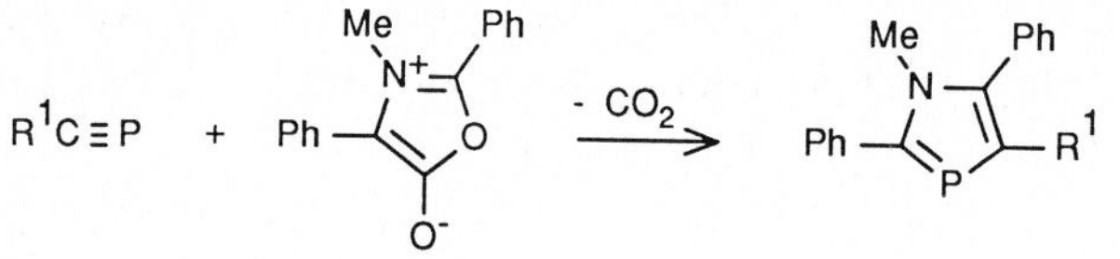

Münchnones and phosphaalkynes correspondingly give 1,3-azaphospholes (R^1 = H, *t*Bu; refs. 55, 61, 71).

Cycloaddition of diphenyl nitrile ylid, obtained by irradiating 2,3-diphenyl-2*H*-azirin, and a phosphaalkene of the second type results in an N-unsubstituted 1,3-azaphosphole (ref. 63):

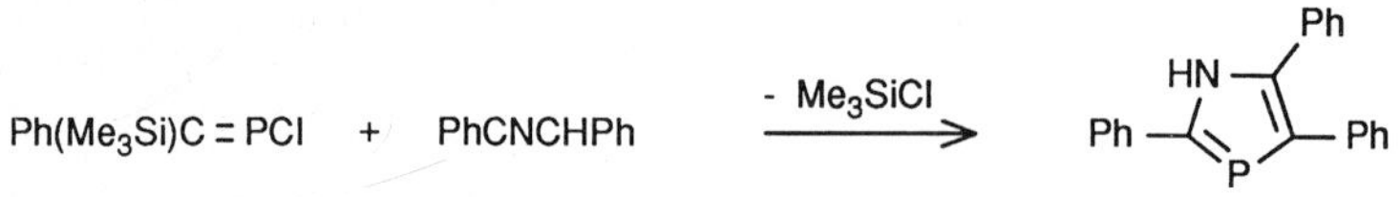

2.7 Conversion of Azaphospholes

In some cases azaphospholes have been converted to others with less nitrogen in the ring. For 1,3,2-diazaphospholes (R = Me, $SiMe_3$; ref. 72) and 1,2,4,3-triazaphospholes (R/R^1 = Me/Ph, $SiMe_3/NMe_2$; ref. 73) this was achieved by reaction with acetylenes. In case of the methyl diazaphosphole a wide variety of acetylenes was used (R^1/R^2 = H/*n*Pr, H/*n*Bu, H/CH_2NMe_2, H/Ph, H/CO_2Et, Me/Me, Me/CO_2Me, Ph/CO_2Et, Ph/OCOPh, CO_2Me/CO_2Me).

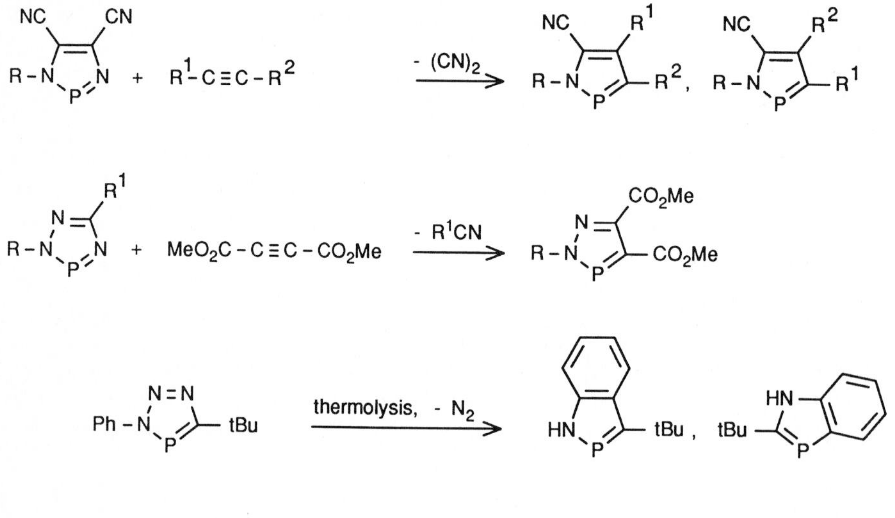

By pyrolysis (540°C) a 3-phenyl 1,2,3,4-triazaphosphole is converted to benzo-1,2- and -1,3-azaphospholes. The same is true for the 2-naphthyl derivative (ref. 60).

3. SPECTRA AND STRUCTURE

Besides the spectra discussed in **3.1** and **3.2** below Raman (refs. 30, 74, 75) and IR spectra (refs. 30, 74 - 77) of 1,2,3-diazaphospholes and IR spectra of 1,2,4-diazaphospholes (refs. 58, 66) as well as mass spectra of various azaphospholes (refs. 12, 22, 78) are reported and discussed in the literature.

3.1 NMR Spectra

A low field chemical shift is characteristic for the azaphosphole phosphorus. Its value depends mainly from the ring members adjacent to phosphorus, but is also influenced by other factors. Even neglecting single extreme values thus only broad and overlapping ranges result (ref. 8):

Ring members adjacent to P	$\delta^{31}P$ Ranges in azaphospholes
C,C	50 - 180
C,N	160 - 270
N,N	220 - 300

The influence of substituents and the influence of ring protonation have also been shown (ref. 8). Correlations of all available data demonstrate, that $\delta^{31}P$ in an azaphosphole is influenced very much the same way as $\delta^{13}C$ or $\delta^{15}N$ is in the corresponding position of an azole (refs. 8, 79). This suggests a rather congruent bonding situation in azoles and azaphospholes.

For many azaphospholes $\delta^{13}C$ values are given and discussed, for some also $\delta^{14,15}N$ values (refs. 79 - 81).

One bond $^{31}P^{13}C$ coupling in azaphospholes is negative (refs. 74, 82) and $^{31}P^{15}N$ coupling is presumably positive (ref. 79). The characteristic range for the one is $^{1}J_{PC}$ = 30 - 60 Hz. The PC coupling is thus larger than in phosphines, e.g. Ph_3P $^{1}J_{PC}$ = -12.5 Hz, and it extends to considerably higher values than in phosphinines (refs. 83, 84). Examples for the other are $^1J_{PN}$ = 24 and 78 Hz (2*H*-1,2,3-diazaphospholes, refs. 79, 81) and 92 Hz (1*H*-1,2,3-diazaphospholes, ref. 79). As in α-unsubstituted phosphinines (ref. 85) and as in phosphaalkenes with a hydrogen trans to the phosphorus substituent (ref. 86) the two-bond $^{31}P^1H$ coupling is large, $^2J_{PCH}$ = 29 - 57 Hz, and positive (refs. 74, 87). Three bond phosphorus coupling to a ring hydrogen is in the range of $^3J_{PH}$ = 4 - 8 Hz.

3.2 UV/VIS Spectra

The known azaphospholes are colourless or pale yellow, except the 4,6-diamino-1,3,5-triaza-2-phosphapentalenes, which are deep red. Their colour like that of other azapentalenes is due to a CT band (ref. 23).

For many azaphospholes UV spectra are reported (refs. 22, 23, 30, 39, 45, 46, 48, 54, 75, 88). The exchange of N in an azole for P in an azaphosphole is generally accompanied by a bathochromic shift of λ_{max} (refs. 45, 46, 75). As compared to pyrrol, imidazol and oxazole 1,3-azaphospholes like thiazoles show an additional $\pi \rightarrow \pi^*$ band (ref. 45).

3.3 X-Ray Structure Investigations

Of 16 azaphospholes including one hydrochloride and some complexes X-ray structures are available. They all show a planar ring with averaged bond distances in accord with a delocalized bond system. The bond distances within the ring and the endocyclic angle at phosphorus are given in Figs. 2 and 3. The angle at P ranges from 88 to 100°. It clearly increases with the number n of two-coordinate nitrogen atoms adjacent to the phosphorus:

	n	angle
1,3-azaphosphole 1,2,4-diazaphosphole	0	88 - 92°
2*H*-1,2,3-diazaphospholes 1*H*-1,2,3-diazaphospholes	1	92 - 95°
2*H*-1,2,4,3-triazaphosphole 1*H*-1,2,4,3-triazaphospholes 1,3,2-diazaphospholate	2	97 - 100°

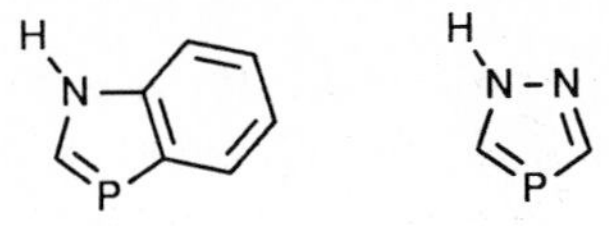

d CN, d NN	137.1		134.0
d CN, d CC	135.1	142.2	132.0
d PC	169.5	180.7	173.4
CPC	88.2°		87.5°

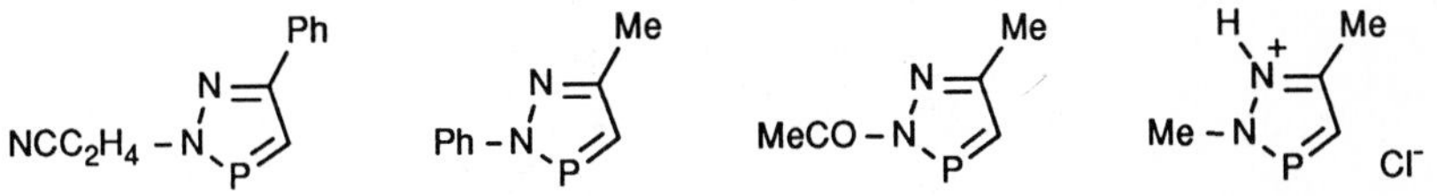

d CN	134.3		132		134		132.6	
d NN, d CC	134.4	140.7	132	142	134	144	134.3	141.5
d PN, d PC	167.6	170.2	165	175	168	175	166.1	174.6
NPC	88.8°		92°		89°		88.2°	

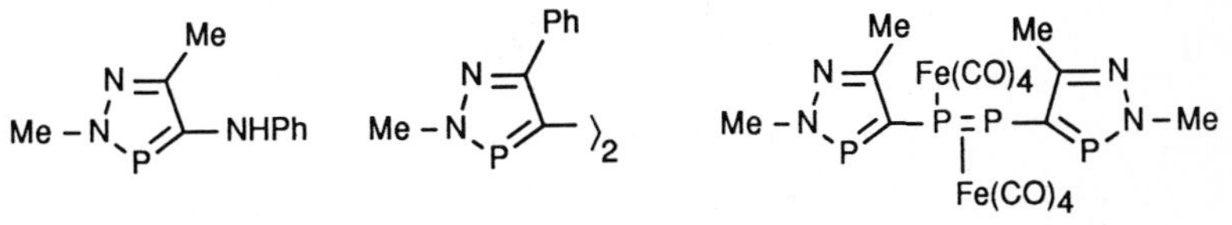

d CN	132.9		132.5		132.0		132.2	
d NN, d CC	134.7	141.6	135.4	142.6	135.4	142.3	141.0	135.1
d PN, d PC	167.4	168.8	168.5	172.6		171.3	171.4	
NPC	90.4°		89.5°		88.7°		89.3°	

Fig. 2. Molecular geometries of 1,3-azaphospholes, 1,2,4- and 2*H*-1,2,3-diazaphospholes (refs. in the order of the above formulas 91, 92, 93, 75, 74, 34, 94, 95, 96). The bond distances are given in pm and arranged to correspond with the position of the bond in the formula.

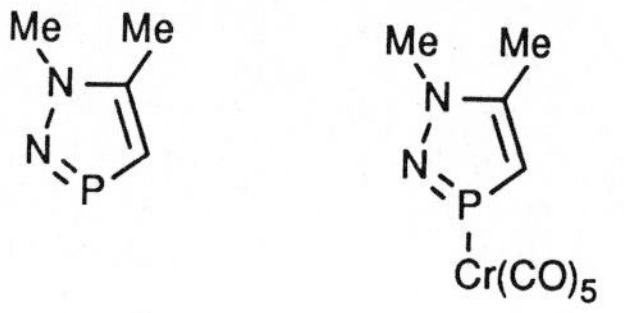

d CN	135.6		136.2	
d NN, d CC	133.0	138.1	135.4	138.2
d PN, d PC	165.4	172.3	163.7	172.9
NPC	92.9 o		95.2 o	

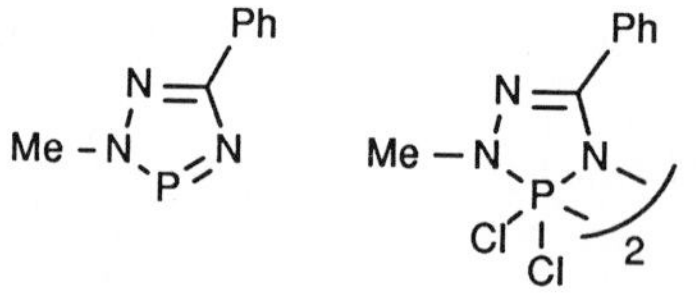

d CN	132.8		128.8	
d NN, d CN	135.4	135.7	141.1	140.1
d PN	166.8	161.6	167.7	168.4
NPN	92.4 o		85.4 o	

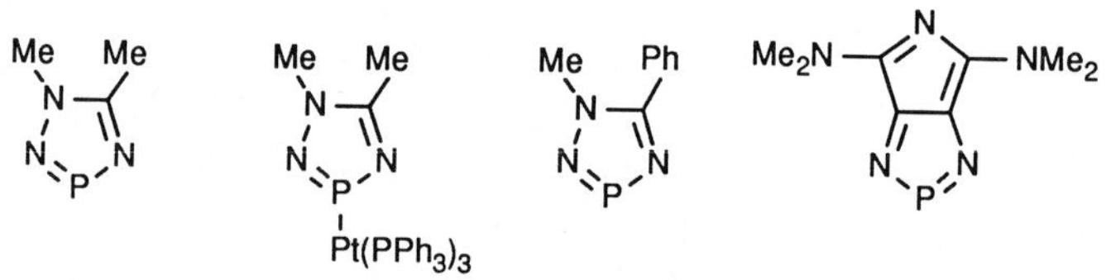

d CN, d CC	134.2		132		135.8		137.6	
d NN, d CN	133.2	134.1	139	131	135.1	132.0	134.5	134.9
d PN	163.8	163.6	164	163	162.7	165.4	166.5	166.1
NPN	96.9 o		96.5 o		97.0 o		99.8 o	

Fig. 3. Molecular geometries of 1*H*-1,2,3-diazaphospholes, 2*H*- and 1*H*-1,2,4,3-triazaphospholes and a pentalene containing a 1,3,2-diazaphosphole ring (refs. in the order of the above formulas 35, 97, 98, 99, 100, 101, 102, 23). For the dichloride mean values of the two triazaphosphole rings of the dimer are given.

PC bond lengths in azaphospholes range from 170 to 175 pm as compared to mean values of 167 pm and 185 pm for the localized double and single bond in phosphaalkenes (ref. 89). The only exception is the long PC bond in the benzo-1,3-azaphosphole, a compound which also shows an exceptional reactivity among azaphospholes (see **4.2**). PN bonds in azaphospholes are 162 to 168 pm long as compared to mean values of 156 pm and 167 pm for the more or less localized double and single bonds in amino-iminophosphines (ref. 90). CC bonds are 141 to 144 pm long in 2*H*-1,2,3-diazaphospholes and only 138 pm long in the 1*H*-isomers.

The molecules of the 1,2,4-diazaphosphole (second formula in Fig. 2) are connected to each other in the crystal by NH···N hydrogen bonding (ref. 92). Three of them form one spire of a helix along a three-fold screw axis (Fig. 4). Pyrazole molecules are connected in the same way forming an 8-spiral however (ref. 103).

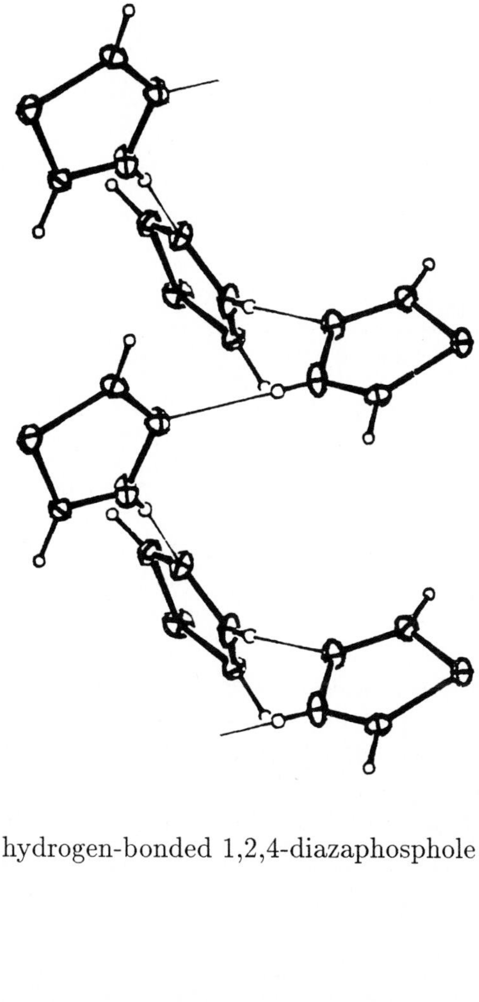

Fig. 4. Helix of hydrogen-bonded 1,2,4-diazaphosphole (ref. 92).

4. REACTIONS

An azaphosphole can mainly undergo three types of reactions (see Fig. 5):

reaction as a P- or N-donor (P- versus N-coordination),
addition to a P=C or P=N bond (1,2-addition),
P-oxidation (oxidative or 1,1-addition).

Two of these possibilities may assist each other or compete with each other. Important combinations are 1,2-addition plus P-coordination and 1,1-addition plus 1,2-addition. The combination of 1,2-addition and 1,2-elimination can result in a substitution at a carbon atom next to phosphorus. In certain cases a double 1,1-addition is found. N-unsubstituted azaphospholes can in addition undergo N-substitution reactions.

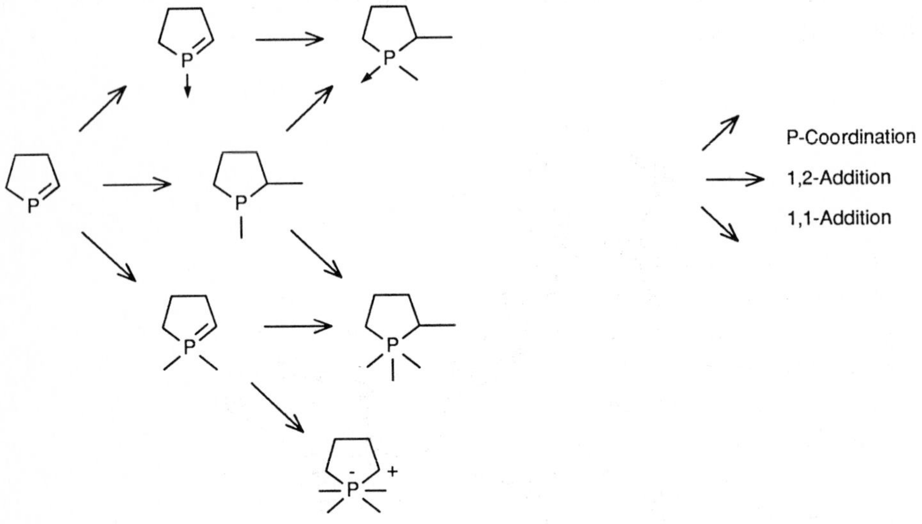

Fig. 5. Reaction modes of azaphospholes involving its phosphorus.

4.1 N-Substitution

Azoles with a two-coordinate (pyridinic) nitrogen like pyrazole and imidazole form NH· · ·N bonded polymers and exchange protons between the two (or more) nitrogen ring members. The same is true for N-unsubstituted di- and triazaphospholes (see **3.3** and ref. 12 in respect to the hydrogen bonding). Depending on the ring system the proton exchange is more or less rapid and the exchange equilibrium is more or less at the side of a certain tautomer (refs. 37, 104 for 1,2,3-diazaphospholes, refs. 49, 61, 64 for 1,2,4-diazaphospholes, ref. 42 for 1,3,4-diazaphospholes, ref. 55 for 1,2,3,4-triazaphospholes and ref. 13 for 1,2,4,3-triazaphospholes). The two-coordinate phosphorus in no case however takes part in the proton exchange (refs. 12, 79).

In substitution reactions of N-unsubstituted azaphospholes the substituent is introduced always at nitrogen again. Except for the cases mentioned below this is also true for two-step substitutions, i.e. if the substituent is introduced into the anion resulting from azaphosphole deprotonation. Table 1 lists the derivatives which were prepared this way.

TABLE 1. Azaphosphole derivatives prepared by N-substitution.

	1,2-aza-	1,2,3-diaza-	1,2,4-diaza-	1,3,2-diaza-	1,2,4,3-triazaphospholes
boryl			ref.105		refs.9,10
alkyl			refs.71,105,106	refs.23,73	
acyl	ref.47	ref.81	ref.105		
silyl	ref.45		refs.65,105	ref.73	refs.12,38,107
germyl			ref.65		
stannyl					ref.12
phosphino	ref.47	ref.72	ref.65		ref.12
phosphinyl					ref.12
arsino			ref.65		
stibino		ref.65			
thiyl	ref.47				

Exceptions are the 1-lithio-1,3-azaphosphole (ref. 45) and 1-lithio-benzo-1,3-azaphosphole (ref. 47) which are alkylated at phosphorus; silyl and acyl substituents even here become N-bonded however.

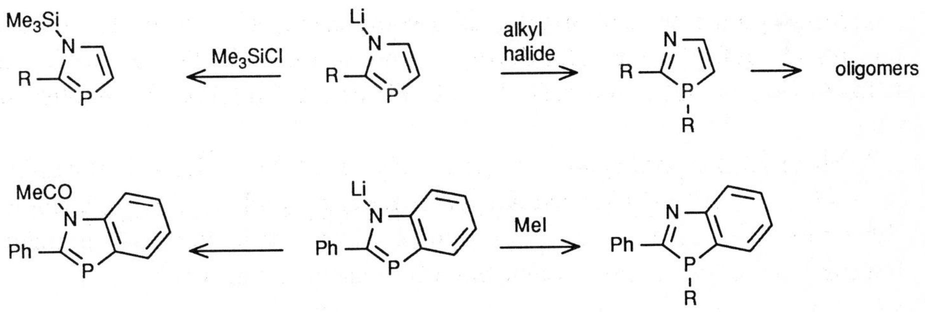

A 2-acetyl-1,2,3-diazaphosphole has been deacylated by phenyl hydrazine (ref. 104). Likewise 1-acyl, 1-silyl and 1-phosphoryl 1,2,4-diazaphospholes (refs. 64, 68) as well as a 2-silyl-1,2,3,4-triazaphosphole (ref. 60) are converted to the NH-compound by methanolysis or hydrolysis.

4.2 N-Coordination

The pyridinic nitrogen in azoles can be alkylated and it coordinates to non-metal and metal acceptors. The same is true again for the di- and triazaphospholes. And surprisingly even here the two-coordinate phosphorus does not very effectively compete with the nitrogen. Where there is a choice alkylation always takes place at the nitrogen (ref. 36 for 1,2,3-diazaphospholes, ref. 105 for 1,2,4-diazaphospholes and ref. 42 for 1,3,4-diazaphospholes). Methylation and demethylation of a 1,2,3-diazaphosphole specificly converts the 1-methyl into the 2-methyl derivative (ref. 36).

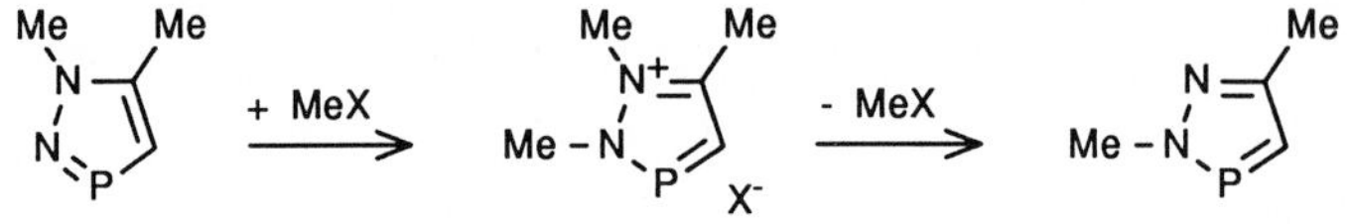

The only case where an alkylation of phosphorus is observed is again the benzo-1,3-azaphosphole (see **4.1**):

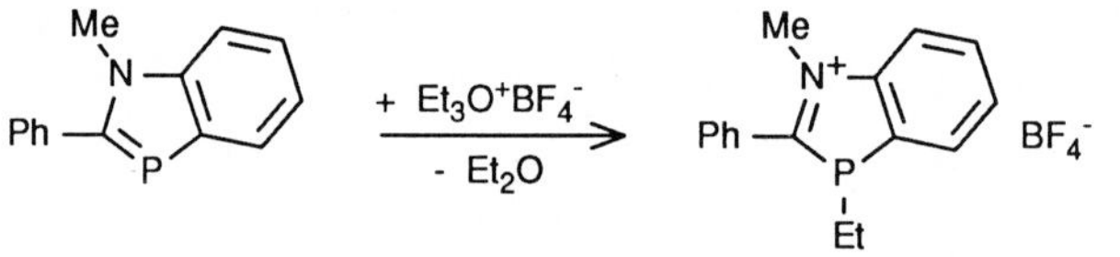

Azaphospholes form N-bonded BF_3-complexes (ref. 108 for 1,2,3-diazaphospholes, ref. 105 for 1,2,4-diazaphospholes, refs. 18, 25, 109 for benzo-1,3,2-diazaphospholes and refs. 11, 14, 15 for 1,2,4,3-triazaphospholes; see **4.4**).

1-Boryl-1,2,4-diazaphospholes (ref. 105) and 2-boryl-1,2,4,3-triazaphospholes (refs. 9, 10) via N-coordination form dimers and adducts with the unsubstituted or metallated azaphosphole. The latter like the pyrazolyl borates are potential monovalent bidentate chelating agents.

TABLE 2. Monomeric P- and N-coordinated azaphosphole transition metal complexes.

Ligand	Mode	Complexed Unit	
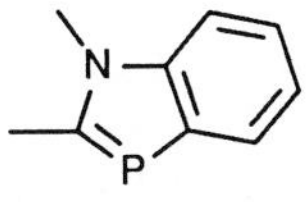	P	$Cr(CO)_5$	ref. 47
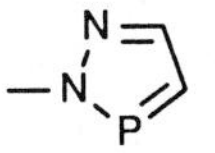	P	$Cr(CO)_5$	refs. 36, 97
	P	$W(CO)_5$	ref. 97
	P	$Mn(CO)_2C_5H_4Me$	ref. 97
	P	$Fe(CO)_4$	ref. 97
	P,N	$AuMe_2Cl$	ref. 116
	P	$Pt(PPh_3)_3$	ref. 101
	P	$Pt(PPh_3)_2/2$	ref. 101
	N	$PdCl_2PEt_3$	ref. 117
	P	$PtCl_2PEt_3$	refs. 117, 118
	P,N	$PtBr_2PEt_3$	ref. 117
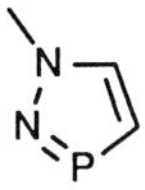	P	$Cr(CO)_5$	ref. 97
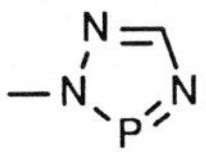	P	$Cr(CO)_5$	ref. 97
	P	$Mo(CO)_5$	ref. 97
	N-4	$PdCl_2PEt_3$	ref. 117
	P, N-4	$PtBr_2PEt_3$	ref. 117
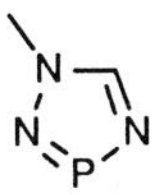	P	$Cr(CO)_5$	ref. 97
	P	$Mo(CO)_5$	ref. 97
	N-2, N-4	$AuMe_2Cl$	ref. 116
	P	$Pt(PPh_3)_3$	ref. 101
	P	$Pt(PPh_3)_2/2$	ref. 101
	N-4	$PdCl_2PEt_3$	ref. 117
	P, N-4	$PtBr_2PEt_3$	ref. 117

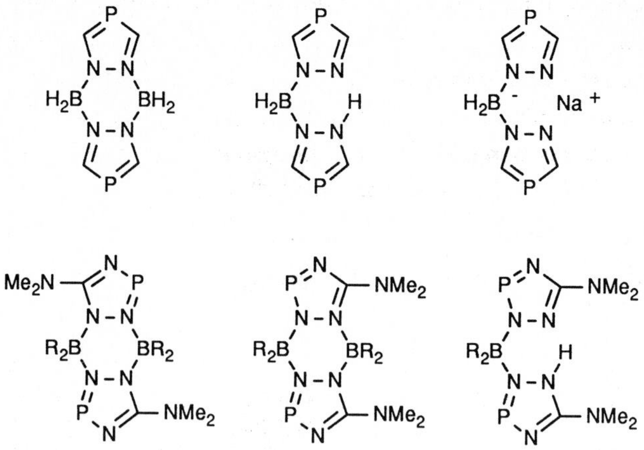

Even transition metal acceptors sometimes prefer N-coordination over P-coordination (see **4.3**).

4.3 P-Coordination

Certainly the most striking feature of azaphospholes (like of phosphinines, ref. 1) is their very low phosphorus nucleophilicity. No P-alkylation has been observed except for the 1,3-azaphospholes mentioned in **4.1** and **4.2**. In most azaphospholes the phosphorus does not react with oxygen or sulfur (ref. 2; 1,2,3,5-diazadiphospholes however are sensitive to oxygen, ref. 43). Though no azaphosphole P-sulfides are known, P-imine (ref. 110) and P-alkylidene derivatives (refs. 111 - 115) are postulated as intermediates from the reaction of azides or diazoalkanes, respectively.

P-coordination of azaphospholes is found in low-valent transition metal complexes (Table 2). Azaphosphole metal(0) complexes invariably are P-coordinated, those of higher oxidation states may be P- or N-coordinated.

Bonding in azaphosphole complexes is discussed in refs. 97, 101. They may be compared to the complexes of phosphaalkenes and iminophosphines (refs. 5, 119). Of two P-coordinated azaphosphole complexes the crystal structure has been investigated (see **3.3**, Fig. 3 and Fig. 6). In both cases the metal atom lies strictly in the plane of the ring and the ring structure has not significantly changed.

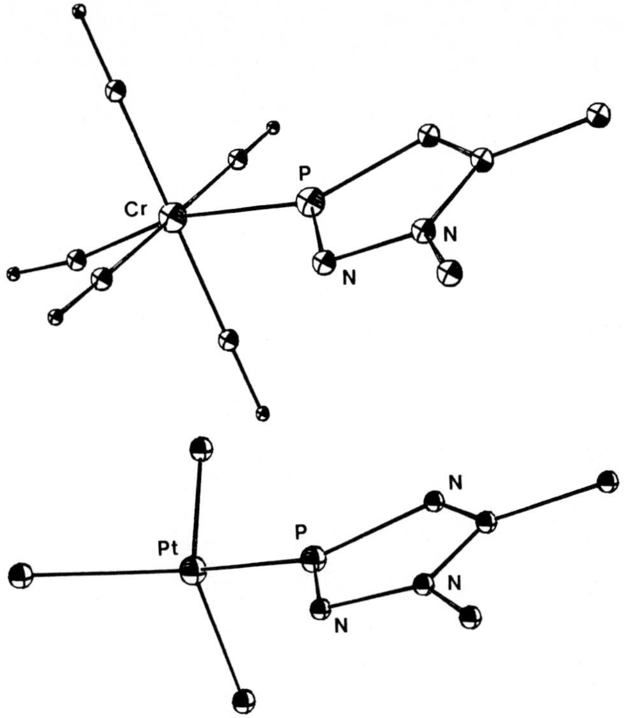

Fig. 6. 1,5-Dimethyl-1,2,3-diazaphosphole $Cr(CO)_5$ complex (ref. 97) and 1,5-Dimethyl-1,2,4,3-triazaphosphole $(Ph_3P)_3Pt$ complex (ref. 101, of the phosphine ligands just the phosphorus atoms are shown). As expected from the different coordination numbers the azaphosphole-platinum distance is considerably shorter (222.7 pm) than the phosphine-platinum distance (233.4 pm).

As mentioned before (see Fig. 5) P-coordination may induce a 1,2-addition. One such example, in which a triazaphosphole tetramer is formed, will be discussed in **4.4**. The dimeric Et_3PPtCl_2-complexes of a 1*H*- and a 2*H*-triazaphosphole (ref. 117) provide two more examples. Here the 1,2-addition is accompanied by a Cl/N-exchange between Pt and P in one or both halfes of the dimer.

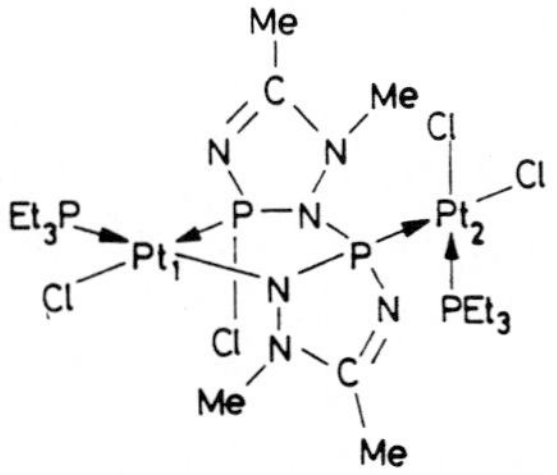

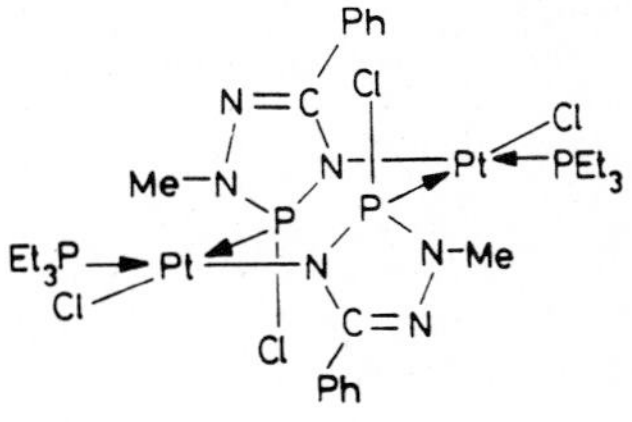

4.4 1,2- (and 1,4-)Addition

The P=C bond and the P=N bond as part of an azaphosphole ring are definitely less inclined to give 1,2-additions than the acyclic and isolated bonds. Nevertheless are such 1,2-additions important in azaphosphole chemistry. Due to the more or less balanced equilibrium situation particularly interesting aspects arise from the 1,2-additions and the 1,2-eliminations to and from the azaphosphole systems.

Azaphospholes normally are stable as monomers. Exceptions are the benzo-1,3,2-diazaphospholes (ref. 25) and the 4*H*-1,2,4,3-triazaphospholes (refs. 11, 120). They form tetramers. Heating converts them to monomers (refs. 25, 120; in one case however it is claimed that heating on the contrary causes a monomeric 2*H*-1,2,4,3-triazaphosphole to oligomerize, ref. 15). The PN-bonds connecting the rings to the tetramer may be understood to arise from a nucleophilic interaction of one monomer's nitrogen with the phosphorus of the other. A hard acceptor such as BF_3 effectively competes for the nitrogen and accordingly degrades the oligomer (refs. 11, 15, 25, 120).

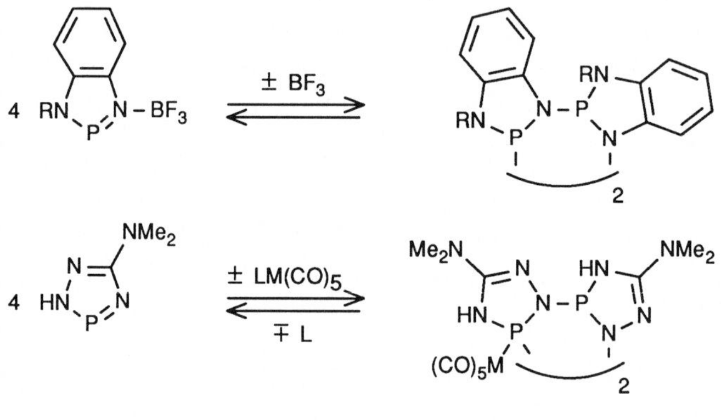

A soft and therefore P-coordinated acceptor can on the contrary induce the formation of an oligomer. This has indeed been observed for the 2:4 chromium and tungsten pentacarbonyl complex of a triazaphosphole (ref. 121).

Protic reagents add to the P=C or P=N bond of an azaphosphole always in such a way that a CH or NH bond is formed. If the reagent is only weakly acidic as in the case of a secondary amine the addition equilibrium often remains well at the left side. As a consequence azaphosphole syntheses with $P(NR_2)_3$ can make use of the spontaneous elimination of R_2NH (see **2.1**). A strong acid such as HCl on the other hand may give the covalent 1,2-adduct or its ionic isomer resulting from N-protonation (see **2.1**).

Additions to the P=C bond of 2*H*-1,2,3-diazaphospholes have been observed with HCl and HBr (refs. 30, 122, 123) and with alcohols (refs. 28, 30, 37, 104, 122); an excess alcohol degrades to $P(OR)_3$ (refs. 28, 30). Amines do not add (refs. 28, 30, 80). 1,3,4-Diazaphospholes add methanol to the P=C bond (ref. 42) and 1,2,3,5-diazadiphospholes add alcohol at the same time to both the P=C and P=N bond.

HX-Addition to the P=N bond has been observed for 1,3,2-diazaphospholes, X = Cl (excess HCl degrades to PCl_3), OR (ref. 20), for benzo-1,3,2-diazaphospholes, X = OR, NR_2 (ref. 25; the intermediate 1,1-addition of butanol has been observed in one case), and for 2*H*-1,2,4,3-triazaphospholes, X = OR (refs. 14, 24, 25; excess ROH degrades to $P(OR)_3$, ref. 124), NH_2, NHR, NR_2 (refs. 13 - 15, 125). The tendency of the three 1,2,4,3-triazaphosphole isomers to form adducts is remarkable different and decreases in the order 4*H*>2*H*>1*H*. 1*H*-1,2,4,3-triazaphospholes are even said not to react with alcohols (ref. 14), in fact however they give a balanced addition equilibrium and are degraded to $P(OR)_3$ by excess ROH (ref. 126).

Where the addition equilibrium of a protic reagent stays at the left side or stops at the ionic form, the covalent adduct may nevertheless participate in a very small concentration. Moreover, as it owns the usual phosphine reactivity it will easily react with sulfur, selenium or some other oxidizing agent. A combination of the two reagents will thus often result in a quantitative 1,2-addition, as in this example (ref. 123):

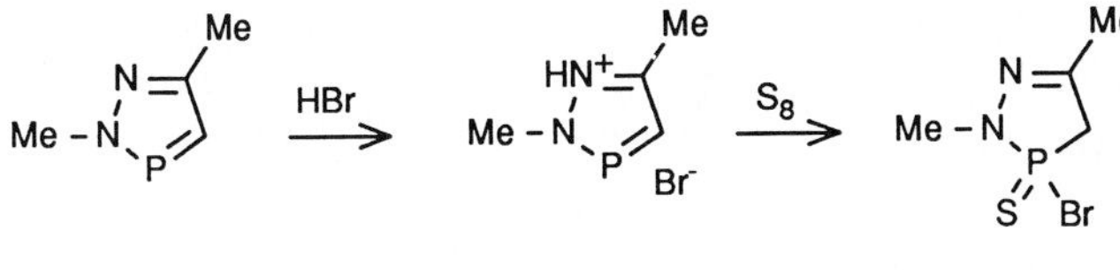

Many more examples are known (e.g. refs. 13, 42). 1,2-Azaphospholes (ref. 73) and 1,2,4-diazaphospholes (ref. 106) are stable against the combination MeOH plus S_8.

There is also an intramolecular version of this type of reaction. The two reagents are here so to speak combined in one. Addition of 1,2-diols yields PH-spirophosphoranes. This has been shown for 1,3,2-diazaphospholes (ref. 20), benzo-1,3,2-diazaphospholes (ref. 25), 2*H*-1,2,4,3-triazaphospholes (refs. 13 - 15, 125) and 2*H*-1,2,3-diazaphospholes (ref. 37). 2*H*-1,2,4,3-triazaphospholes and hydroxy compounds with an additional oxidizing group in 2-position such as 2-hydroxyphenyl heterodienes (ref. 127, just one of many examples is given below) or 2-azidoalcohols (ref. 128) also yield phosphoranes. In these products the phosphorus acts as bridgehead of the azaphosphole ring and two more condensed rings.

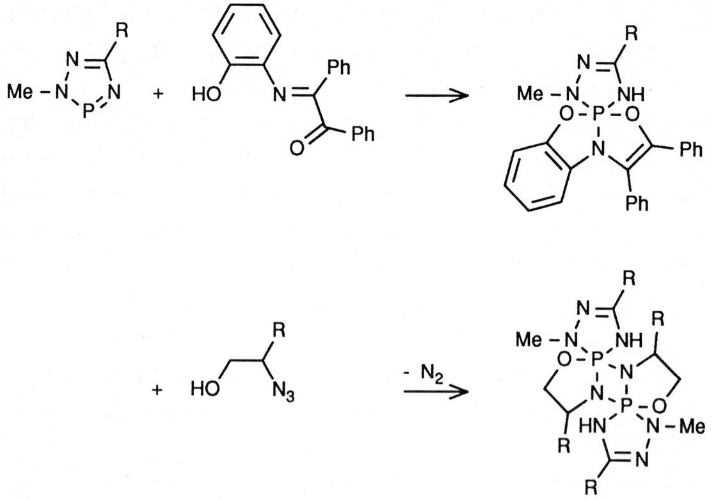

1,2,3-Diaza- and 1,2,4,3-triazaphospholes add one equivalent of water. An excess of water degrades them to phosphorous acid (refs. 28, 30, 129 and 120, 124).

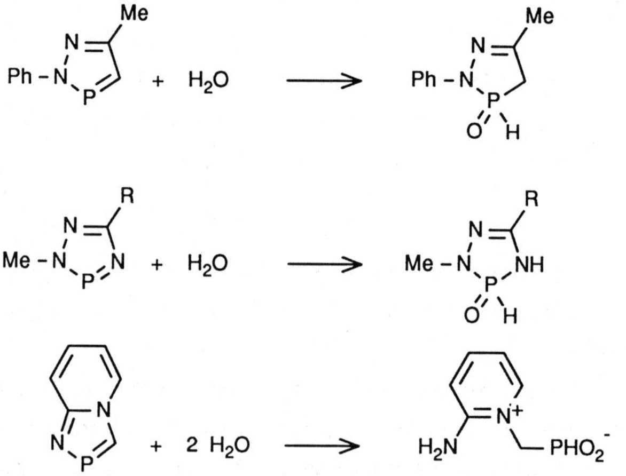

With 1,3,4-diazaphospholes only a 2:1-addition of water is found; it yields a ring opened product (ref. 42). 1,2,4-Diazaphospholes are stable to hydrolysis both at high and low pH (ref. 49). Benzo-1,3-azaphospholes by hydrolysis are reconverted to o-aminophenylphosphine (ref. 47).

$2H$-1,2,4,3-triazaphospholes give 1,2-additions also with phosphoryl halides (ref. 9; more examples occur in the course of their bromination, see **4.5**) and with Grignard reagents (ref. 124).

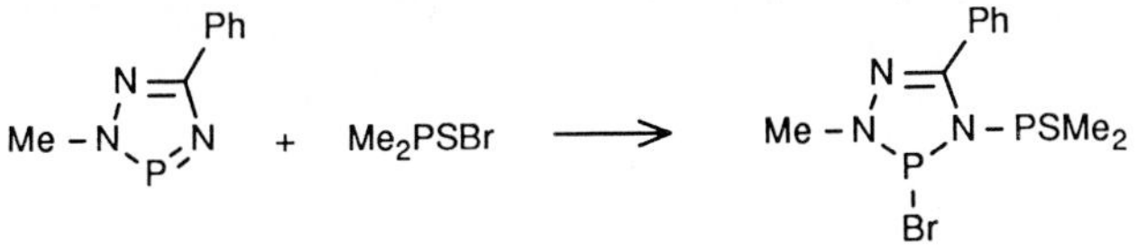

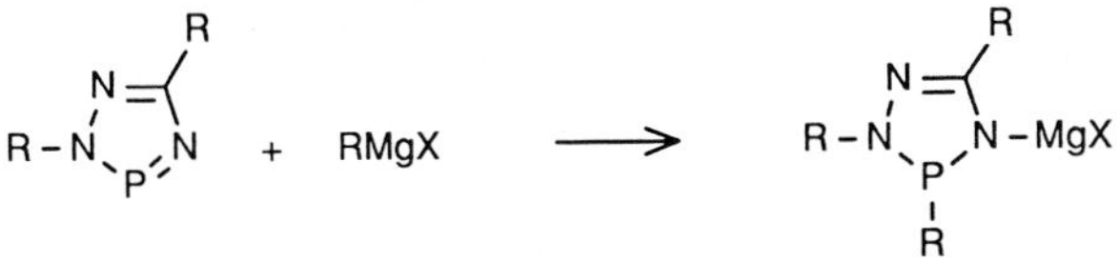

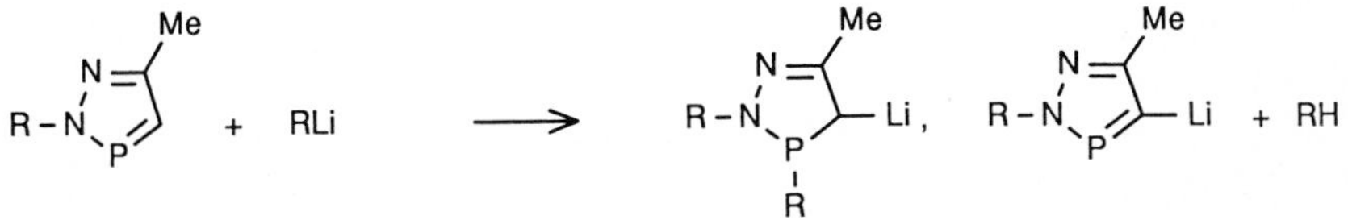

The addition of alkyl lithium to 1,2,3-diazaphospholes competes with α-metallation (refs. 80, 130).

Azaphospholes give 1,2-cycloadditions with dienes and 1,3-dipoles; these are reviewed in ref. 115. A 1,4-cycloaddition to acetylenes is mentioned in **2.7**.

4.5 α-Substitution

While practically all other P(III)-compounds including phosphinines (ref. 1) and azaphospholes with N,N adjacent to phosphorus (see **4.6**) are oxidized by halogen, $2H$-1,2,3-diazaphospholes are substituted instead at carbon in the 4-position. The reaction involves a 1,2-addition of the halogen and a 1,2-elimination of the hydrogen halide (ref. 123). Y = Br has been introduced by Br_2 (refs. 104, 123), BrCN and N-bromo succinimide (ref. 77), Y = Cl, I by ICl, Y = Cl also by SO_2Cl_2 (ref. 77).

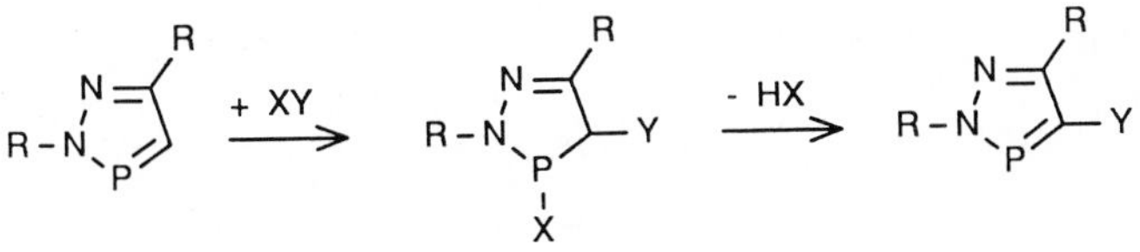

Obviously the phosphorus oxidation is repressed here in order to preserve the azaphosphole 6π system. When protic reagents add to the P=C bond of the 4-bromo derivative however the phosphorus becomes oxidized afterwards and the C-substitution is taken back however (ref. 2):

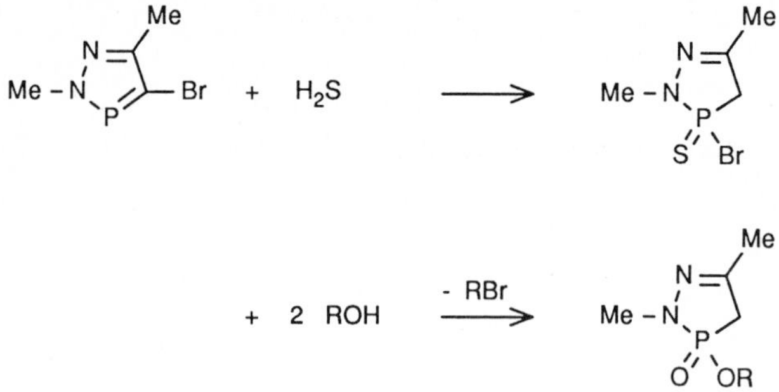

Deuteration (Y = D) of a 2*H*-1,2,3-diazaphosphole by deuterium chloride also follows the above equation and most probably also the same mechanism (ref. 78). Other substituents introduced in this way by means of the corresponding chloro, bromo or cyano compound are Y = SR (refs. 77, 123), PCl_2, PRCl, PR_2, $P(CN)_2$, $PSCl_2$, $PSBr_2$ (refs. 33, 36). A 1*H*-1,2,3-diazaphosphole has also been PCl_2-substituted in 4-position by its reaction with PCl_3. From the same reaction of a bicyclic 1,3,4-diazaphosphole all the substitution products with n = 1, 2, 3 are obtained (ref. 42):

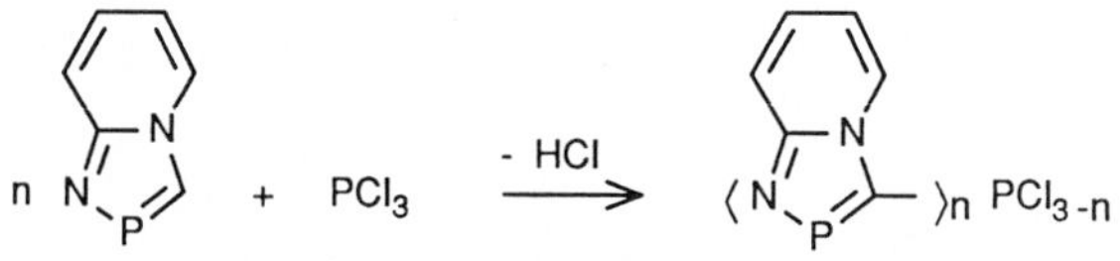

Carbon-bonded 4-substituents have been introduced in 2*H*-1,2,3-diazaphospholes by a formal insertion of a carbonyl group (refs. 2, 129) or of an isocyanate (ref. 76) into the CH-bond. These reactions probably proceed via a cycloaddition and a subsequent proton shift. By a similar sequence and an additional nitrile elimination an anilino derivative is yielded from the addition of a nitrile imine (ref. 94):

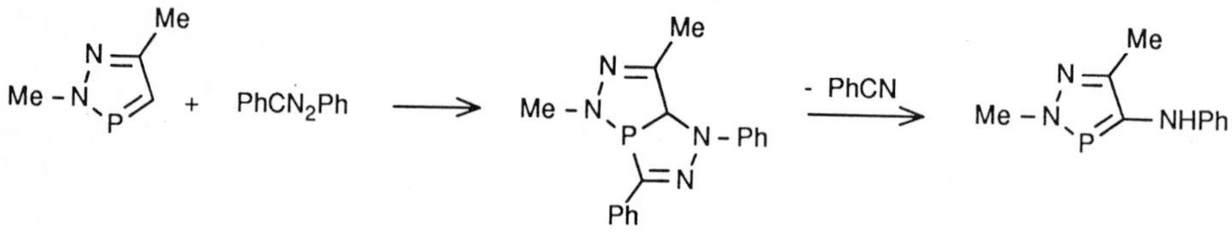

Metallation of an azaphosphole and subsequent reaction with an electrophile is another way of substitution. A 4-methyl group has been introduced in this way into a 1,2,3-diazaphosphole (ref. 80), methyl and stannyl groups into a 1,2,4-diazaphosphole (ref. 80) and silyl, carboxyl and methylthio groups into benzo-1,3-azaphospholes (refs. 48, 131). The 3,5-deuteration of a 1,2,4-diazaphosphole catalyzed by KO*t*Bu (ref. 80) must also be mentioned here.

Trimethylsilyl groups in 3- and 3,5-position of 1,2,4-diazaphospholes can be replaced by hydrogen with KF in dimethylformamide (refs. 61, 68).

4.6 1,1-Additions

As compared to phosphaalkenes $RP{=}CR_2$ iminophosphines $RP{=}NR$ tend to give 1,1-additions instead of 1,2-additions (ref. 132). A parallel trend is observed for azaphospholes. In contrast to 1,2,3-diazaphospholes (see **4.5**) 1,2,4,3-triazaphospholes are oxidized by chlorine (ref. 133; PCl_5 and Cl_3CSCl have also been used, ref. 16) and bromine (refs. 10, 15). The 1,1-addition of halogen is accompanied by a dimerization.

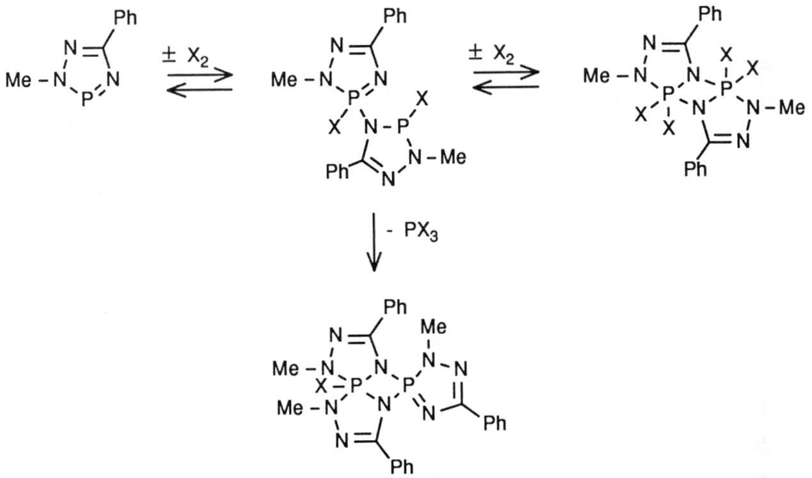

If less than one equivalent halogen is used or if the triazaphosphole is mixed with its dihalide products of intermediate oxidation states are formed of which just one representative is shown here. They involve both a 1,1- and 1,2-addition and they slowly disproportionate yielding polycyclic triazaphospholes of which again just one representative is shown (ref. 133).

The structure of a 3,3-dichloro-1,2,4,3-triazaphosphole dimer is shown in Fig. 7. Its phosphorus is tbp-coordinated with the rings joined by axial PN-bonds. Fig. 3 offers structural data of the triazaphosphole ring in the dimer for comparison with the monomeric ring before oxidation.

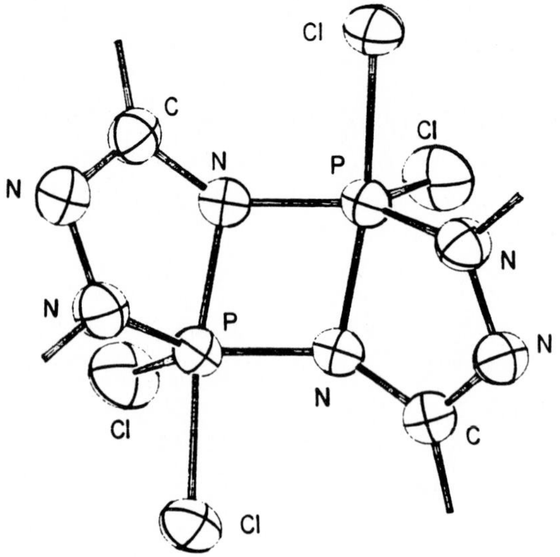

Fig. 7. Molecular structure of the 3,3-dichloro-2-methyl-5-phenyl-1,2,4,3-triazaphosphole dimer (ref. 99, methyl and phenyl groups are not shown).

A special type of 1,1-additions are the [4+1]-cycloadditions of heterodienes. They were observed for 1,2,3-diazaphospholes with α-diimines (ref. 120), for benzo-1,3,2-diazaphospholes with α-diimines (ref. 120) and α-diketones (ref. 134), for a pentalene type 1,3,2-diazaphosphole with azodicarboxylic esters (ref. 135) and for $2H$-1,2,4,3-triazaphospholes with α-diimines (refs. 14, 120), α-diketones (ref. 134) and azodicarboxylic esters (ref. 136). In the last two cases the products are mostly dimers.

Even double 1,1-additions (see Fig. 5) have been observed of 1,3-azaphospholes with quinones (ref. 41), of pentalene type 1,3,2-diazaphospholes with azodicarboxylic esters (ref. 135) and of 1,2,4,3-triazaphospholes with α-diketones (ref. 134). These double additions result in zwitterionic products with the formal negative charge located at the spirotricyclic six-coordinate phosphorus and the positive charge at some nitrogen of the azaphosphole ring.

The azaphosphole phosphorus seems in fact to prefer either low (i.e. two-) coordination or high (i.e. five- or six-)coordination. This may result from the small angle at phosphorus necessitated by the five-membered azaphosphole ring which fits well to the small bond angle associated to both of these situations (ref. 2).

REFERENCES

1. K. Dimroth, Top. Curr. Chem. 38 (1973) 1; Chapter 1.17 in Comprehensive Heterocyclic Chemistry, A. R. Katritzky and Ch. W. Rees eds., Pergamon, Oxford 1984, 508.
2. A. Schmidpeter and K. Karaghiosoff, Nachr. Chem. Tech. Lab. 33 (1985) 793.
3. G. Becker, W. Becker, R. Knebl, H. Schmidt, U. Weeber and M. Westerhausen, Nova Acta Leopoldina 59 (1985) 55.
4. M. Baudler, Angew. Chem. 99 (1987) 429; Angew. Chem. Int. Ed. Engl. 26 (1987) 419.
5. J. F. Nixon, Chem. Rev. 88 (1988) 1327.
6. F. Mathey, Chem. Rev. 88 (1988) 429.
7. W. Güth, E. Niecke and H. Westermann, Phosphorus Sulfur 30 (1987) 798.
8. K. Karaghiosoff and A. Schmidpeter, Phosphorus Sulfur 36 (1988) 217.
9. H. Tautz, Ph. D. Thesis, Universität München, 1980.
10. A. Schmidpeter, Phosphorus Sulfur 28 (1986) 71.
11. A. Schmidpeter and F. Steinmüller, unpublished results; see also ref. 120.
12. A. Schmidpeter and H. Tautz, Z. Naturforsch. B 35 (1980) 1222.
13. L. Lopez, J.-P. Majoral, A. Meriem, Th. N'Gando M'Pondo, J. Navech and J. Barrans, J. Chem. Soc., Chem. Commun. 1984, 183.
14. Y. Charbonnel and J. Barrans, Tetrahedron 32 (1976) 2039.
15. Th. N'Gando M'Pondo, C. Malavaud, L. Lopez and J. Barrans, Tetrahedron Lett. 28 (1987) 6049.
16. A. Schmidpeter, H. Tautz and F. Schreiber, Z. Anorg. Allg. Chem. 475 (1981) 211.
17. A. Schmidpeter, J. Luber and H. Tautz, Angew. Chem. 89 (1977) 554; Angew. Chem. Int. Ed. Engl. 16 (1977) 546.
18. J.-P. Majoral, R. Kraemer, Th. N'Gando M'Pondo and J. Navech, Tetrahedron Lett. 21 (1980) 1307.

19. K. Barlos, H. Nöth and B. Wrackmeyer, J. Chem. Soc., Dalton Trans. 1979, 801.
20. K. Karaghiosoff, J.-P. Majoral, A. Meriem, J. Navech and A. Schmidpeter, Tetrahedron Lett. 24 (1983) 2137.
21. A. Schmidpeter and K. Karaghiosoff, unpublished results.
22. A. Schmidpeter and K. Karaghiosoff, Z. Naturforsch. B 36 (1981) 1273.
23. K. Karaghiosoff, W. S. Sheldrick and A. Schmidpeter, Chem. Ber. 119 (1986) 3213.
24. C. Malavaud, M. T. Boisdon, Y. Charbonnel and J. Barrans, Tetrahedron Lett. 20 (1979) 447.
25. C. Malavaud, Th. N'Gando M'Pondo, L. Lopez, J. Barrans and J.-P. Legros, Can. J. Chem. 62 (1984) 43.
26. K. Pilgram and F. Korte, Tetrahedron 19 (1963) 1037.
27. T. A. Khwaja and H. Pande, Nucl. Acid Res. 7 (1979) 251.
28. N. P. Ignatova, N. N. Melnikov and N. I. Shvetsov-Shilovskii, Chem. Heterocycl. Compds. 3 (1967) 601.
29. N. I. Shvetsov-Shilovskii, N. P. Ignatova and N. N. Melnikov, Zh. Obshch. Khim. 40 (1970) 1501; J. Gen. Chem. USSR 40 (1970) 1488.
30. N. I. Shvetsov-Shilovskii, N. P. Ignatova, R. G. Bobkova, V. Ya. Manyukhina and N. N. Melnikov, Zh. Obshch. Khim. 42 (1972) 1939; J. Gen. Chem. USSR 42 (1972) 1932.
31. A. F. Vasilev, L. V. Vilkov, N. P. Ignatova, N. N. Melnikov, V. V. Negrebetskii, N. I. Shvetsov-Shilovskii and L. S. Khaikin, J. Prakt. Chem. 314 (1972) 806.
32. J. Luber and A. Schmidpeter, Angew. Chem. 88 (1976) 91; Angew. Chem. Int. Ed. Engl. 15 (1976) 111.
33. J. Luber and A. Schmidpeter, J. Chem. Soc., Chem. Commun. 1976, 887.
34. P. Friedrich, G. Huttner, J. Luber and A. Schmidpeter, Chem. Ber. 111 (1978) 1558.
35. J. H. Weinmaier, J. Luber, A. Schmidpeter and S. Pohl, Angew. Chem. 91 (1979) 442; Angew. Chem. Int. Ed. Engl. 18 (1979) 411.
36. J. H. Weinmaier, G. Brunnhuber and A. Schmidpeter, Chem. Ber. 113 (1980) 2278.
37. N. Ayed, B. Baccar, F. Mathis and R. Mathis, Phosphorus Sulfur 21 (1984) 335.
38. M. Hesse and U. Klingebiel, Z. Anorg. Allg. Chem. 501 (1983) 57.
39. L. Dulog, F. Nierlich and A. Verhelst, Phosphorus 4 (1974) 197.
40. G. Baccolini, R. Dalpozzo, V. Mele and E. Mezzina, Phosphorus Sulfur 39 (1988) 179.
41. R. K. Bansal, K. Karaghiosoff and A. Schmidpeter, IUPAC Conference, Stockholm 1989.

42. K. Karaghiosoff, C. Cleve and A. Schmidpeter, Phosphorus Sulfur 28 (1986) 289.

43. A. Schmidpeter, Ch. Leyh and K. Karaghiosoff, Angew. Chem. 97 (1985) 127; Angew. Chem. Int. Ed. Engl. 24 (1985) 124.

44. J. Heinicke and A. Tzschach, Z. Chem. 26 (1986) 407.

45. J. Heinicke, Tetrahedron Lett. 27 (1986) 5699.

46. K. Issleib, R. Vollmer, H. Oehme and H. Meyer, Tetrahedron Lett. 19 (1978) 441.

47. K. Issleib and R. Vollmer, Z. Anorg. Allg. Chem. 481 (1981) 22.

48. J. Heinicke and A. Tzschach, Tetrahedron Lett. 23 (1982) 3643.

49. A. Schmidpeter and A. Willhalm, Angew. Chem. 96 (1984) 901; Angew. Chem. Int. Ed. Engl. 23 (1984) 903.

50. G. Märkl and S. Pflaum, Tetrahedron Lett. 27 (1986) 4415.

51. G. Märkl ans S. Pflaum, Tetrahedron Lett. 28 (1987) 1511.

52. G. Märkl and G. Dorfmeister, Tetrahedron Lett. 27 (1986) 4419.

53. G. Märkl and S. Pflaum, Tetrahedron Lett. 29 (1988) 3387.

54. G. Märkl and G. Dorfmeister, Tetrahedron Lett. 28 (1987) 1089.

55. M. Regitz and P. Binger, Angew. Chem. 100 (1988) 1541; Angew. Chem. Int. Ed. Engl. 27 (1988) 1484.

56. W. Rösch and M. Regitz, Angew. Chem. 96 (1984) 898; Angew. Chem. Int. Ed. Engl. 23 (1984) 900.

57. Y. Y. C. Yeung Lam Ko, R. Carrie, A. Muench and G. Becker, J. Chem. Soc., Chem. Commun. 1984, 1634.

58. T. Allspach, M. Regitz, G. Becker and W. Becker, Synthesis 1986, 31.

59. W. Rösch, U. Vogelbacher, T. Allspach and M. Regitz, J. Organomet. Chem. 306 (1986) 39.

60. W. Rösch, Th. Facklam and M. Regitz, Tetrahedron 43 (1987) 3247.

61. E. O. P. Fuchs, M. Hermesdorf, W. Schnurr, W. Rösch, H. Heydt, M. Regitz and P. Binger, J. Organomet. Chem. 338 (1988) 329.

62. Y. Y. C. Yeung Lam Ko and R. Carrie, J. Chem. Soc., Chem. Commun. 1984, 1640.

63. G. Märkl, I. Troetsch-Schaller and W. Hölzl, Tetrahedron Lett. 29 (1988) 785.

64. W. Rösch, U. Hees and M. Regitz, Chem. Ber. 120 (1987) 1645.

65. H. Keller and M. Regitz, Tetrahedron Lett. 29 (1988) 925.

66. F. Zurmühlen, W. Rösch and M. Regitz, Z. Naturforsch. B 40 (1985) 1077.

67. G. Märkl and I. Troetsch, Angew. Chem. 96 (1984) 899; Angew. Chem. Int. Ed. Engl. 23 (1984) 901.

68. W. Schnurr and M. Regitz, Z. Naturforsch. B 43 (1988) 1285.

69. P. Pellon and J. Hamelin, Tetrahedron Lett. 27 (1986) 5611.
70. W. Rösch and M. Regitz, Synthesis 1987, 689.
71. W. Rösch, H. Richter and M. Regitz, Chem. Ber. 120 (1987) 1809.
72. A. Schmidpeter and H. Klehr, Z. Naturforsch. B 38 (1983) 1484.
73. K. Karaghiosoff, H. Klehr and A. Schmidpeter, Chem. Ber. 119 (1986) 410.
74. L. V. Vilkov, L. S. Khaikin, A. F. Vasilev, N. P. Ignatova, N. N. Melnikov, V. V. Negrebetskii and N. I. Shvetsov-Shilovskii, Dokl. Akad. Nauk SSSR 197 (1971) 1081.
75. A. F. Vasilev, L. V. Vilkov, N. P. Ignatova, N. N. Melnikov, V. V. Negrebetskii, N. I. Shvetsov-Shilovskii and L. S. Khaikin, Dokl. Akad. Nauk SSSR 183 (1968) 95.
76. B. A. Arbuzov, E. N. Dianova and E. Ya. Zabotina, Izv. Akad. Nauk SSSR, Ser. Khim. 1984, 1182.
77. W. Rösch and M. Regitz, Synthesis 1984, 591.
78. R. G. Kostyanovskii, V. G. Plekhanov, N. P. Ignatova, R. G. Bobkova and N. I. Shvetsov-Shilovskii, Izv. Akad. Nauk SSSR, Ser. Khim. 1971, 2611.
79. A. Schmidpeter and B. Wrackmeyer, Z. Naturforsch. B 41 (1986) 553.
80. S. Kerschl, B. Wrackmeyer, A. Willhalm and A. Schmidpeter, J. Organomet. Chem. 319 (1987) 49.
81. V. V. Negrebetskii, L. Ya. Bogelfer, A. V. Vasilev, R. G. Bobkova, N. P. Ignatova, N. I. Shvetsov-Shilovskii and N. N. Melnikov, Izv. Akad. Nauk SSSR, Ser. Khim. 1977, 532.
82. V. V. Negrebetskii, L. Ya. Bogelfer, R. G. Bobkova, N. P. Ignatova and N. I. Shvetsov-Shilovskii, Zh. Strukt. Khim. 19 (1978) 64; J. Struct. Chem. USSR 19 (1978) 52.
83. A. J. Ashe, R. R. Sharp and J. W. Tolan, J. Am. Chem. Soc. 98 (1976) 5451.
84. V. Galasso, J. Magn. Reson. 34 (1979) 199.
85. G. Märkl and K. Hock, Tetrahedron Lett. 24 (1983) 2645; G. Märkl, Angew. Chem. 78 (1966) 907; Angew. Chem. Int. Ed. Engl. 5 (1966) 846.
86. A. A. Prishchenko, A. V. Gromov, Yu. N. Luzikov, A. A. Borisenko, E. I. Lazhko, K. Klaus and I. F. Lutsenko, Zh. Obshch. Khim. 54 (1984) 1520; J. Gen. Chem. USSR 54 (1984) 1354.
87. V. V. Negrebetskii, A. V. Kessenikh, A. F. Vasilev, N. P. Ignatova, N. I. Shvetsov-Shilovskii and N. N. Melnikov, Zh. Strukt. Khim. 12 (1971) 798.
88. G. Märkl, E. Eckl, U. Jakobs, M. L. Ziegler and B. Nuber, Tetrahedron Lett. 28 (1987) 2119.
89. L. N. Markovskii and V. D. Romanenko, Zh. Obshch. Khim. 56 (1986) 2433; J. Gen. Chem. USSR 56 (1986) 2153.

90. A. N. Chernega, A. A. Korkin, M. Yu. Antipin and Yu. T. Struchkov, Zh. Obshch. Khim. 58 (1988) 2045.

91. G. Becker, W. Massa, O. Mundt, R. E. Schmidt and C. Witthauer, Z. Anorg. Allg. Chem. 540/541 (1986) 336.

92. G. A. Jeffrey, unpublished results.

93. V. G. Andrianov, Yu. T. Struchkov, N. I. Shvetsov-Shilovskii, N. P. Ignatova, R. G. Bobkova and N. N. Melnikov, Dokl. Akad. Nauk SSSR 211 (1973) 1101.

94. J. Högel, A. Schmidpeter and W. S. Sheldrick, Chem. Ber. 116 (1983) 549.

95. B. A. Arbuzov, E. N. Dianova, E. Ya. Zabotina, I. A. Litvinov and V. A. Naumov, Izv. Akad. Nauk SSSR, Ser. Khim. 1988, 150.

96. A. G. del Pozo, A.-M. Caminade, F. Dahan, J.-P. Majoral and R. Mathieu, J. Chem. Soc., Chem. Commun. 1988, 574.

97. J. H. Weinmaier, H. Tautz, A. Schmidpeter and S. Pohl, J. Organomet. Chem. 185 (1980) 53.

98. J.-P. Legros, Y. Charbonnel and J. Barrans, C. R. Acad. Sci. Paris, Ser. C 291 (1980) 271.

99. R. O. Day, R. R. Holmes, H. Tautz, J. H. Weinmaier and A. Schmidpeter, Inorg. Chem. 20 (1981) 1222.

100. S. Pohl, Chem. Ber. 112 (1979) 3159.

101. J. G. Kraaijkamp, G. van Koten, K. Vrieze, D. M. Grove, E. A. Klop, A. L. Spek and A. Schmidpeter, J. Organomet. Chem. 256 (1983) 375.

102. J.-P. Legros, Y. Charbonnel, J. Barrans and J. Galy, C. R. Acad. Sci., Ser. C 286 (1978) 319.

103. F. K. Larsen, M. S. Lehmann, I. Sotofte and S. E. Rasmussen, Acta Chem. Scand. 24 (1970) 3248; T. la Cour and S. E. Rasmussen, Acta Chem. Scand. 27 (1973) 1845.

104. R. G. Bobkova, N. P. Ignatova, N. I. Shvetsov-Shilovskii, N. N. Melnikov, V. V. Negrebetskii, L. Ya. Bogelfer, S. F. Dymova and A. F. Vasilev, Zh. Obshch. Khim. 47 (1977) 576; J. Gen. Chem. USSR 47 (1977) 527.

105. A. Willhalm, Ph. D. Thesis, Universität München, 1987.

106. A. Schmidpeter and A. Willhalm, unpublished results.

107. A. Schmidpeter and H. Klehr, unpublished results.

108. N. Ayed, R. Mathis, F. Mathis and B. Baccar, C. R. Acad. Sci., Ser. C 292 (1981) 187.

109. C. Malavaud, L. Lopez, Th. N'Gando M'Pondo, M. T. Boisdon, Y. Charbonnel and J. Barrans, ACS Symp. Ser. 171 (1981) 413.

110. B. A. Arbuzov, E. N. Dianova and S. M. Sharipova, Izv. Akad. Nauk SSSR, Ser. Khim. 1981, 1600.

111. B. A. Arbuzov, E. N. Dianova and I. E. Galeeva, Izv. Akad. Nauk SSSR, Ser. Khim. 1982, 1196.

112. B. A. Arbuzov, E. N. Dianova and Yu. Yu. Samitov, Dokl. Akad. Nauk SSSR 244 (1979) 1117.

113. A. V. Ilyasov, A. N. Chernega, A. A. Mafikova, I. Z. Galeeva, E. N. Dianova and B. A. Arbuzov, Zh. Obshch. Khim. 54 (1984) 1511; J. Gen. Chem. USSR 54 (1984) 1346.

114. B. A. Arbuzov, E. N. Dianova, I. Z. Galeeva I. A. Litvinov, Yu. T. Struchkov, A. N. Chernev and A. V. Ilyasov, Zh. Obshch. Khim. 55 (1985) 3; J. Gen. Chem. USSR 55 (1985) 1.

115. B. A. Arbuzov and E. N. Dianova, Phosphorus Sulfur 26 (1986) 203; more recent contribution: B. A. Arbuzov, E. N. Dianova, R. T. Galiaskarova and A. Schmidpeter, Chem. Ber. 120 (1987) 597.

116. K. C. Dash, H. Schmidbaur and A. Schmidpeter, Inorg. Chim. Acta 41 (1980) 167.

117. J. G. Kraaijkamp, D. M. Grove, G. van Koten and A. Schmidpeter, Inorg. Chem. 27 (1988) 2612.

118. J. G. Kraaijkamp, G. van Koten, D. M. Grove, G. Abbel, C. H. Stam and A. Schmidpeter, Inorg. Chim. Acta 85 (1984) L33.

119. O. J. Scherer, Angew. Chem. 97 (1985) 905; Angew. Chem. Int. Ed. Engl. 24 (1985) 924.

120. O. Diallo, M. T. Boisdon, L. Lopez, C. Malavaud and J. Barrans, Tetrahedron. Lett. 27 (1986) 2971.

121. A. Schmidpeter, H. Tautz, J. von Seyerl and G. Huttner, Angew. Chem. 93 (1981) 420; Angew. Chem. Int. Ed. Engl. 20 (1981) 408.

122. R. G. Bobkova, N. P. Ignatova, N. I. Shvetsov-Shilovskii, V. V. Negrebetskii and A. F. Vasilev, Zh. Obshch. Khim. 46 (1976) 590; J. Gen. Chem. USSR 46 (1976) 588.

123. J. Högel and A. Schmidpeter, Chem. Ber. 118 (1985) 1621.

124. M. Haddad, Th. N'Gando M'Pondo, C. Malavaud, L. Lopez and J. Barrans, Phosphorus Sulfur 20 (1984) 333.

125. Y. Charbonnel and J. Barrans, C. R. Acad. Sci. Paris, Ser. C 278 (1974) 355.

126. A. Schmidpeter and P. Mayer, unpublished results.

127. A. Schmidpeter, M. Junius, J. H. Weinmaier, J. Barrans and Y. Charbonnel, Z. Naturforsch B 32 (1977) 841.

128. M. R. Marre, M. T. Boisdon and M. Sanchez, Tetrahedron Lett. 23 (1982) 853.

129. B. A. Arbuzov, E. N. Dianova, E. Ya. Zabotina, R. L. Korshunov and R. Z. Musin, Zh. Obshch. Khim. 55 (1985) 1464; J. Gen. Chem. USSR 55 (1985) 1304.

130. W. Rösch and M. Regitz, Phosphorus Sulfur 21 (1984) 97.

131. J. Heinicke, A. Petrasch and A. Tzschach, J. Organomet. Chem. 258 (1983) 257.

132. W. W. Schoeller and E. Niecke, J. Chem. Soc., Chem. Commun. 1982, 569; E. Niecke, D. Gudat, W. W. Schoeller and P. Rademacher, J. Chem. Soc., Chem. Commun. 1985, 1050.

133. R. O. Day, A. Schmidpeter and R. R. Holmes, Inorg. Chem. 22 (1983) 3696.

134. O. Diallo, M. T. Boisdon, C. Malavaud, L. Lopez, M. Haddad and J. Barrans, Tetrahedron Lett. 25 (1984) 5521.

135. K. Karaghiosoff, W. S. Sheldrick and A. Schmidpeter, Phosphorus Sulfur 30 (1987) 780.

136. H. Tautz and A. Schmidpeter, Chem. Ber. 114 (1981) 825.

MULTIPLE BONDS BETWEEN TRANSITION METALS AND MAIN-GROUP ELEMENT ATOMS

WOLFGANG A. HERRMANN
Anorganisch-chemisches Institut, Technische Universität München, Lichtenbergstraße 4, D-8046 Garching (Germany)

1. INTRODUCTION

Early in the history of organic chemistry, multiple bonds between carbon atoms were recognized as centers of enhanced reactivity. Alkynes and alkenes thus became the major chemical feedstock for the production of large-volume organic chemicals in industry, especially since various metal compounds had unfolded their capability of once more improving both the reactivity of these functionalities and the selectivity of their transformations towards value-added chemicals. Important landmarks in this context were posted by Walter Reppe at the Badische Anilin- & Sodafabrik and by Karl Ziegler at the Max Planck-Institut für Kohlenforschung. These developments occurred during the forties and early fifties (refs. 1-3). Multiple bonds between transition metal atoms (refs. 4-6) were only recognized around the year 1960 when Cotton et al. examined the structure of the so-called tetrachlororhenate(III) anion which turned out to be dimeric, having the formula $[Re_2Cl_8]^{2-}$. An unprecedented quadruple bond between the rhenium atoms had to be assigned to this particular inorganic species from structural and theoretical considerations. A plethora of inorganic and organometallic compounds exhibiting double, triple, and quadruple bonds has appeared in the literature ever since. It is obvious by now that such entities are among the most reactive ones in chemistry. This work was paralleled by the investigation of metal-to-carbon double and triple bonds that had been established as carbene and carbyne complexes in the laboratory of E. O. Fischer at the Technische Universität München in 1964 and 1973, respectively (ref. 7).

All aspects of the investigations of metal—metal and metal—carbon multiple bonds also apply to a particular class of compounds that has, however, enjoyed little limelight and thus deserves the present review: coordination and organometallic compounds with multiple bonds between transition metals and substituent-free ('bare' or 'naked') main group elements (ref. 8). Although based mostly on accidental discoveries, the few noteworthy examples are now beginning to unfold general concepts of syntheses that are capable of being extended and thus deserving of exploitation in preparative chemistry. The availability of further structural patterns exhibiting multiple bonds between transition metals and ligand-free main group elements might enable preparative organometallic chemistry to expand in a completely new direction (for instance by the stabilizing or activation of small molecules at the metal complex). It is the aim of the present article to summarize some recent developments in this virgin field of chemical research, with the emphasis being restricted to structures that exhibit multiply bonded 'naked' atoms of carbon- and oxygen group elements in bridging positions between transition metal atoms.

2. CARBON GROUP ELEMENT ATOMS CONNECTING TRANSITION METALS

'Bare' carbon atoms have been encountered in a number of clusters and interstitial compounds (refs. 5, 8-12). In these molecules, the carbon atom is most commonly surrounded by five,

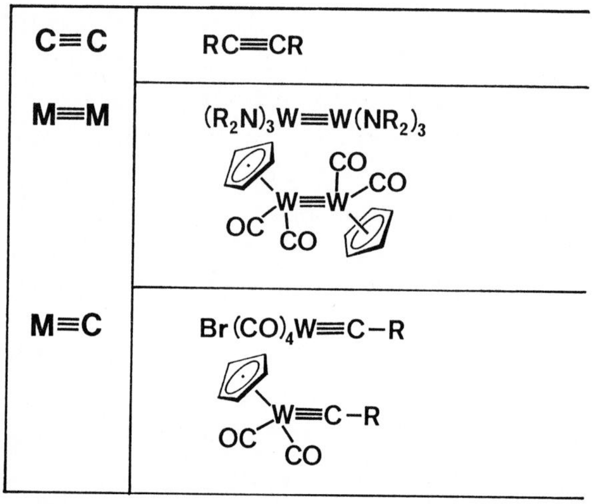

Fig. 1. A selection of carbon—carbon-, carbon—metal-, and metal—metal multiple bonds.

six or eight metal atoms, with the famous textbook examples $Fe_5C(CO)_{15}$ and $Ru_6C(CO)_{17}$ to be mentioned in this context. Carbido clusters of this type can be prepared by routine methods although the reactivity of the carbon atom in its metal 'cage' is as yet difficult to characterize or to classify. Multiple bonding between the central carbon atom and the transition metals cannot be seriously considered here. In addition, the interstitial atom is stericly hardly accessible so that reactions can only occur after one or several vertices have been removed from the cage structures.

Lowering the degree of aggregation generally leads to enhanced reactivity of chemical structures. One way of structural degradation is engagement of two or more atoms in multiple bonding. This line of reasoning may be applied to the simple set of compounds A, B and C shown in Fig. 2. The Fischer-type carbene complexes A (ref. 7) exhibit one double bond between a transition metal and a carbon atom. Replacement of the two carbene substituents R by another 16e fragment M yields linear μ-carbido complexes of type B, which compounds are isolobal (ref. 13) with the allene molecule B' in the same way as carbenes A are isolobal with an olefin A' (Fig. 1). An example belonging to type B is the porphyrine complex of formula (μ-C)[Fe(TPP)]$_2$ (1, TPP = tetraphenylporphyrine) but

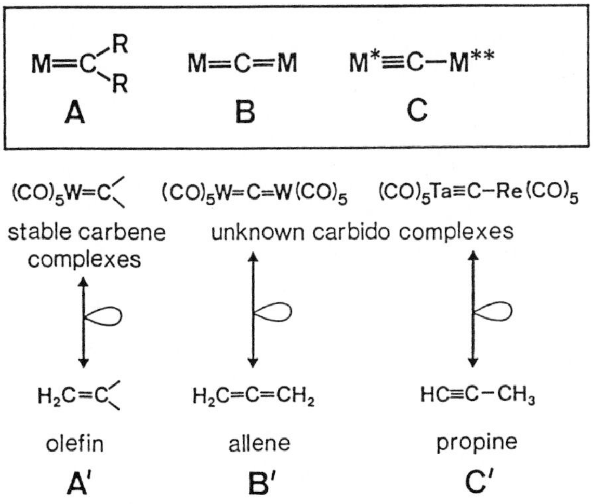

Fig. 2. 'Naked' carbon atoms as multiply bonded bridging ligands in organometallic compounds

structural details have not yet become available (ref. 14). In a previous review we have forecasted the existence of type-C compounds because they are just metal-substituted carbyne complexes, isoelectronic with compounds B, and isolobal with the propyne molecule, provided M* is a 15e- and M** is a 17e organometal fragment (ref. 8). While the tantalum/rhenium compound of Fig. 2, in spite of the correct electron count, has remained unknown up to the present, the tungsten/ruthenium complex synthesized by Selegue et al. (ref. 15) proved for the first time the correctness of our prediction (Fig. 3). The field of multiply bonded carbon atom organometallic chemistry is now open.

The hypothetical μ-carbido complex 2, once again isolobal with allene (and carbon dioxide), has been subjected to molecular orbital calculations. The original literature may be consulted for details (ref. 16), but as a summary it should be stated here that

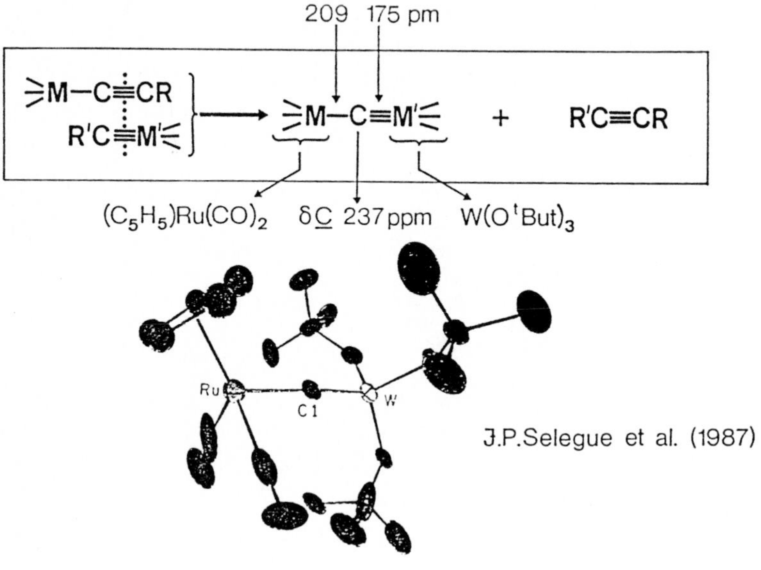

Fig. 3. The first example of a 'carbyne' type bare carbon atom ligand

the metal-centered orbitals 2a'' and 3a' are of very similar energy and that the π-electron system of the three framework atoms possesses an approximate cylindrical symmetry (Fig. 4). The analogy with allenes is thus overly simplistic and would be associated

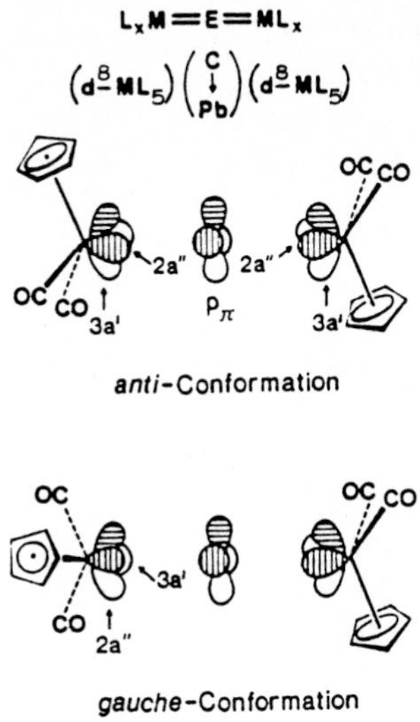

Fig. 4. Molecular orbital scheme of 'bare' group-IV element complexes according to ref. 16

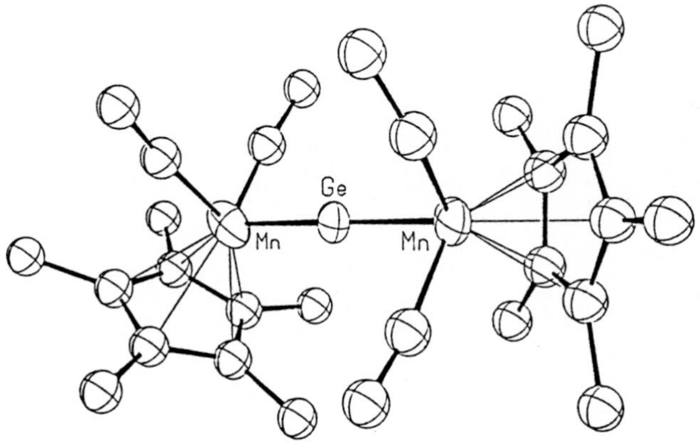

Fig. 5. Structure of the germanium-manganese complex (μ-Ge)[(η^5-C_5Me_5)Mn(CO)$_2$]$_2$ (Me = CH_3) (ref. 17)

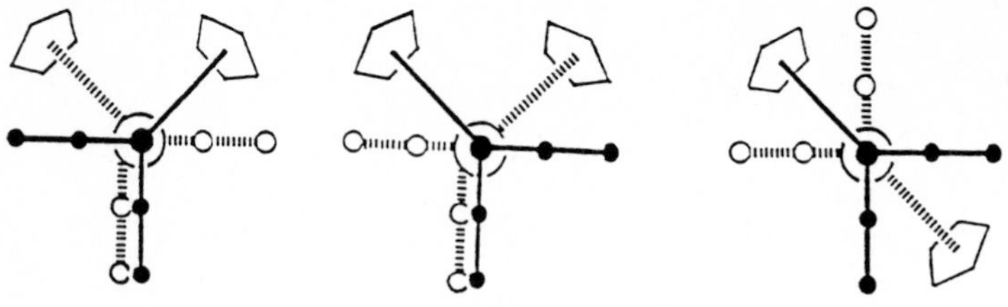

Fig. 6. Conformations of 'cumulene' structures $(\mu\text{-E})[(\eta^5\text{-}C_5R_5)Mn(CO)_2]_2$

with a rotational barrier of ca. 70 kcal/mol. As a matter of fact, infrared spectra of the known germanium, tin, and lead compounds 3a-c rather indicate the presence of several conformers in solution, and both the anticlinal conformers D and E were detected in the crystal structure of compound 4 (Fig. 5; ref. 17). Note that the antiperiplanar conformation F is not observed here while it represents the only rotamer to occur in the crystal structure of the unsubstituted derivative 3a (Fig. 6; ref. 18).

In spite of the fact that the isolobal formalism does not appear to represent correctly the electronic structure of these compounds, it is nevertheless a useful guide to new reactions and structures (refs. 8, 13). For example, a methylene group adds across one of the germanium—manganese double bonds of 3a to give the spirane 5 (diazomethane as methylene precursor compound) according to eqn. 1 (ref. 19). This transformation parallels the formal analogy of 3a with allene on the one, and of 5 with methylene cyclopropane on the other hand. The methylene building block of compound 5 is easily lost, however, even under mild conditions (above 50 °C in solution), with the precursor compound 3a being recovered quantitatively (refs. 19, 20). This same analogy is evident from the **Mn_3Ge** compound 6 resulting from treatment of the **Mn_2Ge** precursor species 3a with the thf-complex of the 16e species $(\eta^5\text{-}C_5H_5)Mn(CO)_2$ (refs. 18, 20). Note that the permethylated derivative of 6 (C_5Me_5 instead of C_5H_5) does not exist; steric effects are likely to be responsible for this. Tetracoordinated germanium is seen in the metallaspirane 7 that results from

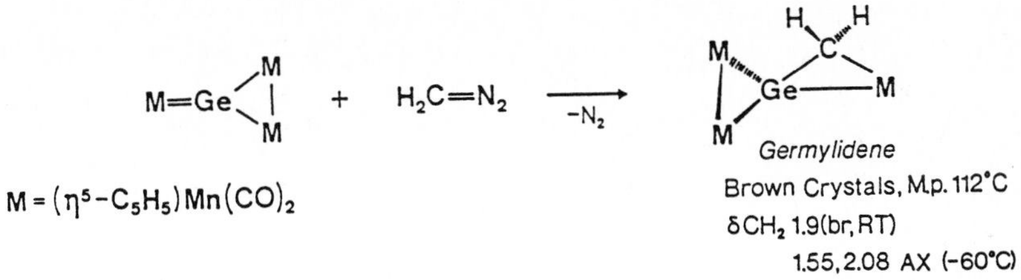

Equation 1

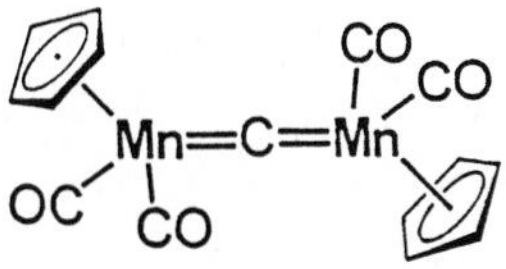

2

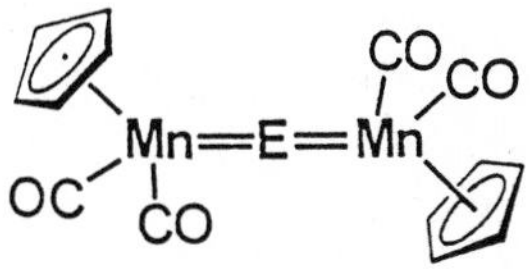

3 a–c

a : E = Ge
b : E = Sn
c : E = Pb

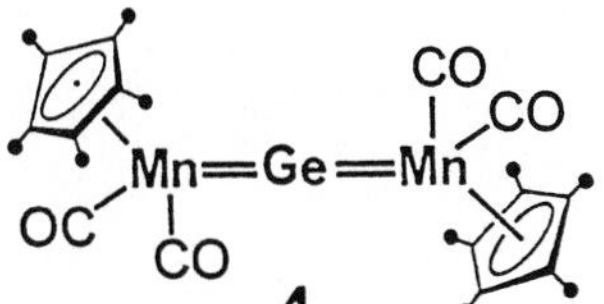

4

(•=CH_3)

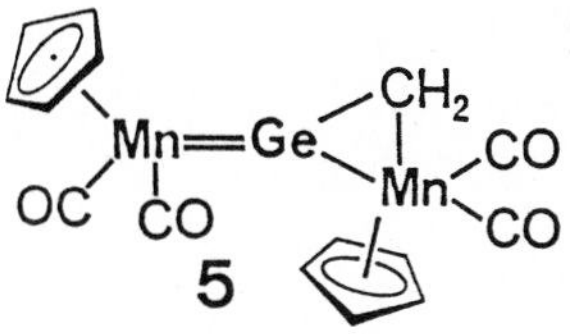

5

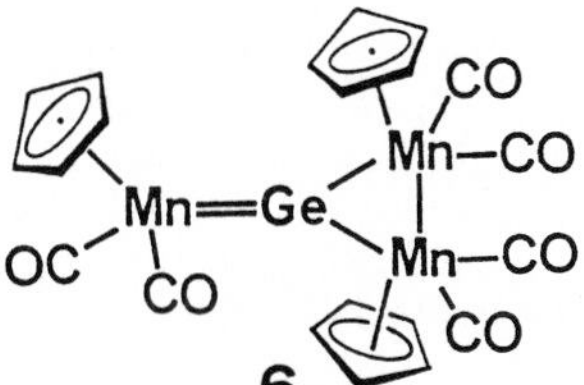

6

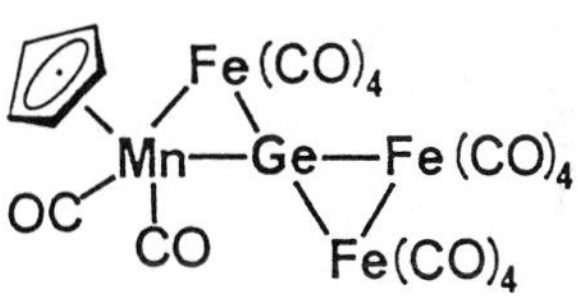

7

$Fe_2(CO)_9$ and 3a in a metathetical reaction (ref. 21). The germanium central atom adopts a distorted tetrahedral geometry, with the bond lengths covering a range typical of single bonds [GeMn 243.1(1) pm; GeFe 241 pm (av.)].

3. SYNTHETIC ASPECTS

Before we present and discuss further examples of multiply bonded combinations between transition metals and main group element atoms, a brief survey of general synthetic methods should be given here. Three main strategies are being employed for the introduction of bare main group elements as bridging ligands in organometallic compounds (Fig. 7).

i) Probably the most straightforward synthetic method is the **'element route'** starting out from the bulk elements. Clean formation of **MEM** core structures is achieved when the organometallic precursor compound contains a labile metal—metal single bond prone to homolytic dissociation. Sulphur- and particularly selenium complexes of iron, chromium, molybdenum, and tungsten present excellent examples how compounds of constitution I and IV (Fig. 7) can be made along these lines (ref. 8). The method relies of course on the availability of labile metal—metal single bonds that can insert a main group element atom E.

ii) The **'hydride route'** usually starts at mononuclear complexes to react with the binary hydrides EH_n. The applicability of this method relies on two prerequisites: oxidative addition of the hydride to a (coordinatively and electronically unsaturated) organometal species (e.g., $(\eta^5\text{-}C_5H_5)Mn(CO)_2$) and further hydrogen loss from the resulting intermediates of simplified formula $L_xM(H)(EH_{n-1})$. A first step of oxidative addition has recently been trapped in the case of tellurium hydride (ref. 22). Structural types I, II, and III are preferentially accessible by this method, with E representing germanium, tin, lead, selenium, and tellurium (ref. 8).

iii) **Reduction of the halides EX_n** may also yield bare main group element atoms E in bridging positions between two or more transition metals. The normally low-valent organometallic precursor compound L_xM (neutral or anionic) sometimes acts as the reducing agent, with one prominent example being the heterocumulene of composition $(\mu\text{-}Pb)[(\eta^5\text{-}C_5H_5)Mn(CO)_2]_2$ (structural type II of Fig. 7; ref. 23). For typical element combinations so far achieved along these three routes, the reader is referred to Fig. 7 and a previous review article (ref. 8).

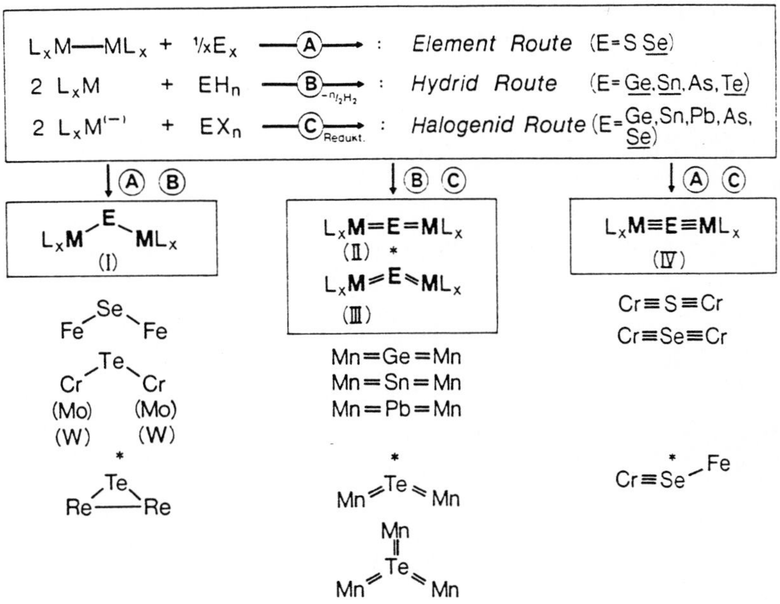

Fig. 7. Synthetic methods for multiple bonds between 'bare' main group element atoms and organotransition metal fragments, including few examples

4. STRUCTURAL ASPECTS

Let us focus on some representative examples of structural types I-IV (Fig. 7) that all contain monoatomic, bare main group element bridges, and compare their structural features with each other. The bent type I is most common for the more electron-rich elements. Complete sets of compounds exhibiting two-coordinate oxygen, sulphur, selenium, and tellurium are known. They are formal derivatives of the corresponding hydrides EH_2, with bond lengths being typical of single bonds. Higher bond orders are generated by successive elimination of terminal ligands L. At the same time, the distances between E and M are shortened by 10-20 pm upon going from I to II/III, and from II/III to IV (Fig. 7, 8). No significant differences in bond lengths are observed upon conside-

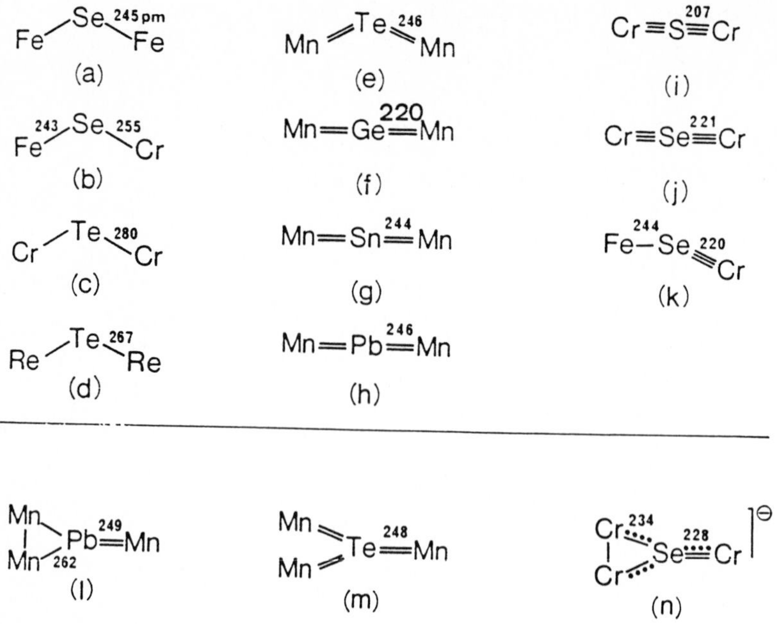

Fig. 8. Typical examples of single, double, and triple bonds between 'bare' main group element atoms and organotransiton metal fragments

ring either linear (type II) or bent core geometries (type III). This statement is supported by comparison of the two- and three-coordinate germanium compounds <u>3a</u> (220.4(1) pm) and <u>6</u> (225.0(1) pm; Fig. 9), respectively (ref. 8a). Germanium—manganese single bonds are in the order of 236 pm. Furthermore, practically the same tellurium—manganese bond lengths (double bonds) have been recorded for the **Mn_2Te**- (245.9(2) pm) and the **Mn_3Te**- (246-251 pm) compounds (e) and (m), respectively, of Fig. 8 (one cyclopentadienyl- and two carbonyl ligands at each manganese atom omitted for clarity). From the data given in Fig. 8, it becomes evident that it is not so much the number of transition metal atoms surrounding the central bridge ligand but rather the bonding situation that dictates the distance between the framework atoms. Thus the manganese—lead double bond has a length of 246-249 pm in both the two- and the three-coordinate lead of compounds (h) and

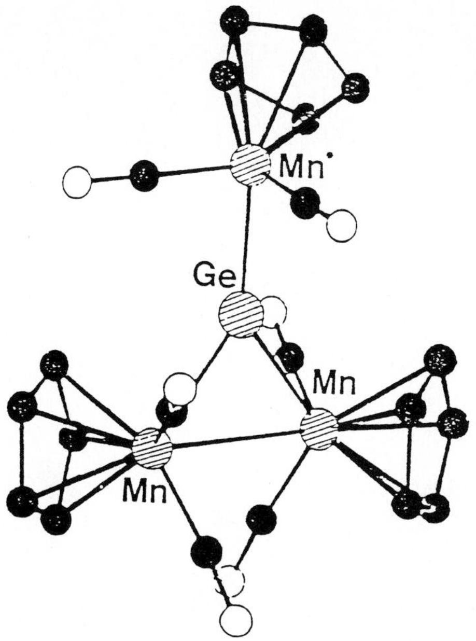

Fig. 9. Structure of the Mn_3Ge complex
(μ_3-Ge)[(η^5-C_5H_5)Mn(CO)$_2$]$_3$

(1), resp., while the manganese—lead single bonds fall in the range of ca. 262 pm. Finally, homologous, isostructural compounds such as (i) and (j) reflect the sizes of their central atoms (sulphur, selenium) so that one expects a metal-to-selenium bond being ca. 15 pm longer than the corresponding metal-to-sulphur bond, no matter whether we are dealing with single or multiple bonds.

Bridging atoms participating in single-bond structures of type I (Fig. 7) represent electron-rich centers. By way of contrast, the bridging atoms E (sulphur, selenium) of the heterocumulenes IV (triple bonds) are electron-deficient. Therefore, compounds I should act as nucleophiles, while compounds IV should rather display electrophilic behaviour. The doubly bonded core structures of II and III (linear and bent, respectively; Fig. 7) should react either weakly electrophilic or weakly nucleophilic, depending on the nature of the central atom E. These simple predictions have been verified experimentally. Thus, the selenium and tellurium bridge atoms of compounds (a)-(c) of Fig. 8 (one cyclopentadienyl and two (Fe) or three (Cr) carbonyl groups at the transition metals omitted for clarity) can be protonated by Brønsted acids and methylated by electrophilic reagents such as $CF_3SO_3CH_3$. On the other hand, compound (m) (Fig. 10) does not at all react with these reagents but rather gives the anionic species $[(\mu_3\text{-}TeCH_3)\text{-}\{(\eta^5\text{-}C_5H_5)Mn(CO)_2\}_3]^-$ upon treatment with the strongly <u>nucleophilic</u> methyllithium. The structure of this anion is shown in Fig. 11.

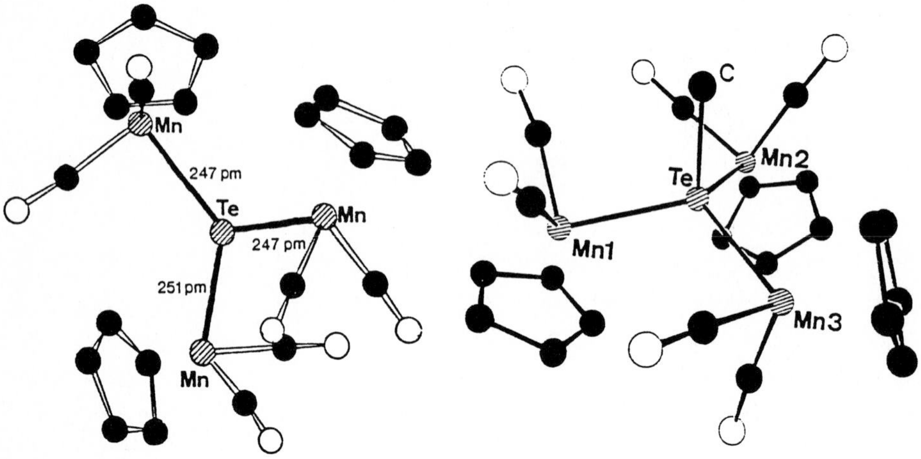

Fig. 10 (left)/Fig. 11 (right). Structures of the **Mn_3Te** complex $(\mu_3\text{-}Te)[(\eta^5\text{-}C_5H_5)Mn(CO)_2]_3$ and the methylation product $[(\mu_3\text{-}TeCH_3)\{(\eta^5\text{-}C_5H_5)Mn(CO)_2\}_3]^-$

The cumulated multiple bond systems II (double) and IV (triple) have sterically accessible **M_2E** core structures since no bridging ligands are present. These linear geometries should allow for various cycloaddition reactions, and in spite of few known examples there is a rich area of chemistry still to come. The special advantage of these compounds is the steric accessibility of reactive, multiply bonded M-E-M framework structures. Neither is a bridging ligand present nor do the stabilizing ligands L (e.g., cyclopentadienyl, carbon monoxide) prevent the attack by smaller molecules. The transition metal atoms M are 210-230 pm apart from each other. Since the covalent radii of most transition metals do not differ too much, the metal—metal distances can be adjusted to the most part by the substituent-free bridging atoms E since the atomic radii of the latter vary by approximately 70 pm within a given group of the periodic table (e.g., C → Si → Ge → Sn → Pb).

An evaluation of the structural data for various combinations between M and E in M_2E- and M_3E type compounds has yielded the following list of covalent bond radii (refs. 8, 17-40).

	Ge	Sn	Pb	S	Se	Te
Single Bonds	122	140	147	104	117	137
Double Bonds	112	130	137	94	107	127
Triple Bonds				87	100	

5. BARE OXYGEN GROUP ELEMENTS AS BRIDGING LIGANDS

While terminal as well as bridging nitrogen atoms form a great number of compounds with multiple bonding to transition metals (ref. 41), this feature is of decreasing significance as one proceeds to the more electron-rich elements of main-group V. It is particularly true that arsenic, antimony, and bismuth tend to forming homopolyions (Zintl phases), including cage structures (ref. 42). Bismuth has not yet been encountered as a multiply bonded bridging atom in organometallic chemistry while the group-IV neighbour lead has at least yielded the two prototypal compounds

$(\mu$-Pb)[(η^5-C_5H_5)Mn(CO)$_2$]$_2$ (ref. 23)
$(\mu_3$-Pb)[(η^5-C_5H_5)Mn(CO)$_2$]$_3$ (ref. 43

On the other hand, all group-IV elements do accomplish such multiply bonded structures. As a matter of fact, the μ-sulfido complexes

(np_3)Co=S=Co(np_3) [np_3 = $N(CH_2CH_2PPh_2)_3$]
[(p_3)Ni=S=Ni(p_3)][BPh_4]$_2$ [p_3 = $H_3CC(CH_2PPh_2)_3$]

of cobalt(I) and nickel(II), respectively, were the first species for which cumulated **M_2S** structures were recognized by Sacconi et al. back in 1978 (ref. 44). The M_2S backbones are strictly linear and display strikingly short bonds (Fig. 12). To quote the authors from their original publication: 'Noteworthy is the Ni-S bond distance (2.034(2) Å), which is the shortest ever found for such a linkage in transition metal complexes ... Since the two individual d^8 entities would be paramagnetic, it seems probable that a superexchange mechanism through the sulphur atom, utilizing

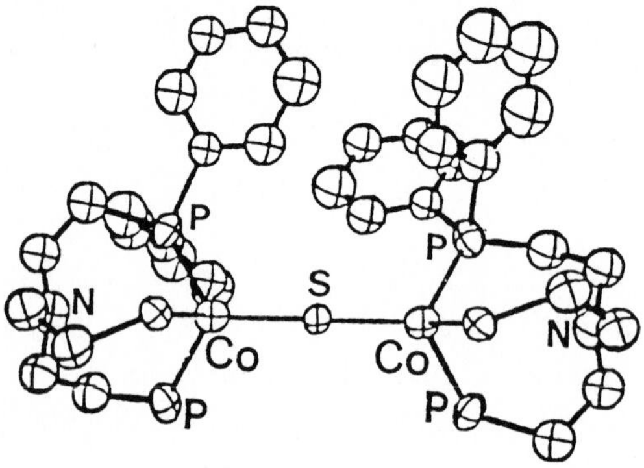

Fig. 12. Structure of Sacconi's μ-sulfido cobalt complex (see text)

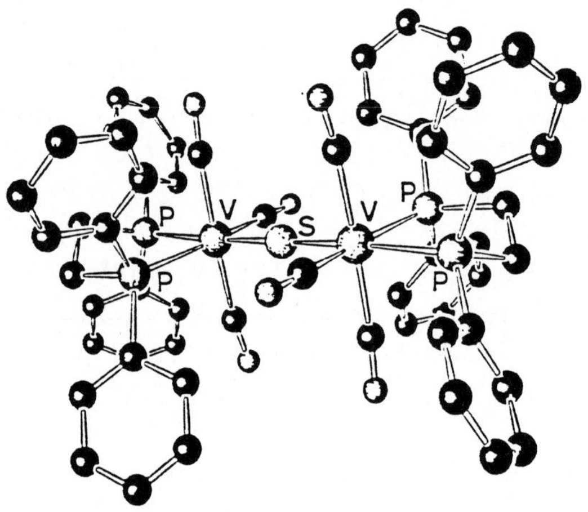

Fig. 13. Structure of Weiss' μ-sulfido vanadium complex (see text)

p_{π} orbitals of the sulphur and orbitals of higher energy of the metal (symmetry D_{3d} or D_{3h}) is involved. This interaction (which must be introduced to account for the diamagnetism of the complexes) will be favoured by the linearity of the sulphur bridge and by the relatively low electronegativity of the sulphur atom' (ref. 44).

The structurally analogous carbonylvanadium(I) complexes of composition

(μ-E)[(dppe)V(CO)$_3$]$_2$ [dppe = $Ph_2PCH_2CH_2PPh_2$; E = S, Se, Te]

reported from the laboratory of E. Weiss (ref. 45) represent examples of linear cumulated triple bond structures (Fig. 13). The bond lengths vary from 217.2(1) pm (sulphur), 229.8(2) pm (selenium), and 251.4(3)/252.(3) pm (tellurium). The shortest bond in which tellurium has been engaged to date was recorded for the bent **Mn_2Te** structure of the compound (μ-Te)[(η^5-C_5Me_5)Mn(CO)$_2$]$_2$ (Me = CH_3), an isolobal analogue of tellurium dioxide (Fig. 14). The manganese—tellurium double bond length amounts to 245.9(2) pm (ref. 46).

Although bent **S_1** bridges are so common that their mention here would depart from the scope of this review, the discovery and correct interpretation of the **Cr_2Se** compound (μ-Se)[(η^5-C_5H_5)Cr(CO)$_2$]$_2$ by Legzdins et al. in 1979 may be regarded another landmark in organometallic chemistry (ref. 47). These authors obtained this particular compound accidentally by reaction of trithiazyl trichloride, $S_3N_3Cl_3$, with the complex ion [(η^5-C_5H_5)Cr(CO)$_3$]$^-$ (Fig. 15). Cumulated triple bond systems of this type have since been obtained from various strategies of synthesis as summarized in Fig. 16 (ref. 8). A necessary condition for the existence of linear M*≡E≡M* complexes is that the organometallic fragment M* must be 15e systems, capable of taking up all six valence electrons of the bare bridging ligand (Fig. 16).

While this μ-sulfido complex of chromium is chemically rather inert, the analogous selenium compound (Fig. 17a and 17b) undergoes several addition reactions, e.g., with diazoalkanes and elemental selenium. We do not yet know the true reasons for this significant a reactivity difference, but better steric accessibility of the selenium in contrast to the sulphur compounds may be part of an explanation. Note that the chromium atoms are by 13 pm further apart from each other in the **Cr_2Se** compound as com-compared with the isostructural sulfur compound (ref. 48).

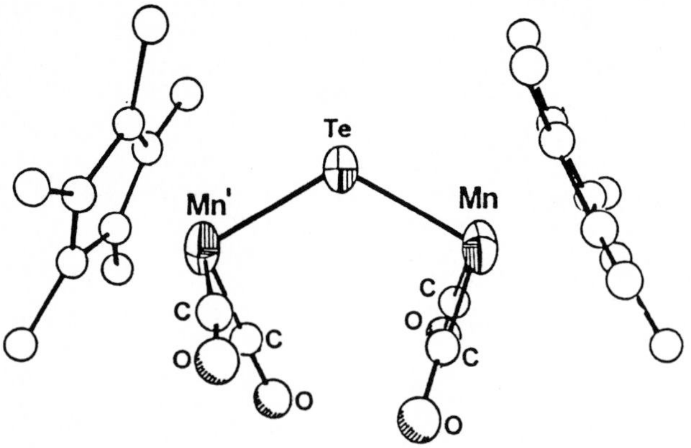

Fig. 14. Structure of the **Mn_2Te** complex (μ-Te)[(η^5-C_5Me_5)Mn(CO)$_2$]$_2$

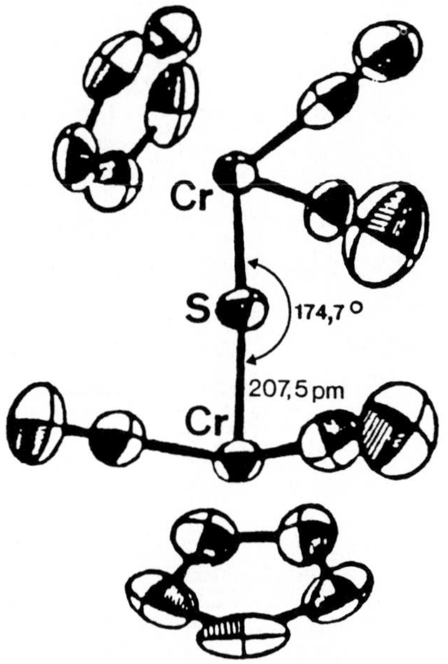

Fig. 15. Structure of Legzdin's μ-sulfido complex (μ-S)[(η^5-C_5H_5)Cr(CO)$_2$]$_2$ according to ref. 48

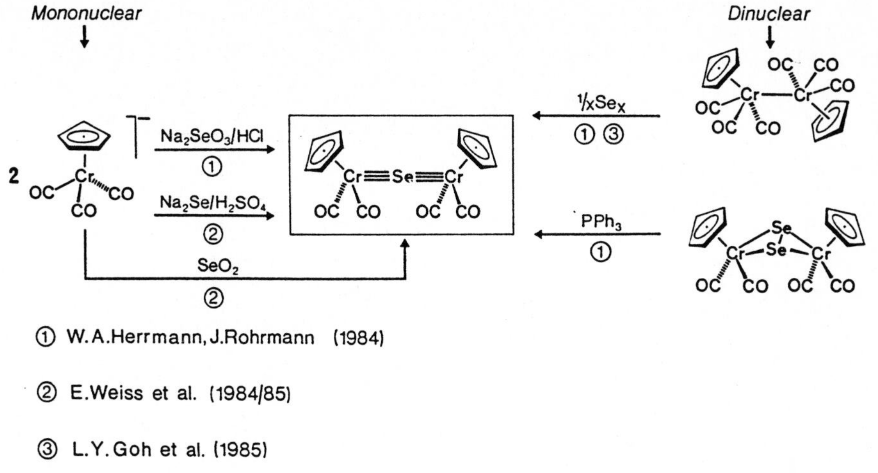

Fig. 16. Synthetic methods towards organochromium compounds that exhibit bare selenium atom ligands. - References: Herrmann/Rohrmann, 49c,d,f; Goh, 49e; Weiss, 45. - See also the review article listed in reference 8a.

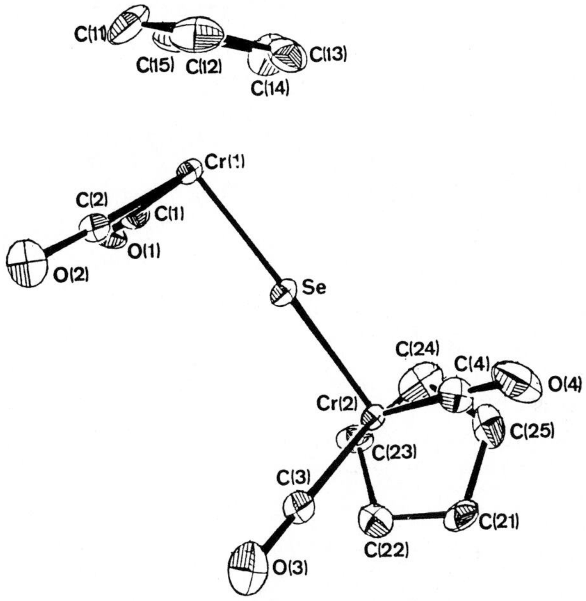

Fig. 17a. Structure of the selenium complex $(\mu\text{-Se})[(\eta^5\text{-}C_5H_5)Cr(CO)_2]_2$ according to ref. 49d.

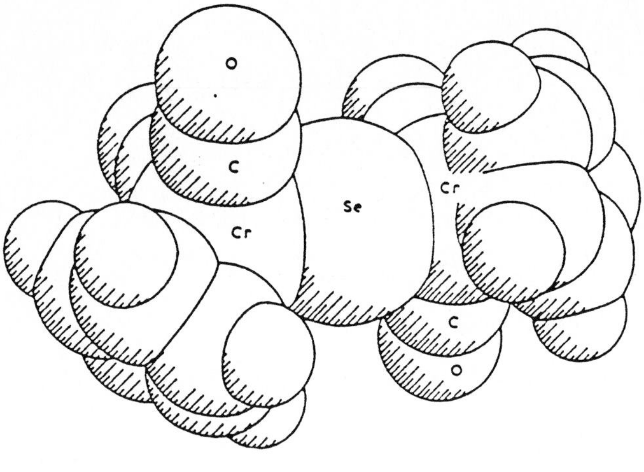

Fig. 17b. Space-filling model of the selenium complex (μ-Se)[(η^5-C_5H_5)Cr(CO)$_2$]$_2$ showing the steric accessibility of the bare selenium atom ligand

6. REDOX CHEMISTRY

The unsaturation of the **Cr_2S–**, **Cr_2Se–**, **Mn_3Te–**, and **Mn_2Te** systems discussed above, with their delocalized π-electron systems, manifests itself not only in addition reactions but also in their electrochemical behavior. The cyclovoltamograms of all the complexes display two reversible (or at least quasi-reversible) one-electron reduction steps. Fig. 18 shows the selenium complex (μ-Se)[(η^5-C_5H_5)Cr(CO)$_2$]$_2$ as a representative example. According to the electrochemical series of Fig. 19, these complexes should also undergo reduction reactions on a preparative scale, provided suitable reducing agents can be found. (The observable oxidation steps in the region +0.46 to −0.87 V are irreversible.) The quoted selenium complex will again serve as an example here. The potentials −0.72 V (radical anion)/−1.40 V (dianion) would suggest that reducing agents with potentials between these values might make the radical anion accessible. The reaction with cobaltocene ($E_{1/2}$[Cp_2Co^+/Cp_2Co] −0.77 V, N,N-dimethylformamide) in tetrahydrofuran at −30 °C slowly gives a black, crystalline, ionic complex of composition [(η^5-C_5H_5)$_2$Co]$^+$[(η^5-C_5H_5)$_3$Cr$_3$(CO)$_6$Se]$^-$ (refs. 50–52). This compound is extremely oxygen-sensitive and on

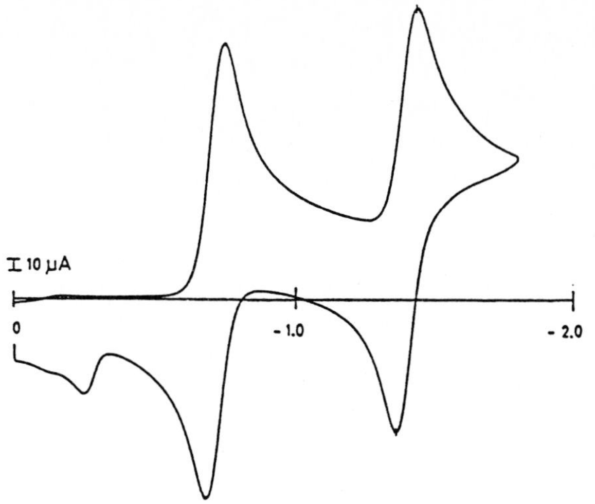

Fig. 18. Cyclovoltamogram of the Cr≡Se≡Cr complex (µ-Se)[(η^5-C_5H_5)Cr(CO)$_2$]$_2$, recorded in N,N-dimethylformamide/ 0.1 M TBAP vs. SCE (100 mV/sec; 28 °C). The two reduction steps are reversible ($E_{1/2}$ = -0.72 V, $E'_{1/2}$ = -1.40 V).

contact with air immediately reverts to the starting material, although not quantitatively. The complex anion of this compound has a planar (!) **Cr_3Se** backbone containing a **Cr_2Se** three-membered ring (Fig. 20). The exo- and endocyclic CrSe bonds (227.9(2) and 233.2(2)/234.3(2) pm, respectively) have different bond orders; the CrCr bond, with a length of 308.(2) pm, compares with a single bond (ref.50). The linear **Cr_2Se** complex (µ-Se)[(η^5-C_5H_5)Cr(CO)$_2$]$_2$ is isoelectronic with the **Mn_2Ge** compound (µ-Ge)- [(η^5-C_5Me_5)Mn(CO)$_2$]$_2$, and both systems are considered as having identical bonding systems, whereas the **Cr_3Se** complex anion of Fig. 20 is isolelectronic with the neutral **Mn_3Ge** compound 6. The central bare selenium atom contributes all six valence electrons for bonding to the three [(η^5-C_5H_5)Cr(CO)$_2$] fragments, which are isolobal with CH and CH_2^+ fragments. Hence, for the singly charged complex anion of fig. 20, as for cpd. 6, there is an isolobal relationship to methylene-cyclopropane because of Se ⟵⟶ C^{2-} ⟵⟶ CH^-, Ge ⟵⟶ C, and (η^5-C_6H_5)Mn(CO)$_2$ ⟵⟶ CH_2.

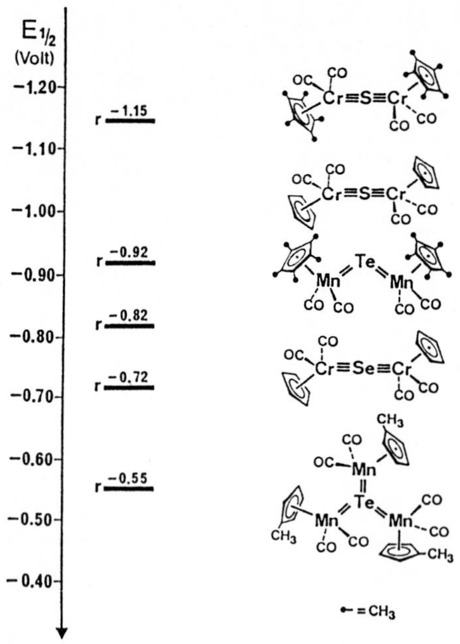

Fig. 19. Electrochemical series for selected complexes that contain multiple bonds between transition metals and bare main group elements. The first reduction potentials are given; for further data see ref. 8.

7. PROSPECTS

In this essay we have met unsubstituted ('bare') atoms of the carbon and oxygen groups as distinctive structural subunits in structurally defined multiple bond systems. Although there is still not the rich variety of compounds that we might have expected on the basis of our two decades' experience of metal—metal multiple bonds, it is to be hoped that preparative methods will prove capable of development (refs. 8, 52). For the heavier, more metallic main group elements, the high reactivity of the elemental modifications themselves and of the binary hydrides on reaction with organometallic complexes is striking. The relationship to the chemistry of the transition metals is very close, since the atomic radii and the electronic structures are broadly

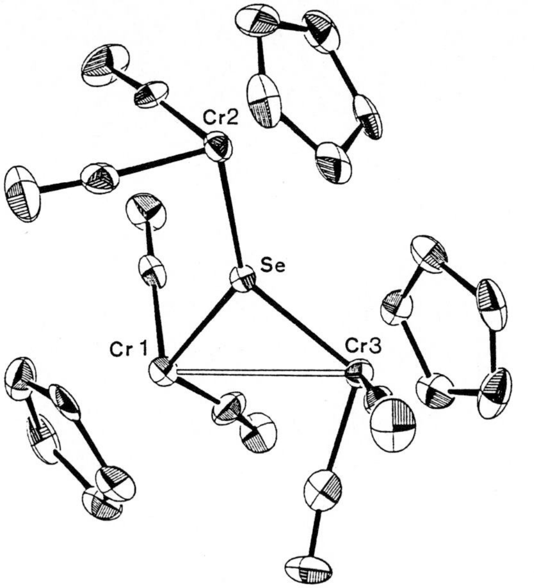

Fig. 20. Structure of the complex anion $[(\mu_3\text{-Se})\{(\eta^5\text{-}C_5H_5)Cr(CO)_2\}_3]^-$ (ref. 50)

comparable. Structural similarities with classical inorganic solid state chemistry have become quite obvious by now, and it is to be expected that iso- and heteropolyanions will soon form part of the organometallic chemist's arsenal. Catalytic applications of M_xE multiple bond systems are obvious to result from such studies.

ACKNOWLEDGEMENT

The author greatly appreciates the contributions of his Ph.D. and post-doctoral students whose names appear in the references. Generous support by the Deutsche Forschungsgemeinschaft, the Fonds der Chemischen Industrie, the Merck'sche Gesellschaft für Kunst und Wissenschaft e.V., and the Alexander von Humboldt Foundation is also acknowledged. Thanks are due to Miss Juliane Geisler for preparing the typoscript of this article.

REFERENCES

1 W. A. Herrmann, Comments Inorg. Chem., (1988), in press.
2 W. A. Herrmann, KONTAKTE (Darmstadt), 1988 (no. 1), pg. 3.
3 G. W. Parshall, Organometallics, 6 (1987) 687.
4 a) "Comprehensive Organometallic Chemistry" (G. Wilkinson, F. G. A. Stone, and E. W. Abel, eds.), vol. 1-9, Pergamon Press, Oxford 1984. - b) "Dictionary of Organometallic Compounds" (J. Buckingham, ed.), Chapman and Hall, London 1984.
5 Textbooks: a) N. N. Greenwood, A. Earnshaw: Chemistry of the Elements, Pergamon Press, Oxford 1984. - b) F. A. Cotton, G. Wilkinson: Advanced Inorganic Chemistry - A Comprehensive Text, 4th Edition, John Wiley & Sons, New York 1980. - c) A. Yamamoto: Organotransition Metal Chemistry - Fundamental Concepts and Applications, Wiley, New York 1986. - d) Ch. Elschenbroich, A. Salzer: Organometallchemie, 2nd Edition, Teubner-Verlag, Stuttgart 1988. - e) J. P. Collman, L. S. Hegedus, J. R. Norton, R. G. Finke: Principles and Applications of Organotransition Metal Chemistry, 2nd Edition, University Science Books, Mill Valley/California (USA) 1987.
6 Monographs and review articles: a) F. A. Cotton, R. A. Walton: Multiple Bonds Between Metal Atoms, Wiley & Sons, New York 1982. - b) M. H. Chisholm (Ed.), Reactivity of Metal-Metal Bonds, ACS Symposium Series, vol. 155, American Chemical Society, Washington D.C. (USA) 1981. - c) M. H. Chisholm, I. P. Rothwell, Progr. Inorg. Chem., 29 (1982) 1. - d) M. H. Chisholm, Angew. Chem., 98 (1986) 21; Angew. Chem. Internat. Edit. Engl., 25 (1986) 21.
7 Review articles on metal carbene and metal carbyne complexes: a) E. O. Fischer (Nobel Lecture), Angew. Chem., 86 (1974) 651; Advan. Organometal. Chem., 14 (1976) 1. - b) J. E. Hahn, Progr. Inorg. Chem., 31 (1984) 205. - c) K. Dötz (Ed.): Transition Metal Carbene Complexes, Verlag Chemie, Weinheim 1983. - d) E. O. Fischer, U. Schubert, J. Organometal. Chem., 100 (1975) 59.
8 a) W. A. Herrmann, Angew. Chem., 98 (1986) 57; Angew. Chem. Internat. Edit. Engl., 25 (1986) 56. - b) K. H. Whitmire, J. Coord. Chem., Series B, 17 (1988) 95.
9 Reviews on carbido clusters: a) P. Chini, G. Longoni, V. G. Albano, Advan. Organometal. Chem., 14 (1976) 285. - b) M. Tachikawa, E. L. Muetterties, Progr. Inorg. Chem., 28 (1981) 203. - c) J. S. Bradley, Advan. Organometal. Chem., 22 (1983) 1.
10 E. H. Braye, L. F. Dahl, W. Hübel, D. L. Wampler, J. Amer. Chem. Soc., 84 (1962) 4633.
11 J. S. Bradley, G. B. Ansell, M. E. Leonowicz, E. W. Hill, J. Amer. Chem. Soc., 103 (1981) 4968.
12 a) J. S. Bradley, E. W. Hill, G. B. Ansell, M. A. Modrick, Organometallics, 1 (1982) 1634. - b) J. S. Bradley, G. B. Ansell, E. W. Hill, J. Amer. Chem. Soc., 101 (1979) 7417. - c) M. R. Churchill, J. Wormald, J. Knight, M. J. Mays, J. Amer. Chem. Soc., 93 (1971) 3073.
13 a) R. Hoffmann (Nobel Lecture), Angew. Chem., 94 (1982) 725; Angew. Chem. Internat. Edit. Engl., 21 (1982) 711. - b) F. G. A. Stone, Angew. Chem., 96 (1984) 85; Angew. Chem. Internat.

Edit. Engl., 23 (1984) 85.
14 D. Mansuy, J.-P. Lecomte, J.-C. Chottard, J. F. Bartolini, Inorg. Chem., 20 (1981) 3119.
15 S. L. Latesky, J. P. Selegue, J. Amer. Chem. Soc., 109 (1987) 4731.
16 N. M. Kostic, R. F. Fenske, J. Organometal. Chem., 233 (1982) 337.
17 J. D. Korp, I. Bernal, R. Hörlein, R. Serrano, W. A. Herrmann, Chem. Ber., 118 (1985) 340.
18 a) W. Gäde, E. Weiss, J. Organometal. Chem., 213 (1981) 451. - b) W. Gäde, E. Weiss, Chem. Ber., 114 (1981) 2399; see also: D. Melzer, E. Weiss, Chem. Ber. 117 (1984) 2464. - c) W. Gäde, E. Weiss, Angew. Chem., 93 (1981) 796; Angew. Chem. Internat. Edit. Engl. 20 (1981) 803.
19 W. A. Herrmann, J. Weichmann, U. Küsthardt, A. Schäfer, R. Hörlein, Ch. Hecht, E. Voss, R. Serrano, Angew. Chem., 95 (1983) 1019; Angew. Chem. Internat. Edit. Engl., 22 (1983) 979; Angew. Chem. Suppl., 1983, 1543.
20 J. Weichmann, Ph.D. Thesis, Universität Frankfurt/Main (Germany), 1983.
21 a) D. Melzer, E. Weiss, J. Organometal. Chem., 255 (1983) 335. - b) D. Melzer, E. Weiss, J. Organometal. Chem., 263 (1984) 67.
22 W. A. Herrmann, Ch. Hecht, E. Herdtweck, H.-J. Kneuper, Angew. Chem., 99 (1987) 158; Angew. Chem. Internat. Edit. Engl., 26 (1987) 132 (oxidative addition of TeH_2 to a 16e organorhenium complex).
23 a) W. A. Herrmann, H.-J. Kneuper, E. Herdtweck, Angew. Chem., 97 (1985) 1060; Angew. Chem. Internat. Edit., 24 (1985) 1062. - b) W. A. Herrmann, H.-J. Kneuper, E. Herdtweck, Chem. Ber., in press.
24 W. A. Herrmann, H.-J. Kneuper, E. Herdtweck, Chem. Ber., (1988), in press; cf., H.-J. Kneuper, Ph.D. Thesis, Technische Universität München 1987.
25 W. A. Herrmann, J. Rohrmann, M. L. Ziegler, Th. Zahn, J. Organometal. Chem., 273 (1984) 221.
26 W. A. Herrmann, Ch. Hecht, M. L. Ziegler, Th. Zahn, J. Organometal. Chem., 273 (1984) 323.
27 W. A. Herrmann, J. Rohrmann, A. Schäfer, J. Organometal. Chem., 265 (1984) C 1.
28 W. A. Herrmann, Ch. Hecht, M. L. Ziegler, B. Balbach, J. Chem. Soc. Chem. Commun., 1984, 686.
29 W. A. Herrmann, Ch. Hecht, M. L. Ziegler, Th. Zahn, J. Organometal. Chem., 273 (1984) 323.
30 W. A. Herrmann, J. Rohrmann, Ch. Hecht, J. Organometal. Chem., 290 (1985) 53.
31 W. A. Herrmann, J. Rohrmann, M. L. Ziegler, Th. Zahn, J. Organometal. Chem., 295 (1985) 175.
32 M. Herberhold, B. Schmidkonz, U. Thewalt, A. Razavi, H. Schöllhorn, W. A. Herrmann, Ch. Hecht, J. Organometal. Chem., 299 (1986) 213.
33 W. A. Herrmann, J. Rohrmann, Chem. Ber., 119 (1986) 1437.
34 J. Rohrmann, W. A. Herrmann, E. Herdtweck, J. Riede, M. L. Ziegler, G. Serguson, Chem. Ber., 119 (1986) 3544.
35 W. A. Herrmann, J. Rohrmann, E. Herdtweck, Ch. Hecht, M. L. Ziegler, O. Serhadli, J. Organometal. Chem., 314 (1986) 295.
36 Ch. Hecht, E. Herdtweck, J. Rohrmann, W. A. Herrmann, W. Beck, P. M. Fritz, J. Organometal. Chem., 330 (1987) 389.
37 W. A. Herrmann, Ch. Hecht, E. Herdtweck, J. Organometal. Chem., 331 (1987) 309.

38 W. A. Herrmann, H.-J. Kneuper, J. Organometal. Chem., (1988), in press.
40 R. Hörlein, Diploma Thesis, Universität Frankfurt/Main (Germany), 1983.
41 K. Dehnicke, J. Strähle, Angew. Chem., 93 (1981) 451; Angew. Chem. Internat. Edit. Engl., 20 (1981) 413.
42 Reviews: a) H. G. von Schnering, Angew. Chem., 93 (1981) 44; Angew. Chem. Internat. Edit. Engl., 20 (1981) 33. - b) H. Schäfer, B. Eisenmann, W. Müller, Angew. Chem., 85 (1973) 742; Angew. Chem. Internat. Edit. Engl., 12 (1973) 694. - c) J. D. Corbett, Progr. Inorg. Chem., 21 (1976) 129.
43 H.-J. Kneuper, E. Herdtweck, W. A. Herrmann, J. Amer. Chem. Soc., 109 (1987) 2508.
44 C. Mealli, S. Midollini, L. Sacconi, Inorg. Chem., 17 (1978) 632.
45 a) J. Schiemann, P. Hübener, E. Weiss, Angew. Chem., 95 (1983) 1021; Angew. Chem. Internat. Edit. Engl., 22 (1983) 980. - b) N. Albrecht, P. Hübener, U. Behrens, E. Weiss, Chem. Ber., 118 (1985) 4059.
46 W. A. Herrmann, Ch. Hecht, M. L. Ziegler, B. Balbach, J. Chem. Soc. Chem. Commun., 1984, 686.
47 a) M. Herberhold, D. Reiner, D. Neugebauer, Angew. Chem., 95 (1983) 46; Angew. Chem. Internat. Edit. Engl., 22 (1983) 59. - b) W. A. Herrmann, Ch. Bauer, J. Weichmann, J. Organomet. Chem., 243 (1985) C 57.
48 T. J. Greenhough, B. W. S. Kolthammer, P. Legzdins, J. Trotter, Inorg. Chem., 18 (1979) 3543.
49 a) L. Y. Goh, T. W. Hambley, G. B. Robertson, J. Chem. Soc. Chem. Commun., (1983) 1458. - b) M. Herberhold, D. Reiner, B. Zimmer-Gasser, U. Schubert, Z. Naturforsch., 35b (1980) 1281. - c) W. A. Herrmann, J. Rohrmann, A. Schäfer, J. Organomet. Chem., 265 (1984) C 1. - d) W. A. Herrmann, J. Rohrmann, H. Nöth, Ch. K. Narula, I. Bernal, M. Draux, J. Organomet. Chem., 284 (1985) 189. - e) L. Y. Goh, Ch. Wei, E. Sinn, J. Chem. Soc., Chem. Commun., 1985, 462. - f) W. A. Herrmann, J. Rohrmann, Ch. Hecht, J. Organomet. Chem., 290 (1985) 53.
50 W. A. Herrmann, J. Rohrmann, E. Herdtweck, H. Bock, A. Veltmann, J. Amer. Chem. Soc., 108 (1986) 3134.
51 H. Bock, A. Veltmann, personal communication.
52 Recent examples: a) K. H. Whitmire, C. B. Lagrone, M. R. Churchill, J. C. Fettinger, L. V. Biondi, Inorg. Chem., 23 (1984) 4227. - b) M. H. Churchill, J. C. Fettinger, K. H. Whitmire, J. Organometal. Chem., 284 (1985) 13. - c) J. M. Cassidy, K. H. Whitmire, J. Chem. Soc. Chem. Commun., (1988), in press.

UNSATURATED FOUR-, SIX- AND EIGHT-MEMBERED METALLAHETEROCYCLES AND METAL-CONTAINING POLYMERS

HERBERT W. ROESKY
Institut für Anorganische Chemie der Universität Göttingen,
Tammannstraße 4, 3400 Göttingen, Federal Republic of Germany

1. INTRODUCTION

In organic chemistry unsaturated four-, six- or eight-membered rings are well known. They have a broad and well-established chemistry. Prominent examples are the six-membered benzene and the eight-membered cyclo-octatetraene ring. These are only two examples of a countless number of unsaturated cyclic compounds.

In inorganic chemistry the origins of unsaturated heterocycles can be traced back to the early 1800s, when the first reports were published on compounds with formulae S_4N_4, $(PNCl_2)_3$ and $(PNCl_2)_4$.

Since that time many derivatives have been developed in which the ring atoms are partly replaced by other main group elements. Concurrently, the variety of chemical transformations and structures found for these heterocycles have attracted both experimental and theoretical interest.

The pyrolysis of $NH_3 \cdot BH_3$ produces, with evolution of hydrogen, the inorganic analog of benzene, $(HBNH)_3$, commonly called borazine. In contrast to benzene, borazine is much more reactive. This is generally attributed to the more localized nature of borazine π-electrons.

In contrast to well-established cyclic compounds of main group elements, derivatives with transition metals in the ring skeleton have been investigated only recently.

This survey begins with a discussion of metallacyclobutadienes and 1,3-dimetallacyclobutadienes and compounds obtained by replacement of a CH group in benzene by a transition metal. There follows a detailed account of transition-metal-containing phosphorus-nitrogen and sulfur-nitrogen compounds and also of metal-containing polymers.

2. FOUR-MEMBERED RINGS WITH MULTIPLE CARBON-METAL BONDS

2.1 METALLACYCLOBUTADIENES

The first triple bonds between a transition metal and a monosubstituted carbon atom were found by Fischer et al. (ref. 1) in chromium, molybdenum and tungsten complexes of the type $X(CO)_4M{\equiv}CR$. Several years later Schrock et al. (refs. 2-4) discovered alkylidyne complexes, mainly of molybdenum and tungsten, of the type $X_3M{\equiv}CR$. In contrast to the Fischer complexes, the $X_3M{\equiv}CR$ complexes have metal atoms in their highest possible oxidation states.

Very recently metallacyclobutadienes have been implicated as active intermediates in alkyne metathesis reactions (refs. 5,6).

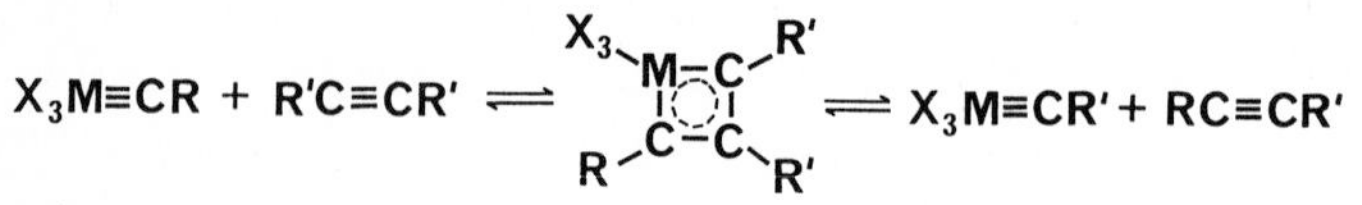

Schrock and co-workers have actually isolated and structurally characterized such compounds, of composition $WC_3Et_3(O\text{-}2,6\text{-}C_6H_3\text{-}i\text{-}Pr_2)_3$ (ref. 7) and $WC_3Et_3[OCH(CF_3)_2]_3$ (ref. 8).

An X-ray structural study of $WC_3Et_3(O\text{-}2,6\text{-}C_6H_3\text{-}i\text{-}Pr_2)_3$ Fig. 1 shows it to be a crowded, distorted trigonal bipyramidal molecule containing a strictly planar, almost symmetrical WC_3 ring.

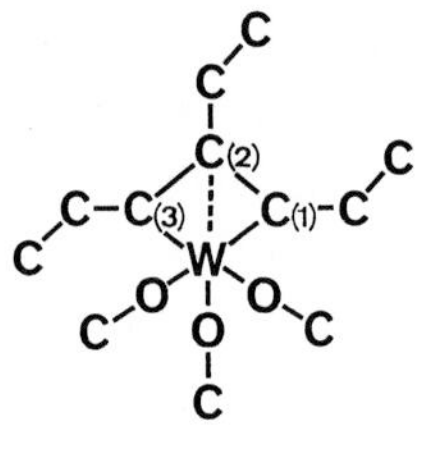

Fig. 1. Molecular structure of $WC_3Et_3(O\text{-}2,6\text{-}C_6H_3\text{-}i\text{-}Pr_2)_3$. The hydrogen atoms are omitted for clarity.

The most interesting features of the ring system are the relatively short W-C(1) (194.9(9)) and W-C(3) (188.3(10)pm) bonds.

By comparison with characteristic carbon-tungsten double (190-200 pm) and triple (175-180 pm) bond lengths, the bond order of W-C(1) and W-C(3) is roughly 2.5. The W-C(2) bond lengths of 215.9(10) correspond to a carbon-tungsten single bond (215-220 pm). The short W-C(2) distance has been ascribed to overlap of a d orbital with the totally symmetric MO of the C_3-fragments (ref. 4). The slight distortion of the WC_3-ring suggests that 3-hexyne is in the process of leaving the ring. The structure of $WC_3Et_3[OCH(CF_3)_2]_3$ is similar to $WC_3Et_3(O\text{-}2,6\text{-}C_6H_3\text{-}i\text{-}Pr_2)_3$ except that the WC_3-ring is more symmetrically arranged.

However, the analogous molybdenum complex $MoC_3Et_3(O\text{-}2,6\text{-}C_6H_3\text{-}i\text{-}Pr_2)_3$ is only stable at low temperatures. It decomposes at -25°C with formation of 3-hexyne (ref. 4).

$$MoC_3Et_3(O\text{-}2,6\text{-}C_6H_3\text{-}i\text{-}Pr_2)_3 \rightarrow EtC{\equiv}Mo(O\text{-}2,6\text{-}C_6H_3\text{-}i\text{-}Pr_2)_3 + EtC{\equiv}CEt$$

Structural data for a number of metallacyclobutadienes with different ligands are summarized in Table 1.

The common feature of all structurally investigated tungstacyclobutadienes is the short W-C_α bond. These bonds are far shorter than one would expect for bonds of order ca. 1.5. However, the W-C_β distances are shorter than typical W-C single bonds.

The 1,2,3-triphenylpropenylium-iridium fragment in the $[IrC_3Ph_3Cl(CO)(PMe_3)_2]^+$ cation shows some remarkable structural similarities to the 4-chloro-1,2,3,4-tetraphenylcyclobutenium cation, $[C_4Ph_4Cl]^+$, in which the 4-carbon atom can be considered equivalent to the iridium atom. Substantial multiple bonding between the three carbon atoms in the ring is indicated by the average value of 138.5 pm determined for the propenylium carbon-carbon distances. This may be compared with the value of 140(6) pm in $[C_4Ph_4Cl]^+$.

Schrock and co-workers have shown that addition of a nitrogen base can convert a tungstacyclobutadiene (1) to a cyclopropenium complex (2) (ref. 13).

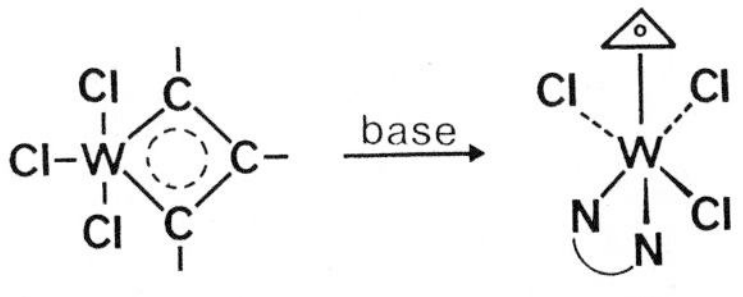

(1) (2)

TABLE 1 Structural data of metallacyclobutadienes

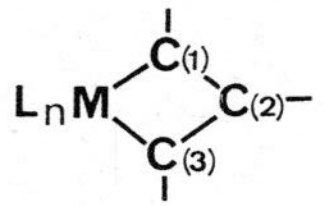

Compound	M-C(1)	M-C(3)	M-C(2)	C(1)-C(2)	C(2)-C(3)	Reference
$W(\eta^5-C_5H_5)[C_3(CMe_3)_2]Cl$	191.9(8)	192.9(6)	204.9(8)	139.9(11)	131.1(21)	(9)
$W[C-t-BuCMeCMe]Cl_3$	186.1(9)	186.4(8)	211.5(8)	145.5(13)	147.8(12)	(10)
$WC_3Et_3[OCH(CF_3)_2]_3$	190.2(16)	186.4(14)	209.3(14)	142.9(18)	143.7(2)	(8)
$W(C-t-BuCHC-t-Bu)-[OCH(CF_3)_2]_3$	192.1(10)	188.9(9)	210.3(9)	141.8(13)	149.9(14)	(11)
$WC_3Et_3(O-2,6-C_6H_3-i-Pr_2)_3$	194.9(9)	188.3(10)	215.9(10)	146.7(14)	143.3(14)	(7)
$[IrC_3Ph_3Cl(CO)(PMe_3)_2]BF_4$	209.9(15)	199.0(15)	261.0(15)	141(2)	136(2)	(12)

2.2. 1.3-DIMETALLACYCLOBUTADIENES

In an elegant paper, Chisholm and Heppert reviewed the chemistry of 1.3-ditungstacyclobutadienes (ref. 14). The 1.3-dimetallacyclobutadienes (3)

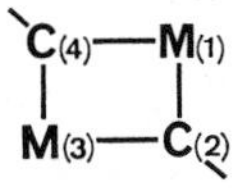

(3)

have been prepared by Chisholm, Cotton, Schmidbaur, Wilkinson and their co-workers. Compounds of type (3) are known for titanium (ref. 18), niobium (ref. 15), tantalum (ref. 15), tungsten (refs. 16,19-23) and rhenium (ref. 17). Several compounds have been fully characterized by X-ray structural investigations. The structural data are compiled in Table 2.

TABLE 2 Structural data of 1.3-dimetallacyclobutadienes

Compound	M-C av	M-M av	C(4)-M(1)-C(2)	Reference
$Nb_2(SiMe_3)_2(CH_2SiMe_3)_4$ [a]	199.5(9)	289.7(2)	85.6(4)	(15)
$W_2(CPh)_2(OCMe_3)_4$	193.8(10)	266.5(1)	93.1(4)	(23)
$W_2(CSiMe_3)_2(CH_2SiMe_3)_4$ [a]	191(2)	253.5(2)	96(1)	(16)
$W_2(CSiMe_3)_2(O\text{-}i\text{-}Pr)_4$ [a]	195(2)	262.0(2)	95.6(7)	(22)
$Re_2(CSiMe_3)_2(CH_2SiMe_3)_4$ [a]	192.9(6)	255.7(1)	97.0(4)	(17)

a two crystallographically independent molecules

The four-membered rings are almost planar. There is some buckling at the bridging carbon in the niobium compound. These distortions may well be the result of mutual repulsion of the bulky ligand groups.

The only important difference between the niobium, tungsten and rhenium structures are associated with the presence of a metal-metal bond in the tungsten and rhenium compounds.

In $W_2(CPh)_2(OCMe_3)_4$ there is structural evidence (ref. 23) that the W-O bonds have remarkable $O \rightarrow W\pi$ character, and it is proposed that this is responsible for the rather long W-W bond of 266.5(1) pm.

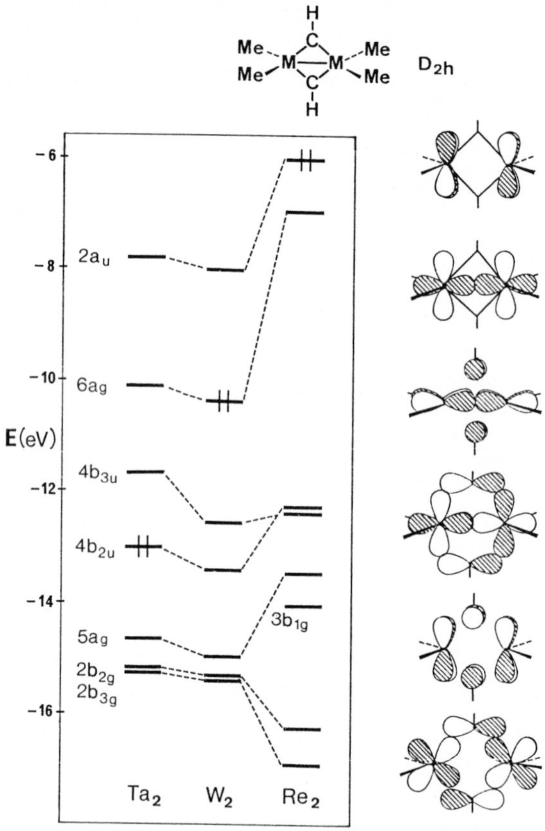

Fig. 2. Molecular orbital energy level diagram of the $Me_4M_2(CH)_2$ molecules employing the method of Fenske and Hall, and showing the orbital population of the HOMO: M-C and M-Mπ, $4b_{3u}$ for M = Ta; M-Mσ, $6a_g$ for M = W; and M-Mσ*, $2a_u$ for M = Re.

The assignment of metal-metal bond order on the basis of bond length is still open to question. In $Re_2(CSiMe_3)_2(CH_2SiMe_3)_4$ the Re-Re distance of 255.7 pm is rather long, but is still considered compatible with a metal-metal double bond.

The results of EHMO and Fenske-Hall calculations, Fig. 2 (ref. 14) are consistent with the M-M bond distance trend, showing that the electronic configurations of the M-M bonds in the Ta, W and Re complexes are $\sigma^{\circ}\delta^{*\circ}$, $\sigma^{2}\delta^{*\circ}$ and $\sigma^{2}\delta^{*2}$, respectively.

$W_2(CSiMe_3)_2(CH_2SiMe_3)_4$ forms adducts with alkynes. The structure of $W_2(CSiMe_3)_2(CH_2SiMe_3)_4$ PhC_2Me was subjected to an X-ray structure analysis (ref. 22). The alkyne is bonded only to one metal atom in an η^2 manner. The W-W contact of 291.5 pm can be considered non-bonding and is comparable to that in the d°-d° complex $Nb_2(CSiMe_3)_2(CH_2SiMe_3)_4$.

$W_2(CSiMe_3)_2(CH_2SiMe_3)_4$ adds chlorine or bromine (ref. 24) to yield compound (4) (X = Cl, Br).

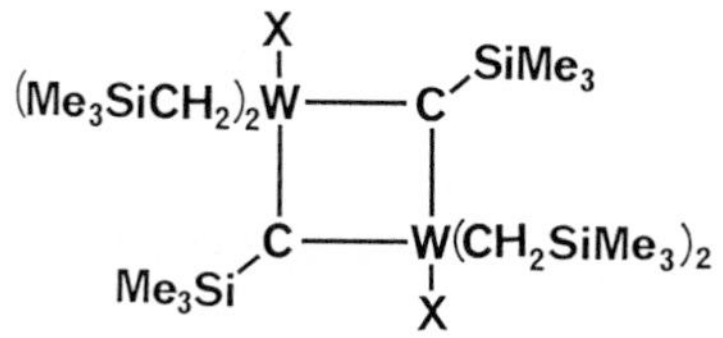

(4)

The W-C distances within the W_2C_2 ring are essentially identical (200 pm), and the W-W bond lengths are 275(1) pm. Each tungsten atom is in a distorted bipyramidal environment joined along a common axial-equatorial edge by the $CSiMe_3$ ligands.

The reactions of d^1-d^1 $W_2(CSiMe_3)_2$-containing compounds with alkynes, halogens, allenes, CO, RNC and Ph_2CN_2 appear to proceed with cleavage of the W-W bond (ref. 14).

2.3. TETRAMETALLACYCLOBUTADIENES

Chisholm and co-workers (ref. 25) were the first to observe that $W_2(O\text{-}i\text{-}Pr)_6$ forms a dimer of type (5)

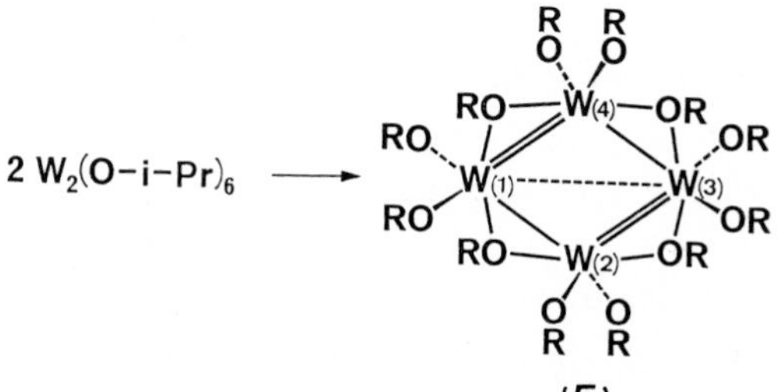

(5)

The X-ray crystal structure of (5) exhibits W-W bond distances W(1)-W(2) 273.3(1), W(1)-W(4) 250.2(1) and W(1)-W(3) 280.7(2) pm. The W_4-cluster forms a rhombus with two shorter and two longer W-W bond distances. The structure of (5) is completed by four edge-bridging OR and eight terminal OR groups. The oxygen atoms of the bridging OR groups form a plane together with the four tungsten atoms.

On the basis of nmr investigations, it was shown that, in solution and depending on the temperature, an equilibrium between (5) and the monomer existed. At 65°C only the monomer was observed. There is an appreciable energy barrier between the monomer and the dimer. It was shown that the corresponding molybdenum compound, $Mo_2(O\text{-}i\text{-}Pr)_6$, does not form a dimer.

3. FIVE-MEMBERED RINGS

3.1. METALLACYCLOPENTADIENES

Five-membered transition metal metallacycles are common products in reactions of transition metal complexes with acetylenes. For a wide range of metals and ligands the structure determination of such compounds show, with some exceptions, bond lengths consistent with localized double bonds of the following type (6).

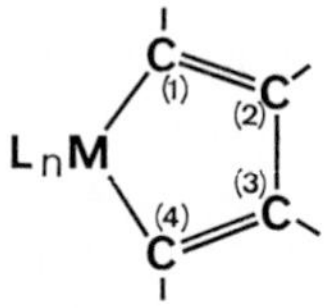

(6)

Structural data of some metallacyclopentadienes are summarized in Table 3.

As a comparison, butadiene has an average short (134 pm) and a long (148 pm) bond. However, there is no appreciable multiple bonding between the metal and the carbon atoms.

During recent years, those compounds have been the subject of considerable research. They play an important role in a number of catalytic reactions. In an excellent paper by Chappell and Cole-Hamilton (ref. 27) the chemistry of metallacyclic compounds has been reviewed. The literature was covered up to the end

TABLE 3 Structural data on some metallacyclopentadienes (ref. 26)

Compound	C(1)-C(2)	C(3)-C(4)	C(2)-C(3)	M-C[a]	References
$RhCl(H_2O)(AsMe_3)_2C_4(CF_3)_4$	131.1(24)	134.6(23)	143.3(26)	202	(28)
$RhCl(PPh_3)_2C_{24}H_{14}O_2$	135	135	145	200	(29)
$CpRh(PPh_3)C_4(C_6F_5)_4$	134.3(16)	135.4(15)	145.7(16)	206.4	(30)
$CpCoPPh_3C_4(C_6F_5)_4$	132.6(15)	133.5(15)	146.7(15)	199.4	(31)
$(Cp)_2TiC_4Ph_4$	136.9(6)	137.0(6)	149.5(6)	215.7	(32)

a average

of 1981. However, such five-membered rings do not fall within the scope of this review.

4. UNSATURATED SIX-MEMBERED RINGS

4.1. METALLABENZENES

Replacement of a CH group in benzene by N, Si, P, As, Sb or Bi leads to the known molecules pyridine, silabenzene, phosphabenzene, arsabenzene, stibabenzene or bismabenzene (refs. 33,34). Some of them are extremely unstable and can be handled only at low temperatures. The replacement of a CH group by a transition metal leads to metallacyclohexatriene or metallabenzene. Octahedral complexes of the type (7) or (8) or a square-pyramidal d^8 complex (9) have been considered by Thorn and Hoffmann (ref. 26). Those complexes should show some delocalization and aromatic character in the C-C bond lengths.

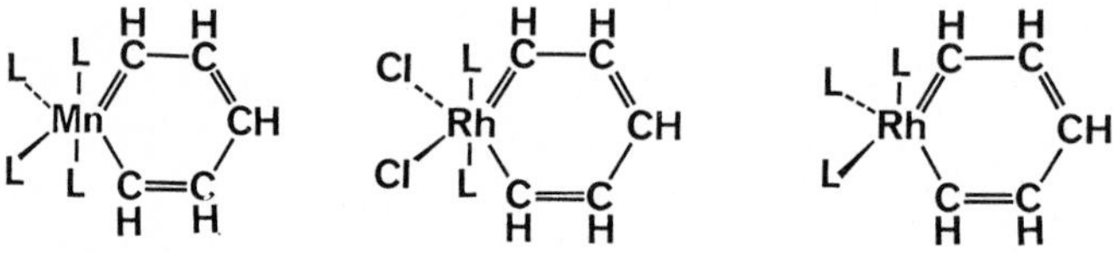

(7) (8) (9)

They suggested that the complexes (7) and (8) would be more stable if the other ligands around the metal were donors instead of acceptors. However, π-donor substituents on the carbon ring, especially in the ortho and para position to the metal, should also help to stabilize the system.

Roper et al. (ref. 35) reported the first osmabenzene derivatives. When they reacted $[Os(CO)(CS)(PPh_3)_3]$ with ethyne in benzene, they obtained a 30 percent yield of brown crystals of composition $[Os(CSCHCHCHCH)(CO)(PPh_3)_2]$ (10). The compound was characterized by X-ray crystal structure determination (Fig. 3).

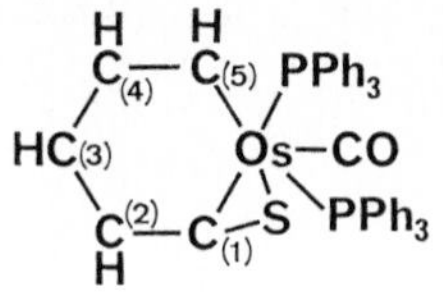

(10)

Fig. 3. Molecular structure of Os(CSCHCHCHCH)(CO)$(PPh_3)_2$

The C-C bond distances within the six-membered ring have the following values: C(1)-C(2) 136(2), C(2)-C(3) 138(2), C(3)-C(4) 142(2), C(4)-C(5) 139(2) pm.

The C(1)-Os and C(5)-Os distances are equal at 200(1) pm. The C-C bond distances within the ring, together with its planar nature, imply aromatic character.

Reaction of [Os(CSCHCHCHCH)(CO)$(PPh_3)_2$] with CO opened the η^2-CS function to yield (11). It forms bronze crystals which exhibit CO stretching frequencies at 2020 and 1945 cm^{-1}. When (11) is treated with methyl iodide, it is converted to (12). Reaction of (10) with hydrogen chloride gave the blue complex (13).

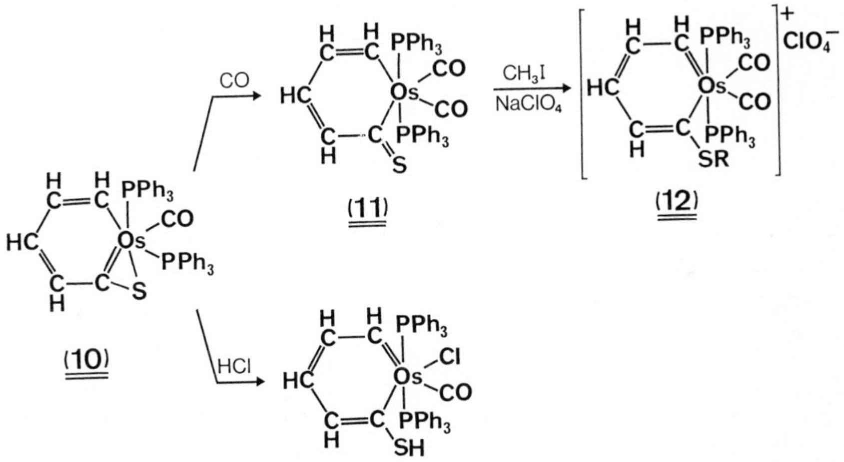

The described structure of compound (10) is intermediate between the suggested possibilities (7) and (8).

5. CYCLOPHOSPHAZENES

5.1 POLYDICHLOROPHOSPHAZENES

Polydichlorophosphazenes with general formula $(PNCl_2)_n$ were first discovered in 1834 by Rose (ref. 36) and Liebig (ref. 37) in very low yield from the reaction of ammonia with PCl_5. Later Stokes (ref. 38), in the reaction of NH_4Cl with PCl_5, produced and isolated various $(PNCl_2)_n$ ranging from the trimer to the hexamer and an impure heptamer in addition to a polymeric residue. Stokes also found that the trimer, when heated in a sealed tube at 250°C formed a polymer, the first inorganic rubber.

The synthesis of $(PNCl_2)_n$ was much improved by Schenk and Römer (ref. 39) in 1924, and their method remains the basis for present-day production on both the industrial and laboratory scale.

During the last few decades, the search for a new inorganic type of polymer provided a strong impetus for the study of polyphosphazenes. Excellent reviews on this subject are by Allcock (refs. 40-43) and Krishnamurthy (ref. 44).

Polyphosphazene chemistry continues to be an important area of research in inorganic chemistry. In particular, the synthesis of transition-metal-bound phosphazenes, phosphazene derivatives with biologically active side groups, organofunctional phosphazene polymers and cyclometalla-phosphazenes are some noteworthy recent advances in this field.

5.2 PHYSICAL PROPERTIES OF CYCLOPHOSPHAZENES

All cyclic phosphazenes contain the formally unsaturated $-N{=}PX_2-$group with nitrogen of coordination number 2 and phosphorus of coordination number 4. The following physical facts have been experimentally demonstrated:

(i) NPX_2, though isoelectronic to OCX_2, is not stable as a monomer; the trimer and tetramer are very stable,

(ii) The P-N distances within the ring skeleton are equal, in the range 158 $\pm$ 2 pm; this represents a bond order of approximately 1.5,

(iii) the nitrogen atoms are weakly basic and can be protonated or form coordination complexes,

(iv) in contrast to aromatic compounds, cyclophosphazenes form no charge-transfer complexes,

(v) in contrast to aromatic compounds they are hard to reduce electrochemically.

The bonding in cyclophosphazenes differs from aromatic σ-π systems, in which there is extensive electron delocalisation via pπ-pπ bonding. Each nitrogen of a cyclophosphazene bears a lone pair of electrons. Two electrons of nitrogen and four of phosphorus are used for the σ-bonds. The remaining electrons, one from nitrogen and one from phosphorus, are believed to form a P-N-P three-center two-electron π-bond. One of the three-center two-electron bonding systems is depicted in Fig. 4.

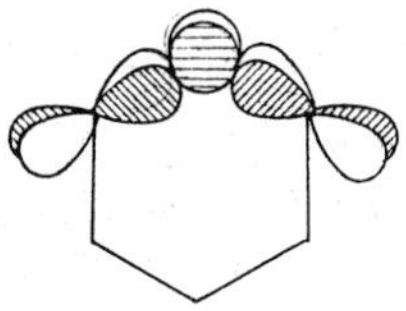

Fig. 4. The Dewar model for a three-centre two-electron bonding system.

These localized three-center bonds are interrupted at each phosphorus atom in a way that permits an equalization of bond lengths around the ring. The possibility of pπ-pπ bonding has been considered by many authors (refs. 40,45,46). In a recent paper Haddon (ref. 47) reported theoretical calculations employing the Hartree-Fock method showing that phosphorus d orbitals are significant participants in the ring bonds. However, extended Hückel calculations by Hoffmann et al. (ref. 48) showed that d orbitals are not essential in phosphazene bonding.

5.3. TRANSITION-METAL-BOUND PHOSPHAZENES

A comprehensive review was recently devoted to the organometallic chemistry of phosphazenes by Allcock et al. (ref. 43).

There are four possibilities for a transition metal to interact with a cyclophosphazene: (i) the lone pair of the ring nitrogen functions as a Lewis base towards the transition metal, (ii) an ionic complex between the transition metal and the cyclophosphazene is formed, (iii) an unsaturated side chain attached to phosphorus may function as a π-ligand towards the transition metal, (iv) a direct covalent bond is formed between a phosphorus atom and a transition metal.

Cyclophosphazenes can act as Lewis base to form complexes such as $N_3P_3Me_6$-$TiCl_4$ (ref. 49) or function as proton acceptors to yield a complex such as (14). The structure of (14) was unambiguously assigned by an X-ray structure investigation (ref. 50).

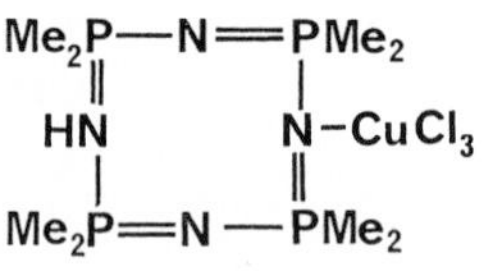

(14)

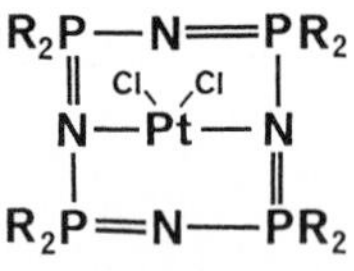

(15)

However, the reaction of $PtCl_2$ with $N_4P_4(NHMe)_8$ or $N_4P_4Me_8$ yields complexes of type (15). The platinum atom in (15) is coordinated to two chlorine ligands and to two nitrogen atoms from the phosphazene ring (refs. 51-53).

An interesting example of a cyclophosphazene acting as a multidentate macrocyclic ligand was observed when $N_6P_6(NMe_2)_{12}$ was reacted with equal amounts of CuCl and $CuCl_2$. The $[N_6P_6(NMe_2)_{12}CuCl]^+$ cation is shown in Fig. 5 (refs. 54, 55).

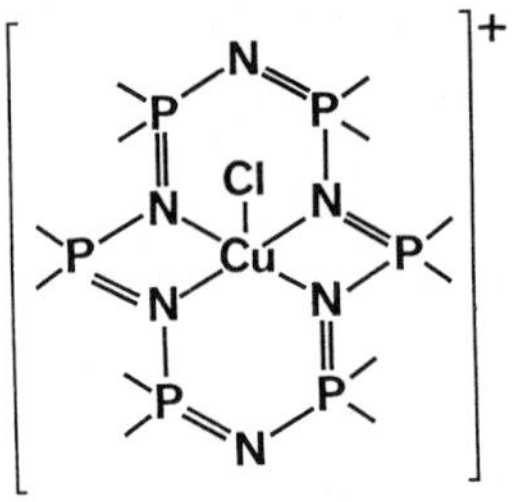

Fig. 5. Structure of the complex cation $[N_6P_6(NMe_2)_{12}CuCl]^+$. The NMe_2 groups are omitted for clarity.

The coordination with Cu^{2+} tightens the ring compared with the free ligand $N_6P_6(NMe_2)_{12}$, the mean angles at phosphorus being reduced from 120.0° to 107.5°, and the mean angles at nitrogen being reduced from 147.5° to 133.6°.

Metal carbonyl complexes of aminophosphazenes are known; products from the reaction of $Mo(CO)_6$ and $W(CO)_6$ with $N_4P_4(NMe_2)_8$ and $N_4P_4Me_8$ have been isolated (refs. 56-59).

As indicated by the molecular structure of (16), $N_4P_4(NMe_2)_8$ functions as a bidentate ligand towards tungsten.

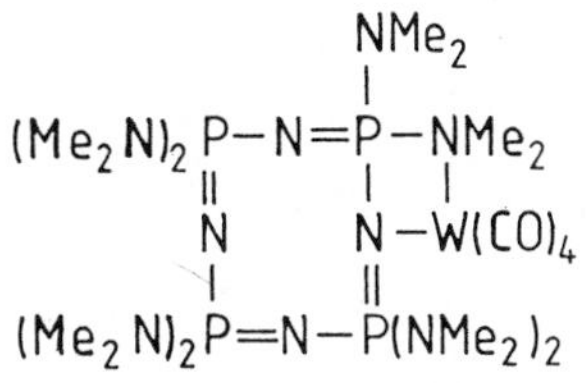

(16)

The reactions between dihalogenophosphazenes and organometallic reagents have been studied by Allcock and co-workers (ref. 43). Some of the beautiful results are summarized in Scheme 1.

$P_3N_3F_3$ (17) reacts with $K[FeCp(CO)_2]$ to yield the monosubstituted product (18). Further treatment of (18) with $K[FeCp(CO)_2]$ results in the formation of the geminally substituted derivative (19). On photolysis, (19) eliminates carbon monoxide to give (20) with an iron-iron bond and a CO group in a bridging position (refs. 60-62).

In contrast, the reaction of lithioferrocene with (17) yields the nongeminal complex (22) (refs. 63,64). It should also be noted that dilithioferrocene reacts with (17) to give the transannular-linked product (23)

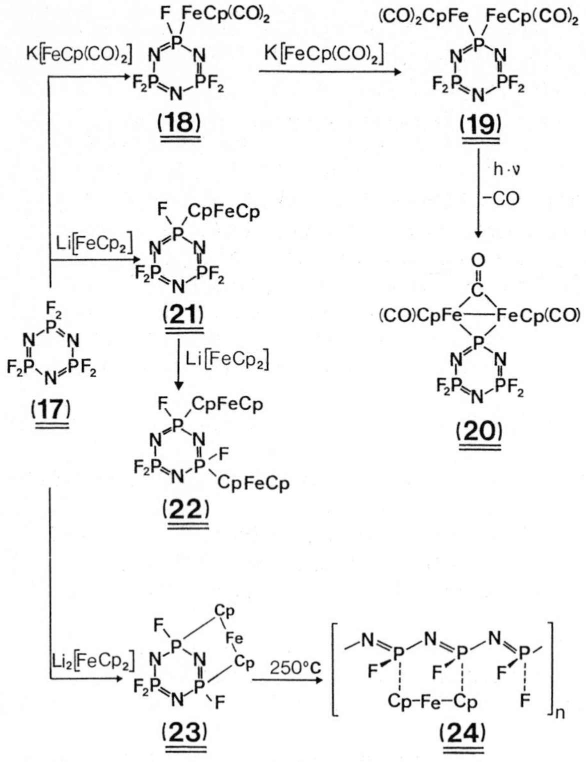

Scheme 1. Reactions of $N_3P_3F_6$ with $K[FeCp(CO)_2]$ and lithiated ferrocenes.

Most of the structures have been confirmed by crystallographic studies.

(23) undergoes ring opening at 250°C to yield a linear polymer (24) with ferrocenyl side groups. The fluorine atoms in (24) can be replaced by 1,1,1-trifluoroethoxy groups. The resulting product is a water- and heat-stable polymer (ref. 65).

Another high-polymeric metal-containing phosphazene similar to (25) was obtained when (18) was heated with $N_3P_3Cl_6$ at 250°C (ref. 43).

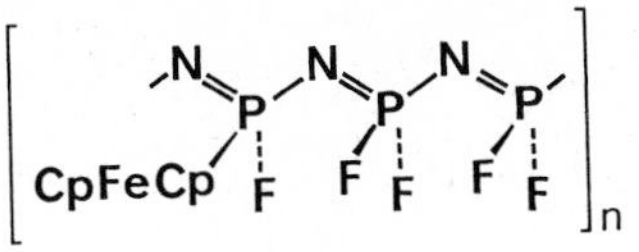

(25)

Similar reactions are reported for the cyclic trimer $P_3N_3Cl_3$ and cyclic tetramer $P_4N_4Cl_8$ (refs. 64,66-68).

6. CYCLOMETALLAPHOSPHAZENES

6.1 PREPARATION

Cyclometallaphosphazenes are derivatives of the well-known class of cyclophosphazenes where one or several of the ring phosphorus atoms are replaced by a transition metal. The versatility of possible compounds is illustrated by selecting derivatives of the six- and eight-membered phosphazene rings (26-31). Compounds of type (26) and (27) may be considered as benzene analogues.

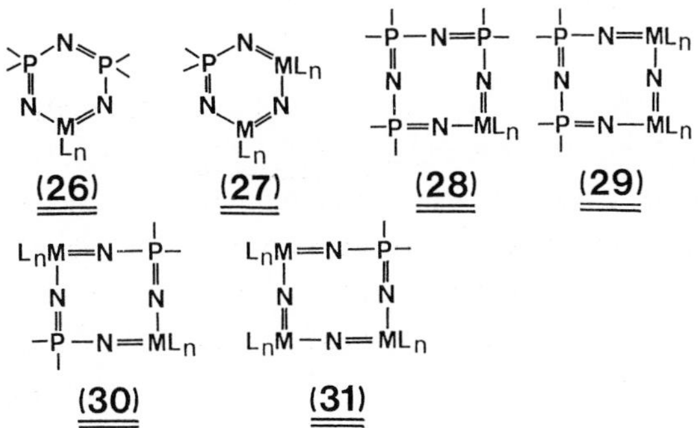

Recently, new synthetic strategies have been developed in our laboratories that led to the preparation of cyclometallaphosphazenes of type (26) and (30).

Reactions of the phosphonium salt (32) with metal halides, oxihalides and nitrides afford compounds of type (26) (Scheme 2).

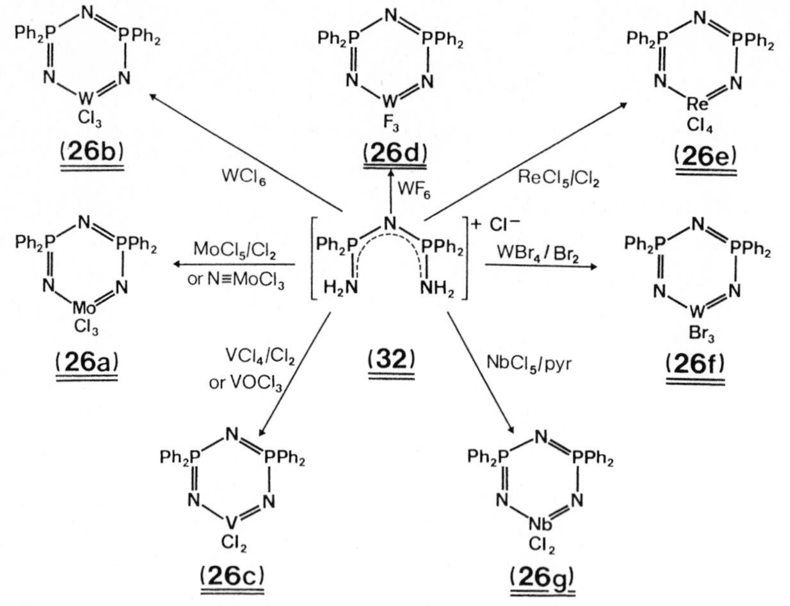

Scheme 2. Reactions of $[H_2NPPh_2NPPh_2NH_2]^+Cl^-$ with transition metal halides.

6.2 MECHANISM OF THE FORMATION OF CYCLOMETALLAPHOSPHAZENES

All reactions of the phosphonium salt (32) with transition metal halides proceed via the elimination of four moles of hydrogen halide. The interaction of the transition metal halides with (32) appears to be a sequential process. The first step is the formation of an acyclic intermediate, followed by the subsequent cyclisation step via elimination of HCl. The acyclic intermediate could not be isolated; however, the progress of the reaction of the various metal halides with (32) was monitored by $^{31}P\{^1H\}$-NMR spectroscopy.

In contrast to the reactions of the transition metal halides, $N{\equiv}MoCl_3$ reacts with insertion of the $N{\equiv}Mo\lessdot$ unit into the phosphazene skeleton. The reaction probably proceeds via a 1,4-cycloaddition step with (32). Ammonium chloride is formed quantitatively in this reaction. Compared with the reaction with $MoCl_5$, this synthetic route is a significant improvement in terms of both yield and purity of (26a).

6.3. PROPERTIES

(**26a**), (**26b**), (**26d**) and (**26g**) are pale yellow, while (**26e**) is pink, (**26c**) red and (**26f**) dark brown.

The compounds mentioned in Scheme 2 are solids showing in the mass spectrum (with one exception) the molecular ion with high relative intensity. These compounds could be sublimed without appreciable decomposition.

Compounds (**26**) provide examples of stable species of vanadium, niobium, molybdenum, tungsten and rhenium in their highest oxidation state. They are soluble in acetonitrile and moderately soluble in chloroform, THF or diethyl ether.

6.4 NMR INVESTIGATIONS

The high resolution ^{31}P-NMR spectra data (Table 4) indicate that all cyclometallaphosphazenes show a significant downfield chemical shift compared to the non-metallated cyclophosphazene analogues. This pronounced deshielding effect is probably a consequence of the electron-withdrawing property of the metal atom.

TABLE 4 ^{31}P-NMR spectra data

Compound		δ ppm	References
$\overline{NPPh_2NPPh_2NVCl_2}$	(**26c**)	43.7	(71)
$\overline{NPPh_2NPPh_2NNbCl_2}$	(**26g**)	40.1	(70)
$\overline{NPPh_2NPPh_2NMoCl_3}$	(**26a**)	42.4	(70)
$\overline{NPPh_2NPPh_2NWF_3}$	(**26d**)	22.5	(71)
$\overline{NPPh_2NPPh_2NWCl_3}$	(**26b**)	39.2	(69)
$\overline{NPPh_2NPPh_2NWBr_3}$	(**26f**)	43.8	(71)
$\overline{NPPh_2NPPh_2NReCl_4}$	(**26e**)	46.2	(71)
$N_3P_3Ph_4Cl_2$		17.2	(72)
$N_3P_3Ph_6$		14.3	(72)

Furthermore, a downfield chemical shift is observed in the series of the fluoro, chloro and bromo compounds of tungsten.

6.5 X-RAY SINGLE CRYSTAL STRUCTURE INVESTIGATION

Crystals suitable for X-ray structure analysis were obtained for (26a) and (26b) by recrystallisation from acetonitrile (Fig. 6)

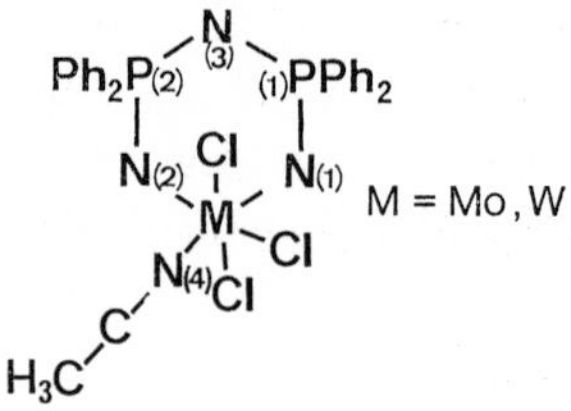

Fig. 6. Molecular strcuture of $\overline{NPPh_2NPPh_2NMo}Cl_3 \cdot CH_3CN$

The crystal structure of this molybdenum compound, which is very similar to that of the corresponding tungsten derivative, contains a second, non interacting acetonitrile within the asymmetric unit; this is probably the reason for obtaining suitable crystals from this solvent.

The central six-membered ring is nearly planar but may best be described as a shallow envelope with the Mo atom 13 pm above the plane formed by the P and N atoms. Its geometry is further characterized by pairwise similar bond lengths and angles (Table 5) and differs from that of non-metallated analogues (ref. 73) mainly in the considerably larger angles at the N-atoms adjacent to the metal, as a consequence of the increased size of the metal atom. The latter is coordinated octahedrally with one ligand position occupied by a weakly bonded acetonitrile molecule.

TABLE 5 Selected bond lengths (pm) and angles (°) of $\overline{NPPh_2NPPh_2NMo}Cl_3 \cdot CH_3CN$ and $\overline{NPPh_2NPPh_2NW}Cl_3 \cdot CH_3CN$

	$\overline{NPPh_2NPPh_2NMo}Cl_3 \cdot CH_3CN$ (ref. 70)	$\overline{NPPh_2NPPh_2NW}Cl_3 \cdot CH_3CN$ (ref. 69)
M-N1	176.6(6)	177.3(5)
M-N2	178.9(6)	179.8(6)
M-N4	239.0(6)	236.4(6)
N1-P1	166.7(6)	166.3(7)
N2-P2	163.3(6)	164.0(7)
P1-N3	158.1(7)	158.7(7)
P2-N3	158.5(6)	159.6(6)
N1-M-N2	96.7(3)	97.3(3)
P1-N1-M	134.7(4)	134.5(4)
P2-N2-M	134.3(4)	133.1(3)
N1-P1-N3	122.4(3)	113.4(3)

In these compounds the main group unit does not function as a chelating ligand but forms multiple bonds towards the metal.

6.6 REACTIONS

In addition to the synthesis of $\overline{NPPh_2NPPh_2NW}Cl_3$ (26b), this reaction can also be used to prepare a cyclophosphazene containing a metal-metal bond within the ring skeleton. $HCl_5N_6P_4Ph_8W_2$ (33) is a crystalline red solid formed in a 5-6 percent yield (ref. 74).

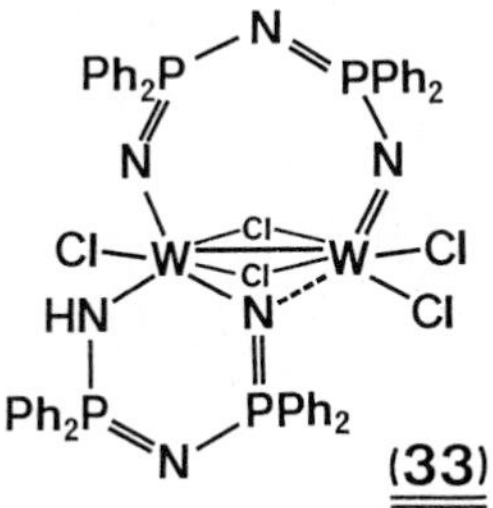

(33)

(33) was characterized by an X-ray structural study. The six-membered ring resembles that observed in (26b) except that the nitrogen adjacent to the tungsten bears a hydrogen. The metal-

metal bond functions as a bridge between the six- and the seven-membered ring. The W-W bond (269.2 pm) falls in the range of comparable d^1-d^1 W-W single bond (252 - 288 pm) lengths. It is noteworthy that the geometry around the tungsten atoms is different.

Spirocyclic metallaphosphazenes of composition (34) and (35) have been obtained either by reacting $[OPPh_2NPPh_2O]Na$ with MoO_2Cl_2 or treating $NPP_2NPPh_2NMoCl_3$ with t-butanol. Both compounds are crystalline solids which have been characterized by X-ray structural investigations (refs. 75,76).

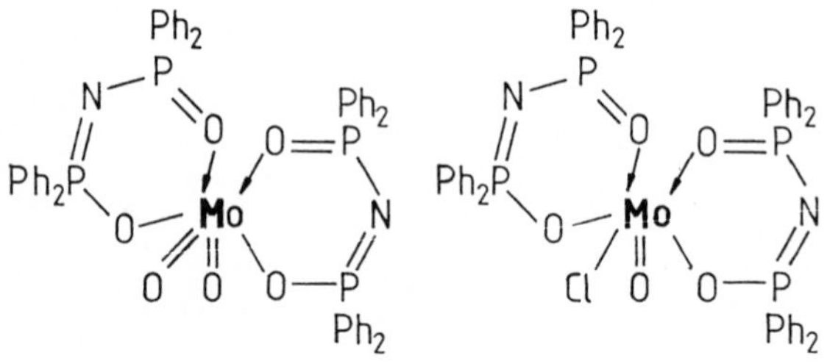

(34) (35)

6.7 POLYMERS

It has been shown that (26a) and (26b) can be converted to polymeric materials (36) at temperatures between 120°C and 250°C. Depending on the temperatures, the polymers have wax- or glass-like properties. These new materials might be used as electrical conductors, polymer-bound catalysts or metal-containing pre-ceramic materials.

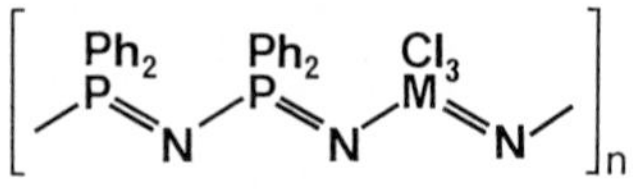

(36)

Clearly, more studies of the properties of these polymers are needed.

6.8 PREDICTIONS

Extended Hückel calculations by Hoffmann et al. (ref. 48) showed that metal-nitrogen distances in d° transition metal compounds parallel C-C bond lengths in linear and cyclic hydrocarbon polyenes. This leads to predicted stability for some as yet unsynthesized $(L_nMN)_n$ oligomers such as the benzene analogue (37).

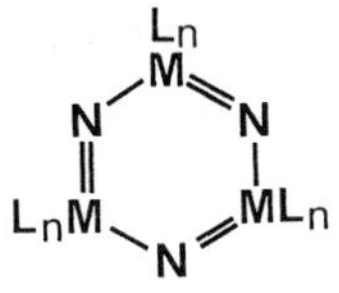

(37)

Obviously, it is an interesting synthetic goal to prepare compounds of type (27) and (37) and also to incorporate other heteroatoms such as boron, germanium, arsenic or sulfur.

7. EIGHT-MEMBERED DIMETALLACYCLOPHOSPHAZENES

7.1 PREPARATION

An eight-membered ring of type (28) has been prepared by two different routes. $VOCl_3$ reacts with $Me_3SiNPPh_2N(SiMe_3)_2$ in dichloromethane with elimination of Me_3SiCl to give $[ClV(OSiMe_3)N_2PPh_2]_2$ (38) (ref. 77).

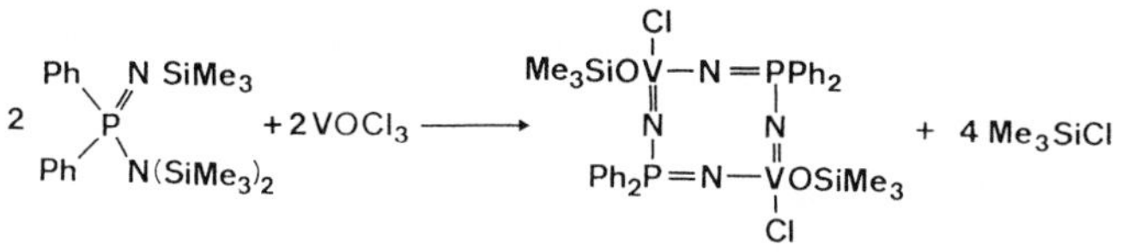

(38)

A derivative of (38) is readily prepared (ref. 76) by reaction of $Me_3SiNVCl_3$ (ref. 79) and $ClPPh_2NSiMe_3$ (ref. 80)

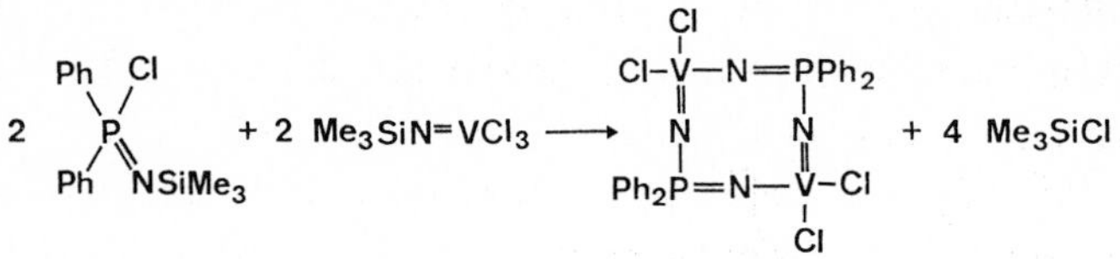

(38a)

(38) and (38a) are dark red solids, soluble in dichloromethane.

7.2 STRUCTURAL INVESTIGATIONS

The X-ray crystal structure of (38) reveals that the molecule possesses an eight-membered planar ring with V-N bond lengths of 167.9(5) and 166.0(5). The short V-N bond lengths indicate multiple bonding between the nitrogen and vanadium atoms.

A tetramer of composition $[(solvent)Cl_3MN]_4$ (39), where M is molybdenum or tungsten, has been characterized (ref. 81). The metal nitrogen bonding in (39) is highly asymmetric with strongly alternating M-N bonds (167 and 217 pm).

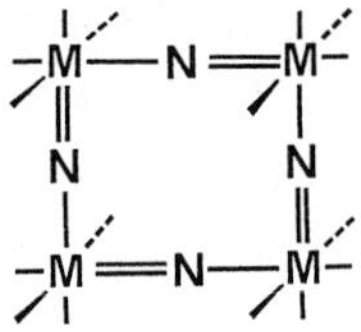

(39)

In these compounds, the pattern of short and long bonds can be attributed to single and triple bonds (ref. 81).

8. ELECTROCHEMICAL INVESTIGATIONS

It is well known that cyclophosphazenes, in contrast to aromatic compounds, are hard to reduce electrochemically. The metallacyclophosphazenes (26a) and (26b) are easily reduced; however, the cyclic voltammograms show no reversible waves. To study the influence of a N=P-unit at a metal center, model compounds of composition $F_5WN{=}PPh_3$ (40) and $F_4W(N{=}PPh_3)_2$ (41) were synthesized (ref. 82). The electrochemical reduction has been investigated using cyclic voltammetry and potentiostatic methods (ref. 83). The cyclic voltammograms of (40) and (41) in CH_2Cl_2 show one-electron reversible waves, and they consist of E_f = -0.45 and -1.45 V (vs. Ag^+-Ag°), respectively. It may be noted that the value of E_f reported by Sharp and co-workers (ref. 84) for the WF_6/WF_6^- couple is 0.51 V. These results indicate that the substitution of a fluorine by a Ph_3P=N-substituent causes a cathodic shift in the E_f value of about 1 V. This considerable cathodic shift for $F_4W(N{=}PPh_3)_2$ compared to $F_5W(N{=}PPh_3)$ and WF_6, respectively, has the following chemical implications:

(i) it is difficult to substitute the third fluorine atom from $F_4W(N{=}PPh_3)_2$ by a Ph_3P=N-substituent and

(ii) $F_4W(N{=}PPh_3)_2$ is extremely stable towards hydrolysis compared to WF_6.

9. CARBON CONTAINING CYCLOMETALLAPHOSPHAZENES

9.1 PREPARATION

For the preparation of carbon-containing cyclometallaphosphazenes the bis-silylated phosphorane $Me_3SiNPPh_2CH_2Ph_2PNSiMe_3$ (42) was used as a starting material. (42) reacts with WF_6, WCl_6 and OsO_4 to yield the heterocycles $NPPh_2CH_2Ph_2PNWF_4$ (43), $NPPh_2CH_2Ph_2NWCl_4$ (44) and $NPPh_2CH_2Ph_2PNOsO_2(OSiMe_3)_2$ (45) (refs. 85,86).

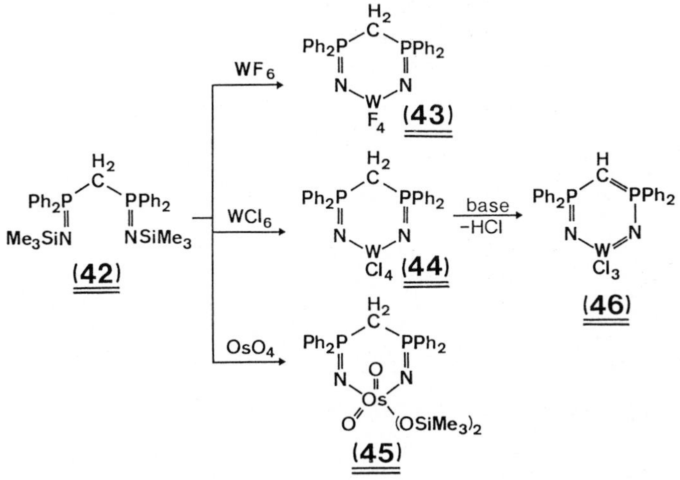

9.2 CHARACTERIZATION

(42) - (44) were characterized by NMR spectroscopy. (43) exhibits two triplets centered at -0.7 and -26.0 in the ^{19}F NMR spectrum. The acyclic compound, $F_4W(N{=}PPh_3)_2$ (41) showed similar ^{19}F NMR spectroscopic patterns as (43) and it was confirmed to be a cis isomer by X-ray crystallography (ref. 82).

The unsaturated ring (44) can be converted into a conjugated heterocyclic compound (46) by intramolecular abstraction of HCl from (44).

The ^{31}P NMR spectrum of (46) consisted of a singlet at 49.2 with ^{183}W satellites ($^2J\ ^{31}P^{183}W$ = 70 Hz). This chemical shift can be compared to that of the corresponding cyclometallaphosphazene, $\overline{NPPh_2NPPh_2NW}Cl_3$ (26b), which resonates at 39.2 (ref. 85).

9.3 HETEROCYCLES CONTAINING METALS IN DIFFERENT OXIDATION STATES

A ferrocenyl and tungsten-containing heterocycle (47) was prepared according to the following reaction steps (ref. 87).

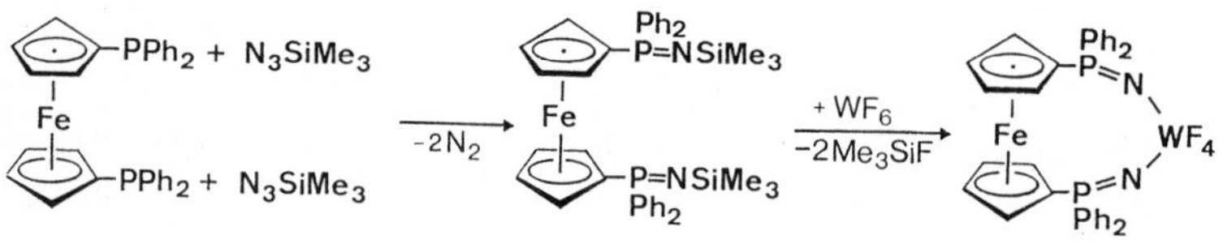

(47)

(47) is a light-yellow non crystalline solid, which was characterized by NMR and mass spectra. In the field desorption mass spectrum the molecular ion was observed at m/z 842 with 100 percent relative intensity.

Compounds of type (47) might have interesting catalytic properties. They contain two metals of very different oxidation states exhibiting both donor and acceptor properties.

10. METAL-CONTAINING SULFUR-NITROGEN RINGS

The chemistry of five-membered metal-containing sulfur-nitrogen compounds dates back to the 50s and has been the subject of several reviews (refs. 88-92).

Five-membered rings of type (48), (49), (50) and (51) have

been isolated and characterized. The known complexes of various metals are summarized:

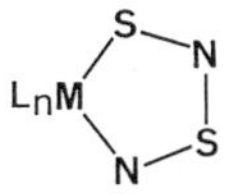

M:Sn,Pb,Co,Ni, Pd,Ir,Pt

(48)

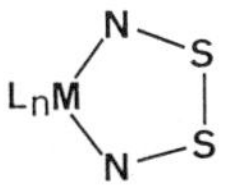

M:Re

(49)

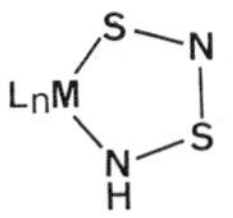

M:Co,Ni,Pd, Pt

(50)

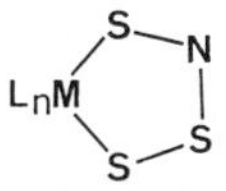

M:Cu,Ag,Au, Co,Ni,Pd,Pt

(51)

Very recent publications of Woollins et al. (ref. 93-95) and Weiss (ref. 96) used $[Me_2SnS_2N_2]_2$(ref. 97,98), S_2N_2, NaS_3N_3 or S_7NH for the preparation of compounds with the composition (48) and (50). Usually they are prepared using S_4N_4 as starting material.

Compounds of type (49) have been characterized by Dehnicke, Müller and their co-workers. Until now, only compounds with Re have been isolated (refs. 114,115). However, the preparation and chemistry is not discussed here in detail, because five-membered rings do not fall within the scope of this review.

In contrast, six-membered cyclometalladithiatriazenes were reported for the first time in 1983 (ref. 99).

10.1 CYCLOMETALLADITHIATRIAZENES, MN_3S_2

10.1.1. Preparation

The cyclometalladithiatriazenes of vanadium, molybdenum and tungsten have been prepared. Various synthetic routes for these compounds are summarized in Scheme 3.

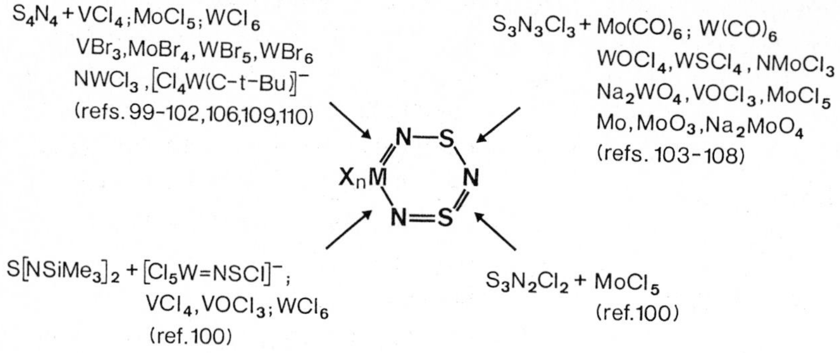

Scheme 3. Reactions for the preparation of $X_nMN_3S_2$ compounds

10.1.2 Reactions

Compounds of molybdenum and tungsten, in particular, have been isolated as adducts with various donor molecules such as CH_3CN, pyridine, 1,2-dipyridyl, HF, S_2N_2 or NSCl. They can also be converted by chloride, bromide or azide to the corresponding anions. The anions are frequently more tractable than their natural precursors and easier to characterize by X-ray structural studies.

Anionic crystalline complexes have been prepared by adding a bulky phosphonium or arsonium salt, such Ph_4PCl, $(PPh_3)_2NCl$, Ph_4AsCl or with sulfur-nitrogen cations such as $[S_5N_5]^+$ or $[S_4N_3]^+$ to yield (52) and (53) (refs. 100,104,105,108,113).

$$\frac{1}{n}(Cl_3MN_3S_2)_n + Ph_4PCl \rightarrow [Ph_4P][Cl_4MN_3S_2] \qquad M = Mo,\ W$$

(52)

$$\frac{2}{n}(Cl_2VN_3S_2)_n + 2Ph_4AsCl \rightarrow [Ph_4As]_2[(Cl_3VN_3S_2)_2]$$

(53)

Silver azide can be used to convert (53) to the azido complex $[Ph_4As]_2[((N_3)_3VN_3S_2)_2]$ (ref. 104).

The reaction of $[Ph_4P][Cl_4MoN_3S_2]$ with $GaCl_3$ affords the uncomplexed $(Cl_3MoN_3S_2)_n$ and $[Ph_4P][GaCl_4]$ (ref. 108).

Partial hydrolysis of $[Ph_4As][Cl_4WN_3S_2]$ yields $[Ph_4As][W(O)Cl_3(HN_3S_2)]$. The structure contains distinct $[Ph_4As]^+$-cations and $[W(O)Cl_3(HN_3S_2)]^-$ anions. The tungsten atom is surrounded by one oxo-ligand, three Cl and two N atoms forming a distorted octahedron (ref. 111).

$[Ph_4As][Cl_4W_2(N_3S_2)_3]\cdot CCl_4$ (54) is obtained by treatment of $[Ph_4As][Cl_4WN_3S_2]$ with H_2S (ref. 112).

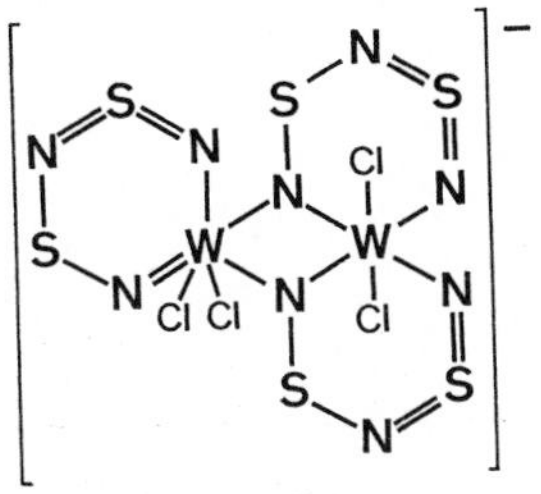

(54)

A paramagnetic molybdenum(V) compound (56) is formed when (55) is reacted with Ph_4PCl. Obviously, S_2N_2 itself functions as a reducing agent (ref. 106).

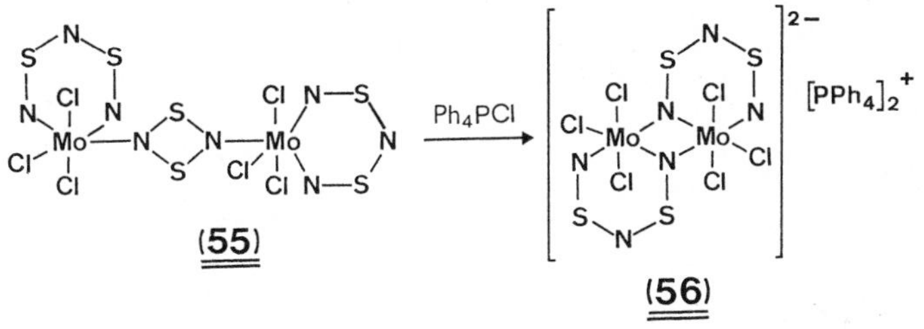

The reaction of $[Ph_4As][Cl_4WC\text{-}t\text{-}Bu]$ with S_4N_4 in dichloromethane leads to $[Ph_4As][Cl_4WN_3S_2]$ and, if traces of water are present, to the salt $[Ph_4As][Cl_3W(O)ON_2S_2]$ (ref. 116). The structure consists of $[Ph_4As]^+$ cations and $[Cl_3W(O)ON_2S_2]^-$ anions (57). The coordination gemometry around the metal may be regarded as distorted octahedral. A weak trans-effect of the NSNSO ligand is indicated by the different axial and equatorial metal-chlorine bond lengths. The metal atom and the NSNSO ligand form a six-membered ring, in which the NSNSO group is planar within experimental error.

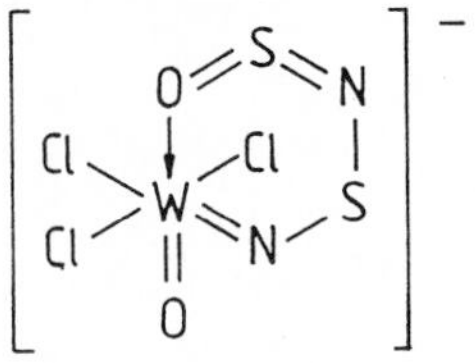

(57)

By reduction of $Cl_2VN_3S_2$ with H_2S, zinc or PPh_3 in dichloromethane/pyridine, $[ClVN_3S_2(Pyr)_2]_2 \cdot 2CH_2Cl_2$(58) is obtained. In this vanadium(IV) compound the metal atoms are part of planar VN_3S_2 units. The two vanadium atoms of the centrosymmetric dimer are linked through one of the nitrogen atoms of each of the VN_3S_2 rings (ref. 117).

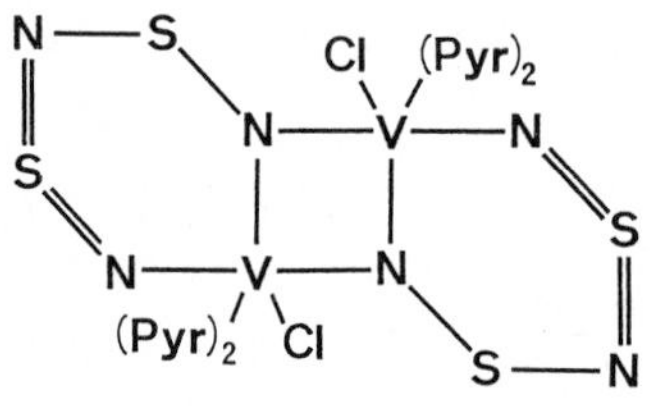

(58)

Treatment of $Cl_2VN_3S_2$ with Me_3SiBr gives $Br_2VN_3S_2$. The structure of $Br_2VN_3S_2$ is similar to that of $Cl_2VN_3S_2$ (ref. 118).

10.1.3 Structural investigations of $L_nMN_3S_2$ compounds

The structures of several $L_nMN_3S_2$ ring systems have been determined by X-ray crystallography. The bond lengths and some bond angles are summarized in Table 6.

The X-ray structural investigations demonstrated that the MN_3S_2 six-membered ring was observed as a monomer (59), as a dimer in (60), (61), (62), as a spirocyclic dimer (63) and as a polymer.

TABLE 6 Selected bond lengths (pm) and bond angles (°) of

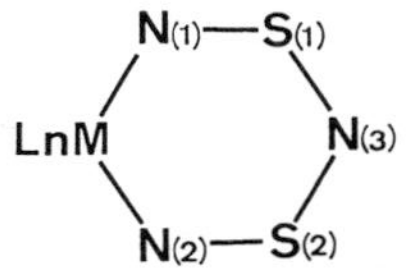

Compound	M-N(1)	M-N(2)	N(1)-S(1)	S(1)-N(3)	N(1)-M-N(2)	M-N(2)-S(2)	Reference
α - $Cl_2VN_3S_2$	184.1(3)	171.4(3)	158.0(3)	157.8(3)	96.3(1)	140.9(2)	99
ß-Cl_2 VN_3S_2	182.8(4)	172.9(4)	158.7(4)	157.5(4)	96.5(2)	139.4(3)	117
$Br_2VN_3S_2$	185(1)	172(1)	157(1)	156(1)	96.1(5)	140.0(7)	118
$[Cl_3VN_3S_2]^-$	179(2)	175(1)	158(2)	158(2)	103.9(10)	130.8(12)	113
$Cl_2VN_3S_2 \cdot NC_5H_5$ [a]	174.8(3)	172.8(3)	157.4(4)	157.7(4)	101.8(2)	132.7(2)	119
$Cl_2VN_3S_2 \cdot NC_5H_5$ [a]	173.6(3)	174	158.5(4)	158.5(4)	100.9(2)	132.6(2)	119
$[(N_3)_3VN_3S_2]^-$	189(1)	172(1)	157(1)	157(1)		132.0(6)	103
$[Cl_4MoN_3S_2]^-$	180.7(11)	186.5(11)	158.1(14)	159.7(13)	92.6(5)		101
$[Cl_3MoN_3S_2]_2^{2-}$	197.1(6)	177.4(7)	156.5(7)	162.5(8)	93.6(3)	142.7(4)	106
$Cl_3WN_3S_2 \cdot THF$	184.3(7)	179.7(8)	158.1(8)	158.9(10)	95.2(3)	138.0(5)	120
$Cl_3WN_3S_2 \cdot CH_3CN$	182.7(14)	182.0(15)	158.5(15)	157.1(17)	94.7(7)		101
$[Cl_4WN_3S_2]^-$	182(1)	188(1)	159(1)	161(2)	93.0(6)	136.1(8)	104
$[(N_3)_2WN_3S_2(O)]_2^{2-}$	222(1)	186(2)	155(2)	157(2)	88.3(6)	143.8(9)	105

a two independent molecules

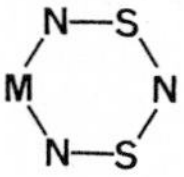

(59)

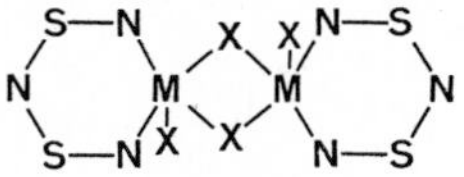

(60)

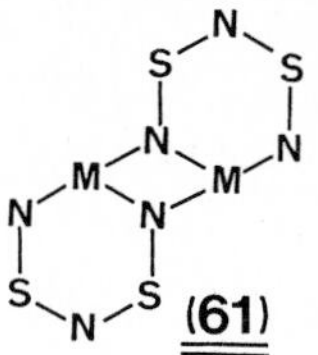

(61)

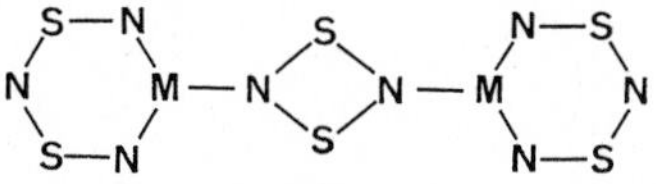

(62)

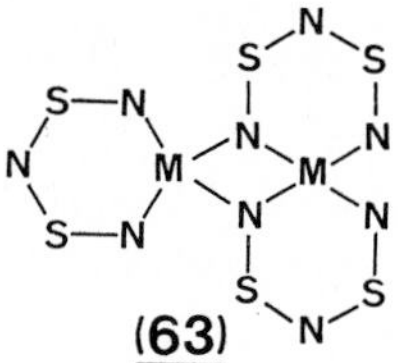

(63)

The polymeric structure has been observed in α- and ß-$Cl_2VN_3S_2$. Both modifications consist of polymeric chains in which the vanadium atoms are linked alternately by chloro and nitrogen bridges. The packing of chains is similar, the only difference being that in ß-$Cl_2VN_3S_2$ the chains are rotated by 11° compared to the orientation in the α-modification. The bridging V-Cl distances (244.5(1) and 246.9(1)) are longer than the terminal V-Cl distance (266.6(1) pm). A distance of 231.2(3) pm was found for the bridging V-N interaction.

The four S-N bond lengths in the six-membered VN_3S_2 ring are approximately equal, and the ring is effectively planar.

In general, the bonding properties of the MN_3S_2 compounds can be described by the resonance structure (64); they indicate multiple bonding between the nitrogen and the metal and approximately equal bond orders for the four S-N bonds.

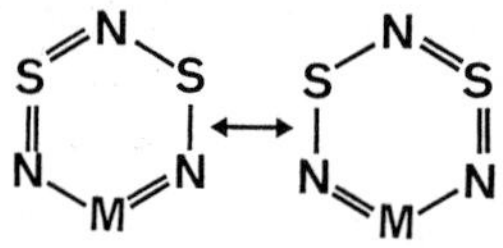

(64) (64)

The endocyclic bond angles at the metal in the MN_3S_2 are in the range 88 - 104°, reflecting a distorted octahedral geometry at the metal center. Consequently, the angles of the metal bonded nitrogens (132 - 144°) are substantially larger than in a regular six-membered ring.

10.1.4 Electrical properties of $L_nMN_3S_2$ compounds

It was observed that $Cl_2VN_3S_2$ in the crystalline state is a semi-conductor at room temperature ($\alpha = 10^{-4}\Omega^{-1}\ cm^{-1}$). However, in the amorphous state, this compound is non-conducting. The N_3S_2 unit of the molecule can be considered as the donor and the vanadium as the acceptor part. In addition the intermolecular interactions lead to stack formations, and, as such, would represent a new type of semi-conductor.

10.2 SIX- AND EIGHT-MEMBERED METAL-SULFUR-NITROGEN RINGS WITH THE N_2S_3-, N_4S_2- AND N_4S_3-LIGANDS

The reaction of $(Cp)_2Ti(CO)_2$ with S_4N_4 produces two titanium-sulfur-nitrogen heterocycles, $(Cp)_2TiN_4S_3$ (65), an eight-membered TiNSNSNSN ring system, and $(Cp)_2TiN_2S_3$ (66), a six-membered ring structure (refs. 120,122).

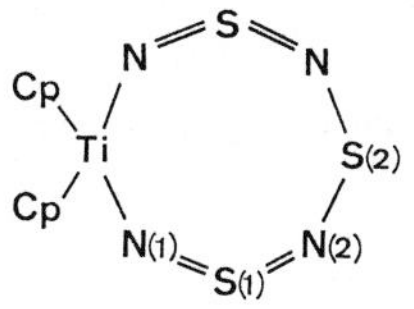

(65)

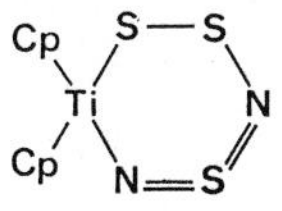

(66)

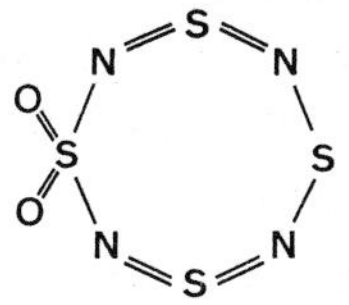

(67)

$(Cp)_2TiN_4S_3$ is structurally related to $S_4N_4O_2$ (67) (refs. 123, 124). The three crystallographically independent molecules in (65) all exhibit the same structure and conformational characteristics. The seven-atom NSNSNSN sequences are planar, with the titanium atoms lying 61.9 - 82.4(1) pm above these planes.

The S-N bond distances vary considerably around the rings. The mean bond lengths of chemically distinct bonds in (65) and (67) are given in Table 7 (ref. 122).

TABLE 7 Mean bond lengths of selected bonds in $(Cp)_2TiN_4S_3$ (65) and $S_4N_4O_2$ (67)

Compound	N(1)-S(1)	S(1)-N(2)	N(2)-S(2)
$(Cp)_2TiN_4S_3$ (65)	148	157.0	159.4
$S_4N_4O_2$ (67)	156.5	155.0	158.4

The variations in the endocyclic bond lengths of (65) are related to the nature of π- interactions between the N_4S_3 fragment and the titanium d orbitals. In contrast, the structure of (67) suggests only limited involvement of sulfur d orbitals with the delocalized π-system of the N_4S_3 chain.

It is surprising that during the reaction of $(Cp)_2Ti(CO)_2$ with S_4N_4 the isomer (68) of $(Cp)_2TiN_2S_3$ is not formed, although compounds such as $(Cp)_2TiS_5$ (69) have been known for quite a long time (ref. 125).

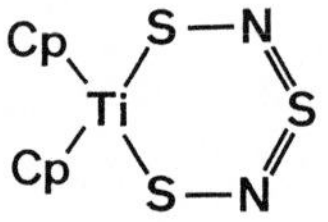

(68)

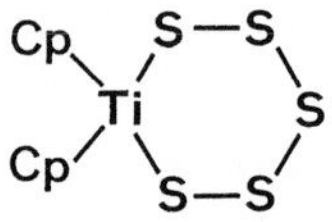

(69)

However, the reaction of S_4N_4 with $PdCl_2$ resulted in the formation of $(S_3N)_2Pd_2S_3N_2$ (70) (refs. 126-128).

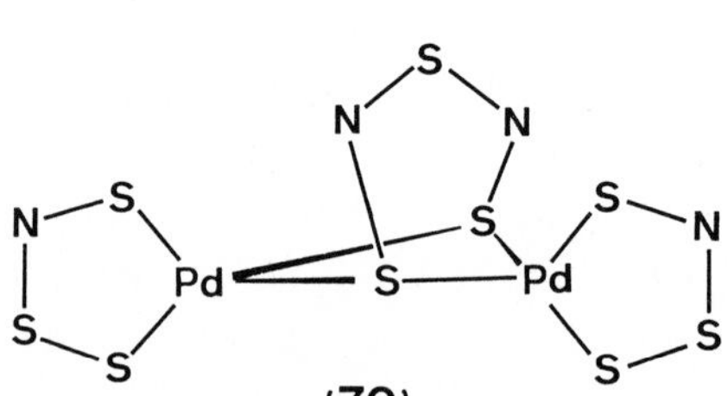

(70)

This molecule consists of two $PdNS_3$ rings bridged by two terminal sulfur atoms of a planar N_2S_3 ring. The small Pd-S-Pd angles (77.1(1)°) and the Pd-Pd distance of 292.1(2) pm (275.1 pm in metallic palladium) indicate a significant Pd-Pd interaction. An eight-membered ring of composition $(H_3N)_2Hg_2(NSN)_2$ (71) has been isolated by the reaction of S_4N_4 with HgI_2.

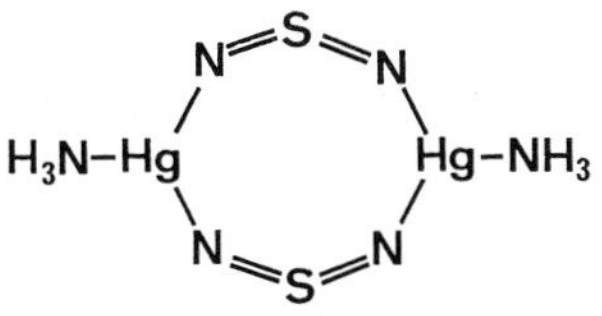

(71)

The X-ray crystal structure revealed that this molecule consists of a planar eight-membered ring, where two Hg atoms are bridged by two NSN-groups. Each Hg atom is bonded to an NH_3 molecule. In addition, each Hg atom is linked to an adjacent ring to give a two-dimensional network (ref. 129).

An insertion of the $[IrCl(CO)PPh_3]$ fragment into a sulfur-nitrogen bond of S_4N_4 is observed (72) when the Vaska complex, $IrCl(CO)(PPh_3)_2$, is treated with S_4N_4 (ref. 130).

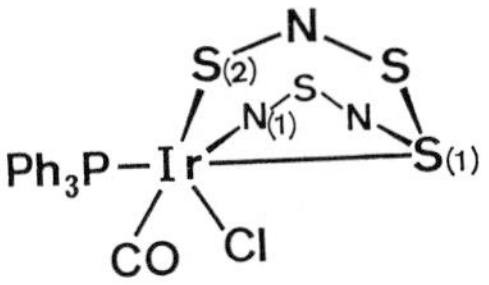

(72)

In (72) the S_4N_4 molecule coordinates through sulfur (S(1), S(2)) and one nitrogen (N(1)) towards the iridium to yield a bicyclic six- and five-membered ring system. A characteristic of (72) is that the two coordinating sulfur atoms have different coordination numbers; this results in different bond distances Ir-S(1) 239.1(3) and Ir-S(2) 233.5(3) pm.

Acknowledgement

Our own work was supported by Deutsche Forschungsgemeinschaft, Fonds der Chemischen Industrie and Volkswagenstiftung. Thanks are due to my colleagues and co-workers whose names are given in the references.

REFERENCES

1 E.O. Fischer and U. Schubert, J. Organomet. Chem., 100 (1975) 59.
2 D.N. Clark and R.R. Schrock, J. Am. Chem. Soc., 100 (1978) 6774.
3 R.R. Schrock, Acc. Chem. Res., 12 (1979) 98.
4 R.R. Schrock, Acc. Chem. Res., 19 (1986) 342.
5 J. Sancho and R.R. Schrock, J. Mol. Catal., 15 (1982) 75.
6 R.R. Schrock, ACS. Symp. Ser., 211 (1983) 369.
7 M.R. Churchill, J.W. Ziller, J.H. Freudenberger and R.R. Schrock, Organometallics, 3 (1984) 1554.
8 J.H. Freudenberger, R.R. Schrock, M.R. Churchill, A.L. Rheingold and J.W. Ziller, Organometallics, 3 (1984) 1563.
9 L.G. McCullough, M.L. Listemann, R.R. Schrock, M.R. Churchill and J.W. Ziller, J. Am. Chem. Soc., 105 (1983) 6729.
10 S.F. Pedersen, R.R. Schrock, M.R. Churchill and J.J. Wasserman, J. Am. Chem. Soc., 104 (1982) 6906.
11 M.R. Churchill and J.W. Ziller, J. Organomet. Chem., 286 (1985) 27.
12 R.M. Tuggle and D.L. Weaver, Inorg. Chem., 11 (1972) 2237.
13 M.R. Churchill, J.W. Ziller, S.F. Pedersen and R.R. Schrock, J. Chem. Soc., Chem. Commun., 1984, 485.
14 M.H. Chisholm and J. A Heppert, Adv. Organomet. Chem., 26 (1986) 97.
15 F. Hug, W. Mowat, A.C. Skapski and G. Wilkinson, J. Chem. Soc., Chem. Commun., 1971, 1477; W. Mowat and G. Wilkinson, J. Chem. Soc., Dalton Trans., 1973, 1120.
16 M.H. Chisholm, F.A. Cotton, M.W. Extine and C.A. Murillo, Inorg. Chem., 17 (1978) 696.
17 M. Bochmann, G. Wilkinson, A.M.R. Galas, M.B. Hursthouse and K.M.A. Malik, J. Chem. Soc., Dalton Trans., 1980, 1797.
18 H. Schmidbaur, W. Scharf and J.J. Füller, Z. Naturforsch., B 32 (1977) 858.
19 M.H. Chisholm, F.A. Cotton, M. Extine and B.R. Stults, Inorg. Chem., 15 (1976) 2252.
20 D.N. Clark and R.R. Schrock, J. Am. Chem. Soc., 100 (1978) 6774.
21 R.A. Andersen, A.L. Galyer and G. Wilkinson, Angew. Chem., 88 (1976) 692; Angew. Chem. Int. Ed. Engl., 15 (1976) 609.
22 M.H. Chisholm, J.C. Huffman and J.A. Heppert, J. Am. Chem. Soc., 107 (1985) 5116.
23 F.A. Cotton, W. Schwotzer and E.S. Shamshoum, Organometallics, 2 (1983) 1167.
24 M.H. Chisholm, J.A. Heppert, J. C. Huffman and P. Thornton, J. Chem. Soc., Chem. Commun., 1985, 1466.
25 M.H. Chisholm, D.L. Clark, K. Folting and J.C. Huffman, Angew. Chem., 98 (1986) 1021; Angew. Dhem. Int. Ed. Engl., 25 (1986) 1014. M.H. Chisholm, C.E. Hammond, M. Hampden-Smith,

J.C. Huffman and W.G. Van Der Sluys, Angew. Chem., 99 (1984) 937. Angew. Chem. Int. Ed. Engl., 26 (1987) 904.
26 D.L. Thorn and R. Hoffmann, Nouv. J. Chem., 3 (1979) 39.
27 S.D. Chappell and D.J. Cole-Hamilton, Polyhedron, 1 (1982) 739.
28 J.T. Mague, Inorg. Chem., 12 (1973) 2649.
29 E. Müller, E. Langer, H. Jäkle, H. Muhm, W. Hoppe, R. Graziani, A. Gieren and F. Brandl, Z. Naturforsch., B 26 (1971) 305.
30 R.G. Gastinger, M.D. Rausch, D.A. Sullivan and G.J. Palenik, J. Organomet. Chem., 117 (1976) 355.
31 R.G. Gastinger, M.D. Rausch, D.A. Sullivan and G.J. Palenik, J. Am. Chem. Soc., 98 (1976) 719.
32 J.L. Atwood, W.E. Hunter, H. Alt and M.D. Rausch, J. Am. Chem. Soc., 98 (1976) 2454.
33 A.J. Ashe III, Topics Current Chem., 105 (1982) 125. G. Märkl and P. Hofmeister, Angew. Chem., 91 (1971) 863; Angew. Chem. Int. Ed. Engl., 18 (1979) 789.
34 G. Raabe and J. Michl, Chem. Rev., 85 (1985) 419.
35 G.P. Elliot, W.R. Roper and J.M. Waters, J. Chem. Soc., Chem. Comm., 1982, 811.
36 H. Rose, Ann. Chem., 11 (1834) 131.
37 J. Liebig, Ann. Chem., 11 (1834) 139.
38 H.N. Stokes, Am. Chem. J., 19 (1897) 782.
39 R. Schenk and G. Römer, Ber., 57 (1924) 1343.
40 H.R. Allcock, Phosphorus Nitrogen Compounds, Academic Press, New York, 1972.
41 H.R. Allcock, Polymer, 21 (1980) 673.
42 H.R. Allcock, Acc. Chem. Res., 12 (1979) 351.
43 H.R. Allcock, J.L. Desorcie and G.H. Riding, Polyhedron, 6 (1987) 119.
44 S.S. Krishnamurthy, A.C. Sau and M. Woods, Adv. Inorg. Chem., Radiochem., 21 (1978) 41.
45 N.L. Paddock, Q. Rev. Chem. Soc., (London) 18 (1964) 168.
46 M.J.S. Dewar, E.A.C. Lucken and M.A. Whitehead, J. Chem. Soc., 1960, 2423.
47 R.C. Haddon, Chem. Phys. Lett., 120 (1985) 372.
48 R.A. Wheeler, R. Hoffmann and J. Strähle, J. Am. Chem. Soc., 108 (1986) 5381.
49 M.F. Lappert and G. Srivastava, J. Chem. Soc., A 1966, 210.
50 J. Trotter and S.A. Whitlow, J. Chem. Soc., A 1970, 455.
51 H.R. Allcock, R.W. Allen and J.P. O'Brien, J. Am. Chem. Soc., 99 (1977) 3984.
52 R.W. Allen, J.P. O'Brien and H.R. Allcock, J. Am. Chem. Soc., 99 (1977) 3987.
53 J.P. O'Brien, R.W. Allen and H.R. Allcock, Inorg. Chem., 18 (1979) 2230.
54 W.C. Marsh, N.L. Paddock, C.J. Stewart and J. Trotter, J. Chem. Soc., Chem. Commun., 1970, 1190.
55 W. Harrison and J. Trotter, J. Chem. Soc., Dalton Trans., 1973, 61.
56 H.P. Calhoun, N.L. Paddock, J. Trotter and J.N. Wingfield, J. Chem. Soc., Chem. Commun., 1972, 875.
57 N.L. Paddock, T.N. Ranganathan and J.N. Wingfield, J. Chem. Soc., Dalton Trans., 1973, 1578.
58 H.P. Calhoun, N.L. Paddock and J. Trotter, J. Chem. Soc., Dalton Trans., 1973, 2708.
59 F.A. Cotton, G.A. Rusholme and A. Shaver, J. Coord. Chem., 3 (1973) 99.

60 P.P. Greigger and H.R. Allcock, J. Am. Chem. Soc., 101 (1979) 2492.
61 H.R. Allcock, L.J. Wagner and M.L. Levin, J. Am. Chem. Soc., 105 (1983) 1321.
62 H.R. Allcock, P.P. Greigger, L.J. Wagner and M.Y. Bernheim, Inorg. Chem.,20 (1981) 716.
63 P.R. Suszko, R.R. Whittle and H.R. Allcock, J. Chem. Soc., Chem. Comm., 1982, 960.
64 H.R. Allcock, K.D. Lavin, G.H. Riding. P.R. Suszco and R.R. Whittle, J. Am. Chem. Soc., 106 (1984) 2337.
65 H.R. Allcok, K.D. Lavin and G.H. Riding, Macromolecules, 18 (1985) 1340.
66 H.R. Allcock, J.J. Fuller and T.L. Evans, Macromolecules, 13 (1980) 1325.
67 H.R. Allcock, P.R. Suszco, L.J. Wagner, R.R. Whittle and B. Boso, J. Am. Chem. Soc., 106 (1984) 4966.
68 H.R. Allcock, P.R. Suszco, L.J. Wagner, R.R. Whittle and B. Boso, Organometallics, 4 (1985) 446.
69 H.W. Roesky, K.V. Katti, U. Seseke, M. Witt, E. Egert, R. Herbst and G.M. Sheldrick, Angew. Chem., 88 (1986) 447; Angew. Chem. Int. Ed. Engl., 25 (1986) 477.
70 H.W. Roesky, K.V. Katti, U.Seseke, H.-G. Schmidt, E. Egert, R. Herbst and G.M. Sheldrick, J. Chem. Soc., Dalton Trans. 1987, 847.
71 K.V. Katti, H.W. Roesky and M. Rietzel, Inorg. Chem., 26 (1987) 4032.
72 H.P. Latscha, Z. Anorg. Allg. Chem., 362 (1968) 7.
73 N.V. Mani, F.R. Ahmed and W.H. Barnes, Acta Crystallogr., 21 (1966) 375.
74 R. Herbst, K.V. Katti, H.W. Roesky and G.M. Sheldrick, Z. Naturforsch., B 42 (1987) 1387.
75 H.W. Roesky, K.V. Katti and B. Krebs, unpublished results.
76 H.W. Roesky, M. Rietzel, M. Noltemeyer and G.M. Sheldrick, unpublished results.
77 M. Witt, H.W. Roesky, M. Noltemeyer and G.M. Sheldrick, Angew. Chem., in press.
78 M. Witt and H.W. Roesky, unpublished results.
79 E. Schweda, K.D. Scherfise and K. Dehnicke, Z. Anorg. Allg. Chem., 528 (1985) 117.
80 R.R. Ford, M.A. Goodman, R.H. Neilson, A.K. Roy, U.G. Wettermark and P. Wisian-Neilson, Inorg. Chem., 23 (1984) 2063.
81 K. Dehnicke and J. Strähle, Angew. Chem, 93 (1981) 451; Angew. Chem. Int. Ed. Engl., 20 (1981) 413.
82 H.W. Roesky, U. Seseke, M. Noltemeyer, P.G. Jones and G.M. Sheldrick, J. Chem. Soc., Dalton Trans., 1986, 1309.
83 H.W. Roesky, T. Tojo, M. Ilemann and D. Westhoff, Z. Naturforsch., B 42 (1987) 877.
84 G.A. Anderson, J. Igbal, D.W.A. Sharp, J.M. Winfield, J.H. Cameron and A.G. McLeod, J. Fluorine Chem., 24 (1984) 303.
85 K.V. Katti, U. Seseke and H.W. Roesky, Inorg. Chem., 26 (1987) 814.
86 K.V. Katti, H.W. Roesky and M. Rietzel, Z. Anorg. Allg. Chem., 553 (1987) 123.
87 H.W. Roesky and K. Swarat, unpublished results.
88 J. Weiss, Fortschr. Chem. Forsch., 5 (1965) 635.
89 Gmelin Handbook, Sulfur-Nitrogen Compounds Part 2, 1985.
90 H.W. Roesky, Chem. Soc. Rev., 15 (1986) 309.
91 T. Chivers and F. Edelmann, Polyhedron, 5 (1986) 1661.
92 P.F. Kelly and J.D. Woollins, Polyhedron, 5 (1986) 607.
93 R. Jones, C.P. Warrens, D.J. Williams and J.D. Woollins, J. Chem. Soc., Dalton Trans., 1987, 907.

94 J.D. Woollins, Polyhedron, 6 (1987) 939.
95 P.A. Bates, M.B. Hursthouse, P.F. Kelly and J.D. Woollins, J. Chem. Soc., Dalton Trans., 1986, 2367.
96 J. Weiss, Z. Anorg. Allg. Chem., 542 (1986) 137.
97 H.W. Roesky, and H. Wiezer, Angew. Chem., 85 (1973) 722; Angew. Chem. Int. Ed. Engl., 12 (1973) 674.
98 H.W. Roesky, Z. Naturforsch., B 31 (1976) 680.
99 H.W. Roesky, J. Anhaus, H.-G. Schmidt, G.M. Sheldrick and M. Noltemeyer, J. Chem. Soc., Dalton Trans., 1983, 1207.
100 J. Anhaus, Z.A. Siddiqi, J. Schimkowiak, H.W. Roesky and H. Lueken, Z. Naturforsch., B 39 (1984) 1722.
101 J. Anhaus, P.G. Jones, M. Noltemeyer, W. Pinkert, H.W. Roesky and G.M. Sheldrick, Inorg. Chim. Acta, 97 (1985) L 7.
102 A. Berg, E. Conradi, U. Müller and K. Dehnicke, Chemiker-Ztg., 108 (1984) 292.
103 J. Hanich, M. Krestel, U. Müller, K. Dehnicke and D. Rehder, Z. Naturforsch., B 39 (1984) 1686.
104 U. Kynast, E. Conradi, U. Müller and K. Dehnicke, Z. Naturforsch., B 39 (1984) 1680.
105 H. Wadle, E. Conradi, U. Müller and K. Dehnicke, Z. Naturforsch., B 40 (1985) 1626.
106 A. Berg, E. Conradi, U. Müller and K. Dehnicke, Z. Anorg. Allg. Chem., 529 (1985) 74.
107 K. Völp, W. Willing, U. Müller and K. Dehnicke, Z. Naturforsch., B 41 (1986) 1196.
108 H. Wadle, E. Conradi, U. Müller and K. Dehnicke; Z. Naturforsch., B 41 (1986) 429.
109 H.W. Roesky, J. Schimkowiak and F. Walther, Z. Naturforsch., B 41 (1986) 393.
110 H.W. Roesky, J. Schimkowiak, M. Noltemeyer and G.M. Sheldrick, Z. Naturforsch., B 41 (1986) 175.
111 E. Conradi, H. Wadle, U. Müller and K. Dehnicke, Z. Naturforsch., B 41 (1986) 46.
112 P. Klingelhöfer, H. Wadle, U. Müller and K. Dehnicke, Z. Anorg. Allg. Chem., 544 (1987) 115.
113 A. El-Kholi, R. Christophersen, U. Müller and K. Dehnicke, Z. Naturforsch., B 42 (1987) 410.
114 W. Hiller, J. Mohy la, J. Strähle, H.G. Hauck and K. Dehnicke, Z. Anorg. Allg. Chem., 514 (1984) 72.
115 E. Conradi, H.G. Hauck, U. Müller and K. Dehnicke, Z. Anorg. Allg. Chem., 539 (1986) 39.
116 J. Anhaus, Z.A. Siddiqi, H.W. Roesky and J.W. Bats, J. Chem. Soc., Dalton Trans., 1985, 2453.
117 R. Christophersen, W. Willing, U. Müller and K. Dehnicke, Z. Naturforsch., B 41 (1986) 1420.
118 J. Hanich, W. Willing, U. Müller, K. Dehnicke and D. Rehder, Z. Naturforsch., B 40 (1985) 1457.
119 R. Christophersen, P. Klingelhöfer, U. Müller, K. Dehnicke and D. Rehder, Z. Naturforsch., B 40 (1985) 1631.
120 A. Khabou, W. Willing, U. Müller and K.Dehnicke, Z. Naturforsch., B 42 (1987) 943.
121 C.G. Marcellus, R.T. Oakley, A.W. Cordes and W.T. Pennington, J. Chem. Soc., Chem. Commun., 1983, 1451.
122 C.G. Marcellus, R.T. Oakley, W.T. Pennington and A.W. Cordes, Organometallics, 5, 1986, 1395.
123 H.W. Roesky , W. Schaper, O. Petersen and T. Müller, Chem. Ber., 110 (1977) 2695.
124 P.G. Jones, W. Pinkert and G.M. Sheldrick, Acta Crystallogr., Sect. C: Cryst. Struct. Commun., C 39 (1983) 827.

125 M. Schmidt and W. Wilhelm, J. Chem. Soc., Chem. Commun., 1970, 111.
126 U. Thewalt, Z. Naturforsch., B 37 (1982) 276.
127 J. Weiss and U. Thewalt, Z. Anorg. Allg. Chem., 346 (1966) 234.
128 J.D. Woollins, R. Grinter, M.K. Johnson and A.J. Thomson, J. Chem. Soc., Dalton Trans., 1980, 1911.
129 H. Martan and J. Weiss, Z. Anorg. Allg. Chem., 515 (1984) 225.
130 F. Edelmann, H.W. Roesky, C. Spang, M. Noltemeyer and G.M. Sheldrick, Angew. Chem., 98 (1986) 908; Angew. Chem. Int. Ed. Engl., 25 (1986) 931.

ORGANOMETALLIC π SYSTEMS

G. HUTTNER and H. LANG
Anorganisch-Chemisches Institut der Universität Heidelberg
Im Neuenheimer Feld 270, 6900 Heidelberg, F.R.G.

Dedicated to Professor E.O. Fischer on the occasion of his 70th birthday

1 INTRODUCTION

It has been known for quite a while that organometallic fragments ML_n may act as building blocks of π systems.

Alkene like carbene compounds $L_nM{=}CRR'$ and alkyne like carbyne species $L_nM{\equiv}CR$, both pioneered by E.O. Fischer and his group (refs. 1-2), are early examples for this statement. Both series of compounds may, at the same time, serve to illustrate the importance of such π systems for organic (refs. 3-4) and organometallic (ref. 5) synthesis and even for technical chemistry (refs. 6-7).

While multiple bond interactions between organometallic fragments and the lighter main group elements such as carbon are well established, π systems built from organometallic fragments and the heavier main group elements have not been so well documented.

2 OPEN CHAIN π SYSTEMS

2.1 THREE-CENTER 4π SYSTEMS

The allylic anion $CH_2CHCH_2^-$ is the basic model for any three-center 4π type system. Organometallic analogues of this system should have two 16-electron L_nM fragments - isolobal to

CH_2 - at the terminal positions and a main group sextet species - isoelectronic to CH^- - at the center. Compounds $[L_nM]_2XR$ conforming to this scheme are known as phosphinidene- (X = P), arsinidene- (X = As), stibinidene- (X = Sb) and bismuthinidene- (X = Bi) species. The bonding in these $[L_nM]_2XR$ entities may be rationalized in terms of σ and π interactions: The σ contribution may be attributed to the interaction of the two lone pairs at the sextet center $-\overline{\underline{X}}$ with the σ acceptor orbitals at the coordinatively unsaturated metal centers L_nM.

Unlike the situation in common $R_3X|$ species, there is no lone pair left at the main goup center X. Therefore the coordination around this center will be trigonally planar. A π system will be built up by the interaction of metal atom d-type donor orbitals with the vacant valence orbital of the trigonally planar coordinated X. Since there are two filled orbitals at the metals to interact with the empty valence p-orbital of the trigonally planar coordinated main group center, the three-center 4π system, as shown in Fig. 1, will appropriately describe the situation.

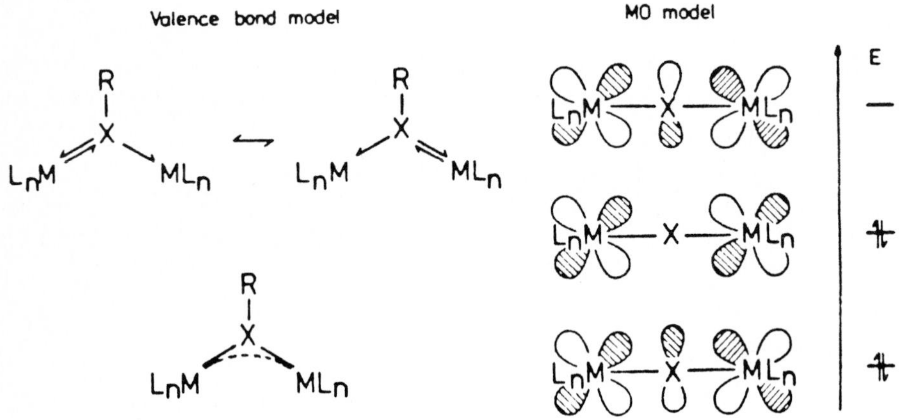

Fig. 1. Bonding in $[L_nM]_2XR$ compounds (X = P ... Bi; L_nM = 16-electron fragment).

Since two reviews (refs. 8-9) deal with the chemistry of such species, only the properties most specific of these compounds will be discussed in this article as far as 5th main group bridging RX entities are concerned. More detailed information will be presented for analogous compounds with RY^+ groups (Y = S, Se, Te) in the same bonding situation.

2.1.1 Phosphinidene complexes and their higher homologues

The specific properties of "inidene" type compounds $[L_nM]_2XR$ (L_nM = 16-electron fragment, X = 5th main group atom) are intimately connected to the presence of the π system.

2.1.1.1 Geometry

Consistent with the bonding scheme lined out in Fig. 1 binuclear phosphinidene complexes $[L_nM]_2XR$ (X = P) show a trigonal planar environment around the phosphorus atom and a short M-P bond length, as a consequence of the phosphorus-metal π interaction (Fig. 1). Equivalent statements apply to arsinidene (X = As) as well as stibinidene (X = Sb) compounds (Fig. 2).

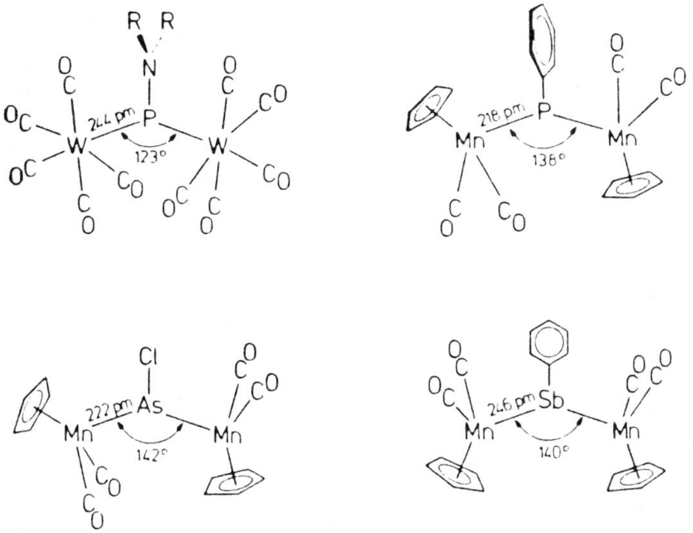

Fig. 2. Structures of $[L_nM]_2XR$ compounds.

2.1.1.2 Spectroscopy

The literally eyecatching property of "inidene" type compounds $[L_nM]_2XR$ (X = P, As, Sb, Bi) is their bright and brilliant color. The visible spectra of these compounds are characterized by an intense long-wavelength absorption which gives rise to bright colors of their solutions and, because of a high specific reflectivity, to a metallic lustre of the crystalline solids.

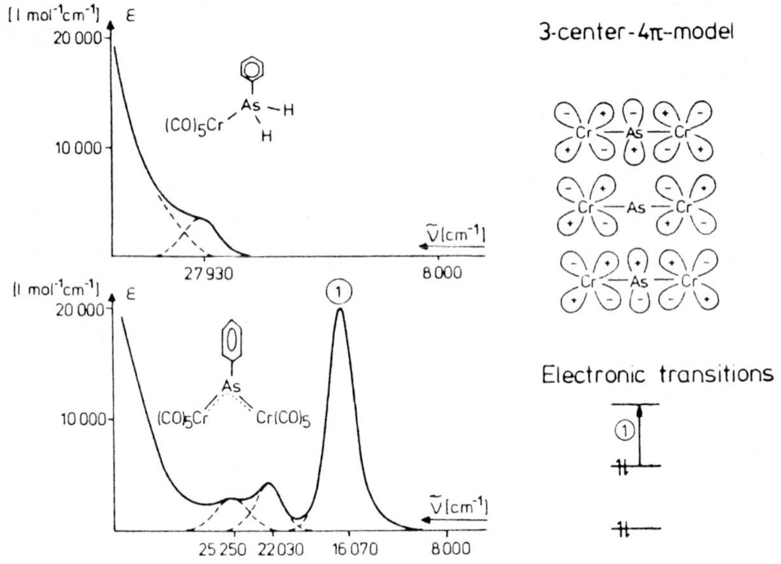

Fig. 3. UV-Vis spectrum and band assignment of $[(CO)_5Cr]_2AsPh$ (the spectrum of $(CO)_5CrAsPhH_2$ is given for comparison).

The spectrum of $[(CO)_5Cr]_2AsPh$ may serve as an example (Fig. 3). The top half of Fig. 3 shows the spectrum of $(CO)_5CrAsPhH_2$ which is typical for a $(CO)_5CrL$ compound where L is spectroscopically innocent, i.e. not a chromophoric system in itself. The characteristic feature of such a compound $(CO)_5CrL$ is the complete absence of any absorption in the long-wavelength visible part. "Inidene" type compounds $[L_nM]_2XR$, in contrast, are characterized by a prominent absorption in this very long-wavelength region (Fig. 3). The interpretation of this band as being due to an electronic transition from the nonbonding metal centered π orbital to the antibonding π^* orbital of the three-center 4π system is within the framework of such simple single electron models (refs. 8-9) beyond doubt.

The long-wavelength absorptions, characteristic for "inidene" type compounds, show that the π interactions in these species are not very strong: There is a relatively small HOMO-LUMO gap! This has some bearing to the explanation of the fact that the ^{31}P-NMR-spectra of phosphinidene complexes show by far the highest paramagnetic ^{31}P-NMR shifts ever observed (ref. 10): The paramagnetic contribution to the NMR shift is dominated by the admixture of low lying excited states to the ground state electronic configuration, where the magnitude σ_{para} of the deshielding effect is proportional to the inverse of the energy difference ΔE between the ground state and the excited state configuration, i.e. $\sigma_{para} \propto 1/\Delta E$ (ref. 11). It is therefore tempting to assume that it is the low lying π^* LUMO of the three-center 4π system (Fig. 3) which gives rise to the observed extraordinary deshielding.

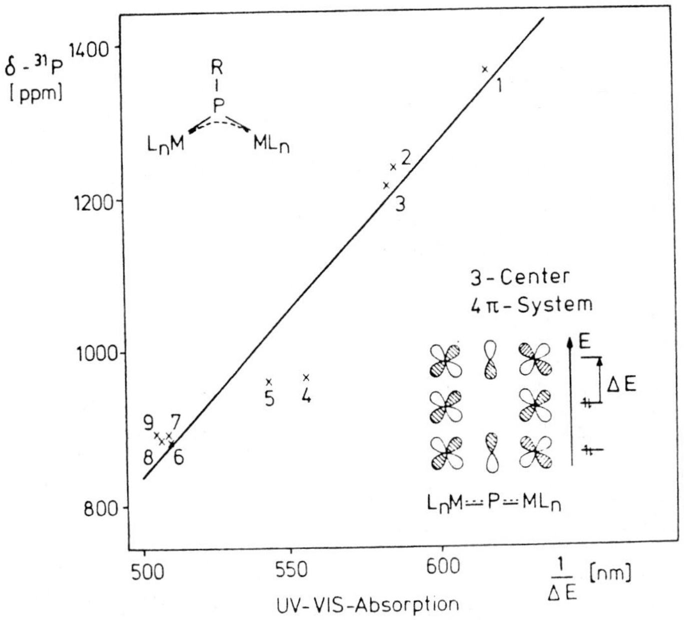

Fig. 4. Correlation between $\delta(^{31}P)$ and λ_{max}-Vis for phosphinidene complexes (numbers correspond to Table 1).

TABLE 1

^{31}P-NMR shifts and $\pi\pi$*-absorptions of $[L_nM]_2PR$ (numbers refer to Fig. 4).

Nr.	RP	L_nM	$\delta(^{31}P)$ (ppm)	$\bar{\nu}(\pi\pi^*)^c$ (cm^{-1})
1	^tBuP	$(CO)_5Cr$	1362	16200
2	PipP [a]	$(CO)_5Cr$	1239	17060
3	MesP [b]	$(CO)_5Cr$	1216	17120
4	MesP [b]	$(CO)_5W$	961	17950
5	PipP [a]	$(CO)_5W$	959	18380
6	PhP	$Cp(CO)_2Mn$	884	19650
7	$4-BrC_6H_4P$	$Cp(CO)_2Mn$	887	19690
8	PhP	$Cp'(CO)_2Mn$	887	19760
9	$4-CH_3OC_6H_4P$	$Cp(CO)_2Mn$	896	19800

[a] Pip = 2,2',6,6'-Tetramethylpiperidyl. [b] Mes = 2,4,6-Trimethylphenyl. [c] Solvens toluene

According to this simple model a correlation between the HOMO-LUMO energy difference of the π system and the ^{31}P-NMR shifts can be expected. Taking the absorption wavelength of the long-wavelength band in these compounds (c.f., Fig. 3) as a rough measure for this energy difference, a plot of wavelength (proportional to $1/\Delta E$) versus $\delta(^{31}P)$ should be monotonic as it is in fact observed for the range of compounds so far studied (Fig. 4, Table 1, ref. 10).

2.1.1.3 Synthesis

There are many routes to synthesize "inidene" compounds. A collection of possible syntheses is shown in Fig. 5; explanations are given in great detail in a recent review on this subject (ref. 9).

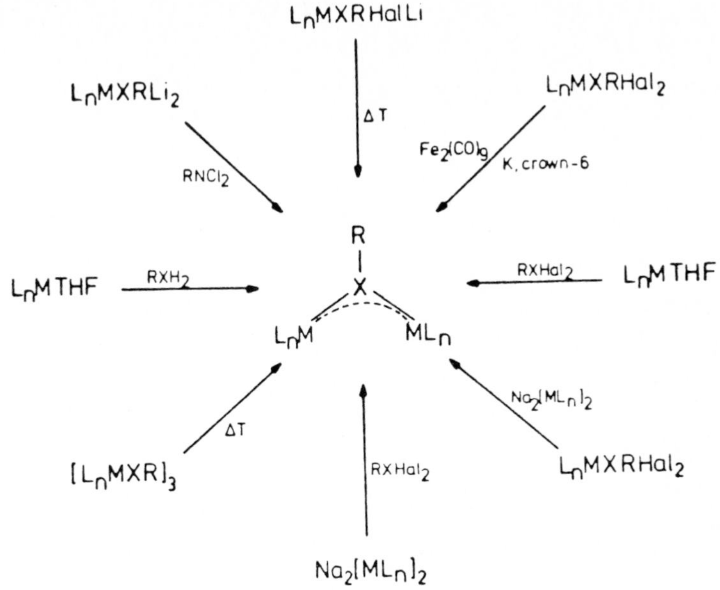

Fig. 5. Collection of synthetic routes to "inidene" complexes.

From all the routes so far reported the ones of the "rocksalt-synthesis-type" appear to be the most efficient ones. Synthesis from $Na_2[ML_n]_2$ and $RXHal_2$ (bottom Fig. 5) rely upon the availability of the organometallic dianionic species $[ML_n]_2^{2-}$ or related anionic compounds which means that this synthetic strategy has been confined to ML_n = $Fe(CO)_4$, $Co(CO)_4$, CpCo(CO) and $M(CO)_5$ (M = Cr, Mo, W) (ref. 9) and has not been applicable to the synthesis of the broad range of known $[Cp(CO)_2Mn]_2XR$ compounds. As shown in Fig. 5 the later may be obtained by different strategies. With the recent synthesis of $[(Cp(CO)_2Mn)_2H]^-$ (ref. 12) an equivalent of $[ML_n]_2^{2-}$ is now also available for the synthesis of $[Cp(CO)_2Mn]_2XR$ compounds:

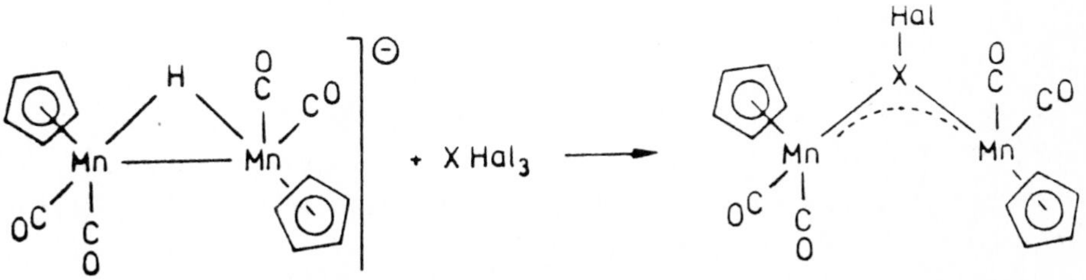

The reactions are simple to carry out and generally give quite good yields (ref. 12).

2.1.1.4 <u>Reactivity</u>

Some of the reaction chemistry of "inidene" compounds is schematically shown in Fig. 6 for the series of chloro-arsinidene species (refs. 8-9). Much of the reactivity of "inidene" compounds is determined by the relatively weak three-center 4π system (Fig. 1) which gives rise to a low lying LUMO with a significant contribution of the main group center X.

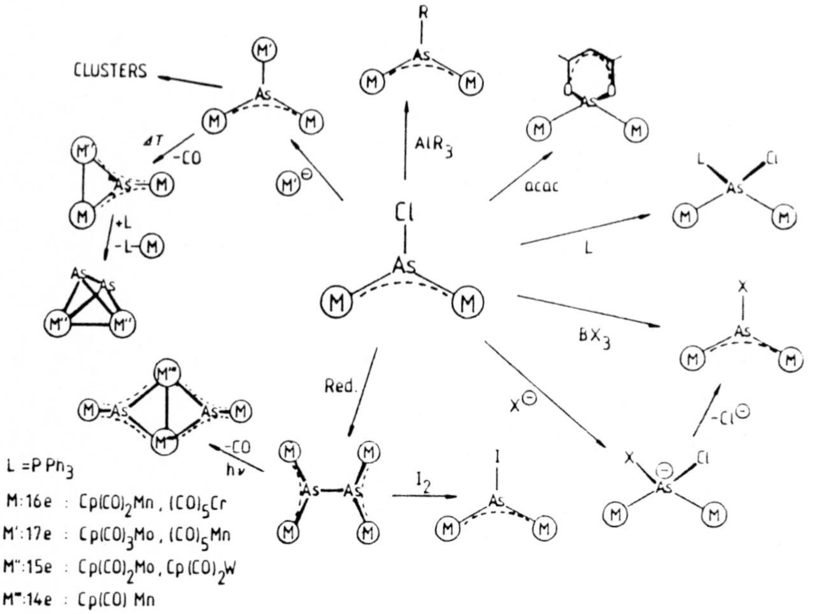

Fig. 6. Reactivity of chloroarsinidene compounds.

This is equivalent to saying that the main group center X is "coordinatively unsaturated" in a trigonal planar environment and will hence be prone to add Lewis bases |B, which results in a tetrahedral type coordination of X for the base adducts formed (Fig. 7).

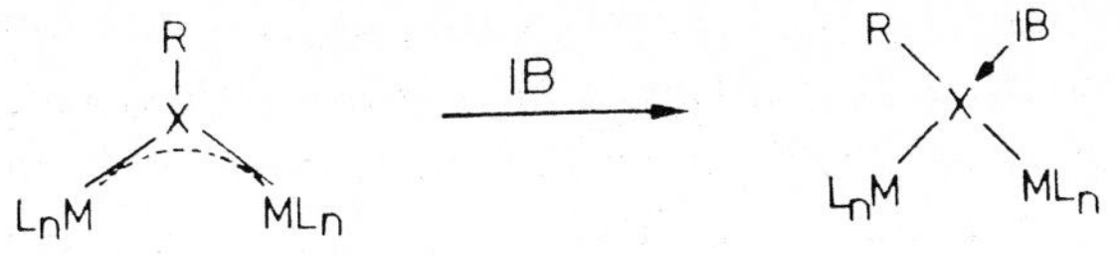

Fig. 7. Lewis base addition to "inidene" type compounds.

Anionic bases give anionic adducts which, in those cases, where the arsenic bonded residue is a nucleophugic group, like Cl^-, may eliminate this group to yield substitution products (bottom right Fig. 6). Substitution of X bonded halides may also be obtained by electrophilic agents such as $BHal_3$ or AlR_3 (right middle and top Fig. 6). Substitution and addition at the same time is observed with a wide variety of monovalent chelating ligands like acetylacetonate shown in the top right corner of Fig. 6.

Organometallic nucleophiles give organometallic substitution products which contain a main group center in a trigonal planar μ_3-bridging position (top left Fig. 6). Using this approach, a wide variety of such compounds with "naked" 5th main group elements in a trigonal planar periphery has been synthesized (Fig. 8, refs. 8-9, 13).

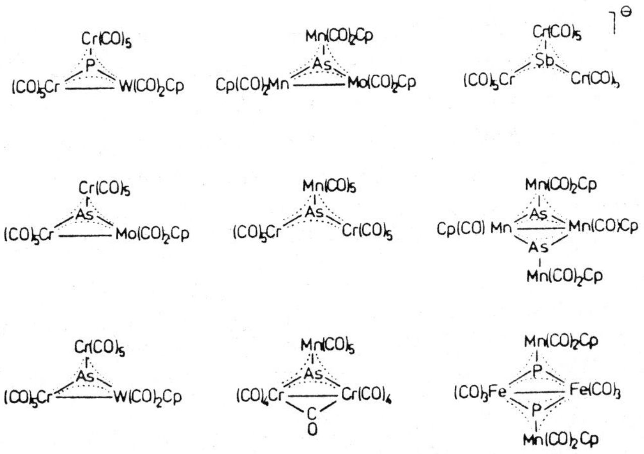

Fig. 8. Structurally characterized μ_3-X systems with a 5th main group element in a trigonally planar environment.

Reductive coupling of $[L_nM]_2AsCl$ to $[L_nM]_4As_2$ (bottom line Fig. 6) is one possible way from "inidene" chemistry into the chemistry of coordinated X_2 species. Side-on coordinated 4-electron donor X_2 ligands (middle left Fig. 6) are as well accessible.

The connection between the chemistry of "inidene" compounds and the one of triply side on coordinated 6-electron donor X_2 ligands is shown in Fig. 9: $[(CO)_5W]_2XCl$ (X = As, Sb, Bi) must be precursors of $[(CO)_5W]_3X_2$, since, in the presence of a Lewis base |B, $[(CO)_5W]_2X(Cl)(B)$ is obtained and the formation of $[(CO)_5W]_3X_2$ is completely suppressed by |B.

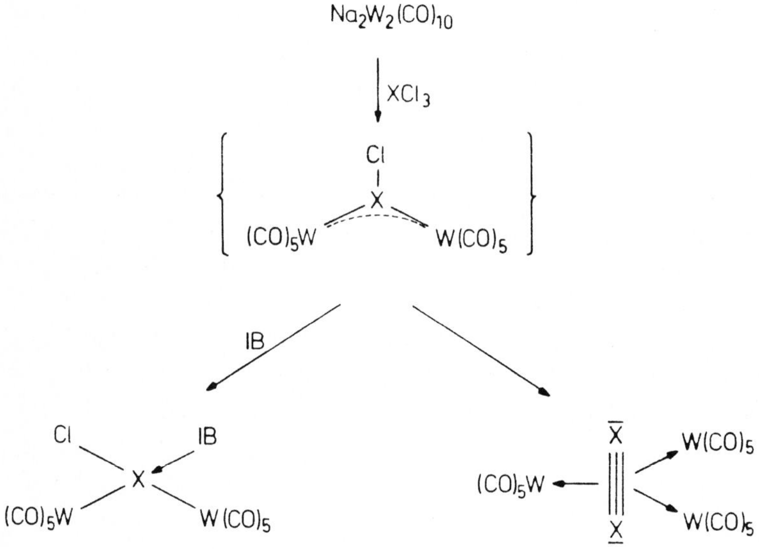

Fig. 9. $[(CO)_5W]_2XCl$ as a precursor to $[(CO)_5W]_3X_2$ (X = As, Sb).

The different interrelations between the chemistry of $[L_nM]_2XR$ species and the one of X_2 and RXXR derivatives have recently been reviewed (refs. 8-9).

While in general "inidene" compounds behave as Lewis acids which add Lewis bases to yield stable adducts (see above, Fig. 6, 7) there are some reactions which indicate that "inidene" compounds undergo valence tautomerism (Fig. 10).

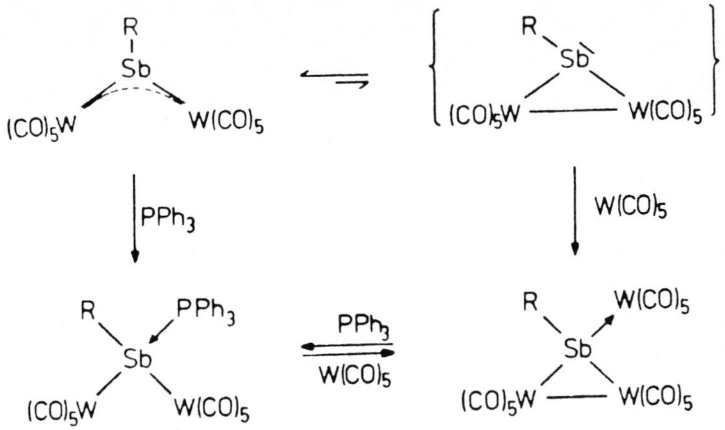

Fig. 10. Valence tautomerism of $[(CO)_5W]_2SbR$.

This is inferred by the observation that $[(CO)_5W]_2SbR$ compounds, while giving the expected Lewis base adduct with PPh_3, will also add the Lewis acid $(CO)_5W$. The formation of the $(CO)_5W$ adduct implies a valence tautomeric equilibrium between the three-center 4π "inidene" type form of the compound and a ring closed valence tautomer in which two of the former 4π electrons are now localized in a metal-metal bond and the other two as an antimony lone pair. The Lewis acid $(CO)_5W$ will trap this valence isomer by blocking the lone pair (ref. 14). This model is substantiated by observations made for analogous 6th main group species (see below).

2.1.2 Sulfanylium ion compounds and their 6th main group homologues

With RS^+ being isoelectronic to RP the relative stability of phosphinidene compounds $[L_nM]_2PR$ might indicate the possible existence of analogous sulfanylium species $[L_nM]_2SR^+$. The characterization of such species has in fact been achieved starting from $Cp(CO)_2MnSR^{\cdot}$ radical compounds (refs. 17-20):

2.1.2.1 Synthesis

When $Cp(CO)_2Mn(Thf)$ is reacted with RSH the sulfane hydrogen is lost even under mild conditions, and stable, intensively colored radical compounds are formed.

$$Cp(CO)_2Mn(Thf) + RSH \xrightarrow[-[H]]{-Thf} Cp(CO)_2MnSR^{\bullet}$$

The relative stability of these radical compounds may be explained in terms of the same simple MO approach as has first been proposed for $[Cp(CO)_2Mn]N(H)(R)^{\bullet}$ radical species by D. Sellmann, P. Hofmann et al. (ref. 15): Once a σ bond is formed between a sulfur donor orbital and the appropriate manganese acceptor orbital, filled π type orbitals of the RSH ligand and the $Cp(CO)_2Mn$ fragment will be forced to interact in a two-center 4π destabilizing way. Loss of hydrogen - presumably by intermediate oxidative addition to give $L_nMS(R)(H)$ (ref. 16) - serves to remove one of the two antibonding electrons (Fig. 11, refs. 17-18).

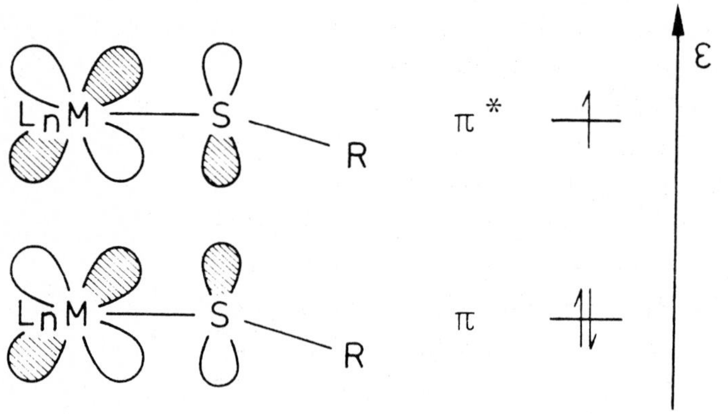

Fig. 11. Two-center π interaction in $L_nMSR^{\bullet}$ radicals.

2.1.2.2 Redox behaviour, spectroscopy and structure

By oxidation or reduction of $L_nMSR^{\bullet}$ the antibonding π^* SOMO may reversibly be transformed in either the LUMO or the HOMO of the corresponding L_nMSR^+ cations or L_nMSR^- anions, respectively (Fig. 12).

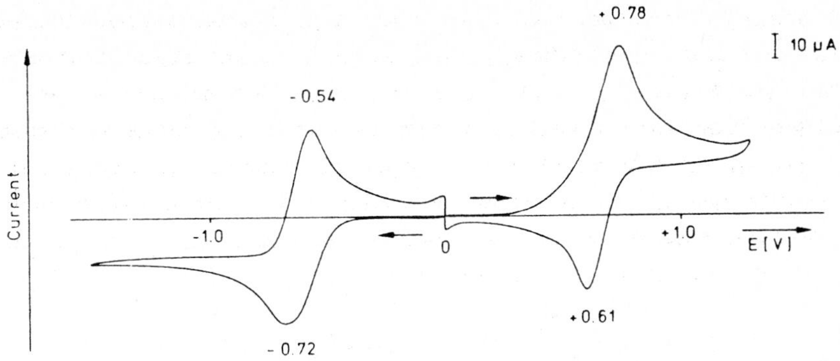

Fig. 12. Reversible one electron oxidation and reduction of $Cp(CO)_2MnSPh^{\bullet}$ (ref. 19).

While the reduction product may be isolated as such (ref. 18), chemical oxidation in $Cp(CO)_2MnSPh^{\bullet}$ inevitably leads to the formation of dinuclear compounds.

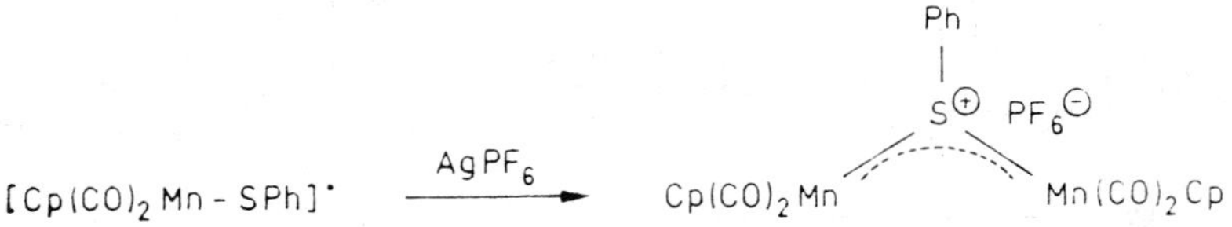

The structure of $[Cp(CO)_2Mn]_2SPh^+$ (ref. 20) is completely analogous to the structure of the isoelectronic phosphinidene compound $[Cp(CO)_2Mn]_2PPh$ (ref. 21); the sulfur is in a trigonal planar environment; the manganese-sulfur bonds are relatively short (Fig. 13).

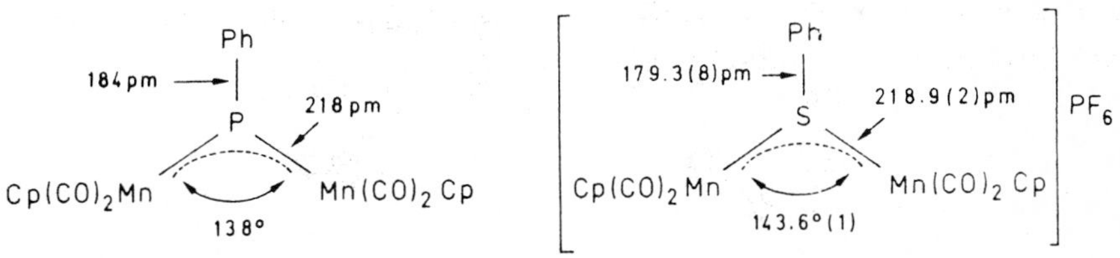

Fig. 13. Comparison of the structures of $[Cp(CO)_2Mn]_2PPh$ (left) and $[(Cp(CO)_2Mn)_2SPh]PF_6$ (right).

The analogy of bonding in the two isoelectronic species $[Cp(CO)_2Mn]_2PPh$ and $[Cp(CO)_2Mn]_2SPh^+$ is also evident from the UV-Vis spectra (Fig. 14): The sulfur compound shows the same prominent long-wavelength absorption as it is characteristic of 5th main group "inidene" type compounds (see above, compare Fig. 3); the long-wavelength band may hence be rationalized by the arguments given above as a $\pi\pi^*$ HOMO-LUMO transition in the three-center 4π system (Fig. 14, ref. 20).

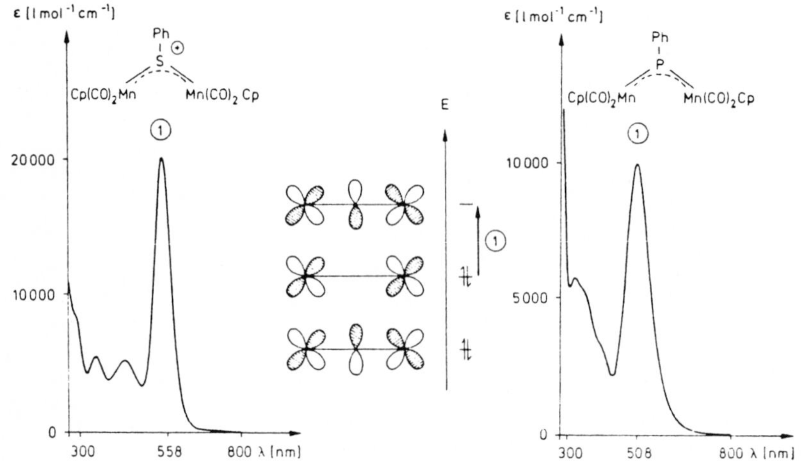

Fig. 14. Characteristic $\pi\pi^*$-absorptions ($[Cp(CO)_2Mn]_2SPh^+$ (left) and $[Cp(CO)_2Mn]_2PPh$ (right)).

The open "inidene" type structure of $[Cp(CO)_2Mn]_2SPh^+$ is in contrast to the ring closed structure of $[Cp'(CO)_2Mn]_2SEt^+$, which had been characterized by R.J. Haines et al. (ref. 22). An isoelectronic species again with a closed structure has also been reported for tellurium (Fig. 15, ref. 23).

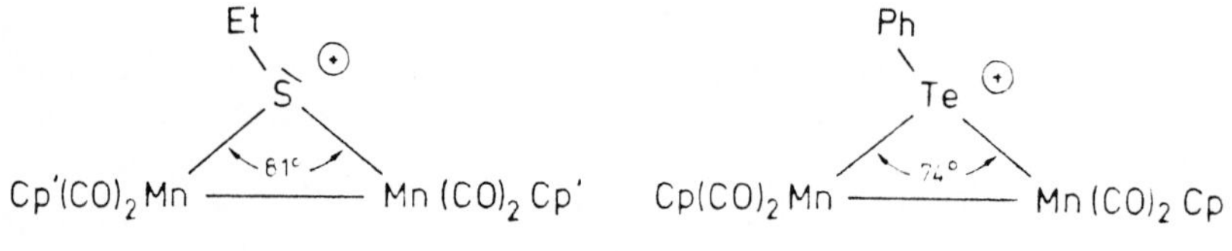

Fig. 15. M-M bond closed compounds $[L_nM]_2YR^+$.

2.1.2.3 Valence isomerism

The above structural observations lead one to suspect that under appropriate circumstances a dynamic equilibrium between bond opened and bond closed forms of $[Cp(CO)_2Mn]_2YR^+$ (Y = S, Se, Te) might be established.

This is in fact the case for $[Cp(CO)_2Mn]_2SePh^+$ (ref. 24):

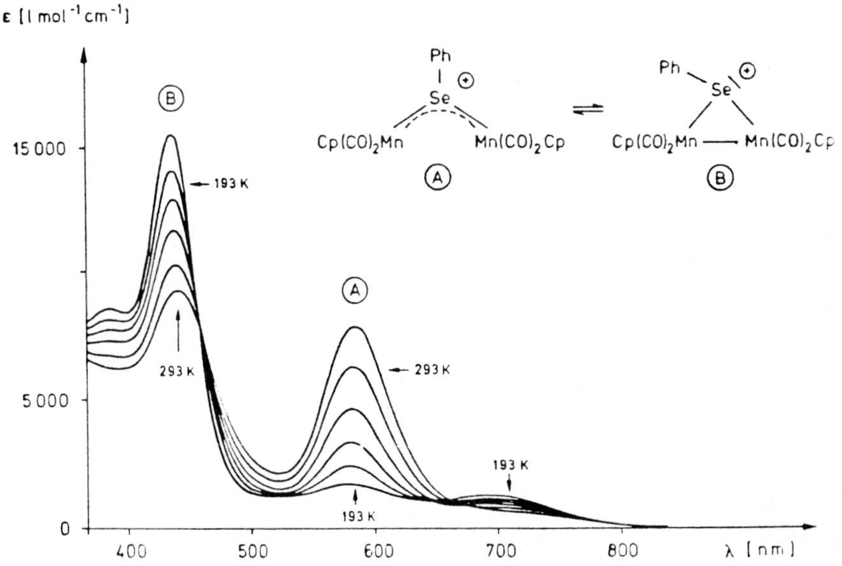

Fig. 16. Valence isomerism of $[Cp(CO)_2Mn]_2SePh^+$ as evidenced by UV-Vis spectra.

Fig. 16 shows that upon cooling a solution of this compound, the long-wavelength band, characteristic for the bond opened structure with trigonally planar coordinated selenium, is reversibly quenched, indicating a valence tautomeric equilibrium between bond opened and bond closed forms. Quantitative analysis of this temperature dependence (Fig. 17) and NMR analysis of reaction dynamics allow a simplified reaction profile to be constructed as shown in the insert of Fig. 17.

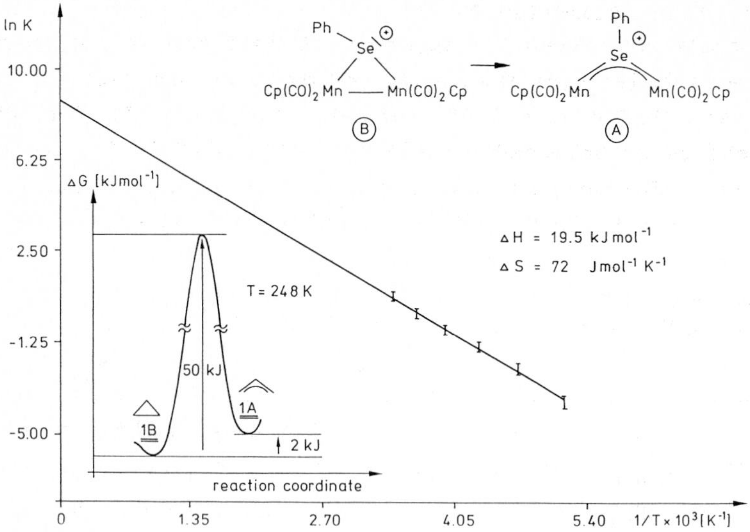

Fig. 17. Thermodynamic (ln K vs. 1/T from UV-Vis, Fig. 16) and kinetic (NMR) analysis of the valence isomerism of $[Cp(CO)_2Mn]_2SePh^+$.

Valence tautomerism of the observed kind (Fig. 18) is well established by scavenging experiments for organic three-center 4π systems especially by the work of R. Huisgen et al. (ref. 25). For stibinidene compounds this type of behaviour has been inferred from trapping the bond closed species (c.f. 2.1.1.4., refs. 8-9, 14).

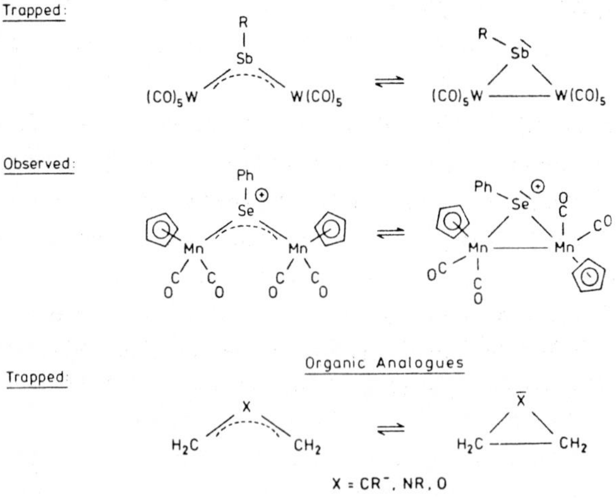

Fig. 18. Valence isomerism of three-center 4π systems.

The case of $[Cp(CO)_2Mn]_2SePh^+$ is unique, since the valence tautomeric equilibrium between the bond opened and closed forms is directly observed in this case.

2.2 CUMULENE COMPOUNDS

Linear twofold coordination of the heavier main group elements has, in molecular inorganic chemistry, first been observed by L. Sacconi et al. for the bridging sulfur in $[(p_3)NiSNi(p_3)]^{2+}$ and $[(p_3)CoSCo(p_3)]$ (p_3 = tripod ligands) (ref. 26). Organometallic analogues of this type of bonding have also been described (refs. 27-28).

4th main group elements in an analogous linear multiply bonded environment have first been described by E. Weiss et al. for $[Cp'(CO)_2Mn]_2Ge$ (ref. 29); a lead analogue has been reported by W.A. Herrmann et al. (ref. 30).

No 5th main group analogues have so far been characterized. However, by dehalogenation of chloroarsinidene complexes such species have now become available (ref. 31).

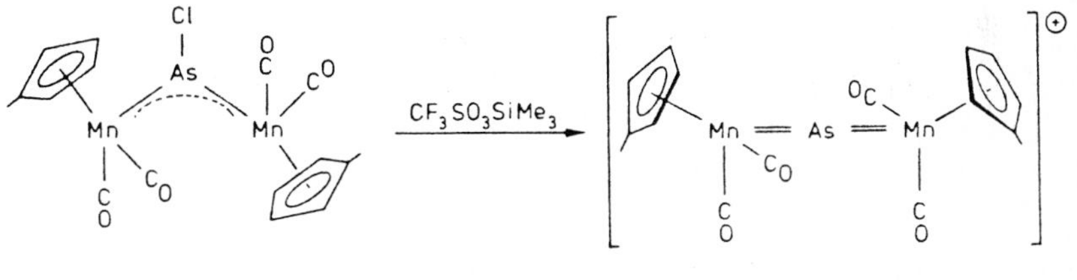

The structure of the cationic arsenic compound is analogous to the one of the isoelectronic 4th main group compound $[Cp'(CO)_2Mn]_2Ge$ (ref. 29).

The linear coordination of arsenic and the short arsenic manganese distances (Fig. 19) - in fact the shortest contacts of this type ever observed - are nicely consistent with the bonding models proposed for these cumulene type compounds (ref. 31).

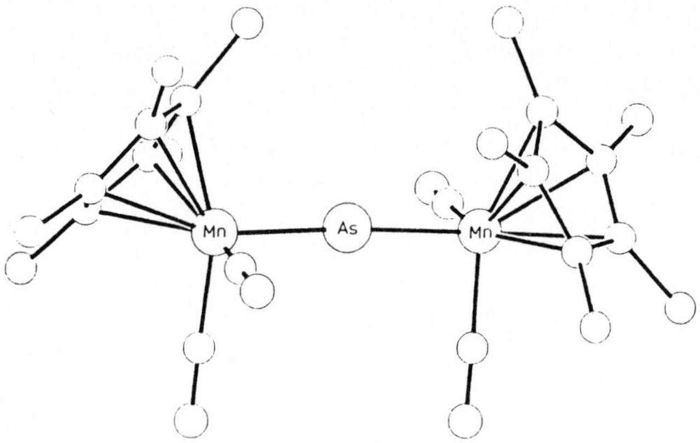

Fig. 19. Structure of $[Cp^*(CO)_2Mn=As=Mn(CO)_2Cp^*]^+BF_4^-$, d_{MnAs} 214.2(2), 215.1(2) pm; ∢ MnAsMn 176.3(1)°.

Since chloro-"inidene" type species $[L_nM]_2XHal$ are available for all the heavier 5th main group elements (X = P...Bi) (see above) it may be expected that the whole series of cationic 5th main group cumulenes $[L_nM=X=ML_n]^+$ should be accessible. These cumulene type cations are, as demonstrated for the arsenic species (ref. 32), highly electrophilic compounds which react with a broad range of nucleophiles to the corresponding "inidene" compounds (refs. 8-9).

3 CYCLIC π SYSTEMS

As a heuristic model the idea of considering clusters as constructed from organometallic π ligands and side-on coordinated L_nM fragments is quite productive. This idea dates back to early observations of structural peculiarities found for an $Fe(CO)_3$ derivative of an $Fe(CO)_4$-carbene (ref. 33), acting as a ligand and has been proposed by L.F. Dahl et al. for the rationalization of observed electron-counts in trigonal bipyramidal clusters like the one shown in Fig. 20 (ref. 34). Additional emphasis of this model came from the observation that phosphinidene type three-center 4π systems may side-on coordinate to $Fe(CO)_3$ fragments quite like allylic anions (ref. 35).

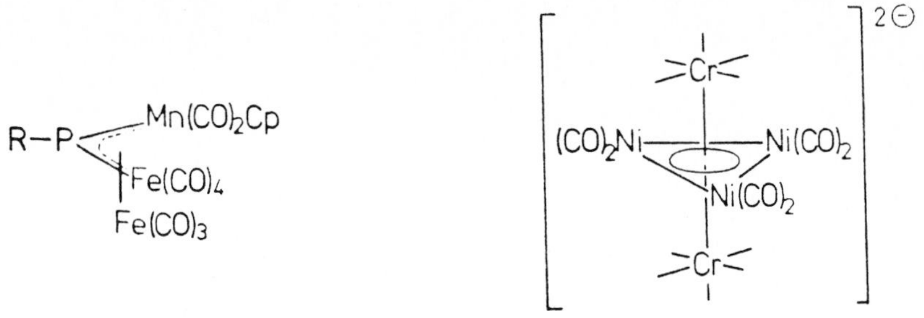

Fig. 20. Organometallic π systems as ligands (refs. 34-35).

A more detailed and more general formulation of this model is possible since R. Hoffmann (ref. 36) coined the concept of isolobality and, in fact, properties of clusters have been rationalized along these lines in quite some detail by J.K. Burdett et al. (ref. 37). The next few paragraphs may be considered as a probe for the heuristic application of this type of model in cluster chemistry.

3.1 THREE-CENTER 2π CYCLES

Organometallic three-center 2π cycles, even while being electron precise within the Hückel framework, are not generally stable in the free state because of their inherent unsaturation. The idea of generating such species as intermediates is a productive one: Organometallic 2 π cycles should be quite reactive because of their unsaturation; at the same time they should be long lived enough to be trapped in binuclear reactions because of their aromatic stabilization: Reductive dehalogenation of the (μ_2-Cl)(μ_2-RPCl) bridged $Fe_2(CO)_6$ derivative (Fig. 21) yields what appears to be a 2π cycle as an intermediate (refs. 38-39). While the unsaturated cyclic 2π species cannot be isolated as such, trapping experiments support the idea of its intermediacy: If no extra trapping agent is added, selftrapping which yields $Fe_4(CO)_{11}(\mu_4\text{-}RP)_2$ is observed. In fact, this type of reaction is the preparatively most satisfying route to synthesize this type of unsaturated cluster species, so well studied by H. Vahrenkamp et al. (ref. 40).

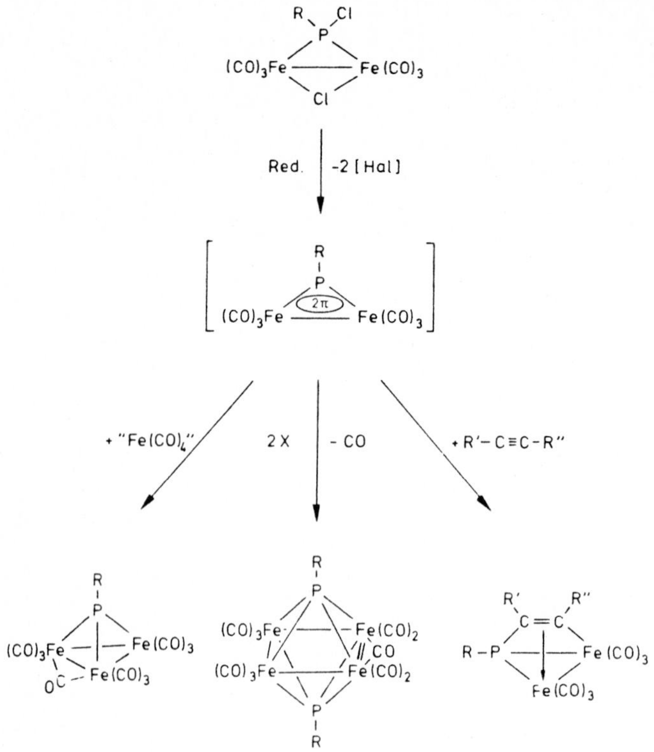

Fig. 21. Generation and trapping of the organometallic 2 π cycle $Fe_2(CO)_6(PR)$.

Adding $Fe_2(CO)_9$ as a source for "$Fe(CO)_4$" as the trapping agent results in high yield formation of the RP-bridged trinuclear clusters $Fe_3(CO)_{10}(\mu_3\text{-RP})$ (ref. 38). Quenching with alkynes R'CCR'' leads to adducts of the type $[\eta^4\text{-}((CO)_3Fe \cdots P(R) \cdots C(R') \cdots C(R''))\text{-}Fe(CO)_3]$ (ref. 41) which themselves may be considered as $Fe(CO)_3$ derivatives of four-membered 4π cycles (see below).

3.2 FOUR-CENTER 4π CYCLES

4π cycles, while destabilized by their antiaromaticity in the free state are, just for the same reason, good ligands for 14-electron ML_n species. The prototype verification of this statement is, of course, the elusive cyclobutadiene species and its many stable and well characterized half-sandwich type derivatives (Fig. 22).

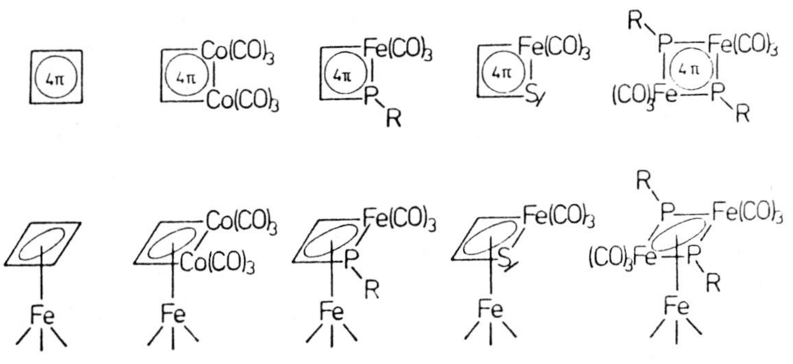

Fig. 22. Nido-octahedral clusters as derived from 4π organometallic π systems.

The cluster fragments described as 4π cycles in Fig. 22 are, not unexpectedly, unstable in the free state. The only exception is $Fe_2(CO)_6(\mu_2\text{-}RP)_2$ which P.P. Power et al. have obtained as such by steric shielding whith bulky groups R (ref. 42). In contrast the $Fe(CO)_3$-derivatives of all these 4π moieties (Fig. 22) are well characterized crystalline compounds (ref. 39).

The main justification for dissecting these clusters into 4π cycles and side-on coordinated $Fe(CO)_3$ fragments stems from structural observations: The bond lengths within the "π systems" are generally shorter than single bonds and well balanced within the cycles. Even more compelling is the observation that the bonds radiating from the carbon and phosphorus centers of these cycles are almost coplanar with the ring planes.

3.3 FIVE-CENTER 4π CYCLES

The same kind of arguments which favor the view of nido octahedral clusters as composed from organometallic four-membered 4π cycles and side-on coordinated 14-electron L_nM fragments (c.f. 3.2) makes it attractive to consider a pentagonal pyramidal cluster as composed of a five-membered organometallic 4π cycle with a side-on coordinated 14-electron fragment. To illustrate this idea some selected examples are given in Fig. 23 (refs. 39, 43-46).

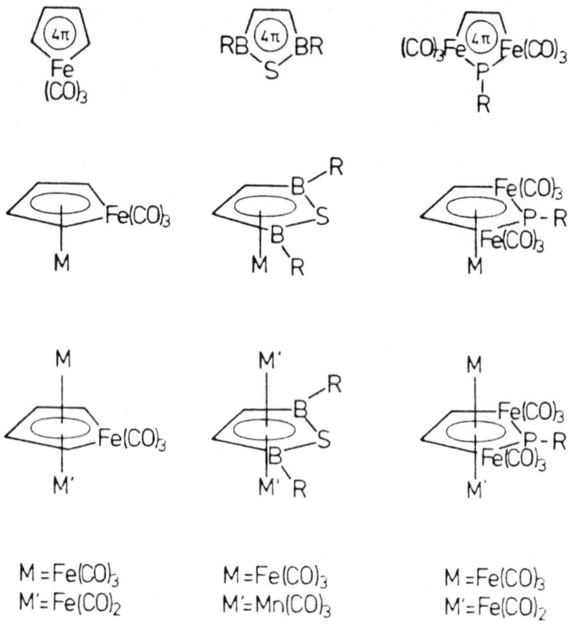

Fig. 23. Five-center 4π cycles and their relation to clusters.

By this way of reasoning W. Hübel's ferracyclopentadiene compounds (Fig. 23, ref. 45) are derived from the organometallic ferracyclopentadiene 4π cycle and W. Siebert's thiadiborolene half-sandwiches are derived from the corresponding organoborene 4π ligand (Fig. 23 middle, ref. 46). A remarkable property of the half-sandwich type compounds is their propensity to expand to triple-decker compounds when treated with the appropriate L_nM fragment (Fig. 23).

Quite the same type of behaviour is found for the diferraphospha 4π cycles (Fig. 23 right) which form stable half-sandwich type derivatives as well as triple-decker compounds (ref. 39,

47). The concept of isolobality relates the different compounds along each horizontal line in Fig. 23 and allows for an explanation of their inherent similarity.

3.4 SKELETAL REARRANGEMENT PROCESSES

The idea put forward in the preceding three paragraphs is in accord with the statement that an organometallic π cycle, given the appropriate electron-count, has some added stability either as a free species (2π) or as a coordinatively stabilized one (4π). This would mean that from the many structural isomers possible for a given set of cluster constituents the isomers conforming to this model would be in the lowest energy moulds on the corresponding hypersurface. It does not mean, however, that there could not be other isomers within the accessible energy range. The situation is similar to the analysis of possible isomers for C_6H_6: Benzene is the most stable isomer for sure. There do exist, however, other isomers like Dewar-benzene, prismane etc. which are less stable than benzene but still accessible. Within a cluster-framework skeletal reorganization is far less energy consuming than within benzene: Breaking metal to carbon, metal to phosphorus and phosphorus to carbon bonds is energetically less demanding than splitting carbon-carbon bonds and, even more, there may be no need to split individual bonds, if concerted pathways exist as they generally do in cluster compounds. It is perhaps not too bad an idea to think of a cluster as a system where the individual constituents behave like being confined to the surface of a sphere and relatively free to change place at this surface. Theoretical justification of this - of course oversimplified idea - comes from the spherical harmonics approach of D.M.P. Mingos et al. (ref. 48). Holding these views in mind it is not astonishing that dynamic isomerization processes have been observed for half sandwich type derivatives of 4π cycles like some of the ones shown in Fig. 22 (refs. 49-52).

A detailed study concerning the framework reorganization of pentagonal pyramidal compounds derived from five-membered 4π compounds is available (refs. 39, 44, 53): This study has been solicitated by the observation that such compounds (Fig. 23) may exist in three well characterized isomeric forms (Fig. 24).

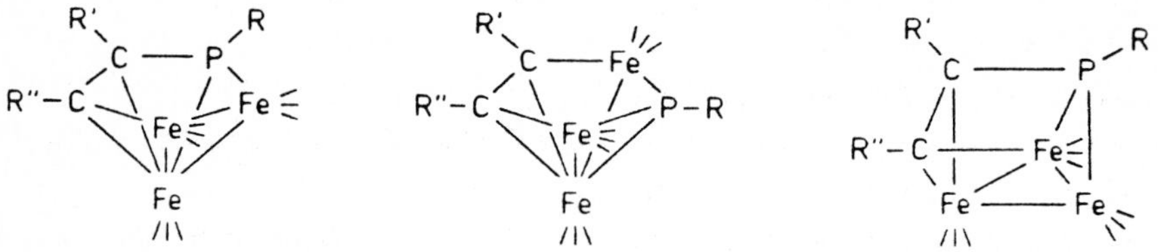

Fig. 24. Three isomeric forms of $Fe_3(CO)_9(PR)(R'CCR'')$.

Two of them are of pentagonal pyramidal shape, differing only in the sequence of ring constituents; a third one is of distorted trigonal prismatic geometry. Depending on the kind of R, R' and R'' the different isomers may be transformed into oneanother in smooth and selective reactions (refs. 44, 53).

The three isomers differ quite remarkably in the kind and number of contacts which the individual constituents form with each other. The fact that the isomerization reactions occur at low temperature and without detectable degradation lends some credit to the hypothesis of a concerted mechanism for these processes. A mechanism, which allows to interpret all the experimental observations (refs. 39, 43-44, 47, 54) in a concise way, is shown in Fig. 25.

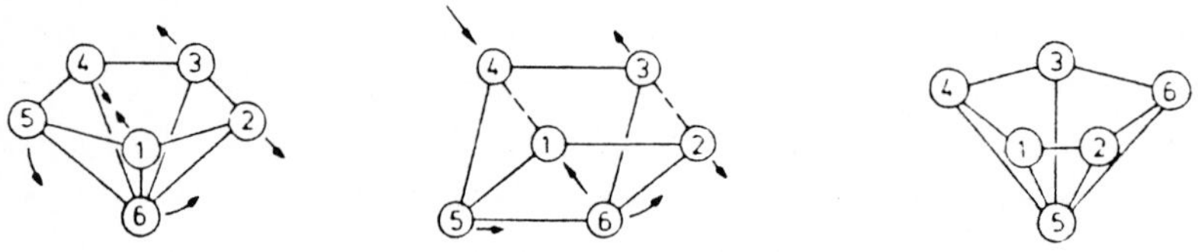

Fig. 25. "Tandem" mechanism for the isomerization of pentagonal pyramids.

The assumed "tandem" pathway implies the distortion of a pentagonal pyramid in such a way, as to give a distorted trigonal prism as an intermediate, which by further distortion along the same coordinate, transforms into an isomerized pentagonal pyramid. For the clusters studied experimentally (ref. 44) the assumption of a trigonally prismatic intermediate is a compelling one, because one of the characterized isomers is in fact of trigonal prismatic shape (Fig. 24). The "tandem" process may be of more general importance for pentagonal pyramidal clusters: As an isolobal model the hypothetical $B_6H_6^{4-}$ has been studied within the extended Hückel framework (ref. 53). It is found that the "tandem" pathway is a quantummechanically allowed one (Fig. 26).

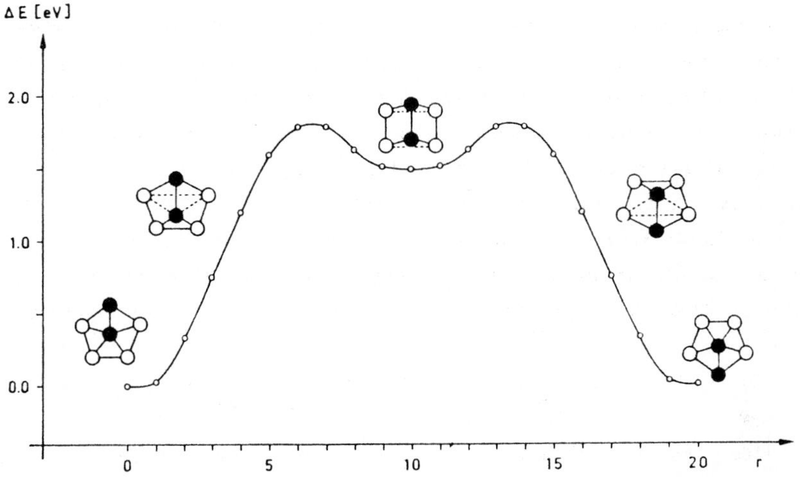

Fig. 26. EHT analysis of the "tandem" isomerization for the $B_6H_6^{4-}$ model.

The energy barrier obtained for this model calculation has, of course, to be far higher than the one expected for the real cluster compounds because boron-boron interactions are stronger than Fe-Fe, Fe-C or C-P and Fe-P contacts in the observed clusters (ref. 53).

4 CONCLUDING REMARKS

The concept of isolobal analogy is a general rational unifying the scarce and individual older approaches towards a development of a chemistry of organometallic π systems. "Inidene" type compounds $[L_nM]_2XR$, with L_nM a 16-electron metal fragment and RX an electron-sextet species, are analogues of well known organic three-center 4π systems like the allylic anion. Their reactions and properties are dominated by their π system which governs the HOMO as well as the LUMO of these compounds. There are many similarities to organic three-center 4π systems: "Inidene" compounds may act as 4π ligands like allylic anions will do; the valence isomerization between the open three-center 4π form and the ring closed form - corresponding to the cyclopropylanion isomer of the allylic anion - is directly observable for "inidene" compounds.

Cumulenes, another well established class of unsaturated organic compounds, have their organometallic analogues. $Cp(CO)_2Mn{=}X{=}Mn(CO)_2Cp$ (X = Ge, Pb) have been known for a while; the field is extending now to analogous isoelectronic compounds with 5th main group cations in the place of a neutral 4th main group atom, e.g. $[Cp^*(CO)_2Mn{=}As{=}Mn(CO)_2Cp^*]^+$.

In the field of cluster chemistry the view of clusters as composed of organometallic π ligands and side-on coordinated L_nM fragments is a productive one. It helps to device and rationalize selective synthetic strategies and it helps to rationalize the structural peculiarities of a broad subset of clusters.

In making use of a model one has to be well aware of its limitation: Quite like the Hückel-model, which does predict benzene as the most stable isomer of C_6H_6, but does not exclude the possible existence of other isomers - which are in fact well known - the organometallic π ligand model for clusters might be capable of predicting the most stable isomer for a given set of cluster constituents, but would not exclude a bunch of other possible and energetically accessible isomers. Observed isomerization processes in clusters may well include partial desintegration and consequent reintegration of the individual building blocks; there are, however, isomerization processes observed where a concerted mechanism, like the "tandem" process, is the appropriate explanation for the experimental facts. The field of organometallic π systems is just emerging from the turbulent surface of organometallic chemistry and the ongoing development will be an exciting and productive one.

Acknowledgement

The creative and stimulating contributions of the following enthusiastic co-workers are gratefully acknowledged:

Dr. H.D. Müller, Dr. H.G. Schmid, Dr. G. Mohr, Dr. H. Willenberg, Dr. J. von Seyerl, Dr. L. Zsolnai, Dr. J. Schneider, Dr. B. Pritzlaff, Dr. A. Winter, Dr. B. Sigwarth, Dr. U. Weber, Dr. O. Scheidsteger, Dr. I. Jibril, Dr. J. Borm, Dr. K. Knoll, Dipl. Chem. K. Plößl, Dipl. Chem. D. Buchholz, Dipl. Chem. A. Strube, Dipl. Chem. P. Lau, Dipl. Chem. T. Feng, Dipl. Chem. F. Ettel, Mr. Th. Fässler, Mr. H. Braunwarth, Mrs. Zufall and Dipl. Chem. Fleiß.

The highly valuable technical assistance of Mrs. M. Gottlieb, S. Fiedler and Mr. D. Günauer is gratefully acknowledged.

Without the outstanding organization and unbelievable tolerance of Mrs. I. Mang, M. Scholz and K. Giessman this work would not have been possible.

The financial support from the University of Heidelberg, from the Fonds der Chemischen Industrie and from the Deutsche Forschungsgemeinschaft has been essential throughout.

REFERENCES

1 E.O. Fischer and A. Maasböl, Angew. Chem., 76 (1964) 645; Angew. Chem. Int. Ed. Engl., 3 (1964) 580.
2 E.O. Fischer, G. Kreis, C.G. Kreiter, J. Müller, G. Huttner and H. Lorenz, Angew. Chem., 85 (1973) 618-620; Angew. Chem. Int. Ed. Engl., 12 (1973) 564.
3 K.H. Dötz, H. Fischer, P. Hofmann, F.R. Kreissl, U. Schubert and K. Weiss, Transition Metal Carbene Complexes, Verlag Chemie, Weinheim, 1983.
4 K.H. Dötz, Angew. Chem., 96 (1984) 573-594; Angew. Chem. Int. Ed. Engl., 23 (1984) 587.
5 G.P. Elliott, J.A.K. Howard, T. Mise, Ch.M. Nunn and F.G.A. Stone, Angew. Chem., 98 (1986) 183-185; Angew. Chem. Int. Ed. Engl., 25 (1986) 190.
F.G.A. Stone, Angew. Chem., 96 (1984) 85-96; Angew. Chem. Int. Ed. Engl., 23 (1984) 89.
6 R.H. Grubbs, Comprehensive Organometallic Chemistry, Vol. 8, Chapter 54 in G. Wilkinson, F.G.A. Stone, E.W. Abel (Editors), Pergamon Press, 1982.
7 R.R. Schrock, J. Organomet. Chem., 300 (1986) 249-262.
R.R. Schrock, Science, 219 (1983) 13-18.
M.E. Thompson, S.M. Baxter, A.R. Bulls, B.J. Burger, M.C. Nolan, B.D. Santariero, W.P. Schäfer and J.E. Bercaw, J. Am. Chem. Soc., 109 (1987) 203-219.
8 G. Huttner, Pure & Appl. Chem., 58 (1986) 585-596.
9 G. Huttner and K. Evertz, Acc. Chem. Res., 19 (1986) 406-413.
10 G. Huttner, J. Organomet. Chem., 308 (1986) C11-C13.

11 R.S. Drago, Physical Methods in Chemistry, Saunders Golden Sunburst Series, W.B. Saunders Company, Philadelphia, 1977.
12 G. Huttner, K. Plößl and L. Zsolnai, unpublished.
13 G. Huttner, B. Sigwarth, O. Scheidsteger, L. Zsolnai and O. Orama, Organometallics, 3 (1985) 326-332.
14 U. Weber, G. Huttner, O. Scheidsteger and L. Zsolnai, J. Organomet. Chem., 289 (1985) 357-366.
15 D. Sellmann, J. Müller and P. Hofmann, Angew. Chem., 94 (1982) 708-709; Angew. Chem. Int. Ed. Engl., 21 (1982) 691.
D. Sellmann and J. Müller, J. Organomet. Chem., 281 (1985) 249-262.
16 R. Ugo, G. La Monica, S. Cenini, A. Segre and F. Conti, J. Chem. Soc. (A), (1971) 522-528.
D. Morelli, A. Segre, R. Ugo, G. La Monica, S. Cenini, F. Conti and F. Bonati, J. Chem. Soc., Chem. Commun., (1967) 524-526.
17 A. Winter, G. Huttner, L. Zsolnai, P. Kroneck and M. Gottlieb, Angew. Chem., 96 (1984) 986-987; Angew. Chem. Int. Ed. Engl., 23 (1984) 975-976.
18 A. Winter, G. Huttner, M. Gottlieb and I. Jibril, J. Organomet. Chem., 286 (1985) 317-327.
19 G. Huttner and H. Braunwarth, unpublished.
20 H. Braunwarth, G. Huttner and L. Zsolnai, Angew. Chem., 100 (1988) 731-732; Angew. Chem. Int. Ed. Engl., 27 (1988) 698-699.
21 G. Huttner, H.D. Müller, A. Frank and H. Lorenz, Angew. Chem., 87 (1975) 714; Angew. Chem. Int. Ed. Engl., 14 (1975) 705.
22 J.C.T.R. Burckett-St.Laurent, M.R. Caira, R.B. English, R.J. Haines and L.R. Nassimbeni, J. Chem. Soc. Dalton Trans., (1977) 1077-1081.
23 G. Huttner, S. Schuler, L. Zsolnai, M. Gottlieb, H. Braunwarth and M. Minelli, J. Organomet. Chem., 299 (1986) C4-C6.
24 H. Braunwarth, F. Ettel and G. Huttner, J. Organomet. Chem., in press.
25 R. Huisgen, Angew. Chem., 89 (1977) 589-602; Angew. Chem. Int. Ed. Engl., 16 (1977) 572.
26 C. Mealli, S. Midollini and L. Sacconi, Inorg. Chem., 17 (1978) 632-637.
M.D. Vaira and L. Sacconi, Angew. Chem., 94 (1982) 338-351; Angew. Chem. Int. Ed. Engl., 21 (1982) 330.
27 T.J. Greenhough, B.W.S. Kolthammer, P. Legzdins and J. Trotter, Inorg. Chem., 18 (1979) 3543-3548.
28 J. Schiemann, P. Hübener and E. Weiss, Angew. Chem., 95 (1983) 1021; Angew. Chem. Int. Ed. Engl., 22 (1983) 980.
29 W. Gäde and E. Weiss, J. Organomet. Chem., 213 (1981) 451-460.
30 W.A. Herrmann, H.J. Kneuper and E. Herdtweck, Angew. Chem., 97 (1985) 1060-1061; Angew. Chem. Int. Ed. Engl., 24 (1985) 1062.
31 A. Strube, G. Huttner and L. Zsolnai, Angew. Chem., in press.
32 G. Huttner and A. Strube, unpublished.
33 G. Huttner and D. Regler, Chem. Ber., 105 (1972) 2726-2737.
34 J.K. Ruff, R.P. White Jr. and L.F. Dahl, J. Am. Chem. Soc., 93 (1971) 2159-2176.
35 G. Huttner, G. Mohr and A. Frank, Angew. Chem., 88 (1976) 719; Angew. Chem. Int. Ed. Engl., 15 (1976) 682.
36 R. Hoffmann, Angew. Chem., 94 (1982) 725-739; Angew. Chem. Int. Ed. Engl., 21 (1982) 711.
37 Th.A. Albright, J.K. Burdett and M.H. Whangbo, Orbital Interactions in Chemistry, John Wiley & Sons, New York, 1985.

38 H. Lang, L. Zsolnai and G. Huttner, J. Organomet. Chem., 282 (1985) 23-51.
39 G. Huttner and K. Knoll, Angew. Chem., 99 (1987) 765-783; Angew. Chem. Int. Ed. Engl., 26 (1987) 743-760.
40 H. Vahrenkamp, Adv. Organomet. Chem., 22 (1983) 169-208. T. Jaeger, S. Aime and H. Vahrenkamp, Organometallics, 5 (1986) 245- 252.
41 H. Lang, L. Zsolnai and G. Huttner, Chem. Ber., 118 (1985) 4426-4432.
42 K.M. Flynn, R.A. Bartlett, M.M. Olmstead and P.P. Power, Organometallics, 5 (1986) 813-815.
43 K. Knoll, G. Huttner and L. Zsolnai, J. Organomet. Chem., 312 (1986) C57-C60.
44 K. Knoll, G. Huttner and L. Zsolnai, J. Organomet. Chem., 332 (1987) 175-199.
45 W. Hübel and E.H. Braye, J. Inorg. Nucl. Chem., 10 (1959) 250-268.
46 W. Siebert, Adv. Organomet. Chem., 18 (1980) 301-337. W. Siebert, Angew. Chem., 97 (1985) 924-939; Angew. Chem. Int. Ed. Engl., 24 (1985) 943.
47 K. Knoll, Th. Fässler and G. Huttner, J. Organomet. Chem., 332 (1987) 309-320.
48 R.L. Johnsten and D.M.P. Mingos, Polyhedron, 5 (1986) 2059-2061.
49 G. Huttner and K. Evertz, unpublished.
50 G. Huttner and Th. Fässler, unpublished.
51 G.N. Schrauzer, H.N.Rabinowitz, J.A.K. Frank and I.C. Paul, J. Am. Chem. Soc., 92 (1970) 213-214.
52 J.P. Hickey, J.C. Huffman and L.J. Todd, Inorg. Chim. Acta, 28 (1978) 77-78.
53 G. Huttner, K. Knoll, Th. Fässler and H. Berke, J. Organomet. Chem., 340 (1988) 223-226.
54 K. Knoll, G. Huttner, Th. Fässler und L. Zsolnai, J. Organomet. Chem., 327 (1987) 255-267.

POLYNUCLEAR TRANSITION METAL COMPLEXES WITH SULFUR LIGANDS

B. KREBS[1] and G. HENKEL[2]

[1]Anorganisch-Chemisches Institut der Universität, Wilhelm-Klemm-Strasse 8, D-4400 Münster (Federal Republic of Germany)

[2]Fachgebiet Anorganische Chemie der Universität (GH), Lotharstrasse 1, D-4100 Duisburg 1 (Federal Republic of Germany)

1. INTRODUCTION

The inorganic and organometallic chemistry of transition metal complexes with sulfur-containing ligands has experienced an exciting and expansive development in recent years with a clear tendency of further significant growth. A large number of novel classes of coordination compounds with interesting electronic properties and with a wide field of potential applications have been prepared and characterized. The ligand types in these systems reach from sulfides, monofunctional thiolates, thioethers, thiocarbonyl compounds such as thioureas, thiocyanates, and sulfur-containing amino acids to a variety of di- and polyfunctional ligands such as elemental sulfur, polysulfides, derivatives of thiocarbonic, thiophosphoric and thiophosphonic acids, multifunctional thiolates, and sulfur analogues of the crown-ethers and cryptands, and the degree of molecular complexity extends from mononuclear through various types of polynuclear association to polymeric species.

Especially the chemistry of mono- and polynuclear sulfide and thiolate complexes has received much attention and was stimulated greatly by its relevance to structure, bonding, and function of biologically active reaction centers in metalloproteins and metalloenzymes such as rubredoxins, ferredoxins, nitrogenases, blue copper proteins, or in metallothioneins. The RS^- thiolate group can act as a close electronic model to the metal-binding sulfur-containing amino acid cysteine in metallo-biomolecules. Besides the interest in general catalytic properties it was this bio-inorganic aspect which led to an impressive development of this field.

Within this context, the present article will concentrate, out of the vast field of metal-sulfur ligand complexes, on some aspects of the very important class of homoleptic and heteroleptic polynuclear transition metal complexes with thiolate or mixed sulfide-thiolate ligands. Complexes with 1,1-dithiolates, 1,2-dithiolenes, and other abiotic dithio ligands as well as complexes with partial chalcogen ligand spheres and complexes containing oxo ligands have generally been excluded. These aspects have been subject of a large number of reviews (see for example refs. 1-14).

2 THIOLATES AND SULFIDE-THIOLATES OF OPEN-SHELL TRANSITION METALS

2.1 Zirconium

The coordination chemistry of titanium and zirconium with complete sulfur coordination is at present largely limited to mononuclear dithiolene, dithiocarbamate, and aromatic dithiolate complexes (ref. 15 and references cited therein), among them the tris(benzene-1,2-dithiolato)zirconate(IV) anion in $[Me_4N]_2[Zr(S_2C_6H_4)_3]$ (ref. 16).

The first example for a polynuclear zirconium thiolate was prepared from a remarkable simple reaction of $Zr(CH_2Ph)_4$ with t-BuSH in toluene. Yellow $[Zr_3S(S\text{-}t\text{-}Bu)_{10}]$ is formed as a new type of trinuclear complex in which the t-butanethiolate ligands are present in three different coordination modes (ref. 17). The structure shown in Fig. 1 contains the $[Zr_3(\mu_3\text{-}S)(\mu_3\text{-}SR)(\mu\text{-}SR)_3]$ core. It can be described as a hexagonal bipyramid where three μ-thiolates bridge the Zr atoms in a puckered equatorial plane, and a μ_3-S^{2-} and a μ_3-thiolate ligand serve as 'capping' ligands on the axial positions. The molecular framework can also be described as three face-sharing ZrS_6 octahedra. The structure presents an opportunity to study the bonding of a thiolate ligand in three different coordination modes within the same molecule. The differen-

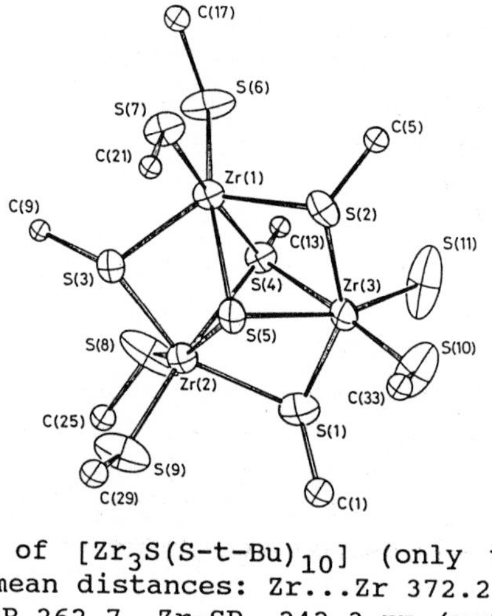

Fig. 1. Structure of $[Zr_3S(S\text{-}t\text{-}Bu)_{10}]$ (only the tertiary carbon atoms are shown); mean distances: Zr...Zr 372.2, Zr-μ_3-S 260.1, Zr-μ_3-SR 276.5, Zr-μ-SR 262.7, Zr-SR_t 242.3 pm (ref. 17).

ces are clearly reflected in the Zr-S distances (Fig. 1). The structural data also clearly show the Zr...Zr distances to be non-bonding, in accordance with the d^0 configuration of the metals.

It is interesting to note that the central Zr_3S_5 core of the $[Zr_3S(S\text{-}t\text{-}Bu)_{10}]$ type is stabilized in slightly different forms in two novel mixed-ligand tri- and hexanuclear zirconium complexes with predominant sulfide-thiolate coordination. Yellow $[Zr_3S_3(t\text{-}BuS)_2(BH_4)_4(THF)_2]$ is obtained from THF solutions according to the reaction

$$3\ Zr(BH_4)_4 + 13\ t\text{-}BuSH + 2\ THF \longrightarrow [Zr_3S_3(S\text{-}t\text{-}Bu)_2(BH_4)_4(THF)_2] + 8/3\ (BH_2S\text{-}t\text{-}Bu)_3 + 13\ H_2 + 3\ t\text{-}BuH \qquad \text{Eq. (1)}$$

and, following abstraction of a THF molecule, it dimerizes in solution to give the novel hexanuclear $[Zr_6S_6(S\text{-}t\text{-}Bu)_4(BH_4)_8(THF)_2]$ (ref. 15). The structures of both are shown in Fig. 2; they are closely related and contain hexagonal bipyramidal $[Zr_3(\mu_3\text{-}S)_2(\mu\text{-}SR)_2(\mu\text{-}S)]$ cores with puckered $[Zr_3(\mu\text{-}SR)_2(\mu\text{-}S)]$ equatorial planes. The BH_4^- ligands are bidentate and tridentate (not shown in the Figure), so that Zr(1) and Zr(3) are eight-coordinate, and Zr(2) is nine-coordinate in both compounds. Due to sterical requirements, the non-bonded Zr...Zr distances are much shorter than in $[Zr_3S(S\text{-}t\text{-}Bu)_{10}]$ (Figs. 1 and 2).

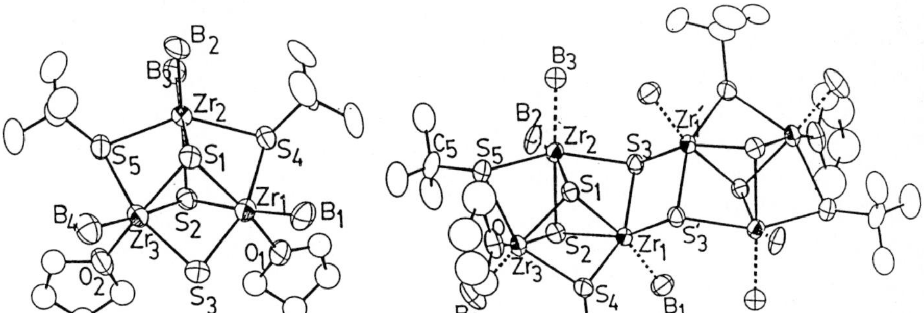

Fig. 2. Structures of $[Zr_3S_3(S\text{-}t\text{-}Bu)_2(BH_4)_4(THF)_2]$ (**1**) (left) and of $[Zr_6S_6(S\text{-}t\text{-}Bu)_4(BH_4)_8(THF)_2]$ (**2**) (right); mean distances in the cores of 1 and 2: Zr...Zr 347, 348.3, Zr-μ_3-S 259.4, 259.4, Zr(2)-SR 263.2, 259.6, Zr(1,3)-SR 265.0, 264.5 pm (ref. 15).

2.2 <u>Vanadium</u>

The chemistry of polynuclear homo- and heteroleptic sulfur ligand complexes of vanadium is of special current interest because of increasing evidence for the important role of vanadium- sulfur compounds in various chemical, biological, and catalytically active industrial systems. Examples are the interesting magnetic and electrical properties of binary and ternary vanadium sulfides together with their structural relations to molecular vanadium cluster compounds (refs. 18,19), the nature and properties of vanadium species in crude mineral oils as well as their role in poisoning of hydrodesulfurization catalysts (refs. 20-23) and, most importantly, the recently reported existence and partial structural characterization of vanadium nitrogenase (refs. 24-28). The interest in these systems has stimulated significant research into thiolate and thiolate/sulfide model compounds.

The preparative routes towards vanadium thiolate and sulfide-thiolate anions start with the reaction of VCl_3 and sodium thiolate in alcohols or acetonitrile as solvents, yielding vanadium-(III) compounds. Mixed-ligand sulfide-thiolate compounds with higher-valent vanadium are obtained by addition of appropriate amounts of elemental sulfur which produces sulfide by redox reactions of the type

$$1/8\ S_8 + edt^{2-} \longrightarrow S^{2-} + 1/2\ (edt)_2 \qquad \text{Eq. (2)}$$

(edt^{2-}: ethane-1,2-dithiolate) and which is also able to oxidize the metal (see ref. 29).

Dinuclear $[V_2(SCH_2CH_2S)_4]^{2-}$ with trivalent metal centers was the first homoleptic vanadium thiolate and the first oligomeric vanadium complex with complete sulfur coordination. It was reported independently in 1983 by three groups (refs. 30-32) as the only product from 1:2 to 1:3 VCl_3/ethanedithiolate reaction mixtures. During its formation, no redox reaction takes place, and it was isolated as the crystalline salts $[Ph_4P]_2[V_2(SCH_2CH_2S)_4]$-.4MeOH (ref. 30), $[Ph_4P]_2[V_2(SCH_2CH_2S)_4]$ (ref. 31), and $[Et_4N]_2$-$[V_2(SCH_2CH_2S)_4]$ (ref. 32). The structure (Fig. 3) differs fundamentally from those of the $[Fe_2(SCH_2CH_2S)_4]^{2-}$ and $[Mn_2(SCH_2CH_2$-$S)_4]^{2-}$ analogues (refs. 33-36): Besides two terminal edt^{2-} ligands there are two bridging ones; each μ-sulfur atom of the bridging edt^{2-} is bonded to both vanadium atoms, resulting in a remarkable quadruply bonded V(III) dimer. The overall symmetry of the chiral complex is close to D_2 (in the $[Et_4N]^+$ salt: exactly D_2), the V-S-V bridges are significantly asymmetric (Fig. 3), and the VS_6 coordination can be described as a twisted trigonal prism having one edge rotated by ca. 30° - 45° from the D_{3h} arrangement. According to extended Hückel MO calculations (ref. 32) the molecular diamagnetism of the complex is ascribed to a V-V single bond and antiferromagnetic coupling of the remaining two electrons transmitted through bridging sulfur atoms; for a more detailed discussion see ref. 37.

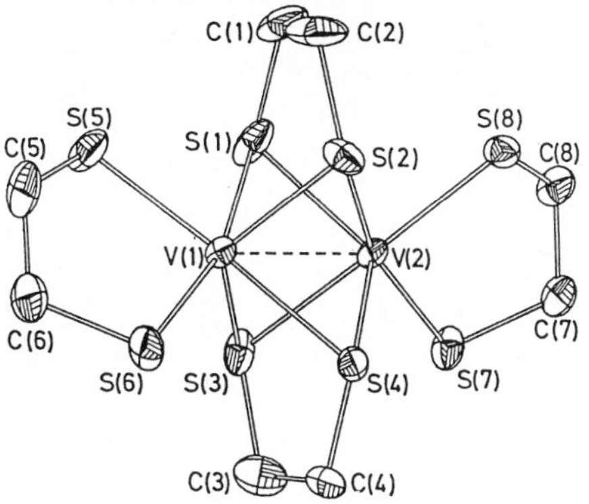

Fig. 3. Structure of $[V_2(SCH_2CH_2S)_4]^{2-}$; distances: V(1)...V(2) 257.5, V(1)-S(2) 236.8, V(1)-S(3) 239.3, V(1)-S(1) 248.5, V(1)-S(4) 249.3, V(1)-S(5) 238.4, V(1)-S(6) 238.0 pm (ref. 30).

As the first example of this quadruply bridged dimer with aromatic chelating thiolate ligands, different salts of the anion $[V_2(tdt)_4]^{2-}$ (tdt^{2-} : toluene-3,4-dithiolate) were prepared and structurally characterized recently (ref. 38). The core structure and electronic properties are very similar to the edt^{2-} species. A similar quadruply bridged $[V_2(edt)_2]^{2+}$ core has also been found in $[Cp_2V_2(edt)_2]$ which has a V...V distance of 254.2 pm (ref. 39). Quite remarkably, the structural topology of the central quadruply bridged V_2S_4 unit is also present in various electronically diverse organometallic systems and solid state materials with sulfide, disulfide or disulfur, 1,1-dithiolates, or dithiolenes as donating ligands, such as in the $[V(S_2)_2]_n$ chains of the mineral patronite, VS_4, (refs. 40-42), in $[Cp_2V_2S_2\{S_2C_2(CF_3)_2\}]$ (ref. 43), or in $[V_2(S_2)_2\{S_2CN(C_4H_9)_2\}_4]$ (ref. 42). For a general discussion on the bonding in this type of dinuclear systems: see ref. 44.

If sulfide formation and oxidation of vanadium is allowed in the VCl_3/thiolate reaction system by addition of sulfur, and by adjusting the $VCl_3/S/Na_2edt$ ratio to 1:1.5:2, the trinuclear mixed-valence complex $[V_3S_4(edt)_3]^{3-}$ is formed. It was isolated as black $[Et_4N]_3[V_3S_4(edt)_3].2MeCN$ (ref. 29). The molecular geometry of the complex anion of approximate C_{3v} symmetry (Fig. 4), with one central capping μ_3-S atom, with a μ-S atom bridging each edge, and with approximate trigonal bipyramidal coordination of vanadium, is isostructural (but not isoelectronic) with similar trinuclear $[M_3S_4]$ cores with M = Fe, Co, Ni, Zr, and especially Mo (see below). There are no indications of a trapped valence assignment of the mixed-valence trivanadium(III,2xIV) system, and the anion is best described as electronically delocalized with an average metal oxidation number of 11/3. Cyclic voltammetry indicates three one-electron processes with two of them reversible which shows $[V_3S_4(edt)_3]^{3-}$ to be a member of the four-component electron transfer series:

$$[V_3S_4(edt)_3]^{4-} \underset{}{\overset{-1.51\ V}{\rightleftharpoons}} [V_3S_4(edt)_3]^{3-} \overset{-0.63\ V}{\rightleftharpoons} [V_3S_4(edt)_3]^{2-} \overset{0.00\ V}{\longrightarrow} [V_3S_4(edt)_3]^{-} \qquad \text{Eq. (3)}$$

A dinuclear species with the multiply bonded VS^{2+} unit is obtained in another redox reaction from the same system when the $VCl_3/S/Na_2edt$ ratio is chosen to be 1:2:2.5. The remarkable mixed valence $[V_2S_5(edt)]^{3-}$ anion, which was isolated as black $[Et_4N]_2$-$Na[V_2S_5(edt)]$ (**1**), contains one thiovanadyl-V(IV) besides one V(V) (ref. 29). According to the crystal structure of the homologous purple $[Et_4N]_2Na[V_2OS_4(edt)]$, which is formed in a mixture together with **1** if the above ratio is fixed at 1:2:2, and according to spectroscopic evidence $[V_2S_5(edt)]^{3-}$ very probably has the structure of the $[V_2OS_4(edt)]^{3-}$ anion as shown in Fig. 5. Localized valencies are to be assumed with the tetrahedral V(2) being typical for V(V) as in the well-known $[VS_4]^{3-}$ anion (refs. 45,46) and the square pyramidal $[O{=}VS_4]$ (or $[S{=}VS_4]$) core being characteristic for V(IV). The $[S{=}VS_4]$ unit has been structurally characterized in the mononuclear model compound $[S{=}V(edt)_2]^{2-}$ (refs. 47,48). A weak V-V bonding interaction may be present within the central nonplanar $[V_2S_2]$ ring of $[V_2OS_4(edt)]^{3-}$ (Fig. 5). It is interesting to

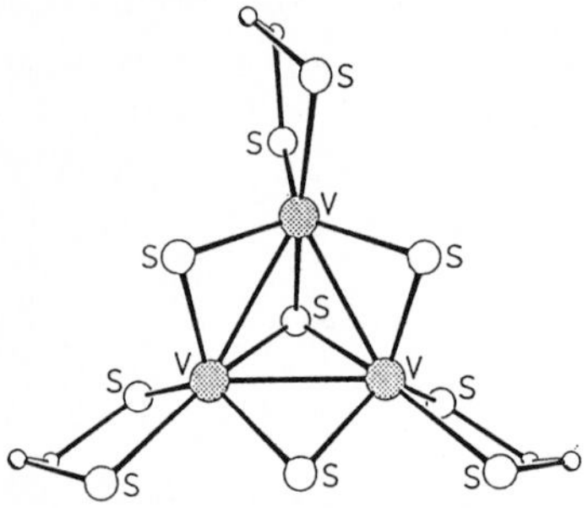

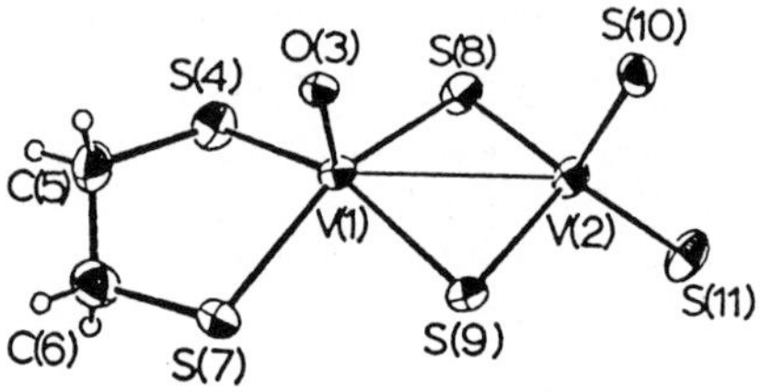

Fig. 4. The $[V_3S_4(SCH_2CH_2S)_3]^{3-}$ anion in $[Et_4N]_3[V_3S_4(SCH_2CH_2$-$S)_3]\cdot 2MeCN$; distances: V...V 288.1-291.5, mean values: V-(μ_3-S) 233.9, V-(μ-S) 225.2, V-S(thiolate) 232.2, 237.4 pm (ref. 29).
Fig. 5. Structure of $[V_2OS_4(edt)]^{3-}$ (the analogous $[V_2S_5(edt)]^{3-}$ anion with a V=S group instead of the V=O one is isostructural); distances: V...V 297.7, V(1)-O 162.6, V(1)-S(b) 237.5, V(1)-S(t) 237.0, V(2)-S(b) 220.7, V(2)-S(t) 211.6 pm (refs. 49,29)

note that $[V_2S_5(edt)]^{3-}$ as well as $[V_2OS_4(edt)]^{3-}$ show a reversible one-electron reduction with similar potential, suggesting that the identity of the multiply bonded atom has a negligible effect on the acceptor orbital (ref. 29).

The novel tetranuclear $[V_4S_2(edt)_6]^{2-}$ anion (ref. 50) may serve as an example for an especially close relationship in structure and electronic properties between a solid state material (here: the phase Li_xVS_2, refs. 51,18) and a soluble building block of it (here: the tetranuclear thiolate). Black crystals of $[Et_4N]_2[V_4S_2(edt)_6].2MeCN$ are formed by the reaction of VCl_3, Li_2S, Na_2edt, and $[Et_4N]Br$ in a 3:4:3:6 ratio in MeCN (ref. 50). In the structure of $[V_4S_2(edt)_6]^{2-}$ which is shown in Fig. 6(a) there are, beside the central $[V_4S_2]$ core, three different types of edt^{2-} ligands: (a) with both S atoms terminal, S(12) and S(15), (b) with both bridging, S(4) and S(7), and (c) with one bridging, S(8), and one terminal, S(11). The structural data and the magnetic moment (2.43 μ_B/V_4 unit) suggest an electronically delocalized system with partial spin coupling and with an average formal metal oxidation state of +3.5.

The $[V_4S_{14}]$ core of the $[V_4S_2(edt)_6]^{2-}$ anion can be regarded (ref. 50) as a fragment of the CdI_2-type VS_2 layers within the structure of the Li_xVS_2 phases (variable numbers of Li^+ ions are located in the interlayer spaces, ref. 51). As shown in Fig. 6 (b), all thiolate S atoms are essentially in positions occupied by S^{2-} atoms in the VS_2 layers (with the exception of S(7) which replaces two S^{2-}). Remarkably, $[V_4S_2(edt)_6]^{2-}$ also shows strong similarities to the Li_xVS_2 phases by its variable electron content. CV measurements of the complex anion, which corresponds to $Li_{0.5}VS_2$, show a series of partially reversible one-electron transfer reactions towards other oxidation levels of the Li_xVS_2 phases. This is indicated in Eq. (4) which shows, besides the measured potentials vs. NHE, the mean oxidation numbers of V:

$$\begin{array}{ccccccc} [V_4S_2(edt)_6]^{0} & \xleftarrow{?} & [V_4S_2(edt)_6]^{-} & \underset{}{\overset{-0.56\ V}{\rightleftharpoons}} & [V_4S_2(edt)_6]^{2-} & \overset{-1.01\ V}{\rightleftharpoons} & \\ 4.00 & & 3.75 & & 3.50 & & \\ VS_2 & \rightleftharpoons & Li_{0.25}VS_2 & \rightleftharpoons & Li_{0.5}VS_2 & \rightleftharpoons & \end{array}$$

$$\begin{array}{ccc} [V_4S_2(edt)_6]^{3-} & \xrightarrow{-1.82\ V} & [V_4S_2(edt)_6]^{4-} \\ 3.25 & & 3.00 \\ Li_{0.75}VS_2 & \rightleftharpoons & LiVS_2 \end{array} \qquad \text{Eq. (4)}$$

Thus, in the present system the molecular fragment of the polymeric material can serve not only as a structural but also as an electronic model for the practically relevant properties of the material.

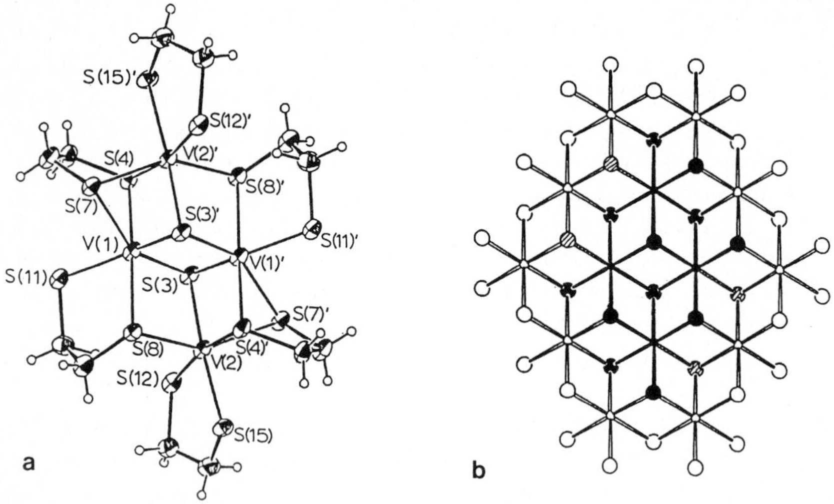

Fig. 6. (a) Structure of $[V_4S_2(SCH_2CH_2S)_6]^{2-}$: V(1)...V(1)' 303.9, V(1)...V(2) 330.0, V(1)...V(2)' 277.1, V(1)-S(3) 231.5, V(1)-S(3)' 238.7 pm (ref. 50). (b) Portion of the two-dimensional extended lattice of the Li_xVS_2 phases. V and S corresponding to those in $[V_4S_2(edt)_6]^{2-}$ are indicated in black.

Considerable progress has also been obtained recently in the field of polynuclear vanadium thiolates and sulfide-thiolates. From anaerobic reactions of tetraalkylammonium salts of $[Fe(SPh)_4]^{2-}$ and $[Fe_4(SPh)_{10}]^{2-}$ or of $[FeCl_4]^{2-}$ with $[NH_4]_3[VS_4]$ red-black salts of $[VFe_2S_4X_4]^{3-}$ complex anions (X = SPh, Cl) were synthesized (refs. 45,52). The thiolate complex has the structure $[(PhS)_2FeS_2VS_2Fe(SPh)_2]^{3-}$ and belongs to the series of linear trinuclear core systems consisting of three edge-sharing tetrahedra. The magnetic moment clearly supports the valence formulation 2 high-spin Fe(II) + 1 V(V). The crystal structure is known for the chlorine analogue (refs. 45,52).

A series of highly interesting novel heterometallic vanadium-iron-sulfur clusters containing the cubane-type $[VFe_3S_4]^{2+}$ core (which is isoelectronic to the important $[MoFe_3S_4]^{3+}$ core) and novel double-cubane systems have been prepared and characterized recently (refs. 53-56). They shed some new light on the redox properties of heteronuclear metal-sulfur cubanes as well as on the understanding of the nitrogenase problem, especially if compared to the analogous molybdenum-iron systems, and they open new ways for the synthesis of other heterocubanes.

The linear trinuclear cluster system $[VFe_2S_4X_4]^{3-}$ serves as starting material for assembly reactions involving electron transfer towards the synthesis of cubane-type clusters with the novel $[VFe_3S_4]$ framework. If the linear trinuclear system with X = Cl is reacted with $FeCl_2$ in DMF, V(V) is reduced by iron(II), and cubane-like $[VFe_3S_4Cl_3(DMF)_3]^-$ (**1**) with VS_3O_3 coordination of vanadium is formed (refs. 53-55). This and similar solvated complexes are reactive towards substitution by a variety of ligands. By reaction of **1** with ethane-1,2-dithiolate (edt^{2-}) the interesting novel double-cubane cluster $[V_2Fe_6S_8Cl_4(edt)_2]^{4-}$ is synthesized in which vanadium is trigonal-bipyramidally coordinated by sulfide and thiolate sulfur atoms. The centrosymmetric complex is shown in Fig. 7. The bridge structure is cleaved by excess thiolate RS^- to give the single-cubane trianion $[VFe_3S_4(SR)_3(edt)]^{3-}$ which does not bind additional ligands, probably because of steric constraints (ref. 54). The electronically delocalized $[VFe_3S_4]^{2+}$ type clusters show reversible one-electron oxidations (refs. 54-56). The formation of the isoelectronic cores $[VFe_3S_4]^{2+}$, $[MoFe_3S_4]^{3+}$ and $[WFe_3S_4]^{3+}$ by cluster assembly reactions indicates a special stability of these 51e, S = 3/2 configurations.

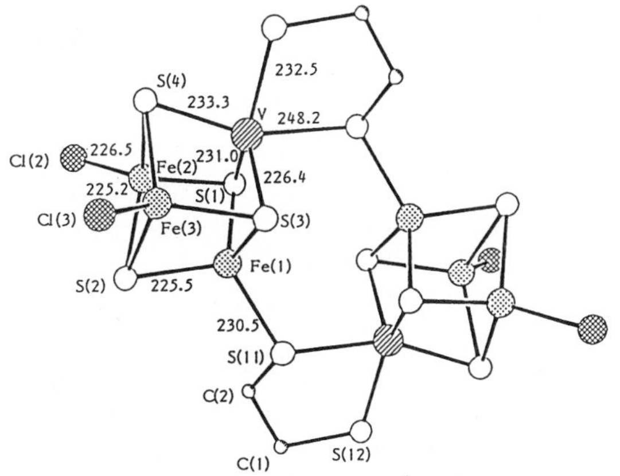

Fig. 7. The $[V_2Fe_6S_8Cl_4(SCH_2CH_2S)_2]^{4-}$ double cubane anion in solid $[Me_4N]_2[Et_4N]_2[V_2Fe_6S_8Cl_4(SCH_2CH_2S)_2]$. (ref. 54).

2.3 Niobium and Tantalum

Especially little is known about fully sulfur-coordinated thiolate and sulfide/thiolate cluster systems of the heavy early transition metals. For niobium and tantalum this chemistry differs considerably from the corresponding one of vanadium, and at present it is dominated by mononuclear $[Nb^VL_3]^-$ and $[Ta^VL_3]^-$ tris-chelate species with L = $(SCH_2CH_2S)^{2-}$ (ref. 57), $(SCH_2CH_2CH_2S)^{2-}$, $(SCH{=}CHS)^{2-}$ (ref. 58), and benzene-1,2-dithiolate (refs. 59,60). Among the mononuclear species, the formation of $[NbS(SCH_2CH_2S)$-

$(SCH_2CH_2SCH_2CH_2S)]^-$ from $[Nb(SCH_2CH_2S)_3]^-$ by spontaneous isomerization with C-S bond disruption and concomitant formation of a sulfide (as a Nb=S bond) is expected to be a general reaction particularly for early-transition-metal thiolates and their cluster formation, and it may be highly relevant to the mechanism of hydrodesulfurization processes (ref. 61).

In an unusual sulfide-mediated coupling reaction of two dimers to a tetramer, the interesting tetranuclear $[Nb_4S_2(SPh)_{12}]^{4-}$ cluster (Fig. 8) was obtained directly from the double Nb=Nb-bonded $[Nb_2Cl_6(Me_2S)_3]$ with LiSPh.THF in toluene (ref. 62). The sulfide present in the product must have been formed by C-S bond cleavage within PhS^-, the overall reaction being

$$2\ [Nb_2Cl_6(Me_2S)_3] + 16\ LiSPh \longrightarrow Li_4[Nb_4S_2(SPh)_{12}] + 12\ LiCl + 6\ Me_2S + Ph\text{-}Ph + PhSSPh \qquad \text{Eq. (5)}$$

The centrosymmetric anion contains a planar square Nb_4 unit; each edge is bridged by two μ-SPh^- groups, with one terminal SPh^- on each Nb. Two μ_4-S atoms, one above and one below the Nb_4 plane, complete seven-coordination for each Nb. The symmetry of the $[Nb_4S_{14}]$ core is close to D_{4h}. In the diamagnetic anion niobium(III) makes available eight d-electrons for Nb-Nb bonding, leading to a bond order assignment of one. The cluster is surprisingly stable against oxidation.

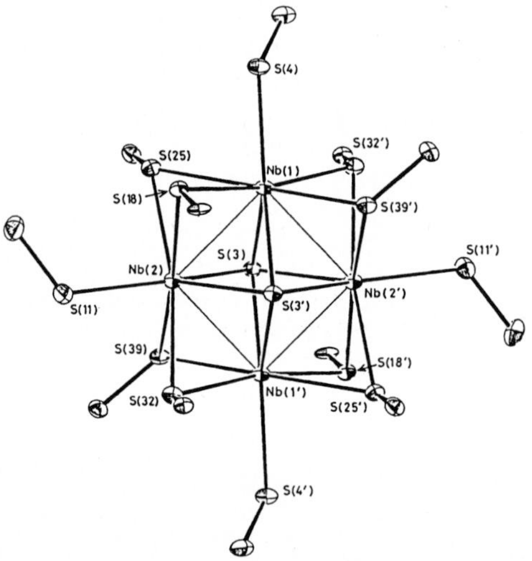

Fig. 8. Molecular framework (phenyl rings omitted) of the $[Nb_4S_2(SPh)_{12}]^{4-}$ cluster anion; selected distances: Nb(1)-Nb(2) 282.6, Nb(1)-Nb(2') 282.9, Nb-μ_4-S 246.2-251.6, Nb(1)-S(4) 253.4, Nb(2)-S(11) 265.0, Nb-μ_2-S 255.7-266.5 pm (ref. 62).

The only other examples of oligomeric Nb or Ta thiolates are two diamagnetic dinuclear complexes of niobium(IV). They were obtained by self-reduction of the mononuclear propanedithiolate complex and subsequent reaction with methanol:

$$[Ph_4P][Nb(SCH_2CH_2CH_2S)_3] \xrightarrow{30d, MeCN} [Ph_4P]_2[Nb_2(SCH_2CH_2CH_2S)_5] \xrightarrow{MeOH}$$
(1) **(2)**
$$[Ph_4P][Nb_2(OMe)_3(SCH_2CH_2CH_2S)_3] \qquad \text{Eq. (6)}$$
(3)

The structure of **2** is not yet known, **3** with predominant thiolate ligation is a rare example of a triply bridged Nb(IV) dimer, with a Nb-Nb single bond at 290.1 pm. It is related to the known dinuclear niobium sulfide halide complexes, and its structure is shown in Fig. 9 (ref. 238).

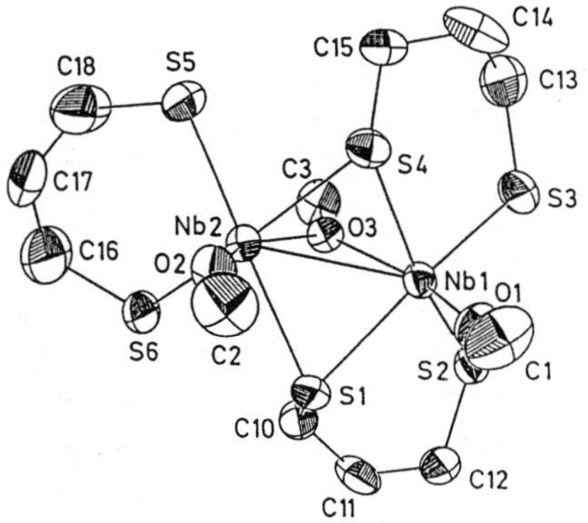

Fig. 9. Structure of $[Nb_2(OMe)_3(SCH_2CH_2CH_2S)_3]^-$ (ref. 238).

2.4 Molybdenum and Tungsten

Research on the coordination chemistry of sulfur-rich molybdenum complexes has been stimulated by the large current interest in molybdenum enzymes (e.g. nitrogenase, nitrate reductase, xanthine oxidase, sulfite oxidase) and in molybdenum desulfurization catalysts. A number of oxidation-reduction reactions which are catalyzed by molybdoenzymes occur at sites where the molybdenum is coordinated by one or more S atoms. Especially, the biological relevance of iron-molybdenum-sulfur systems in enzymes such as nitrogenase has initiated considerable activity in the synthesis and characterization of heterometallic mixed-ligand molybdenum-sulfur clusters in general. Much of this research is centered on tetranuclear cubane-like clusters with $[MoFe_3S_4]$ cores and with various types of external ligands other than thiolates for which a variety of investigations are reported.

In the field of poly-homonuclear molybdenum thiolates and sulfide-thiolates with complete MoS_n coordination some di- and trinuclear species have been reported. An especially stable diamagnetic dinuclear di-μ-sulfido bridged Mo(V) species with a $[Mo_2O_2S_2]^{2+}$ or $[Mo_2S_4]^{2+}$ core in which each Mo atom is also bonded to a terminal oxo or sulfido group can be prepared by a variety of reactions (e.g. refs. 63-66, and lit. cited in refs. 67,68). Among them, the unique pair of isomeric *syn*- and *anti*-forms of the redox-active red-violet $[Mo_2S_4(SCH_2CH_2S)_2]^{2-}$ anion (Fig. 10) were obtained as the $[Et_4N]^+$ salt from the system $MoCl_3$/NaHS/Na_2edt/$[Et_4N]$Br in MeOH (refs. 69,67). $[Mo_2S_4(SCH_2CH_2S)_2]^{2-}$ is also produced from interesting redox reactions of $[NH_4]_2[Mo_2(S_2)_6]$ (ref. 70), of $[NH_4]_2[MoS_4]$ or of $[MoO_2(MeCHOCHOHMe)_2]$ with a large excess of ethane-1,2-dithiolate (or similarly with o-aminobenzenethiolate) which acts both as a reducing agent and as a ligand (refs. 71,72,68). $[Mo_2S_4(SPh)_4]^{2-}$ was prepared in a similar way (ref. 71).

In both isomers of $[Mo_2S_4(SCH_2CH_2S)_2]^{2-}$ (Fig. 10) two tetragonal pyramidal $[MoS_5]$ units with a terminal sulfido ligand in the axial position are connected via two μ-sulfido bridges, with nearly equal Mo...Mo distances of 286.3 and 287.8 pm for the *syn* and *anti*-isomer. Both isomers show reversible one-electron reductions. From molecular orbital calculations significant Mo-Mo bonding interaction is found to account for the diamagnetism of both isomers (ref. 73).

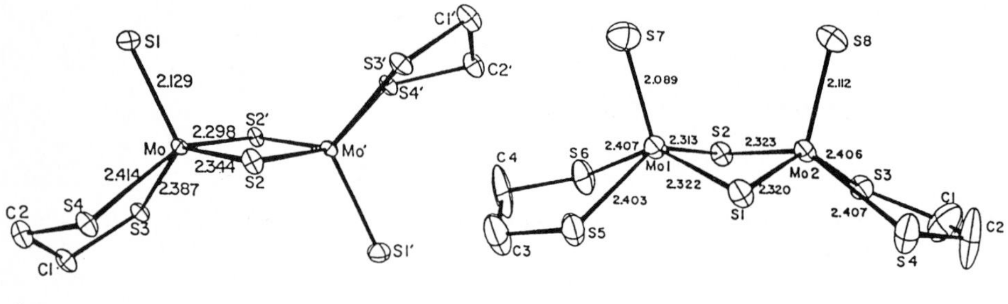

Fig. 10. Structures of the *anti*- (left) and *syn*-isomer (right) of $[Mo_2S_4(SCH_2CH_2S)_2]^{2-}$ in the $[Et_4N]^+$ salts (ref. 67).

Analogous dinuclear tungsten compounds with pentavalent metal centers, $[W_2S_4(SCH_2CH_2S)_2]^{2-}$ and $[W_2S_4(SC_6H_4NH)_2]^{2-}$, were successfully prepared from a very similar redox reaction of $[NH_4]_2[WS_4]$ with a large excess of ethane-1,2-dithiolate or 2-aminobenzenethiol (ref. 74). They represent rare examples of oligomeric tungsten thiolate complexes which remain a field to be explored (see also ref. 75). Structure and bonding in the diamagnetic $[W_2S_4(SCH_2CH_2S)_2]^{2-}$ anion which has the *syn*-configuration are very similar to the Mo analogue (Fig. 10, right). The W-W distance is 286.2 pm, indicating a single bond as in the Mo compound (refs. 73,74).

The first trinuclear molybdenum thiolate/sulfide clusters were recently obtained by substitution reactions of $[Mo_3S(S_2)_6]^{2-}$ with difunctional saturated thiolates. If solid $[NH_4]_2[Mo_3S(S_2)_6]$ (ref. 76) is reacted with a tenfold excess of $Na_2(SCH_2CH_2S)$ in MeOH or MeCN, the three terminal disulfide ligands are replaced by ethane-1,2-dithiolate, and dark brown $[Et_4N]_2[Mo_3S(S_2)_3(SCH_2CH_2S)_3]$.MeOH (**1**) is formed upon addition of NH_4Br (refs. 77,78). Using a twentyfold excess of thiolate, reductive elimination of the remaining three disulfides and replacement by S^{2-} lead to deep red-black $[Et_4N]_2[Mo_3S_4(SCH_2CH_2S)_3]$ (**2**) (refs. 77-79). **1** is also formed from **2** by reaction with 3 equiv. of S (as S_8) or dibenzyltrisulfide, and treatment of **1** with ethane-1,2-dithiolate, PPh_3, or CN^- in MeCN regenerates **2**; an excess of S produces again $[Mo_3S_{13}]^{2-}$ (ref. 79). Both novel Mo(IV) anions, $[Mo_3(\mu_3\text{-}S)(\mu\text{-}S_2)_3(SCH_2CH_2S)_3]^{2-}$ (**1a**), and $[Mo_3(\mu_3\text{-}S)(\mu\text{-}S)_3(SCH_2CH_2S)_3]^{2-}$ (**2a**), which are shown in Fig. 11, belong to a large class of trinuclear transition metal (M) complexes with $[M_3X_7]$ and $[M_3X_4]$ cores (X = O, S, halogen, for a review see ref. 80). The $[Mo_3S_7]^{4+}$ core of **1a** has also been characterized in $[(R_2PS_2)_3Mo_3S_7][R_2PS_2]$ (ref. 75) and in $Mo_3S_7Cl_4$ (ref. 81), the $[Mo_3S_4]^{4+}$ core in **2a** is very similar to that in $[Mo_3S_4(CN)_9]$ (ref. 82) and in $[Mo_3S_4(Cp)_3]^+$ (ref. 83).

The facile redox interconversion between disulfide and sulfide in **1a** and **2a** without net formal charge in metal oxidation state or

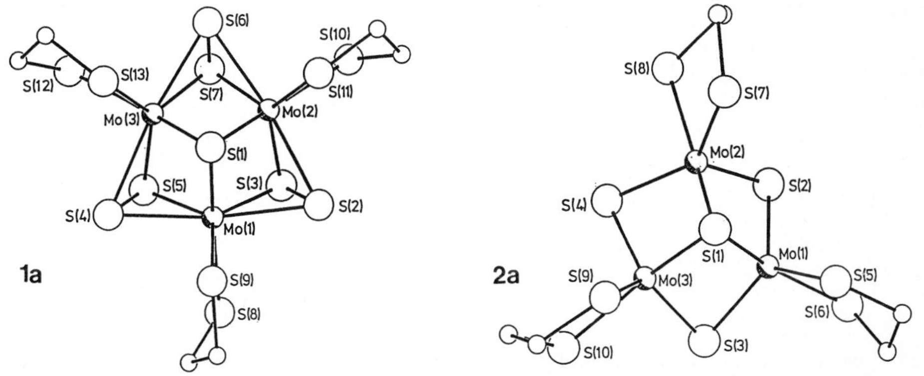

Fig. 11. Structures of $[Mo_3S_7(SCH_2CH_2S)_3]^{2-}$ (**1a**) (refs. 77,78) and $[Mo_3S_4(SCH_2CH_2S)_3]^{2-}$ (**2a**) (refs. 77-79); **1a**: mean Mo...Mo 276.8, Mo(1,2,3)-S(2,4,6) 250.2, Mo(1,2,3)-S(3,5,7) 240.9, Mo(1,2,3)-S(1) 237.0, Mo(1,2,3)-S(9,11,13) 244.6, Mo(1,2,3)-S(8,10,12) 250.6 pm. **2a**: similar, except mean Mo(1,2,3)-S(2,3,4) 229.4 pm (ref. 77).

in overall charge (see also ref. 84) leaves a vacant coordination position on each metal in **2a.** It makes this system an important model within the concept of sulfide "vacancies" being sites of catalytic activity in transition-metal sulfide-based heterogeneous catalysts. Also, based on the resonance Raman spectrum, **2a** is discussed as a hetero-analogues structural model for the 3Fe ferredoxin centers in *Desulfovibrio gigas* (ref. 79).

In spite of the continuous current interest in the chemistry of dinuclear complexes with metal-metal multiple bonds, there is still a lack of sulfur compounds in this class. As the first examples of thiolate analogues within the group of homoleptic d^3-molybdenum(III) compounds of formula $[X_3Mo{=}MoX_3]$ (X = bulky ligands such as ß-elimination-stabilized alkyls or bulky alkoxide groups), the derivatives with X = 2,4,6-trimethylbenzenethiolate and 2,4,6-triisopropylbenzenethiolate were reported (refs. 85,86). Orange-red $[Mo_2(SC_6H_2Me_3)_6]$ was at first prepared (a) by the reaction of $[1,2\text{-}Mo_2(S\text{-}t\text{-}Bu)_2(NMe_2)_4]$ with $C_6H_2Me_3SH$ (ref. 85), and then (b) with better yield from $MoCl_4$ and the thiolate in 1,2-dimethoxyethane as a solvent. Red $[Mo_2(SC_6H_2\text{-}i\text{-}Pr_3)_6]$ (Fig. 12) was obtained in a similar way in a mixture with $[MoO(SC_6H_2\text{-}i\text{-}Pr_3)_4]$ (ref. 86). The structures of both very stable hexathiolates contain well-shielded short Mo-Mo triple bonds of 222.8 and 223.9 pm length. They undergo a reversible one-electron reduction at ca. -0.85 V and a further irreversible one-electron reduction accompanied by thiolate anion loss (ref. 86).

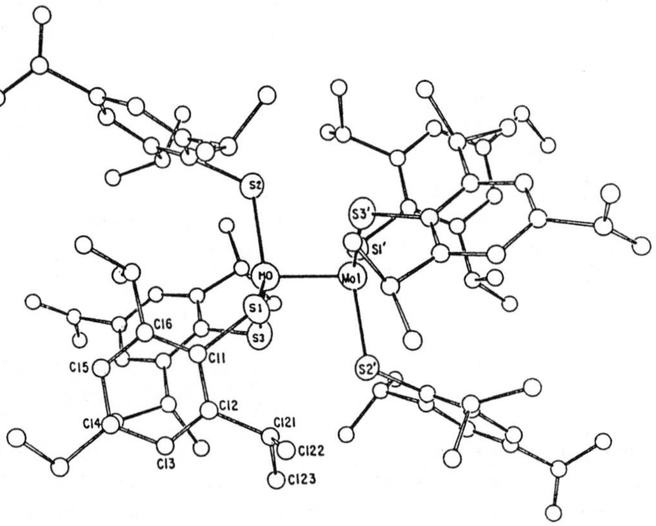

Fig. 12. Structure of $[Mo_2(SC_6H_2\text{-}i\text{-}Pr_3)_6]$. Bond lengths: Mo-Mo 223.9, Mo-S(1) 231.3, Mo-S(2) 234.8, Mo-S(3) 230.6 pm (ref. 86).

2.5 Molybdenum-Iron and Tungsten-Iron

The only second and third row transition metals with biological functions are molybdenum and tungsten, respectively. The search for models which are able to reflect characteristic properties of the molybdenum-iron-sulfur centers (M-centers) of nitrogenases, or of a cofactor (FeMo-co) which can be removed from the enzyme, in different laboratories has produced a highly advanced molybdenum-iron chemistry with biologically relevant sulfur ligands. Despite this fact, the coordination properties of molybdenum and iron in these biomolecules could neither be completely defined by synthetic models nor by investigations using native enzyme samples.

Nitrogenases are strictly limited to procariotic organisms and have been isolated from a number of different bacteria. Investigations of the enzyme complex have shown it to consist of the Fe protein responsible for electron transport and of the FeMo protein which contains the M-centers believed to be the catalytic sites for binding and reduction of substrates such as N_2, H^+ and C_2H_2 (see e.g. ref. 87 and references cited therein).

A convenient approach towards the Mo-Fe-S centers of nitrogenases started from 1 equivalent of $[Et_4N]_2[MoS_4]$, 3 equivalents of $FeCl_3$ and 9 equivalents each of EtSH and NaOMe in methanol. This work was conducted in 1978. The product obtained was recrystallized from MeCN/THF, resulting in black crystals of $[Et_4N]_3[Mo_2Fe_6S_9(SEt)_8]$ (ref. 88). The metal-sulfur frame of the $[Mo_2Fe_6S_9(SEt)_8]^{3-}$ anion is given in Fig. 13. At the same time the closely related complex salt $[(n\text{-}Bu)_4N]_3[Mo_2Fe_6S_8(SEt)_9]$ was independently reported and subsequently characterized by X-ray diffraction (refs. 89-91).

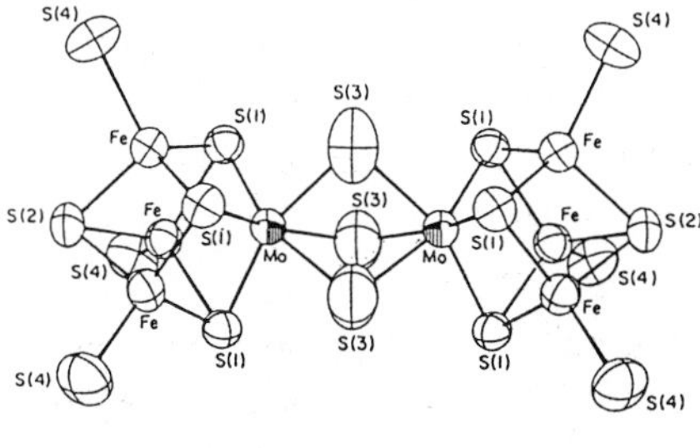

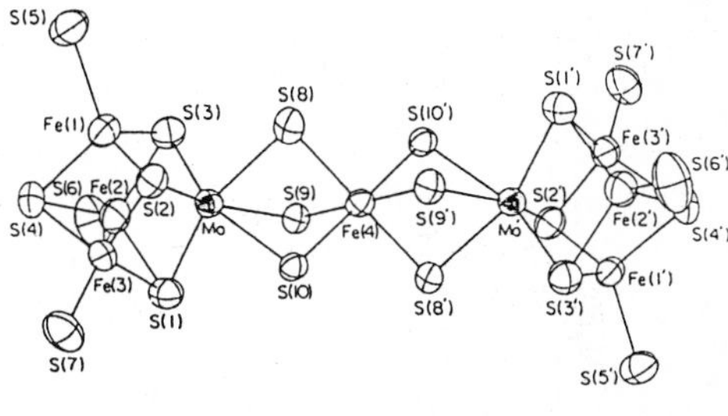

Fig. 13 (left). The M-S frame of $[Mo_2Fe_6S_9(SEt)_8]^{3-}$ (ref. 92).
Fig. 14 (right). The M-S frame of $[Mo_2Fe_7S_8(SEt)_{12}]^{3-}$ (ref. 95).

Both the $[Mo_2Fe_6S_8(SEt)_9]^{3-}$ (see also ref. 92) and the $[Mo_2Fe_6S_9(SEt)_8]^{3-}$ complex anion have been considered as models of the molybdenum sites in nitrogenases. They contain two $[Fe_3MoS_4]$ subunits with cubane-like geometries familiar from the $[Fe_4S_4]$ clusters (see section 2.8) which are connected via three sulfur bridges. In $[Mo_2Fe_6S_8(SEt)_9]^{3-}$, these bridges are thiolate functions exclusively, whereas in $[Mo_2Fe_6S_9(SEt)_8]^{3-}$ one thiolate bridge is replaced by a S^{2-} ion giving rise to positional disorder in the crystals. Due to their characteristic subunits, these compounds have been termed 'double-cubane' cluster complexes (ref. 92). At that time, direct structural information concerning the first and second set of molybdenum neighbours in the biological Mo-Fe-S centers was available from X-ray absorption spectroscopy (refs. 93, 94), and the Mo EXAFS characteristics of FeMo proteins from *Azotobacter vinelandii* and *Clostridium pasteurianum* are strikingly similar to that ones determined for $[Et_4N]_3[Mo_2Fe_6S_9(SEt)_8]$ (refs. 88,92).

An interesting extension of the $[Mo_2Fe_6S_8(SEt)_9]^{n-}$ double-cubane-like structure has been observed in the anions $[Mo_2Fe_7S_8(SEt)_{12}]^{3-}$ and $[M_2Fe_7S_8(SCH_2Ph)_{12}]^{4-}$ (M = Mo, W), the structure of the former one being depicted in Fig. 14. In all these cases the molybdenum (or tungsten) atoms of both subunits are not directly fused via three bridging functions but are bonded to a central iron atom each which is surrounded by an octahedron of sulfur atoms (refs. 95,96).

The material which has emerged in this field up to mid-1981 has been extensively reviewed elsewhere (ref. 97) and is therefore not discussed in complete detail here. Since that time numerous attempts have been performed to stabilize the $[MoFe_3S_4]$ subunits as isolated complexes in their own chemical environment, and a variety of other double-cubane-type complexes and derivatives have been synthesized and investigated by various methods (see e.g. refs.

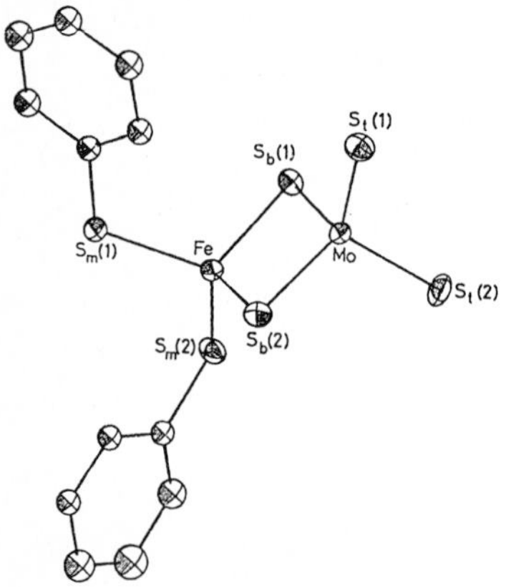

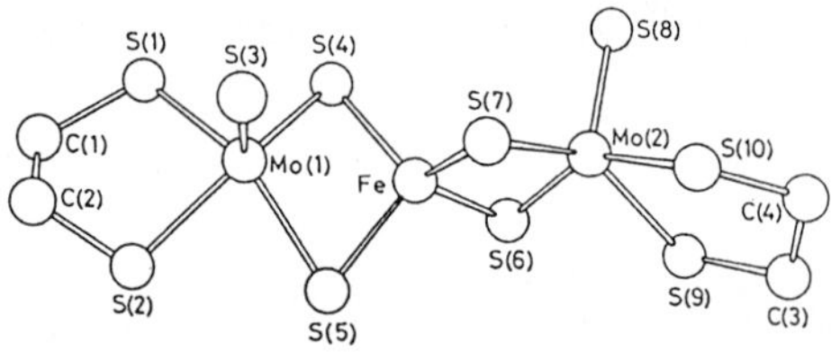

Fig. 15 (left). The structure of $[(PhS)_2FeS_2MoS_2]^{2-}$; distances: Mo...Fe 275.0, Mo-S_t 214.8 and 215.9, Mo-S_b 224.2 and 224.5, Fe-S_b 224.2 and 225.8, Fe-SPh 230.3 and 232.0 pm (ref. 147).
Fig. 16 (right). The structure of $[(SCH_2CH_2S)Mo(S)S_2FeS_2Mo(S)(SCH_2CH_2S)]^{3-}$ (ref. 155).

98-132). In the case of the single-cubane-type $[MoFe_3S_4]$ complexes, the molybdenum atoms are at least partly bonded to oxygen or phosphorous atoms (see e.g. refs. 133-146) and these clusters are therefore not discussed in detail here.

Another class of heteronuclear mixed sulfide-thiolate complexes is defined by the open-chain anions $[(PhS)_2FeS_2MS_2]^{2-}$ (M = Mo, refs. 147-150 and W, ref. 149) which have been isolated as tetraethylammonium and tetraphenylphosphonium salts, respectively. The binuclear complex $[(PhS)_2FeS_2MoS_2]^{2-}$ is shown in Fig. 15. Its central $[FeS_2Mo]$ frame with tetrahedrally coordinated metal centers has been discussed as a minimum structural basis for the Mo site in nitrogenase. The terminal ligands are sulfido groups with respect to Mo and thiophenolate groups with respect to Fe.

For complexes of general type $[(PhS)_nMS_2MoS_2]^{2-}$ it is not only possible to introduce Fe (M = Fe, n = 2) but also Cu (M = Cu, n = 1). The CuS_2Mo frame can be further extended by CuSPh synthones resulting in the formation of $[(PhS)CuS_2MoS_2Cu(SPh)]^{2-}$ (refs. 151,152) and similar complexes of higher nuclearity.

It is possible to identify these complexes by ^{95}Mo NMR spectroscopy (ref. 153). No direct copper counterpart is known for the mixed Ag-Mo complex $[Mo_2Ag(SCH_2CH_2S)_4]^+$, which has complete sulfur ligand spheres, too (ref. 154).

A related complex with an extended M-S frame but with iron acting as heteroatom has also been described. The trinuclear anion $[(SCH_2CH_2S)Mo(S)S_2FeS_2Mo(S)(SCH_2CH_2S)]^{3-}$ has been obtained by reaction of a mixture of $[NH_4]_2[MoS_4]$ and $Na_2(SCH_2CH_2S)$ in DMF/MeCN and precipitated as tetramethylammonium salt (ref. 155; see also refs. 156-160). The structure of the anion is shown in Fig. 16. The UV-Vis spectrum in DMF is similar to those observed for the binuclear Fe-Mo complexes discussed in the last section. The two Mo atoms are surrounded by five S atoms each two of which are thiolate functions, and the Fe atom is the center of a tetrahedron defined by bridging sulfide ions.

Though all Mo-Fe complexes discussed in this paragraph and certain other ones mimic certain structural features of the Mo sites of the M-centers of nitrogenase or of FeMo-co, none of them is a true analogue with respect to the second or perhaps to the third set of metal neighbours. Different models for the complete Mo-Fe-S cluster core have been proposed in the literature (for a review see ref. 161) but all of them suffer from substantial inconsistencies with respect to one or more important properties of the biological centers. The double- and single-cubane clusters containing the $[MoFe_3S_4]$ unit are perhaps the closest chemical approaches towards the Mo-Fe-S centers of nitrogenases, but these complexes have Fe/Mo/S ratios (3:1:4) significantly different from the values found in FeMo-co (6-8:1:4-6, see e.g. refs. 162,163). Moreover, the number of nearest iron neighbours determined for the iron atoms of FeMo-co by means of X-ray absorption spectroscopy (EXAFS region, ref. 363) is 0.4±0.1 and does not match with the corresponding properties in the synthetic complexes. For example, the number of nearest iron neighbours is exactly 1 in complexes with $[MoFe_3S_4]$ units. For these reasons we (refs. 164,165) and others (ref. 166) have proposed a model for the Mo-Fe-S centers of nitrogenases substantially different from other ones which is not only fully consistent with analytical data but also with both Fe and Mo EXAFS properties. This model has been discussed in section 2.8 and is depicted in Fig. 38.

2.6 Manganese

In spite of its diversity with respect to its oxidation states, the oxidation potentials of the higher ones do not allow stable thiolato and/or sulfido complexes, so that the known thiolates all contain Mn(II) and/or Mn(III). As there is a growing number of known manganese-containing metallobiomolecules (e.g. refs. 167-171 and lit. cited in refs. 36,172) with possible binding to thiolate sulfur (and/or, e.g., imidazole, carboxylate, phenoxide) from amino acid side-chains, interest in thiolates as model systems is increasing. Besides mononuclear compounds such as $[Mn(SCH_2CH_2S)_2]^{2-}$ (ref. 35), $[Mn(SPh)_4]^{2-}$ (refs. 173,174), $[Mn(S_2C_6H_3Me)_2]^{2-}$ and $[Mn(S_2C_6H_3Me)_2]^{-}$ ($S_2C_6H_3Me^{2-}$: toluene-3,4-dithiolate) (ref. 175), a number of di-, tri-, and tetranuclear species have been prepared.

If monofunctional ethanethiolate is employed as a ligand, $[Et_4N]_2[Mn_2(SEt)_6]$ with Mn(II) can be prepared by heterogeneous reaction of a 3:1:1 molar mixture of $NaSEt/MnCl_2/[Et_4N]Cl$ in acetonitrile (ref. 176). In its structure, which is shown in Fig. 17, the dinuclear complex is very similar to the isomorphous zinc(II) and cadmium(II) compounds: The centrosymmetric $[Mn_2S_6]$ core consists of two edge-sharing distorted tetrahedra; the complex has *anti*-configuration. Similar to the Zn and Cd isomorphs (however in contrast to the *anti* forms of the Fe(II) and Co(II) analogues, ref. 177), the planar $Mn_2(\mu\text{-}S)_2$ bridge units show slight but significant differences in the two independent $Mn\text{-}S_b$ distances (Fig. 17).

In a slightly distorted manner, the tetrahedral edge-shared $[Mn_2S_6]$ core can be stabilized with o-xylene-α,α'-dithiolate ($(S_2\text{-o-xyl})^{2-}$, $(SCH_2C_6H_4CH_2S)^{2-}$) as a ligand. The bite distance of this ligand matches the requirements for bidentate ligation at tetrahedral centers and can also bridge dinuclear systems. $[Mn_2(SCH_2C_6H_4CH_2S)_3]^{2-}$ (Fig. 17) was prepared in the system $Na_2(S_2$-o-xyl)/$MnCl_2.4H_2O/[Me_4N]Cl$ in methanol and was isolated as pink $[Me_4N]_2[Mn_2(SCH_2C_6H_4CH_2S)_3].3MeOH.1/2H_2O$ (refs. 77,178). From the three possible isomers of this type of complex, the one with a "basket-handle" dithiolate (S(5),S(6)), bridging two metals, is realized here. The other two ligands donate one sulfur each (S(1), S(3)) to the central $[Mn(\mu\text{-}S)_2Mn]$ unit, whereas the second sulfur (S(2), S(4)) of each ligand coordinates Mn terminally. Due to the steric strain of the S(5)-S(6) bite the central $[Mn(\mu\text{-}S)_2Mn]$ ring is folded about the S(1)...S(3) vector, yielding the *syn-exo* configuration for the molecule (refs. 77,178).

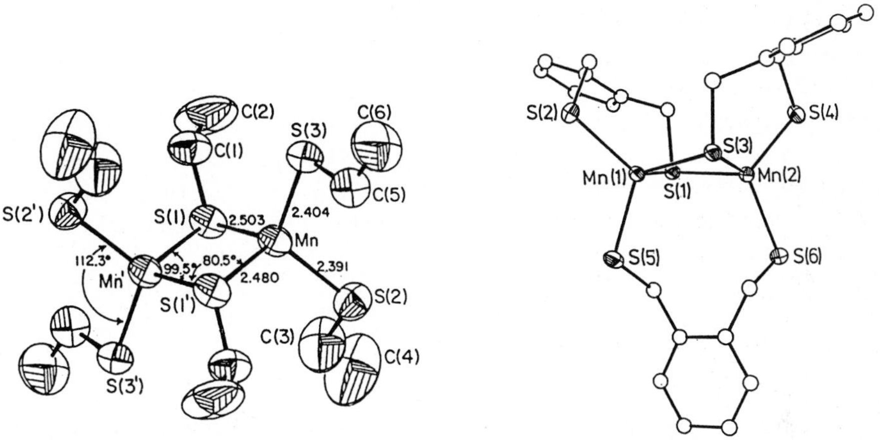

Fig. 17. $[Mn_2(SEt)_6]^{2-}$ (left). Mn...Mn 322.0 pm (ref. 176) and $[Mn_2(SCH_2C_6H_4CH_2S)_3]^{2-}$ (right) in $[Me_4N]_2[Mn_2(SCH_2C_6H_4CH_2S)_3].3MeOH.1/2H_2O$. Selected distances: Mn(1)...Mn(2) 312.4, Mn(1,2)-S_{br} 244.9-248.1, mean Mn(1,2)-S_t 238.5 pm (refs. 77,178).

A different reaction pattern results if EtS^-, as an aliphatic thiolate, is replaced in the system $MnCl_2.4H_2O/NaSR/[R_4N]^+$ by an aromatic one like benzenethiolate. At a Mn^{2+}/SR^- molar ratio of 1:2.5, adamantane-like $[Mn_4(SPh)_{10}]^{2-}$ forms as brown-red $[Me_4N]^+$ or $[Et_4N]^+$ salts (ref. 35). This tetranuclear anion is the largest polynuclear thiolato-manganate known up to now, and it is a member of the large class of transition metal thiolates known e.g. for Fe(II), Co(II), Zn(II), or Cd(II) with adamantane-like cage stereochemistry (reviewed in ref. 35). A view of a member of this class is given in Fig. 35.

A series of interesting dinuclear mixed-ligand thiolate complexes of manganese(II) is obtained if the primary polymeric product in the system Mn(II)/o-xylene-α,α'-dithiolate/MeOH is reacted with other sulfur donor ligands such as S^{2-}, monofunctional thiolates, or dithiocarbamates (e.g. $(Et_2NCS_2)^-$) (refs. 77,178). If benzenethiolate is used as the co-ligand, the resulting mixed-ligand anion $[Mn_2(SCH_2C_6H_4CH_2S)_2(SPh)_2]^{2-}$, which is shown in Fig. 18, derives from $[Mn_2(SCH_2C_6H_4CH_2S)_3]^{2-}$ (Fig. 17) by replacement of the "basket-handle" bridging $(S_2\text{-o-xyl})^{2-}$ by two monofunctional terminal thiophenolate ligands. However, as Fig. 18 shows, the mixed-ligand complex has *anti* configuration with approximate D_{2h} symmetry of the $[Mn_2S_6]$ molecular core, which is more stable in the absence of the steric requirements of the replaced bidentate ligand. In spite of the planar central $[Mn(\mu\text{-}S)_2Mn]$ ring, the Mn...Mn distance is very similar to the value in the folded ring of $[Mn_2(SCH_2C_6H_4CH_2S)_3]^{2-}$ (Fig. 17).

A similar dinuclear framework with an expansion of the manganese coordination spheres to trigonal-bipyramidal five-coordinate is formed if dithiocarbamate, $(Et_2NCS_2)^-$, is added to the above reaction system. In the resulting $[Mn_2(SCH_2C_6H_4CH_2S)_2(S_2CNEt_2)_2]^{2-}$ anion (Fig. 18) all interatomic distances in the core (Fig. 18) are longer than in the tetrahedral parent tris-dithiolate complex (Fig. 17). Magnetic measurements indicate antiferromagnetic coupling of the two high-spin d^5 centers (2.87 μ_B/Mn at 90 K) (refs. 77,178).

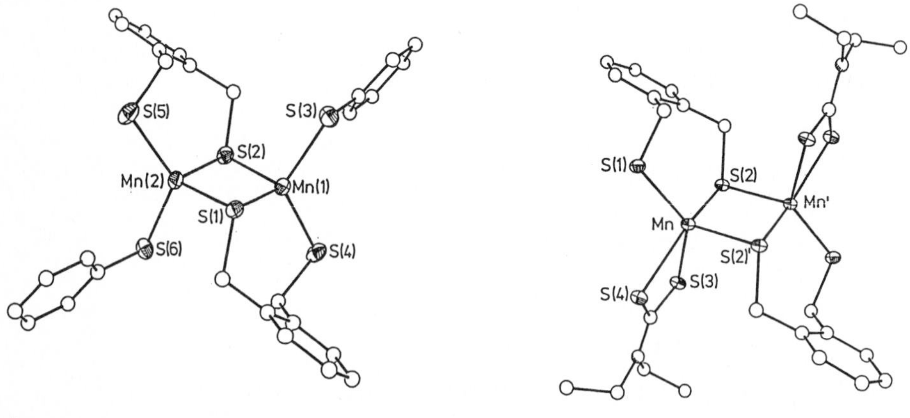

Fig. 18. Structures of $[Mn_2(S_2\text{-o-xyl})_2(SPh)_2]^{2-}$ (left) and $[Mn_2(S_2\text{-o-xyl})_2(S_2CNEt_2)_2]^{2-}$ (right) in the $[Et_4N]^+$ salts; distances: Mn(1)...Mn(2) 308.5, mean Mn(1,2)-S_t 239.6, Mn(1,2)-S_b 246.9 pm; Mn...Mn' 335.3, Mn-S(1) 241.3, Mn-S(2) 255.9, Mn-S(2)' 249.7, Mn-S(3) 255.9, Mn-S(4) 271.1 pm (refs. 77,178).

A number of interesting novel mono-, di-, and trinuclear thiolates of manganese(III) have recently been prepared by controlled air oxidation of Mn(II) thiolates. They are important as their existence supports the possible occurrence of Mn-S bonds in manganese(III)-containing biological systems.

Deep-green $[Mn_2(SCH_2CH_2S)_4]^{2-}$ was obtained from exposure of a solution of $[Mn^{II}(SCH_2CH_2S)_2]^{2-}$ to air with subsequent isolation as $[Ph_4P]^+$ and $[Et_4N]^+$ salts (refs. 35,36). The centrosymmetric anion, as shown in Fig. 19, has a dimer structure in which two bis-chelate monomers are joined laterally by two rather long Mn-S bonds. The five-coordination of Mn is between trigonal-bipyramidal and square-pyramidal. The Mn...Mn distance excludes direct Mn-Mn bonding. Magnetic and spectral data indicate dissociation of the dimer to solvated monomers in solution with some antiferromagnetic coupling in the solid (3.96 μ_B/Mn, ref. 35; J = -18.7 cm^{-1}, ref. 172) and a limiting d^4 full-spin value of 5.06 μ_B in solution. The large solvent-dependent separation between one-electron reduction and re-oxidation potential of ca. 0.5-0.7 V supports the existence of dissociation-solvation equilibria (ref. 35).

Interestingly, a reaction analogous to the above oxidation towards $[Mn_2(SCH_2CH_2S)_4]^{2-}$, but starting with the toluene-3,4-dithiolate ($(S_2C_6H_3Me)^{2-}$, tdt^{2-}) complex $[Mn^{II}(S_2C_6H_3Me)_2]^{2-}$ in MeOH, yields upon controlled air oxidation the mononuclear black thiolates $[Mn^{III}(S_2C_6H_3Me)_2]^-$ and $[Mn^{III}(S_2C_6H_3Me)_2(MeOH)]^-$ (refs. 175,179). They were isolated as the $[Ph_4P]^+$ double salt and are square-planar with methanol as the fifth axial ligand in the latter anion. Both are shown in Fig. 19 for comparison. Their existence strongly supports the reported model of dissociation and solvation of $[Mn_2(SCH_2CH_2S)_4]^{2-}$, and they show the d^4 high-spin value of 5.2 μ_B (ref. 175).

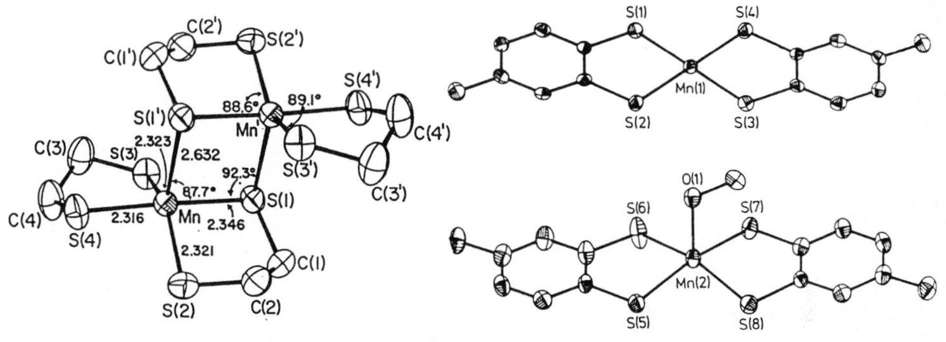

Fig. 19. The $[Mn_2(SCH_2CH_2S)_4]^{2-}$ anion (left); Mn...Mn 359.6 pm (refs. 35,36) in comparison to the mononuclear analogues $[Mn(S_2C_6H_3Me)_2]^-$ (mean Mn-S 228.2 pm) and $[Mn(S_2C_6H_3Me)_2(MeOH)]^-$ (mean Mn-S 231.0 pm) in the $[Ph_4P]^+$ salts (right) (ref. 175).

The stability of the manganese(III) complexes with ethanedithiolate and toluene-3,4-dithiolate is remarkable, especially towards the usual redox reaction:

$$Mn^{3+} + RS^- \longrightarrow Mn^{2+} + 1/2\ RSSR \qquad \text{Eq. (7)}$$

It is suggested (see also ref. 36) that this stability relative to the tetrahedrally coordinated Mn(II) is largely a function of the special nature of the chelating ligands. The ligand bites of both are unfavourably small for bidentate coordination of tetrahedral centers (this is well-known from many examples), but they match perfectly into five-membered chelate rings with bond angles of ca. 90° such as in the square-planar chelate systems of manganese(III). Thus, the usual redox instability of Mn(III) towards readily oxidizable thiolate functions is suppressed by use of appropriate dithiolate ligands (see also ref. 172).

A dinuclear complex with an unusual tetradentate thiolate ligand has recently been prepared from reaction of 1:2:4 mixtures in the system $MnCl_2/Na_3(ptt)/[PhCH_2Et_3N]Br/EtOH$ (ptt^{3-}: propane-1,2,3-trithiolate) and subsequent treatment with air (ref. 172). The anion, $[Mn_2(SCH_2CH(S)CH_2SSCH_2CH(S)CH_2S)_2]^{2-}$ (Fig. 20), has a core structure remarkably similar to $[Mn_2(SCH_2CH_2S)_4]^{2-}$. In the tetradentate ligand which has been formed as an oxidation product during complexation, two adjacent thiolate S atoms behave like edt^{2-} groups. The two mononuclear sub-units of the $[Mn_2S_8]$ core are connected through the weak Mn(1)-S(8) and Mn(1')-S(8') bonds (Fig. 20). According to the magnetic moments there is significant temperature-dependent antiferromagnetic coupling with a fit value of J = -18.7 cm^{-1} for the exchange parameter (ref. 172).

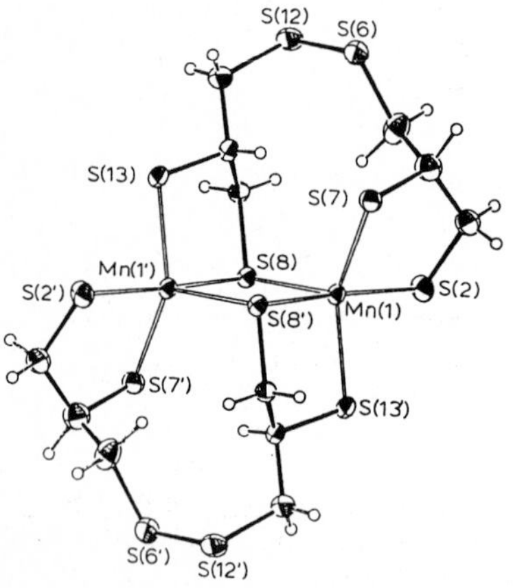

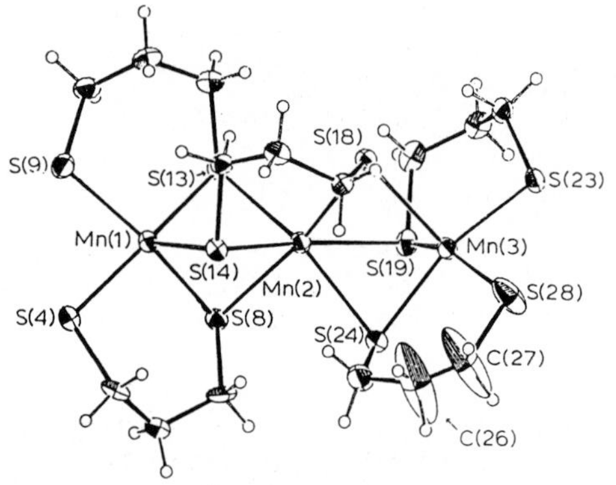

Fig. 20 (left). The $[Mn_2(SCH_2CH(S)CH_2SSCH_2CH(S)CH_2S)_2]^{2-}$ anion: Mn...Mn 359.8, Mn-S(8) 265.5, Mn(1)-S(8') 235.3 pm (ref. 172).
Fig. 21 (right). The trinuclear Mn(II,III) anion $[Mn_3(SCH_2CH_2CH_2S)_5]^{2-}$: Mn(1)...Mn(2) 312.3, Mn(2)...Mn(3) 310.1, Mn(1,3)-S_t 231.5-232.7, Mn(2)-S 255.4-267.7 pm (ref. 172).

Of very special interest is the recently reported trinuclear mixed-valence Mn(II,III) propane-1,3-dithiolate (pdt^{2-}) complex of composition $[Mn_3(SCH_2CH_2CH_2S)_5]^{2-}$ (ref. 172). It is the first example of a mixed-valence homoleptic metal thiolate and was obtained from air-oxidized solutions of $MnCl_2/Na_2pdt/[Ph_4P]Br$ (1:2:1) in the very same way as the above Mn(III) thiolates. In the structure, as shown in Fig. 21, the central Mn is assigned to be the Mn(II) atom; its six-coordinate thiolate environment is unique. The outer Mn(III) atoms in the linear 3Mn arrangement are five-coordinate with a geometry intermediate between trigonal-bipyramidal and square-pyramidal. The metal centers are also high-spin, the magnetic moments of 6.75 μ_B at 300 K decreasing to ca. 3.9 μ_B below 25 K indicates again antiferromagnetic coupling with exchange parameters $J_{12} = J_{23} = -18.3\ cm^{-1}$.

2.7 Technetium and Rhenium

Besides the interest in the general coordination and electronic properties, the growing relevance of technetium compounds as ^{99m}Tc radiopharmaceuticals in the field of nuclear medicine has greatly stimulated recent research in saturated and unsaturated thiolates of ^{99}technetium (see ref. 180). The chemistry of these compounds is largely restricted to various mononuclear complexes; no polynuclear systems are known yet, however, some dinuclear species could have been prepared in the field recently.

By reaction of ammonium pertechnetate with benzene-1,2-dithiol (H_2bdt) in aqueous ethanol/HCl and subsequent extraction with chloroform, wine-red crystals of $[Tc_2(bdt)_4].CHCl_3$ are obtained (refs. 181,182). The binuclear complex which is shown in Fig. 22 represents the first characterized example of a Tc(IV) dithiolate complex. Each Tc atom is coordinated to a rather regular novel type of trigonal-prismatic array of six sulfur atoms. The arrays are connected through a sharing quadrilateral face defined by the four bridging S atoms of two bdt^{2-} ligands to give a $[Tc_2S_8]$ core of approximate D_{2h} symmetry. The structure of $[Tc_2(bdt)_4]$ is characteristically different from the formally analogous $[M_2(edt)_4]^{2-}$ (M = Mn, Fe, V) anion structures, with the Mn (ref. 36) and Fe (ref. 34) compounds having five-coordinate distorted trigonal pyramidal geometries and the more closely related vanadium isomer (refs. 30-32) being severely distorted from trigonal prismatic coordination (see also ref. 37). The Tc-Tc bond length of 259.1 pm is similar to the V-V distance of 257.5 pm in $[V_2(SCH_2CH_2S)_4]^{2-}$ (ref. 30), but it is considerably longer than in systems such as $[Tc_2Cl_8]^{3-}$ (ref. 183) which have been assigned bond orders of 3.5 and above (for a discussion of the electronic properties relevant for the trigonal-prismatic coordination: see ref. 182).

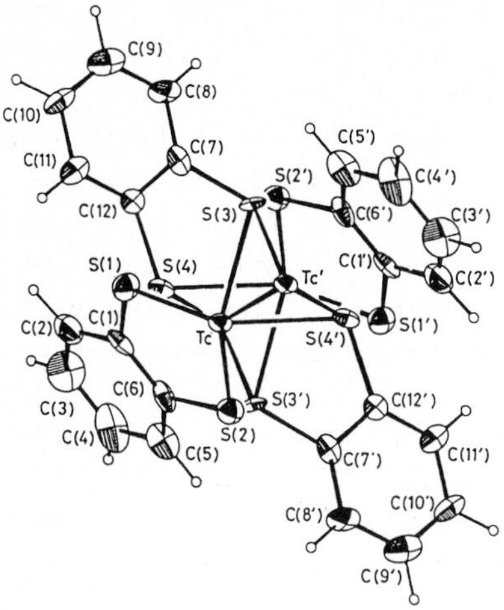

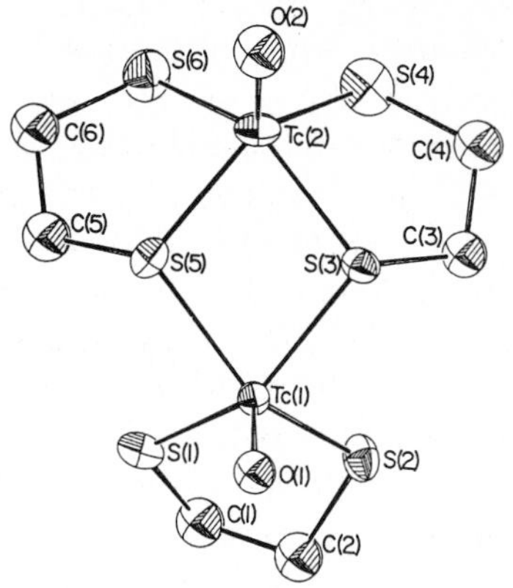

Fig. 22 (left). Structure of $[Tc_2(SC_6H_4S)_4]$: Tc-Tc 259.1, mean Tc-S(1,2) 229.5, mean Tc-S(3,4,3',4') 240.9 pm (refs. 181,182).
Fig. 23 (right). Structure of $[(TcO)_2(SCH_2CH_2S)_3]$: Tc(1)...Tc(2) 365.4, mean Tc(1)-S(1,2) 225.8, Tc(1)-S(3,5) 241.1, Tc(1)-O(1) 166.5, Tc(2)-S(4,6) 228.0, Tc(2)-S(3,5) 236.8, Tc(2)-O(2) 166.1 pm (ref. 184).

Another unique dinuclear technetium complex with predominant thiolate ligation, $[(TcO)_2(SCH_2CH_2S)_3]$ with pentavalent technetium (low-spin d^2), was synthesized as an orange solid by the reaction of $[TcOCl_4]^-$ with either an excess of bis(acetamidomethyl)ethane-1,2-dithiol or with 1.5 equiv. of ethane-1,2-dithiol (ref. 184). A by-product of the reaction is the known $[TcO(SCH_2CH_2S)_2]^-$ ion (refs. 185-187). Both Tc(V) in $[(TcO)_2(SCH_2CH_2S)_3]$ are in square-pyramidal coordination with four sulfur atoms in the base and a tightly bonded oxygen atom typical for Tc(V) in the apex. The structure as shown in Fig. 23 has four distinct types of Tc-S bonds with the shortest ones to the unique edt^{2-} ligand indicating a stronger trans influence in the Tc(1) coordination (ref. 184). The two square-pyramidal units are arranged in the *syn*-configuration. The Tc(1)...Tc(2) distance (Fig. 23) excludes any metal-metal bonding.

The nature of the reaction products of $[TcOCl_4]^-$ with various thiols and dithiols is largely governed by steric factors (refs. 184,188). With 2,3,5,6-tetramethylbenzenethiol (Htmbt) only mononuclear $[TcO(tmbt)_4]^-$ is formed; reaction of $[TcCl_6]^{2-}$ with Htmbt/Zn in MeCN/MeOH gives $[Tc^{III}(tmbt)_3(MeCN)_2]$ which is reactive toward CO and CNR (ref. 189) and which is reacted with pyridine-N-oxide to dinuclear $[(TcO)_2(tmbt)_6]$ (ref. 188). The bridging function of such sterically hindered thiolate is unexpected.

Among the few examples of oligomeric thiolate complexes of rhenium are the remarkable dinuclear $[Re_2(SPh)_7(NO)_2]^-$ and $[Re_2(SC_6H_4Me\text{-}4)_7(NO)_2]^-$ anions (ref. 190). They are prepared from the reaction of $[ReCl_2(OMe)(NO)(PPh_3)_2]$ with thiophenol or 4-methylthiophenol in the presence of triethylamine, in contrast to sterically more demanding o-substituted thiophenols which lead to mononuclear complexes. The very similar $[Re_2(SPh)_7(NNPh)_2]^-$ with diazenido coligands (which are of intrinsic interest as intermediates in dinitrogen assimilation) was obtained in an analogous way starting from $[ReCl(PPh_3)_2(NNPh)_2]$ (ref. 191).

The core of this type of complex (Fig. 24) consists of two octahedra sharing a face defined by three bridging thiolate donors. The asymmetric structure of the triple thiolate bridge implies the presence of a strong metal-metal bond which is confirmed by the short Re-Re distance (Fig. 24) and the observed diamagnetism. If the NO and NNPh ligands are considered to function as three-electron donors, an 18-electron count is attained by the formation of the Re-Re bond. Qualitative MO considerations in comparison to other triply bridged 34 and 36 electron systems support this picture (refs. 191,247). NMR data suggest a rapid exchange of the six equatorial thiolate ligands about a rigid Re(1)-S(2)-Re(2) core. The complexes show irreversible one-electron oxidations and reductions.

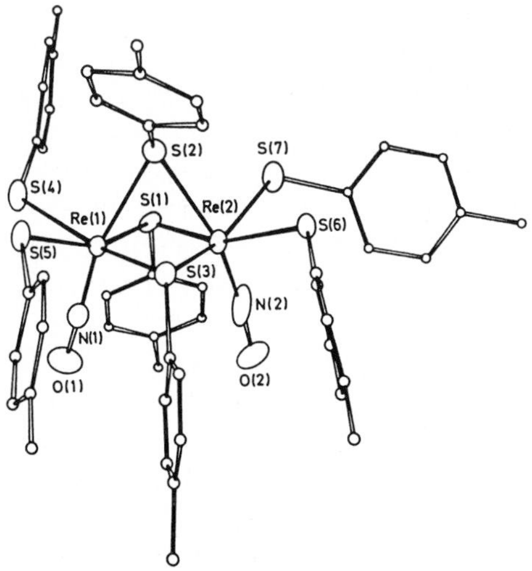

Fig. 24. Structure of $[Re_2(SC_6H_4Me\text{-}4)_7(NO)_2]^-$: Re(1)-Re(2) 278.3, mean Re(1,2)-S(1,3) 240.0, Re(1,2)-S(2) 256.4, Re(1,2)-S(4,5,6,7) 244.5, Re(1)-N(1) 172, Re(2)-N(2) 164 pm (ref. 190).

Whereas no tetranuclear rhenium cluster with complete thiolate/sulfide coordination has been reported yet, cubane-like [Re-

$(CO)_3SMe]_4$ is known as a member of the series of tetranuclear organothio-tricarbonyls $[M(CO)_3SR]_4$. In the very stable $[Re_4(SMe)_4]$ core with Re and S occupying alternating corners of the cube, low-valent Re is in octahedral coordination of three bridging thiolate-S (Re-S 249-253 pm) and three terminal CO ligands (Re-C 185-193 pm). There are no Re-Re bonding interactions in the highly symmetric tetramer (Re...Re 387.3-395.7 pm) (ref. 192 and lit. cited therein).

In addition, there is a number of mononuclear thiolate complexes of rhenium outside the scope of this review, including the oxo-complexes $[ReO(SCH_2CH_2S)_2]^-$ (ref. 187) and $[ReO(SPh)_4]^-$ (ref. 193), the neutral complex $[Re(SPh)_3(MeCN)(PPh_3)]$ (ref. 194), the trigonal prismatic trithiolate complex $[Re\{(SCH_2)_3CCH_3\}_2]^-$ (ref. 86), and the square-pyramidal sulfide-thiolate $[ReS(SCH_2CH_2S)_2]^-$ (ref. 195).

2.8 Iron

The most important transition metal in living matter and presumably also in the inanimate nature is iron. Due to the fact that the natural occurrence of iron is at least 10 times larger than that of any other transition element, all organisms known so far have developed specific iron binding sites during biological evolution and thus are able to make use of the unique coordinating properties of this metal in a variety of life processes such as transport, storage, electron transfer and detoxication. On the other hand, these properties have put their stamp on the technological evolution as well. The development of a synthetic iron chemistry with simple sulfur-containing ligands such as thiolate and sulfide ions during the past two decades can be traced back to the importance of biologically active iron-sulfur coordination sites which have been discovered in a variety of metalloenzymes and metalloproteins.

Biomolecules containing iron-sulfur entities include simple iron-sulfur proteins such as rubredoxins, ferredoxins and high-potential iron proteins (HIPIPs) and more complex metalloenzymes (refs. 197-198). Members of the latter group contain additional prosthetic groups such as nickel (e.g. certain hydrogenases and CO-dehydrogenases), molybdenum or vanadium (e.g. nitrogenases), flavin (e.g. xanthine dehydrogenase) or heme (e.g. UQ-cytochrome c reductase).

Up to 1980, three distinct iron-sulfur coordination sites (Fig. 25) have been identified as prosthetic groups in simple iron-sulfur proteins and characterized by X-ray crystallography (ref. 199).

They contain one, two, and four iron atoms, respectively, and are characteristic constituents of rubredoxins (type a, mononuclear 1Fe centers), 2Fe-ferredoxins (type b, binuclear [2Fe-2S] centers) and 8Fe-ferredoxins and high potential iron proteins (HIPIPs) (type c, tetranuclear [4Fe-4S] centers).

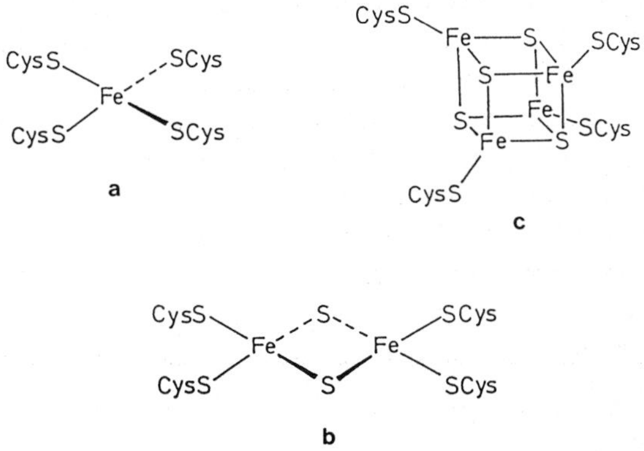

Fig. 25. The 'classic' types of iron-sulfur coordination sites in simple iron-sulfur proteins with terminal cysteine ligands; (a): 1Fe center of rubredoxins; (b): [2Fe-2S] center of chloroplast-type ferredoxins; (c): [4Fe-4S] centers of bacterial-type ferredoxins and HIPIPs.

The overall structural principles of these units including rubredoxin-type $Fe(S\text{-}Cys)_4$ coordination sites can be summarized as follows:

- each iron atom is located at the center of a more or less regular tetrahedron defined by four sulfur atoms
- in the polynuclear complexes (type b and type c) the FeS_4 tetrahedra share common edges resulting in short Fe...Fe distances of ca. 270 - 280 pm
- the bridging ligands are inorganic sulfide ions
- the other (non-bridging) functional groups are thiolate functions of cysteinyl residues (S-Cys) of the polypeptide chain

In the case of the [2Fe-2S] proteins, two biologically active core oxidation levels do exist, namely $[Fe_2S_2]^{2+}$ and $[Fe_2S_2]^{+}$; the situation is slightly more complicated in the [4Fe-4S] proteins where three different oxidation levels can be activated by organisms. The ferredoxins utilize the $[4Fe\text{-}4S]^{2+}/[4Fe\text{-}4S]^{+}$ redox couple, while the HIPIPs depend on the $[4Fe\text{-}4S]^{3+}/[4Fe\text{-}4S]^{2+}$ oxidation states (refs. 196,200,201).

A combined e.p.r. and Mößbauer study of a 7Fe-ferredoxin from *Azotobacter vinelandii* (*Av* FdI, ref. 202) as well as of a ferredoxin from

Desulfovibrio gigas (*Dg* FdII, ref. 203) in 1980 revealed for the first time the existence of trinuclear coordination sites in iron-sulfur proteins. However, the X-ray structure determination of crystalline *Av* FdI was interpreted in terms of unusual $[Fe_3S_3(S\text{-}Cys)_5L]$ protein centers (Fig. 26(a), refs. 204,205) which have been proven by resonance Raman spectroscopy not to be identical with the corresponding coordination sites of the native enzyme (ref. 206) and other ones in different 3Fe proteins.

In due course, EXAFS investigations of *Desulfovibrio gigas* ferredoxin II (*Dg* FdII, ref. 207) and beef heart aconitase (ref. 208) gave evidence for a biologically significant [3Fe-4S] coordination unit which derives from the cubane-like [4Fe-4S] cluster core of the tetranuclear protein centers simply by removal of one iron atom. A recent reinvestigation of the structure of *Av* FdI (ref. 209) confirmed the structural model which has been derived from suitable model complexes (Fig. 26(b), refs. 210,165).

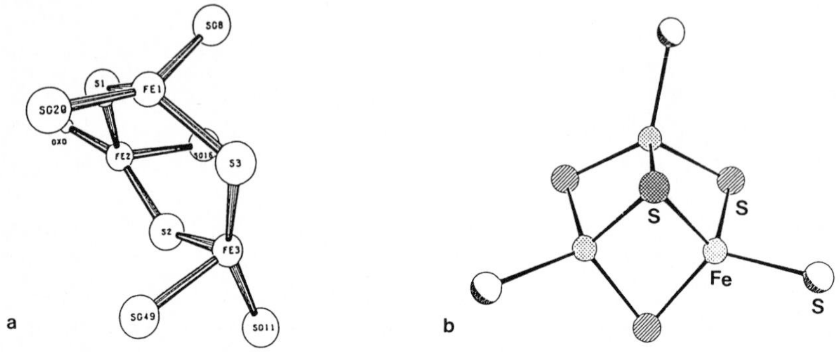

Fig. 26 (a) The $[Fe_3S_3(S\text{-}Cys)_5L]$ coordination unit derived from the X-ray diffraction pattern of a single crystal of *Av* FdI (ref. 6); (b) structural model of the [3Fe-4S] site of 3Fe proteins as derived from synthetic complexes (ref. 165).

Efforts to model the active centers of iron-sulfur proteins resulted in the successful syntheses and characterization of a large body of true [2Fe-2S] and [4Fe-4S] site analogues in different core oxidation states. These compounds have been reviewed elsewhere (refs. 197,198) and are therefore not discussed in complete detail here. It is interesting to note, however, that up to now no direct chemical and/or electronic model for the [3Fe-4S] protein sites could be synthesized. The closest synthetic approach towards these centers mimics the oxidation state and chemical definition, but is a condensation product of two trinuclear subunits (refs. 211,212,215,164,165,214), whereas a simple trinuclear sulfide-thiolate complex lacks chemical identity, but reflects important topological features (refs. 210,215, see below).

(i) Binuclear Complexes. The first binuclear iron-sulfide-thiolate complex was reported in 1973. $[Fe_2S_2(SCH_2C_6H_4CH_2S)_2]^{2-}$ (Fig. 27) was obtained by reaction of one equivalent of $FeCl_3$ with two equivalents of $HSCH_2C_6H_4CH_2SH$ and NaOMe in methanol, followed by addition of one equivalent of a methanolic solution of sodium methoxide/sodium hydrosulfide. The anion, which contains trivalent iron centers, has been proven to be the first well-characterized synthetic analogue of the 2Fe proteins and was isolated as its red-black $[Et_4N]^+$ and $[Ph_4As]^+$ salt, respectively (refs. 216,217). The $[Fe_2S_6]$ frame with tetrahedral FeS_4 coordination sites includes a central Fe_2S_2 unit and represents a binuclear fraction of the infinite linear chain of condensed FeS_4 tetrahedra present in natural occurring minerals such as erdite ($NaFeS_2.2H_2O$, ref. 218) with terminal thiolate ligands and bridging sulfide ions.

Several other members of the $[Fe_2X_2(YR)_4]^{2-}/[Fe_2X_2(YR)_4]^{3-}$ complex family (X, Y = S, Se) have been described (e.g. refs. 213,219-223) and investigated by Mößbauer spectroscopy (e.g. refs. 224-226), electrochemical and magnetic measurements (e.g. refs. 224,225,227), e.p.r. spectroscopy (e.g. refs. 225,226,228,229), vibrational spectroscopy (e.g. ref. 230), EXAFS spectroscopy (e.g. ref. 231) and ^1H- (e.g. ref. 219) as well as ^{13}C-NMR spectroscopy (e.g. ref. 232). In this context it is worth to note that pure inorganic derivatives in this biologically important class of complexes such as $[Fe_2S_2X_4]^{2-}$ (X = Cl, Br, I) (refs. 233,234), $[Fe_2S_2(S_5)_2]^{2-}$ (ref. 235) and $[Fe_2Se_2(Se_5)_2]^{2-}$ (Fig. 27, ref. 236) do exist as well. A convenient route to $[Fe_2S_2(SR)_4]^{2-}$ complexes by reaction of $[Fe(SR)_4]^-$ with elemental sulfur has recently been

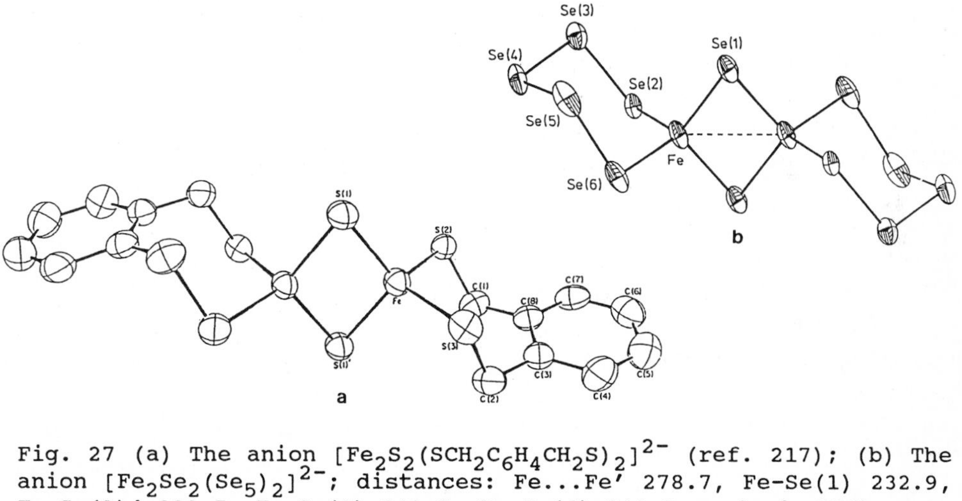

Fig. 27 (a) The anion $[Fe_2S_2(SCH_2C_6H_4CH_2S)_2]^{2-}$ (ref. 217); (b) The anion $[Fe_2Se_2(Se_5)_2]^{2-}$; distances: Fe...Fe' 278.7, Fe-Se(1) 232.9, Fe-Se(1)' 231.7, Fe-Se(2) 242.8, Fe-Se(6) 242.1 pm (ref. 236).

described (ref. 237).

Another important family of binuclear complexes is defined by homo- and heteroleptic iron thiolates all but one of general formula $[Fe_2(SR)_6]^{2-}$. Within this group, all members known so far contain two FeS_4 tetrahedra which share a common edge. In this respect, there are similarities with the binuclear mixed sulfide-thiolate complexes. The main differences between both classes of compounds arise from the chemical nature of the bridging ligands which are μ-thiolate functions here in contrast to unsubstituted sulfide ions associated with the ferredoxin-type centers. Compared with doubly charged sulfide ions, monovalent thiolate bridges introduce less negative charges thus allowing for reduced oxidation states of the metal. It is therefore not surprising that the most stable oxidation state of iron in these complexes with the coordination number of 4 is +2, and oxidized species have not been isolated yet. The only binuclear iron thiolate in a higher oxidation state known so far is the aforementioned exception of formula $[Fe_2(SCH_2CH_2S)_4]^{2-}$, which contains trivalent iron centers surrounded by five sulfur atoms each (refs. 33,34,239). Its structure is depicted in Fig. 28 showing the anion to consist of two monomeric subunits with slightly distorted square-planar FeS_4 coordination sites (mean Fe-S bond length 224.2 pm). Both subunits are associated via long axial Fe-S bonds (250.3 pm) to give an overall coordination number of 5 for each metal site.

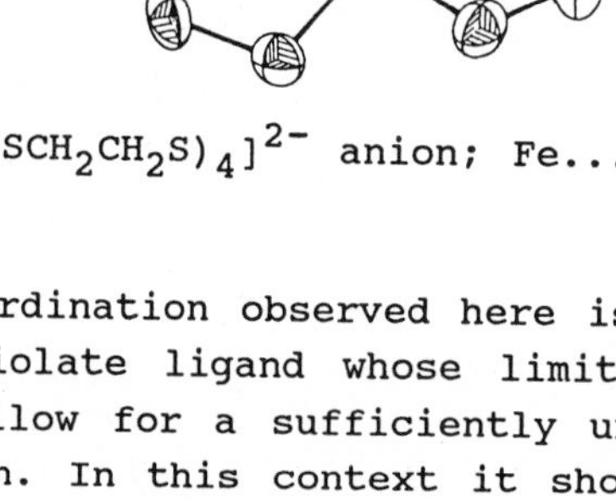

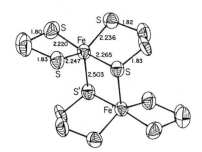

Fig. 28. The $[Fe_2(SCH_2CH_2S)_4]^{2-}$ anion; Fe...Fe distance: 341.0 pm (ref. 239).

The unusual coordination observed here is certainly introduced by the ethanedithiolate ligand whose limited S...S bite is not large enough to allow for a sufficiently undistorted tetrahedral sulfur coordination. In this context it should be mentioned that the related benzene-1,2-dithiolate ligand and its methylated derivative toluene-3,4-dithiolate form binuclear complexes of coordination number 5 as well (refs. 240,241).

The first binuclear thiolate complex with tetrahedral FeS_4 sites has been observed as an unexpected product during initial attempts to synthesize analogues for the 1Fe protein sites and identified as $[Fe_2(SCH_2C_6H_4CH_2S)_3]^{2-}$ (refs. 242,243). Later its structure in the complex salt $[Et_4N]_2[Fe_2(SCH_2C_6H_4CH_2S)_3]$.2MeCN has been proven to be of type IIa (Fig. 29) (refs. 244,177). The bifunctional thiolate ligands lead to an *anti*-configuration of the central (slightly nonplanar) $[Fe_2S_2]$ unit with a Fe...Fe distance of 303.4 pm. The binuclear complex $[Fe_2(SEt)_6]^{2-}$ with a similar (more regular) iron-sulfur framework but with monofunctional thiolate ligands has also been described (ref. 177). Due to a crystallographically imposed inversion center the central $[Fe_2S_2]$ unit is perfectly planar (Fe...Fe 297.8 pm).

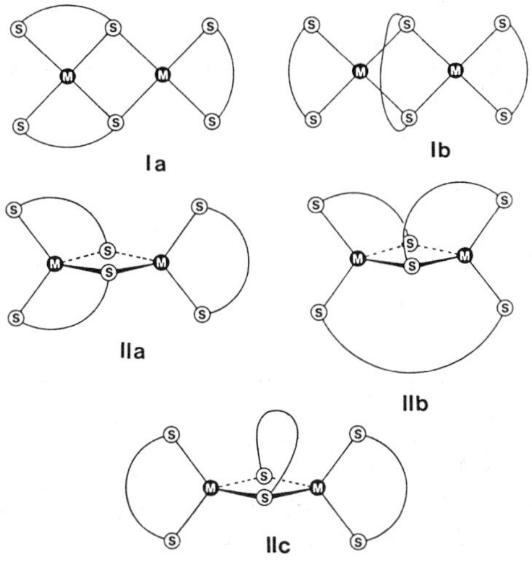

Fig. 29. Possible structures of dinuclear tetracoordinated metal-(II) dithiolates $[M_2(SRS)_3]^{2-}$: Ia,b square planar; IIa-c tetrahedral metal coordination (ref. 245).

At the same time isomer IIb of $[Fe_2(SCH_2C_6H_4CH_2S)_3]^{2-}$ has been established in crystals of $[Me_4N]_2[Fe_2(SCH_2C_6H_4CH_2S)_3]$.5MeOH (ref. 246). Its structure is depicted in Fig. 30(a). The unusual coordination of one o-xylene-α,α'-dithiolate ligand which does not act as a chelating group but occupies terminal coordination positions on both metal sites introduces considerable steric strain (S...S bite distance 425.3 pm). As one of the results, the central $[Fe_2S_2]$ unit is no longer planar but folded along its S...S vector resulting in a *syn-exo*-configuration which has been observed here for the first time with respect to metal-thiolate complexes. Due to this folding, the Fe...Fe distance is diminished to 287.9 pm. Similar complexes containing the 'unconventional' o-xylene-α,α'-dithiolate ligand have been synthesized in the manganese-thiolate and cobalt-thiolate reaction systems (see sections 2.6 and 2.9).

If the dark brown precipitate which is formed in the reaction of $FeCl_2$ and equimolar amounts of $Na_2(SCH_2C_6H_4CH_2S)$ in methanolic solution is treated with thiophenolate, it dissolves readily resulting in the formation of $[Fe_2(SCH_2C_6H_4CH_2S)_2(SPh)_2]^{2-}$. This anion, which is the first heteroleptic iron thiolate, exhibits the more usual *anti*-configuration (Fig. 30(b)) and has been precipitated as $[Me_4N]^+$ salt of formula $[Me_4N]_2[Fe_2(SCH_2C_6H_4CH_2S)_2(SPh)_2]$-.1/3MeCN.2/3$C_5H_{10}O$ from a mixture of MeCN/$C_5H_{10}O$/EtOH (ref. 246).

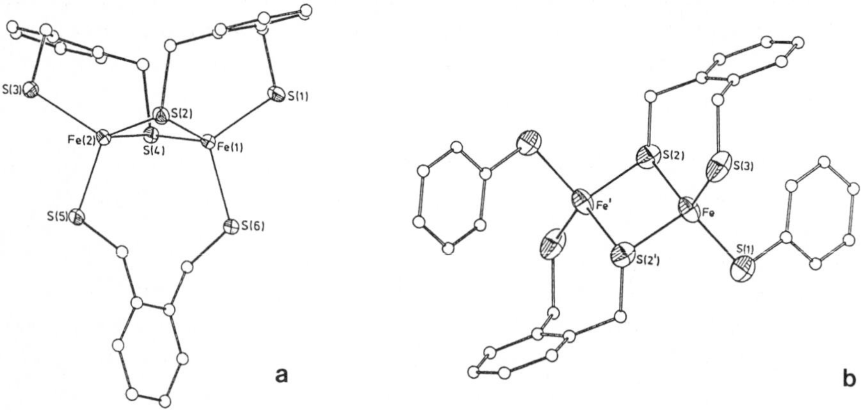

Fig. 30. (a) Structure of $[Fe_2(SCH_2C_6H_4CH_2S)_3]^{2-}$ (isomer IIb): Fe...Fe 287.9, S(1)...S(3) 671.5, S(5)...S(6) 425.3, Fe-S_{br} 235.1 - 236.9(4x), Fe-S_t 229.1 - 231.4 (4x) pm, S-Fe-S_{av} 109.47 (12x), Fe-S-Fe 75.06 and 75.19° (ref. 246); (b) Structure of $[Fe_2(SCH_2$-$C_6H_4CH_2S)_2(SPh)_2]^{2-}$: Fe...Fe 297.7, Fe-$S_{br}$ 230.3 (4x), Fe-S_t 236.3 - 236.8 (4x) pm, S-Fe-S_{av} 109.14 (12x), Fe-S-Fe 77.97 (2x)° (ref. 246).

It is interesting to note here that complexes of formula $[Fe_2(SR)_6]^{2-}$ with aromatic thiolates in bridging positions are not known. The most likely explanation is a significant contribution of sp^2 hybridisation to the valence state of sulfur in thiophenolate type ligands which prevents the orbitals of sulfur to overlap with corresponding valence orbitals of iron efficiently. Aliphatic thiolates, however, are much better suited to realize acute Fe-S-Fe valence angles in the region of 73 - 80° which are typical for cyclic $[Fe_2S_2]$ portions in structures assembled from condensed FeS_4 tetrahedra due to their hybrid functions predominantly sp^3 in character.

The monofunctional thiophenolate ligands of $[Fe_2(SCH_2C_6H_4$-$CH_2S)_2(SPh)_2]^{2-}$ can be replaced by bifunctional 1,1-dithiolates in appropriate ligand substitution reactions or by direct synthetic approaches (ref. 248). As an example, the reaction of polymeric

$[Fe(SCH_2C_6H_4CH_2S)]_n$ with dimethyldithiocarbamate groups leads to the binuclear compound $[Fe_2(SCH_2C_6H_4CH_2S)_2(S_2CNMe_2)_2]^{2-}$ with expanded coordination spheres of iron (ref. 241). The realisation of coordination number 5 produces larger Fe...Fe separations (316.3 pm) and less acute Fe-S-Fe valence angles (81.08°). The magnetic behaviour of the binuclear iron thiolate complexes including $[Fe_2(SCH_2C_6H_4CH_2S)_2(S_2CNMe_2)_2]^{2-}$ is characterized by significant antiferromagnetic coupling of electron spins which can be traced back either to superexchange mechanisms mediated by bridging ligand functions or to direct metal-metal interactions (ref. 249). The heteroleptic complexes discussed in the previous section find their counterparts in the manganese/thiolate systems (for a discussion of the corresponding $[Mn_2(SCH_2C_6H_4CH_2S)_2(SPh)_2]^{2-}$ and $[Mn_2(SCH_2C_6H_4CH_2S)_2(S_2CNEt_2)_2]^{2-}$ complexes see section 2.6).

(ii) Trinuclear Complexes. This series of compounds consists of sulfide-thiolate complexes only; they can be classified into trinuclear cyclic complexes of divalent iron and trinuclear linear-chain complex anions of trivalent iron, respectively. Trinuclear homo- and heteroleptic thiolates are not known.

The first synthetic trinuclear iron-sulfide-thiolate complex was reported in 1981. The exceptional anion $[Fe_3S(SCH_2C_6H_2Me_2CH_2S)_3]^{2-}$ (Fig. 31) has initially been identified from the products obtained by reaction of 1,2-bis(mercaptomethyl)-4,5-dimethylbenzene, sodium methanolate, ferric chloride and p-thiocresol in methanolic solution and could be isolated as its $[Et_4N]_2[Fe_3S(SCH_2C_6H_2Me_2CH_2S)_3]$.MeOH complex salt (ref. 210). In subsequent investigations, more rational synthetic routes towards trinuclear complexes of general formula $[M_3S(SR)_6]^{2-}$ utilizing NaHS (M = Fe, Co and Ni, refs. 246,250,251) or Li_2S (M = Fe, Co, ref. 215) have been developed, and the selenium-containing derivative $[Fe_3Se(SCH_2C_6H_4CH_2S)_3]^{2-}$ has also been reported (ref. 246).

The $[Fe_3S_7]$ framework of $[Fe_3S(SCH_2C_6H_2Me_2CH_2S)_3]^{2-}$ is a structural model for the [3Fe-4S] centers of 3Fe proteins (Fig. 26(b), refs. 207-209) and can be considered to be a condensation product of three FeS_4 tetrahedra each of which shares one edge with each of the remaining ones. As a result, an incomplete cubane-type structure is formed which lacks one FeX unit of the complete $[Fe_4X_8]$ portion depicted in Fig. 33(b). The triangular 3Fe unit is bridged by a μ_3-S ion which belongs to all three FeS_4 tetrahedra simultaneously. Additional μ-$S_{thiolate}$ bridges, which are unique features of $[Fe_3S(SR)_6]^{2-}$ complexes in iron-sulfide-thiolate chemistry, allow for a metal oxidation state of +2, another unique feature of these complexes. In this respect, the trinuclear compound lacks chemical identity with the [3Fe-4S] protein sites which have four sulfur bridges exclusively defined by sulfide ions. Consequently, the oxidation state of iron exceeds +2 in the latter case, and a mixed-valent behaviour (mean oxidation state 2.67) is observed (ref. 252).

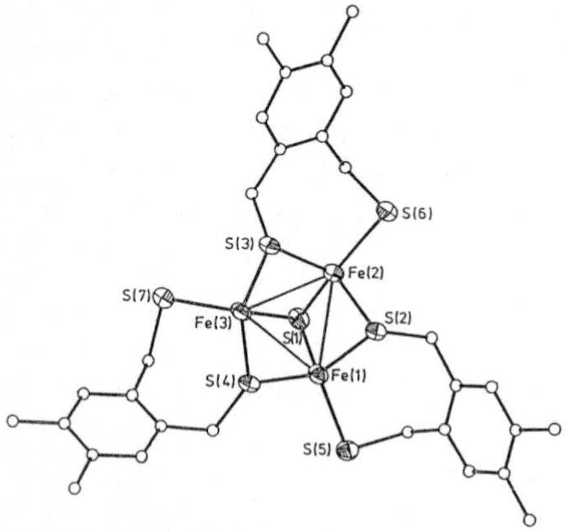

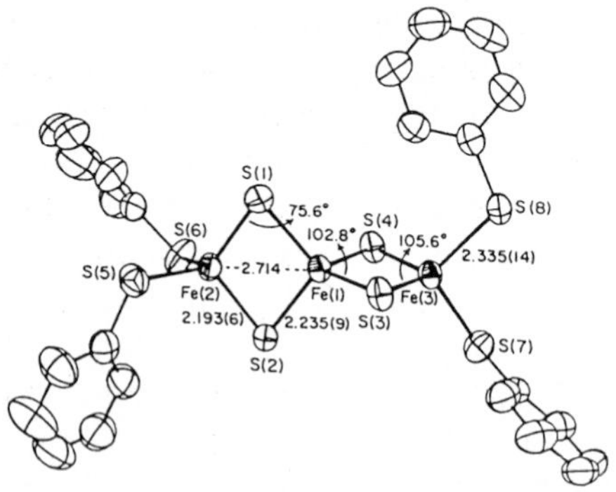

Fig. 31 (left). Structure of the cyclic trinuclear anion $[Fe_3S(SCH_2C_6H_2Me_2CH_2S)_3]^{2-}$: Fe...Fe 278.0 - 281.1, Fe-(μ_3-S) 230.2 - 230.4, Fe-(μ-S) 233.2 - 236.0, Fe-S_t 228.4 - 230.4 pm, Fe-Fe-Fe 59.39 - 60.47, Fe-(μ_3-S)-Fe 74.3 - 75.2, Fe-(μ-S)-Fe 72.3 - 73.7, (μ_3-S)-Fe-(μ-S) 102.6 - 104.0, (μ_3-S)-Fe-S_t 119.1 - 121.3, (μ-S)-Fe-(μ-S) 107.2 - 109.3, (μ-S)-Fe-S_t 107.7 - 112.9 ° (ref. 210).
Fig. 32 (right). The linear trinuclear anion $[Fe_3S_4(SPh)_4]^{3-}$ (ref. 244).

The reaction of $[Et_4N]_2[Fe(SEt)_4]$ with elemental sulfur in dry acetone afforded the trinuclear linear sulfide-thiolate complex $[Fe_3S_4(SEt)_4]^{3-}$ which was isolated as $[Et_4N]^+$ salt (ref. 244). Its $[Fe_3S_8]$ framework represents a trinuclear fraction of the infinite linear chain of condensed FeS_4 tetrahedra present in natural occurring minerals such as erdite ($NaFeS_2 \cdot 2H_2O$, ref. 218) and is thus an extension of the $[Fe_2S_6]$ core observed in binuclear sulfide-thiolate complexes (see above). The complex has been claimed to be a structural isomer of the $[Fe_3S_4]^+$ unit in iron-sulfur proteins (ref. 253) but lacks the triply bridging sulfide ion which has to be present in the corresponding biomolecule due to steric requirements provided the metal atoms form a regular triangle with an edge length of ca. 270 pm. The iron atoms are antiferromagnetically coupled, and an S = 5/2 ground state has been determined (ref. 253). A derivative containing thiophenolate ligands (Fig. 32) is also known (refs. 244,213).

(iii) <u>Tetranuclear Complexes</u>. $[Fe_4S_4(SCH_2Ph)_4]^{2-}$ was the first synthetic iron-sulfide-thiolate complex reported in the literature and introduced as a synthetic analogue of bacterial iron-sulfur proteins (refs. 254,255). The complex anion has been prepared by reaction of $FeCl_3$, NaOMe, NaSH and $HSCH_2Ph$ in methanol and isolated as tetraethylammonium salt. Up to now a large body of derivatives of general formula $[Fe_4S_4(SR)_4]^{z-}$ with z = 1 (refs. 256-261), z = 2 (refs. 262-305,220,231,257), and z = 3 (refs. 306-321, 275, 276, 220, 256) have been investigated by numerous techniques. It should be pointed out that several closely related tetranuclear systems with oxygen, selenium and tellurium ligands could be prepared as well, e.g. $[Fe_4Se_4(SR)_4]^{2-}$ (refs. 323,324,275,220) and $[Fe_4Se_4(SR)_4]^{3-}$ (refs. 275,220). Structurally characterized examples include $[Fe_4S_4(OR)_4]^{2-}$ (ref. 325), $[Fe_4Se_4(SPh)_4]^{2-}$ (ref. 323), $[Fe_4Se_4(OR)_4]^{2-}$, $[Fe_4S_4(SePh)_4]^{2-}$, $[Fe_4S_4(TePh)_4]^{2-}$, $[Fe_4Se_4(SePh)_4]^{2-/3-}$, $[Fe_4Se_4(TePh)_4]^{3-}$, $[Fe_4Te_4(SPh)_4]^{2-/3-}$, $[Fe_4Te_4(SePh)_4]^{3-}$ (refs. 326-328) and $[Fe_4Te_4(TePh)_4]^{3-}$ (Fig. 33(a), refs. 326-329).

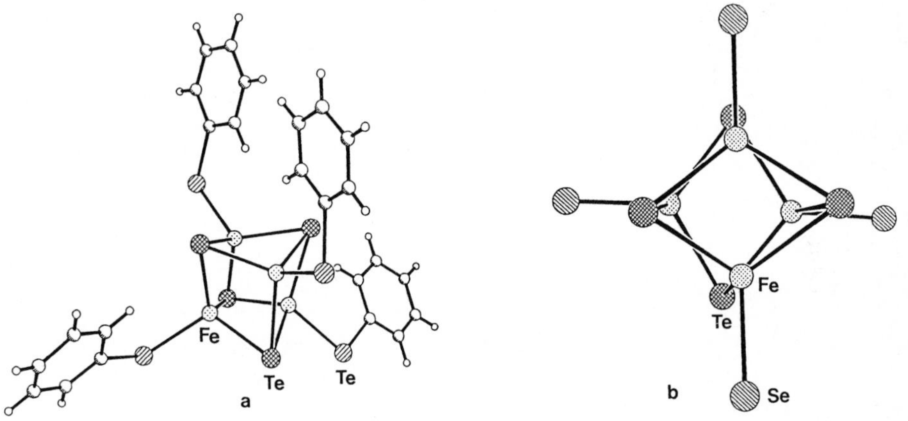

Fig. 33. (a) The anion $[Fe_4Te_4(TePh)_4]^{3-}$ in crystals of $[Et_4N]_3[Fe_4Te_4(TePh)_4]$.2MeCN; selected distances: Fe...Fe 262.4 and 263.9 (2x), 278.1 - 282.4 (4x), Fe-Te 256.3 - 258.7 (4x), 259.7 - 264.6 (8x) (ref. 329); (b) The $[Fe_4Te_4Se_4]$ framework of $[Fe_4Te_4(SePh)_4]^{3-}$ (ref. 328).

These compounds, together with the halide series $[Fe_4S_4X_4]^{2-}$ (X = Cl, Br, I, ref. 233), define a family of complexes with (distorted) cubanoidal $[Fe_4X_4]$ cages and tetrahedral iron coordination. The rigid metal-chalcogen framework is composed of two interpenetrating tetrahedra, a smaller one defined by iron and a larger one defined by chalcogen atoms, which share a common centroid. Fig. 34 shows the Fe_4 tetrahedron in the structure of the simplest member in the $[Fe_4S_4(SR)_4]^{2-/3-}$ complex family, namely $[Fe_4S_4(SMe)_4]^{2-}$, which has been determined by neutron diffraction (ref. 362). Both tetrahedra are oriented with respect to each other in such a way that the chalcogen atoms reside above the centers of the triangular metal faces and vice versa. A third tetrahedron with faces roughly parallel to that of the iron cage but ca. 12 times as large in volume if thiolate ligands are considered is established by the terminal donor functions. The $[Fe_4Te_4]$ core of $[Fe_4Te_4(SePh)_4]^{3-}$ with terminally bonded selenolate functions as viewed along an idealized molecular S_4 axis is given in Fig. 33(b) (ref. 328).

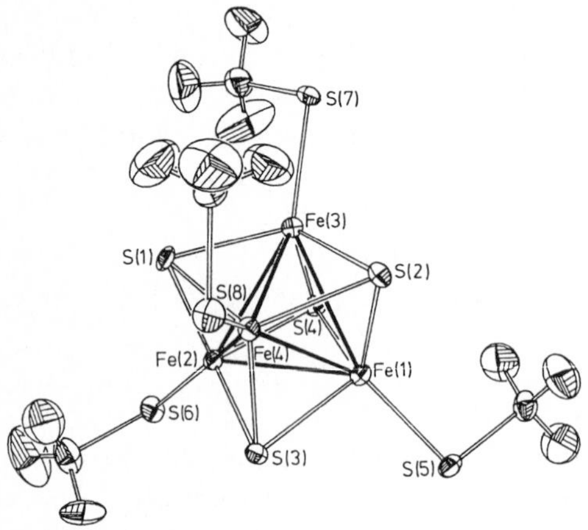

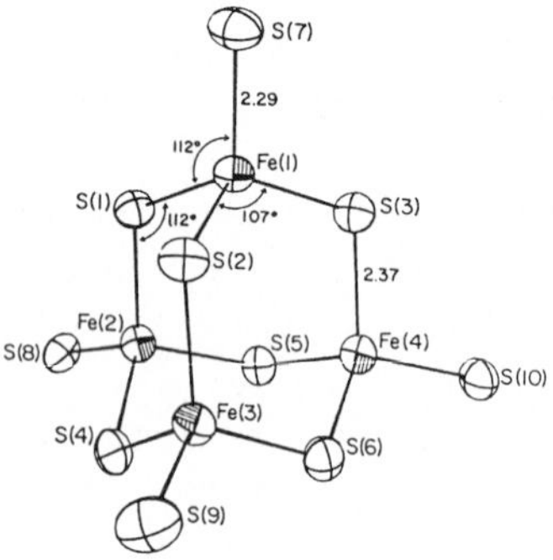

Fig. 34 (left). Structure of $[Fe_4S_4(SMe)_4]^{2-}$ with vibrational ellipsoids (methyl groups without labels) from a single crystal neutron diffraction study (ref. 362).
Fig. 35 (right). The $[Fe_4(SPh)_{10}]^{2-}$ anion without phenyl groups; given in the picture are mean values of independent bond distances and angles under idealized T_d symmetry (ref. 331).

In the case of the $[Fe_4S_4(SPh)_4]^{2-}/[Fe_4S_4(SPh)_4]^{3-}$ redox couple, the volumes of the interpenetrating Fe_4 and S_4 tetrahedra $[(pm.10^{-2})^3]$ are calculated to be 2.41/2.40 and 5.54/5.75, respectively (refs. 266, 318). These values increase steadily in the series $[Fe_4X_4(SPh)_4]^{2-}/[Fe_4X_4(SPh)_4]^{3-}$ on going from X = S via Se to Te, individual volumes being 2.54/2.54 and 6.72/6.99 for Se and 2.62/2.61 and 8.73/9.43 for Te, respectively (refs. 326, 327).

Virtually all members of the cubane-type complexes which have been characterized crystallographically do not have the ideal T_d symmetry with respect to their $[Fe_4X_4]$ cores but are significantly distorted instead. These distortions have been subject of several discussions concerning their origin. While it is believed that the electronic ground state (S = 0) of the dianionic species $[Fe_4S_4(SR)_4]^{2-}$ favours the tetragonal compression of the $[Fe_4X_4]^{2+}$ cluster core resulting in idealized (compressed) D_{2d} symmetry, the picture which emerges from the crystal structures of about nine reduced trianionic species $[Fe_4S_4(SR)_4]^{3-}$ is somewhat confusing showing three of them to be tetragonally elongated, while four others are distorted in an opposite sense and appear to be compressed. Of the remaining two species not included in the two sets, one has a more regular structure, while the other is irregular in shape. This structural diversity, which has been termed 'plasticity' (ref. 320), may be related with different ground state properties of the trianions, which have been shown to exist either with pure $S = 1/2$ or $S = 3/2$ ground states or as 'spin-admixed' species with properties not determined by simply adding pure $S = 1/2$ and $S = 3/2$ ground states (refs. 318-320).

The situation is even more complicated inasmuch as there is no clear criterion for the choice of the molecular reference axis. A compression of the $[Fe_4X_4]$ cubane core with respect to an axis which passes through the midpoints of opposite faces manifests itself by 4 short (and roughly parallel) Fe-X bonds and 8 longer ones which are roughly perpendicular to the former set. From another point of view, however, simply by changing the reference axis, the former compression looks like an elongation if mean Fe-X bond distances are considered (for a discussion see ref. 329).

Though the group of tetranuclear complexes is dominated by sulfide-thiolate compounds with cubanoidal $[Fe_4S_4]$ frameworks, there exists a family of homoleptic thiolates with adamantanoidal $[M_4S_6]$ cages. Two discrete members of this family are known which contain iron coordination sites, namely $[Fe_4(SPh)_{10}]^{2-}$ (Fig. 35, refs. 330,331) and $[Fe_4(SC_2H_5)_{10}]^{2-}$ (ref. 177). In contrast to all other multinuclear iron complexes with tetrahedral FeS_4 coordination, both molecules are assembled from FeS_4 tetrahedra which share common corners instead of edges. The adamantane-like $[Fe_4(\mu\text{-}S_6)]$ cage consists of four Fe_3S_3 cycles in the ideal chair conformation resulting in Fe...Fe distances as long as ca. 394 pm. The six sulfur atoms are arranged at the corners of a regular octahedron, while the iron atoms are located above the centers of each second trigonal face thus defining a regular tetrahedron. They complete their coordination spheres by binding a terminal thiolate function each.

(iv) <u>Complexes of higher nuclearity</u>. Various investigations of diverse iron/sulfide/thiolate reaction systems in different laboratories were directed towards the syntheses of multinuclear complexes which are able to mimic structural, chemical and/or electronic properties of the metal coordination sites of [3Fe-4S] proteins as well as of the Mo-Fe-S centers of nitrogenase. These efforts resulted in the isolation and characterisation of members of two different compound families both with a significant degree of relevance. One of them, namely that which comprises the trinuclear sulfide-thiolate complexes, has been introduced as suitable structural models for the [3Fe-4S] protein sites and discussed in an earlier section. The other one comprises sulfur-rich hexanuclear species of general formula $[Fe_6(\mu_4\text{-}S)(\mu_3\text{-}S)_2(\mu\text{-}S)_6(SR)_2]^{4-}$ which are related to the [3Fe-4S] protein sites with respect to chemical identity and electronic structure. Moreover, from the unprecedented coordination properties of the central $[Fe_4(\mu_4\text{-}S)(\mu_3\text{-}S)_2(\mu\text{-}S)_6]$ core portion a novel structural model for the Mo-Fe-S centers could be derived.

The structure of $[Fe_6S_9(SCH_2Ph)_2]^{4-}$ is depicted in Fig. 36 (refs. 164,165,332).

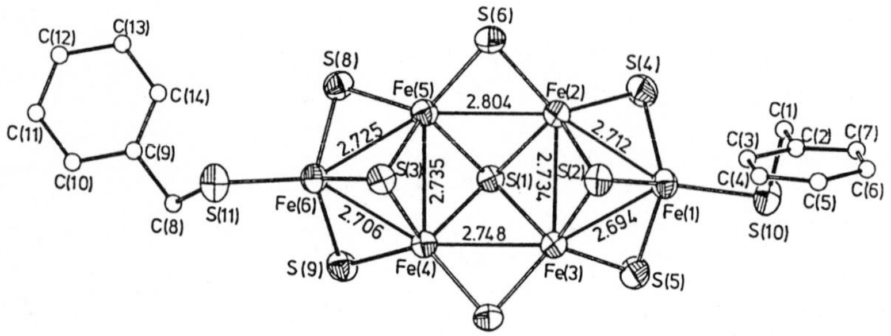

Fig. 36. The anion $[Fe_6S_9(SCH_2Ph)_2]^{4-}$ in crystals of $[Et_4N]_4$-$[Fe_6S_9(SCH_2Ph)_2].H_2O$ (ref. 164).

The complex is formed by reaction of $FeCl_3$ with $NaSCH_2Ph$ and Na_2S_2 in methanolic solution and has been precipitated as black tetraethylammonium salt. The $[Fe_6S_{11}]$ framework can be described in terms of six FeS_4 tetrahedra. They are coupled via common edges and corners to give a planar metal arrangement that consists of two triangles participating in a central metal square. The sulfur atoms are involved in four different coordination modes reflecting the number of FeS_4 tetrahedra they are belonging to. These numbers are one for the terminally bonded thiolate groups S(10) and S(11), two for the doubly bridging atoms S(4) - S(9), three for the triply bridging atoms S(2) and S(3) and four for the quadruply bridging unique atom S(1).

Similar complexes with tert.-butanethiolate (refs. 211,212) and ethanethiolate ligands which have been synthesized by a different procedure (ref. 213) have also been described. An interesting mixed-metal variant consisting of two of these complex anions hold together via sodium-sulfur interactions has been observed in crystals of $[Et_4N]_6[\{Fe_6S_9(SMe)_2\}_2Na_2]$ (ref. 165). The metal-sulfur framework is depicted schematically in Fig. 37 with emphasis placed on two significantly different core portions. The related selenium derivative $[Fe_6Se_9(SMe)_2]^{4-}$ with the corresponding Fe-Se/S core portions has also been characterized (refs. 322,214).

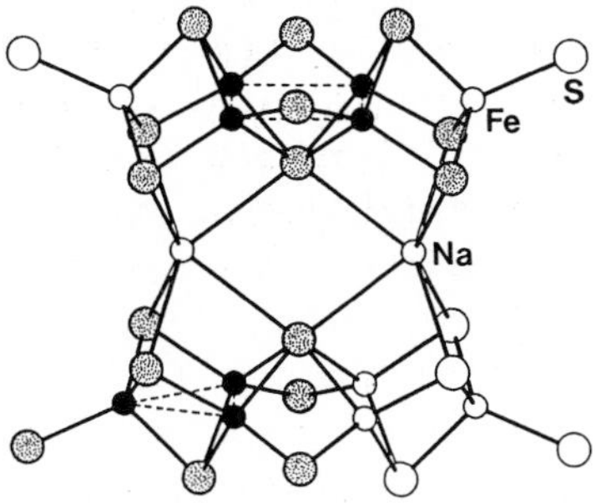

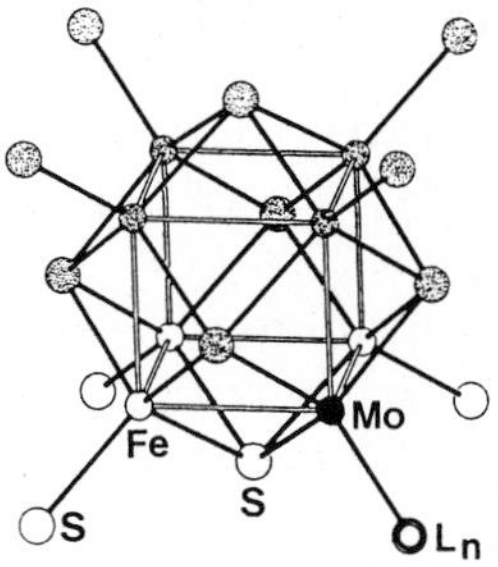

Fig. 37 (left). The $[(Fe_6S_{11})_2Na_2]$ framework of the $[\{Fe_6S_9(SMe)_2\}_2Na_2]^{6-}$ entity in crystals of $[Et_4N]_6[\{Fe_6S_9(SMe)_2\}_2Na_2]$ (large circles, S; small circles, Fe and Na; Na is located at the equatorial line); (a) the $[Fe_4S_9]$ core portion as fraction of a model for the Mo-Fe-S cluster of nitrogenase (top) and the $[Fe_3S_7]$ core portion as a model for the [3Fe-4S] centers of 3Fe proteins (bottom); (b) the mixed sodium-iron-sulfur cubane cluster with terminal sulfur ligands (ref. 165).
Fig. 38 (right). The $[Fe_7MoS_{13}L_n]$ frame of the $[Fe_7MoS_6(SR)_7L_n]^{z-}$ cluster (n = 1, 2 or 3) proposed as a model for the Mo-Fe-S center of nitrogenase. The dotted atoms indicate the $[Fe_4S_9]$ unit observed in $[\{Fe_6S_9(SMe)_2\}_2Na_2]^{6-}$ (ref. 165, see Fig. 37 (a)).

The $[Fe_4S_9]$ unit shown at the top of Fig. 37 has been used to design a possible structural model for the Mo-Fe-S cluster (and of the MoFe-cofactor which after isolation has different terminal ligands) of nitrogenase (Fig. 38, refs. 164,165). This model has the advantage that it does not suffer from substantial inconsistencies as do most of the other ones proposed in the literature (ref. 161). A similar model has been derived from the architecture of $[Co_8S_6(SPh)_8]^{4-}$ (ref. 166, see section 2.9). In this structure type each iron atom has four sulfur neighbours, whereas molybdenum is coordinated to three sulfide ions allowing for the activation of up to three additional bonds for substrate binding.

2.9 Cobalt

In sharp contrast to iron, the sulfide-thiolate chemistry of cobalt is determined predominantly by the metal in the oxidation state +2, and more reduced species are also known. Probably for this reason no tetranuclear complexes of general formula $[M_4S_4(SR)_4]^{2-/3-}$ are known if M = Co though this complex type is by far the most stable in synthetic iron-sulfide-thiolate chemistry.

The first complex containing sulfide and thiolate ligands simultaneously was reported in 1982 and subsequently in 1985 (refs. 333,166). $[Co_8S_6(SPh)_8]^{4-}$ has been synthesized by reaction of $CoCl_2$, Li_2S and LiSPh in methanol which contained an excess of LiOMe. The metal-sulfur frame of the anion is given in Fig. 39. The rigid $[Co_8S_6]$ cage consists of an octahedron of sulfur atoms which interpenetrates a cube of cobalt atoms. A third cube with faces roughly parallel to that of the cobalt frame but much larger in volume is defined by the thiolate sulfur functions. Each metal atom thus reaches a tetrahedral sulfur environment. In this respect and with reference to nearest metal neighbour distances the $[M_8S_6]$ frame with a metal cube has coordination properties very similar to that of the mixed-atom $[M_4S_4]$ cubane frame observed in $[Fe_4S_4(SR)_4]^{2-/3-}$ complexes (see section 2.8). $[Co_8S_6(SPh)_8]^{4-}$ can be reduced by one electron with sodium acenaphthylenide to give $[Co_8S_6(SPh)_8]^{5-}$ with slightly larger core dimensions (ref. 166).

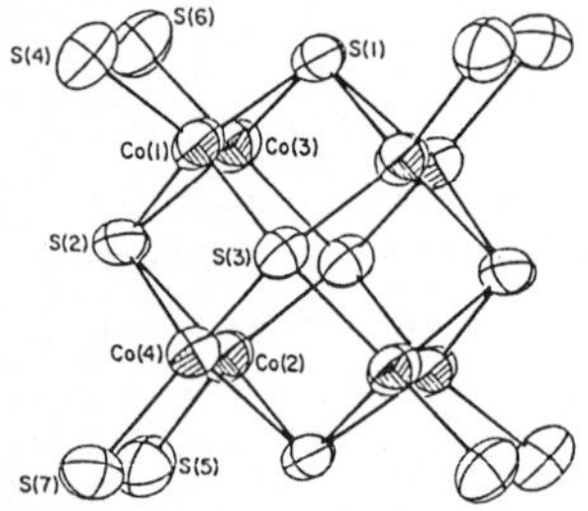

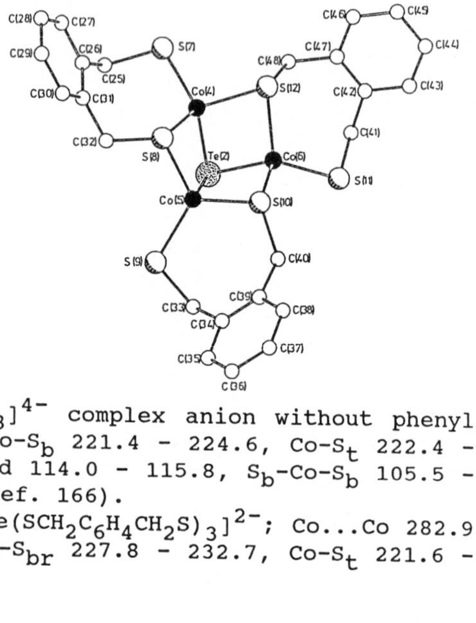

Fig. 39 (left). The $[Co_8S_6(SPh)_8]^{4-}$ complex anion without phenyl groups. Co...Co 263.4 - 268.2, $Co\text{-}S_b$ 221.4 - 224.6, $Co\text{-}S_t$ 222.4 - 225.0 pm, Co-S-Co 72.2 - 74.1 and 114.0 - 115.8, $S_b\text{-}Co\text{-}S_b$ 105.5 - 107.6, $S_b\text{-}Co\text{-}S_t$ 100.3 - 119.9° (ref. 166).
Fig. 40 (right). The anion $[Co_3Te(SCH_2C_6H_4CH_2S)_3]^{2-}$; Co...Co 282.9 - 289.5, Co-Te 256.5 - 259.9, $Co\text{-}S_{br}$ 227.8 - 232.7, $Co\text{-}S_t$ 221.6 - 223.8 pm (ref. 334).

Sulfide-thiolate complexes of the type $[M_3S(SRS)_3]^{2-}$ with M = Co have also been described (refs. 215,250). They were introduced in the corresponding section of the chemistry of iron (M = Fe, Fig. 31) and are the sole example for similarities in the behaviour of cobalt and iron towards sulfide/thiolate ligands.

The triply bridging sulfur atoms in $[Co_3S(SCH_2C_6H_4CH_2S)_3]^{2-}$ can be replaced by selenium and tellurium atoms, respectively, if NaHSe and NaHTe is used instead of NaHS in the synthetic procedures. The corresponding complex anions $[Co_3Se(SCH_2C_6H_4CH_2S)_3]^{2-}$

and $[Co_3Te(SCH_2C_6H_4CH_2S)_3]^{2-}$ could be isolated as tetraethylammonium salt and characterized by X-ray crystallography (ref. 334). The structure of $[Co_3Te(SCH_2C_6H_4CH_2S)_3]^{2-}$ is shown in Fig. 40. The telluride ion occupies the apex of a pyramid whose basal plane is defined by Co atoms. Similar coordination principles are realized in the mixed selenide-selenolate complex $[Co_3Se(SeCH_2C_6H_4CH_2Se)_3]^{2-}$ as well as in the mixed sulfide-selenolate system $[Co_3S(SeCH_2C_6H_4CH_2Se)_3]^{2-}$ (ref. 334).

Besides these sulfide-thiolate complexes and their selenium and tellurium analogues several polynuclear homoleptic thiolates are known in which cobalt is invariantly divalent. All of them find their counterparts in the corresponding chemistry of iron, and general synthetic and structural principles have been discussed in chapter 2.8. Examples include $[Co_4(SPh)_{10}]^{2-}$ (refs. 331,335,336), $[Co_2(SPh)_6]^{2-}$ (ref. 177) and $[Co_2(SCH_2C_6H_4CH_2S)_3]^{2-}$ (ref. 337). The latter complex has the isomer IIb-type structure (see Figs. 29 and 30(a)) with a Co...Co distance of 278.6 pm.

2.10 Nickel

Since the discovery of nickel-sulfur interactions in the Ni-containing hydrogenases from *Methanobacterium thermoautotrophicum* (ref. 338) and *Desulfovibrio gigas* (ref. 339) and in CO dehydrogenase of *Clostridium thermoaceticum* (ref. 340) the nickel chemistry with sulfur containing ligands has become one of the most attractive research fields in the area of bioinorganic chemistry (ref. 341). Another interesting aspect stimulating the exploration of nickel thiolate chemistry is given by the molecular-condensed state analogy in the light of a plethora of sulfide minerals and synthetic phases with a variety of structures not yet encountered in discrete molecular units (refs. 342,343).

All nickel complexes with thiolate and mixed sulfide-thiolate ligand spheres known so far contain the metal in its divalent form

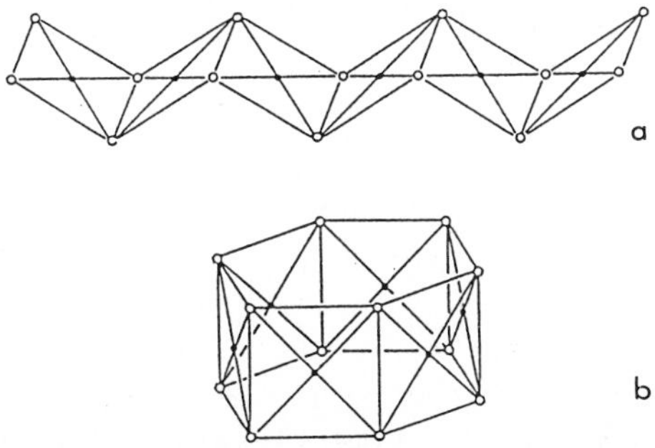

Fig. 41. Stereochemically distinct patterns containing square-planar NiS_4 subunits; (a) folded one-dimensional chain, trans-condensation; (b) cyclic arrangement, cis-condensation (ref. 245).

(see below). In contrast to iron and cobalt, the chemistry of divalent nickel with aliphatic thiolate ligands is determined by square-planar metal-sulfur coordination units. These units show a pronounced tendency to condensate via opposite edges in protic and aprotic media (ref. 344). The $[Ni_2S_2]$ heterocycles which are formed in the condensation steps are in all but one cases not planar but folded along their S...S diagonals. The Ni-S frameworks of polynuclear nickel thiolates should therefore be fragments of the extended one-dimensional $[NiS_2]_\infty$ chain (all-trans condensation, Fig. 41(a)) or of cyclic arrangements such as $[NiS_2]_6$ (all-cis condensation, Fig. 41(b)) or should exhibit mixtures of both condensation types.

All three principles have been recognized in synthetic nickel-thiolate chemistry. With the sole exception of $[Ni_2(SEt)_6]^{2-}$ and $[Ni_3(SEt)_8]^{2-}$ (ref. 176), all thiolates adopting an open-chain

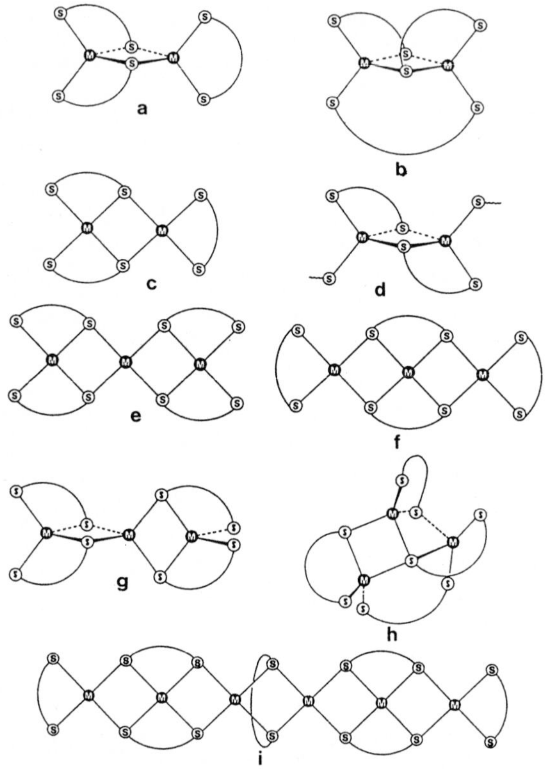

Fig. 42. Structures of tetracoordinated polynuclear metal(II) dithiolates that have been crystallographically established for the following metals: (a) Fe, Zn; (b) Mn, Fe, Co, Cd; (c) Ni; (d) Hg; (e) Ni; (f) Ni; (g) Hg; (h) Hg; (i) Ni (for a more complete specification see ref. 245).

configuration are formed by difunctional ligands. Fig. 42 gives an idea of the structure types which have been established in metal(II) dithiolate complexes with coordination number 4 of which the tetrahedrally coordinated species have been addressed in the corresponding sections.

The nickel species define the group of complexes with planar metal coordination (c-, e-, f- and i-type structures). Structure c with a non-planar central $[Ni_2S_2]$ cycle has been observed in $[Ni_2(SCH_2CH_2S)_3]^{2-}$ (refs. 345,346). The related complex $[Ni_2(SEt)_6]^{2-}$ (ref. 176) with monofunctional thiolate ligands is planar in this respect (see above). Extension of structure c by one NiS_2 unit leads to complexes e and f, realized in $[Ni_3(SCH_2CH_2S)_4]^{2-}$ (refs. 347,245) and $[Ni_3(SCH_2C_6H_4CH_2S)_4]^{2-}$ (refs. 337,245), respectively (Figs. 43 and 44).

Both complexes differ not only in the arrangement of the difunctional thiolate ligands but also in condensation type (see Fig. 41): trans-condensation is observed in $[Ni_3(SCH_2CH_2S)_4]^{2-}$, cis-condensation in $[Ni_3(SCH_2C_6H_4CH_2S)_4]^{2-}$. Further extension of structures e or f by three NiS_2 units leads to structure i which could be realized in $[Ni_6(SCH_2CH_2CH_2S)_7]^{2-}$ (Fig. 45, ref. 245).

$[Ni_6(SCH_2CH_2CH_2S)_7]^{2-}$ is especially interesting in the sense that its $[Ni_6S_{14}]$ framework has a mixed cis/trans configuration. The architecture can best be described on the basis of two trinuclear subunits adopting the cis configuration, which are condensed in a trans-like manner.

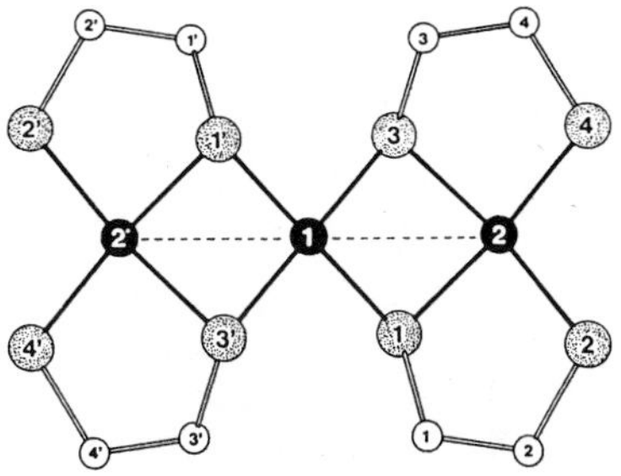

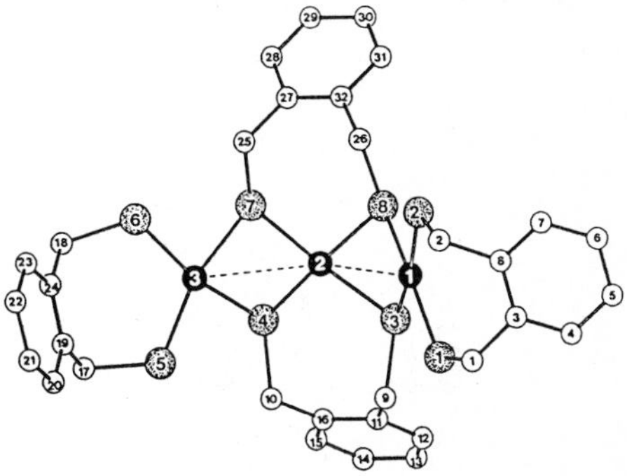

Fig. 43 (left). The anion $[Ni_3(SCH_2CH_2S)_4]^{2-}$ in crystals of $[Ph_4P]_2[Ni_3(SCH_2CH_2S)_4]$ (trans configuration); distances: Ni...Ni 285.6, Ni-S_{br} 218.9 - 221.0, NiS_t 217.4 - 219.6 pm (ref. 245).
Fig. 44 (right). The anion $[Ni_3(SCH_2C_6H_4CH_2S)_4]^{2-}$ (cis configuration); selected distances: Ni...Ni 301.6, 313.1, Ni-S_{br} 219.2 - 222.9, Ni-S_t 217.7 - 219.3 pm (ref. 245).

Hexanuclear nickel thiolates with an overall cis geometry must of course be cyclic. Examples are $[Ni_6(SR)_{12}]$ with R = Et (refs. 348-350), R = C_2H_4OH (ref. 351) and $[Ni_6(SC_3H_6NHMe_2)_{12}]^{12+}$ (ref. 352). The smallest cyclic members proven to be stable are $[Ni_4(SR)_8]$ with R = C_5H_9NMe (ref. 353), C_6H_{11} (Fig. 46 (a), ref. 354) and C_3H_7 (ref. 355), and the octanuclear complex $[Ni_8(SCH_2\text{-}COOEt)_{16}]$ is also known (ref. 356). Besides these cyclic thiolates containing even numbers of nickel atoms the pentanuclear forms $[Ni_5(SEt)_{10}]$ (Fig. 46 (b), ref. 354) and $[Ni_5(SC_2SiMe_3)_{10}]$ (ref. 357) with an odd number of nickel centers do exist as well.

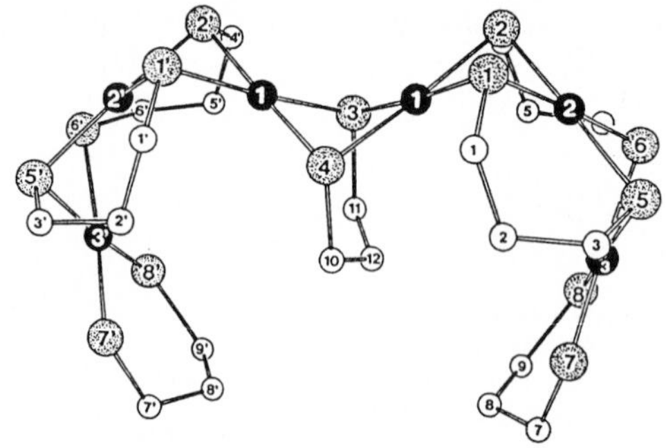

Fig. 45. The anion $[Ni_6(SCH_2CH_2CH_2S)_7]^{2-}$ (mixed cis/trans configuration); selected distances: Ni...Ni 275.0 - 284.3, Ni-S_{br} 217.5 - 224.7, Ni-S_t 215.3 and 216.1 pm (ref. 245).

Condensation of three square-planar NiS_4 coordination sites is not only possible via opposite edges (as observed in structure e or f) but also via adjacent ones. As a result, one sulfur atom has

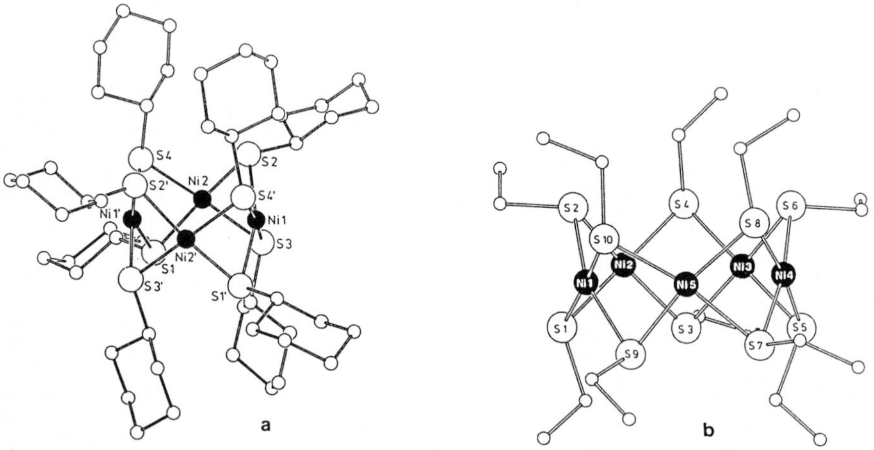

Fig. 46. Cyclic nickel thiolates with even and odd numbers of Ni centers; (a) $[Ni_4(SC_6H_{11})_8]$; (b) $[Ni_5(SEt)_{10}]$ (ref. 354).

to belong to three NiS_4 units simultaneously. Trinuclear complexes of nickel which obey this rule are known indeed, but they are not pure thiolates but mixed sulfide-thiolates instead due to sulfide ions which occupy the triply bridging positions. One example is $[Ni_3S(SCH_2C_6H_4CH_2S)_3]^{2-}$. The complex anion is formed by reaction of $NiCl_2$ with $Na_2(SCH_2C_6H_4CH_2S)$ and NaSH in methanol and contains bifunctional thiolate groups (ref. 251). $[Ni_3S(SMe)_6]^{2-}$ (Fig. 47, ref. 358) and $[Ni_3S(SPh)_6]^{2-}$ (ref. 359), mixed sulfide-thiolate derivatives with monofunctional thiolate ligands, are also known.

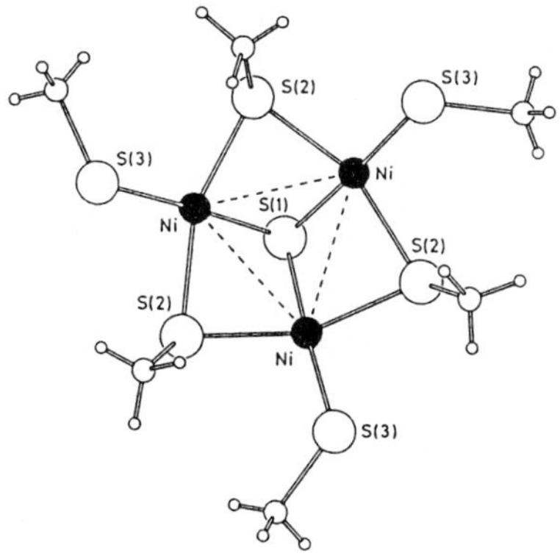

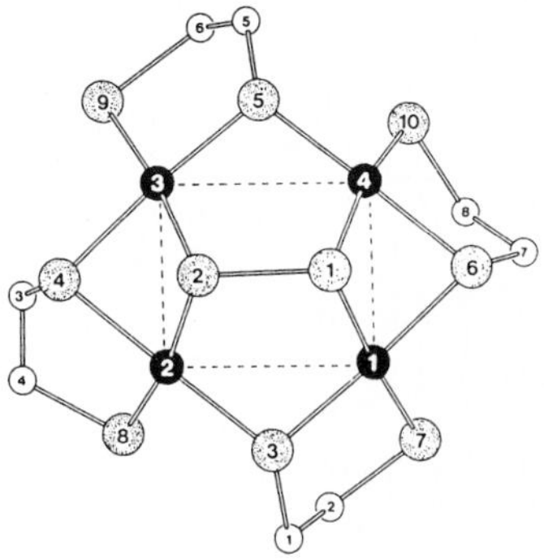

Fig. 47 (left). The anion $[Ni_3S(SMe)_6]^{2-}$; distances and angles: Ni...Ni 285.5, Ni-(μ_3-S) 218.6, Ni-(μ-S) 218.2 and 218.7, Ni-S_t 219.4 pm, (μ_3-S)-Ni-(μ-S) 81.0 and 81.1, (μ_3-S)-Ni-S_t 175.8, (μ-S)-Ni-(μ-S) 158.2, (μ-S)-Ni-S_t 97.1 and 101.6° (ref. 358).
Fig. 48 (right). The anion $[Ni_4(S_2)(SCH_2CH_2S)_4]^{2-}$; distances: Ni(1)-Ni(2) 332.9, Ni(1)-Ni(4) 282.2, Ni(2)-Ni(3) 286.9, Ni(3)-Ni(4) 334.4, Ni-(S_2) 220.6 - 222.4, Ni-S_{br} 212.1 - 222.8, Ni-S_t 215.2 - 219.4, S-S 208.4 pm (ref. 360).

Besides these 'classical' complexes with mixed sulfide-thiolate ligand spheres of nickel a related disulfide-thiolate species is also known. The tetranuclear anion $[Ni_4(S_2)(SCH_2CH_2S)_4]^{2-}$ (Fig. 48) is formed by reaction of $NaSCH_2CH_2SNa$ and $NiCl_2$ with Na_2S_2 in methanol and could be precipitated as brown $[Ph_4P]_2[Ni_4(S_2)(SCH_2CH_2S)_4]$.MeOH (refs. 347,360). Compounds containing polysulfide and thiolate groups coordinated to one metal site simultaneously are most uncommon due to the delicate redox balance of the couple S_n^{2-}/SR^- which favours oxidation of the thiolate component, the only other structurally characterized examples being $[Mo_3S(S_2)_3(SCH_2CH_2S)]^{2-}$ (refs. 77,78) and $[V(S_2)S_2(SPh)]^{2-}$ (ref. 361).

Very recently the 'conventional' thiolate chemistry of nickel which is determined by d^8-configurated divalent metal sites has been for the first time extended towards reduced species capable of forming direct metal-metal bonds. This unexpected development opens not only novel perspectives in inorganic coordination chemistry but also in bioinorganic chemistry inasmuch as biologically important nickel centers have been proven to cycle between various metal oxidation states (see e.g. ref. 341 and references cited therein). Initial evidence for the existence of novel nickel-sulfide-thiolate clusters came from the observation that the $[Ni_8S(SC_4H_9)_9]^-$ anion (Fig. 49) which contains eight metal atoms in two different (formal) oxidation states is a chemically stable entity (ref. 355). The compound was prepared by reaction of $NiCl_2$ with $NaSC_4H_9$ in methanol and isolated as black needles of $[Ph_4P][Ni_8S(SC_4H_9)_9].MeOH$. Six Ni atoms form a trigonal prism, whose triangular faces are capped by two further Ni atoms. The average oxidation state of the metal centers is +1.25 and can formally be ascribed to six Ni(I) and two Ni(II). The diamagnetic behaviour indicates a delocalized electron system. Nickel-nickel bonds are established between the capping Ni atoms (Ni^k) and their neighbours (Ni^p, average 247.1 pm) and between Ni atoms of opposite triangular faces of the metal prism (average 244.4 pm). The Ni-Ni bonds are supported by thiolate bridges each. Within the triangular metal faces the Ni-Ni distance has an average value of 312.2 pm; this excludes additional metal-metal bonding. With a total of 114 skeletal electrons in the Ni_8S_{10} framework, a 'magic count' is achieved.

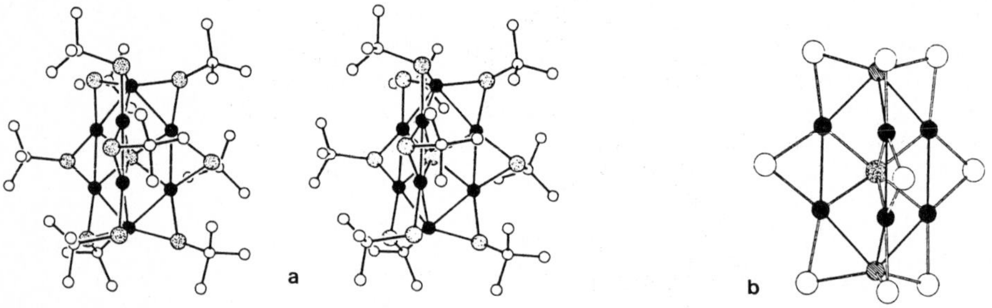

Fig. 49. The $[Ni_8S(SC_4H_9)_9]^-$ anion; (a) Stereoscopic view; (b) Schematic drawing of the $[Ni_8S_{10}]$ framework (ref. 355).

The center of the nickel prism is occupied by a unique sulfide ion which is bonded to six Ni atoms simultaneously. This $(\mu_6\text{-S})Ni_6$ fragment has been observed for the first time in isolated complexes and represents a slightly distorted portion of the nickel-arsenide type structure of hexagonal NiS. The average $(\mu_6\text{-S})$-Ni bond length of 217.9 pm is even slightly shorter than the average (RS)-Ni^k bond length (219.6 pm). Just as unusual as the oxidation state of the Ni centers is their coordination. The ligands are arranged in such a way that each Ni atom is surrounded by three S atoms in a roughly trigonal-planar fashion.

3 THIOLATES AND SULFIDE-THIOLATES OF CLOSED-SHELL TRANSITION METALS

3.1 Copper, Silver, and Gold

The chemistry and structural properties of the thiolates and sulfide-thiolates of the d^9 and d^{10} transition metals are addressed only briefly. The interest in fully or partially sulfur-coordinated biologically active centers of copper, zinc, cadmium, and mercury such as the blue copper proteins (e.g. plastocyanins, azurins) or different varieties of metallothioneins have again stimulated research in the field of metal thiolate complexes.

In TABLE 1 a listing of the presently known oligonuclear fully sulfur-coordinated thiolates of the copper group is given.

Copper is univalent in its thiolates, its normal coordination mode is trigonal-planar, though some linearly coordinated examples are known. In the oligonuclear complexes, which are formed from Cu(I) or Cu(II) salts with thiolates under different conditions in solvents such as MeOH, the ligands have mostly bridging functions.

Copper reacts with simple saturated thiolates to give tetranuclear complexes of the type $[Cu_4(SR)_6]^{2-}$ containing a cage-like tetrahedro-Cu_4-octahedro-(μ-S) moiety (refs. 364-371) (Fig. 50). They are known with monofunctional as well as with difunctional ligands (TABLE 1). $[Cu_3(SCH_2CH_2S)_3]^{3-}$ (ref.239) contains a nonplanar $[Cu_3(\mu\text{-}S)_3]$ ring with trigonal Cu coordination. With thiophenolate pentanuclear $[Cu_5(SPh_7)]^{2-}$ is formed under different reaction conditions. Its structure derives from that of $[Cu_4(SPh)_6]^{2-}$ by replacement of a SR^- ligand by a linear PhS-Cu-SPh unit (refs. 375,376). A different 5Cu cage anion, $[Cu_5(S\text{-}t\text{-}Bu)_6]^-$, where two Cu atoms are three-coordinate and three Cu atoms are two-coordinate, was obtained with the sterically demanding $S\text{-}t\text{-}Bu^-$ ligand. As in other d^{10} metal thiolates, there are strong structural indications for attractive metal-metal interactions in this

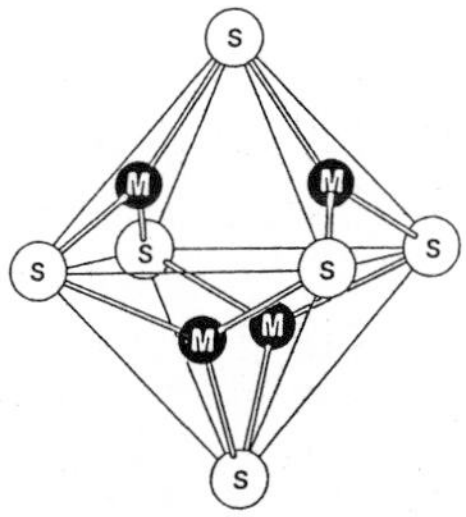

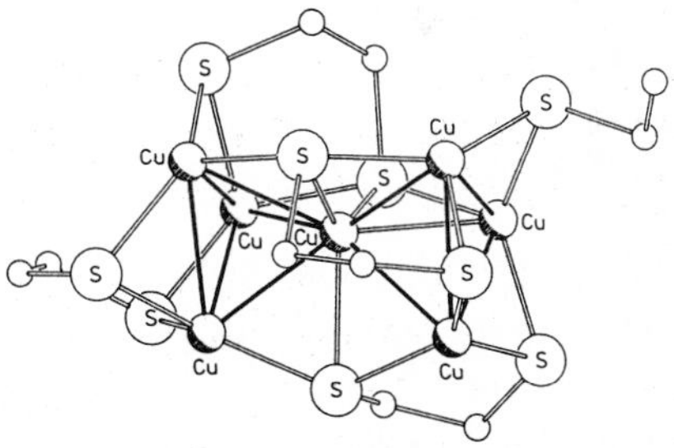

Fig. 50 (left). The $[M_4S_6]$ frame of the $[Cu_4(SR)_6]^{2-}$ complexes.
Fig. 51 (right). $[Cu_7(SCH_2CH_2S)_4(SEt)]^{2-}$ (refs. 369,371,389).

complex (refs. 372-374). The remarkable novel $[Cu_7(SCH_2CH_2S)_4(SEt)]^{2-}$ anion contains both difunctional and monofunctional thiolate ligands, and it has a 7Cu core consisting of two corner-sharing Cu_4 tetrahedra (refs. 369,371,389) (Fig. 51).

The recently reported neutral 1:1 copper thiolate molecules, $[Cu_8\{SC_6H_2(i\text{-}Pr)_3\}_8]$ (**1**) (ref. 377) and $[Cu_{12}\{SC_6H_4(SiMe_3)\}_{12}]$ (**2**) (ref. 378), are also highly interesting. **1** exhibits a twisted 16-membered cyclic aggregate of alternating Cu and S atoms; the sulfur atoms are doubly bridging and each copper atom has linear two-coordination. The sterically hindered ligands in **2** afford a dodecametallic cluster with a "paddle-wheel" $[Cu_{12}S_{12}]$ core.

A number of thiolates with similar preparation methods and similar cage structures have also been reported for silver. They range from the tetranuclear $[Ag_4S_6]$ frameworks in $[Ag_4(SCH_2C_6H_4CH_2S)_3]^{2-}$ (ref. 379) (Fig. 53) and in $[Ag_4(SC_6H_3(Me)S)_3]^{2-}$ (refs. 380,381) over the $[Ag_5S_6]$ and $[Ag_5S_7]$ cores (refs. 374,376,383) to the neutral $[Ag_{12}S_{12}]$ core (refs. 385,386) (TABLE 1). Other novel species are $[Ag_4\{SSi(O\text{-}t\text{-}Bu)_3\}_4]$ with a central planar $[Ag_4S_4]$ square (ref. 382), $[Ag_6(SPh)_8]^{2-}$ (ref. 388), and $[Ag_9(SCH_2CH_2S)_6]^{3-}$ (ref. 368). A series of novel mixed-metal species could be prepared with $[Au_{6-n}M_n]$ (M = Ag, n = 4, M = Cu, n = 3,4) and $[Ag_{9-n}Cu_n]$ cores derived from the hexa- and nonanuclear complexes (refs. 368,369,371,379,389) (Fig. 52).

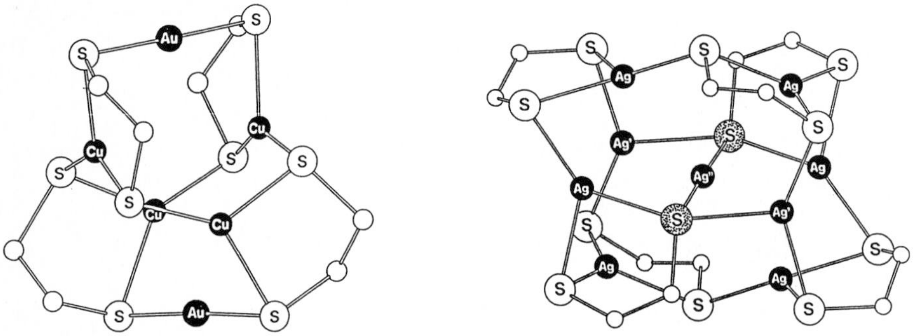

Fig. 52. Structures of $[Au_2Cu_4(SCH_2CH_2S)_4]^{2-}$ (left) and $[Ag_9(SCH_2CH_2S)_6]^{3-}$ (right) (ref. 368).

The chemistry of gold(I) thiolates is strikingly different from that of the homologous copper and silver complexes because of its preference for linear two-coordination. This has already been shown in the above-mentioned mixed-metal complexes with Ag(I) and Cu(I), and also the structures and reactivities of the homonuclear complexes are unique. In contrast to copper and silver, gold is able to form stable thiolate complexes in its trivalent form.

TABLE 1
Polynuclear thiolates of copper, silver and gold

	Refs.		Refs.
Copper:			
$[Cu_3(SCH_2CH_2S)_3]^{3-}$	239	$[Cu_5(S\text{-}t\text{-}Bu)_6]^-$	372-374
$[Cu_4(SMe)_6]^{2-}$	364	$[Cu_5(SPh)_7]^{2-}$	375,376
$[Cu_4(SPh)_6]^{2-}$	364-367	$[Cu_7(SCH_2CH_2S)_4(SEt)]^{2-}$	369,389
$[Cu_4(SCH_2CH_2S)_3]^{2-}$	368	$[Cu_8\{SC_6H_2(i\text{-}Pr)_3\}_8]$	377
$[Cu_4(SCH_2C_6H_4CH_2S)_3]^{2-}$	369-371	$[Cu_{12}\{SC_6H_4(SiMe_3)\}_{12}]$	378
Silver:			
$[Ag_4(SCH_2C_6H_4CH_2S)_3]^{2-}$	379	$[Ag_5(SCH_2CH_2CH_2NH_{0.5}Me_2)_6]^{2+}$	384
$[Ag_4\{SC_6H_3(Me)S\}_3]^{2-}$	380,381	$[Ag_9(SCH_2CH_2S)_6]^{3-}$	368
$[Ag_4\{SSi(O\text{-}t\text{-}Bu)_3\}_4]$	382	$[Ag_{12}(SC_6H_{11})_{12}]$	385,386
$[Ag_5(S\text{-}t\text{-}Bu)_6]^-$	374,383	$[Ag(SCMeEt_2)]_n$	387
$[Ag_5(SPh)_7]^{2-}$	376	$[Ag_6(SPh)_8]_n^{2n-}$	388
Gold:			
$[Au_2(SCH_2CH_2CH_2S)_2]^{2-}$	380,381	$[Au_2Se_2(SCH_2CH_2S)_2]^{2-}$	380,381
$[Au_2(SCH_2C_6H_4CH_2S)_2]^{2-}$	380,381		
Mixed metal:			
$[Au_2Cu_4(SCH_2CH_2S)_4]^{2-}$	368	$[Au_2Ag_4(SCH_2CH_2S)_4]^{2-}$	368
$[Au_3Cu_3(SCH_2CH_2S)_4]^{2-}$	368	$[Ag_{9-n}Cu_n(SCH_2CH_2S)_6]^{3-}$	368,389 371,379

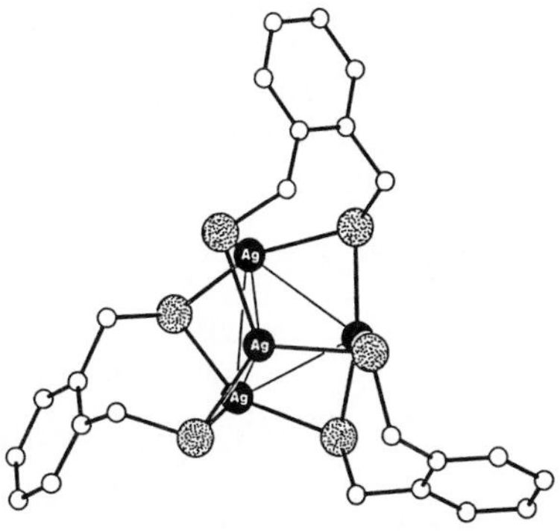

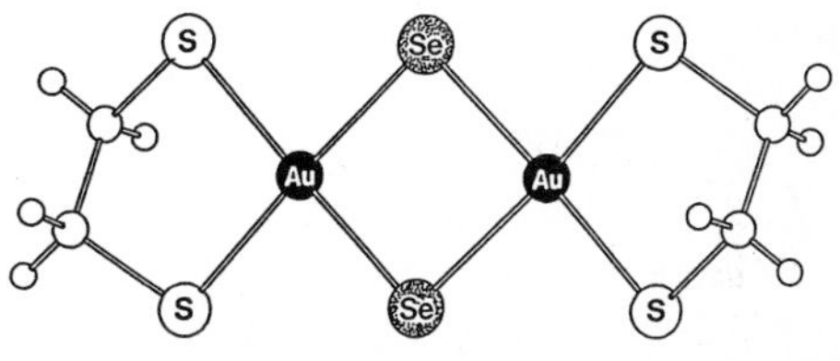

Fig. 53. The anion $[Ag_4(SCH_2C_6H_4CH_2S)_3]^{2-}$ in crystals of $[Ph_4P]_2$-$[Ag_4(SCH_2C_6H_4CH_2S)_3]$ (ref. 379).
Fig. 54. The structure of $[Au_2Se_2(SCH_2CH_2S)_2]^{2-}$ (refs. 380,381).

With saturated thiolates only low-nuclearity species were obtained up to now. For d^8-Au(III), a mononuclear $[Au(SCH_2CH_2S)_2]^-$ anion with square-planar configuration and a remarkable novel selenide-thiolate, $[Au_2Se_2(SCH_2CH_2S)_2]^{2-}$, could be prepared (Fig. 54, refs. 380,381). The latter was synthesized in the system $NaHSe/Se/BH_4^-/[Ph_4P][AuCl_3Br]$/ethane-1,2-dithiolate in MeOH.

Two dinuclear gold(I) complexes were prepared on reaction of $[AuCl_4]^-$ with 1,3-propanedithiolate and o-xylene-α,α-dithiolate, using the thiolates as reductants. These complexes, $[Au_2(SCH_2CH_2CH_2S)_2]^{2-}$ (Fig. 55) and $[Au_2(SCH_2C_6H_4CH_2S)_2]^{2-}$ (Fig. 55), contain interesting 12- and 14-membered metallacycles with linearly coordinated gold (refs. 380,381).

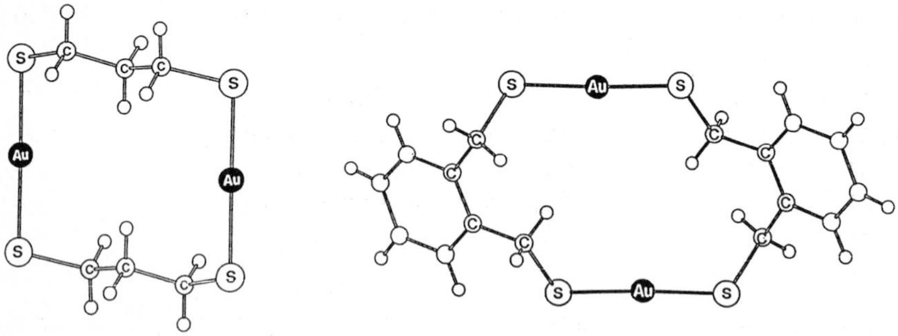

Fig. 55. The structures of $[Au_2(SCH_2CH_2CH_2S)^{2-}$ (left) and $[Au_2(SCH_2C_6H_4CH_2S)_2]^{2-}$ (right) (refs. 380,381).

3.2 <u>Zinc, Cadmium, and Mercury</u>

In the search for structural models of the metal-sulfur bonding in metallothioneins considerable research activity is centered on the thiolates of zinc(II), cadmium(II), mercury(II). A comprehensive review and dicussion of the structural aspects is given in ref. 10. The preparative methods for their synthesis are similar to those for the thiolates of the copper group and involve typically the reaction of thiolate anions with the dihalides in solvents like methanol with subsequent precipitation as salts with large quaternary cations.

In TABLE 2 the known oligomeric complexes of the three metals with complete thiolate coordination are listed. Especially for zinc and cadmium there is a large group of additional mixed-ligand thiolates with, e.g., halides or solvent molecules such as THF and a number of polymeric species structurally related to the compounds discussed (ref. 10).

Binuclear $[M_2(SR)_6]^{2-}$ complexes with R = Et or Me have been reported for all three metals (refs. 176,395,397). They contain the $[M_2S_6]$ core composed of two edge-sharing tetrahedra which is also present in analogous Mn(II) and Fe(II) species as well as in the related Zn(II) and Cd(II) anions of composition $[Zn_2(SPh)_6]^{2-}$ (ref. 390) and $[M_2(SCH_2C_6H_4CH_2S)_3]^{2-}$ (M = Zn, Cd) (refs. 391,

395). A very characteristic structural feature of the polynuclear Zn and Cd thiolate and sulfide systems is their tendency to form

TABLE 2
Polynuclear thiolates of zinc, cadmium, and mercury

	Refs.		Refs.
Zinc:			
$[Zn_2(SEt)_6]^{2-}$	176	$[Zn_4(SPh)_{10}]^{2-}$	392,393
$[Zn_2(SPh)_6]^{2-}$	390	$[Zn_{10}S_4(SPh)_{16}]^{4-}$	394
$[Zn_2(SCH_2C_6H_4CH_2S)_3]^{2-}$	391	$[Zn_{10}S_4(SCH_2C_6H_4CH_2S)_8]^{4-}$	391
Cadmium:			
$[Cd_2(SMe)_6]^{2-}$	395	$[Cd_4(SPh)_{10}]^{2-}$	393,396
$[Cd_2(SEt)_6]^{2-}$	176	$[Cd_{10}S_4(SPh)_{16}]^{4-}$	394
$[Cd_2(SCH_2C_6H_4CH_2S)_3]^{2-}$	395	$[Cd_{17}S_4(SPh)_{28}]^{2-}$	402
Mercury:			
$[Hg_2(SMe)_6]^{2-}$	397	$[Hg_2(SCH_2CH_2S)_3]_n^{2n-}$	398
$[Hg_3(SCH_2CH_2S)_4]^{2-}$	398	$[Hg_2(SCH_2CH_2CH_2S)_3]_n^{2n-}$	369,371
$[Hg_3(SCH_2C_6H_4CH_2S)_4]^{2-}$	399	$[Hg(S\text{-}t\text{-}Bu)_2]_n$	400
$[Hg_5(SCH_2CH_2S)_4(SEt)_4]^{2-}$	369,371	$[Hg(SeMe)_2]_n$	401

large cages consisting of parallel $[MS_4]$ cores in a zincblende-like fashion with ultimate formation of polymeric products (e.g. ref. 10). Examples for the first stage of the condensation process are the adamantane-like thiophenolate complexes of composition

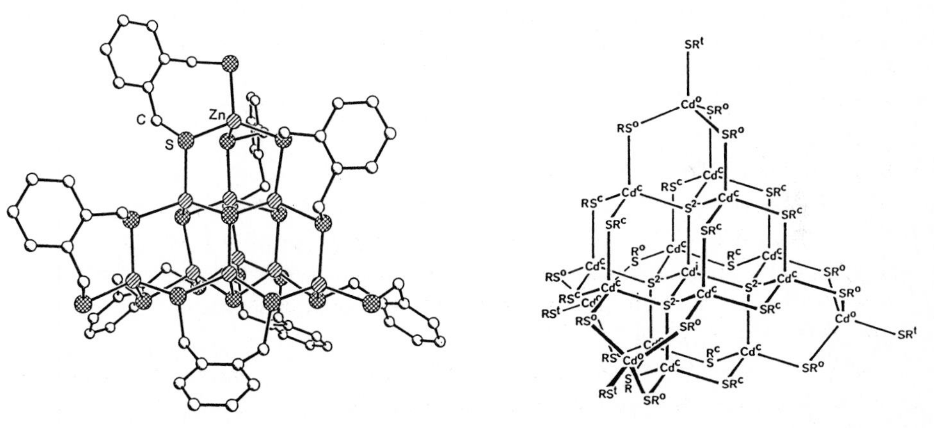

Fig. 56. The structures of $[Zn_{10}S_4(SCH_2C_6H_4CH_2S)_8]^{4-}$ (left, ref. 391) and $[Cd_{17}S_4(SPh)_{28}]^{2-}$ (right, ref. 402).

$[M_4(SPh)_{10}]^{4-}$ (M = Zn, Cd) (refs. 392,393,396). Remarkable molecular examples for products of even higher nuclearity are $[Zn_{10}S_4(SPh)_{16}]^{4-}$ (ref. 394), $[Zn_{10}S_4(SCH_2C_6H_4CH_2S)_8]^{4-}$ (Fig. 56, left) (ref. 391), $[Cd_{10}S_4(SPh)_{16}]^{4-}$ (ref. 394), and the recently prepared novel macro-anion $[Cd_{17}S_4(SPh)_{28}]^{2-}$ (Fig. 56, right) (ref. 402). In all of them the thiolates are bridging or terminal, whereas the sulfides are triply or quadruply bridging.

The trinuclear $[Hg_3(SCH_2CH_2S)_4]^{2-}$ and $[Hg_3(SCH_2C_6H_4CH_2S)_4]^{2-}$ anions (Fig. 57) (refs. 398,399) are remarkable examples of polynuclear mercury thiolates, and the pentanuclear mixed-ligand species $[Hg_5(SCH_2CH_2S)_4(SEt)_4]^{2-}$ is the first mixed-ligand compound in this series (refs. 369,371). Other interesting mercury thiolates are polymeric (see TABLE 2).

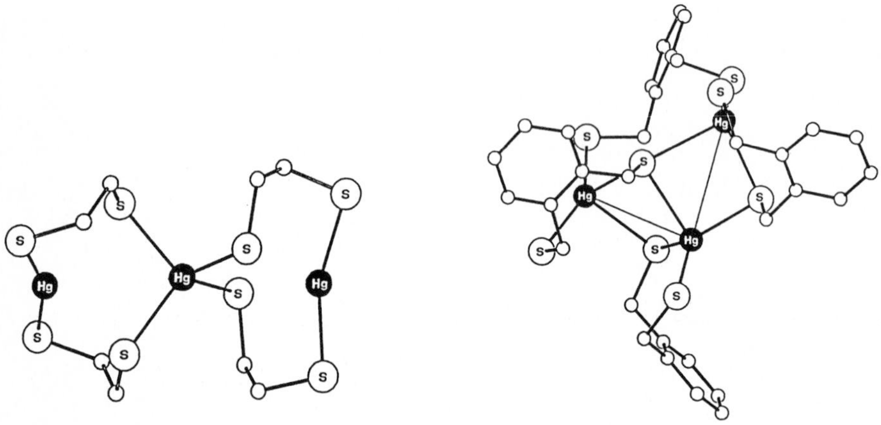

Fig. 57. The structures of $[Hg_3(SCH_2CH_2S)_4]^{2-}$ (left, ref. 398) and $[Hg_3(SCH_2C_6H_4CH_2S)_4]^{2-}$ (right, ref. 399).

ACKNOWLEDGMENT

The authors wish to acknowledge the invaluable collaboration of the students whose names appear in the articles cited. Special thanks are due to those who gave their assistance in the preparation of this article.

Work at the Universities of Münster and Duisburg is sponsored by the Deutsche Forschungsgemeinschaft, the Minister for Science and Research NRW and the German Federal Minister for Science and Technology. We also thank the Fonds der Chemischen Industrie for generous support.

REFERENCES

1 J.A. McCleverty, Prog. Inorg. Chem., 10 (1969) 49-221.
2 D. Coucouvanis, Prog. Inorg. Chem., 11 (1970) 233-371; 26 (1979) 301-469.
3 R. Eisenberg, Prog. Inorg. Chem., 12 (1970) 295-369.
4 J. Willemse, J.A. Cras, J.J. Steggerda and C.P. Keijzers, Struct. Bonding, 28 (1976) 83-126.
5 R.P. Burns and C.A. McAuliffe, Adv. Inorg. Chem. Radiochem., 22 (1979) 303-48.
6 A. Müller and B. Krebs (Editors), Sulfur - Its Significance for Chemistry, for the Geo-, Bio- and Cosmosphere and Technology, Elsevier, Amsterdam, 1984.
7 R.P. Burns, F.P. McCullough and C.A. McAuliffe, Adv. Inorg. Chem. Radiochem., 23 (1980) 211-80.
8 D. Coucouvanis, Acc. Chem. Res., 14 (1981) 201-9.
9 M. Draganjac and T.B. Rauchfuss, Angew. Chem., 97 (1985) 745-60; Angew. Chem. Int. Ed. Engl., 24 (1985) 742-57.
10 I.G. Dance, Polyhedron, 5 (1986) 1037-104.
11 A. Müller and E. Diemann, Adv. Inorg. Chem., 31 (1987) 89-122.
12 P.J. Blower and J.R. Dilworth, Coord. Chem. Rev., 76 (1987) 121-85.
13 A. Müller, Polyhedron, 5 (1986) 323-40.
14 D. Coucouvanis, A. Hadjikyriacou, M. Draganjac, M.G. Kanatzides and O. Ileperuma, Polyhedron, 5 (1986) 349-56.
15 D. Coucouvanis, R.K. Lester, M.G. Kanatzidis and D. Kessisoglou, J. Am. Chem. Soc., 107 (1985) 8279-80.
16 M. Cowie and M.J. Bennet, Inorg. Chem., 15 (1976) 1595-603.
17 D. Coucouvanis, A. Hadjikyriacou and M.G. Kanatzidis, J. Chem. Soc., Chem. Commun., (1985) 1224-5.
18 J. Rouxel and R. Brec, Annu. Rev. Mater Sci., 16 (1986) 137-62.
19 D.W. Murphy and P.A. Christian, Science, 205 (1979) 651-6.
20 L.A. Rankel and L.D. Rollman, Fuel, 62 (1983) 44-6.
21 R. Galiasso, R. Blanco, C. Gonzalez and N. Quinteros, Fuel, 62 (1983) 817-22.
22 B.G. Silbernagel, J. Catal., 56 (1979) 315-20.
23 J.G. Reynolds, W. Biggs and J.C. Fetzer, Liq. Fuels Technol., 3 (1985) 423-48.
24 R.L. Robson, R.R. Eady, T.H. Richardson, R.W. Miller, M. Hawkins and J.R. Postgate, Nature, 322 (1986) 388-90.
25 B.J. Hales, E.E. Case, J.E. Morningstar, M.F. Dzeda and L.A. Mauterer, Biochemistry, 25 (1986) 7251-5.
26 J.M. Arber, B.R. Dobson, R.R. Eady, P. Stevens, S.S. Hasnain, C.D. Garner and B.E. Smith, Nature, 325 (1987) 372-4.
27 J.E. Morningstar, M.K. Johnson, E.E. Case and B.J. Hales, Biochemistry, 26 (1987) 1795-1800.
28 J.E. Morningstar and B.J. Hales, J. Am Chem. Soc., 109 (1987) 6854-5.
29 J.K. Money, J.C. Huffman and G. Christou, Inorg. Chem., 27 (1988) 507-14.
30 D. Szeymies, B. Krebs and G. Henkel, Angew. Chem., 95 (1983) 903-4; Angew. Chem. Int. Ed. Engl., 22 (1983) 885-6; Angew. Chem. Suppl. (1983), 1176-82.
31 R.W. Wiggins, J.C. Huffman and G. Christou, J. Chem. Soc., Chem. Commun., (1983) 1313-5.
32 J.R. Dorfman and R.H. Holm, Inorg. Chem., 22 (1983) 3179-81.
33 M.R. Snow and J.A. Ibers, Inorg. Chem., 12 (1973) 249-54.
34 T. Herskovitz, B.V. DePamphilis, W.O. Gillum and R.H. Holm, Inorg. Chem., 14 (1975) 1426-9.
35 T. Costa, J.R. Dorfman, K.S. Hagen and R.H. Holm, Inorg. Chem., 22 (1983) 4091-9.
36 G. Christou and J.C. Huffman, J. Chem. Soc., Chem. Commun., (1983) 558-60.
37 Ch.P. Rao, J.R. Dorfman and R.H. Holm, Inorg. Chem., 25 (1986) 428-39.
38 D. Szeymies, Doctoral Thesis, Univ. of Münster 1986; G. Henkel, D. Szeymies and B. Krebs, to be published.

39 O.A. Rajan, M. McKenna, J. Noordik, R.C. Haltiwanger and R.C. Rakowski-DuBois, Organometallics, 3 (1984) 831-40.
40 R. Allmann, I. Baumann, A. Kutoglu, H. Rosch and E. Hellner, Naturwissenschaften, 51 (1964) 263-4.
41 W. Klemm and H.G. von Schnering, Naturwissenschaften, 52 (1965) 12.
42 T.R. Halbert, L.L. Hutchings, R. Rhodes and E.I. Stiefel, J. Am. Chem. Soc., 108 (1986) 6437-8.
43 C.M. Bolinger, T.B. Rauchfuss and A.L. Rheingold, J. Am. Chem. Soc., 105 (1983) 6321-3.
44 W. Tremel, R. Hoffmann and E.D. Jemmis, Inorg. Chem., 28 (1989) 1213-24.
45 Y. Do, E.D. Simhon and R.H. Holm, J. Am. Chem. Soc., 105 (1983) 6731-2.
46 H. Schäfer, P. Moritz and A. Weiss, Z. Naturforsch., B20 (1965) 603.
47 D. Szeymies, B. Krebs and G. Henkel, Angew. Chem., 96 (1984) 797-8; Angew. Chem. Int. Ed. Engl., 23 (1984) 804-5.
48 J.K.Money, J.C. Huffman and G. Christou, Inorg.Chem., 24 (1985) 3297-302.
49 J.K. Money, J.R. Nicholson, J.C. Huffman and G. Christou, Inorg. Chem., 25 (1986) 4072-4.
50 J.K. Money, J.C. Huffman and G. Christou, J. Am. Chem. Soc., 109 (1987) 2210-1.
51 D.W. Murphy, C. Cros, F.J. DiSalvo and J.V. Waszczak, Inorg. Chem., 16 (1977) 3027-31.
52 Y. Do, E.D. Simhon and R.H. Holm, Inorg. Chem., 24 (1985) 4635-42.
53 J.A. Kovacs and R.H. Holm, J. Am. Chem. Soc., 108 (1986) 340-1.
54 J.A. Kovacs and R.H. Holm, Inorg. Chem., 26 (1987) 702-11.
55 J.A. Kovacs and R.H. Holm, Inorg. Chem., 26 (1987) 711-8.
56 M.J. Carney, J.A. Kovacs, Y.P. Zhang, G.C. Papaefthymiou, K. Spartalian, R.B. Frankel and R.H. Holm, Inorg. Chem., 26 (1987) 719-24.
57 K. Tatsumi, Y. Sekiguchi, A. Nakamura, R.E. Cramer and J.J. Rupp, Angew. Chem., 98 (1986) 95-6; Angew. Chem. Int. Ed. Engl.,25 (1986) 86-7.
58 K. Tatsumi, I. Matsubara, Y. Sekiguchi, A. Nakamura and C. Mealli, Inorg. Chem., 28 (1989) 773-80.
59 J.L. Martin and J. Takats, Inorg. Chem., 14 (1975) 1358-64.
60 M. Cowie and M.J. Bennett, Inorg. Chem., 15 (1976) 1589-95.
61 K. Tatsumi, Y. Sekiguchi, A. Nakamura, R.E. Cramer and J.J. Rupp, J. Am. Chem. Soc., 108 (1986) 1358-9.
62 J.L. Seela, J.C. Huffman and G. Christou, J. Chem. Soc. Chem. Commun., (1985) 58-60.
63 E.I. Stiefel, Prog. Inorg. Chem., 22 (1977) 1-233.
64 B. Spivack and Z. Dori, Coord. Chem. Rev., 17 (1975) 99-136; B. Spivack, Z. Dori and E.I. Stiefel, Inorg. Nucl. Chem. Lett., 11 (1975) 501-3.
65 J.I. Gelder and H. Enemark, Inorg. Chem., 15 (1976) 1839-43.
66 W.E. Newton, G.J.-J. Chen and J.W. McDonald, J. Am. Chem. Soc., 98 (1976) 5387-8.
67 G. Bunzey and J.H. Enemark, Inorg. Chem., 17 (1978) 682-8.
68 T.C. Hsieh, K. Gebreyes and J. Zubieta, Transition Met. Chem., 10 (1985) 81-4.
69 G. Bunzey, J.H. Enemark, J.K. Howie and D.T. Sawyer, J. Am. Chem. Soc., 99 (1977) 4168-70.
70 A. Müller, W.O. Nolte and B. Krebs, Angew. Chem., 90 (1978) 286-7; Angew. Chem. Int. Ed. Engl., 17 (1978) 279-80; A. Müller, W.O. Nolte and B. Krebs, Inorg. Chem., 19 (1980) 2835-6.
71 K.F. Miller, A.E. Bruce, J.L. Corbin, S. Wherland and E.I. Stiefel, J. Am. Chem. Soc., 102 (1980) 5102-4.
72 W.H. Pan, M.E. Leonowicz and E.I. Stiefel, Inorg. Chem., 22 (1983) 672-8.
73 T. Chandler, D.L. Lichtenberger and J.H. Enemark, Inorg. Chem., 20 (1981) 75-7.

74 W.H. Pan, T. Chandler, J.H. Enemark and E.I. Stiefel, Inorg. Chem., 23 (1984) 4265-9.
75 H. Keck, W. Kuchen, J. Mathow, B. Meyer, D. Mootz and H. Wunderlich, Angew. Chem., 93 (1981) 1019-20; Angew. Chem. Int. Ed. Engl., 20 (1981) 975-6.
76 A. Müller, R.G. Bhattacharyya and B. Pfefferkorn, Chem. Ber., 112 (1979) 778-80, A. Müller, S. Sarkar, R.G. Bhattacharyya, S. Pohl and M. Dartmann, Angew. Chem., 90 (1978) 564-5; Angew. Chem. Int. Ed. Engl., 17 (1978) 535-6.
77 K. Greiwe, Doctoral Thesis, Univ. of Münster 1985.
78 G. Henkel, K. Greiwe and B. Krebs, publication in preparation.
79 T.R. Halbert, K. McGauley, W.H. Pan, R.S. Czernuszewicz and E.I. Stiefel, J. Am. Chem. Soc., 106 (1984) 1849-51.
80 A. Müller, R. Jostes and F.A. Cotton, Angew. Chem., 92 (1980) 921-9; Angew. Chem. Int. Ed. Engl., 19 (1980) 875-82.
81 J. Marcoll, A. Rabenau, D. Mootz and H. Wunderlich, Rev. Chim. Miner., 11 (1974) 607-15.
82 A. Müller and U. Reinsch, Angew. Chem., 92 (1980) 69-70; Angew. Chem. Int. Ed. Engl., 19 (1980) 72-3.
83 P.J. Vergamini, H. Vahrenkamp and L.F. Dahl, J. Am. Chem. Soc., 93 (1971) 6327-9.
84 H. Keck, W. Kuchen, J. Mathow and H. Wunderlich, Angew. Chem., 94 (1982) 927-8; Angew. Chem. Int. Ed. Engl., 21 (1982) 929-30; Angew. Chem. Suppl. (1982) 1962-66.
85 M.H. Chisholm, J.F. Corning and J.C. Huffman, J. Am. Chem. Soc., 105 (1983) 5924-5.
86 P.J. Blower, J.R. Dilworth, J.P. Hutchinson and J. Zubieta, Transition Met. Chem., 7 (1982) 353.
87 C. Veeger and W.E. Newton, Advances in Nitrogen Fixation Research, Mertinus Nijhoff, The Hague, 1984.
88 T.E. Wolff, J.M. Berg, C. Warrick, K.O. Hodgson, R.H. Holm and R.B. Frankel, J. Am. Chem. Soc., 100 (1978) 4630-2.
89 G. Christou, C.D. Garner and F.E. Mabbs, Inorg. Chim. Acta, 29 (1978) L189-L190.
90 G. Christou, C.D. Garner, F.E. Mabbs and T.J. King, J. Chem. Soc., Chem. Commun., (1978) 740-1.
91 S.R. Acott, G. Christou, C.D. Garner, T.J. King, F.E. Mabbs and R.M. Miller, Inorg. Chim. Acta, 35 (1979) L337-L338.
92 T.E. Wolff, J.M. Berg, K.O. Hodgson, R.B. Frankel and R.H. Holm, J. Am. Chem. Soc., 101 (1979) 4140-50.
93 S.P. Cramer, K.O Hodgson, W.O. Gillum and L.E. Mortenson, J. Am. Chem. Soc., 100 (1978) 3398-407.
94 S.P. Cramer, W.O. Gillum, K.O. Hodgson, L.E. Mortenson, E.I. Stiefel, J.R. Chisnell, W.J. Brill and V.K. Shah, J. Am. Chem. Soc., 100 (1978) 3814-9.
95 T.E. Wolff, J.M. Berg, P.P. Power, K.O. Hodgson and R.H. Holm, Inorg. Chem., 19 (1980) 430-7.
96 T.E. Wolff, J.M. Berg, P.P. Power, K.O. Hodgson, R.H. Holm and R.B. Frankel, J. Am. Chem. Soc., 101 (1979) 5454-6.
97 R.H. Holm, Chem. Soc. Rev., 10 (1981) 455-490.
98 R.E. Palermo, P.P. Power and R.H. Holm, Inorg. Chem., 21 (1982) 173-81.
99 G. Christou, P.K. Mascharak, W.H. Armstrong, G.C. Papaefthymiou, R.B. Frankel and R.H. Holm, J. Am. Chem. Soc., 104 (1982) 2820-31.
100 M. Tanaka, K. Tanaka and T. Tanaka, Chem. Lett., (1982) 767-70.
101 K. Tanaka, Y. Imasaka, M. Tanaka, M. Honjo and T. Tanaka, J. Am. Chem. Soc., 104 (1982) 4258-60.
102 K. Tanaka, Y. Hozumi and T. Tanaka, Chem. Lett., (1982) 1203-6.
103 D. Collison and F.E. Mabbs, J. Chem. Soc., Dalton Trans., (1982) 1575-85.
104 G. Christou, D. Collison, C.D. Garner, S.R. Acott, F.E. Mabbs and V. Petrouleas, J. Chem. Soc., Dalton Trans., (1982) 1575-85.
105 B.K. Teo, M.R. Antonio, R.H. Tieckelmann, H.C. Silvis and B.A. Averill, J. Am. Chem. Soc., 104 (1982) 6126-9.

106 T. Yamamura, G. Christou and R.H. Holm, Inorg. Chem., 22 (1983) 939-49.
107 P.K. Mascharak, M.C. Smith, W.H. Armstrong, B.K. Burgess and R.H. Holm, Proc. Natl. Acad. Sci. U. S. A., 79 (1982) 7056-60.
108 M.R. Antonio, B.K. Teo, W.E. Cleland and B.A. Averill, J. Am. Chem. Soc., 105 (1983) 3477-84.
109 B.K. Teo, M.R. Antonio and B.A. Averill, J. Am. Chem. Soc., 105 (1983) 3751-62.
110 Y. Hozumi, Y. Imasaka, K. Tanaka and T. Tanaka, Chem. Lett., (1983) 897-900.
111 K. Tanaka, M. Honjo and T. Tanaka, J. Inorg. Biochem., 22 (1984) 187-99.
112 J.A. Kovacs, J.K. Bashkin and R.H. Holm, J. Am. Chem. Soc., 107 (1985) 1784-6.
113 S. Kuwabata, Y. Hozumi, K. Tanaka and T. Tanaka, Chem. Lett., (1985) 401-4.
114 J. Cia and B. Kang, Jiegou Huaxue, 3 (1984) 143-5.
115 S.D. Conradson, B.K. Burgess, W.E. Newton, J.W. McDonald, J.F. Rubinson, S.F. Gheller, L.E. Mortenson, M.W.W. Adams, P.K. Mascharak et al., J. Am. Chem. Soc., 107 (1985) 7935-40.
116 A.M. Flank, M. Weininger, L.E. Mortenson and S.P. Cramer, J. Am. Chem. Soc., 108 (1986) 1049-55.
117 K. Tanaka, M. Moriya and T. Tanaka, Inorg. Chem., 25 (1986) 835-8.
118 S. Kuwabata, S. Uezumu, K. Tanaka and T. Tanaka, J. Chem. Soc., Chem. Commun., (1986) 135-6.
119 S. Kuwabata, K. Tanaka and T. Tanaka, Inorg. Chem., 25 (1986) 1691-7.
120 D.J. Rutstrom and A. Robbat, Jr., J. Electroanal. Chem. Interfacial Electrochem., 200 (1986) 193-203.
121 W.E. Newton, S.F. Gheller, B. Hedman, K.O. Hodgson, S.M. Lough and J.W. McDonald, Eur. J. Biochem., 159 (1986) 111-5.
122 B. Kang, J. Cai, C. Cheng and J. Lu, Huaxue Xuebao, 44 (1986) 781-6.
123 J. Cai and B. Kang, Jiegou Huaxue, 4 (1985) 82-5.
124 Y. Cai and Z. Wang, Huaxue Tongbao, (1986) 35-7.
125 J. Cai and C. Cheng, Jiegou Huaxue, 3 (1984) 33-5.
126 Z. Zhang, S. Li and L. Xu, Jilin Daxue Ziran Kexue Xuebao, (1986) 105-8.
127 J. Xu, X. Liu, L. Wang and S. Li, Jilin Daxue Ziran Kexue Xuebao, (1986) 121-4.
128 K. Tanaka, M. Nakamoto, M. Tsunomori and T. Tanaka, Chem. Lett., (1987) 613-6.
129 J.M. Arber, A.C. Flood, C.D. Garner, S.S. Hasnain and B.E. Smith, J. Phys., Colloq. C8, (1986) 1159-63.
130 H. Liu, D. Wu and B. Kang, Bopuxue Zazhi, 4 (1987) 13-9.
131 P.A. Lindahl, B.K. Teo and W.H. Orme-Johnson, Inorg. Chem., 26 (1987) 3912-6.
132 B. Hedman and P. Frank, J. Am. Chem. Soc., 110 (1988) 3798-805.
133 T.E. Wolff, J.M. Berg and R.H. Holm, Inorg. Chem., 20 (1981) 174-80.
134 W.H. Armstrong and R.H. Holm, J. Am. Chem. Soc., 103 (1981) 6246-8.
135 W.H. Armstrong, P.K. Mascharak and R.H. Holm, Inorg. Chem., 21 (1982) 1699-701.
136 W.H. Armstrong, P.K. Mascharak and R.H. Holm, J. Am. Chem. Soc., 104 (1982) 4373-83.
137 P.K. Mascharak, W.H. Armstrong, Y. Mizobe and R.H. Holm, J. Am. Chem. Soc., 105 (1983) 475-83.
138 R.E. Palermo and R.H. Holm, J. Am. Chem. Soc., 105 (1983) 4310-18.
139 P.K. Mascharak, G.C. Papaefthymiou, W.H. Armstrong, S. Foner, R.B. Frankel and R.H. Holm, Inorg. Chem., 22 (1983) 2851-8.
140 Y. Mizobe, P.K. Mascharak, R.E. Palermo and R.H. Holm, Inorg. Chim. Acta, 80 (1983) L65-L67.
141 R.E. Palermo, R. Singh, J.K. Bashkin and R.H. Holm, J. Am. Chem. Soc., 106 (1984) 2600-12.

142 M.J. Carney, J.A. Kovacs, Y.-P. Zhang, G.C. Papaefthymiou, K. Spartalian, R.B. Frankel and R.H. Holm, Inorg. Chem., 26 (1987) 719-24.
143 Y.P. Zhang, J.K. Bashkin and R.H. Holm, Inorg. Chem., 26 (1987) 694-702.
144 S. Ciurli and R.H. Holm, Inorg. Chem., 28 (1989) 1685-90.
145 K. Tanaka, M. Moriya and T. Tanaka, Chem. Lett., (1987) 373-6.
146 K. Tanaka, M. Moriya, S. Uezumi and T. Tanaka, Inorg. Chem., 27 (1988) 137-43.
147 D. Coucouvanis, E.D. Simhon, D. Swenson and N.C. Baenziger, J. Chem. Soc. Chem. Commun., 1979, 361-2.
148 R.H. Tieckelmann, H.C. Silvis, T.A. Kent, B.H. Huynh, J.V. Waszczak, B.-K. Teo and B.A. Averill, J. Am. Chem. Soc., 102 (1980) 5550-9.
149 D. Coucouvanis, P. Stremple, E.D. Simhon, D. Swenson, N.C. Baenziger, M. Draganjac, L.T. Chan, A. Simopoulos, V. Papaefthymiou, A. Kostikas and V. Petrouleas, Inorg. Chem., 22 (1983) 293-308.
150 L. He, L. Zhang and J. Lu, Huaxue Xuebao, 45 (1987) 676-81.
151 S.R. Acott, C.D. Garner, J.R. Nicholson and W. Clegg, J. Chem. Soc., Dalton Trans., (1983) 713-9.
152 R.J.H. Clark, S. Joss, M. Zvagulis, C.D. Garner and J.R. Nicholson, J. Chem. Soc., Dalton Trans., (1986) 1595-601.
153 M. Minelli, J.H. Enemark, J.R. Nicholson and C.D. Garner, Inorg. Chem., 23 (1984) 4384-6.
154 J. Hyde, J. Zubieta and N. Seeman, Inorg. Chim. Acta, 54 (1981) L137-L139.
155 P.L. Dahlstrom, S. Kumar and J. Zubieta, J. Chem. Soc., Chem. Commun., (1981) 411-2.
156 F. Wang, Z. Zhang, Y. Fan and C. Shen, Jilin Daxue Ziran Kexue Xuebao, (1984) 94-8.
157 Z. Zhang, F. Wang and Y. Fan, Kexue Tongbao (Foreign Lang. Ed.), 29(11) (1984) 1486-9
158 Z. Zhang, F. Wang, Y. Lin and Y. Fan, Gaodeng Xuexiao Huaxue Xuebao, 6 (1985) 544-8.
159 S. Niu, M. Pan, Z. Zhang and S. Li, Kexue Tongbao, 32 (1987) 427.
160 Z. Zhang, C. Guo, Y. Fan and F. Wang, Wuji Huaxue, 2 (1986) 26-39.
161 B.A. Averill, Struct. Bond., 53 (1983) 59-103.
162 D.M. Kurtz, Jr., R.S. McMillan, B.K. Burgess, L.E. Mortenson and R.H. Holm, Proc. Natl. Acad. Sci. U.S.A., 76 (1979) 4986-9.
163 B.K. Burgess, D.B. Jacobs and E.I. Stiefel, Biophys. Acta, 614 (1980) 196-209.
164 G. Henkel, H. Strasdeit and B. Krebs, Angew. Chem., 94 (1982) 204-5; Angew. Chem. Int. Ed. Engl., 21 (1982) 201; Angew. Chem. Suppl. (1982) 489-98.
165 H. Strasdeit, B. Krebs and G. Henkel, Inorg. Chem., 23 (1984) 1816-25.
166 G. Christou, K.S. Hagen and R.H. Holm, J. Am. Chem. Soc., 104 (1982) 1744-5.
167 G.D. Lawrence and D.T. Sawyer, Coord Chem. Rev., 27 (1978) 173.
168 Y. Sugiura, H. Kawabe and H.Tanaka, J. Am. Chem. Soc., 102 (1980) 6581
169 Y. Sugiura, H. Kawabe, H. Tanaka, S. Fujimoto and A. Ohara, J. Am. Chem. Soc., 103 (1981) 963-4
170 Y. Sugiura, H. Kawabe, H. Tanaka, S. Fujimoto and A. Ohara, J. Biol. Chem., 256 (1981) 10664-70
171 H. Kawabe, Y. Sugiura, M. Terachi and H. Tanaka, Biochim. Biophys. Acta, 784 (1984) 81-8.
172 J.L. Seela, K. Folting, R.J. Wang, J.C. Huffman, G. Christou, H.R. Chang and D.N. Hendrickson, Inorg. Chem., 24 (1985) 4454-6.
173 D. Swenson, N.C. Baenziger and D. Coucouvanis, J. Am. Chem. Soc., 100 (1978) 1932-4.
174 D.G. Holah and D. Coucouvanis, J. Am. Chem. Soc., 97 (1975) 6917-9.

175 G. Henkel, K. Greiwe and B. Krebs, Angew. Chem., 97 (1985) 113-4; Angew. Chem. Int. Ed. Engl., 24 (1985) 117-8.
176 A.D. Watson, C.P. Rao, J.R. Dorfman and R.H. Holm, Inorg. Chem., 24 (1985) 2820-6.
177 K.S. Hagen and R.H. Holm, Inorg. Chem., 23 (1984) 418-27.
178 G. Henkel, K. Greiwe and B. Krebs, publ. in preparation.
179 G. Henkel, K. Greiwe and B.Krebs, Acta Crystallogr., A40 (1984) C-306.
180 E. Deutsch, K. Libson, S. Jurisson and L.F. Lindoy, Progr. Inorg. Chem., 30 (1983) 75-139.
181 S.F. Colmanet and M.F. Mackay, J. Chem. Soc., Chem. Commun., (1987) 705-6.
182 S.F. Colmanet and M.F. Mackay, Aust. J. Chem., 41 (1988) 269-77.
183 W.K. Bratton and F.A. Cotton, Inorg. Chem., 9 (1970) 789-793; F.A. Cotton and L.W. Shive, Inorg. Chem., 14 (1975) 2032-5.
184 A. Davison, B.V. DePamphilis, R. Faggiani, A.G. Jones, C.J.L. Lock and C. Orvig, Can. J. Chem., 63 (1985) 319-23.
185 J.E. Smith, E.F. Byrne, F.A. Cotton and J.C. Sekutowski, J. Am. Chem. Soc., 100 (1978) 5571-2.
186 E.F. Byrne and J.E. Smith, Inorg. Chem., 18 (1979) 1832-5.
187 A. Davison, C. Orvic, H.S. Trop, M. Sohn, B.V. De Pamphilis and A.G. Jones, Inorg. Chem., 19 (1980) 1988-92.
188 A. Davison, N. De Vries, J. Dewan and A. Jones, Inorg. Chim. Acta, 120 (1986) L15-L16.
189 N. de Vries, J.C. Dewan, A.G. Jones and A. Davison, Inorg. Chem., 27 (1988) 1574-80.
190 P.J. Blower, J.R. Dilworth, J.P. Hutchinson and J.A. Zubieta, Transition Met. Chem., 7 (1982) 354-5; P.J. Blower, J.R. Dilworth, J.P. Hutchinson and J.A. Zubieta, J. Chem. Soc., Dalton Trans. (1985) 1533-41.
191 T.-C. Hsieh, T. Nicholson and J. Zubieta, Inorg. Chem., 27 (1988) 241-50.
192 E.W. Abel, W. Harrison, R.A.N. McLean, W.C. Marsh and J. Trotter, J. Chem. Soc. Chem. Commun., (1970) 1531-3.
193 A.C. McDonnel, T.W. Hambley, M.R. Snow and A.G. Wedd, Aust. J. Chem., 36 (1983) 253-8.
194 J.R. Dilworth, B.D. Neaves, J.P. Hutchinson and J.A. Zubieta, Inorg. Chim. Acta, 65 (1982) L223-L224.
195 P.J. Blower, J.R. Dilworth, J.P. Hutchinson and J.A. Zubieta, Inorg. Chim. Acta, 65 (1982) L225-L226.
196 A.J. Thomson, in P. Harrison (Editor), Metalloproteins; Part 1: Metal Proteins with Redox Roles, VCH Verlagsgesellschaft, Weinheim, 1985, pp. 79-120.
197 R.H. Holm and J.A. Ibers, in W. Lovenberg (Editor), Iron-Sulfur Proteins, Vol. III, Academic Press, New York, 1977, pp. 205-81.
198 J.M. Berg and R.H. Holm, in T.G. Spiro (Editor), Iron-Sulfur Proteins, John Wiley & Sons, New York, 1982, pp. 1-66.
199 C.D. Stout, in T.G. Spiro (Editor), Iron-Sulfur Proteins, John Wiley & Sons, New York, 1982, pp. 97-146.
200 W. Lovenberg (Editor), Iron-Sulfur Proteins, Academic Press, New York, 1973 (Vol. I and II), 1977 (Vol. III).
201 T.G. Spiro (Editor), Iron-Sulfur Proteins, John Wiley & Sons, New York, 1982.
202 M.H. Emptage, T.A. Kent, B.H. Huynh, J. Rawlings, W.H. Orme-Johnson and E. Münck, J. Biol. Chem., 255 (1980) 1793-6.
203 B.H. Huynh, J.J.G. Moura, I. Moura, T.G. Kent, J. LeGall, A.V. Xavier and E. Münck, J. Biol. Chem., 255 (1980) 3242-4.
204 C.D. Stout, V. Pattabhi and A.H. Robins, J. Biol. Chem., 255 (1980) 1797-800.
205 D. Ghosh, W. Furey, Jr., S. O'Donnell and C.D. Stout, J. Biol. Chem., 256 (1981) 4185-92.
206 M.K. Johnson, R.S. Czernuszewicz, T.G. Spiro, J.A. Fee and M.V. Sweeney, J. Am. Chem. Soc., 105 (1983) 6671-8.
207 M.R. Antonio, B.A. Averill, I, Moura, J.J.G. Moura, W.H. Orme-Johnson, B.-K. Teo and A.V. Xavier, J. Biol. Chem., 257 (1982) 6646-9.

208 H. Beinert, M.H. Emptage, J.-L. Dreyer, R.A. Scott, J.E. Hahn, K.O. Hodgson and A.J. Thomson, Proc. Natl. Acad. Sci. U.S.A., 80 (1983) 393-6.
209 G.H. Stout, S. Turley, L.C. Sieker and L.H. Jensen, Proc. Natl. Acad. Sci. U.S.A., 85 (1988) 1020-2.
210 G. Henkel, W. Tremel and B. Krebs, Angew. Chem., 93 (1981) 1072-3; Angew. Chem. Int. Ed. Engl., 20 (1981) 1033.
211 G. Christou, R.H. Holm, M. Sabat and J.A. Ibers, J. Am. Chem. Soc., 103 (1981) 6269-71.
212 G. Christou, M. Sabat, J.A. Ibers and R.H. Holm, Inorg. Chem., 21 (1982) 3518-26.
213 K.S. Hagen, A.D. Watson and R.H. Holm, J. Am. Chem. Soc., 105 (1983) 3905-13.
214 H. Strasdeit, B. Krebs and G. Henkel, Z. Naturforsch., B: Chem. Sci., 42 (1987) 565-72.
215 K.S. Hagen, G. Christou and R.H. Holm, Inorg. Chem., 22 (1983) 309-14.
216 J.J. Mayerle, R.B. Frankel, R.H. Holm, J.A. Ibers, W.D. Phillips and J.F. Weiher, Proc. Nat. Acad. Sci. U. S. A., 70 (1973) 2429-33.
217 J.J. Mayerle, S.E. Denmark, B.V. DePamphilis, J.A. Ibers and R.H. Holm, J. Am. Chem. Soc., 97 (1975) 1032-45.
218 J.A. Konnert and H.T. Evans, Jr., Am. Mineral., 65 (1980) 516-21.
219 J.G. Reynolds and R.H. Holm, Inorg. Chem., 19 (1980) 3257-60.
220 J.G. Reynolds and R.H. Holm, Inorg. Chem., 20 (1981) 1873-8.
221 S. Ueno, N. Ueyama, A. Nakamura and T. Tukihara, Inorg. Chem., 25 (1986) 1000-5.
222 J. Cai and C. Cheng, Jiegou Huaxue, 4 (1985) 199-202.
223 W. Cen and H. Liu, Jiegou Huaxue, 5 (1986) 203-8.
224 W.O. Gillum, R.B. Frankel, S. Foner and R.H. Holm, Inorg. Chem., 15 (1976) 1095-100.
225 P.K. Mascharak, G.C. Papaefthymiou, R.B. Frankel and R.H. Holm, J. Am. Chem. Soc., 103 (1981) 6110-6.
226 P. Beardwood, J.F. Gibson, C.E. Johnson and J.D. Rush, J. Chem. Soc., Dalton Trans., (1982) 2015-20.
227 N. Ueyama, A. Kajiwara, T. Terakawa, S. Ueno and A. Nakamura, Inorg. Chem., 24 (1985) 4700-4.
228 P. Beardwood, J.F. Gibson, P. Bertrand and J.P. Gayda, Biochim. Biophys. Acta, 742 (1983) 426-33.
229 P. Beardwood and J.F. Gibson, J. Chem. Soc., Dalton Trans., (1983) 737-48.
230 P. Beardwood and J.F. Gibson, J. Chem. Soc., Dalton Trans., (1984) 1507-16.
231 B. Teo, R.G. Shulman, G.S. Brown and A.E. Meixner, J. Am. Chem. Soc., 101 (1979) 5624-31.
232 T.J. Ollerenshaw, S. Bristow, B.N. Anand and C.D. Garner, J. Chem. Soc., Dalton Trans., (1986) 2013-5.
233 G.B. Wong, M.A. Bobrik and R.H. Holm, Inorg. Chem., 17 (1978) 578-84.
234 Y. Do, E.D. Simhon and R.H. Holm, Inorg. Chem., 22 (1983) 3809-12.
235 D. Coucouvanis, D. Swenson, P. Stremple and N.C. Baenziger, J. Am. Chem. Soc., 101 (1979) 3392-4.
236 H. Strasdeit, B. Krebs and G. Henkel, Inorg. Chim. Acta, 89 (1984) L11-L13.
237 S. Han, R.S. Czernuszewicz and T.G. Spiro, Inorg. Chem., 25 (1986) 2276-7.
238 K. Tatsumi, Y. Sekiguchi, M. Sebata, A. Nakamura, R.E. Cramer and T.Chung, Angew.Chem., 101 (1989) 83-4; Angew. Chem. Int. Ed. Engl., 28 (1989) 98-9.
239 C.P. Rao, J.R. Dorfman and R.H. Holm, Inorg. Chem., 25 (1986) 428-39.
240 D.T. Sawyer, G.S. Srivatsa, M.E. Bodini, W.P. Schaefer, R.M. Wing, J. Am. Chem. Soc., 108 (1986) 936-42.
241 G. Henkel, W. Tremel, U. Kuhlmann and B. Krebs, Proc. Int. Conf. Coord. Chem., 21 (1980) 351.
242 R.W. Lane, J.A. Ibers, R.B. Frankel and R.H. Holm, Proc. Natl. Acad. Sci. U. S. A., 72 (1975) 2868-72.

243 R.W. Lane, J.A. Ibers, R.B. Frankel, G.C. Papaefthymiou and R.H. Holm, J. Am. Chem. Soc., 99 (1977) 84-98.
244 K.S. Hagen and R.H. Holm, J. Am. Chem. Soc., 104 (1982) 5496-7.
245 W. Tremel, M. Kriege, B. Krebs and G. Henkel, Inorg. Chem., 27 (1988) 3886-95.
246 G. Henkel, W. Tremel and B. Krebs, Angew. Chem., 95 (1983) 317-18; Angew. Chem. Int. Ed. Engl., 22 (1983) 319; Angew. Chem. Suppl. (1983) 323-46.
247 F.A. Cotton and D.A. Ucko, Inorg. Chim. Acta., 6 (1972) 161-72.
248 G. Henkel, W. Tremel and B. Krebs, in preparation.
249 G. Henkel, W. Tremel and B. Krebs, unpublished results.
250 G. Henkel, W. Tremel and B. Krebs, Angew. Chem., 95 (1983) 314; Angew. Chem. Int. Ed. Engl., 22 (1983) 318; Angew. Chem. Suppl. (1983), 307-22.
251 W. Tremel, B. Krebs and G. Henkel, Inorg. Chim. Acta, 80 (1983) L31-L32.
252 H. Beinert and A.J. Thomson, Arch. Biochem. Biophys., 222 (1983) 333-61.
253 J.J. Girerd, G.C. Papaefthymiou, A.D. Watson, E. Gamp, K.S. Hagen, N. Edelstein, R.B. Frankel and R.H. Holm, J. Am. Chem. Soc., 106 (1984) 5941-7.
254 T. Herskovitz, B.A. Averill, R.H. Holm, J.A. Ibers, W.D. Phillips and J.F. Weiher, Proc. Nat. Acad. Sci. U. S. A., 69 (1972) 2437-41.
255 B.A. Averill, T. Herskovitz, R.H. Holm and J.A. Ibers, J. Amer. Chem. Soc., 95 (1973) 3523-34.
256 J. Gloux, P. Gloux, B. Lamotte and G. Rius, Phys. Rev. Lett., 54 (1985) 599-602.
257 P.K. Mascharak, K.S. Hagen, J.T. Spence and R.H. Holm, Inorg. Chim. Acta, 80 (1983) 157-70.
258 T. O'Sullivan and M.M. Millar, J. Am. Chem. Soc., 107 (1985) 4096-7.
259 C.J. Pickett, J. Chem. Soc., Chem. Commun., (1985) 323-6.
260 V. Papaefthymiou, M.M. Millar and E. Muenck, Inorg. Chem., 25 (1986) 3010-4.
261 J. Gloux, P. Gloux, H. Hendriks and G. Rius, J. Am. Chem. Soc., 109 (1987) 3220-4.
262 M.A. Bobrik, L. Que, Jr. and R.H. Holm, J. Amer. Chem. Soc., 96 (1974) 285-7.
263 R.H. Holm, W.D. Phillips, B.A. Averill, J.J. Mayerle and T. Herskovitz, J. Amer. Chem. Soc., 96 (1974) 2109-17.
264 R.H. Holm, B.A. Averill, T. Herskovitz, R.B. Frankel, H.B. Gray, O. Siiman and F.J. Grunthaner, J. Amer. Chem. Soc., 96 (1974) 2644-6.
265 B. V. DePamphilis, B. A. Averill, T. Herskovitz, L. Jr. Que, R. H. Holm, J. Amer. Chem. Soc., 96 (1974) 4159-67.
266 L. Que, Jr., M.A. Bobrik, J.A. Ibers and R.H. Holm, J. Amer. Chem. Soc., 96 (1974) 4168-78.
267 L. Que, Jr., J.R. Anglin, M.A. Bobrik, A. Davison and R.H. Holm, J. Am. Chem. Soc., 96 (1974) 6042-8.
268 S.P.W. Tang, T.G. Spiro, C. Antanaitis, T.H. Moss, R.H. Holm, T. Herskovitz and L.E. Mortensen, Biochem. Biophys. Res. Commun., 62 (1975) 1-6.
269 G.R. Dukes and R.H. Holm, J. Am. Chem. Soc., 97 (1975) 528-33.
270 W.O. Gillum, L.E. Mortenson, J.S. Chen and R.H. Holm, J. Am. Chem. Soc., 99 (1977) 584-95.
271 R.W. Johnson and R.H. Holm, J. Am. Chem. Soc., 100 (1978) 5338-44.
272 K.S. Hagen, J.G. Reynolds and R.H. Holm, J. Am. Chem. Soc., 103 (1981) 4054-63.
273 T.D.P. Stack and R.H. Holm, J. Am. Chem. Soc., 109 (1987) 2546-7.
274 T.D.P. Stack and R.H. Holm, J. Am. Chem. Soc., 110 (1988) 2484-94.
275 J.G. Reynolds, C.L. Coyle and R.H. Holm, J. Am. Chem. Soc., 102 (1980) 4350-5.

276 G.C. Papaefthymiou, E.J. Laskowski, S. Frota-Pessoa, R.B. Frankel and R.H. Holm, Inorg. Chem., 21 (1982) 1723-8.
277 K.W. Browall, T. Bursh, L.V. Interrante and J.S. Kasper, Inorg. Chem., 11 (1972) 1800-6.
278 J. Cambray, R.W. Lane, A.G. Wedd, R.W. Johnson and R.H. Holm, Inorg. Chem., 16 (1977) 2565-71.
279 B.A. Averill, J.R. Bale and W.H. Orme-Johnson, J. Am. Chem. Soc., 100 (1978) 3034-43.
280 P.J. Stephens, A.J. Thomson, T.A. Keiderling, J. Rawlings, K.K. Rao and D.O. Hall, Proc. Natl. Acad. Sci. U. S. A., 75 (1978) 5273-5.
281 M.W.W. Adams, K.K. Rao, D.O. Hall, G. Christou and C.D. Garner, Biochim. Biophys. Acta, 589 (1980) 1-9.
282 T. Nagano, K. Yoshikawa and M. Hirobe, Tetrahedron Lett., 21 (1980) 297-300.
283 T. Itoh, T. Nagano and M. Hirobe, Tetrahedron Lett., 21 (1980) 1343-6.
284 M. Tezuka, T. Yajima, A. Tsuchiya, Y. Matsumoto, Y. Uchida and M. Hidai, J. Am. Chem. Soc., 104 (1982) 6834-6.
285 D.B. Beach, J.L. Hoskins, W.L. Jolly, S.P. Smit and S.F. Xiang, J. Electron Spectrosc. Relat. Phenom., 28 (1983) 299-302.
286 K. Tano and G.N. Schrauzer, J. Am. Chem. Soc., 97 (1975) 5404-8.
287 R.A. Henderson and A.G. Sykes, Inorg. Chem., 19 (1980) 3103-5.
288 K. Tanaka, T. Tanaka and I. Kawafune, Inorg. Chem., 23 (1984) 516-8.
289 D.M. Kurtz, Jr. and W.C. Stevens, J. Am. Chem. Soc., 106 (1984) 1523-4.
290 J.M. Moulis, J. Meyer and M. Lutz, Biochemistry, 23 (1984) 6605-13.
291 N. Ueyama, T. Sugawara, M. Fuji, A. Nakamura and N. Yasuoka, Chem. Lett., (1985) 175-8.
292 H. Tsai, W.V. Sweeney and C.L. Coyle, Inorg. Chem., 24 (1985) 2796-8.
293 R.J. Burt, B. Ridge and H.N. Rydon, J. Chem. Soc., Dalton Trans., (1980) 1228-35.
294 R. Maskiewicz and T.C. Bruice, J. Chem. Soc., Chem. Commun., (1978) 703-4.
295 R.B. Frankel, W.M. Reiff, I. Bernal and M.L. Good, Inorg. Chem., 13 (1974) 493-4.
296 W. C. Stevens and D.M. Kurtz, Jr., Inorg. Chem., 24 (1985) 3444-9.
297 F. Bonomi, M.T. Werth and D.M. Kurtz, Jr., Inorg. Chem., 24 (1985) 4331-5.
298 M. Nakazawa, Y. Mizobe, Y. Matsumoto, Y. Uchida, M. Tezuka and M. Hidai, Bull. Chem. Soc. Jpn., 59 (1986) 809-14.
299 K. Tanaka, M. Masanaga and T. Tanaka, J. Am. Chem. Soc., 108 (1986) 5448-52.
300 H. Inoue and T. Nagata, J. Chem. Soc., Chem. Commun., (1986) 1177-8.
301 G. Lin, H. Zhang, S. Hu and T.C.W. Mak, Acta Crystallogr., Sect. C: Cryst. Struct. Commun., C43 (1987) 352-3.
302 T. Itoh, T. Nagano and M. Hirobe, Chem. Pharm. Bull., 34 (1986) 2013-7.
303 Y. Okuno, K. Uoto, Y. Sasaki, O. Yonemitsu and T. Tomohiro, J. Chem. Soc., Chem. Commun., (1987) 874-6.
304 Y. Okuno, K. Uoto, O. Yonemitsu and T. Tomohiro, J. Chem. Soc., Chem. Commun., (1987) 1018-20.
305 D.J. Evans, G.J. Leigh, A. Houlton and J. Silver, Inorg. Chim. Acta, 146 (1988) 5.
306 R.B. Frankel, T. Herskovitz, B.A. Averill, R.H. Holm, P.J. Krusic and W.D. Phillips, Biochem. Biophys. Res. Commun., 58 (1974) 974-82.
307 C.L. Hill, R.H. Holm and L.E. Mortenson, J. Am. Chem. Soc., 99 (1977) 2549-57.
308 R.W. Lane, A.G. Wedd, W.O. Gillum, E.J. Laskowski, R.H. Holm, R.B. Frankel and G.C. Papaefthymiou, J. Am. Chem. Soc., 99 (1977) 2350-2.

309 J.G. Reynolds, E.J. Laskowski and R.H. Holm, J. Am. Chem. Soc., 100 (1978) 5315-22.
310 E.J. Laskowski, R.B. Frankel, W.O. Gillum, G.C. Papaefthymiou, J. Renaud, J.A. Ibers and R.H. Holm, J. Am. Chem. Soc., 100 (1978) 5322-37.
311 J.M. Berg, K.O. Hodgson and R.H. Holm, J. Am. Chem. Soc., 101 (1979) 4586-93.
312 R.S. McMillan, J. Renaud, J.G. Reynolds and R.H. Holm, J. Inorg. Biochem., 11 (1979) 213-27.
313 E.J. Laskowski, J.G. Reynolds, R.B. Frankel, S. Foner, G.C. Papaefthymiou and R.H. Holm, J. Am. Chem. Soc., 101 (1979) 6562-70.
314 B.A. Averill and W.H. Orme-Johnson, J. Am. Chem. Soc., 100 (1978), 5234-6
315 D.W. Stephan, G.C. Papaefthymiou, R.B. Frankel and R.H. Holm, Inorg. Chem., 22 (1983) 1550-7.
316 K.S. Hagen, A.D. Watson and R.H. Holm, Inorg. Chem., 23 (1984) 2984-90.
317 M.J. Carney, R.H. Holm, G.C. Papaefthymiou and R.B. Frankel, J. Am. Chem. Soc., 108 (1986) 3519-21.
318 M.J. Carney, G.C. Papaefthymiou, M.A. Whitener, K. Spartalian, R.B. Frankel and R.H. Holm, Inorg.Chem., 27 (1988) 346-52.
319 M.J. Carney, G.C. Papaefthymiou, K. Spartalian, R.B. Frankel and R.H. Holm, J. Am. Chem. Soc., 110 (1988) 6084-95.
320 M.J. Carney, G.C. Papaefthymiou, R.B. Frankel and R.H. Holm, Inorg. Chem., 28 (1989) 1497-503.
321 T. Mashino, T. Nagano and M. Hirobe, Tetrahedron Lett., 24 (1983) 5113-6.
322 H. Strasdeit, B. Krebs and G. Henkel, Proc. Int. Conf. Coord. Chem., 23 (1984) 576.
323 M.A. Bobrik, E.J. Laskowski, R.W. Johnson, W.O. Gillum, J.M. Berg, K.O. Hodgson and R.H. Holm, Inorg. Chem., 17 (1978) 1402-10.
324 G. Christou, B. Ridge and H.N. Rydon, J. Chem. Soc., Dalton Trans., (1978) 1423-5.
325 W.E. Cleland, D.A. Holtman, M. Sabat, J.A. Ibers, G.C. DeFotis and B.A. Averill, J. Am. Chem. Soc., 105 (1983) 6021-31.
326 G. Henkel, W. Simon, A. Wilk, B. Krebs, Z. Kristallogr., 92 (1987) 178.
327 G. Henkel, W. Simon and B. Krebs, Z. Kristallogr., (1989) in press.
328 G. Henkel, A. Wilk and W. Simon in Chemiedozententagung, Selbstverlag, Göttingen, 1987, p. 60.
329 W. Simon, A. Wilk, B. Krebs and G. Henkel, Angew. Chem., 99 (1987) 1039-40; Angew. Chem. Int. Ed. Engl., 26 (1987) 1009-10.
330 K.S. Hagen, J.M. Berg and R.H. Holm, Inorg. Chim. Acta, 45 (1980) L17-L18.
331 K.S. Hagen, D.W. Stephan and R.H. Holm, Inorg. Chem., 21 (1982) 3928-36.
332 G. Henkel, H. Strasdeit and B. Krebs, Proc. Int. Conf. Coord. Chem., 22 (1982) 618.
333 G. Christou, K.S. Hagen, J.K. Bashkin and R.H. Holm, Inorg. Chem., 24 (1985) 1010-8.
334 G. Henkel and F. Hardinghaus, in preparation
335 I.G. Dance and J.C. Calabrese, J. Chem. Soc., Chem. Commun., (1975) 762-3.
336 I.G. Dance, J. Am. Chem. Soc., 101 (1979) 6264-73.
337 W. Tremel, B. Krebs and G. Henkel, Angew. Chem., 96 (1984) 604-5; Angew. Chem. Int. Ed. Engl., 23 (1984) 634.
338 P.A. Lindahl, N. Kojima, R.P. Hausinger, J.A. Fox, B.K. Teo, C.T. Walsh and W.H. Orme-Johnson, J. Am. Chem. Soc., 106 (1984) 3062-4.
339 R.A. Scott, S.A. Wallin, M. Czechowski, D.V. DerVartanian, J. LeGall, H.D. Peck, Jr. and I. Moura, J. Am. Chem. Soc., 106 (1984) 6864-5.
340 S.P. Cramer, W.-H. Pan, M.K. Eidsness, T. Morton, S.W. Ragsdale, D.V. DerVartanian, L.G. Ljungdahl and R.A. Scott, Inorg. Chem., 26 (1987) 2477-9.

341 J.R. Lancaster, Jr. (Editor), The Bioinorganic Chemistry of Nickel, VCH Verlagsgesellschaft, Weinheim, 1988.
342 A.F. Wells, Structural Inorganic Chemistry, 5th ed., Oxford University Press, Oxford, 1986.
343 D.J. Vaughn and J.R. Craig, Mineral Chemistry of Metal Sulfides, Cambridge University Press, New York, 1978.
344 T. Yamamura, H. Miyamae, Y. Katayama and Y. Sasaki, Chem. Lett., (1985) 269-72.
345 B.S. Snyder, C.P. Rao and R.H. Holm, Aust. J. Chem., 39 (1986) 963-74.
346 J.R. Nicholson, G. Christou, J.C. Huffman and K. Folting, Polyhedron, 6 (1987) 863-70.
347 W. Tremel, B. Krebs and G. Henkel, J. Chem. Soc., Chem. Commun., (1986) 1527-9.
348 P. Woodward, L.F. Dahl, E.W. Abel and B.C. Crosse, J. Am. Chem. Soc., 87 (1965) 5251-3.
349 H. Miyamae and T. Yamamura, Acta Crystallogr., Sect. C: Cryst. Struct. Commun., C44 (1988) 606-9.
350 T. Yamamura, Bull. Chem. Soc. Jpn., 61 (1988) 1975-8.
351 R.O. Gould and M.M. Harding, J. Chem. Soc. A, 1970, 875-86.
352 H. Barrera, J.C. Bayon, J. Suades, C. Germain, J.P. Declerq, Polyhedron, 3 (1984) 969-75.
353 W. Gaete, J. Ros, X. Solans, M. Font-Altaba, J.L. Brianso, Inorg. Chem., 23 (1984) 39-43.23 (1984) 39-43.
354 M. Kriege and G. Henkel, Z. Naturforsch., B: Chem. Sci., 42 (1987) 1121-8.
355 T. Krüger, B. Krebs and G. Henkel, Angew. Chem., 101 (1989) 54; Angew. Chem. Int. Ed. Engl., 28 (1989) 61-2.
356 I.G. Dance, M.L. Scudder and R. Secomb, Inorg. Chem., 24 (1985) 1201-8.
357 B.K. Koo, E. Block, H. Kang, S. Liu, J. Zubieta, Polyhydron, 7 (1988) 1397-9.
358 G. Henkel, M. Kriege and K. Matsumoto, J. Chem. Soc., Dalton Trans., (1988) 657-9.
359 K. Matsumoto, H. Nakano and S. Ooi, Chem. Lett., (1988) 823-6.
360 W. Tremel and G. Henkel, Inorg. Chem., 27 (1988) 3896-9.
361 J.K. Money, J.R. Nicholson, J.C. Huffmann and G. Christou, Inorg. Chem., 25 (1986) 4072-4.
362 G. Henkel, W. Tremel, B. Krebs, T.F. Koetzle and P. Coppens, Proc. Eur. Crystallogr. Meeting, 7 (1982) 223.
363 M.R. Antonio, B.-K. Teo, W.H. Orme-Johnson, M.S. Nelson, S.E. Groh, P.A. Lindahl and S.M. Kauzlarich, J. Am. Chem. Soc., 104 (1982) 4703-5.
364 I.G. Dance, G.A. Bowmaker, G.R. Clark and J.K. Seadon, Polyhedron, 2 (1983) 1031-43.
365 D. Coucouvanis, C.N. Murphy and S.K. Kanodia, Inorg. Chem., 19 (1980) 2993-8.
366 I.G. Dance and J.C. Calabrese, Inorg. Chim. Acta, 19 (1976) L41-L42.
367 M. Baumgartner, W. Bensch, P. Hug and E. Dubler, Inorg. Chim. Acta, 136 (1987) 139-47.
368 G. Henkel, B. Krebs, P. Betz, H. Fietz and K. Saatkamp, Angew. Chem., 100 (1988) 1373-5; Angew. Chem. Int. Ed. Engl., 27 (1988) 1326-9.
369 P. Betz, Doctoral Thesis, Univ. of Münster 1986.
370 J.R. Nicholson, I.L. Abrahams, W. Clegg and C.D. Garner, Inorg. Chem., 24 (1985) 1092-6.
371 G. Henkel, P. Betz and B. Krebs, publ. in preparation.
372 G.A. Bowmaker, G.R. Clark, J.K. Seadon and I.G. Dance, Polyhedron, 3 (1984) 535-44.
373 I.G. Dance, J. Chem. Soc., Chem. Commun., (1976) 68-9.
374 G.A. Bowmaker and L.C. Tan, Aust. J. Chem., 32 (1979) 1443-52.
375 I.G. Dance, J. Chem. Soc., Chem. Commun., (1976) 103-4.
376 I.G. Dance, Aust. J. Chem., 31 (1978) 2195-206.
377 Q. Yang, K. Tang, H. Liao, Y. Han, Z. Chen and Y. Tang, J. Chem. Soc., Chem. Commun., (1987) 1076-7.
378 E. Block, M. Gernon, H. Kang, S. Lu and J. Zubieta, J. Chem. Soc., Chem. Commun., (1988) 1031-3.

379 G. Henkel, P. Betz and B. Krebs, Angew. Chem., 99 (1987) 131-2; Angew. Chem. Int. Ed. Engl., 26 (1987) 145-6.
380 K. Saatkamp, Doctoral Thesis, Univ. of Münster 1988.
381 G. Henkel, K. Saatkamp and B. Krebs, publ. in preparation.
382 W. Wojnowski, M. Wojnowski, K. Peters, E.-M. Peters and H.G. v. Schnering, Z. Anorg. Allg. Chem., 530 (1985) 79-88.
383 I.G. Dance, L.J. Fitzpatrick and M.L. Scudder, Inorg. Chem., 23 (1984) 2276-81.
384 P. Gonzalez-Duarte, J. Sola, J. Vives and X. Solans, J. Chem. Soc., Chem. Commun., (1987) 1641-2.
385 S.-H. Hong, Å. Olin and R. Hesse, Acta Chem. Scand., A29 (1975) 583-9.
386 I.G. Dance, Inorg. Chim. Acta, 25 (1977) L17-L18.
387 I.G. Dance, L.J. Fitzpatrick, A.D. Rae and M.L. Scudder, Inorg. Chem., 22 (1983) 3785-8.
388 I.G. Dance, Inorg. Chem., 20 (1981) 1487-92.
389 G. Henkel, P. Betz and B. Krebs, Proc. Int. Conf. Coord. Chem., 24 (1986) 737 (Chemika Chronika New Series, Special Issue, August 1986).
390 I.L. Abrahams, C.D. Garner and W. Clegg, J. Chem. Soc., Dalton Trans., (1987) 1577-9.
391 G. Henkel, J. Bremer, W. Tremel and B. Krebs, publ. in preparation.
392 J.L. Hencher, M.A. Khan, F.F. Said and D.G. Tuck, Inorg. Nucl. Chem. Lett., 17 (1981) 287-90.
393 J.L. Hencher, M.A. Khan, F.F. Said and D.G. Tuck, Polyhedron 4 (1985) 1263-7.
394 I.G. Dance, A. Choy and M.L. Scudder, J. Am. Chem. Soc., 106 (1984) 6436-7.
395 B. Krebs, W. Puls and G. Henkel, publ. in preparation.
396 K.S. Hagen and R.H. Holm, Inorg. Chem., 22 (1983) 3171-4.
397 G.A. Bowmaker, I.G. Dance, C.B. Dobson and D.A. Rogers, Aust. J. Chem., 37 (1984) 1607-18.
398 G. Henkel, P. Betz and B. Krebs, J. Chem. Soc., Chem. Commun. (1985) 1498-9.
399 G. Henkel, P. Betz and B. Krebs, Inorg. Chim. Acta, 134 (1987) 195-6.
400 D.C. Bradley and N.R. Kunchur, Can. J. Chem., 43 (1965) 2786-92.
401 A.P. Arnold, A.J. Canty, B.W. Skelton and A.H. White, J. Chem. Soc., Dalton Trans., (1982) 607-13.
402 G.S.H. Lee, D.C.Craig, I. Ma, M.L. Scudder, T.D. Bailey and I.G. Dance, J. Am. Chem. Soc., 110 (1988) 4863-4.

CLUSTERS OF METALS AND NONMETALS

KENTON H. WHITMIRE
Department of Chemistry, Rice University, P.O. Box 1892, Houston, Texas (USA)

1 INTRODUCTION

Cluster compounds which contain both main group elements and transition metals lie at the interface between homogeneous, solution phase chemistry and heterogeneous, solid state chemistry. They are important hybrids of two distinct types of cluster molecules - the traditional metal carbonyls and the main group clusters known as Zintl ions (refs. 1-2). Many examples of this class of compound have been structurally characterized, and practically every main group III to VI element has been incorporated into a transition metal cluster. The syntheses are not always clean or rational, and this has hindered understanding of the chemistry and reactivity of these molecules. Significant progress towards this end has been made in the past few years, however, and synthetic methods are becoming more systematic.

Transition metal clusters in general have generated a considerable interest as models for heterogeneous catalysts and more recently have been examined for their relationship to solid state systems and materials synthesis (refs. 3-9). The hybrid main group/transition metal systems offer many possibilities in regard to understanding heterogeneous systems. Main group elements are often added to heterogeneous catalyst systems and may have a dramatic effect either as promoters or as inhibitors. The mixed element compounds have stoichiometries and geometries which possess an intriguing potential as building blocks for extended solid arrays.

Due to limited space, this chapter cannot be an exhaustive review of these compounds, but rather it is hoped that the salient features of their chemistry will be presented. The discussion will be focused on complexes in which the main group atom is bonded only to other main group elements or transition metals. A large number of related clusters which contain organo-main group functionalities and have the similar structural and reactivity patterns are known but will not be discussed. A more thorough treatment of this class of cluster compound is available as are several recent, related reviews (refs. 10-16).

2 STRUCTURAL FEATURES

Main group atoms show a tremendous flexibility in their bonding in transition metal clusters. Trends in bonding are apparent and depend on size and charge, although these relationships are not simple. The smaller, naked main group atoms such as boron, carbon and nitrogen tend to prefer geometries which are considered interstitial while the heavier and larger members of these groups do not so readily fit inside such cavities. Most commonly, the larger main group atoms are bonded to three or four metal atoms, with the most observed structure type containing an EM_3 tetrahedron and derivatives thereof formed by breakage of M-M bonds. Increasingly, examples of more complex structures are being reported which show a larger number of main group vertices in unusual bonding modes.

2.1 Interstitial Main Group Atoms

In a number of solid state phases, main group atoms are considered to occupy interstitial sites in a transition metal lattice. Classic examples are the metal carbide phases such as NbC and WC where carbon is seen to occupy either octahedral or trigonal prismatic vacancies in a metal lattice (ref. 17). The square antiprismatic interstitial geometry is rare in the solid state but has been observed for $Cr_{23}C_6$ (ref. 18).

Geometries of discrete cluster molecules mimic the interstitial sites found in the solid state. By conventional bonding theories (refs. 19-35), the presence of seven skeletal electron pairs produces an octahedral complex while nine skeletal pairs is most often observed for the trigonal prismatic molecules. For the interstitial situation, the main group atom is considered to donate all of its electrons to the cluster skeletal pair count. Several octahedral rhenium compounds (refs. 36-38) structurally similar to the parent $[H_2Re_6(CO)_{18}C]^{2-}$ (ref. 39) have been prepared and are isoelectronic to a series of octahedral iron, ruthenium and osmium clusters (refs. 40-63) whose prototype is $Ru_6(CO)_{17}C$ (Fig. 1, ref. 64). Rhenium, iron, ruthenium and osmium compounds seem to favor the formation of octahedral carbides and nitrides, but one example of these metals with a trigonal prismatic carbide has been structurally characterized, that being the large cluster complex $[PPh_3Me][Os_{11}(CO)_{27}\{Cu(NCMe)\}C]$ (ref. 65).

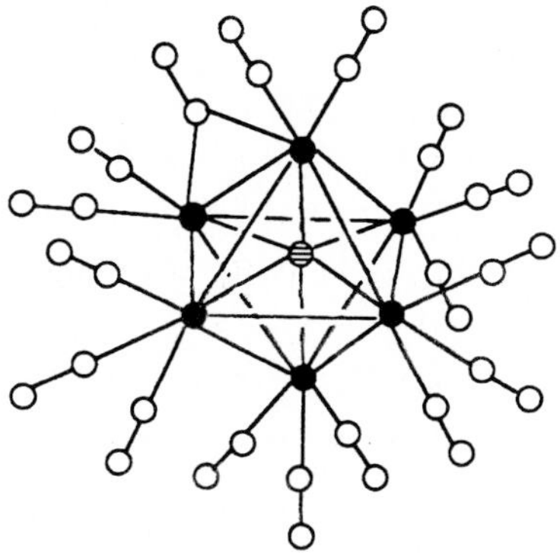

Fig. 1. $Ru_6(CO)_{17}C$

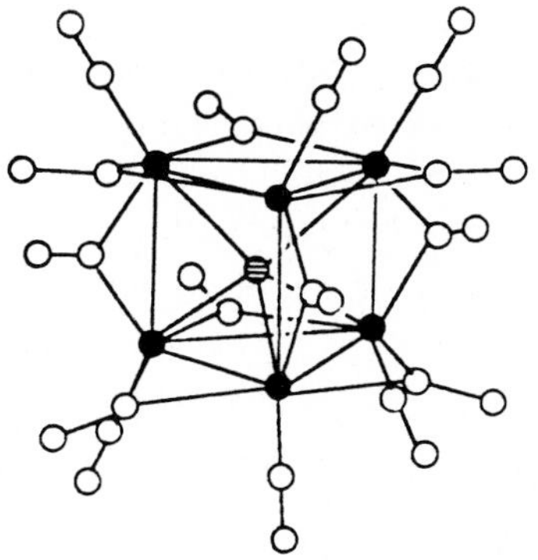

Fig. 2. $[Co_6(CO)_{15}C]^-$

On the other hand, cobalt and rhodium show a preference for trigonal prismatic geometries (refs. 66-80). The prototype is the $[Co_6(CO)_{15}C]^-$ ion (Fig. 2, ref. 66). These clusters show nine skeletal electron pairs which corresponds to the number of M-M bonds. Three exceptions have been characterized which have an octahedral metal array: the anion $[Co_6(CO)_{13}C]^{2-}$ (ref. 81), the related paramagnetic $[Co_6(CO)_{14}C]^-$ (ref. 82), and $[H_3O][Rh_{15}(CO)_{28}(C)_2]$ (ref. 83). In the $[Co_8(CO)_{18}C]^{2-}$ ion, the carbon occupies a square antiprismatic cavity (ref. 84). Other examples of carbides in square anti-prismatic holes include the anions $[Ni_8(CO)_{16}C]^{2-}$ (ref. 85), $[[Ni_9(CO)_{17}C]^{2-}$ (ref. 85) and $[Co_3Ni_9(CO)_{20}C]^{3-}$ (ref. 86).

These cavities have different sizes which can be seen in the metal to carbon distances (Table 2.1). Octahedral cavities are the smallest followed by trigonal prismatic and then square anti-prismatic. For completely idealized geometries in which all metal-metal bond distances are equal and all angles are regular, the metal-carbon bond distances should vary directly with the metal-metal distances and increases in the order octahedron < trigonal prism < square antiprism. If the metal-metal bond distance is X, then the metal-carbon distances should be ideally 0.7071 X for an octahedron, 0.7638 X for a trigonal prism and 0.8226 X for a square antiprism. Of course, the metal-metal bonding framework is very flexible and can allow for a range of metal-metal and metal-carbon distances. Sometimes, these parameters are very distorted from idealized values.

TABLE 2.1 Comparative Data for Cavity Sizes in the Carbido Clusters.

Compound	Geometry[a]	M-C(ave)pm	Ref.
$[NEt_4]_2[Co_6(CO)_{13}C]$	O	187	81
$[NMe_4][Co_6(CO)_{14}C]$	O	190	82
$[NMe_3CH_2Ph]_2[Co_6(CO)_{15}C]$	TP	195	66
$[PPN][Co_6(CO)_{15}N]$	TP	193.8	67
$[NMe_3CH_2Ph]_2[Co_8(CO)_{18}C]$	SA	207	84
$[H_3O][Rh_{15}(CO)_{28}C_2]$	O	204	83
$[PPN]_2[Rh_{12}(CO)_{24}C_2]$	TP	212	76
$[NPr_4]_3[Rh_{12}(CO)_{23}C_2]$	TP	212.5	77
$[NPr_4]_4[Rh_{12}(CO)_{23}C_2]$	TP	213	78
$[PPN]_2[Os_{10}(CO)_{24}C]$	O	204	56, 61
$[PPN][HOs_{10}(CO)_{24}C]$	O	204	55
$[PPh_3Me][Os_{11}(CO)_{27}\{CuNCMe\}C]$	TP	217	65

a. O = octahedral, TP = trigonal prismatic, SA = square anti-prismatic

Carbides and nitrides are the most common interstitial atoms but boron, silicon, phosphorus, arsenic, antimony and sulfur have also been found in encapsulated configurations. The structure of the one boron example $BCo_6(CO)_{18}$ is unknown (ref. 87). As the main group atom becomes larger, its steric requirements are more demanding and the cavities reflect this increase in size with a pronounced tendency to give metal arrays which have larger cavities. Thus, square antiprismatic interstitial geometries but not octahedral or trigonal prismatic are observed, examples being $[Co_9(CO)_{21}Si]^{2-}$ (ref. 88), $[Rh_9(CO)_{22}P]^{3-}$ (ref. 89), $[Rh_9(CO)_{21}P]^{2-}$

(ref. 90) and $[Rh_{10}(CO)_{22}As]^{3-}$ (ref. 91). A distorted icosahedrally coordinated interstitial antimony atom is found in $[Rh_{12}(CO)_{27}Sb]^{3-}$ (ref. 92) while two seven coordinate, encapsulated sulfur atoms are seen in $[Rh_{17}(CO)_{32}(S)_2]^{3-}$ (ref. 93). The anion $[Co_6(CO)_{15}P]^-$ resembles an octahedral cluster which has been splayed open by the size of the P atom, but it also has one more skeletal electron pair than required for the octahedral configuration (ref. 94).

Encapsulated dicarbides have been reported. These are different from the two carbide-containing rhodium clusters given in Table 2.1 in which the carbides are well-separated. The dicarbides show a range of C-C bond lengths as indicated in Table 2.2. For comparison, the C-C separation in Cr_3C_2 is 166pm (ref. 16).

TABLE 2.2 C-C Bond Distances in Encapsulated Dicarbide Clusters

compound	C-C Bond Distance, pm	Ref.
$[NMe_3CH_2Ph]_3[Co_{11}(CO)_{22}C_2]$	162	95
$[NEt_4]_2[Co_6Ni_2(CO)_{16}C_2]$	149.4	96
$[Co_3Ni_7(CO)_{15}C_2]^{3-}$	141	97
$[PPh_4]_2[Co_3Ni_7(CO)_{16}C_2]$	148	97
$[AsPh_4]_2[Ni_{10}(CO)_{16}C_2]$	140	98
$[NMe_4]_4[Ni_{16}(CO)_{23}(C_2)_2]$	138	99
$Rh_{12}(CO)_{25}C_2$	148	100

2.2 <u>Partially Encapsulated Main Group Atoms</u>

In addition to the completely encapsulated atom geometries listed above, partially encapsulated structures are known which can be viewed as fragments of the larger structures. This is best illustrated by the octahedral cluster carbides and nitrides where one or two metal vertices may be lost. The resultant structures are the five metal vertex square-based pyramid, the bridged-butterfly (derived from the square pyramid by addition of an electron pair and the breakage of an apical to basal M-M bond) and the butterfly geometry which can be viewed as a square pyramid with a basal metal vertex removed.

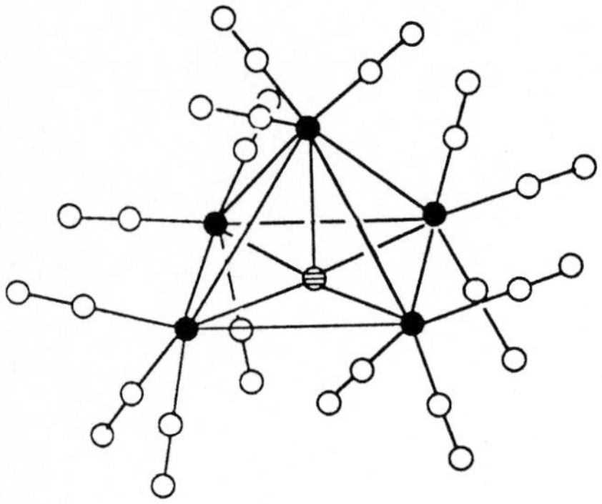

Fig. 3. $Fe_5(CO)_{15}C$

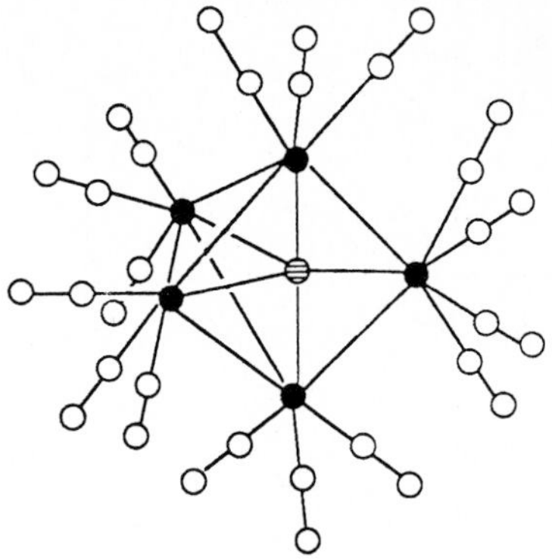

Fig. 4. $Os_6(CO)_{16}C$

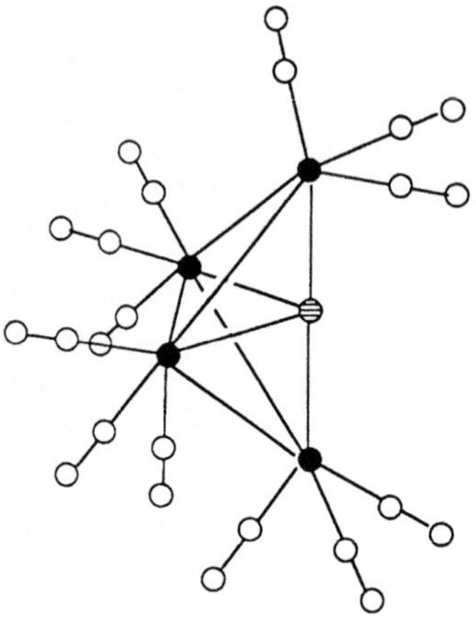

Fig. 5. $[Fe_4(CO)_{12}C]^{2-}$

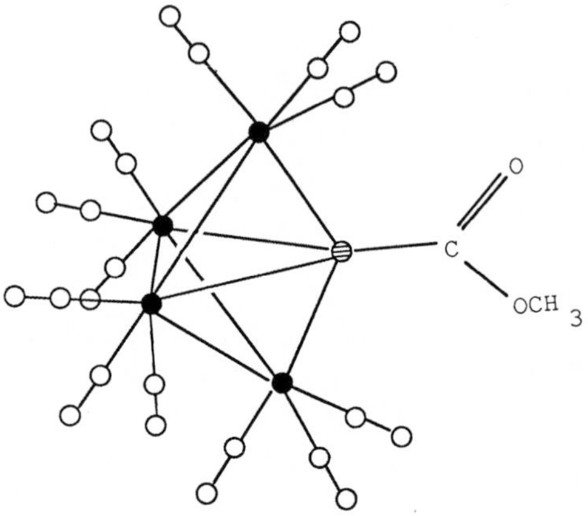

Fig. 6. $[Fe_4(CO)_{12}CCO_2Me]^-$

Figures 3-5 illustrate how the metal cores are derived from the octahedron. This also has chemical significance. The lower nuclearity iron carbides can be synthesized from the octahedral parent $[Fe_6(CO)_{16}C]^{2-}$ by a series of oxidative, vertex removal reactions (refs. 101-107) and the ruthenium cluster $Ru_5(CO)_{15}C$ can be obtained from $Ru_6(CO)_{17}C$ by treatment with CO (ref. 108). The chemistry for the corresponding nitride complexes is similar and has been recently reviewed (ref. 109). The obvious structural relationship between the bridged-butterfly and square pyramidal geometries is seen in the synthetic routes. The bridged-butterfly configuration is most often observed when ligands are added, or HX or MX bonds oxidatively added, to the square pyramid with a net increase of two electrons to the skeletal electron count. This is well-established for the $Ru_5(CO)_{15}C$ cluster which adds MeCN, HX (X = Cl, Br, I) and $Au(PPh_3)X$ (refs. 108, 110-111). In simple ligand addition of MeCN, the ligand is found attached to the unique bridging metal atom. In the HX or MX addition, the X group attaches to that bridging metal, but the hydride or metal group ends up bridging the Ru-Ru hinge bond.

The main group atoms generally sit somewhat below the base of the square pyramid. Values for this parameter are given in Table 2.3. The data indicates that this is favored by increase in the cluster charge. In spite of this, neither the pentametal carbides nor nitrides have shown reactivity at the main group atom. A theoretical discussion of this phenomenon has been published (ref. 112). Bridged-butterfly clusters have also been structurally characterized. Examples are given in Table 2.4.

TABLE 2.3 Distances of the Main Group Atom Below the Square Base Plane in Selected M_5E Clusters

Cluster	Distance, Å	ref.
$Fe_5(CO)_{15}C$	0.09	113, 114
$[NBu_4]_2[Fe_5(CO)_{14}C]$	0.18	62
$[PPN][Fe_5(CO)_{14}N]$	0.11	114
$HFe_5(CO)_{14}N$	0.093	115
$Ru_5(CO)_{15}C$	0.11	108
$Ru_5(CO)_{14}(PPh_3)C$	0.19	108
$Ru_5(CO)_{13}(PPh_3)_2C$	0.23	108
$[NEt_3CH_2Ph][Ru_5(CO)_{14}N]$	0.21	116
$Os_5(CO)_{15}C$	0.12	117
$[PPN]_2[Os_5(CO)_{14}C]$	0.21	118

TABLE 2.4 Examples of Bridged-Butterfly M_5E Clusters

Cluster	Ref.	Cluster	Ref.
$[NEt_4]_2[Fe_5(CO)_{12}(Br)_2C]$	104	$HFe_4(CO)_{12}(AuPPh_3)C$	119
$Ru_5(CO)_{15}(MeCN)C$	108	$Ru_5(CO)_{15}Cl(AuPPh_3)C$	110
$Ru_5(CO)_{14}(\mu\text{-}Br)(AuPPh_3)C$	110	$HRu_4(CO)_{12}(AuPPh_3)C$	120
$Ru_4(CO)_{12}(AuPMe_2Ph)_2C$	120	$Ru_5(CO)_{13}Cp(AuPPh_3)C$	111
$Os_5(CO)_{16}C$	118	$Os_5(CO)_{15}(diphos)C$	121
$[PPN][Os_5(CO)_{15}(I)C]$	117		

The chemistry of the tetranuclear series, however, is more extensive. Reactivity at the carbide atom is seen depending upon the complex being studied. For example, methyl triflate will attack the carbide in $[Fe_4(CO)_{12}C]^{2-}$ giving rise to the μ_3-ethylidyne complex, $[Fe_4(CO)_{12}(\mu_3\text{-}CMe)]^-$ (refs. 122-123). Reactivity at carbon is also exemplified by $Fe_4(CO)_{13}C$ which reacts with MeOH to give $[Fe_4(CO)_{12}(CC(O)Me)]^-$ (ref. 104). This reaction can be reversed upon protonation. In this latter case the carbide acts very similarly to a metal vertex, accepting a CO ligand, and $[Fe_4(CO)_{12}(CC(O)Me)]^-$ (Fig. 6) has only six skeletal electron pairs suggesting that it is indeed better viewed as a trigonal bipyramid with N+1 skeletal electron pairs (N = number of vertices) rather than as a fragment of an octahedral core. Most of the carbide butterflies have seven electron pairs and their structural parameters are more consistent with the interstitial nature of the carbon atom. This can be seen in the apical metal-carbon-apical metal bond angle which is very close to 180° for the seven skeletal pair clusters and quite different from that for the examples with only six skeletal pairs (Table 2.5). The net effect is that the lower electron count clusters have a carbide that sticks out considerably from the metal core. MO calculations for the carbide clusters indicate some multiple bonding between the carbon and the wing-tip metals in the butterfly configuration (ref. 112). A fundamentally different type of cluster carbide showing the same butterfly geometry is $W_4(O\text{-}iPr)_{12}(NMe)C$ arising from the reaction of i-PrOH and $W_2(NMe_2)_6$ (ref. 124).

The butterfly nitrides show less reactivity at the nitrogen atom than carbon, but data are available that show the nitride can be protonated (ref. 109). A series of butterfly borides (refs. 125-127) and a butterfly oxide cluster (ref. 128) have been recently prepared. Key differences in the reactivity in these isostructural, isoelectronic molecules may arise from the charge differences across the series, e.g. $[Fe_4(CO)_{12}B]^{3-}$, $[Fe_4(CO)_{12}C]^{2-}$, $[Fe_4(CO)_{12}N]^{-}$.

TABLE 2.5 Apical M-E-Apical M Angles in the Butterfly Clusters

Cluster	Angle deg	Skeletal Electron Pairs	Ref.
$[NEt_4][Fe_4(CO)_{12}C(CO_2Me)]$	148	6	106
$Fe_4(CO)_{12}(C{=}C(OMe)_2$	149	6	107
$HFe_4(CO)_{12}BH_4$	162	7	127
$Fe_4(CO)_{13}C$	175	7	105
$Fe_4(CO)_{10}(PMe_3)_3C$	174	7	107
$HFe_4(CO)_{12}CH$	170.5	7	129
$[PPN][HFe_4(CO)_{12}C]$	174	7	130
$[Zn(NH_3)_4][Fe_4(CO)_{12}C]$	177.6	7	131
$[PPN][RhFe_3(CO)_{12}C]$	173.2	7	132
$[PPN][Fe_4(CO)_{12}N]$	179.0	7	133,134
$HFe_4(CO)_{12}N$	178.4	7	115
$HRu_4(CO)_{10}(P(OMe)_3)_2N$	173.2	7	135
$[NEt_4][FeRu_3(CO)_{10}(P(OMe)_3)N]$	172.0	7	136
$H_3Ru_4(CO)_{11}N$	174.0	7	137
$Ru_4(CO)_{12}(NO)N$	173.0	7	138
$Ru_4(CO)_{12}(NCO)N$	173.5	7	138
$[PPN][Os_4(CO)_{12}N]$	-----	7	137

2.3 Metallated Main Group Atoms and Without Transition Metal-Transition Metal Bonds

Metal moieties can act as simple replacement groups for halide or organic functionalities attached to a central main group atom with planar, pyramidal and tetrahedral configurations (Figs. 7-8, Table 2.6). It has been demonstrated that planar, six e^- $E[Co(CO)_4]_3$ (E = In, Tl) can function as Lewis acids, adding

$[Co(CO)_4]^-$ to give tetrahedral, eight electron $[E\{Co(CO)_4\}_4]^-$ ions (ref. 139). Some planar EM_3 complexes are believed to have main group-transition metal multiple bonding (ref. 11). This phenomenon is more thoroughly discussed in the chapter by Herrmann. In a fundamentally different reaction, $Bi[Co(CO)_4]_3$ will also add $[Co(CO)_4]^-$ to produce $[Bi\{Co(CO)_4\}_4]^-$ but in that ion the bismuth is a ten electron center (ref. 140-141). This gives rise to extremely long Bi-Co bonds and paramagnetic behavior (two unpaired electrons). The ability of the main group atom to accept an extra pair of electrons may be a very important aspect of the chemistry of these metallated complexes.

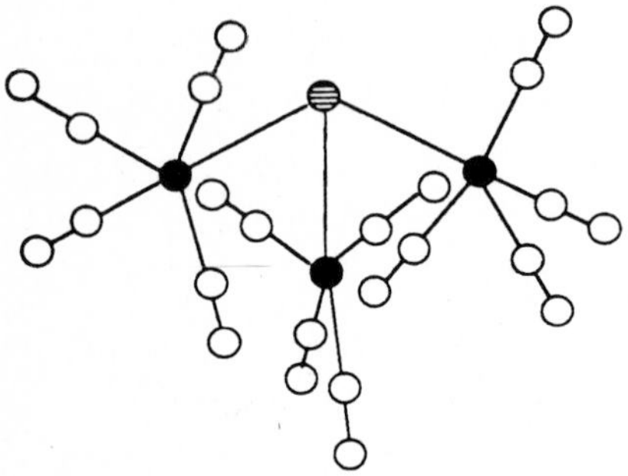

Fig. 7. $BiCo_3(CO)_{12}$

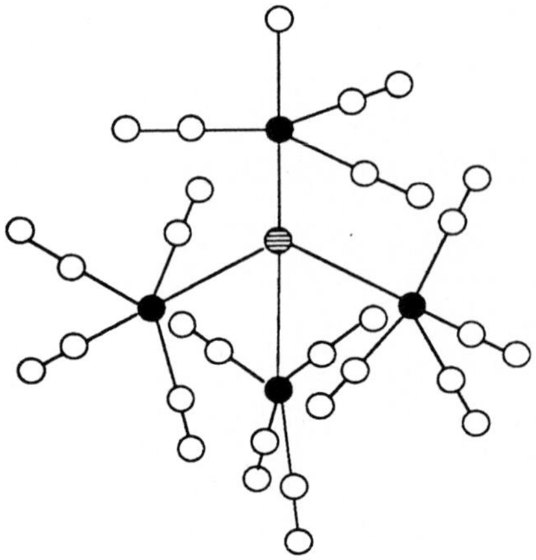

Fig. 8. $[Bi\{Fe(CO)_4\}_4]^{3-}$

TABLE 2.6 Metallated Main Group Atoms with No M-M Bonds

Compound	Ref.	Compound	Ref.
$Ga[CpW(CO)_2]_3$	142	$In[Co(CO)_4]_3$	143
$Tl[CpMo(CO)_3]_3$	144	$As[Cr(CO)_5]_2[Mn(CO)_5]$	145
$[Na(THF)_2][Sb\{Cr(CO)_5\}_3]$	145	$Te[CpMn(CO)_3]_3$	146
$Se[CpMn(CO)_2][CpFe(CO)_2]_2$	147	$Bi[Co(CO)_4]_3$	148
$Bi[Mn(CO)_5]_3$	149	$Bi[CpFe(CO)_2]_3$	150
$[NEt_4]_3[Bi\{Fe(CO)_4\}_4]$	151	$[Cp_2Co][Bi\{Co(CO)_4\}_4]$	140,141

2.4 Spirocyclic Main Group Atoms and Related Complexes

An intermediate structural type exists between the completely open form of the compounds discussed in section 2.3 and the closed molecules to be described in the next section. Thus main group atoms attached to three or more transition metals may show varying numbers of transition metal-transition metal bonds. In some cases these bonds may be created or destroyed by electron or ligand loss/addition. For example, $[NEt_4]_2[PbFe_4(CO)_{16}]$ (Fig. 9) is oxidized to spirocyclic $Pb[Fe_2(CO)_8]_2$ (Fig. 10) which is isostructural with analogous Ge and Sn complexes. Planar as well as pyramidal complexes are known for the combination of one main group atom and three transition metals. In the planar molecules with main group V and VI elements some multiple bonding is believed to be present. Examples are found in Tables 2.7 and 2.8.

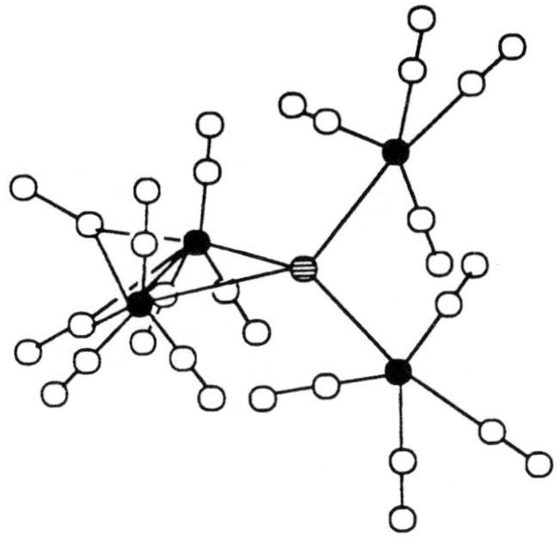

Fig. 9. $[PbFe_4(CO)_{16}]^{2-}$

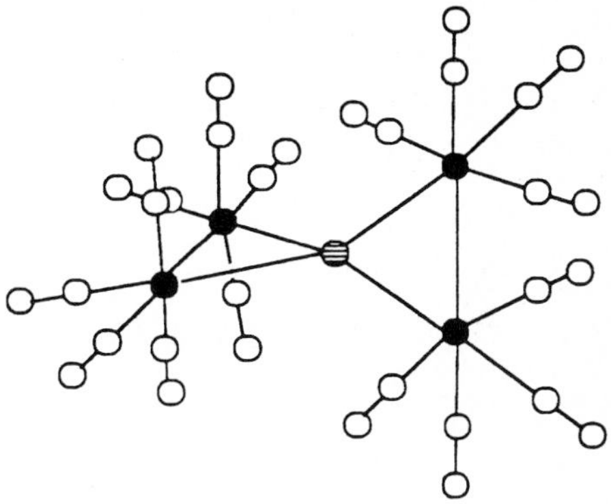

Fig. 10. $Pb[Fe_2(CO)_8]_2$

TABLE 2.7 EM_3 and EM_4 Clusters Showing One Metal-Metal Bond

Compound	Ref.	Compound	Ref.
$Ge[W(CO)_5][W_2(CO)_{10}]$	145	$Sn[W(CO)_5][W_2(CO)_{10}]$	145
$Se[CpMn(CO)_2][Cp_2Fe_2(CO)_3]$	147	$As[Mn(CO)_5][Cr_2(CO)_9]$	145
$[Cp_2Co][Se\{CpCr(CO)_2\}\{Cp_2Cr_2(CO)_4\}]$	152	$P[Cr(CO)_5][CpWCr(CO)_7]$	145
$[NEt_4]_2[Pb\{Fe(CO)_4\}_2\{Fe_2(CO)_8\}]$	153		

TABLE 2.8 Spirocyclic Complexes

Compound	Ref.	Compound	Ref.
$Ge[Fe_2(CO)_8]_2$	154, 155	$Sn[Fe_2(CO)_8]_2$	156
$Pb[Fe_2(CO)_8]_2$	157	$Ge[Co_2(CO)_7]_2$	158
$Ge[Fe_2(CO)_8][C_5H_4MeMnFe(CO)_6]$	154	$Ge_3Co_8(CO)_{26}$	159
$Fe_2(CO)_7[GeCo_2(CO)_7]_2$	159	$P[Fe_2(CO)_8][Fe_2(CO)_6Cl]$	160
$As[Fe_2(CO)_8][Fe_2(CO)_6Cl]$	160	$P[FeW(CO)_9][Fe_2(CO)_6Br]$	161
$As[Cp_2Co_2(CO)_2]_2[BF_4]$	162	$Fe_2(CO)_6[SbFe_2(CO)_8]_2$	163

2.5 Tetrahedra and Related Structures Containing Main Group Atoms

(i) EM_3 Tetrahedra. The whole range of E_xM_{4-x} clusters have been synthesized, but among the main group element-containing clusters the most commonly observed case is where the tetrahedron possesses one main group atom. These molecules most often show six skeletal electron pairs which can be thought of either as fitting a delocalized model based on a *nido*-trigonal bipyramid or a edge-localized scheme. In the counting formalisms the main group atoms retain a lone pair of electrons that is not involved in cluster bonding. EM_3 examples are known for almost all stable main group III through VI elements except gallium, thallium and lead. At one time it was believed that the larger main group atoms would place a strain on the metal-metal bonding framework of the smaller transition metals (i.e., the first period) (ref. 164), but numerous closed complexes are now known (Table 2.9; Fig. 11). A fourth metal atom may be attached to the main group element either as an acceptor of the main group lone pair for sixteen electron metal fragments or via a shared pair as for seventeen electron metal fragments (Table 2.10; Fig. 12) Interesting examples are $Co_3(CO)_9E$ (E = P, As) which do not appear stable unless the lone pairs are complexed to external metals. The phosphorus derivative has been trapped with $Fe(CO)_4$ to give $Co_3(CO)_9(\mu_3\text{-}PFe(CO)_4)$. If trapping is not done then both the P and As-containing molecules form cyclic trimers in which the lone pair of the main group V element coordinates to a cobalt of an adjacent cluster. Cyclic trimerization also occurs for $H_2Ru_3(CO)_9S$. This basicity of the main group element appears to decrease as one descends the periodic table as expected, and the heavier elements often are found with lone pairs of electrons not coordinated while that condition is rare for the lighter elements, expecially in main groups IV and V.

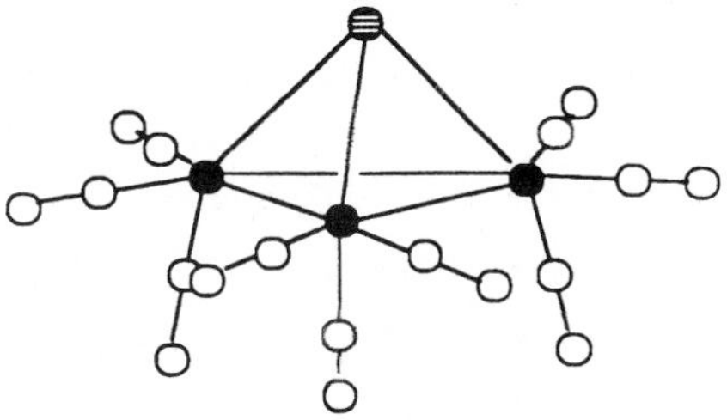

Fig. 11. $BiIr_3(CO)_9$

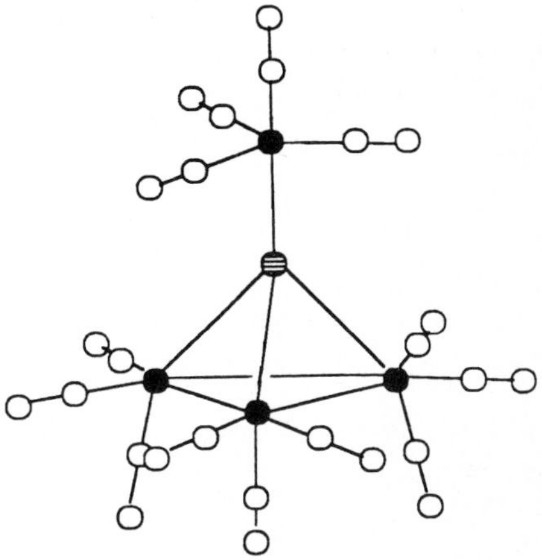

Fig. 12. $Co_3(CO)_9SiCo(CO)_4$

TABLE 2.9 Selected Examples of Complexes Containing a Naked μ_3-Main Group Atom on a Closed Triangle of Transition Metals

Compound	Ref.	Compound	Ref.
$PCo_3(CO)_9$	165, 166	$AsMo_3Cp_3(CO)_6$	167
$AsCo_3(CO)_9$	166	$BiCo_3(CO)_9$	141, 168
$[NEt_4][BiFe_3(CO)_{10}]$	169	$BiFe_3(CO)_9(COMe)$	170
$H_3BiFe_3(CO)_9$	170	$BiCp_3Fe_3(CO)_3$	150
$BiIr_3(CO)_9$	171	$[NEt_4]_2[H_3Re_3(CO)_9O]$	172
$[NMe_3CH_2Ph]_2[Fe_3(CO)_9O]$	173	$Ru_3(CO)_6(Ph_2AsCH_2AsPh_2)O]$	174
$Os_6(CO)_{19}O$	175	$FeCo_2(CO)_9S$	176
$Fe_2Co(CO)_8(NO)S$	177	$[Fe_3(CO)_9S]^{2-}$	177
$Fe_3(CO)_9(AuPPh_3)_2S$	177	$Os_3(CO)_{10}S$	178
$H_2Os_3(CO)_9S$	179	$Os_4(CO)_{12}S$	180
$Os_4(CO)_{13}S$	181	$Co_3(CO)_9Se$	182
$FeCo_2(CO)_9Se$	182	$FeCo_2(CO)_9Te$	182
$Cp^*MnFe_2(CO)_8Te$	183	$Cp_2Mo_2Fe_2(CO)_7Te_2$	184

TABLE 2.10 Selected Examples of Complexes Containing an EML_n Group Attached to a Closed Triangle of Metal Atoms.

Compound	Ref.	Compound	Ref.
$Re_4(CO)_{12}(\mu_3\text{-}InRe(CO)_5)_4$	185	$[NEt_4]_2[Fe_3(CO)_{10}(GeFe(CO)_4)]$	157
$Co_3(CO)_9(SiCo(CO)_4)$	186	$Co_3(CO)_9(GeFeCp(CO)_2)$	187
$CpMoCo_2(CO)_8(GeWCp(CO)_3)$	188	$[NEt_4][Co_3(CO)_9(GeCo_2(CO)_8)]$	189
$Co_3(CO)_9(AsCr(CO)_5)$	190	$[PPh_4]_2[Fe_3(CO)_{10}(PFe(CO)_4)]$	191
$Co_3(CO)_9(AsCo_4(CO)_{11})$	167	$[NEt_4][Fe_3(CO)_{10}(SbFe(CO)_4)]$	192
$FeCo_2(CO)_9(SCr(CO)_5)$	193	$[H_2Ru_3(CO)_8S]_3$	194
$Co_3(CO)_9(PMn(CO)_2Cp)$	190	$Co_3(CO)_7(\{P(OMe)_3\}_2(PMn(CO)_2Cp)$	190

(ii) Trigonal Bipyramidal and Square Pyramidal E_2M_3 Clusters. E_2M_3 clusters which have two main group atoms bridging the same triangle of metal atoms give rise to a trigonal bipyramidal cluster (or capped tetrahedron) as in $As_2Fe_3(CO)_9$ (Fig. 13). Six skeletal electron pairs are also most commonly observed for this series of compounds, and one or both of the main group atoms may be attached to an external metal moiety (Table 2.11).

Adding another electron pair, either directly by reduction or by incorporation of higher electron count atoms (e.g., replacing As with Se) results in the formation of clusters with seven skeletal electron pairs and a nido octahedral configuration (Fig. 14, Table 2.12). The resultant configuration is the well-known square pyramidal geometry where a transition metal-transition metal bond has been broken in the original trigonal bipyramid.

Fig. 13. $As_2Fe_3(CO)_9$

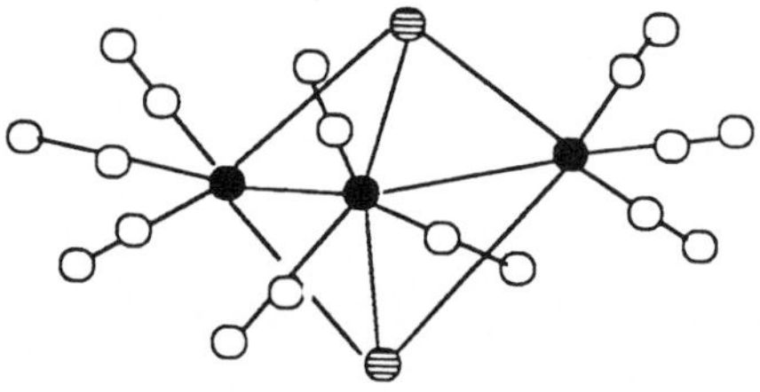

Fig. 14. $S_2Fe_3(CO)_9$

TABLE 2.11 Examples of Trigonal Bipyramidal E_2M_3 Clusters

Compound	Ref.	Compound	Ref.
$Fe_3(CO)_9As_2$	195	$Fe_3(CO)_9Bi_2$	196
$Fe_3(CO)_9(SnFeCp(CO)_2)_2$	197	$Fe_3(CO)_9(PMnCp(CO)_2)(PFe(CO)_4)$	161
$Fe_3(CO)_9(PMnCp(CO)_2)_2$	161	$Fe_3(CO)_9(PMnCp(CO)_2)(PFe_2(CO)_8)$	161

TABLE 2.12 Examples of Square Pyramidal E_2M_3 Clusters

Compound	Ref.	Compound	Ref.
$Fe_3(CO)_9S_2$	198	$Os_3(CO)_9S_2$	199
$Os_3(CO)_8(PMe_2Ph)S_2$	199	$Os_3(CO)_8(CS)S_2$	200
$H_2Os_3(CO)_8S_2$	201	$Fe_3(CO)_9Se_2$	202
$Fe_3(CO)_9Te_2$	203,204	$CrCo_2(CO)_{10}(AsCr(CO)_5)_2$	191
$[NEt_4]_2[Bi_2Fe_4(CO)_{13}]$	205,206	$[NEt_4]_2[Fe_3(CO)_9(SbFe(CO)_4)_2]$	207

(iii) E_2M_2 Tetrahedra. Tetrahedra of two main group atoms and two transition metals are known and are isoelectronic with the EM_3 clusters. The lone pairs on the main group atoms can function as donors to external metal groups (Table 2.13; Fig. 15). The E-E bond distances for the Co_2P_2, Co_2As_2 and Bi_2W_2 clusters are suggestive of multiple bonding, while that in $Fe_2(CO)_6S_2$ lies in the range of other known S-S single-bonded compounds.

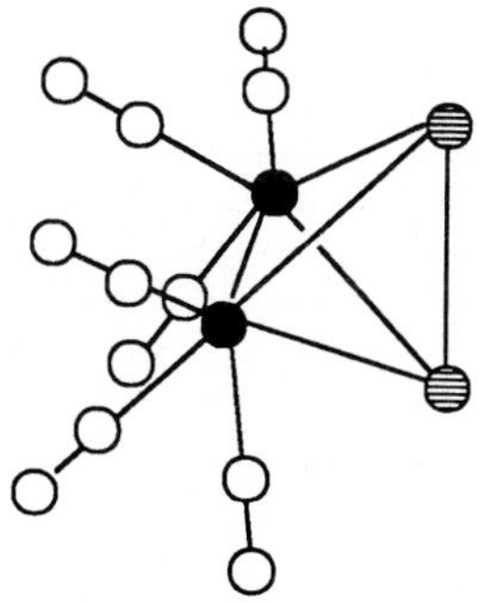

Fig. 15. $As_2Co_2(CO)_6$

TABLE 2.13 Complexes Containing a Tetrahedral E_2M_2 Core Geometry

Compound	E-E Bond Distance, pm	Ref.
$Co_2(CO)_5(PPh_3)P_2$	201.9(9)	208
$Co_2(CO)_5(PPh_3)As_2$	227.3(3)	209, 210
$Fe_2(CO)_6S_2$	200.7(5)	211, 212
$Fe_2(CO)_6Se_2$	229.3(2)	213
$Fe_2(CO)_6Te_2$	---	204
$[FeW(CO)_8Se_2][SbF_6]_2$	228.1(3)	214
$Co_2(CO)_6(PCr(CO)_5)(PW(CO)_5)$	206.1(3)	215
$Fe_2(CO)_7(PCr(CO)_5)_2$	208.7(6)	161
$Cp_2Mo_2(AsCr(CO)_5)_2$	231.0(3)	216
$Bi_2W_2(CO)_8(\mu\text{-}Bi\{W(CO)_5\}Me)$	279.6(1)	217

A very recently discovered cobalt/bismuth compound shows a Bi_2Co_2 tetrahedron capped on the Bi_2Co faces with $Co(CO)_3$ groups. The cluster has a formulation of $[NMe_4][Bi_2Co_4(CO)_{11}]$ and was obtained from the pyrolysis of salts of $[Bi\{Co(CO)_4\}_4]^-$ (ref. 141). We have found that this paramagnetic species may also be obtained via the reduction of $BiCo_3(CO)_9$ with cobaltocene in a 1:1 ratio. The Bi-Bi bond distance is 308.8 pm in both the $[NMe_4]^+$ and $[Cp_2Co]^+$ salts.

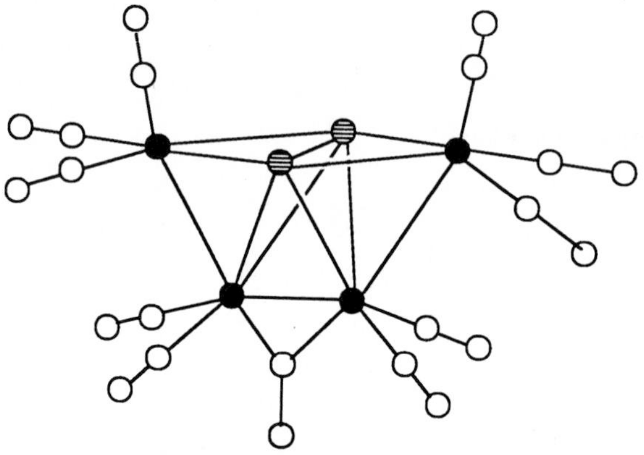

Fig. 16. $[Bi_2Co_4(CO)_{11}]^-$

(iv). E_2M_2 Clusters with a Third Metal Fragment Bridging the E-E Bond. Some more recently discovered molecules add an interesting twist to this structural class by having a third transition metal fragment bridging the main group-main group bond. What is most notable about these molecules is that the main group-main group bond appears to be retained to some degree. This is true in a series of tellurium clusters studied by Rauchfuss and in one example of a bismuth compound (Table 2.14; Fig. 17). The tellurium distances are long and are thought to represent ca. 2/3 of a bond, while the Bi-Bi distance is in good agreement with single bond values.

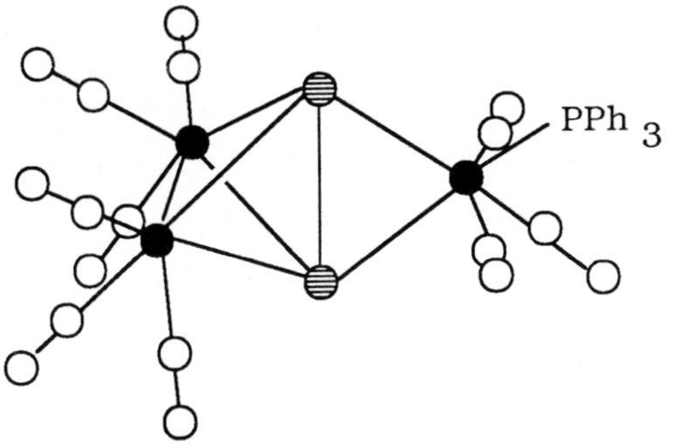

Fig. 17. $Te_2Fe_3(CO)_9(PPh_3)$

TABLE 2.14 Complexes Containing an E_2M_2 Core Geometry with a Third Transition Metal Group Bridging the E-E Bond

Compound	E-E Bond Distance, pm	Ref.
$[NEt_4][Bi_2CoFe_2(CO)_{10}]$	309.2(2)	205, 206
$Te_2Fe_3(CO)_9(PPh_3)$	313.8(1)	218
$Te_2FeMo_2Cp_2(CO)_7$	314.6(1)	184

These complexes have a close structural tie to another series of very unusual compounds which have the E-E bond bridged by three transition metal groups but have no M-M bonds (Fig. 18). These complexes discovered by Huttner and coworkers arise from the reaction of ECl_3 (E = As, Sb, Bi) with salts of $[W_2(CO)_{10}]^{2-}$ (refs. 219-221). They all show very short E-E distances (As: 227.9 pm; Sb, 266.3 pm; Bi, 281.8 pm) which have been attributed to triple bond character. It should be remembered that the fragments As_2, Sb_2 and Bi_2 are isoelectronic with N_2.

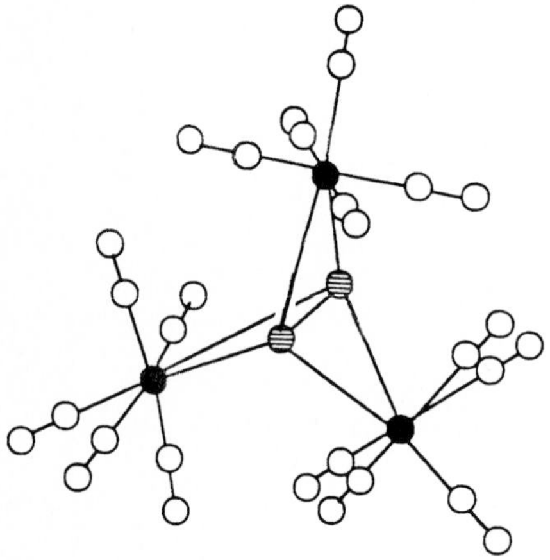

Fig. 18. $E_2[W(CO)_5]_3$, E = As, Sb, Bi

(v) Open E_2M_2 Clusters. Increasing or decreasing the electron count from the six skeletal pairs observed for the tetrahedral E_2M_2 clusters produces compounds in which one of the bonds within the tetrahedron is broken. Usually this bond is the E-E bond and the electron change can be accomplished by change in the main group element or transition metal. Thus a series of $[ML_n]_2[\mu\text{-}EM'L_n]_2$ are known for Ga (refs. 222-223), In (refs. 222-225) and Tl (ref. 226) with ML_n including Fe, Mn and Re carbonyl groups. They have planar E_2M_2 rings with alternating E and M atoms. Most of the known examples also have a metal-metal bond across the ring, but examples are known in which there is no M-M bond and in some an E-E bond is present instead. Examples of the latter are $Te_2(\mu\text{-}M(triphos))_2$ generated from ditelluride and

$M(ClO_4)_2$ (M = Co, Ni) (ref. 227). An interesting series of compounds with no M-M or E-E bonds centers around $[NEt_4][Tl\{Fe(CO)_4\}_2]$ which is a weak dimer in the solid with the E_2M_2 ring structure (Fig. 19). In the ring one set of Fe-Tl distances is short, 263.2 pm compared the the other at 303.8 pm substantiating the weak dimer nature of this compound. It loses CO and redimerizes to give $[NEt_4]_4[Tl_4Fe_8(CO)_{30}]$ (Fig. 20). A further derivative has been characterized. It is $[NEt_4]_6[Tl_6Fe_{10}(CO)_{36}]$ shown in Fig. 21 (ref. 228). All three compounds show the E_2M_2 ring structure with no E-E or M-M bonds across the ring.

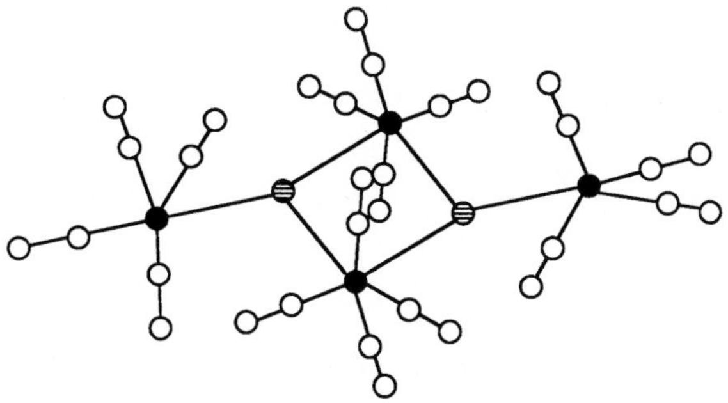

Fig. 19. $[Tl_2Fe_4(CO)_{16}]^{2-}$

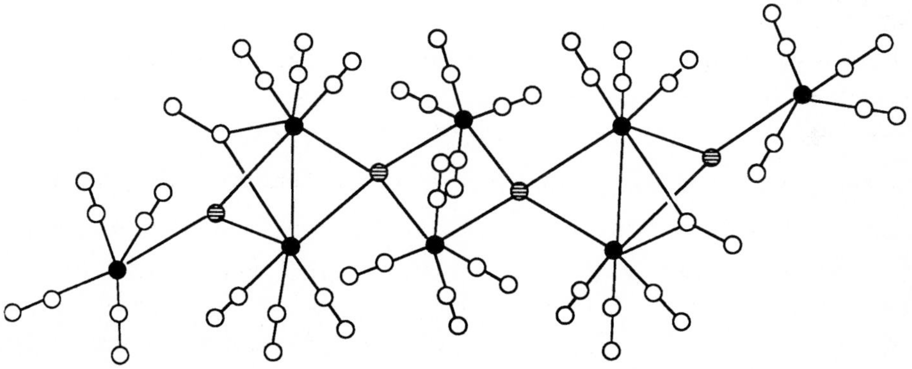

Fig. 20. $[Tl_4Fe_8(CO)_{30}]^{4-}$

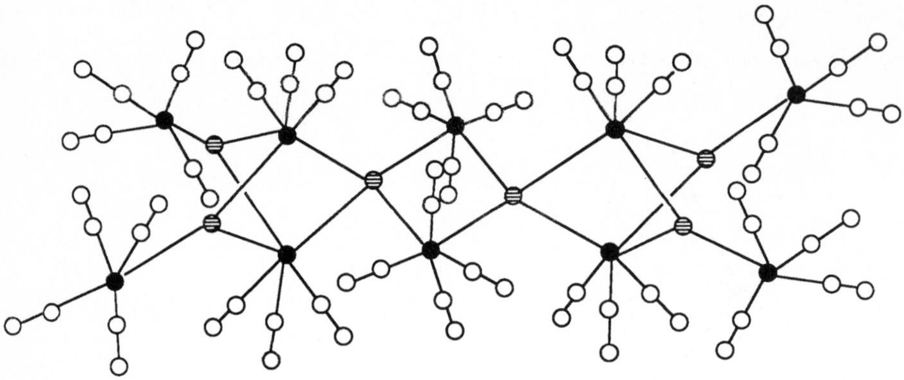

Fig. 21. $[Tl_6Fe_{10}(CO)_{36}]^{6-}$

Butterfly structures are seen for the later main group atoms. It has been shown that reduction of $Fe_2(CO)_6E_2$ (E = S,Se) gives butterfly dianions in which the E-E bonds are broken (ref. 229). The sulfur dianion has been trapped with a number of cationic reagents yielding products having the cation attached at S and no E-E bond (e.g.the reaction with RX gives $Fe_2(CO)_6(\mu\text{-}SR)_2$) (ref. 230). Butterfly clusters with the main group and transition metal groups in the opposite positions are $Cp^*{}_2Mn_2(CO)_4As_2$ (refs. 231-232) $Cp_2Cr_2(CO)_4Se_2$ (ref. 233). The latter has a Se-Se bond distance of 227.7 pm which is thought to be intermediate between a single and double bond,

(vi) Tetrahedra with Three Main Group Atoms and One Transition Metal. This last of the hybrid main group/transition metal tetrahedral prototype compounds is also well-known with examples available for P, As and Te. A recent review of the phosphorus compounds has been published (ref. 234). Structurally characterized examples are P_3Co(triphos) (triphos = 1,1,1-tris(diphenylphosphino methyl)ethane), $[P_3Pd(triphos)][BF_4]$, $[P_3Cu(triphos)]_2Cu_6Br_6$, $P_3Co(np_3)$ (np_3 = tris(2-diphenylphosphinomethyl)ethane), $As_3Co(CO)_3$ (ref. 235), $As_3(CpMo(CO)_2)$ (ref. 236) and $[Te_3W(CO)_4][SbF_6]_2$ (Fig. 22; ref. 237) which is generated when $[Te_4][SbF_6]_2$ is treated with $W(CO)_6$ in liquid SO_2-AsF_3. The Te-Te distances range from 271.8 to 273.6 pm and compare favorably with those found in elemental tellurium (274 pm).

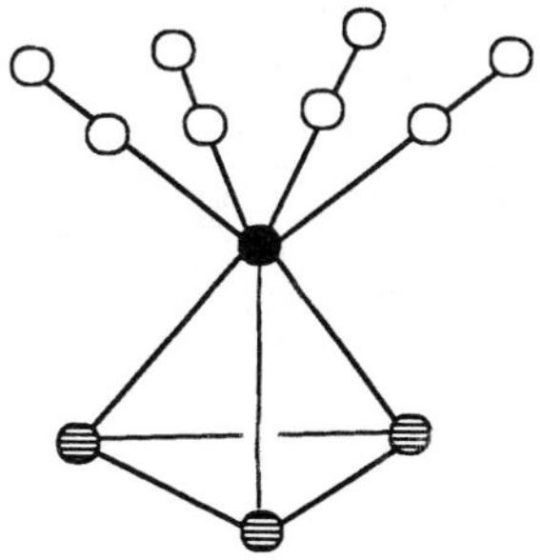

Fig. 22. $[Te_3W(CO)_4]^{2+}$

The clusters $[P_3\{Ni(triphos)\}_2]X_2$, $[P_3CoNi(triphos)_2][BPh_4]_2$, $[P_3Pd_2(triphos)_2][BPh_4]$ and $[As_3Co_2(triphos)_2][BPh_4]_2$ are trigonal bipyramids where the main group atoms occupy the equatorial sites and the transition metals the axial sites (ref. 234).

The most general method of synthesis of the E_3M and E_3M_2 clusters has been reaction of white phosphorus (P_4) or freshly generated, unstable yellow arsenic (As_4) with transition metal cations in the presence of a tridentate ligand such as triphos or np_3. For the np_3 and triphos metal complexes, the arsenic derivatives appear to be more susceptible to oxidation than their phosphorus analogues. The triphos and np_3 complexes of P_3 and As_3 adodpt a variety of oxidation states including 0, +1 and +2. The complexes containing P_3 and As_3 and two transition metals can adopt oxidation states with total electron counts ranging from 30 to 34. The 31 to 33 electron molecules are paramagnetic while the 30 and 34 electron cases are diamagnetic. Mixed metal 32 electron clusters may show diamagnetic behavior (or low residual paramagnetism) which has been attributed to a Jahn-Teller-like distortion of the e orbitals into which electrons 31 to 34 are placed.

As with the other tetrahedral derivatives, the lone pairs on the main group atoms in P_3Co(triphos) can donate to metals as has been demonstrated by the X-ray analysis of the bis($CpMn(CO)_2$) and bis{$Cr(CO)_5$} derivatives (refs. 234, 238).

2.6 Main Group Atoms Bridging Square Faces of Metal Atoms

Bare and metallated main group atoms are capable of bridging square metal arrays. Examples are given in Table 2.15. In addition to those listed in the table, there are a large number of μ_4-PR clusters of this general structure type in which R = alkyl or aryl.

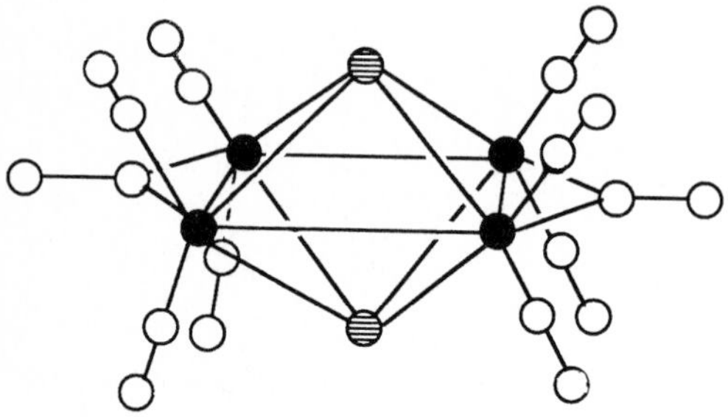

Fig. 23. $Te_2Co_4(CO)_8(\mu\text{-}CO)_2$

TABLE 2.15 Some Clusters with Main Group Atoms Bridging Square Metal Arrays

compound	ref.	Compound	ref.
$Co_4(CO)_{11}(GeCo(CO)_4)_2$	239	$Co_4(CO)_{10}S_2$	240, 241
$Fe_2Co_2(CO)_{11}S_2$	242	$Ru_4(CO)_9(PMe_2Ph)_2S_2$	243
$Os_5(CO)_{15}S$	244	$Os_6(CO)_{17}S$	245
$Os_7(CO)_{19}S$	180	$Os_6(CO)_{16}(\mu_3\text{-}S)(\mu_4\text{-}S)$	246, 247
$Os_6(CO)_{17}S_2$	247	$Co_4(CO)_{10}Te_2$	248

2.7 More Complex Structures

Several reports of unusual cluster geometries with a higher proportion of main group atoms have been reported and emphasize the importance of the main group component of these compounds. Of special note are the pentagonal bipyramidal configuration observed for $As_5(\mu_5\text{-MoCp})_2$ (Fig. 24; ref 249) and the hexagonal bipyramidal one seen for $P_6(\mu_6\text{-MoCp}^*)_2$ (Fig. 25; ref. 250). Analogy can be drawn between these main group ring structures and isoelectronic aromatic hydrocarbons: $[As_5]^-/[Cp]^-$ and P_6/C_6H_6.

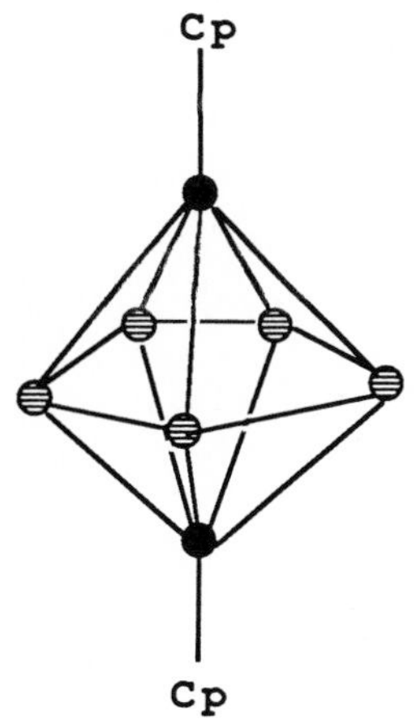

Fig. 24. $(CpMo)_2As_5$

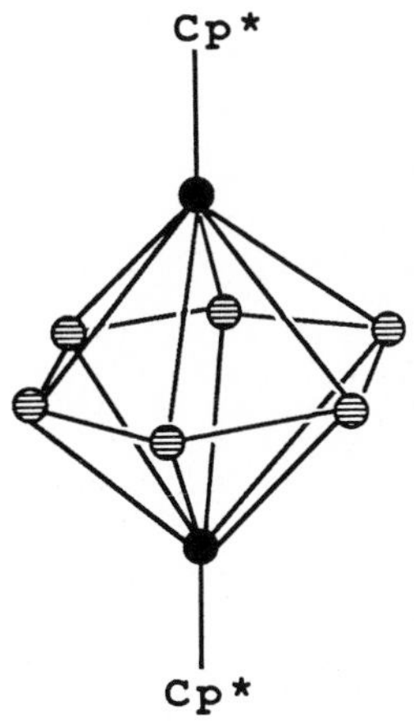

Fig. 25. $(Cp^*Mo)_2P_6$

The complex $[NEt_4]_2[Bi_4Fe_4(CO)_{13}]$ shows an interesting structure similar in several regards to the Zintl ions and Zintl phases (refs. 251-252). The derivation of the structure can be thought of in two ways. One can start with a Bi_4 tetrahedron and cap three of the faces with $Fe(CO)_3$ groups. The lone pair on the Bi which is bonded to all three of these groups also donates its lone pair to an external $Fe(CO)_4$ moiety. Alternatively, the structure can be viewed starting from a $[P_7]^{3-}$ geometry in which the three P's in the base and the apical P are replaced by Bi and the 3 bridging P's by the $Fe(CO)_3$ groups. A simple 60° twist is then required to allow the irons to bond to two rather than just one Bi in the base. The Bi-Bi distances within the cluster fall into two categories. The first are those values between bismuth atoms in the triangular base which are ca. 310 pm while the distances between those bismuth atoms and the apical one are ca. 345 pm. The second value is not believed to represent a direct bonding interaction but the first is consistent with a bond order of one. The two values, however, are very similar to those seen for the two closest neighbor contacts in elemental Bi. Another feature of this molecule which emphasizes its relationship to the solid state is the Bi···Bi interactions between adjacent clusters. Each Bi of the triangular base is on the order of 400 pm from a Bi in a neighboring cluster giving the crystal lattice a semi-polymeric nature. These distances are about the same as those observed between Bi clusters within Zintl phases such as $Ca_{11}Bi_{10}$ (ref. 253).

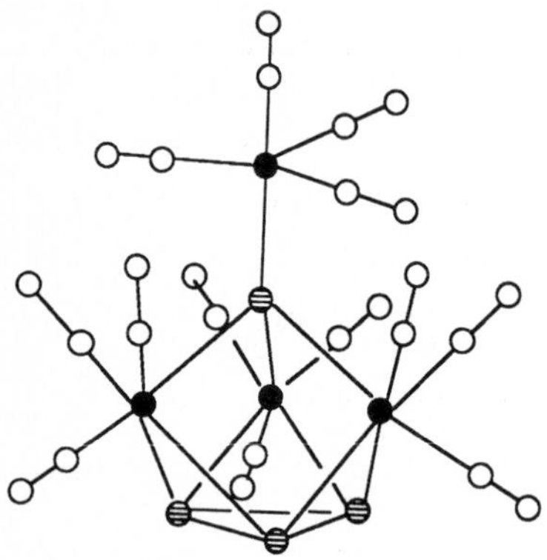

Fig. 26. $[Bi_4Fe_4(CO)_{13}]^{2-}$

A cationic cluster is generated when $[Se_4][SbF_6]_2$ is treated with $W(CO)_6$ in liquid SO_2 (ref. 254). The product $[Se_4\{W(CO)_5\}_2][SbF_6]$ structure shows partial retention of the Se_4^{2+} ring with $W(CO)_5$ groups bridging opposite edges. The Se-Se distance bridged by W is 220.8 pm with an estimated bond order of 1.7, while the non-bridged Se-Se distances are quite long, 301.7 pm. These parameters suggest that the molecule may be considered a weak dimer of $[W(CO)_5Se_2]^+$. Se_4^{2+} is a six π electron system and it is interesting that the metal groups do not coordinate in a π fashion as does benzene. By comparison, the same reaction using Te_4^{2+} in place of Se_4^{2+} produces the $[Te_3\{W(CO)_4\}]^{2+}$ discussed earlier (ref. 237).

3 SYNTHETIC METHODOLOGIES

Many strategies are available for producing low valent clusters containing both main group and transition metal atoms. Almost any main group element-containing compound will react with metal carbonyl complexes under suitable conditions. The product geometries are not always predictable but certain patterns are becoming evident. Some general methods are outlined below with examples of the various reaction types. The list is certainly not exhaustive in terms of methods or examples, but the most common stategies are presented.

3.1 Direct Reaction of Metal Reagents with the Main Group Elements

In several cases, direct reaction of the main group element and the transition metal reagent results in formation of the desired complexes. Sn, Pb and Bi react with $Co_2(CO)_8$ to give $E[Co(CO)_4]_4$ (E = Sn, Pb) and $Bi[Co(CO)_3]_3$ respectively (refs. 148, 255). Bismuth, when treated with $Na_2Fe(CO)_4$, generates $Na_3[Bi\{Fe(CO)_4\}_4]$. It has already been mentioned that a number of the E_3M and E_3M_2 compounds are generated by addition of P_4 or As_4 to various metal cations in the presence of triphos or np_3 (ref. 234).

3.2 Thermal or Photochemical Reactions with Organo Main Group Compounds

This approach is one of the most common methods for producing mixed clusters. The simplest and perhaps oldest route was to pyrolyze organo main group complexes with transition metal carbonyls. The problem with the method is that it often gives rise to mixtures of compounds in low yields. Nevertheless, some complexes have not yet been accesible by other routes. Reaction conditions depend on the ease of cleavage of the E-H or E-C and

M-CO bonds. For the heavier main group elements the E-H and E-C bonds are weak enough that reaction proceeds under very mild conditions. Table 3.1 gives some examples of reagents and products pertinent to this section.

TABLE 3.1 Products Arising from the Photochemical or Thermal Insertion of Metals into E-H, E-Halide or E-C Bonds

Main Group Reagent	Transition Metal Complex(es)	Product	Ref.
AsF_3	$Fe(CO)_5$	$As_2Fe_3(CO)_9$	195
H_2Se	$Co_2(CO)_8$	$Co_3(CO)_9Se$	182
H_2Se	$Co_2(CO)_8$, $Fe_3(CO)_{12}$	$FeCo_2(CO)_9Se$	182
Et_2Te	$Co_2(CO)_8$, $Fe_3(CO)_{12}$	$FeCo_2(CO)_9Te$	182
$[Fe(CO)_4GeH_2]_2$	$Co_2(CO)_8$	$Fe_2Co_2(CO)_{21}Ge_2$	159
$Fe_2(CO)_6B_2H_6$	$Fe_2(CO)_9$	$H_3Fe_4(CO)_{12}B$	127
$(AsMe)_5$	$Co_2(CO)_8$	$As_3Co(CO)_3$	235
GeH_4	$Co_2(CO)_8$	$Ge[Co_2(CO)_7]_2$	158
$Mn(CO)_5GeH_3$	$Co_2(CO)_8$	$Co_3(CO)_9GeMn(CO)_5$	256

One nice application of this methodology is to treat lower nuclearity pre-formed compounds containing main group-hydrogen or halide bonds with metal carbonyls in order to build higher nuclearity and/or heteronuclear complexes. Some examples are found at the end of Table 3.1.

3.3 Reaction of Main Group Element Halides with Metal Carbonyl Anions

This very straightforward ionic displacement strategy appears to work best for the heavier main group elements, and the method is cleanest where the product desired is insoluble in the solvent employed. TABLE 3.2 gives a list of some reagents used and products formed by this method.

TABLE 3.2 Reactions of Main Group Element Halides With Metal Carbonyl Anions

Main Group Halide	Metal anion	Product	Ref.
$InCl_3$	$[Mn(CO)_5]^-$	$In[Mn(CO)_5]_3$	257
$InCl_3$	$[CpMo(CO)_3]^-$	$In[CpMo(CO)_3]_3$	258
$TlCl_3$	$Fe(CO)_5/KOH$	$[NEt_4][Tl\{Fe(CO)_4\}_2]$	228
$TlCl_3$	$[Re(CO)_5]^-$	$Tl[Re(CO)_5]_3$	259
$PbCl_2$, $SnCl_2$	$[Fe_2(CO)_8]^{2-}$	$[E\{Fe_2(CO)_8\}\{Fe(CO)_4\}_2]^{2-}$	153
GeI_2	$[Fe_2(CO)_8]^{2-}$	$[GeFe_4(CO)_{14}]^{2-}$	157
SiI_4	$[Co(CO)_4]^-$	$Co_3(CO)_9SiCo(CO)_4$	186
C_2Cl_4 or C_2Cl_6	$[Ni_6(CO)_{12}]^{2-}$	$[Ni_{10}(CO)_{16}C_2]^{2-}$	98
GeI_4	$[Co(CO)_4]^-$	$Ge[Co_2(CO)_7]_2$	158

3.4 Reaction of Main Group Oxides with Metal Carbonyls

Metal carbonyls are very good reagents for deoxygenating main group element oxides. The primary driving force for this reaction is the oxidation of CO to CO_2 although in some cases it is possible that water is also a product. Internal disproportionation of two carbonyl ligands to give a carbide and a CO_2 is the accepted mechanism of formation of a number of carbide clusters in pyrolysis reactions of the pure metal carbonyl (ref. 260). Nitrite, NO_2^-, and nitrosyl, NO^+, are effective at producing nitrido clusters of iron, ruthenium and osmium (ref. 109). EO_3^{2-} (E = S, Se, Te), are known to react with iron and ruthenium carbonyl anions (refs. 261-264). Examples of compounds obtained in this fashion include $Fe_2(CO)_6S_2$, $Fe_3(CO)_9S_2$, $Fe_3(CO)_9(S)(SO)$, $Fe_2(CO)_6Te_2$, $Fe_3(CO)_9Te_2$ and $H_2Ru_3(CO)_9E$. Bismuthate, BiO_3^-, has been used to produce $[Bi\{Fe(CO)_4\}_4]^{3-}$ from methanolic KOH solutions of $Fe(CO)_5$ (ref. 151).

3.5 Reaction of Main Group Trimethylsilyl Complexes with Metal Halides.

Trimethylsilyl derivatives of the main group elements react

easily with metal halides to eliminate trimethylsilyl halide and form complex cage and cluster compounds. For example, the metal complexes $MCl_2(PPh_3)_2$ (M = Ni, Co) react with $E(SiMe_3)_2$ (E = S, Se) with the formation of $Co_4As_6(PPh_3)_4$, $Co_7S_6(PPh_3)_5Cl_2$, $[Co_6S_8(PPh_3)_6][CoCl_3(THF)]$, $Ni_8S_6(PPh_3)_6Cl_2$, $Ni_8S_5(PPh_3)_7$, $Co_4Se_4(PPh_3)_4$, $Co_6Se_8(PPh_3)_6$, $Co_9Se_{11}(PPh_3)_6$ and $Ni_{34}Se_{22}(PPh_3)_{10}$ among others(refs. 265-268). This method is noteworthy for its generality and for its ability to generate clusters of higher formal metal oxidation states.

3.6 Metal Exchange Reactions.

This method has been employed for the synthesis of main group III element clusters and consists of treating $Hg[ML_n]_2$ with the desired element. Thus, indium and thallium react with $Hg[Co(CO)_4]_2$ to give $In[Co(CO)_4]_3$ and $Tl[Co(CO)_4]$, respectively (refs. 13, 143, 258). Exchanges will also occur among the main group elements as evidenced by the replacement of Tl in $Tl[Mn(CO)_5]_3$ using elemental indium (ref. 257).

4 REACTIVITY PATTERNS

Even though a large number of main group/transition metal complexes have been prepared and structurally characterized, their reactivity is not so well understood, especially that of the heavier elements. Some attention has been focused on this deficiency of late and some progress is being made. This is especially true of the carbido and nitrido clusters. There is a general sentiment that the presence of main group atoms somewhat stabilizes the cluster framework. This issue will need to be examined carefully if these compounds are to find uses in materials synthesis and catalysis applications. While some data on the thermal and photochemical stability of simple metal carbonyl clusters has been reported, the mixed main group-transition metal systems are virtually unexplored.

Much of the chemistry centers around oxidation/reduction reactions where one of several processes may occur. As with the non-derivatized transition metal clusters, simple ligand loss may occur upon reduction. For example, $Fe_5(CO)_{15}C$ is reduced to $[Fe_5(CO)_{14}C]^{2-}$ (ref. 62). Alternatively, M-M, E-M or E-E bonds may be broken/formed as found for the pair $[Pb\{Fe_2(CO)_8\}\{Fe(CO)_4\}_2]^{2-}$ and $Pb[Fe_2(CO)_8]_2$ (refs. 153, 157). The reduction of $Fe_2(CO)_6E_2$ (E = S, Se) has already been mentioned as leading to the breakage of the E-E bond (refs. 229-230). The redox activity of the E_3M and E_3M_2 compounds has already been discussed (ref. 234). This ability to funnel electrons in and out of the cluster bonding framework is an important feature of this chemistry and has implications for catalysis. One related area which has been under-represented is the study of anionic cluster derivatives. Often, work has centered on the neutral compounds which can be purified by chromatographic methods which ares often inappropriate for the charged species.

As found for the simple transition metal clusters, these compounds undergo ligand substitution reactions where CO can be replaced with phosphines. Normally, many substitutions are thought to proceed via CO dissociation but in some main group/transition metal species an associative process may be likely. For example, $Ru_5(CO)_{15}C$ will replace one or two CO's with phosphines, but this may proceed through a bridged-butterfly intermediate similar to that formed when MeCN adds to the cluster (ref. 108). The addition of a ligand such as MeCN adds two electrons to the cluster count and may be formally considered a type of reduction reaction.

The lone pairs of electrons that reside on main group atoms may be active as donors to extra-cluster metal fragments. A number of structurally documented examples have already been mentioned.

An important feature that may be important in determining the chemistry of clusters containing the heavier main group elements is the potential for the main group atom to expand its octet in order to accomodate one or more extra pairs of electrons. This has been demonstrated for tellurium in $CpMoFe(CO)_5Te_2Br$ and $(MeC_5H_4)MoFe(CO)_5Te_2(S_2CNEt_2)$ (ref 269) and for bismuth in $[Cp_2Co][Bi\{Co(CO)_4\}_4]$ (refs. 140-141).

A reaction found for a large number of cases is transition metal substitution. These observations were often an outgrowth of attempted cluster building reactions where it was hoped to introduce another metal. Instead, replacement of one of the cluster metal vertices with another metal was observed. Vahrenkamp has very nicely demonstrated this metal exchange for a series of germanium cobalt clusters (refs. 270-272).

5 SUMMARY

Cluster compounds containing both main group atoms and transition metals constitute a growing and interesting area of study. These compounds have a rich chemistry of their own arising from the synergism of the metal and main group fragments and from the increasing observations of unexpected bonding situations. More and more complex structures are being reported which show relationships to Zintl ion and Zintl phase chemistry. They have properties relevant to catalysis and solid state chemistry and may prove to be convenient building blocks for materials of designed stoichiometry and structure.

Acknowledgments

We gratefully acknowledge the support of the Petroleum Research Fund, the Robert A. Welch Foundation, the National Science Foundation (CHE-8421217) and the Alexander von Humboldt Stiftung for financial support of our research efforts which have contributed to this paper.

REFERENCES

1 J.D. Corbett, Chem. Rev., 85 (1985) 383.
2 J.D. Corbett, Prog. Inorg. Chem., 21 (1976) 129.
3 P. Braunstein, New J. Chem., 10 (1986) 365.
4 E.L. Muetterties and M.J. Krause, Angew. Chem. Int. Ed. Eng., 22 (1983) 135.
5 E.L. Muetterties, T.N. Rhodin, E. Band, C.F. Brucker and W.R. Pretzer, Chem. Rev., 79 (1979) 91.
6 E.L. Muetterties and J. Stein, Chem. Rev., 79 (1979) 479.
7 M. Moskovits, Accts. Chem. Res., 12 (1979) 229.
8 M. Moskovits (editor), Metal Clusters, Wiley-Interscience,New York, 1986.
9 B.F.G. Johnson (editor), Transition Metal Clusters, John Wiley, New York, 1980.
10 K.H. Whitmire, J. Coord. Chem. Rev., in press.
11 W.A. Herrmann, Angew. Chem., 98 (1986) 57; Angew. Chem. Int. Ed. Eng., 25 (1986) 56.
12 J.N. Nicholls, Polyhedron, 3 (1984) 1307.
13 A.T.T. Hsieh, Inorg. Chim. Acta, 14 (1975) 87.

14 N.E. Kolobova, A.B. Antonova and K.N. Anisimov, Russ. Chem. Rev., 38 (1969) 822.
15 W. Petz, Chem. Rev., 86 (1986) 1019.
16 O.Scherer, Angew. Chem., 97 (1985) 905; Angew. Chem. Int. Ed. Eng., 24 (1985) 924.
17 L.E. Toth, Transition Metal Carbides and Nitrides, Academic Press, New York, 1971.
18 A.L. Bowman, P.G. Arnold, E.K. Storms and N.G. Nereson, Acta Cryst., B28 (1972) 3102.
19 K. Wade, Adv. Inorg. Chem. Radiochem., 18 (1976) 1.
20 J.W. Lauher, J. Am. Chem. Soc., 100 (1978) 5305.
21 D.M.P. Mingos, Inorg. Chem., 21 (1982) 464.
22 B.W. Clare and D.L. Keppert, Inorg. Chem., 23 (1984) 1521.
23 L.D. Brown and W.N. Lipscomb, Inorg. Chem., 16 (1977) 2989.
24 D.G. Evans and D.M.P. Mingos, Organometallics, 2 (1983) 435.
25 D.M.P. Mingos and D.G. Evans, J. Organomet. Chem., 251 (1983) C13.
26 M.E. O'Neill and K. Wade, Inorg. Chem. 21 (1982) 461.
27 B.K. Teo, Inorg. Chem., 23 (1984) 1251.
28 B.K. Teo, G. Longoni and F.R.K. Chung, Inorg. Chem., 23 (1984) 1257.
29 D.M.P. Mingos, Inorg. Chem., 24 (1985) 114.
30 B.K. Teo, Inorg. Chem., 24 (1985) 115.
31 B.K. Teo, Inorg. Chem., 24 (1985) 1627.
32 B.K. Teo, Inorg. Chem., 24 (1985) 4209.
33 B.K. Teo and N.J.A. Sloane, Inorg. Chem., 24 (1984) 1289.
34 D.M.P. Mingos, Polyhedron, 3 (1984) 1289.
35 M. McPartlin and D.M.P. Mingos, Polyhedron, 3 (1984) 1321.
36 G. Ciani, G. D'Alfonso, M. Freni, P. Romiti and A. Sironi, J. Chem. Soc., Chem. Comm., (1982) 339.
37 G. Ciani, G. D'Alfonso, M. Freni, P. Romiti and A. Sironi, J. Chem Soc., Chem. Comm., (1982) 705.
38 T. Beringhelli, G. D'Alfonso, M. Freni, G. Ciani and A. Sironi, J. Organomet. Chem., 282 (1985) C39.
39 G. Ciani, G. D'Alfonso, P. Romiti, A. Sironi and M. Freni, J. Organomet. Chem., 244 (1983) C27.
40 M.R. Churchill, J. Wormald, J. Knight and M.J. Mays, J. Am. Chem. Soc., 93 (1971) 3073.
41 A. Gourdon and Y. Jeannin, J. Organomet. Chem., 282 (1985) C39.
42 B.F.G. Johnson, D.A. Kaner, J. Lewis and M. Rosales, J. Organomet. Chem., 238 (1982) C73.
43 J.S. Bradley, G.B. Ansell and E.W. Hill, J. Organomet. Chem., 184 (1980) C33.
44 B.F.G. Johnson, J. Lewis, S.W. Sankey, K. Wong, M. McPartlin and W.J.H. Nelson, J. Organomet. Chem., 191 (1980) C3.
45 J.S. Bradley, R.L. Pruett, E. Will, G.B. Ansell, M.E. Leonowicz and M.A. Modrick, Organometallics, 1 (1982) 748.
46 B.F.G. Johnson, J. Lewis, W.J.H. Nelson, J. Puga, P.R. Raithby, D. Braga, M. McPartlin and W. Clegg, J. Organomet. Chem. 243 (1983) C13.
47 S.C. Brown, J. Evans and M. Webster, J. Chem. Soc., Dalton Trans., (1981) 2263.
48 G.B. Ansell and J.S. Bradley, Acta Crystallogr. 36B (1980) 1930.
49 P.F. Jackson, B.F.G. Johnson, J. Lewis, P.R. Raithby, G.J. Will, M. McPartline and W.J.H. Nelson, J. Chem. Soc., Chem. Comm., (1980) 1190.
50 R. Mason and W.R. Robinson, J. Chem. Soc., Chem. Comm., (1968) 468.

51 P. Gomez-Sal, B.F.G. Johnson, J. Lewis, P.R. Raithby and A.H. Wright, J. Chem. Soc. Chem. Comm., (1985) 1682.
52 R.D. Adams, P. Mathur and B.E. Segmüller, Organometallics, 2 (1983) 1258.
53 S.R. Bunkhall, H.D. Holden, B.F.G. Johnson, J. Lewis, G.N.Pain, P.R. Raithby and M.J. Taylor, J. Chem. Soc., Chem. Comm., (1984) 25.
54 C.-M.T Howard, J.R. Shapley, M.R. Churchill, C. Bueno and A.L. Rheingold, J. Am. Chem. Soc., 104 (1982) 7347.
55 P.F. Jackson, B.F.G. Johnson, J. Lewis, M. McPartlin and W.J.H. Nelson, J. Chem. Soc., Chem. Comm., (1982) 49.
56 P.F. Jackson, B.F.G. Johnson, J. Lewis, M. McPartlin and W.J.H. Nelson, J. Chem. Soc., Chem. Comm., (1980) 224.
57 B.F.G. Johnson, J. Lewis, W.J.H. Nelson, M.D. Vargas, D. Braga, K. Henrick and M. McPartlin, J. Chem. Soc., Dalton Trans., (1986) 975.
58 S.R. Drake, K. Henrick, B.F.G. Johnson, J. Lewis, M. McPartlin and M. Morris, J. Chem. Soc., Chem. Comm., (1986) 928.
59 D. Braga, K. Henrick, B.F.G. Johnson, J. Lewis, M. McPartlin, W.J.H. Nelson and J. Puga, J. Chem. Soc., Chem. Comm., (1982) 1083.
60 D.H. Farrar, P.G. Jackson, B.F.G. Johnson, J. Lewis, W.J.H. Nelson, M.D. Vargas and M. McPartlin, J. Chem. Soc., Chem. Comm., (1981) 1009.
61 P.F. Jackson, B.F.G. Johnson, J. Lewis, W.J.H. Nelson and M. McPartlin, J. Chem. Soc., Dalton Trans., (1982) 2099.
62 V.E. Lopatin, S.P. Gubin, N.M. Mikova, M. Ts. Tsybenov, Yu. L. Slovokhotov and Yu. T. Struchov, J. Organomet. Chem., 292 (1985) 275.
63 M. Tachikawa, A. C. Sievert, E.L. Muetterties, M.R. Thompson, C.S. Day and V. W. Day, J. Am. Chem. Soc., 102 (1980) 1725.
64 A. Sirigu, M. Bianchi and E. Bendetti, J. Chem. Soc., Chem. Comm., (1969) 596.
65 D. Braga, K. Henrick, B.F.G. Johnson, J. Lewis, M. McPartlin, W.J.H. Nelson, A. Sironi and M.D. Vargas, J. Chem. Soc., Chem. Comm., (1983) 1131.
66 S. Martinengo, D. Strumolo, P. Chini, V.G. Albano and D. Braga, J. Chem. Soc., Chem. Comm., (1985) 35.
67 S. Martinengo, G. Ciani, A. Sironi, B.T. Heaton and J. Mason, J. Am. Chem. Soc. 101 (1979) 7095.
68 V.G. Albano, M. Sansoni, P. Chini and S. Martinengo, J. Chem. Soc., Dalton Trans., (1973) 651.
69 R. Bonfichi, G. Ciani, A. Sironi and S. Martinengo, J. Chem. Soc., Dalton Trans., (1983) 253.
70 S. Martinengo, G. Ciani and A. Sironi, J. Chem. Soc., Chem. Comm., (1984) 1577.
71 V.G. Albano, D. Braga, S. Martinengo, P. Chini, M. Sansoni and D. Strumolo, J. Chem. Soc., Chem. Comm., (1980) 52.
72 V.G. Albano, M. Sansoni, P. Chini, S. Martinengo and D. Strumolo, J. Chem. Soc., Dalton Trans., (1975) 305.
73 V.G. Albano, D. Braga, A. Fumagalli and S. Martinengo, J. Chem. Soc., Dalton Trans., (1985) 1137.
74 V.G. Albano, D. Braga, P. Chini, G. Ciani and S. Martinengo, J. Chem. Soc., Dalton Trans., (1982) 645.
75 A. Arigoni, A. Ceriotti, R. Della Pergola, G. Longoni, M. Manassero, N. Masciocchi and M. Sansoni, Angew. Chem. 96 (1984) 290; Angew. Chem. Int. Ed. Eng., 23 (1984) 322.
76 V.G. Albano, D. Braga, P. Chini, D. Strumolo and S. Martinengo, J. Chem. Soc., Dalton Trans., (1983) 249.
77 D. Strumolo, C. Serengi, S. Martinengo, V.G. Albano and D. Braga, J. Organomet. Chem. 252 (1983) C93.

78 V.G. Albano, D. Braga, D. Strumolo, C. Serengi and S. Martinengo, J. Chem. Soc., Dalton Trans., (1985) 1309.
79 B.T. Heaton, L. Strona, S. Martinengo, D. Strumolo, V.G. Albano and D. Braga, J. Chem. Soc., Dalton Trans., (1983) 2175.
80 S. Martinengo, D. Strumolo, P. Chini, V.G. Albano and D. Braga, J. Chem. Soc., Dalton Trans., (1984) 1837.
81 V.G. Albano, D. Braga and S. Martinengo, J. Chem. Soc., Dalton Trans., (1986) 981.
82 V.G. Albano, P. Chini, G. Ciani, M. Sansoni and M. Martinengo, J. Chem. Soc., Dalton Trans., (1980) 163.
83 V.G. Albano, P. Chini, S. Martinengo, M. Sansoni and D. Strumolo, J. Chem. Soc., Chem. Comm., (1974) 299.
84 V.G. Albano, P. Chini, G. Ciani, S. Martinengo and M. Sansoni, J. Chem. Soc., Dalton Trans., (1978) 463.
85 A. Ceriotti, G. Longoni, M. Manassero, M. Perego and M. Sansoni, Inorg. Chem., 24 (1985) 117.
86 A. Ceriotti, R. Della Pergola, G. Longoni, M. Manassero and M. Sansoni, J. Chem. Soc., Dalton Trans., (1984) 1181.
87 G. Schmid, V. Bätzel, G. Etzrodt and R. Pfeil, J. Organomet. Chem., 86 (1975) 257.
88 K. MacKay, B.K. Nicholson, W.T. Robinson and A.W. Sims, J. Chem. Soc., Chem. Comm., (1984) 1276.
89 J.L.Vidal, W.E Walker and R.C Schoening, Inorg Chem., 20 (1981) 238.
90 J.L.Vidal, W.E Walker, R.L. Pruett and R.C Schoening, Inorg Chem., 18 (1979) 129.
91 J.L. Vidal, Inorg Chem., 20 (1981) 243.
92 J.L. Vidal and J.M. Troup, J. Organomet. Chem., 213 (1981) 351.
93 J.L. Vidal, R.A. Fiato, L.A. Cosby and R.L. Pruett, Inorg. Chem., 17 (1978) 2574.
94 P. Chini, G. Ciani, S. Martinengo, A. Sironi, L. Longhetti and B. T. Heaton, J. Chem. Soc., Dalton Trans., (1979) 188.
95 V.G. Albano, D. Braga, G. Ciani and S. Martinengo, J. Organomet. Chem., 213 (1981) 293.
96 A. Arigoni, A. Ceriotti, R. Della Pergola, G. Longoni, M. Manassero, N. Masciocchi and M. Sansoni, Angew. Chem. Int Ed. Eng., 23 (1984) 322.
97 A. Arigoni, A. Ceriotti, R. Della Pergola, G. Longoni, M. Manassero, and M. Sansoni, J.Organomet. Chem., 296 (1985) 243.
98 A. Ceriotti, G. Longoni, M. Manassero, N. Masciocchi, L. Resconi and M. Sansoni, J. Chem. Soc., Chem. Comm., (1985) 181.
99 A. Ceriotti, G. Longoni, M. Manassero, N. Masciocchi, G. Piro, L. Resconi and M. Sansoni, J. Chem. Soc., Chem. Comm., (1985) 1402.
100 V.G. Albano, P. Chini, S. Martinengo and D. Strumolo, J. Chem. Soc., Dalton Trans., (1978) 459.
101 M.R. Churchill, J. Wormald, J. Knight and M.J. Mays, J. Am. Chem. Soc., 93 (1971) 3073.
102 R.P. Stewart, U. Anders and W.A.G. Graham, J. Organomet. Chem., 32 (1971) C49.
103 M. Tachikawa and E.L. Muetterties, J. Am. Chem. Soc., 102 (1980) 4541.
104 J.S. Bradley, E.W. Hill, G.B. Ansell and M. A. Modrick, Organometallics, 1 (1982) 1634.
105 J.S. Bradley, G.B. Ansell, M.E. Leonowicz and E.W. Hill, J. Am. Chem. Soc., 103 (1981) 4968.
106 J.S. Bradley, G.B. Ansell and E.W. Hill, J. Am. Chem. Soc., 101 (1979) 7417.
107 J.S. Bradley, Phil. Trans. R. Soc. Lond. A. 308 (1982) 103.

108 B.F.G. Johnson, J. Lewis, J.N. Nicholls, J. Puga, P.R. Raithby, M.J. Rosales, M. McPartlin and W. Clegg, J. Chem. Soc., Dalton Trans., (1983) 277.
109 W.L. Gladfelter, Adv. Organomet. Chem., 24 (1985) 41.
110 B.F.G. Johnson, J. Lewis, J.N. Nicholls, J. Puga and K.H. Whitmire, J. Chem. Soc., Dalton Trans., (1983) 787.
111 A.G. Cowie, B.F.G. Johnson, J. Lewis, J.N. Nicholls, P.R. Raithby and A.G. Swanson, J. Chem. Soc., Chem. Comm., (1984) 637.
112 S.D. Wijeyesekera, R. Hoffmann and C.N. Wilker, Organometallics, 3 (1984) 962.
113 E.H. Braye, L.F. Dahl, W. Hübel and D.L. Wampler, J. Am. Chem. Soc., 84 (1962) 4633.
114 A. Gourdon and Y. Jeannin, J. Organomet. Chem., 290 (1985) 199.
115 M. Tachikawa, J. Stein, E.L. Muetterties, R.G. Teller, M.A. Beno, E. Gebert and J.M. Williams, J. Am. Chem. Soc., 102 (1980) 6648.
116 M.L. Blohm and W.L. Gladfelter, Organometallics, 4 (1985) 45.
117 P.F. Jackson, B.F.G. Johnson, J. Lewis and J.N. Nicholls, J. Chem. Soc., Chem. Comm., (1980) 564.
118 B.F.G. Johnson, J. Lewis, W.J.H. Nelson, J.N. Nicholls, J. Puga, P.R. Raithby, M.J. Rosales, M. Schröder and M.D. Vargas, J. Chem. Soc., Dalton Trans., (1983) 2447.
119 B.F.G. Johnson, D.A. Kaner, J. Lewis, P.R. Raithby and M.J. Rosales, J. Organomet. Chem., 231 (1982) C59.
120 A.G. Cowie, B.F.G. Johnson, J. Lewis and P.R. Raithby, J. Chem. Soc., Chem. Comm., (1984) 1710.
121 B.F.G. Johnson, J. Lewis, P.R. Raithby, M.J. Rosales and D.A. Welch, J. Chem. Soc., Dalton Trans., (1986) 453.
122 P.L. Bogdan, C.P. Horwitz and D.F. Shriver, J. Chem. Soc., Chem. Comm., (1986) 553.
123 E.M. Holt, K.H. Whitmire and D.F. Shriver, J. Am. Chem. Soc., 104 (1982) 5621.
124 M.H. Chisholm, K. Folting, J.C. Huffman, J. Leonelli, N.S. Marchant, C.A. Smith and L.C.E. Taylor, J. Am. Chem. Soc., 107 (1985) 3722.
125 J.C. Vites, C.E. Housecroft, G.B. Jacobsen and T.P. Fehlner, Organometallics, 3 (1984) 1591.
126 C.E. Housecroft and T.P. Fehlner, Organometallics, 5 (1986) 1279.
127 T.P. Fehlner, C.E. Housecroft, W.R. Scheidt and K.S. Wong, Organometallics, 2 (1983) 825.
128 C.K. Schauer and D.F. Shriver, Angew. Chem., 99 (1987) 275; Angew. Chem. Int. Ed. Eng., 26 (1987) 255.
129 M.A. Beno, J.M. Williams, M. Tachikawa and E.L. Muetterties, J. Am. Chem. Soc., 102 (1980) 4542.
130 E.M. Holt, K.H. Whitmire and D.F. Shriver, J. Organomet. Chem., 213 (1981) 125.
131 J.H. Davis, M.A. Beno, J.M. Williams, J. Zimmie, M. Tachikawa and E.L. Muetterties, Proc. Nat. Acad. Sci. USA, 78 (1981) 668.
132 J.A. Hriljac, P.N. Sweptson and D.F. Shriver, Organometallics, 4 (1985) 158.
133 D.E. Fjare and W.L. Gladfelter, J. Am. Chem. Soc., 103 (1981) 1572.
134 D.E. Fjare and W.L. Gladfelter, Inorg. Chem., 20 (1981) 3533.
135 D. Braga, B.F.G. Johnson, J. Lewis, J. Mace., M. McPartlin, J. Puga, W.J.H. Nelson and K.H. Whitmire, J. Chem. Soc., Chem. Comm., (1982) 1081.
136 M.L. Blohm, D.E. Fjare and W.L. Gladfelter, 108 (1986) 2301.

137 M.A. Collins, B.F.G. Johnson, J.Lewis, J.M. Mace, J. Morris, M. McPartlin, W.J.H. Nelson, J. Puga and P.R. Raithby, J. Chem. Soc., Chem. Comm., (1983) 689.
138 J.P. Attard, B.F.G. Johnson, J. Lewis, J. Mace and P.R.Raithby, J. Chem. Soc., Chem. Comm., (1985) 1526.
139 W.R. Robinson and D.P. Schussler, J. Organomet. Chem., 30 (1971) C5.
140 J.S. Leigh and K.H.Whitmire, Angew. Chem., in press.
141 S. Martinengo and G. Ciani, J. Chem. Soc., Chem. Comm., (1987) 1589.
142 A.J. Conway, P.B. Hitchcock and J.D. Smith, J. Chem. Soc., Dalton Trans., (1975) 1945.
143 W.R. Robinson and D.P. Schussler, Inorg. Chem., 12 (1973) 848.
144 J. Rajaram and J.A. Ibers, Inorg. Chem., 12 (1973) 1313.
145 G. Huttner, U. Weber, B. Sigwarth, O. Scheidsteger, H. Lang and L. Zsolnai, J. Organomet. Chem., 282 (1985) 331.
146 M. Heberhold, D. Reiner and D. Neugebauer, Angew. Chem., 95 (1983) 46; Angew. Chem. Int. Ed. Eng., 22 (1983) 59.
147 W.A. Herrmann, J. Rohrmann, M.L. Ziegler and T. Zahn, J. Organomet. Chem, 295 (1985) 175.
148 G. Etzrodt, R. Boese and G. Schmid, Chem. Ber., 112 (1979) 2574.
149 J.M. Wallis, G. Müller and H. Schmidbaur, Inorg. Chem., 26 (1987) 458.
150 J.M. Wallis, G. Müller and H. Schmidbaur, J. Organomet. Chem., 325 (1987) 159.
151 M.R. Churchill, J.C. Fettinger, K.H. Whitmire and C.B. Lagrone, J. Organomet. Chem., 303 (1986) 99.
152 J. Rohrmann, W.A. Herrmann, E. Herdtweck, J. Riede, M. Ziegler and G. Sergerson, Chem. Ber., 119 (1986) 3544.
153 C.B. Lagrone, K.H. Whitmire, M.R. Churchill and J.C. Fettinger, Inorg. Chem., 25 (1986) 2080.
154 D. Melzer and E. Weiss, J. Organomet. Chem., 255 (1983) 335.
155 A.S. Batsanov, L.V. Rybin, M.I. Rybinskaya, Yu. T. Struchov, I.M. Salimgareeva and N.G. Bogatova, J. Organomet. Chem., 249 (1983) 319.
156 P.F. Lindley and P. Woodward, J. Chem. Soc. (A), (1967) 382.
157 K.H. Whitmire, C.B. Lagrone, M.R. Churchill, J.C. Fettinger and B.H. Robinson, Inorg. Chem., 26 (1987) 3491.
158 R.F. Gerlach, K.M. Mackay, B.K. Nicholson and W.T. Robinson, J. Chem. Soc., Dalton Trans., (1981) 80.
159 S.G. Anema, K.M. Mackay, L.C. McLeod, B.K. Nicholson and J.M. Whittaker, Angew. Chem., 98 (1986) 744.
160 G. Huttner, G. Mohr, B. Pritzlaff, J. von Seyerl and L. Zsolnai, Chem. Ber., 115 (1982) 2044.
161 H. Lang, G. Huttner, L. Zsolnai, G. Mohr, B. Sigwarth, U. Weber, O. Orama and I Jibril, J. Organomet. Chem., 304 (1986) 157.
162 C.F. Campana and L.F. Dahl, J. Organomet. Chem., 127 (1977) 209.
163 A.L. Rheingold, S.J Geib, M. Shieh and K.H. Whitmire, Inorg. Chem., 26 (1987) 463.
164 G. Schmid, Angew. Chem., 90 (1978) 417; Angew. Chem. Int. Ed. Eng., 17 (1978) 392.
165 A. Vizi-Orosz, J. Organomet. Chem., 111 (1976) 61.
166 A. Vizi-Orosz, V. Galamb, G. Pályi, L. Markó, G. Bor and G. Natile, J. Organomet. Chem., 107 (1976) 235.
167 K. Blechsmitt, H. Pfisterer, T. Zahn and M.L. Ziegler, Angew. Chem. Int. Ed. Eng., 24 (1985) 66.
168 K.H. Whitmire, J.S. Leigh and M.E. Gross, J. Chem. Soc., Chem. Comm., (1987) 926.

169 K.H. Whitmire, C.B. Lagrone, M.R. Churchill, J.C. Fettinger and L.V. Biondi, Inorg. Chem., 23 (1984) 4227.
170 K.H. Whitmire, C.B. Lagrone and A.L. Rheingold, Inorg. Chem. 25 (1986) 2472.
171 W. Kruppa, D. Bläser, R. Boese and G. Schmid, Z. Naturforsch., 37B (1982) 209.
172 A. Bertolucci, M. Freni, P. Romiti, G. Ciani, A. Sironi and V.G. Albano, J. Organomet. Chem., 113 (1976) C61.
173 A. Ceriotti, L. Reconi, F. Demartin, G. Longoni, M. Manassero and M. Sansoni, J. Organomet. Chem., 249 (1983) C35.
174 G. Lavigne, N. Lugan and J.-J. Bonnet, Nouv. J. Chem., 5 (1981) 423.
175 R.J. Goudsmit, B.F.G. Johnson, J. Lewis, P.R. Raithby and K.H. Whitmire, J. Chem. Soc., Chem. Comm., (1983) 246.
176 D.L. Stevenson, C.H. Wei and L.F. Dahl, J. Am. Chem. Soc., 93 (1971) 6027.
177 K. Fischer, W. Deck, M. Schwarz and H. Vahrenkamp, Chem. Ber., 118 (1985) 4946.
178 R.D. Adams, I.T. Horváth and H.-S. Kim, Organometallics, 3 (1984) 548.
179 B.F.G. Johnson, J. Lewis, D. Pippard, P.R. Raithby, G.M. Sheldrick and K.D. Rouse, J. Chem. Soc., Dalton Trans., (1979) 616.
180 R.D. Adams, D.F. Foust and P. Mathur, Organometallics, 2 (1983) 990.
181 R.D. Adams, I.T. Horváth, B.E. Segmüller and L.-W. Yang, Organometallics, 2 (1983) 1301.
182 C.E. Strouse and L.F. Dahl, J. Am. Chem. Soc., 93 (1971) 6032.
183 W.A. Herrmann, C. Hecht, M.L.Ziegler and T. Zahn, J. Organomet. Chem., 273 (1984) 323.
184 L.E. Bogan, Jr., T.B. Rauchfuss and A.L. Rheingold, J. Am. Chem. Soc., 107 (1985) 3843.
185 H.-J. Haupt, F. Neumann and H. Preut, J. Organomet. Chem., 99 (1975) 439.
186 G. Schmid, V. Bätzel and G. Etzrodt, J. Organomet. Chem., 112 (1976) 345.
187 P. Gusbeth and H. Vahrenkamp, Chem. Ber., 118 (1985) 1746.
188 P. Gusbeth and H. Vahrenkamp, Chem. Ber., 118 (1985) 1770.
189 R.A. Croft, D.N. Duffy and B.K. Nicholson, J. Chem. Soc., Dalton Trans., (1982) 1023.
190 H. Lang, G. Huttner, B. Sigwarth, I. Jibril, L. Zsolnai and O. Orama, J. Organomet. Chem., 304 (1936) 137.
191 A. Gourdon and Y. Jeannin, J. Organomet. Chem., 304 (1986) C1.
192 A.L. Rheingold, K.H. Whitmire, M.D. Fabiano and J.S. Leigh, unpublished results.
193 F. Richter and H. Vahrenkamp, Angew. Chem. Int. Ed. Eng., 17 (1978) 444.
194 R.D. Adams, D. Männig and B.E. Segmüller, 2 (1932) 149.
195 L.T. Delbaere, L.J. Kruczynski and D.W. McBride, J. Chem. Soc., Dalton Trans., (1973) 307.
196 M.R. Churchill, J.C. Fettinger and K.H. Whitmire, J. Organomet. Chem., 284 (1985) 13.
197 T.J. McNeese, S.S. Wreford, D.L. Tipton and R. Bau, J. Chem. Soc., Chem. Comm., (1977) 390.
198 A.J. Bard, A.H. Cowley, J.K. Leland, J.N. Thomas, N.C. Norman, P. Jutzi, C.P. Morley and E. Schlüter, J. Chem. Soc., Dalton Trans., (1985) 1303.
199 R.D. Adams, I.T. Horváth, B.E. Segmüller and L.-W. Yang, Organometallics, 2 (1983) 144.
200 P.V. Broadhurst, B.F.G. Johnson, J. Lewis and P.R. Raithby, J. Organomet. Chem., 194 (1980) C35.

201 R.D. Adams and I.T. Horváth, J. Am. Chem. Soc., 106 (1984) 1869.
202 L.F. Dahl and P.W. Sutton, Inorg. Chem., 2 (1963) 1067.
203 W. Hieber and J. Gruber, Z. Anorg. Allg. Chem., 296 (1958) 91.
204 D.A. Lesch and T.B. Rauchfuss, Inorg. Chem., 20 (1981) 3583.
205 K.H. Whitmire, K.S. Rhaguveer, M.R. Churchill, J.C. Fettinger and R.F. See, J. Am. Chem. Soc., 108 (1986) 2778.
206 K.H.Whitmire, M. Shieh, C.B. Lagrone, B.H. Robinson, M.R. Churchill, J.C. Fettinger and R.F. See, Inorg. Chem., 26 (1987) 2798.
207 S. Luo and K.H. Whitmire, unpublished results.
208 C.F. Campana, A. Vizi-Orosz, G. Palyi, L. Markó and L.F. Dahl, Inorg. Chem., 18 (1979) 3054.
209 A.S. Foust, M.S. Foster and L.F. Dahl, J. Am. Chem. Soc., 91 (1969) 5633.
210 A.S. Foust, C.F. Campana, D. Sinclair and L.F. Dahl, Inorg. Chem., 18 (1979) 3047.
211 C.H. Wei and L.F. Dahl, Inorg. Chem., 4 (1965) 1.
212 C.H. Wei and L.F. Dahl, Inorg. Chem., 4 (1965) 493.
213 C.F. Campana, F.W.-K. Lo and L.F. Dahl, Inorg. Chem., 18 (1979) 3060.
214 D.J. Jones, T. Makani and J. Rozière, J. Chem. Soc., Chem. Comm., (1986) 1275.
215 A. Vizi-Orosz, G. Pályi, L. Markó, R. Boese and G. Schmid, J. Organomet. Chem., 288 (1985) 179.
216 G. Huttner, B. Sigwarth, O. Scheidsteger, L. Zsolnai and O. Orama, Organometallics, 4 (1985) 326.
217 A.M. Arif, A.H. Cowley, N.C. Norman and M. Pakulski, Inorg. Chem., 25 (1986) 4836.
218 D.A. Lesch and T.B. Rauchfuss, Organometallics, 1 (1982) 499.
219 B. Sigwarth, L. Zsolnai, H. Berke and G. Huttner, J. Organomet. Chem., 226 (1982) C5.
220 G. Huttner, U. Weber, B. Sigwarth and O. Scheidsteger, Angew. Chem. Int. Ed. Eng. 21 (1982) 215.
221 G. Huttner, U. Weber and L. Zsolnai, Z. Naturforsch. 37B (1982) 707.
222 H. Preut and H.-J. Haupt, Chem. Ber., 107 (1974) 2860.
223 H. Preut and H.-J. Haupt, Chem. Ber., 108 (1975) 1447.
224 H.-J. Haupt and F. Neumann, J. Organomet. Chem. 74 (1974) 185.
225 H. Preut and H.-J. Haupt, Acta Crystallogr., 35B (1979) 729.
226 K.H. Whitmire, J. Cassiday, A.L. Rheingold and R.R. Ryan, Inorg. Chem., in press.
227 M. DiVaira, M. Peruzzini and P. Stoppioni, J. Chem. Soc., Chem. Comm., (1986) 374.
228 K.H. Whitmire, R.R. Ryan, H.J. Wasserman, T.A. Albright and S.-K. Kang, J. Am. Chem. Soc., 108 (1986) 6831.
229 T.D. Weatherill, T.B. Rauchfuss and R.A. Scott, Inorg. Chem., 25 (1986) 1466.
230 D. Seyferth and R.S. Henderson, Organometallics, 1 (1982) 125.
231 W.A. Herrmann, B. Koumbouris, T. Zahn and M.L. Ziegler, Angew. Chem. Int. Ed. Eng., 23 (1984) 812.
232 W.A. Herrmann, B. Koubouris, A. Schäfer, T. Zahn and M.L. Ziegler, Chem. Ber. 118 (1985) 2472.
233 L.Y. Goh, C. Wei and E. Sinn, J. Chem. Soc., Chem. Comm., (1985) 462.
234 M. DiVaira and L. Sacconi, Angew. Chem.94 (1982) 338; Angew. Chem. Int. Ed. Eng., 21 (1982) 330, and references therein.
235 A.S. Foust, M.S. Foster and L.F. Dahl, J. Am. Chem. Soc., 91 (1969) 5631.

236 I. Bernal, H. Brunner, W. Meier, H. Pfisterer, J. Wachter and M.L. Ziegler, Angew. Chem., 96 (1984) 428; Angew. Chem. Int. Ed. Eng., 23 (1984) 438.
237 R. Faggiani, R.J. Gillespie, C. Campana and J.W. Kolis, J. Chem. Soc., Chem. Comm., (1987) 485.
238 C.A. Ghilardi, S. Midollini, A. Orlandini and L. Sacconi, Inorg. Chem., 19 (1980) 301.
239 S.P. Foster, K.M. Mackay and B. K. Nicholson, Inorg. Chem. 24 (1985) 909.
240 R.C. Ryan, C.U. Pittman, Jr., J.P. O'Connor and L.F. Dahl, J. Organomet. Chem., 193 (1980) 247.
241 C.H. Wei and L.F. Dahl, Cryst. Struct. Comm., 4 (1975) 583.
242 H. Vahrenkamp and E.J. Wucherer, Angew. Chem., 93 (1981) 715; Angew. Chem. Int. Ed. Eng., 20 (1981) 680.
243 R.D. Adams, J.E. Babin and M. Tasi, Inorg. Chem. 25 (1986) 4514.
244 R.D. Adams, I.T. Horváth, B.E. Segmüller and L.-W. Yang, Organometallics, 2 (1983) 1301.
245 R.D. Adams, I.T. Horváth and P. Mathur, Organometallics, 3 (1984) 623.
246 R.D. Adams and L.-W. Yang, J. Am. Chem. Soc., 104 (1982) 4115.
247 R.D. Adams, I.T. Horváth and L.-W. Yang, J. Am. Chem. Soc., 105 (1983) 1533.
248 R.C. Ryan and L.F. Dahl, J. Am. Chem. Soc., 97 (1975) 6904.
249 A.L. Rheingold, M.J Foley and P.J. Sullivan, J. Am. Chem. Soc., 104 (1982) 4727.
250 O.J. Scherer, H. Sitzmann, and G. Wolmershäuser, Angew. Chem., 97 (1985) 358; Angew. Chem. Int. Ed. Eng., 24 (1985) 351.
251 K.H. Whitmire, M.R. Churchill and J.C. Fettinger, J. Am. Chem. Soc., 107 (1985) 1056.
252 K.H. Whitmire, T.A. Albright, S.-K. Kang, M.R. Churchill and J.C. Fettinger, Inorg. Chem., 25 (1986) 2799.
253 K. Deller and B. Eisenmann, Z. Naturforsch B: Anorg. Chem., Org. Chem., 31B (1976) 29.
254 C. Belin, T. Makani and J. Rozière, J. Chem. Soc., Chem. Comm., (1985) 118.
255 G. Schmid and G. Etzrodt, J. Organomet. Chem., 131 (1977) 477.
256 J.A. Christie, D.N. Duffy, K.M. Mackay and B.K. Nicholson, J. Organomet. Chem., 226 (1982) 165.
257 A.T.T. Hsieh and M.J. Mays, J. Chem. Soc., Dalton Trans., (1972) 516.
258 A.T.T. Hsieh and M.J. Mays, J. Organomet. Chem., 37 (1972) 9.
259 H.-J. Haupt, F. Neumann and B. Schwab, Z. Anorg. Allg. Chem., 485 (1982) 234.
260 C.R. Eady, B.F.G. Johnson and J. Lewis, J. Chem. Soc., Dalton Trans., (1975) 2606.
261 W. Hieber and J. Gruber, Z. Anorg. Allg. Chem. 296 (1958) 91.
262 L. Markó, B. Markó-Monostory, T. Madach and H. Vahrenkamp, Angew. Chem., 92 (1980) 225; Angew. Chem. Int. Ed. Eng., 19 (1980) 226.
263 D.A. Lesch and T.B. Rauchfuss, J. Organomet. Chem., 199 (1980) C6.
264 E. Sappa, O. Gambino and G. Cetini, J. Organomet. Chem., 35 (1972) 375.
265 D. Fenske and J. Hachgenei, Angew. Chem., 98 (1986) 165; Angew. Chem. Int. Ed. Eng., 25 (1986) 175.
266 D. Fenske, J. Hachgenei and J. Ohmer, Angew. Chem., 97 (1985) 684; Angew. Chem. Int. Ed. Eng., 24 (1985) 706.
267 D. Fenske and J. Ohmer, Angew. Chem., 99 (1987) 155; Angew. Chem. Int. Ed. Eng., 26 (1987) 148.

268 D. Fenske, J. Ohmer and J. Hachgenei, Angew. Chem., 97 (1985) 993; Angew. Chem. Int. Ed. Eng., 24 (1985) 993.
269 L.E. Bogan, Jr., T.B. Rauchfuss and A.L. Rheingold, Inorg. Chem., 24 (1985) 3722.
270 P. Gusbeth and H. Vahrenkamp, Chem. Ber., 118 (1985) 1770.
271 P. Gusbeth and H. Vahrenkamp, Chem. Ber., 118 (1985) 1746.
272 P. Gusbeth and H. Vahrenkamp, Chem. Ber., 118 (1985) 1758.

Subject Index

For brevity only a single citation is given for a specific section. The reader is advised to read a few pages before and after the listed citation as the topic of interest is often discussed on several pages.